Chemoresponsive Materials
Stimulation by Chemical and Biological Signals

RSC Smart Materials

Series Editors:
Professor Hans-Jörg Schneider, *Saarland University, Germany*
Professor Mohsen Shahinpoor, *University of Maine, USA*

Titles in this Series:

1: Janus Particle Synthesis, Self-Assembly and Applications
2: Smart Materials for Drug Delivery: Volume 1
3: Smart Materials for Drug Delivery: Volume 2
4: Materials Design Inspired by Nature
5: Responsive Photonic Nanostructures: Smart Nanoscale Optical Materials
6: Magnetorheology: Advances and Applications
7: Functional Nanometer-Sized Clusters of Transition Metals: Synthesis, Properties and Applications
8: Mechanochromic Fluorescent Materials: Phenomena, Materials and Applications
9: Cell Surface Engineering: Fabrication of Functional Nanoshells
10: Biointerfaces: Where Material Meets Biology
11: Semiconductor Nanowires: From Next-Generation Electronics to Sustainable Energy
12: Supramolecular Materials for Opto-Electronics
13: Photocured Materials
14: Chemoresponsive Materials: Stimulation by Chemical and Biological Signals

How to obtain future titles on publication:
A standing order plan is available for this series. A standing order will bring delivery of each new volume immediately on publication.

For further information please contact:
Book Sales Department, Royal Society of Chemistry, Thomas Graham House, Science Park, Milton Road, Cambridge, CB4 0WF, UK
Telephone: +44 (0)1223 420066, Fax: +44 (0)1223 420247
Email: booksales@rsc.org
Visit our website at www.rsc.org/books

Chemoresponsive Materials
Stimulation by Chemical and Biological Signals

Edited by

Hans-Jörg Schneider
FR Organische Chemie der Universität des Saarlandes,
Saarbrücken, Germany
Email: h-j.schneider@mx.uni-saarland.de

ROYAL SOCIETY
OF CHEMISTRY

THE QUEEN'S AWARDS
FOR ENTERPRISE:
INTERNATIONAL TRADE
2013

RSC Smart Materials No. 14

Print ISBN: 978-1-78262-062-4
PDF eISBN: 978-1-78262-242-0
ISSN: 2046-0066

A catalogue record for this book is available from the British Library

Published by The Royal Society of Chemistry,
Thomas Graham House, Science Park, Milton Road,
Cambridge CB4 0WF, UK

Registered Charity Number 207890

For further information see our web site at www.rsc.org

Preface

Life functions depend on chemical communication. It will take decades before man-made systems can approach the regulation complexity of living cells. Only in recent years have scientists started to develop artificial systems that mimic biological functions, and can furthermore interact with living organisms in an intelligent manner. Smart chemical systems that can respond to external signals and can adopt to their environment are based on very different, often soft, materials. These encompass most frequently gels in large or nanoscale dimensions in the form of films, brushes, capsules, vesicles, or membranes. All of these can exhibit swelling or shrinking in response to external signals. Other smart materials are based on gelators, which exhibit solubility changes or a sol–gel phase transition in response to different chemical components. An important extension is particles such as silica microspheres with pores, which can open or close in response to their environment. All these materials can be designed to respond selectively to a chemical input, which can range from changes of pH, of redox conditions, of light and of specific substances. In this book, the first attempt is made to illustrate the development of smart materials, which, unlike those described in related monographs, are not primarily responsive to physical signals such as electric or magnetic fields, but to the chemical environment.

The possible applications of chemically responsive materials span from artificial muscles, actuators for process control, molecular machines, tissue engineering, self-healing surfaces, electronics, to targeted drug delivery. The large variety of different smart materials and the many possible applications would make it difficult to systematically put them together in a single-authored book. The chapters of the present monograph illustrate, on the basis of different chemically-stimulated materials, their great promise. Each chapter was written by experts in their field; the editor is most grateful to all of them for their indefatigable cooperation, but takes responsibility for

RSC Smart Materials No. 14
Chemoresponsive Materials: Stimulation by Chemical and Biological Signals
Edited by Hans-Jörg Schneider

Published by the Royal Society of Chemistry, www.rsc.org

possible shortcomings. The indispensable support by the staff of the RSC is also gratefully acknowledged. It is hoped that the book will stimulate experienced and young scientists to explore the intriguing and virtually endless possibilities of smart materials for applications in many new technologies, and particularly in biomedical fields.

Hans-Jörg Schneider
Saarbrücken, Germany

Contents

RSC Smart Materials No. 14
Chemoresponsive Materials: Stimulation by Chemical and Biological Signals
Edited by Hans-Jörg Schneider

Published by the Royal Society of Chemistry, www.rsc.org

Highlights from Book Chapters

In this section, the contents of the different book chapters are highlighted in order to provide a flavour for the less informed reader. Obviously it is impossible to summarize in a few lines all the essential information from a chapter, which covers a broad range of topics, occasionally with some overlap between the different contributions. Nevertheless it is hoped that with these highlights many readers will be intrigued enough to have a closer look at the various chapters.

Chapter 1 by the editor contains a general introduction and examples of some applications not covered in other chapters, such as molecular machines based on rotaxanes with protonation-controlled stations. Other examples are electroactive polymers in the form of conductive materials, and some mass-sensitive sensors, which can also be triggered by pH or redox changes, as well as by ion recognition.

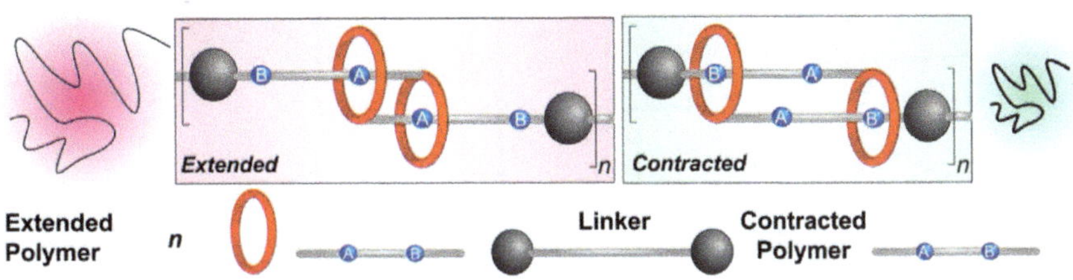

Drug delivery is at present the most actively pursued application of chemoresponsive materials; the topic is introduced in Chapter 2 by Peppas, Caldorera-Moore *et al.*, one of the leading groups in the field, but it is also discussed in several other chapters. The characteristics of often-used

RSC Smart Materials No. 14
Chemoresponsive Materials: Stimulation by Chemical and Biological Signals
Edited by Hans-Jörg Schneider

Published by the Royal Society of Chemistry, www.rsc.org

hydrogels are described in this chapter; they serve as reservoirs for drugs, which can be released at a controlled rate. Release can be triggered by signals from the environment, such as endogenous pH, blood glucose, or inflammation. Thus, in a hydrogel containing immobilized catalase and glucose oxidase, glucose is oxidised to gluconic acid, and insulin is released as the gel swells. It is shown that nanogel particles have particular advantages for targeted delivery; they can circulate in the blood for an extended time and exhibit a high level of accumulation at tumour sites.

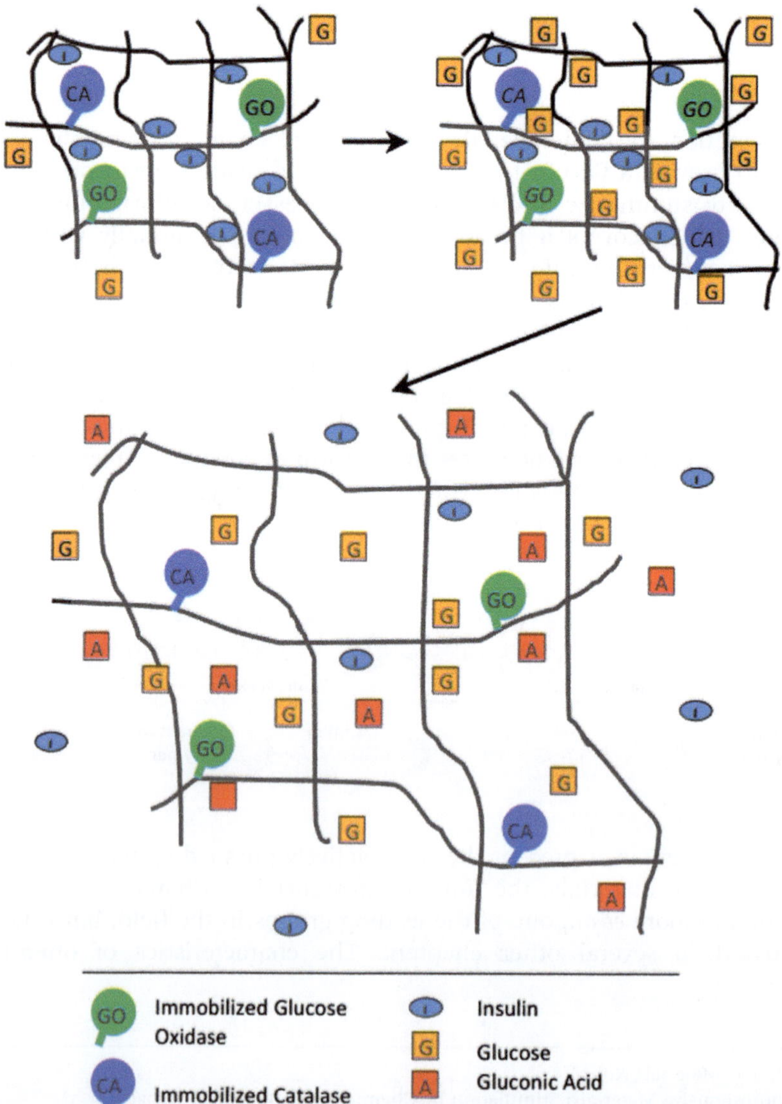

In Chapter 3, Schneider discusses how non-covalent binding functions incorporated into hydrogels lead to chemomechanical swelling or shrinking by water uptake or release as a response to biogenic effectors. These can include peptides if suitable binding sites allow the binding of metal ions simultaneously. Related cooperativity between two effectors in the sense of logical AND gates is seen between, for example, benzoic acid and amino acids. The speed and sensitivity of size changes as a function of the gel particle size and shape are quantified, as well as the selectivity between different stimuli, including chiral compounds. Also discussed are recent cyclodextrin-based materials produced by Harada *et al.* to undergo macroscopic movements triggered by external stimuli.

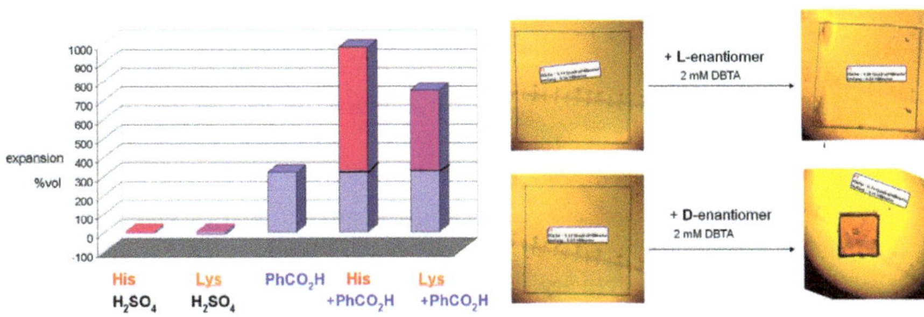

In Chapter 4, Shinkai, Dawn and Roy show how organic gelators can be used to trigger either sol–gel transitions or optical signal changes. For example, the crown ether derivative **A** exhibits with Li, K, and Cs cations a unique allosterism-like sigmoidal response with respect to gel nature and fluorescence signals. In another case, a logic gate with two outputs (UV/vis and fluorescence) and three inputs (acid, base and light) was obtained using a xerogel, which bears amide groups for anion binding and a benzimidazole moiety for anion and proton binding. The gelator **B** can be used for glucose-selective sensing. The chiral organogel from **C** leads to an unprecedented significant enantioselectivity in the photoaddition of asymmetrically substituted anthracene.

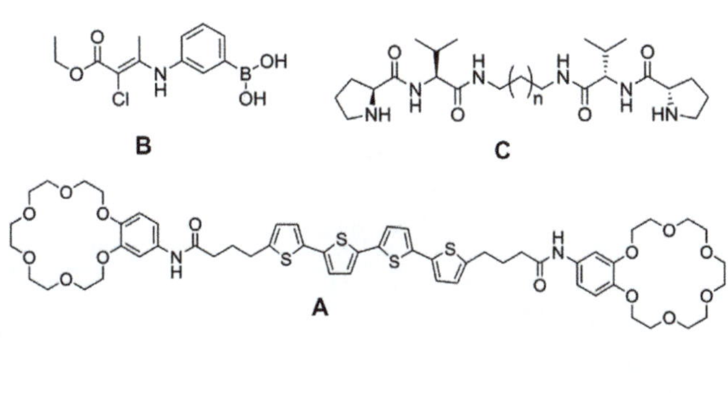

In Chapter 5, Schalley and Qi show intriguing examples for dual response materials including nanocarriers, stimulated, for example, by pH, light, redox condition, or bioactive components. They also highlight the use of fast covalent reactions, as illustrated, for example, with a hydrogel exhibiting reversible gel–sol transitions in response to either pH and redox condition. Logic gates are shown based on multi-stimuli responsive bis-urea gelators, or an example with three enzymes as input. Oscillating reactions including swelling and deswelling of hydrogels also demonstrate the enormous potential of smart soft materials for biomedical applications.

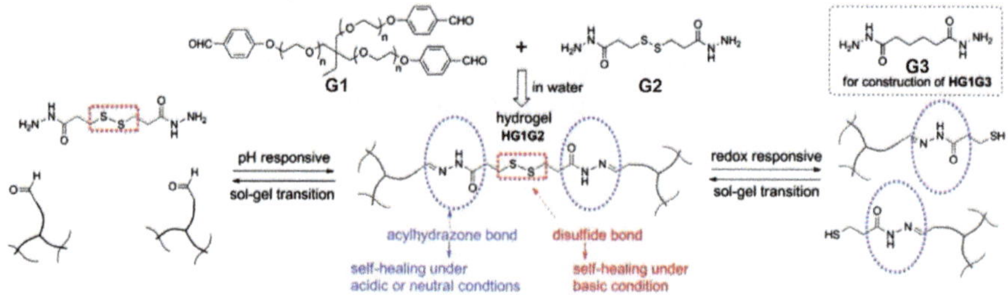

Chapter 6 by Vallet-Regí and Colilla illustrates how mesoporous silica nanoparticles (MSNPs) can be equipped with various binding functions for their use as smart drug delivery systems. Tumour-targeted MSNPs can be engineered to encapsulate, protect, transport and release therapeutic agents to specific diseased tissues. Such devices encompass, for example, glutathione-responsive co-delivery of drugs and single-stranded DNA, aptamer-based ATP-responsive MSNPs, or MSNPs end-capped with glucose oxidase enzyme. Parameters that rule the passive accumulation of MSNPs within solid tumours, such as size, shape and surface functionalisation, are described.

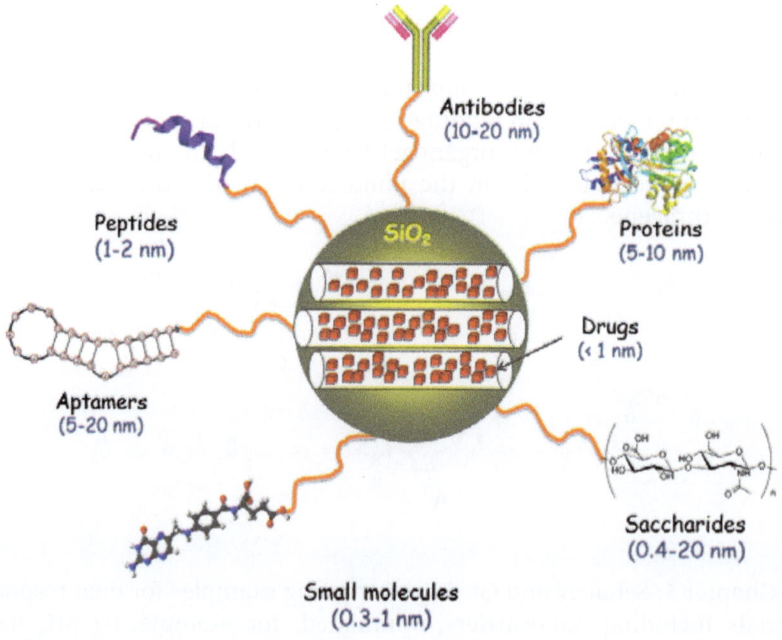

Chapter 7 by Harada and Hashidzume contains impressive examples of the use of cyclodextrins (CDs) for the design of hydrogels and microcapsules for drug release or gene delivery as a function of pH and temperature. Similar applications are possible with responsive sol-to-gel transitions. CD-based polypseudorotaxanes and dendrimers can be used for biomedical applications, showing promise, for example, for photodynamic therapy based on the release of a porphine derivative. The formation of CD-containing macroscopic gel assemblies opens the way to chemoresponsive self-healing materials, as a function of pH, or with ferrocenyl complexes by oxidants. pH-Controlled self-assembly occurs, for example, between a gel possessing CD residues and a gel possessing dansyl residues; only at pH $\geq$4 does the then unprotonated dansyl unit form an inclusion complex with the CD unit.

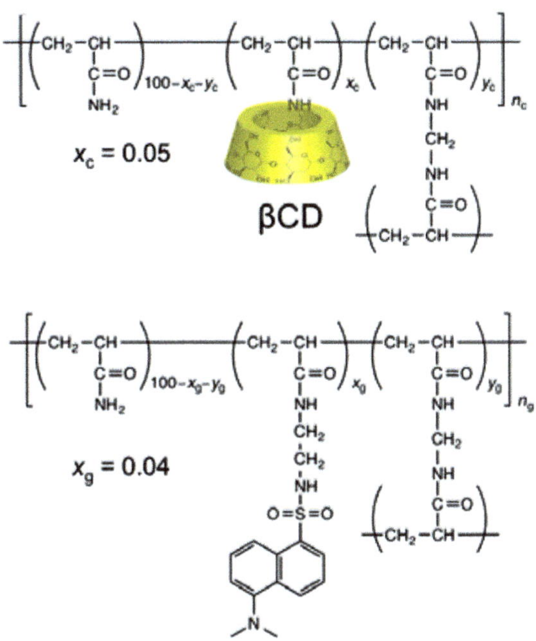

In Chapter 8, Payne, Shi, Xiong, Qu, Liu and Bentley deal with polysaccharides, which have historically been a major source of functional materials for applications in foods, pharmaceuticals and the life sciences (*e.g.*, agar gels in biotechnology). Polysaccharides are also appealing for more recent applications where biocompatibility and sustainability are important. Recent studies show that by precisely controlling the spatiotemporal cues that trigger polysaccharide self-assembly, it is possible to generate hydrogels with complex internal structures. Such studies are inspiring a deeper probing of the structure–property–function relations for polysaccharide and these should reveal even greater possibilities for applying smart polysaccharides.

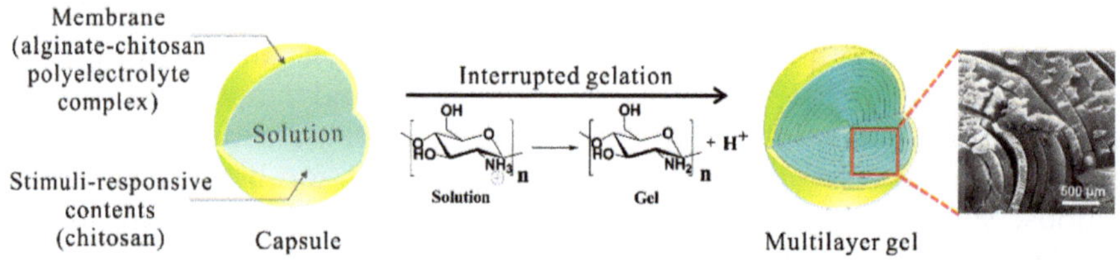

In Chapter 9, Cayre, Biggs and Yow illustrate the use of colloids as environmentally-responsive polymers. Various formulations, whether in the core, on the surface or both, have opened up many novel applications. After an introduction to the different types of particles, examples of the preparation of responsive polymers are given. Applications for these systems range from oil recovery, emulsifiers and sensors to drug delivery. As an example, vesicles containing a basic polymeric membrane can be used for a tunable membrane permeation, regulating the amount of CO_2 introduced, allowing the selective separation and release of active ingredients from the vesicle inside.

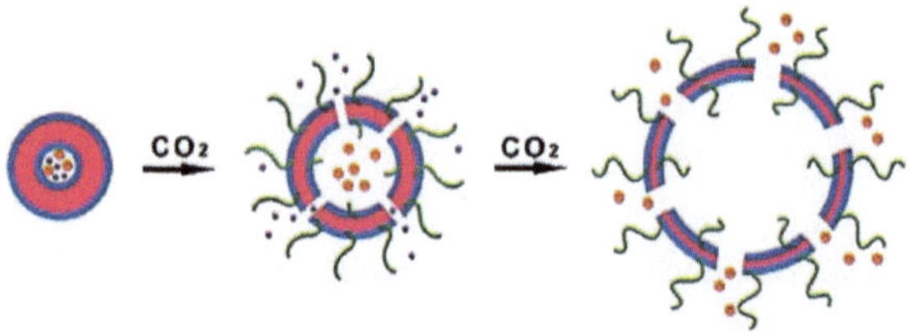

Chapter 10 by Wickramasinghe, Qian and Darvishmanesh gives a brief overview of responsive membranes, which alter their separation properties in response to the presence of external chemical agents. Taking advantage of extensive knowledge in smart polymers and unique membrane separation applications, various responsive membrane separation systems have been developed and engineered. This chapter discusses the separation performances of these intelligent membranes when they are triggered by a change in solution pH, salt and ionic strength as well as biological stimuli. Finally, insights from molecular simulations of responsive polymers are discussed with regard to the design and prediction of separation performances.

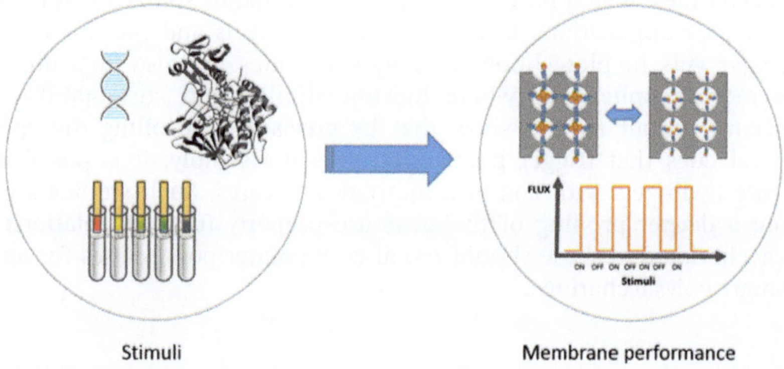

In Chapter 11, Paulusse, Engbersen, Hujaya and Wohl describe chemically and biologically responsive layer-by-layer assemblies. The use of multi-layered thin films in the preparation of smart coatings and capsules is described here. The layer-by-layer assembly process is described, as well as the physicochemical aspects of the resulting films and the importance of specific fabrication techniques and deposition conditions, such as pH and ionic strength, for the responsiveness of the resulting films. Specific biomedical applications of these intriguing systems are highlighted and an outlook on the future of layer-by-layer assemblies is given.

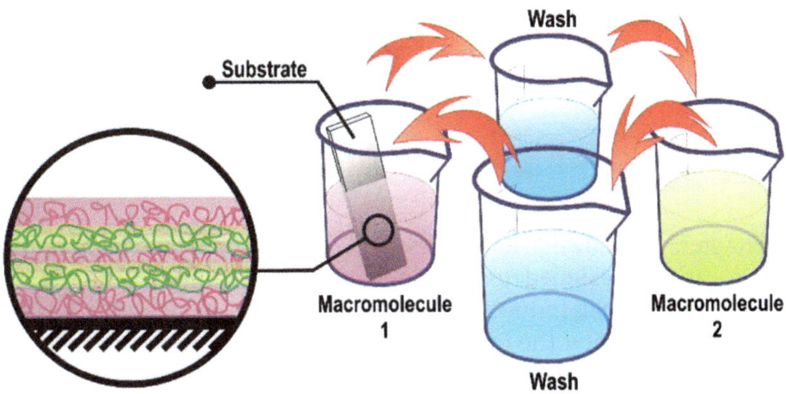

Chapter 12 by Adams and Draper is devoted to gels based on the use of low molecular weight gelators (LMWGs) which, other than cross-linked hydrogelators, exhibit reversible formation from soluble to insoluble material. Here, gel formation from the LMWGs, which are often based on amino acids, is triggered by different stimuli; these range from pH to proteins. Gels incorporating bioactive molecules can be used as sensors, for drug delivery, and for cell culture. The bioactive triggering molecules can be released when the gel dissolves at the target site in the body.

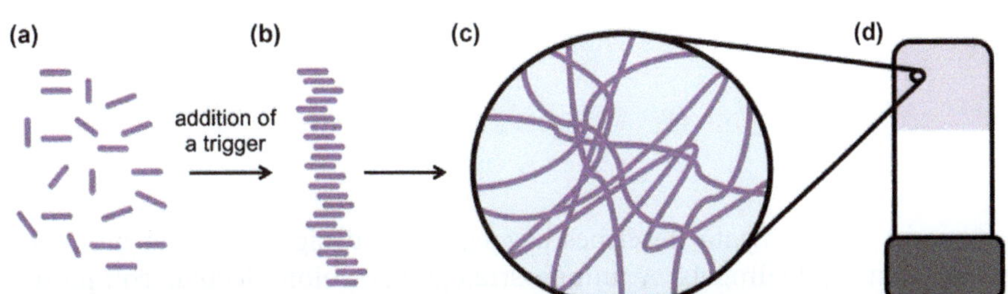

In Chapter 13, Puoci and Parisi discuss how molecularly imprinted polymers (MIPs) can be used to design stimuli-dependent new materials, such as a pH-responsive MIP with pore-filling poly(acrylic acid) gels in a bisphenol-imprinted polyethersulfone. Multi-stimulated MIPs with photo-, thermo- and pH-responsive imprinted microspheres are described, with narrowly dispersed "living" core polymer microspheres. Introduction of polymeric brushes onto these core–shell imprinted microspheres provides a responsive layer with controllable collapse and dissolution of polymer brushes towards external stimuli, resulting in stimuli-responsive template binding or release.

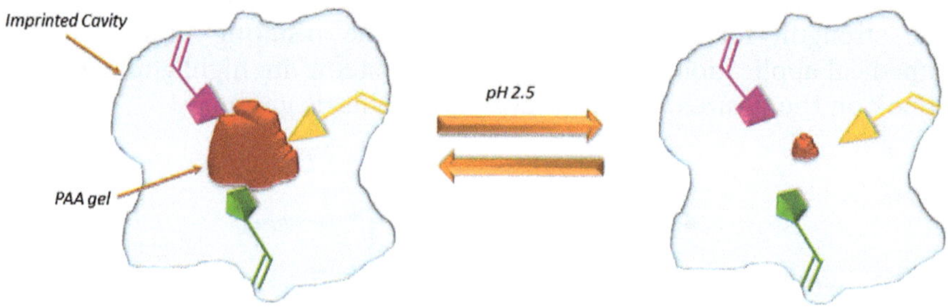

In Chapter 14, Löwik, Ramakers and Guo show how peptide-based materials can be designed to respond to external stimuli. While a great variety of stimuli that induce responsive behaviour in peptides has been reported, in this chapter the focus lies on chemical (pH, ions and reduction–oxidation) and biological signals. The responsiveness of these peptide-based materials is very often due to a conformational transition that can be induced. These transitions can be ascribed to a change in intramolecular interactions such as hydrogen bonding and hydrophobic contacts that stabilise a particular fold. These responsive materials are believed to have potential applications in areas such as drug delivery, tissue engineering and bio-sensing.

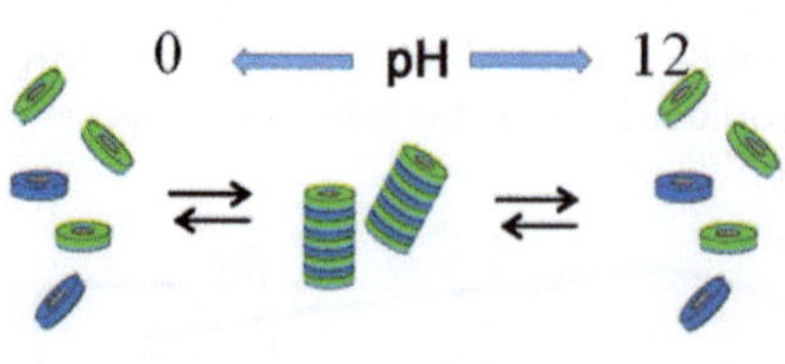

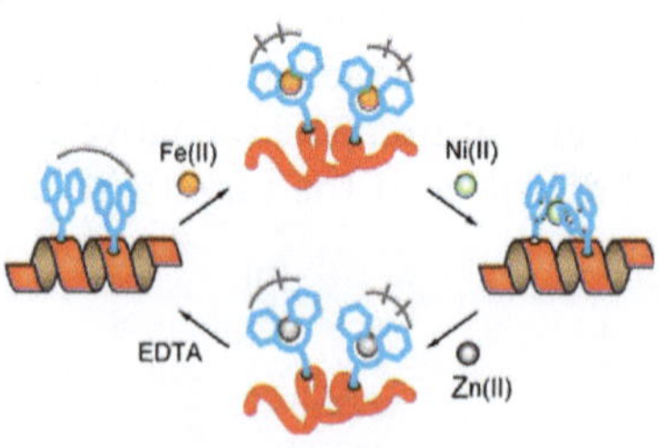

Chapter 15 by Miyata describes strategies for designing target biomolecule-responsive hydrogels. A unique strategy uses biomolecular complexes such as saccharide–lectin and antigen–antibody complexes as dynamic crosslinks. For example, bioconjugated hydrogels with antigen–antibody complexes underwent reversible volume changes in response to changes in antigen concentration, with autonomous drug permeation as a function of antigen concentration. On the other hand, bioconjugated hydrogels with lectins and antibodies as ligands, which were prepared by molecular imprinting, shrank gradually in the presence of a target tumor marker glycoprotein (AFP) but not in the presence of other similar glycoproteins. Such biomolecule-responsive hydrogels have many potential applications as smart biomaterials for drug delivery systems, molecular diagnosis and so on.

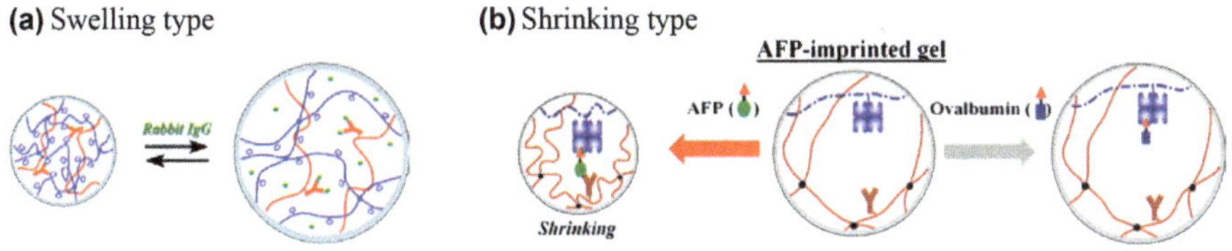

Chapter 16 by Chu and Sidorenko is dedicated to chemo-responsive polymer brushes. These polymer chains, permanently tethered to a solid surface, represent a unique responsive system. Polymer brushes provide dynamic control of surface properties by either a chemical stimulus or a combination of stimuli. The authors review three of the most important cases of chemically-responsive brushes. The polyelectrolyte brush is the most studied system that demonstrates a response to pH and the nature and concentration of ions. The mechanisms of response on molecular and mesoscale levels are briefly discussed. Another interesting group of responsive brushes is composed of polymers bearing RedOx groups, such as ferrocenyl, phenothiazine, and complexes with transition metals. Also reviewed are recent advances in photoresponsive polymer brushes bearing photo-cleavable and photo-crosslinkable groups. Several examples of complex responses and remarks about applications conclude the chapter.

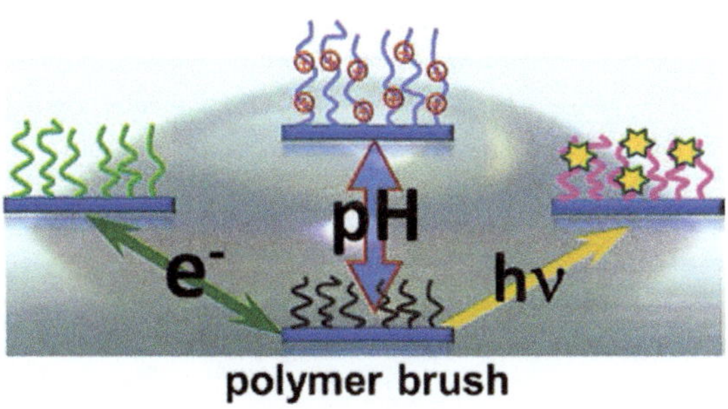

In Chapter 17, Shioi *et al.* show how autonomous motions can be realized in man-made colloidal systems. These can rectify random fluctuations caused by chemical reactions, so that we can see their regulated motions. Moreover, they can convert a microscopic pH gradient into mechanical work and exhibit ion-sensitive autonomous motions. Studies of these moving systems may lead to an understanding of biological motions and also a new type of chemical technology based on biomimetics. Autonomous motion is one of the most outstanding characteristics of biological systems. They can respond to chemical and physical stimuli and exhibit motions with various scales.

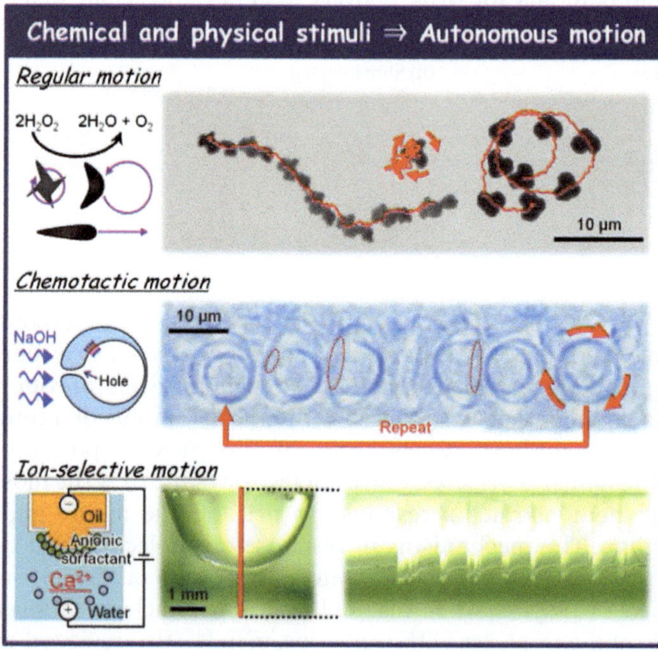

CHAPTER 1

Overview/Strategies for Chemoresponsive Materials

HANS-JÖRG SCHNEIDER

FR Organische Chemie der Universität des Saarlandes, D 66041, Saarbrücken, Germany
Email: h-j.schneider@mx.uni-saarland.de; ch12hs@rz.uni-sb.de

1.1 Introduction

Smart or "intelligent" materials change their properties upon stimulation by external signals. Modern technologies have already led to many materials that can be stimulated by, for example, voltage, light, temperature, pressure, electric or magnetic fields.[1,2] This has opened new ways for a large variety of possible applications, such as sensors, artificial muscles, molecular machines, actuators for process control, tissue engineering, self-healing surfaces, electronics, and drug delivery. In contrast to the situation with many man-made externally controlled smart materials, living systems are mostly regulated and thus maintained by response to **chemical** signals. Countless publications have described the transformation of chemical interactions into macroscopic motions after the first discovery.[3] Hybrid multifunctional self-adapting materials can be integrated into multicompartmental systems, allowing simultaneous control by chemical stimuli, light, temperature, and other environmental factors.

The present book concentrates on often bio-inspired materials stimulated primarily by chemical signals, which may range from pH to, for example, antibodies. The potential of chemoresponsive materials encompasses the same applications as mentioned above, but reaches much beyond them.

RSC Smart Materials No. 14
Chemoresponsive Materials: Stimulation by Chemical and Biological Signals
Edited by Hans-Jörg Schneider

Published by the Royal Society of Chemistry, www.rsc.org

Like in nature, one can hope to realize sensors and actuators within the same material, thus allowing self-regulating systems, or, in the probably most important application, selective drug targeting. The size and shape of the materials can be adopted to any desired dimension; this allows the modification of both sensitivity and speed of the response. In future, such devices could serve as nanorobots, even as injectable or implantable therapeutic systems.

1.2 Basic Structures of Chemoresponsive Materials

A multitude of polymers can be made sensitive to a variety of chemical stimuli, as depicted in Figure 1.1.[2d] The common feature of almost all of these materials, which are by no means restricted to nanosizes, is a volume expansion or constriction triggered by their specific environment. Many of these smart polymers are discussed in separate chapters of this book. The gels that are used most often, based on a large variety of structures, are

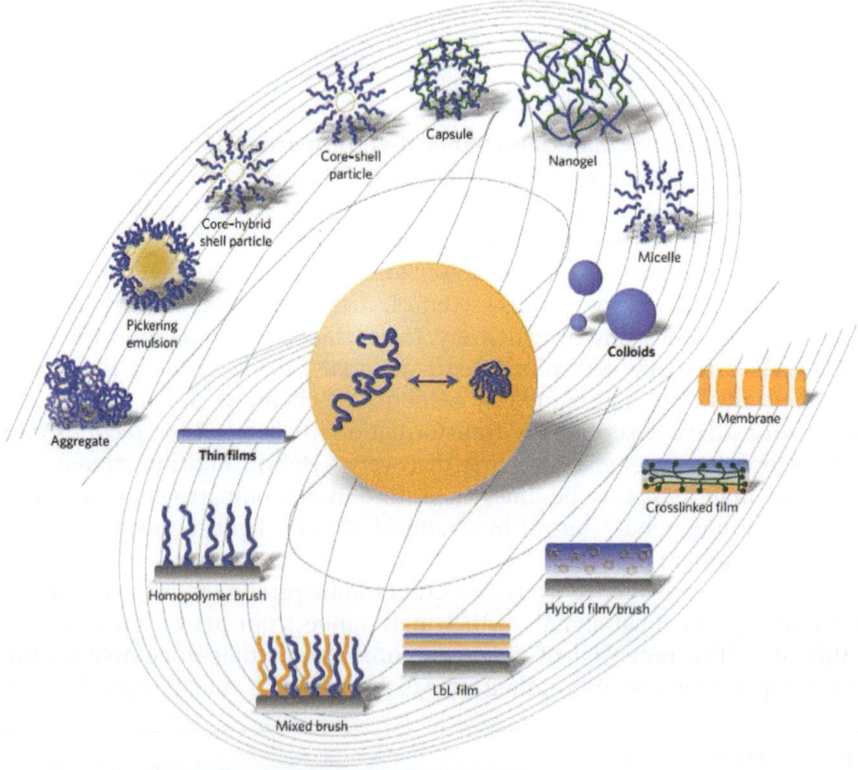

Figure 1.1 Overview of stimulus-responsive materials. Reprinted with permission from ref. 2c.

the topics of Chapters 2–5, 8 and 12–15, colloids in Chapter 9, multilayered films in Chapter 11, responsive membranes in Chapter 10, and polymer brushes in Chapter 16. Chemically-induced surface changes of such materials can involve their permeability, wettability, hydrophobicity, and polarity. They can control their adhesive, adsorptive, mechanical or optical properties, the latter being in particular the basis for sensors. Another strategy for drug delivery materials relies on particles with pores that can open or close in response to external stimuli, as shown with core–shell silica particles in Chapter 6. In terms of applications of smart materials, such as drug delivery or as actuators, their capacity to take up and release effector molecules is the essential feature. In the next sections, some chemoresponsive systems are discussed for which there are no special chapters in this book, and which do not appear in Figure 1.1.

1.3 Electroactive Polymers and Solid State Receptors with Redox- or pH-activation

Technically well advanced chemoresponsive systems are electroactive polymers (EAPs) in the form of conductive materials (CPs).[4–8] They can be obtained, for example, by electropolymerization as polypyrrole, polyaniline, or polythiophene; alternatively, one may use doping of an inactive polymer by introduction of a redox-active agent. Oxidation leads, by an electron hopping mechanism, to their cationic form, which together with the simultaneous import of anions and uptake of ion solvation water leads to significant swelling. This can be used for the construction of artificial muscles, for example in the form of bilayers with an active conductive polypyrrole and an adherent flexible passive polymer tape, with bending resulting from swelling of the active layer (Figure 1.2).[1e] Changes of the pH lead to changes in the backbone structure of CPs bearing acidic or basic elements. This depends on the corresponding pK values; for polypyrrole, these are either between 2 and 4, or between 9 and 11; protonation enhances the conductivity, deprotonation does the opposite. Such CPs can also be used as nanoparticles, for example for drug release.[7] For sensing, one can use the change in conductivity, measuring the resistance, or optical signals such as fluorescence.

Ligands attached to CPs can complex with high selectively and sensitivity analytes; thus a poly(*p*-phenylene ethynylene) CP bearing crown ether (Figure 1.3A) detects by fluorescence spectroscopy K^+-ions in the presence of 1500 fold excess of Li^+ or Na^+ ions. Electron-rich systems such as in Figure 1.3B, can be used for the detection of electron-poor nitroaromatic explosives.[5]

Drug release from EAP- or CP-based hydrogels is mainly due to the movement of charged particles out of the hydrogel; for example protonated dimethyldopamine incorporated into a polypyrrole composite is released anodically. Biodegradable or bioerodible electroactive implantable

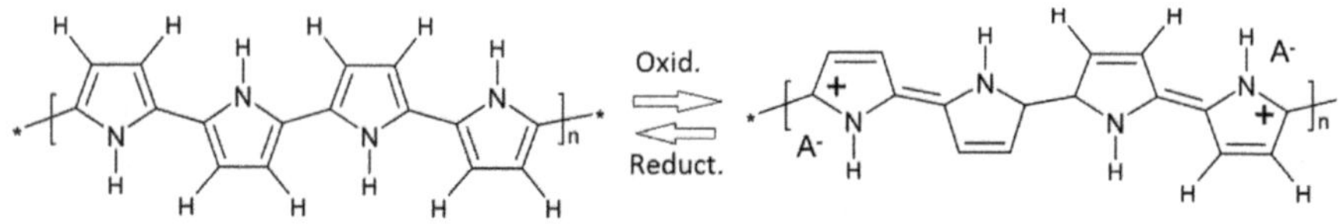

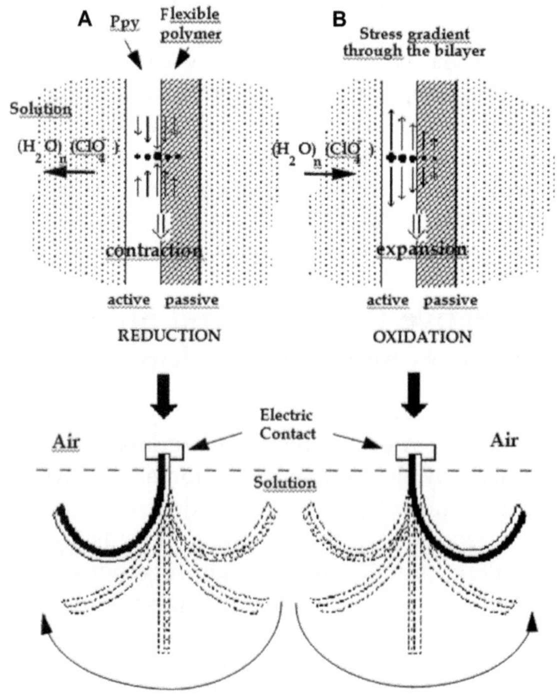

Figure 1.2 Scheme of a bilayer with the active conductive polymer PPy (polypyrrole) and an adherent flexible passive polymer tape, with ionic interchanges between the PPy and the solution during a redox reaction produced by a current; bending movements between both layers induced by the ionic interchanges and concomitant volume changes in the active layer; from ref. 1*e*, p. 159.

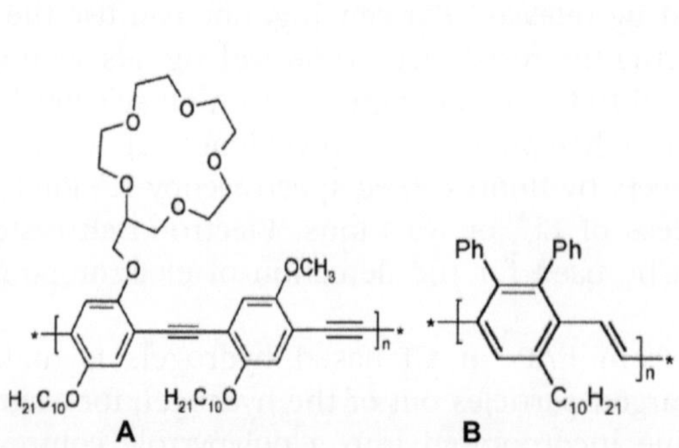

Figure 1.3 Conductive polymers with binding sites for sensing alkali ions (A) or nitroaromatic explosives (B).

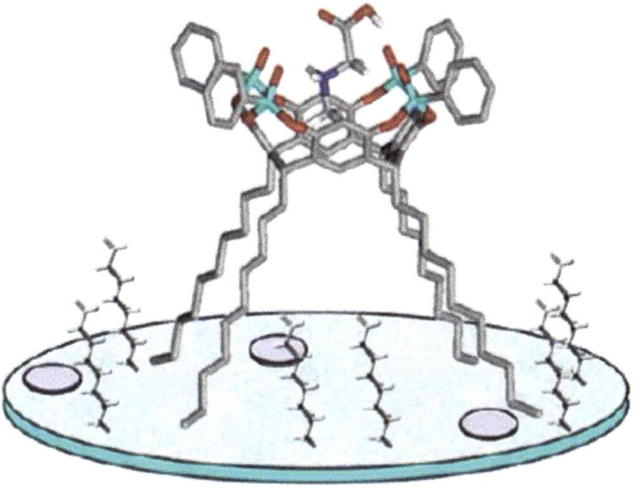

Figure 1.4 Silicon wafer after exposure to a water solution of sarcosine. Reproduced from ref. 9.

polymers, based, for example, on oligoaniline-based electroactive blocks linked to polyethylene glycol, hold particular promise for controlling the chronopharmacology of drugs.[8]

Solid receptors can be used for sensing at gas–solid and liquid–solid interfaces, in the first case for vapor detection, in the second case for measurement of analytes in solution. The technique has been particularly successful with receptors bound to silica wafers, as illustrated in Figure 1.4 with a phosphonate cavitand as a receptor for the determination of sarcosin in urine. Sensitive detection is possible with the aid of a traditional quartz balance, with a microcantilever or by displacement of a fluorescence dye.[9] Recently it has been shown that in crystal polymorphs, guest molecules can leave a memory that can be read by differential scanning calorimetry.[10]

1.4 Incorporation of Supramolecular Binding Sites/ Molecular Machines

The rapid development of supramolecular chemistry in the last decades has paved the way to whole families of chemoresponsive materials by implementation of selective binding sites with suitable polymers. Such responsive host–guest systems are most often based on macrocycles such as cyclodextrins, crown ethers, cyclophanes, calixarenes, and cucurbiturils, or on foldamers.[11] Cyclodextrins play a prominent role here due to their availability and large chemical variability,[12] as shown in particular in Chapter 7 by Harada *et al.* and also in Chapter 3. Several systems for movements triggered by pH or redox reactions have been designed on the basis of rotaxanes, as for example shown in Figure 1.5.

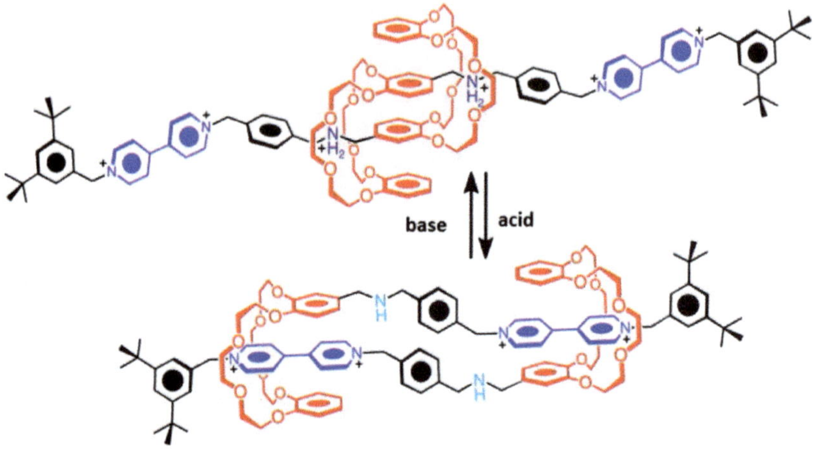

Figure 1.5 Actuation of movement of crown ether shuttles in a rotaxane by pH change.
Adapted with permission from C. J. Bruns, J. F. Stoddart, *Acc. Chem. Res.*, 2014, **47**, 2186–2199. Copyright 2014 American Chemical Society.

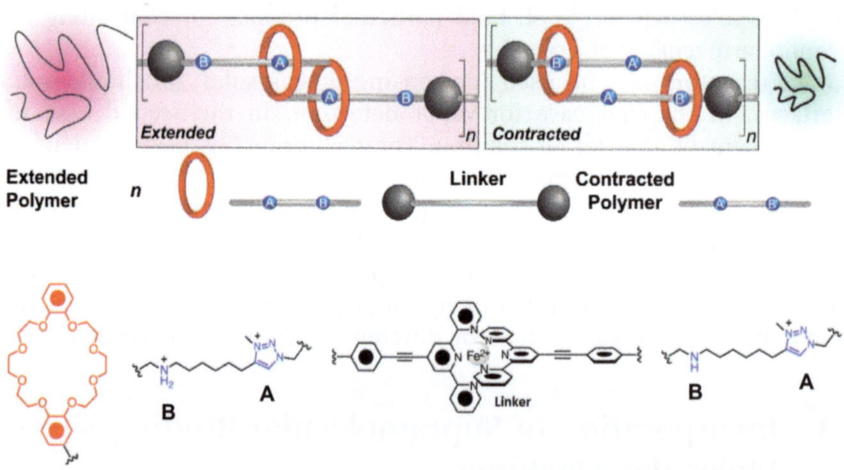

Figure 1.6 Contractile/extensile motions in rotaxane-based molecular muscles. In the polydaisy chains, the crown ether (red) binds at low pH at station B, at higher pH at A.
Adapted with permission from C. J. Bruns, J. F. Stoddart, *Acc. Chem. Res.*, 2014, **47**, 2186–2199. Copyright 2014 American Chemical Society.

Figure 1.6 illustrates how, with ligands that dimerize around a Fe^{2+} complex as a linker, one can obtain linear polymers with an average of ~3000 repeating units with a total length of several nm. The long poly[c2]daisy chains can amplify nanomotions of rotaxanes over several orders of magnitude, as in the single elements the contour lengths with the protonated part at B

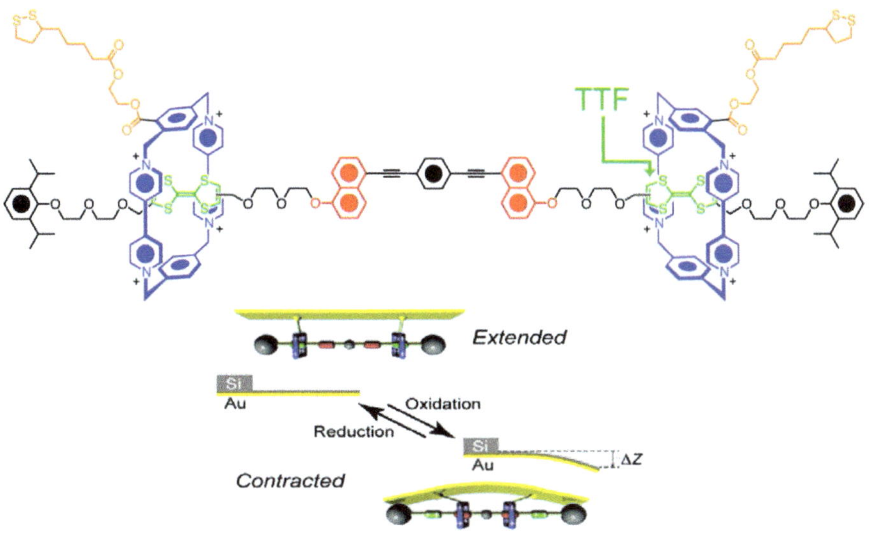

Figure 1.7 Bending of a microcantilever with a redox-switchable palindromic [3]rotaxane. Electrochemical or chemical oxidation generates a positive tetrathiafulvalene station and the cyclophane shuttle moves to the red naphthalene station. The resulting bending can be measured by fixing the shuttles to a gold microcantilever as a monolayer *via* sulfur bridges. Adapted with permission from C. J. Bruns, J. F. Stoddart, *Acc. Chem. Res.*, 2014, **47**, 2186–2199. Copyright 2014 American Chemical Society.

(15.9 μm) and at the unprotonated part at B (9.4 μm) differ by 6.5 μm.[13] Fixation of rotaxane shuttles on a microcantilever allowed the measurement of the mechanical bending force resulting from a redox input (Figure 1.7).

1.5 Outlook

Many applications of smart materials are discussed in the other chapters of this book, which are highlighted in a section of the preface. Not all applications could find a place in this book. Besides drug release applications,[14] which have dominated until now, molecular electronics is a rapidly developing field where such materials can be used;[15] even the cleaning of paintings can be improved with gels that respond to the surface.[16] Most sensors are based on chemoresponsive materials, which encompass traditional systems such as chemically induced optical changes, piezoelectric devices like quartz balances loaded with suitable receptor compounds, or more recently surface plasmon resonance (SPR), or photonic crystals.[17] Such sensor techniques are described in many books and reviews, and are mentioned here only in the context of other applications.[18] Actuators including artificial muscles have until now most often relied on piezoelectrics[19] or in particular on electrical activation.[20] Chemical and light-induced molecular machines have already reached a high degree of sophistication[13,21] (see also Section 1.4), but need much further development for real-life applications.

References

1. (a) M. Wun-Fogle, *Materials for Smart Systems*, Cambridge University Press, 2014; (b) M. R. Aguilar and J. S. Roman, *Smart Polymers and their Applications*, Woodhead Publishing, 2014; (c) Q. Li, *Intelligent Stimuli-Responsive Material*, Wiley, 2013; (d) *Handbook of Stimuli-Responsive Materials*, ed. M. H. Urban, Wiley-VCH, Weinheim, 2011; (e) *Intelligent Materials*, ed. M. Shahinpoor and H.-J. Schneider, Royal Society of Chemistry, Cambridge, UK, 2007.
2. Recent reviews: (a) L. D. Zarzar and J. Aizenberg, *Acc. Chem. Res.*, 2013, **47**, 530; (b) E. G. Kelley, J. N. L. Albert, M. O. Sullivan and T. H. Epps III, *Chem. Soc. Rev.*, 2013, **42**, 7057; (c) A. Grinthal and J. Aizenberg, *Chem. Soc. Rev.*, 2013, **42**, 7072; (d) M. A. C. Stuart, W. T. S. Huck, J. Genzer, M. Müller, C. Ober, M. Stamm, G. B. Sukhorukov, I. Szleifer, V. V. Tsukruk, M. Urban, F. Winnik, S. Zauscher, I. Luzinov and S. Minko, *Nat. Mater.*, 2010, **9**, 101; (e) J. M. Spruell and C. J. Hawker, *Chem. Sci.*, 2011, **2**, 18–26; (f) F. D. Jochum and P. Theato, *Chem. Soc. Rev.*, 2013, **42**, 7468; (g) J. Thevenot, H. Oliveira, O. Sandre and S. Lecommandoux, *Chem. Soc. Rev.*, 2013, **42**, 7099.
3. M. V. Sussman and A. Katchalsky, *Science*, 1970, **167**, 45.
4. T. F. Otero, *J. Mater. Chem.*, 2009, **19**, 681; T. F. Otero, *Polym. Rev.*, 2013, **53**, 211; T. F. Otero, in ref 1*e*, 142.
5. U. Lange, N. V. Roznyatovskaya and V. M. Mirsky, *Anal. Chem. Acta*, 2008, **614**, 1; D. T. McQuade, A. E. Pullen and T. M. Swager, *Chem. Rev.*, 2000, **100**, 2537–2574.
6. V. Pillay, T.-S. Tsai, Y. E. Choonara, L. C. du Toit, P. Kumar, G. Modi, D. Naidoo, L. K. Tomar, C. Tyagi and V. M. K. Ndesendo, *J. Biomed. Mater. Res., Part A*, 2014, **102A**, 2039.
7. (a) D. Park, Y. Cho, S.-H. Goh and Y. Choi, *Chem. Commun.*, 2014, **50**, 15014; (b) E. Shamaeli and A. Naader, *Int. J. Pharm.*, 2014, **472**, 327.
8. J. G. Hardy, D. J. Mouser, N. Arroyo-Currás, S. Geissler, J. K. Chow, L. Nguy, J. M. Kim and C. E. Schmidt, *J. Mater. Chem. B*, 2014, **2**, 6809.
9. R. Pinalli and E. Dalcanale, *Acc. Chem. Res.*, 2013, **46**, 399.
10. G. D. Safina, M. A. Ziganshin, A. T. Gubaidullin and V. V. Gorbatchuk, *Org. Biomol. Chem.*, 2013, **11**, 1318.
11. K. Liu, Y. Kang, Z. Wang and X. Zhang, *Adv. Mater.*, 2013, **25**, 5530; X. Ma and H. Tian, *Acc. Chem. Res.*, 2014, **47**, 1971; Z. Qi and C. A. Schalley, *Acc. Chem. Res.*, 2014, **47**, 2222; S. Dong, B. Zheng, F. Wang and F. Huang, *Acc. Chem. Res.*, 2014, **47**, 1982; J. M. Spruell and C. J. Hawker, *Chem. Sci.*, 2011, **2**, 18; L. Isaacs, Use of cucurbiturils: *Acc. Chem. Res.*, 2014, **47**, 2052; D.-S. Guo and Y. Liu, of calixarenes: *Acc. Chem. Res.*, 2014, **47**, 1925; D.-W. Zhang, X. Zhao and Z. T. Li, of foldamers, *Acc. Chem. Res.*, 2014, **47**, 1961; special issue of *Acc. Chem. Res.*, 2014, **47**(7), http://pubs.acs.org/toc/achre4/47/7.

12. A. Harada, Y. Takashima and M. Nakahata, *Acc. Chem. Res.*, 2014, **47**, 2128; A. Luzardo-Alvarez, E. Sobarzo-Sanchez, J. Blanco-Mendez and F. J. Otero-Espinar, *Curr. Top. Med. Chem.*, 2014, **14**, 542.
13. C. J. Bruns and J. F. Stoddart, *Acc. Chem. Res.*, 2014, **47**, 2186; G. Du, E. Moulin, N. Jouault, E. Buhler and N. Giuseppone, *Angew. Chem., Int. Ed.*, 2012, **51**, 12504.
14. See e.g. *Smart Materials for Drug Delivery*, ed. C. Alvarez-Lorenzo and A. Concheiro, RSC, Cambridge, 2013; R. Tong, L. Tang, L. Ma, C. Tu, R. Baumgartner and J. Cheng, *Chem. Soc. Rev.*, 2014, **43**, 6982, for insulin delivery systems see W. Wu and S. Zhou, *Macromol. Biosci.*, 2013, **64**, 1464.
15. *Supramolecular Materials for Opto-Electronics*, ed. N. Koch, RSC, Cambridge, UK, 2014; W. R. Browne and B. L. Feringa, *Chimia*, 2010, **46**, 64398; A. Coskun, J. M. Spruell, B. Barin, W. R. Dichtel, A. H. Flood, Y. Y. Botros and J. F. Stoddart, *Chem. Soc. Rev.*, 2012, **41**, 4827, Beckmann, D., Müllen, K. in ref. 1e, p. 187ff.
16. E. Carretti, M. Bonini, L. Dei, B. H. Berrie, L. V. Angelova, P. Baglioni and R. G. Weiss, *Acc. Chem. Res.*, 2010, **43**, 751.
17. C. Pacholski, *Sensors*, 2013, **13**, 4694.
18. *Smart Membranes and Sensors*, ed. A. Gugliuazza, Wiley, 2014; M. Schwartz, *Smart Materials*, CRC, 2008; *Smart Sensor Systems*, ed. G. Meijer, Wiley, 2008, 2014; R. N. Dsouza, U. Pischel and W. M. Nau, *Chem. Rev.*, 2011, **11**, 7941.
19. S.-B. Choi and Y. M. Han, *Piezoelectric Actuators: Control Applications of Smart Materials*, CRC Press, 2012.
20. T. Mirfakhrai, J. D. Madden and R. H. Baughman, *Mater. Today*, 2007, **10**, 30; P. Brochu and Q. Pei, *Macromol. Rapid Commun.*, 2010, **31**, 10; Shahinpoor, M. in ref. 1*e*, p. 126f.
21. *Molecular Motors*, ed. M. Schliwa, Wiley-VCH, Weinheim, 2003; M. Von Delius and D. Leigh, *Chem. Soc. Rev.*, 2011, **40**, 3656; A. Coskun, M. Banaszak, R. D. Astumian, J. F. Stoddart and B. Grzybowski, *Chem. Soc. Rev.*, 2012, **41**, 19; E. R. Kay, D. A. Leigh and F. Zerbetto, *Angew. Chem., Int. Ed.*, 2007, **46**, 72; W. R. Browne and B. L. Feringa, *Nat. Nanotechnol.*, 2006, **1**, 25; M. C. Jimenez-Molero, C. Dietrich-Buchecker and J.-P. Sauvage, *Chem. Commun.*, 2003, 1613; V. Balzani, A. Credi, S. Silvi and M. Venturi, *Chem. Soc. Rev.*, 2006, **35**, 1135.

CHAPTER 2

Intelligent, Responsive and Theranostic Hydrogel Systems for Controlled Delivery of Therapeutics

STEVE R. MAREK,[a] MARTIN L. GRAN,[a] NICHOLAS A. PEPPAS[a,b,c,d] AND MARY CALDORERA-MOORE*[e]

[a] McKetta Department of Chemical Engineering, The University of Texas at Austin, Austin, TX 78712, USA; [b] Institute of Biomaterials, Drug Delivery and Regenerative Medicine, The University of Texas at Austin, Austin, TX 78712, USA; [c] Department of Biomedical Engineering, The University of Texas at Austin, Austin, TX 78712, USA; [d] College of Pharmacy, The University of Texas at Austin, Austin, TX 78712, USA; [e] Department of Biomedical Engineering, Louisiana Tech University, Ruston, LA 71270, USA

*Email: mcmoore@latech.edu

2.1 Introduction to Hydrogels

Hydrogels are hydrophilic polymers that have been crosslinked to form a three-dimensional polymer matrix as shown in Figure 2.1. Under the correct environmental conditions, these materials can imbibe significant amounts of water, often over two orders of magnitude more water than polymer by weight. Due to the large water content, hydrogels are currently the synthetic materials that most resemble living tissue.[1] These properties help make

RSC Smart Materials No. 14
Chemoresponsive Materials: Stimulation by Chemical and Biological Signals
Edited by Hans-Jörg Schneider

Published by the Royal Society of Chemistry, www.rsc.org

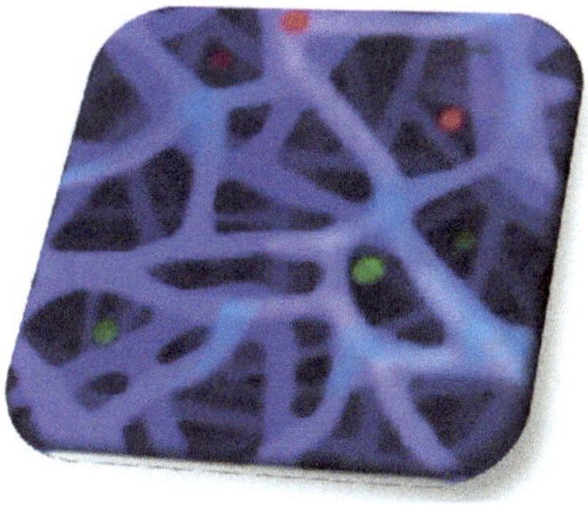

Figure 2.1 Schematic of a polymeric hydrogel network.

hydrogels ideal for use in tissue engineering, drug delivery, surface coatings, contact lenses, wound dressings, and even polluted water treatment.[2–5]

There are several important physical parameters of hydrogels that alter their properties and ultimately can determine their potential uses as biomaterials. One of the most important parameters is the number average molecular weight between crosslinks, $\overline{M_c}$, which indicates the average molecular distance between two consecutive crosslinks along the polymer backbone. However, for diffusion studies, it is more important to know the physical size between the polymer chains, rather than the molecular weight. The term $\overline{M_c}$ can be used to calculate the mesh size, ξ, which essentially indicates the actual average size between crosslinks. For diffusion purposes, this can lead to identifying the size of nanoparticles or molecules that could theoretically diffuse through the polymer network.

$$\xi = Q^{-1/3}\left(C_n l^2 \frac{\overline{M_c}}{M_r}\right)^{1/2} \tag{2.1}$$

In this equation, Q is the volume swelling ratio, C_n is the Flory characteristic ratio, l is the length of a C–C bond on the polymer backbone ($l = 1.54$ A), and M_r is the molecular weight of the repeat unit. As a hydrogel swells, its mesh size increases allowing larger solutes to diffuse through the macromolecular network. This swelling is also characterized by the polymer volume fraction in the swollen state, $\nu_{2,s}$, which is the inverse of the volume swelling ratio.

The study of hydrogel use in drug delivery applications has been in place for decades.[6–9] It was realized early on that reservoir devices yielded zero-order release kinetics, that is, a release rate that was not a function of time. Transport through the membrane shell is the diffusion-limiting step, and so the rate of release can be controlled by changing the thickness or type of membrane. These devices tend to have either a burst-effect or lag-effect, depending on whether or not the drug had time to dissolve in the membrane during storage. Since many diseased states can be treated with a constant, therapeutic level of drug, reservoir devices seemed ideal. However, difficulties in manufacturing as well as problems with membrane rupture were significant hindrances. Also, large molecular weight drugs such as proteins are not able to diffuse at an appreciable rate through the biocompatible polymers used in reservoir devices.

Non-degrading matrix type systems, on the other hand, typically yield first-order release kinetics. A matrix system has the drug homogeneously dispersed or dissolved throughout the polymer. For these devices, the release profile is a function of the square root of time. Thus, the drug is released faster at earlier time points compared to later ones because the drug has to diffuse through the whole polymer system in order to be released, and so the diffusional path length increases with time as the drug concentration decreases. The burst and lag effects tend to not be present, and the manufacturing of these devices can be less intricate than reservoir systems. There is also no need to worry about the polymer rupturing, since the drug is dispersed throughout the whole polymer system and it is less likely to cause a significant overdose. These devices permit site-specific drug delivery and others can even permit tissue targeting.[10]

2.2 Neutral Hydrogels

Depending on their charges, hydrogels can be divided into two types: neutral and ionic. Neutral hydrogels are a class of polymer gels that do not have any ionizable moieties on their side chains, such as poly(vinyl alcohol) (PVA). These polymers swell due to a balance between the elastic retractive forces of the polymer chains and the thermodynamic mixing of the polymer chains with the solvent molecules. These polymers can be used as bioadhesives or in both reservoir and matrix drug delivery systems, among other applications.

Certain hydrogels exhibit environmental sensitivity due to these forces being functions of temperature[11,12] or some other variable. One such polymer, poly(*N*-isopropyl acrylamide), exhibits a lower critical solution temperature (LCST); at temperatures above this LCST, the polymer is hydrophobic and remains in a collapsed state. Below the LCST, the hydrogel is hydrophilic and imbibes large amounts of water. Thus, if a drug is dissolved or suspended in the polymer, it is more easily able to diffuse out when the polymer is swollen. A similar swelling behavior is observed with polymers that exhibit an upper critical solution temperature (UCST), though these systems swell above the threshold temperature and collapse below it Figure 2.2.

Poly(ethylene glycol) (PEG) is a highly biocompatible, non-responsive polymer comprising repeating ethylene glycol units. In an aqueous environment, each ethylene glycol repeat unit is able to bind multiple molecules of water. This phenomenon allows the PEG chains to repel each other and, when tethered to a surface, form a "brush-like" conformation when the appropriate PEG density is achieved.[13] It is believed that the chains form a mushroom morphology when the surface density of PEG is much lower. Thus, a high surface density of PEG grafts is usually desired.

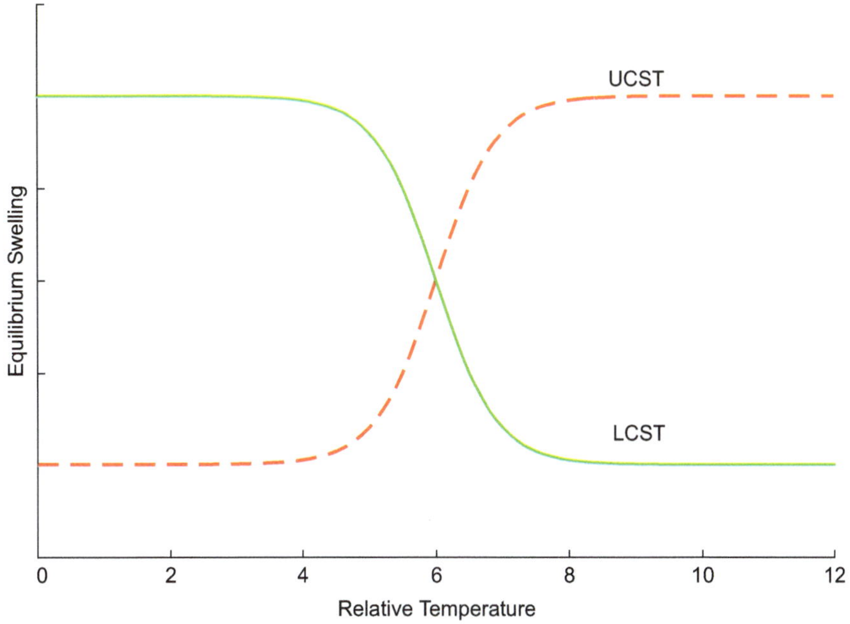

Figure 2.2 Swelling of theoretical temperature-sensitive polymers. The solid line represents a theoretical polymer exhibiting a lower critical solution temperature (LCST), the hashed line represents a theoretical polymer exhibiting an upper critical solution temperature (UCST).

PEG has long been known for its resistance to protein adsorption[14,15] due to its steric repulsion, hydrophilicity, and chain mobility.[16] Two surfaces coated with PEG will develop repulsive forces when brought close enough, due to the steric stabilization of each PEG chain. The interaction between infinitely sized[14] to small sized[14] proteins has been modeled, showing that van der Waals interactions between the PEG and protein are negligible. Nanoparticles decorated with PEG tethers have been shown to decrease opsonization and often complement system activation.[17] These properties can allow the prolonged use of medical devices *in vivo* while preventing coagulation, protein binding, or severe immune response. Thus, these properties have made PEG the most widely used polymer to impart biocompatibility and stealth properties to materials.

2.3 Ionic Hydrogels

One of the interesting types of environmentally-sensitive hydrogels is the ionic hydrogel.[18] These polymers exhibit extensive swelling when in certain media, in part due to electrostatic forces. Changes in environmental conditions such as pH,[19,20] ionic strength,[21] buffer species,[22,23] temperature,[24,25] or other conditions cause swelling of these systems. Anionic pH-sensitive hydrogels are neutral at a pH below their pK_a, where the backbone and/or pendant groups are uncharged. When in this state, hydrogen bonding between moieties is possible, especially when acidic and ethereal groups are present such as methacrylic acid and PEG. It is interesting to note that the pK_a can be a slight function of temperature, thus adding some temperature dependence to pH-sensitive polymers.

Once the pH rises above the pK_a of the pendant groups, they become charged. Usually, a carboxylic acid becomes deprotonated, which imparts a negative charge on the system. Often, hydrogen bonds are broken and electrostatic repulsion ensues. The polymer also tends to become more hydrophilic, thus imbibing more water as it swells. Osmotic forces also play a role in the extent of swelling. Thus, the pH, buffer species, and ionic strength are all important parameters in determining the extent of swelling of a given polymer system. Common anionic polymers consist of methacrylic acid (MAA) or acrylic acid (AA), which have carboxylic acid functional groups that can deprotonate to form carboxylate anions (Figure 2.3). It has also been shown that copolymers of acrylic acid and acrylamide exhibit both a pH and very significant temperature swelling response.[26]

Cationic hydrogels function in the opposite manner to anionic hydrogels. When the pH of the medium is above the pK_a of the polymer, the polymer is in its uncharged, deprotonated state. Many polymers are also

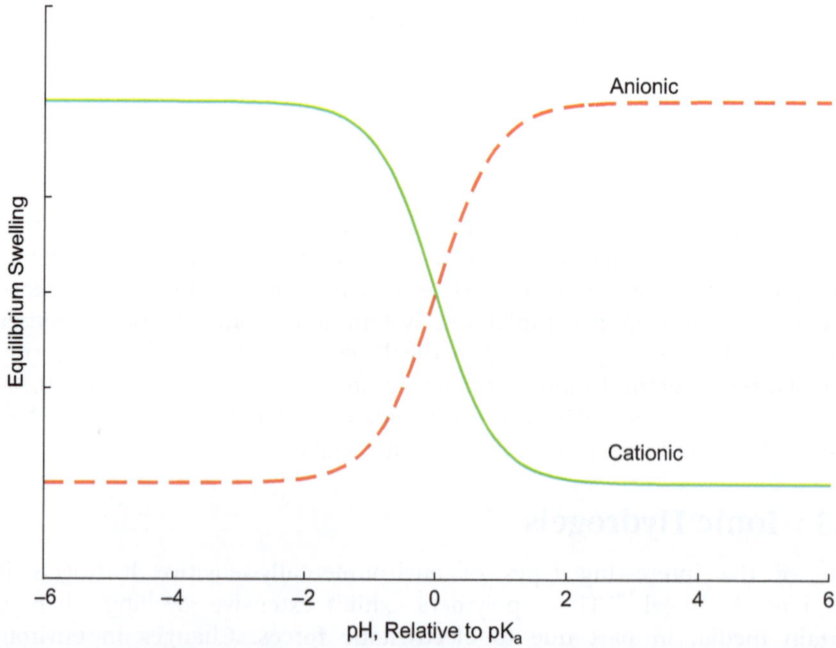

Figure 2.3 Swelling of theoretical pH-sensitive polymers. The solid line represents a theoretical cationic polymer; the hashed line represents a theoretical anionic polymer.

fairly hydrophobic in this state. The polymer is collapsed, excluding a significant amount of water from the interior of the system. Once the pH is dropped below the pK_a, significant portions of the polymer become protonated, thus taking on a positive charge. This positive charge can then induce swelling due to charge repulsion and an increase in hydrophilicity of the system, imbibing significant amounts of water. These hydrogels can then be collapsed again by raising the pH. Many cationic hydrogels are based on amines since the lone electron pair on the nitrogen is able to capture a hydrogen ion, imparting a positive charge to the nitrogen atom (Figure 2.3).

Significant research has been conducted on cationic systems based on either polymers or copolymers of *N*,*N*-diethylaminoethyl methacrylate (DEAEM) or *N*,*N*-dimethylaminoethyl methacrylate (DMAEMA).[21,23,27–31] The pK_a of PDEAEM at physiological temperature is very close to physiological pH, between about 7.0 and 7.3; therefore, a slight drop in local pH from physiological will cause significant swelling of the hydrogel system. This swelling is coupled to an increase in the mesh size of the system, allowing imbibed solutes to more easily diffuse out. One important limitation that Siegel *et al.*[27] observed is that Donnan equilibrium excludes hydrogen ions from the hydrogels, which can be partly overcome by using buffered media. In high ionic strength media, Donnan equilibrium also limits the extent of swelling, due to the partitioning of the dissolved salts. Mallapragada and Anderson have synthesized copolymers of DEAEM and PEG *via* anionic polymerizations.[32] The initiator potassium *tert*-butoxide was used to produce very monodisperse linear pH responsive polymers. Cytocompatibility tests were also conducted. While polymers of pure DEAEM were cytotoxic, polymers containing PEG tethers were shown to be cytocompatible at concentrations as high as 3 mg L^{-1}. Thus, the PEG tethers act as expected and can help decrease cell toxicity.

2.4 Intelligent, Responsive Delivery of Therapeutics with Hydrogels

In recent years, there has been interest in developing more efficient therapeutic systems based on hydrogels that can provide highly effective therapy at a disease site while minimizing systemic effects. The goal is to improve the effectiveness of treatment while reducing any adverse effects associated with a therapeutic or drug, thereby improving overall patient quality of life. Targeted delivery systems involve the use of a carrier system that is capable of delivering a therapeutic in a preferential manner to a disease site. Two notable areas where environmentally responsive hydrogels have been successfully employed have been for oral delivery of proteins—specifically insulin and for the development of externally-controlled drug delivery at tumor sites. Advancements in these fields will be discussed in the following sections.

2.5 Hydrogels in Insulin Delivery

Insulin delivering hydrogels have been heavily researched, though the success of these systems is variable. Ishihara *et al.* characterized the permeation of insulin through membranes made of 2-hydroxyethyl methacrylate (HEMA) with methyl methacrylate (MMA) as a comonomer.[33] These devices exhibited zero order release profiles of insulin, and permeability increased with increasing HEMA content. However, the insulin needs of a patient that has just consumed a meal are not met with a zero order release. They also showed the responsiveness of a poly(amine) membrane to glucose using immobilized glucose oxidase.[33,34] The diffusion of insulin through the membrane was shown to increase as the glucose concentrations increased. This concept was later used to form polymer capsules of 1.5 mm diameter synthesized out of complexed random copolymers of DEAEM, 2-hydroxyethyl acrylate (HEA), and 4-trimethylsilylstyrene (TMS) with entrapped glucose oxidase.[35] These capsules were loaded with insulin when synthesized, and release rates were determined. A cyclic response was observed when the capsules were repeatedly exposed to glucose solutions.

Most hydrogel-based insulin delivery systems rely on glucose oxidase coupled with environmentally sensitive hydrogels to detect glucose. Glucose oxidase (GOx) is approximately a 160 kDa dimeric enzyme for oxidizing glucose into gluconic acid. Dissolved oxygen is reduced to hydrogen peroxide in the process of glucose oxidation (Figure 2.4). Since an acid is produced from glucose, it is possible to use the decrease in pH as a signal for increased glucose levels. Glucose oxidase has been used for decades in the determination of blood glucose levels for diabetic patients. It has also been demonstrated that molecules other than oxygen can be reduced, thus allowing the transfer of charge along a conductive material to directly operate as a sensor. Due to the toxicity of hydrogen peroxide, catalase (CAT), a 250 kDa tetrameric enzyme, has been incorporated in some devices to prevent a toxic buildup (Figure 2.4). Catalase also regenerates the consumed dissolved oxygen, thus helping to eliminate one limitation of the glucose oxidase/glucose sensor. Both enzymes occur naturally in many organisms, though the mold *Aspergillus niger* is a common source for industrial and medical uses.

A graphical representation of how a pH responsive cationic polymer with immobilized glucose oxidase and catalase responds to glucose levels is shown in Figure 2.5. An appropriate hydrogel needs to be selected such that the system is collapsed at normal physiological pH, with a relatively small mesh size. Thus, the pK_a of the hydrogel needs to be at or below about 7.4. As the glucose levels rise, the glucose oxidase catalyzes more oxidations of glucose and therefore the concentration of gluconic acid increases. Locally, this gluconic acid buildup drops the pH. The hydrogel absorbs these excess hydrogen ions, imparting a positive charge along the polymer backbone. The governing thermodynamics force the polymer to swell. Larger quantities of gluconic acid cause larger amounts of swelling. As the hydrogel swells, the

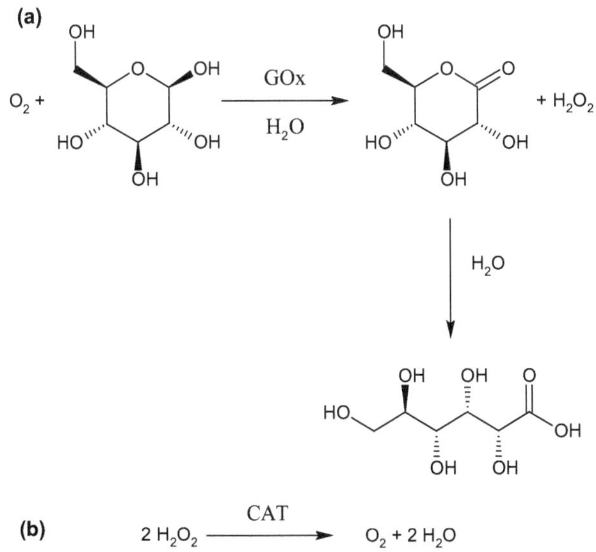

Figure 2.4 Chemical reaction catalyzed by glucose oxidase. (a) β-D-Glucose is first oxidized to β-D-glucono-1,5-lactone while dissolved oxygen is reduced to hydrogen peroxide by the enzyme glucose oxidase (GOx). β-D-Glucono-1,5-lactone is then hydrolyzed to gluconic acid in the presence of water. (b) Hydrogen peroxide is oxidized back into water by catalase (CAT), evolving molecular oxygen in the process.

mesh size increases, thus allowing imbibed insulin to diffuse out of the system. Once the released insulin starts to cause a drop in glucose levels, the concentration of gluconic acid inside the system decreases and the insulin release slows or stops.

Horbett *et al.* designed systems based on crosslinked DMAEMA with HEMA and glucose oxidase.[36] These films were synthesized by first freezing the monomer mixture and then exposing it to radiation using a ^{60}Co source to initiate the polymerization. Glucose oxidase was entrapped in the hydrogel structure, thus imparting glucose sensitivity to the pH sensitive hydrogel. Storage of the systems for 300 days in buffer still retained more than a third of the enzymatic activity, whereas storage in blood for 60 days retained almost half of the enzymatic activity. Thus, long-term usage of such devices may be possible, since the enzyme remains stable over long periods of time. ^{125}I labeled insulin was used for release studies, which were performed *in vitro* as well as *in vivo*. They concluded that insufficient amounts of insulin were being released, however, and that macroporous structures may be necessary. Further work showed that these macroporous structures were able to regulate insulin release at more relevant levels.[37]

Siegel and Firestone[31] also used polybasic gels based on DMAEMA. Oscillatory swelling and deswelling was demonstrated, yielding an ultimate collapsed state. These systems were then used to selectively release caffeine as a model drug.[21] At a very low pH, release continued for almost two hours.

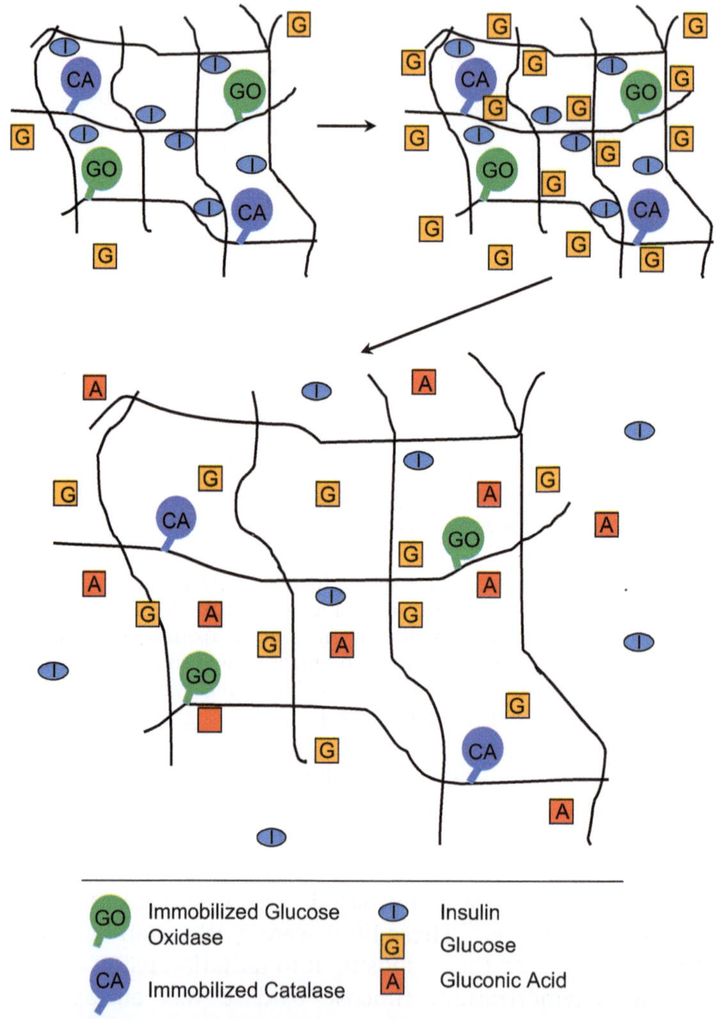

Figure 2.5 Diagram of a cationic hydrogel system response to a glucose stimulus. As glucose concentrations rise, gluconic acid production increases. The system swells and releases insulin.

At slightly acidic and neutral pH, the gels took nearly twice as long to fully release the caffeine. Thus, the collapsed hydrogels were able to trap the small molecular weight drug significantly more than the swollen hydrogels. When they copolymerized DMAEMA with methyl methacrylate,[28] they showed that temperature and ionic strength also affected swelling and deswelling. Significant amounts of water could be imbibed in these systems. As the amount of MAA or the alkyl chain length of *n*-alkyl methacrylate esters increased, swelling volume decreased.[21] This effect was attributed to an increase in the overall hydrophobicity of the system, which decreased the total amount of water able to be imbibed.

Siegel and Firestone also studied the effects of the ionic strength and buffer composition of the swelling media.[23,27,29] They tested how different ions affected swelling extent and rates using a 70 : 30 gel of MMA–DMAEMA. These systems exhibited significant sensitivities to the different ions tested, especially the multivalent ions. Increased ion concentrations decreased overall swelling. Weak electrolytes allowed faster swelling than strong electrolytes. A rigorous application of the ideal Donnan equilibrium theory was unsuccessful in fitting all the data, and so a heuristic approach was taken to qualitatively describe the responses of the systems to different ionic species.[23]

Kinetic experiments were performed using DMAEMA and *n*-alkyl methacrylate esters.[28] The alkyl chain length was altered to vary the overall hydrophobicity of the systems. A biphasic means of swelling was observed in glassy disks, having an initial slow uptake of water followed by a rapid volume expansion. This initial swelling rate was observed to increase with four factors: increased temperature, decreased pH, decreased *n*-alkyl methacrylate content, and decreased alkyl chain length. When the opposite conditions are met, a non-Fickian zero-order behavior is observed. Weak electrolytes were observed to increase swelling kinetics substantially compared to strong electrolytes, which was attributed to the ease of ion transport through the membranes. Initially rubbery disks, on the other hand, did not exhibit these biphasic swelling fronts. Instead, they had monophasic but still non-Fickian sorption of the solvent.

More recently, Siegel has moved away from a hydrogel for releasing insulin towards one that acts as an actuator for a glucose sensing device based on phenylboronic moieties.[38] The glucose bonded reversibly to the phenylboronic groups, imparting a charge to the boron atom. This charge induced swelling in the system similar to the cationic hydrogels. As the gel swelled in the presence of glucose, it pressed against a plate attached to a variable capacitor, thus altering the resonant frequency of an LC resonator. This device had a response time of 90 min, though this should be improved by decreasing the thickness of the responsive hydrogel layer. The group have also been working on a P(NIPAAm-*co*-MAA) membrane that could temporarily shut off glucose transport at low pH.[39] Glucose flux resumed after equilibration at this low pH, albeit at much lower rates. These results were attributed to a collapsed "skin" that formed on the acidic side of the membrane.

Goldraich and Kost[40] studied systems based on DEAEM and HEMA, crosslinked with tetraethylene glycol dimethacrylate (TEGDMA). Glucose oxidase, either immobilized or entrapped, was included in these systems. Swelling as a function of the comonomer and crosslinking ratios was determined, and insulin release was tested. In high pH media, little insulin was released. Unfortunately, the collapse of a gel was much slower than the swelling, and thus the cutoff of insulin release was much slower than the initiation of insulin release. This slowed collapse may be due to Donnan equilibrium effects, which make it more difficult to remove the hydrogen ion from the swollen gel. Other problems noticed were low physiological oxygen concentrations and possible encapsulation after implantation. Since glucose oxidase is dependent on dissolved oxygen as a reactant, low physiological levels significantly hinder the kinetic rate of the glucose oxidation.

Imanishi and associates took a different approach to releasing insulin. Porous cellulose films were grafted with poly(acrylic acid) and glucose oxidase.[41] At neutral pH, the chains were charged and extended due to charge repulsion, thus blocking the pores of the cellulose membrane. Once the glucose levels increased, glucose oxidase catalyzed the formation of gluconic acid. The pH dropped, and the acrylic acid groups became protonated and lost their charge. This decrease in electrostatic repulsion between polymer chains allowed the opening of the pores, releasing insulin. A second method of insulin release consisted of insulin immobilized to a polymer membrane using a disulfide bond.[42] Glucose dehydrogenase (GDH) was used to ultimately reduce the disulfide bond in the presence of glucose, thus releasing insulin. NAD was also incorporated to help with the charge transfer from the GDH to the insulin disulfide bond. Released insulin activity was reported to be the same as untreated insulin.

Heller *et al.* also took a different route to release insulin.[43] A bioerodible poly(ortho ester) was used, which should degrade in response to very slight pH decreases. Thus, with the incorporation of glucose oxidase, it would render the system glucose erodible, since it would degrade much faster in the presence of glucose. This degradation could be coupled to the release of infused insulin. However, it was noticed that the poly(ortho ester) was not very specific for its degradation. The inclusion of tertiary amine components solved this problem by increasing the hydrophilicity of the system at an acidic pH. The increased hydrophilicity allowed higher water penetration, which in turn led to a faster erosion rate. This erosion was coupled to a release of the impregnated insulin. The group demonstrated cyclic responses as well, and these data suggest the possibility of a long-term device that could release insulin in bursts, in response to high glucose concentrations.

Peppas *et al.* have been studying both anionic and cationic hydrogels for the delivery of insulin and other drugs for decades.[44–58] Hariharan and collaborators[44–46] studied copolymers of DEAEM and HEMA, *N*,*N*-diethylaminoethyl acrylate (DEAEA) and HEMA, methacrylaminopropyl ammonium chloride and HEMA, and AA and HEMA. Model drugs such as insulin, oxprenolol, lysozyme, and bovine serum albumin (BSA) were used in the studies. As the crosslinking amount increased, the mesh size decreased and solute release was either decreased or slowed. For the cationic gels, solute release increased with a decrease in pH. Anionic gels exhibited the opposite behavior, where solute release increased with an increase in pH. Release of solutes was by a diffusion-controlled process.

Dorski *et al.* took another approach to insulin release.[48,49] Instead of coupling glucose oxidase to a cationic hydrogel for insulin release in the presence of glucose, an anionic gel was studied. Since the decrease in pH still needed to elicit the release of insulin, a tighter network was used. Thus, upon constriction of the hydrogel, insulin was "squeezed" out of the system. Copolymers of MAA and ethylene glycol (EG) were synthesized, containing immobilized glucose oxidase and catalase. The system exhibited strong complexation at acidic pH, thus providing a means of insulin release in the presence of glucose.

There is great promise in DEAEM-based devices, as shown by Podual *et al.*[50–56] Hydrogels were synthesized with DEAEM and PEG grafts, and crosslinked with TEGDMA. Glucose oxidase and catalase were chemically immobilized in the systems using acryloyl chloride. Both films and microparticles were synthesized, both of which showed pH and glucose sensitivity. The films were made *via* UV initiated polymerization between two glass plates. The particles were synthesized *via* suspension polymerization in silicone oil with redox initiators, and different sized particles were formed depending on the speed of mixing and the amount of surfactant used. Smaller particles swelled faster, whereas larger particles and disks were slower to swell. The films were determined to have too slow a response time for physiological needs; however, the microparticles did have an adequately fast response time.

Other parameters were also changed, such as crosslinking ratio and PEG graft length. Varying the crosslinking ratio was shown to alter the swelling rates and maxima. It was found that a crosslinking ratio of 3 mol% gave the highest ratio of diffusion of insulin out of the swollen hydrogel to the collapsed hydrogel. Systems with longer PEG grafts were determined to have swelling responses dominated by relaxation of chains, whereas shorter PEG grafted systems had diffusion-controlled swelling rates. Modeling of these systems was performed, which provided a means of estimating insulin release profiles. It was determined that a reasonable amount of polymer particles could be loaded and delivered to a patient.

Hydrogels have also been used in the oral delivery of proteins. Lowman *et al.* showed that copolymers of MAA and PEG form significant complexation in their collapsed states.[57,58] Hydrogen bonding at low pH between the hydrogen on the MAA group to an ether bond on the PEG unit was observed. This complexation was broken once the pH rose above the pK_a of the material and the acid moiety became deprotonated and negatively charged. These systems were also very efficient at loading insulin. Since these hydrogels were collapsed at a low pH and swollen at a neutral pH, they were ideal for oral insulin delivery to the small intestines. The stomach has a low pH as well as peptidases that are designed to destroy proteins. Thus, when insulin was encapsulated in the collapsed hydrogel at low pH values, it was protected from the harsh environment of the stomach. The environment in the upper small intestine is much more hospitable to insulin, with fewer peptidases and a much higher pH. Thus, after being protected in the stomach, the insulin was released in the neutral environment of the small intestine when the complexation of the MAA and PEG broke down.

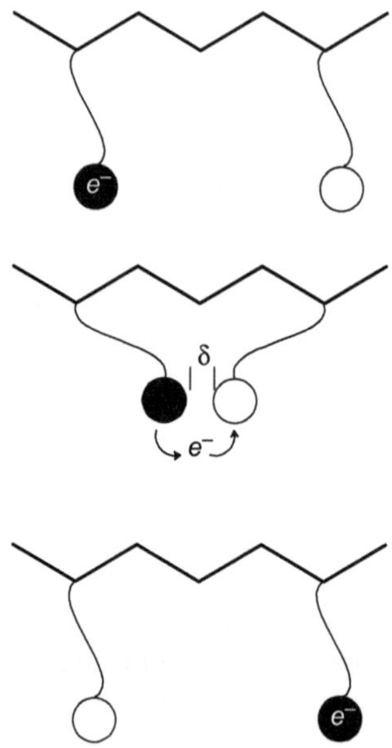

Figure 2.6 Electron diffusion in redox hydrogels. Electrons diffuse in the redox hydrogels by collisions between mobile redox centers tethered to the backbone of the crosslinked and hydrated polymer. When the tethers are long enough, the reduced (black) and the oxidized (white) members of the couple can come within the Marcus theory-defined distance δ that the electrons can cross (adapted from Heller, 2006).[59]

One last example worth mentioning, though not for insulin delivery, is the detection of glucose levels using glucose oxidase embedded in a conductive polymer.[59] Instead of coupling glucose oxidase with acid production to generate a signal, the reduction of redox hydrogels was used to capture electrons from the enzymatic reaction and pass them along the polymer to an electrode. The polymer backbone itself was not conductive, contrary to standard conductive polymers such as polyaniline. Rather, long flexible pendant chains off the polymer backbone had a functional group susceptible to a redox reaction, which allowed the charge to diffuse by swinging from pendant chain to pendant chain towards the embedded electrode, as shown in Figure 2.6. This arrangement allowed the enzyme to be throughout the entire hydrogel as opposed to monolayers coating an electrode.

2.6 Externally-controlled Drug Delivery Systems

Externally-controlled, sometimes referred to as remotely-triggered delivery systems, have the potential to go one step beyond standard targeted systems. In these types of systems, a therapeutic can be triggered to release at a specified time and location using a non-invasive external stimulus. Past work has shown that ultrasound, radio frequency pulses, magnetic fields, and light may be used to trigger delivery.[60–64] This is accomplished with the incorporation of metallic nanoparticles into polymeric hydrogel systems.[65] Along with being able to induce an externally-triggered response, metallic particles, such as gold nanoparticles (AuNPs), also serve as an imaging moiety for the detection and localization of diseased cells. Following this methodology, multifunctional carriers that could target, detect, and treat diseased cells can be developed. The term "theranostics" has been created to describe this auspicious area of research that focuses on the combination of diagnostics detection agents with therapeutic drug delivery carriers. Figure 2.7 summarizes the different hydrogel particle architectures that can be used for the development of "theranostics" carriers.

Here we focus on nanoscale externally-controlled, responsive hydrogel-based therapeutic systems for the delivery of chemotherapeutics for cancer treatment. These systems respond to exposure to a near infrared (NIR) light source (thus the name "responsive") to release an entrapped therapeutic agent. Light in the NIR range is of interest because of its ability to penetrate deeply through tissue compared to light in other wavelength ranges. Gold nanorods that are capable of adsorbing light in this region and converting it to heat will serve as a transducer to convert light to a drug release signal.

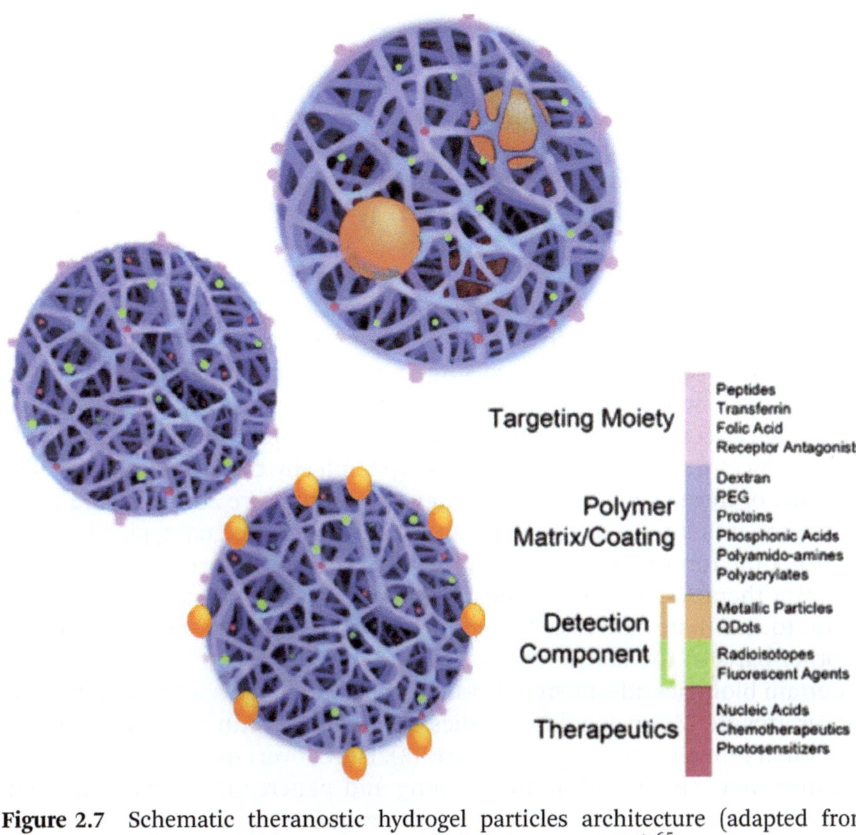

Figure 2.7 Schematic theranostic hydrogel particles architecture (adapted from reference Caldorera-Moore, Liechty, Peppas, 2011).[65]

The gold nanorods heat an adjacent thermally responsive polymer nanoparticle that undergoes a volume swelling change upon heating in an aqueous environment. This volume change leads to a subsequent release of the entrapped therapeutic. In this manner, a light signal can be applied externally to release the therapeutic at a desired time and location.

It is well known that cancer leads to uncontrolled, rapid growth of cells in the body typically leading to formation of a tumor often invading surrounding tissues and in some cases metastasizing, spreading to different areas of the body through blood or the lymphatic system. Overall, cancer is the second leading cause of death following cardiovascular disease. Approximately 13% of all deaths worldwide each year are caused by various types of cancer.[66,67]

The extent of the growth and spread of cancer is typically quantified by classification into one of five different stages. Determination of the malignant stage is based on the TNM classification system: size of tumor (T), involvement of lymph nodes (N), and distant metastasis (M). Stage 0 is a precancerous disease or marker such as an *in situ* carcinoma. Stage I is an early stage cancer referring to a small localized invasive tumor, typically less than 2 cm in diameter in the case of breast cancer. Cancers classified as Stage II are larger and more advanced. Breast cancers classified as Stage II are generally at least 2–5 cm in diameter and/or there is spread of cancerous cells to the lymph nodes. Stage III cancers are larger still and have in all cases spread to the lymph nodes. Stage IV cancers are metastatic meaning cancerous cells have spread to other organs of the body. Stage I–III cancers are considered early stage cancers and potentially curable through treatment, while metastasized Stage IV cancer is typically terminal but may be treated.[68]

Primary methods of treatment of Stage I–III cancers include surgery, radiation therapy, and systemic therapies such as chemotherapy, biologic treatments, and hormone treatments. Most individuals who are diagnosed with an invasive cancer will have surgery to remove the cancer if possible accompanied by one or more of the other therapies to shrink the tumor, eliminate cancerous tissue following surgery, or to prevent recurrence.[68–70]

Anti-cancer drugs are often delivered systemically through the oral delivery route or intravenous injection. When these drugs are administered prior to surgery, it is known as neoadjuvant therapy, with the goal typically being to shrink the tumor enough to allow for easier removal during lumpectomy. Adjuvant therapy consists of drugs administered following the removal of a tumor to eliminate cancer cells that may have migrated to the lymph nodes or other parts of the body.[70]

Certain biologics administered systemically have been shown to minimize cancer growth. Monoclonal antibodies such as herceptin target the human epidermal growth factor receptor 2 (HER2). HER2 overexpression in a variety of carcinoma cells including breast, lung and pancreatic cancers has been shown to be associated with faster-growing, more aggressive tumors. The drugs can be used to minimize the HER2 effect.[71] Another monoclonal antibody, bevacizumab, blocks vascular endothelial growth factor A thereby minimizing growth of new blood vessels and slowing tumor growth.[72]

Chemotherapeutic drugs are a broad class of therapeutics that target fast-dividing cells, by either impairing cell division or causing actual cell death. Neoadjuvant treatments help to shrink the tumor prior to resection, while adjuvant treatments can treat remaining cancer cells and prevent resection. In later stage cancers that cannot be entirely removed or in the case of metastatic cancers, chemotherapies are often applied to minimize further spread of the tumor with the goal of extending lifespan.[73–75]

Chemotherapy is one of the primary treatments used among cancer patients and is important for both pre- and postoperative therapies. Chemotherapeutic agents that are typically delivered systemically *via* either an oral or intravenous route most commonly work by disrupting cell growth in rapidly dividing cells. The systemic delivery of these agents leads to a number of adverse effects including nausea, vomiting, hair loss, cardiotoxicity, and immunosuppression.[69,76]

One promising approach for effective theranostic drug delivery systems includes nanoparticles of temperature-responsive polymeric hydrogels with gold nanorods grafted to the polymer nanoparticles that are capable of absorbing light and transmitting heat to the polymer. The heating triggers a swelling response in the carrier and an associated release of a therapeutic. Nanoscale systems are of interest because they could be administered intravenously, localize at a disease site because of the enhanced permeability and retention (EPR) effect and have extended circulation half-lives.

2.7 Targeted Drug Delivery

Biomaterials research has sought to develop systems that can increase effective dosing while simultaneously decreasing adverse side effects that occur as a result of therapeutic administration. Targeted drug delivery has been widely studied as a method of delivering a therapeutic to a localized disease site, such as a tumor or a site of inflammation, while minimizing the distribution of the drug to other parts of the body. The ability to concentrate therapeutic delivery at the disease site is advantageous over traditional systemic delivery because it may increase therapy at the disease site while decreasing the overall necessary dose and lessening the severity of adverse side effects.

Externally-controlled or triggered delivery systems have been developed to specify both the time and location of therapeutic delivery. Externally-controlled therapeutic systems are devices that can be triggered to release a therapeutic at a desired time and location, by external means. These systems release a therapeutic agent in response to triggering mechanisms such as ultrasound waves,[60,64] light,[63] a magnetic field,[61,62] or other methods that can be used to transmit a signal to a delivery system *in vivo*. Metal nanoparticles can act as the signal receiver in light-triggered systems because of their ability to convert light into heat. When coupled with a temperature sensitive polymer, metal nanoparticles can trigger release of a drug by causing a swelling transition in the polymer through heating.

Nanoscale devices are desirable for drug delivery systems because they localize at tumor sites more readily as a result of the EPR effect. Nanoparticles have improved circulation half-lives in the bloodstream and collect in tumor microvasculature with poor lymphatic drainage.[77] Nanoscale materials are essential for the development of injectable targeted and externally-triggered delivery systems.

2.7.1 Chemotherapeutics and Targeted Drug Delivery

Anticancer drugs known as chemotherapeutics are administered systemically to around 75% of all cancer patients *via* either an oral or intravenous route. Chemotherapeutics may be used in all stages of various cancers and delivered either preoperatively or postoperatively.[69] Preoperative or neoadjuvant delivery can help to shrink the tumor allowing for easier and more total removal of tumors while adjuvant therapy following operation is used to prevent recurrence and to eliminate cancer cells that may be in the lymph nodes or other parts of the body. There are a number of different chemotherapeutic drugs that are administered to cancer patients, and frequently these drugs are administered together in combinations of 2–4 of the different types of drugs. The most common classes of chemotherapeutic drugs used are alkylating agents, antimetabolites, anthracyclines, and taxanes.[74]

Alkylating agents such as cyclophosphamide work by effectively cross-linking DNA strands thereby preventing cell division and tumor growth. The alkylating agents alkylate guanine bases in DNA, and because of the fast-dividing nature of cancer cells, they are susceptible to this modification. Side effects for cyclophosphamide tend to be less severe than for many other chemotherapeutics, but some side effects do commonly occur including nausea and vomiting, hair loss, diarrhea, and joint pain.

Antimetabolites used in chemotherapy typically mimic the structure of either purine or pyrimidine. The pyrimidine analog fluorouracil (5-FU) is the most common antimetabolite used to treat breast cancer. 5-FU blocks DNA replication by acting as a thymidylate synthase inhibitor preventing synthesis of pyrimidine thymidine, which is necessary for replication.[78] Antimetobolites affect all DNA replication and cell growth, most pronounced in any fast-dividing cells. Because of their far-reaching effects, they may have severe side effects such as bone marrow suppression, digestive tract inflammation, dermatitis, diarrhea, and hair loss. Additionally, individuals with a condition known as dihydropyrimidine dehydrogenase deficiency (DPD deficiency) are unable to metabolize pyrimidine-based drugs and for these patients 5-FU treatment can be lethal.

Anthracyclines are another class of drugs that inhibit the growth of cancer cells. Doxorubicin (adriamycin) is used most commonly in breast cancer chemotherapeutic regimens. Anthracyclines such as doxorubicin are administered intravenously and inhibit replication by intercalating DNA. Doxorubicin is a very versatile drug used to treat a wide range of cancers including breast cancer. The positive effects of doxorubicin are accompanied by a number of adverse effects. Cardiotoxicity is a major concern for patients who are treated with anthracyclines, and high doses can lead to a number of heart-related conditions including heart arrhythmias, cardiomyopathy, heart failure, and the possibility of death. Other common side effects include nausea, vomiting, and total hair loss.[75]

To reduce the side effects of doxorubicin, systems have been developed to enhance localization of the drug at the target cancer site and decrease systemic effects. PEGylated liposomal doxorubicin systems including Doxil and Caelyx have been shown to increase the therapeutic index of doxorubicin. PEGylation inhibits uptake by the reticulo-endothelial system (RES), increasing the half-life of circulation as well as increasing extravasation into the leaky vasculature of tumors. The liposomal systems have been shown to increase concentration of drug or radiotracers at tumor sites. Decreased cardiotoxicity has been observed in patients treated with PEGylated liposomal doxorubicin compared to free drug.[79,80] A non-PEGylated liposomal formulation, Myocet, has shown antitumor activity comparable to free doxorubicin while decreasing cardiotoxicity in the treatment of metastatic breast cancer patients.[81,82] Myocet has been approved for treatment of breast cancer in Europe and Canada, but it has not yet been approved by the FDA in the United States.

The targeting of drug delivery systems, most commonly micro- and nanoscale particles, provides more efficient treatment and improves quality of life for patients by minimizing potential adverse effects. In a perfect targeted system, the entire drug payload of all carriers would be delivered at the preferential site. In this case, the dosage can be tailored to have the most significant effect at the disease site without causing systemic adverse affects. Externally-triggered devices seek to go a step beyond simple targeting as release of drug is only triggered at the site where treatment is desired at a specific point in time.

Nanoparticles are desirable in many targeted delivery systems because their properties allow them to circulate in the bloodstream for extended periods of time and they have a high level of accumulation at tumor sites. Delivery of anticancer drugs to tumor sites is the most commonly investigated use of targeted nanoparticles to date. Intravenously-administered nanoparticles under the size of 200 nm have been shown to circulate much longer than systems larger than this size, because of their ability to escape the body's natural defense mechanisms such as the RES.[17] These carriers are also the right size to take advantage of the EPR effect. Particles smaller than 400 nm have been shown to extravasate leaky tumor vasculature *via* the EPR effect, and those smaller than 200 nm may be even more effective.[83]

While the EPR effect is an example of passive targeting, taking advantage of properties of the drug or carrier to localize at the target site, active targeting involves the incorporation of molecules in the delivery system that are intended to interact with the physiological target. Typically a targeting ligand is adhered to the surface of the nanoparticle or is covalently attached. The addition of targeting ligands has been shown to improve the therapeutic effect of delivery systems both *in vitro* and *in vivo*.[84,85]

Folate receptor is a glycoprotein that has frequently been observed to have elevated expression in tumors while it is not present in most normal tissues making it a very attractive candidate for targeted anticancer delivery.[86] Nanoparticles functionalized with folate—the receptor target—have been shown to have higher therapeutic efficacies in the treatment of tumors than non-targeted nanoparticles in both *in vitro* and *in vivo* models.[84]

Another common ligand used to target delivery systems to tumors is transferrin. The transferrin receptor, TfR, has been shown to be expressed at higher levels on cancer cells than on normal cells, so nanoparticle systems have been designed to target cancer cells by attaching transferrin to the delivery system.[85] Improvements in delivery and associated cancer cell death have been observed using transferrin-functionalized systems.[87] Several studies have shown that the tumor accumulation is similar when comparing systems incorporating transferrin and their non-targeted counterparts.[88] The improved delivery has therefore been attributed to increased tumor cell internalization of the non-targeted nanoparticles.

Several systems have been developed to use monoclonal antibodies to target cancerous tumors. An anti-HER2 monoclonal antibody has been incorporated into doxorubicin-loaded liposomes and shown to increase antitumor efficacy over control doxorubicin formulations.[89] Other studies have demonstrated that targeted liposomes with the anti-HER2 antibody were found at high concentrations in tumor cells while non-targeted liposomes were more commonly found in the extracellular spaces. Additionally, targeted liposomes had a much greater preference for tumor cells over host cells while control systems showed no preference.[90]

Another monoclonal antibody, 2C5, has been shown to increase efficacy when incorporated into liposomal doxorubicin systems. The targeted systems showed increased accumulation in lung cancer models when compared to non-targeted liposomal systems. The tumor reduction measured was also shown to be higher for the targeted systems compared to the control.[91]

2.8 Hydrogel Nanoparticulate Systems

Although particles less than 0.1 μm in diameter are considered as nanoparticles, hydrogel particles smaller than 1 micron are also commonly referred to as nanoparticles. Polymer nanoparticles are of interest for applications in drug delivery because of their potential to target a variety of different specific locations in the body, elongate periods of dosage, protect sensitive therapeutics such as proteins, increase half-life circulation, and reduce systemic effects.[92,93] Polymer nanoparticles with many different types of properties have been developed for use in drug delivery systems including biodegradable systems, hydrogel systems, and self-assembled systems such as micelles, liposomes, and polymersomes.

The use of hydrogel nanoparticles is particularly advantageous in drug delivery because of their inherent potential to avoid the RES and take advantage of the EPR effect.[94,95] Microparticles or large nanoparticles are rapidly recognized by macrophages and cleared by the RES through the liver and spleen. Nanoparticles measuring less than 300 nm in diameter show increased avoidance of the RES particularly in particles that are less than 100 nm in diameter.

This avoidance of the RES leads to a longer circulation half-life and ultimately greater localization at the delivery site. Because of the fast growth of tumors, the vasculature tends to be leaky and lymphatic drainage poor. Nanoparticles are preferentially extravasated and concentrated at these tumor sites while small molecular weight drugs are not retained at such high concentrations.

Primarily, hydrogel nanoparticles are formed by one of two broad methods, either the dispersion of preformed polymers into nanoparticles, or monomer polymerization designed to directly polymerize into the final particle form. Additionally, block copolymers have been synthesized in order to assemble nanoscale polymersomes or micelles. These polymer nanoparticles, along with various block copolymer assemblies, have been extensively investigated for use as drug carriers.[96,97]

2.8.1 Biodegradable Systems

Biodegradable hydrogel nanoparticulate systems are of interest for the ability to combine desirable nanoscale properties with enhanced biocompatibility and controlled release. The most widely investigated biodegradable polymers for nanoparticle delivery are those based on the polyesters poly(lactic acid) (PLA), poly(glycolic acid) (PGA), and their copolymer poly(lactic acid-*co*-glycolic acid) (PLGA). By adjusting the ratio of lactic acid to glycolic acid in PLGA, the rate of degradation can be controlled. Increased lactic acid content extends the degradation time of the polymer. The polymer degrades by hydrolysis to the monomers lactic acid and glycolic acid, which can be metabolized by the body. Because the degradation time of PLGA can be easily tuned and the biocompatibility of the system has led to FDA approval for resorbable sutures, a large number of PLGA nanoparticle delivery systems have been developed.

Additional PLGA nanoparticle synthesis methods employ nanoprecipitation and spray drying. In traditional spray-drying, the polymer is dissolved in a volatile solvent and the solution in sprayed into an apparatus where the particles are formed as the solvent evaporates.[98] Typically, this method yields particles at the upper limit of the nanoscale and on the microscale. A modification of this method known as spray freezing into liquid (SFL) is also used to synthesize polymer micro and nanoparticles. In this case, a solution of the polymer is sprayed through an atomizing nozzle into liquid nitrogen over a solvent such as frozen ethanol.[99] Particles form upon contact with the liquid nitrogen and are collected in the ethanol after evaporation of the liquid nitrogen. Drugs have been incorporated during particle formation by addition to polymer solution or formation of an oil-in-water emulsion prior to atomization.[99,100]

2.8.2 Nanogels

Upon injection into the bloodstream, nanoscale systems have the ability to circulate for extended periods of time and to be taken up by cells. Their dimensional properties, such as high surface area and rapid response to external stimuli, make them attractive candidates for *in vivo* systems.[101] In particular, nanoscale hydrogel systems, known as nanogels, are of interest for use as "intelligent" drug carriers. These intelligent materials respond to external stimuli such as pH, temperature, and ionic strength by imbibing water and swelling under certain conditions.

Hydrogels are hydrophilic but insoluble in aqueous environments. Under specific conditions, the polymer networks imbibe water to swell. Hydrogel carriers have been designed to exhibit stimuli-sensitive behavior, where the degree of swelling is affected by an external stimulus such as pH, temperature, ionic strength, or electric field.[102,103] Because of this swelling behavior in response to their environment, stimuli-sensitive polymers are known as "intelligent" or "smart" polymer systems.

Intelligent hydrogel carriers have been used to encapsulate various therapeutic molecules and subsequently release them following a change in one of these external conditions. Release from these systems has been triggered by both swelling and shrinking of hydrogel carriers.[63,104] In the first case, a therapeutic molecule is encapsulated in a carrier in its collapsed state where the drug is protected and a small mesh size limits diffusion in or out of the polymer matrix.[105] If the polymer encounters an environment that triggers swelling, the mesh size increases, and there is an increase in the rate of diffusion, so that the therapeutic is released into the surrounding environment. In the latter case, a polymer is synthesized so that in its swollen state the properties are such that the polymer can encapsulate and protect a molecule of interest. Upon deswelling or shrinking there is a fast release as some of the encapsulated material is squeezed out of the polymer matrix.[63] As this cycle is repeated, a pulsatile release of the therapeutic may be observed.

Temperature-responsive polymers can exhibit one of two different types of temperature behaviors in aqueous solutions and will be examined separately in this work. Both types of systems—those that exhibit UCST and those that fall into the class of LCST behavior—will be synthesized and investigated here. Hydrogels with UCST behavior have characteristics such that at lower temperatures some polymer–polymer interactions lead to a less swollen state. Upon heating through the UCST these interactions are weakened, so that the polymer effectively becomes miscible with the aqueous phase, but because the polymer is crosslinked, the system imbibes water and swells.

Certain types of interpenetrating polymer networks, IPNs, such as those based on acrylamide (AAm) and AA, demonstrate this type of behavior with transition temperatures near 40 °C.[106] IPNs consist of two separately crosslinked polymer networks that are physically entangled with the first network, but not covalently attached. The PAA network and the PAAm network undergo hydrogen bonding at low temperatures, which breaks down above the UCST and electrostatic and steric interactions cause the polymer to swell.

Polymers demonstrating LCST behavior, most commonly PNIPAAm and related copolymers, have also been investigated for use as temperature-sensitive drug carriers. The presence of separate hydrophilic and hydrophobic groups in the polymer chain generally leads to this behavior. At low temperatures, hydrogen bonding interactions between water and the hydrophilic groups lead to a swollen state, but an increase in temperature beyond the transition temperature leads to a weakening of the water–polymer hydrogen bonding and a strengthening of hydrophobic polymer–polymer interactions.[107] LCST drug carriers have been developed that expel encapsulated therapeutics upon heating above transition temperatures at therapeutically relevant temperatures by what is known as a "squeezing out" method upon collapse of the polymer system.[63]

Hydrogel nanoparticles, or nanogels, have been synthesized using several techniques including dispersion and emulsion polymerization as well as step and flash imprint lithography.[102,108,109] These nanoscale hydrogels show rapid response rates, swelling quickly with small changes in stimuli. Synthetic nanogels have stimuli-sensitive properties similar to their bulk hydrogel counterparts, exhibiting volume phase transitions in response to pH, temperature, electric field, or other environmental stimuli.[110]

Several heterogeneous polymerization techniques have been used to synthesize monomers directly into the desired polymer hydrogel nanoparticles. Two commonly used techniques are emulsion and dispersion polymerizations, and both are used to synthesize the various temperature-sensitive polymers in the proposed therapeutic systems. Oil-in-water emulsions are commonly used to synthesize hydrophobic polymers from similarly hydrophobic monomers. Polymerization occurs in micelles that are formed with the addition of surfactant as monomers diffuse from large droplets into micelles and react with initiator.[111] Similarly, water-in-oil emulsions have been used to synthesize polymers from more water-soluble molecules. Previous work by Landfester featured the use of consecutive water-in-oil microemulsion polymerizations to synthesize IPN nanoparticles, and this technique has been adapted for use in our work.[111] Monomer and crosslinker are dissolved in an aqueous phase and emulsified into a cyclohexane oil phase with surfactant. Initiator is added to start the polymerization in the aqueous micelles.[26]

Dispersion polymerization has been used in free radical polymerizations where the growing polymer is much less soluble in a particular solvent than the monomer. Monomer and initiator are dissolved in a solvent in a homogeneous mixture often with the inclusion of some molecule to act as a steric stabilizer for particle formation. As polymerization begins, growing macroradicals precipitate, and form stabilized particles. Continued polymerization occurs as the particles are slightly swollen in the polymerization medium.[112] An aqueous dispersion polymerization has been used to synthesize temperature-sensitive hydrogel polymers with LCST behavior. Nanoparticles of PNIPAAm and related copolymers have been synthesized by reaction of monomers in aqueous medium at elevated temperature with or without steric stabilizer molecules such as sodium dodecyl sulfate (SDS).[113] The reaction is performed at a temperature well above the LCST of the resultant polymer. Since the temperature is above the LCST, the polymer is immiscible in the aqueous environment and particle nucleation occurs.

2.9 Externally-controlled Delivery Using Hydrogels

Externally-controlled or externally-triggered delivery systems are those in which some source external to the body can be used to create a signal to induce release of a therapeutic at a desired time and location. The external source in most cases is a form of electromagnetic radiation or an acoustic signal. The signal must be able to penetrate through tissue to reach the delivery system. Typically the signal will either induce heating in the delivery vehicle or interact mechanically with the system to trigger release. These remotely-controlled systems can have the same benefits as targeted systems in that they can improve therapeutic index locally while reducing systemic effects. In some cases, externally-controlled systems have been functionalized with targeting ligands to further improve efficacy.

Externally-applied electric fields, ultrasound, and mechanical forces have also been used to trigger a swelling transition and subsequent drug release.[62,114,115] For example, ultrasound microbubbles have been developed for use as externally-triggered delivery and imaging systems.[64] A focused ultrasound pulse can be used to disrupt the microbubbles and locally deliver a drug or contrast agent.

The swelling characteristics of temperature-responsive hydrogels can be combined with a means to heat them *via* an external signal to trigger drug delivery. Heating can be triggered by several types of signals including light and magnetic fields. A magnetic field has been used to induce heating in a PNIPAAm hydrogel film system loaded with superparamagnetic iron oxide particles, which triggers a negative swelling and release of a model drug.[62]

Other systems have been developed to control the release of a therapeutic using light. In one case, gold nanoshells were incorporated into poly(*N*-isopropyl acrylamide-*co*-acrylamide)—P(NIPAAm-*co*-AAm)—films or disks. Upon exposure to certain wavelengths of light, gold nanoshells heat the polymer and trigger a negative swelling transition. This behavior has been used to externally control the opening and closing of a microvalve and to trigger pulsatile release of insulin from a polymer loaded film.[63]

Previous work in the authors' lab has focused on developing a nanoscale system that incorporates light as the trigger to produce a temperature change, swelling transition, and subsequent drug release from UCST nanogel particles. In the IPN systems that have been developed, complexation of the two networks occurs at low temperatures because of hydrogen bonding. As the polymer is heated, hydrogen bonding begins to break down and a positive swelling transition is observed around 38–42 °C.[26] A synthetic procedure has been developed for temperature-sensitive IPNs composed of interpenetrating networks of poly(acrylic acid), PAA, and poly(acrylamide), PAAm.[116] The inverse-microemulsion produces particles between 100 nm and 1 μm in diameter, and there is potential to incorporate gold particles in the polymer matrix when added to the aqueous phase of the water-in-oil emulsion leading to a light-triggered release system.

Finally, a large number of other chemotherapeutic systems based on hydrogels have been studied in our laboratories and have been advanced for applications.[113–116]

2.10 Concluding Remarks and Future Directions

It is apparent that designing a device to release insulin in a physiological manner is highly desirable. Hydrogels seem a logical building block, since their parameters can be precisely tuned to provide specific key features, such as mesh size, response time, and release rates. Also, they can easily be made biocompatible, either by choosing already biocompatible materials or by grafting chains such as PEG to the surface. The environmental sensitivity of the many hydrogels being studied also provides a means of incorporating a drug reservoir, a sensor, and an actuator into one device with minimal components. Thus, designing an optimal insulin-releasing device with hydrogels is quite a promising endeavor.

We believe that nanotechnology that is applicable to biological and pharmaceutical problems, and more specifically to processes and devices based on synthetic and hybrid biomimetic hydrogels, can be used for the next generation of drug, protein and gene delivery devices. The discovery and delivery of drugs to cure chronic diseases have been achieved by the combination of intelligent material design with advances in nanotechnology. In particular, there has been considerable work in preparing nanostructured biomaterials for various applications, such as carriers for controlled and targeted drug delivery, micropatterned devices, and systems for biological recognition. Since many drugs act as protagonists or antagonists to different chemicals in the body, a delivery system that can respond to the concentrations of certain molecules in the body is invaluable. For this purpose, intelligent therapeutics or smart drug delivery calls for the design of the next generation of responsive devices and materials.

In particular, novel nanoscale biomimetic materials, especially polymeric networks, capable of molecular recognition can be prepared by designing interactions between the building blocks of biocompatible networks and the desired specific ligands and by stabilizing these interactions by a three-dimensional structure. These structures are, at the same time, flexible enough to allow for diffusion of solvent and ligand into and out of the networks. Synthetic networks designed to recognize and bind biologically significant molecules are of great importance and influence a number of emerging technologies. These artificial materials can be used as unique systems or incorporated into existing drug delivery technologies that can aid in the removal or delivery of biomolecules and restore the natural profiles of compounds in the body.

In addition, biomimetic methods are now used to build biohybrid systems or even biomimetic materials (mimicking biological recognition) for drug delivery, drug targeting, and tissue engineering. The synthesis and characterization of biomimetic gels and molecularly imprinted drug release and protein delivery systems has been advanced by our work and others. Configurational biomimetic imprinting of an important analyte on an intelligent hydrogel leads to the preparation of new biomaterials that not only recognize the analyte but also act therapeutically by locally or systemically releasing an appropriate drug.

The design of a precise macromolecular chemical architecture that can recognize target molecules from an ensemble of closely related molecules has had a large number of potential applications. Nanoimprinting creates stereo-specific three-dimensional binding cavities based on the template of interest. For example, efforts for the imprinting of large molecules and proteins has led to novel methods of two-dimensional surface imprinting, a method of recognition at a surface rather than within a bulk polymer matrix.

Additionally, micro- and nanofabrication techniques have enabled the development of novel drug delivery devices that can improve the therapeutic effect of a drug, such as micro- and nanoscale needles, pumps, valves, and implantable drug delivery devices. Electronic devices have now reached a stage of dimensions comparable to those of biological macromolecules. Peppas' work has raised exciting possibilities for combining microelectronics and biotechnology to develop new technologies with unprecedented power and versatility. Such devices are developed by platforms with well-led functions at the micro- or nanoscale, and they include nanoparticulate systems, recognitive molecular systems, biosensing devices, and microfabricated and microelectronic devices. Thus, there are a variety of microelectronic devices that have been studied for controlled drug delivery systems, such as microchips capable of the storage and then delivery of multiple drugs in a controlled manner. Solid-state silicon microchips that can provide controlled release of single or multiple chemical substances on demand have been fabricated and demonstrated. Microreservoirs that can be filled with chemicals in solid, liquid, or gel form are released individually *via* electrochemical dissolution of the thin anode membranes covering their opening.

Because of the adverse health effects associated with systemic delivery of therapeutic agents such as anticancer drugs, there is great interest in developing systems that can deliver a therapeutic locally to a disease site. In the case of chemotherapy, the use of nanoscale drug delivery systems is of interest because of their ability to passively target tumors, and in some cases actively target following functionalization of these particulate systems with targeting ligands. Externally-triggered systems can go one step further by precisely controlling the time and place that a drug is released. There is a great need for improved systems that can better maximize therapy while minimizing adverse systemic effects.

List of Abbreviations

5-FU	pyrimidine analog fluorouracil
AA	acrylic acid
AAm	acrylamide
AuNP	gold nanoparticles
BSA	bovine serum albumin
CAT	catalase
DEAEA	*N,N*-diethylaminoethyl acrylate
DEAEM	*N,N*-diethylaminoethyl methacrylate
DMAEMA	*N,N*-dimethylaminoethyl methacrylate
DPD	dihydropyrimidine dehydrogenase
EG	ethylene glycol
EPR	enhanced permeability and retention
GDH	glucose dehydrogenase
GOx	glucose oxidase
HEA	2-hydroxyethyl acrylate
HEMA	2-hydroxyethyl methacrylate
HER2	human epidermal growth factor receptor 2
LCST	lower critical solution temperature
MAA	methacrylic acid
MMA	methyl methacrylate
NIR	near infrared
PAA	poly(acrylic acid)
PAAm	poly(acrylamide)
PEG	poly(ethylene glycol)
PGA	poly(glycolic acid)
PLA	poly(lactic acid)
PLGA	poly(lactic acid-*co*-glycolic acid)
PNIPAAm	poly(*N*-isopropyl acrylamide)
PVA	poly(vinyl alcohol)
RES	reticulo-endothelial system
SDS	sodium dodecyl sulfate
SFL	spray freezing into liquid
TEGDMA	tetraethylene glycol dimethacrylate
TNM	size of tumor (T), involvement of lymph nodes (N), and distant metastasis (M)

TMS 4-trimethylsilylstyrene
UCST upper critical solution temperature

References

1. *Biomaterials Science: An Introduction to Materials in Medicine*, ed. B. D. Ratner, A. S. Hoffman, F. J. Schoen and J. E. Lemons, Elsevier Academic Press, Amsterdam, Boston, 2nd edn, 2004.
2. J. C. Wheeler, J. A. Woods, M. J. Cox, R. W. Cantrell, F. H. Watkins and R. F. Edlich, Evolution of hydrogel polymers as contact lenses, surface coatings, dressings, and drug delivery systems, *J. Long-Term Eff. Med. Implants*, 1996, **6**(3–4), 207–217.
3. J. Chen, Y. Q. Yang, P. B. Qian, Z. T. Ma, W. B. Wu, P. Z. Sung, X. G. Wang and J. H. Li, Drug Carrying Hydrogel Base Wound Dressing, *Radiat. Phys. Chem.*, 1993, **42**(4–6), 915–918.
4. A. S. Hoffman, Hydrogels for Biomedical Applications, *Ann. N. Y. Acad. Sci.*, 2001, **944**(1), 62–73.
5. H. A. Shawky, M. H. El-Sayed, A. E. Ali and M. S. A. Mottaleb, Treatment of polluted water resources using reactive polymeric hydrogel, *J. Appl. Polym. Sci.*, 2006, **100**(5), 3966–3973.
6. H. E. Bodde, E. A. C. Vanaalten and H. E. Junginger, Hydrogel Patches for Transdermal Drug Delivery - Invivo Water Exchange and Skin Compatibility, *J. Pharm. Pharmacol.*, 1989, **41**(3), 152–155.
7. S. W. Kim, Hydrogel Drug Delivery Systems, *Pharm. Int.*, 1983, **4**(7), 182.
8. P. I. Lee, Kinetics of drug release from hydrogel matrices, *J. Controlled Release*, 1985, **2**, 277–288.
9. J. Rosiak, K. Burozak and W. Pekala, Polyacrylamide hydrogels as sustained release drug delivery dressing materials, *Radiat. Phys. Chem.*, 1983, **22**(3–5), 907–915.
10. J. Kopecek and R. Duncan, Targetable polymeric prodrugs, *J. Controlled Release*, 1987, **6**(1), 315–327.
11. C. V. Rice, Phase-transition thermodynamics of N-isopropylacrylamide hydrogels, *Biomacromolecules*, 2006, **7**(10), 2923–2925.
12. S. Hirotsu, Coexistence of Phases and the Nature of 1st-Order Phase-Transition in Poly-N-Isopropylacrylamide Gels, *Adv. Polym. Sci.*, 1993, **110**, 1–26.
13. A. Halperin, Polymer brushes that resist adsorption of model proteins: Design parameters, *Langmuir*, 1999, **15**(7), 2525–2533.
14. S. I. Jeon and J. D. Andrade, Protein–surface interactions in the presence of polyethylene oxide : II. Effect of Protein Size, *J. Colloid Interface Sci.*, 1991, **142**(1), 159–166.
15. L. M. Seifert and R. T. Greer, Evaluation of in vivo adsorption of blood elements onto hydrogel-coated silicone rubber by scanning electron microscopy and fourier transform infrared spectroscopy, *J. Biomed. Mater. Res.*, 1985, **19**(9), 1043–1071.

16. S. Sharma, K. C. Popat and T. A. Desai, Controlling nonspecific protein interactions in silicon biomicrosystems with nanostructured poly(ethylene glycol) films, *Langmuir*, 2002, **18**(23), 8728–8731.
17. D. E. Owens and N. A. Peppas, Opsonization, biodistribution, and pharmacokinetics of polymeric nanoparticles, *Int. J. Pharm.*, 2006, **307**(1), 93–102.
18. N. Peppas, P. Bures, W. Leobandung and H. Ichikawa, Hydrogels in pharmaceutical formulations, *Eur. J. Pharm. Biopharm.*, 2000, **50**(1), 27–46.
19. B. Zhao and J. S. Moore, Fast pH- and ionic strength-responsive hydrogels in microchannels, *Langmuir*, 2001, **17**(16), 4758–4763.
20. J. P. Baker, L. H. Hong, H. W. Blanch and J. M. Prausnitz, Effect of Initial Total Monomer Concentration on the Swelling Behavior of Cationic Acrylamide-Based Hydrogels, *Macromolecules*, 1994, **27**(6), 1446–1454.
21. R. A. Siegel, M. Falamarzian, B. A. Firestone and B. C. Moxley, Ph-Controlled Release from Hydrophobic Poly-Electrolyte Copolymer Hydrogels, *J. Controlled Release*, 1988, **8**(2), 179–182.
22. Y. B. Zhao, W. Y. Chen, Y. J. Yajiang, X. L. Yang and H. B. Xu, Swelling behavior of ionically cross-linked polyampholytic hydrogels in varied salt solutions, *Colloid Polym. Sci.*, 2007, **285**(12), 1395–1400.
23. B. A. Firestone and R. A. Siegel, Ph, Salt, and Buffer Dependent Swelling in Ionizable Copolymer Gels - Tests of the Ideal Donnan Equilibrium-Theory, *J. Biomater. Sci., Polym. Ed.*, 1994, 5(5), 433–450.
24. H. K. Ju, S. Y. Kim and Y. M. Lee, pH/temperature-responsive behaviors of semi-IPN and comb-type graft hydrogels composed of alginate and poly (N-isopropylacrylamide), *Polymer*, 2001, **42**(16), 6851–6857.
25. H. Yu and D. W. Grainger, Thermosensitive Swelling Behavior in Cross-Linked N-Isopropylacrylamide Networks - Cationic, Anionic, and Ampholytic Hydrogels, *J. Appl. Polym. Sci.*, 1993, **49**(9), 1553–1563.
26. D. E. Owens, J. K. Eby, Y. Jian and N. A. Peppas, Temperature-responsive polymer-gold nanocomposites as intelligent therapeutic systems, *J. Biomed. Mater. Res., Part A*, 2007, **83A**(3), 692–695.
27. R. A. Siegel, I. Johannes, C. A. Hunt and B. A. Firestone, Buffer Effects on Swelling Kinetics in Polybasic Gels, *Pharm. Res.*, 1992, **9**(1), 76–81.
28. B. A. Firestone and R. A. Siegel, Kinetics and Mechanisms of Water Sorption in Hydrophobic, Ionizable Copolymer Gels, *J. Appl. Polym. Sci.*, 1991, **43**(5), 901–914.
29. R. A. Siegel and B. A. Firestone, Mechanochemical Approaches to Self-Regulating Insulin Pump Design, *J. Controlled Release*, 1990, **11**(1–3), 181–192.
30. R. A. Siegel and B. A. Firestone, pH-dependent equilibrium swelling properties of hydrophobic polyelectrolyte copolymer gels, *Macromolecules*, 1988, **21**(11), 3254–3259.
31. B. A. Firestone and R. A. Siegel, Dynamic Ph-Dependent Swelling Properties of a Hydrophobic Poly-Electrolyte Gel, *Polym. Commun.*, 1988, **29**(7), 204–208.

32. B. C. Anderson and S. K. Mallapragada, Synthesis and characterization of injectable, water-soluble copolymers of tertiary amine methacrylates and poly(ethylene glycol) containing methacrylates, *Biomaterials*, 2002, **23**(22), 4345–4352.
33. K. Ishihara, M. Kobayashi, N. Ishimaru and I. Shinohara, Glucose-Induced Permeation Control of Insulin through a Complex Membrane Consisting of Immobilized Glucose-Oxidase and a Poly(Amine), *Polym. J.*, 1984, **16**(8), 625–631.
34. K. Ishihara, M. Kobayashi and I. Shionohara, Control of Insulin Permeation through a Polymer Membrane with Responsive Function for Glucose, *Makromol. Chem., Rapid Commun.*, 1983, **4**(5), 327–331.
35. K. Ishihara and K. Matsui, Glucose-Responsive Insulin Release from Polymer Capsule, *J. Polym. Sci., Part C: Polym. Lett.*, 1986, **24**(8), 413–417.
36. J. Kost, T. A. Horbett, B. D. Ratner and M. Singh, Glucose-Sensitive Membranes Containing Glucose-Oxidase - Activity, Swelling, and Permeability Studies, *J. Biomed. Mater. Res.*, 1985, **19**(9), 1117–1133.
37. G. Albin, T. A. Horbett and B. D. Ratner, Glucose sensitive membranes for controlled delivery of insulin: Insulin transport studies, *J. Controlled Release*, 1985, **2**, 153–164.
38. L. Zhu, S. H. Shabbir, M. Gray, V. M. Lynch, S. Sorey and E. V. Anslyn, A structural investigation of the NB interaction in an o-(N, N-dialkylaminomethyl) arylboronate system, *J. Am. Chem. Soc.*, 2006, **128**(4), 1222–1232.
39. A. P. Dhanarajan and R. A. Siegel, Time-dependent Permeabilities of Hydrophobic, pH-Sensitive Hydrogels Exposed to pH Gradients, in *Macromolecular Symposia.* 2005, Wiley Online Library.
40. M. Goldraich and J. Kost, Glucose-sensitive polymeric matrices for controlled drug delivery, *Clin. Mater.*, 1993, **13**(1–4), 135–142.
41. Y. Ito, M. Casolaro, K. Kono and Y. Imanishi, An Insulin-Releasing System That Is Responsive to Glucose, *J. Controlled Release*, 1989, **10**(2), 195–203.
42. D. J. Chung, Y. Ito and Y. Imanishi, An Insulin-Releasing Membrane System on the Basis of Oxidation Reaction of Glucose, *J. Controlled Release*, 1992, **18**(1), 45–53.
43. J. Heller, A. C. Chang, G. Rood and G. M. Grodsky, Release of insulin from pH-sensitive poly(ortho esters), *J. Controlled Release*, 1990, **13**(2–3), 295–302.
44. D. Hariharan and N. A. Peppas, Characterization, dynamic swelling behaviour and solute transport in cationic networks with applications to the development of swelling-controlled release systems, *Polymer*, 1996, **37**(1), 149–161.
45. M. T. Amende, D. Hariharan and N. A. Peppas, Factors Influencing Drug and Protein-Transport and Release from Ionic Hydrogels, *React. Polym.*, 1995, **25**(2–3), 127–137.

46. D. Hariharan, N. A. Peppas, R. Bettini and P. Colombo, Mathematical-Analysis of Drug-Delivery from Swellable Systems with Partial Physical Restrictions or Impermeable Coatings, *Int. J. Pharm.*, 1994, **112**(1), 47–54.
47. C. M. Hassan, F. J. Doyle and N. A. Peppas, Dynamic behavior of glucose-responsive poly(methacrylic acid-g-ethylene glycol) hydrogels, *Macromolecules*, 1997, **30**(20), 6166–6173.
48. C. M. Dorski, F. J. Doyle and N. A. Peppas, Preparation and characterization of glucose-sensitive P(MAA-g-EG) hydrogels, *Abstr. Pap. Am. Chem. Soc.*, 1997, **213**, 172-PMSE.
49. C. M. Dorski, F. J. Doyle and N. A. Peppas, Glucose-responsive, complexation hydrogels, *Abstr. Pap. Am. Chem. Soc.*, 1996, **211**, 417-POLY.
50. K. Podual, *Glucose-sensitive Cationic Hydrogels for Insulin Release*, Purdue University, 1998, p. xviii, 263.
51. K. Podual and N. A. Peppas, Relaxational behavior and swelling-pH master curves of poly[(diethylaminoethyl methacrylate)-graft-(ethylene glycol)] hydrogels, *Polym. Int.*, 2005, **54**(3), 581–593.
52. K. Podual, F. Doyle and N. A. Peppas, Modeling of water transport in and release from glucose-sensitive swelling-controlled release systems based on poly(diethylaminoethyl methacrylate-g-ethylene glycol), *Ind. Eng. Chem. Res.*, 2004, **43**(23), 7500–7512.
53. K. Podual, F. J. Doyle and N. A. Peppas, Dynamic behavior of glucose oxidase-containing microparticles of poly(ethylene glycol)-grafted cationic hydrogels in an environment of changing pH, *Biomaterials*, 2000, **21**(14), 1439–1450.
54. K. Podual, F. J. Doyle and N. A. Peppas, Glucose-sensitivity of glucose oxidase-containing cationic copolymer hydrogels having poly(ethylene glycol) grafts, *J. Controlled Release*, 2000, **67**(1), 9–17.
55. K. Podual, F. J. Doyle and N. A. Peppas, Preparation and dynamic response of cationic copolymer hydrogels containing glucose oxidase, *Polymer*, 2000, **41**(11), 3975–3983.
56. K. Podual, F. J. Doyle and N. A. Peppas, Release of insulin from glucose-sensitive hydrogels, *Abstr. Pap. Am. Chem. Soc.*, 1996, **211**, 221-PMSE.
57. M. Morishita, A. M. Lowman, K. Takayama, T. Nagai and N. A. Peppas, Elucidation of the mechanism of incorporation of insulin in controlled release systems based on complexation polymers, *J. Controlled Release*, 2002, **81**(1–2), 25–32.
58. A. M. Lowman and N. A. Peppas, Molecular analysis of interpolymer complexation in graft copolymer networks, *Polymer*, 2000, **41**(1), 73–80.
59. A. Heller, Electron-conducting redox hydrogels: design, characteristics and synthesis, *Curr. Opin. Chem. Biol.*, 2006, **10**(6), 664–672.
60. I. Lentacker, B. Geers, J. Demeester, S. C. De Smedt and N. N. Sanders, Design and evaluation of doxorubicin-containing microbubbles for ultrasound-triggered doxorubicin delivery: cytotoxicity and mechanisms involved, *Mol. Ther.*, 2009, **18**(1), 101–108.

61. J. Zhang, R. Srivastava and R. Misra, Core-shell magnetite nanoparticles surface encapsulated with smart stimuli-responsive polymer: synthesis, characterization, and LCST of viable drug-targeting delivery system, *Langmuir*, 2007, **23**(11), 6342–6351.
62. N. S. Satarkar and J. Z. Hilt, Magnetic hydrogel nanocomposites for remote controlled pulsatile drug release, *J. Controlled Release*, 2008, **130**(3), 246–251.
63. S. Sershen, N. Halas, and J. West. *Pulsatile release of insulin via photothermally modulated drug delivery. in Engineering in Medicine and Biology, 2002. 24th Annual Conference and the Annual Fall Meeting of the Biomedical Engineering Society EMBS/BMES Conference, 2002. Proceedings of the Second Joint*. 2002. IEEE.
64. K. W. Ferrara, Driving delivery vehicles with ultrasound, *Adv. Drug Delivery Rev.*, 2008, **60**(10), 1097–1102.
65. M. E. Caldorera-Moore, W. B. Liechty and N. A. Peppas, Responsive theranostic systems: integration of diagnostic imaging agents and responsive controlled release drug delivery carriers, *Acc. Chem. Res.*, 2011, **44**(10), 1061–1070.
66. A. Jemal, R. Siegel, E. Ward, Y. Hao, J. Xu, T. Murray and M. J. Thun, Cancer statistics, *Ca-Cancer J. Clin.*, 2008, **58**(2), 71–96.
67. D. M. Parkin, F. Bray, J. Ferlay and P. Pisani, Global cancer statistics, *Ca-Cancer J. Clin.*, 2005, **55**(2), 74–108.
68. J. T. McPhee, J. S. Hill, G. F. Whalen, M. Zayaruzny, D. E. Litwin, M. E. Sullivan, F. A. Anderson and J. F. Tseng, Perioperative mortality for pancreatectomy: a national perspective, *Ann. Surg.*, 2007, **246**(2), 246.
69. S. G. Nayfield, G. C. Bongiovanni, M. H. Alciati, R. A. Fischer and L. Bergner, Statutory requirements for disclosure of breast cancer treatment alternatives, *J. Natl. Cancer Inst.*, 1994, **86**(16), 1202–1208.
70. B. L. Fowble, L. J. Solin, D. J. Schultz and R. L. Goodman, Ten year results of conservative surgery and irradiation for stage I and II breast cancer, *Int. J. Radiat. Oncol., Biol., Phys.*, 1991, **21**(2), 269–277.
71. E. H. Romond, E. A. Perez, J. Bryant, V. J. Suman, C. E. Geyer Jr, N. E. Davidson, E. Tan-Chiu, S. Martino, S. Paik and P. A. Kaufman, Trastuzumab plus adjuvant chemotherapy for operable HER2-positive breast cancer, *N. Engl. J. Med.*, 2005, **353**(16), 1673–1684.
72. K. Miller, M. Wang, J. Gralow, M. Dickler, M. Cobleigh, E. A. Perez, T. Shenkier, D. Cella and N. E. Davidson, Paclitaxel plus bevacizumab versus paclitaxel alone for metastatic breast cancer, *N. Engl. J. Med.*, 2007, **357**(26), 2666–2676.
73. R. I. CTSU, Effects of chemotherapy and hormonal therapy for early breast cancer on recurrence and 15-year survival: an overview of the randomised trials, *Lancet*, 2005, **365**, 1687–1717.
74. O. Abe, R. Abe, K. Enomoto, K. Kikuchi, M. Baum and J. Houghton, Polychemotherapy for early breast cancer: an overview of the randomised trials, *Lancet*, 1998, **352**(9132), 930–942.

75. National Institutes of Health Consensus Development Panel, National Institutes of Health Consensus Development Conference statement: adjuvant therapy for breast cancer, November 1–3, 2000, *J. Natl. Cancer Inst.*, 2001, **93**(13), 979–989.
76. B. K. Edwards, A.-M. Noone, A. B. Mariotto, E. P. Simard, F. P. Boscoe, S. J. Henley, A. Jemal, H. Cho, R. N. Anderson, B. A. Kohler, C. R. Eheman and E. M. Ward, Annual Report to the Nation on the status of cancer, 1975-2010, featuring prevalence of comorbidity and impact on survival among persons with lung, colorectal, breast, or prostate cancer, *Cancer*, 2014, **120**(9), 1290–1314.
77. J. D. Byrne, T. Betancourt and L. Brannon-Peppas, Active targeting schemes for nanoparticle systems in cancer therapeutics, *Adv. Drug Delivery Rev.*, 2008, **60**(15), 1615–1626.
78. D. B. Longley, D. P. Harkin and P. G. Johnston, 5-fluorouracil: mechanisms of action and clinical strategies, *Nat. Rev. Cancer*, 2003, **3**(5), 330–338.
79. A. Gabizon, H. Shmeeda and Y. Barenholz, Pharmacokinetics of pegylated liposomal doxorubicin, *Clin. Pharmacokinet.*, 2003, **42**(5), 419–436.
80. M. L. Immordino, P. Brusa, S. Arpicco, B. Stella, F. Dosio and L. Cattel, Preparation, characterization, cytotoxicity and pharmacokinetics of liposomes containing docetaxel, *J. Controlled Release*, 2003, **91**(3), 417–429.
81. L. Harris, G. Batist, R. Belt, D. Rovira, R. Navari, N. Azarnia, L. Welles and E. Winer, Liposome-encapsulated doxorubicin compared with conventional doxorubicin in a randomized multicenter trial as first-line therapy of metastatic breast carcinoma, *Cancer*, 2002, **94**(1), 25–36.
82. H. S. Kim and I. W. Wainer, Simultaneous analysis of liposomal doxorubicin and doxorubicin using capillary electrophoresis and laser induced fluorescence, *J. Pharm. Biomed. Anal.*, 2010, **52**(3), 372–376.
83. D. Peer, J. M. Karp, S. Hong, O. C. Farokhzad, R. Margalit and R. Langer, Nanocarriers as an emerging platform for cancer therapy, *Nat. Nanotechnol.*, 2007, **2**(12), 751–760.
84. L. Yang, H. Mao, Y. A. Wang, Z. Cao, X. Peng, X. Wang, H. Duan, C. Ni, Q. Yuan and G. Adams, Single chain epidermal growth factor receptor antibody conjugated nanoparticles for in vivo tumor targeting and imaging, *Small*, 2009, **5**(2), 235–243.
85. T. R. Daniels, T. Delgado, G. Helguera and M. L. Penichet, The transferrin receptor part II: targeted delivery of therapeutic agents into cancer cells, *Clin. Immunol.*, 2006, **121**(2), 159–176.
86. J. Sudimack and R. J. Lee, Targeted drug delivery via the folate receptor, *Adv. Drug Delivery Rev.*, 2000, **41**(2), 147–162.
87. S. K. Sahoo and V. Labhasetwar, Enhanced antiproliferative activity of transferrin-conjugated paclitaxel-loaded nanoparticles is mediated via sustained intracellular drug retention, *Mol. Pharmaceutics*, 2005, **2**(5), 373–383.

88. D. W. Bartlett, H. Su, I. J. Hildebrandt, W. A. Weber and M. E. Davis, Impact of tumor-specific targeting on the biodistribution and efficacy of siRNA nanoparticles measured by multimodality in vivo imaging, *Proc. Natl. Acad. Sci.*, 2007, **104**(39), 15549–15554.
89. H. Shmeeda, D. Tzemach, L. Mak and A. Gabizon, Her2-targeted pegylated liposomal doxorubicin: Retention of target-specific binding and cytotoxicity after *in vivo* passage, *J. Controlled Release*, 2009, **136**(2), 155–160.
90. S. Y. Lee, M. Ferrari and P. Decuzzi, Shaping nano-/micro-particles for enhanced vascular interaction in laminar flows, *Nanotechnology*, 2009, **20**(49), 495101.
91. T. A. ElBayoumi and V. P. Torchilin, Tumor-targeted nanomedicines: enhanced antitumor efficacy in vivo of doxorubicin-loaded, long-circulating liposomes modified with cancer-specific monoclonal antibody, *Clin. Cancer Res.*, 2009, **15**(6), 1973–1980.
92. M. Hans and A. Lowman, Biodegradable nanoparticles for drug delivery and targeting, *Curr. Opin. Solid State Mater. Sci.*, 2002, **6**(4), 319–327.
93. R. Singh and J. W. Lillard Jr, Nanoparticle-based targeted drug delivery, *Exp. Mol. Pathol.*, 2009, **86**(3), 215–223.
94. L. Brannon-Peppas and J. O. Blanchette, Nanoparticle and targeted systems for cancer therapy, *Adv. Drug Delivery Rev.*, 2004, **56**(11), 1649–1659.
95. H. Maeda, J. Wu, T. Sawa, Y. Matsumura and K. Hori, Tumor vascular permeability and the EPR effect in macromolecular therapeutics: a review, *J. Controlled Release*, 2000, **65**(1), 271–284.
96. C. Vauthier and K. Bouchemal, Methods for the preparation and manufacture of polymeric nanoparticles, *Pharm. Res.*, 2009, **26**(5), 1025–1058.
97. K. Kataoka, A. Harada and Y. Nagasaki, Block copolymer micelles for drug delivery: design, characterization and biological significance, *Adv. Drug Delivery Rev.*, 2001, **47**(1), 113–131.
98. A. Raffin Pohlmann, V. Weiss, O. Mertins, N. Pesce da Silveira and S.l. Stanisçuaski Guterres, Spray-dried indomethacin-loaded polyester nanocapsules and nanospheres: development, stability evaluation and nanostructure models, *Eur. J. Pharm. Sci.*, 2002, **16**(4), 305–312.
99. T. L. Rogers, A. C. Nelsen, M. Sarkari, T. J. Young, K. P. Johnston and R. O. Williams Iii, Enhanced aqueous dissolution of a poorly water soluble drug by novel particle engineering technology: spray-freezing into liquid with atmospheric freeze-drying, *Pharm. Res.*, 2003, **20**(3), 485–493.
100. J. Hu, K. P. Johnston and R. O. Williams III, Spray freezing into liquid (SFL) particle engineering technology to enhance dissolution of poorly water soluble drugs: organic solvent versus organic/aqueous co-solvent systems, *Eur. J. Pharm. Sci.*, 2003, **20**(3), 295–303.
101. I. Brigger, C. Dubernet and P. Couvreur, Nanoparticles in cancer therapy and diagnosis, *Adv. Drug Delivery Rev.*, 2002, **54**(5), 631–651.

102. S. Ganta, H. Devalapally, A. Shahiwala and M. Amiji, A review of stimuli-responsive nanocarriers for drug and gene delivery, *J. Controlled Release*, 2008, **126**(3), 187–204.
103. Y. Qiu and K. Park, Environment-sensitive hydrogels for drug delivery, *Adv. Drug Delivery Rev.*, 2001, **53**(3), 321–339.
104. N. A. Peppas, P. Bures, W. Leobandung and H. Ichikawa, Hydrogels in pharmaceutical formulations, *Eur. J. Pharm. Biopharm.*, 2000, **50**(1), 27–46.
105. L. Brannon-Peppas and N. A. Peppas, Equilibrium swelling behavior of pH-sensitive hydrogels, *Chem. Eng. Sci.*, 1991, **46**(3), 715–722.
106. T. Okano, Molecular Design of Temperature-responsive Polymers as Intelligent Materials, in *Responsive Gels: Volume Transitions II*, Springer, 1993, pp. 179–197.
107. R. Pelton, Temperature-sensitive aqueous microgels, *Adv. Colloid Interface Sci.*, 2000, **85**(1), 1–33.
108. C. F. Lee, C. C. Lin and W. Y. Chiu, Thermosensitive and control release behavior of poly (N-isopropylacrylamide-co-acrylic acid) latex particles, *J. Polym. Sci., Part A: Polym. Chem.*, 2008, **46**(17), 5734–5741.
109. L. C. Glangchai, M. Caldorera-Moore, L. Shi and K. Roy, Nanoimprint lithography based fabrication of shape-specific, enzymatically-triggered smart nanoparticles, *J. Controlled Release*, 2008, **125**(3), 263–272.
110. K. Raemdonck, J. Demeester and S. De Smedt, Advanced nanogel engineering for drug delivery, *Soft Matter*, 2009, **5**(4), 707–715.
111. K. Landfester, The generation of nanoparticles in miniemulsions, *Adv. Mater.*, 2001, **13**(10), 765–768.
112. M. Murray and M. Snowden, The preparation, characterisation and applications of colloidal microgels, *Adv. Colloid Interface Sci.*, 1995, **54**, 73–91.
113. M. Andersson and S. L. Maunu, Structural studies of poly (N-isopropylacrylamide) microgels: Effect of SDS surfactant concentration in the microgel synthesis, *J. Polym. Sci., Part B: Polym. Phys.*, 2006, **44**(23), 3305–3314.
114. R. Langer, Drug delivery and targeting, *Nature*, 1998, **392**(6679 Suppl), 5–10.
115. R. Langer and N. A. Peppas, Advances in biomaterials, drug delivery, and bionanotechnology, *AIChE J.*, 2003, **49**(12), 2990–3006.
116. P. Bouillot and B. Vincent, A comparison of the swelling behaviour of copolymer and interpenetrating network microgel particles, *Colloid Polym. Sci.*, 2000, **278**(1), 74–79.

CHAPTER 3

Chemomechanical Gels—Actuators and Sensors

HANS-JÖRG SCHNEIDER

FR Organische Chemie der Universität des Saarlandes, D 66041, Saarbrücken, Germany
Email: h-j.schneider@mx.uni-saarland.de; ch12hs@rz.uni-sb.de

3.1 Introduction and Overview

After the first report on chemically-induced changes of macromolecular materials in 1950,[1] many researchers have led the way to the development of these, mostly in the form of gels.[2] Gel-like materials have found widespread applications in the design of smart materials, as described in several chapters of the present book. The most important application of smart hydrogels concerns drug delivery, on which Chapter 2 by Peppas *et al.* or Chapter 5 by Schalley *et al.*, and several recent reviews,[3,4] provide an excellent overview, which allows us to restrict our discussion to principles and some typical examples. In most cases, the response of these gels towards external stimuli is based on swelling, and also on sol–gel transition,[5] whereby, for example, a pre-loaded drug can be set free from inside the particle into the surrounding medium. The present chapter deals with size changes of a gel triggered by the surrounding medium, without losing the character of a not-dissolved material. The size changes, which may reach several hundred percent, can be contraction or swelling, always based on interactions of the solute with recognition groups present in the polymeric gel network. The underlying gels are usually hydrophilic polymers, either

RSC Smart Materials No. 14
Chemoresponsive Materials: Stimulation by Chemical and Biological Signals
Edited by Hans-Jörg Schneider

Published by the Royal Society of Chemistry, www.rsc.org

with functions already present in the chains, or with functions implemented specifically for reaction with the solutes. These functions can be incorporated either in polymers that exist already or before polymerization in corresponding monomers. Modern synthetic chemistry allows the introduction of a large variety of functional entities, which are designed for molecular recognition with practically all kinds of chemical substances in the surroundings. The wide experience now available in the field of supramolecular chemistry, also in aqueous solution, allows the design of suitable host functions. Macrocycle-containing supramolecular gels make use of the very effective recognition with such entities,[5] using, besides crown ethers *etc.*, in particular cyclodextrin- or cucurbituril-based hosts.[6] The intriguing property of chemomechanical systems is that they combine sensor and actuator in the same particle.[7] They can be made in all dimensions, which enables further enhancement of sensitivity and speed of reaction (see Section 3.2).

Generally, a large variety of polymers can be made responsive to electric, thermal, light, magnetic, and solvent/medium stimuli.[8] There are already many systems in which the sensitivity of supramolecular complexation in gels towards external stimuli has been explored.[9] Mechanotransduction phenomena in the narrow sense play an important role in living systems,[10] and, for example, in mechanosensitive polymer brushes (see also Chapter 16).[11] Probe tips used in atomic force microscopy (AFM) allow chemomechanical nanografting on surfaces.[12] Intermolecular interactions with chemically-loaded tips used in chemical force microscopy (CFM) lead directly to microscopic mechanical motions.[13] Microcapsules are capable of selective release of encapsulated chemicals.[14] Nanocomposite hydrogels based on combinations with carbon nanotubes, graphene, metal oxides *etc.* hold particular promise if equipped with stimuli-responsive elements.[15] The formation and properties of hydrogels from low molecular gelators are aptly described in Chapter 12 by Adams *et al.* Click chemistry offers a promising new approach for hydrogels, using, for example, azide–alkene cycloaddition with azide functionalized polyethyleneglycol and alkene-derivatised peptides.[16]

3.2 Mechanisms Involved in Size Changes of Chemomechanical Hydrogels[17,18]

Ionic groups in gels lead to swelling mostly due to a Donnan-type equilibrium with the counterions of the fixed charges inside the gel, and as shown later, in the context of pH-induced changes by salts within the gel. High electrolyte concentrations lead to higher osmotic pressure within the hydrogel, which leads to gel swelling. Repulsions between charged groups also contribute to loosening the network and thus swelling. The degree of swelling of a ionic gel is proportional to the number of ionic side groups per polymer chain in the gel.[19,20]

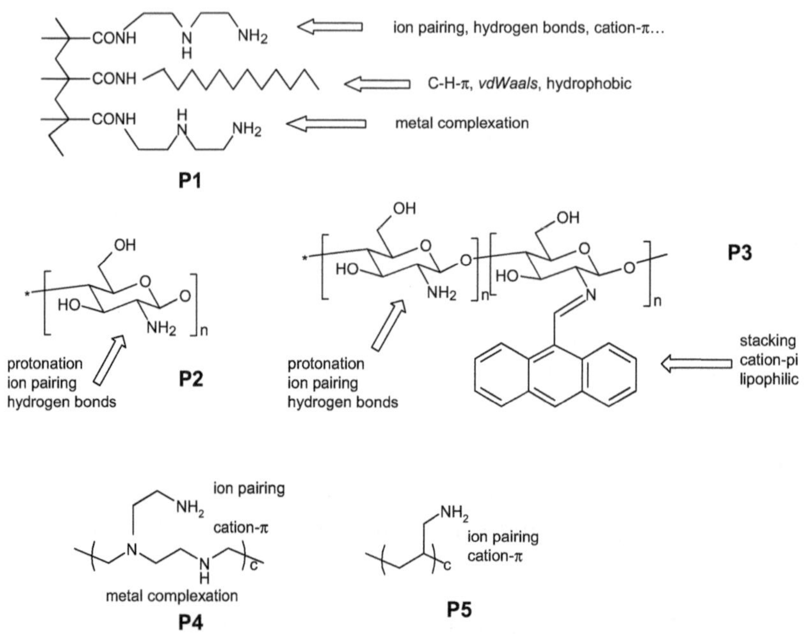

Scheme 3.1 Structural elements in the majority of chemomechanical hydrogels discussed in this chapter, with possible non-covalent interactions. **P1**: polymethyl(methyl)acrylic derivative (contains also free—unreacted—carboxylate groups); **P2**: chitosan; **P3**: chitosan-anthryl derivative; **P4**: polyethyleneimine; **P5**: polyallylamine.

In most cases, size changes in chemomechanical hydrogels are related to non-covalent interactions between the functional groups of the polymer and the solute. Scheme 3.1 shows the structural host elements on which most gel polymers in the following discussion are based. Fast covalent reactions between interacting groups are discussed in Section 3.8 for the recognition of carbohydrates with boronic acid derivatives. Generally it has been found that size changes induced in hydrogels by a large range of solutes are invariably based on changes of the water content.[21] The uptake of water molecules counterbalances the osmotic pressure difference between the outside and inside of the gel. As illustrated in Figure 3.1, the water uptake largely determines the volume of the particles of polymer **P1**, with only small contributions by the solute itself.[22] The water content can, for example, with polymer **P1** increase from about 75% up to 98%, triggered either by pH change, or by solutes such as $Cu(OAc)_2$ or, for example, AMP (adenosine monophosphate). Similar observations were made in the cases of contraction instead of swelling, for example by non-covalent crosslinking (see Section 3.7).

As illustrated in Figure 3.2, there is no indication of distinct stepwise phase changes, other than those often assumed.[23] Instead one observes a

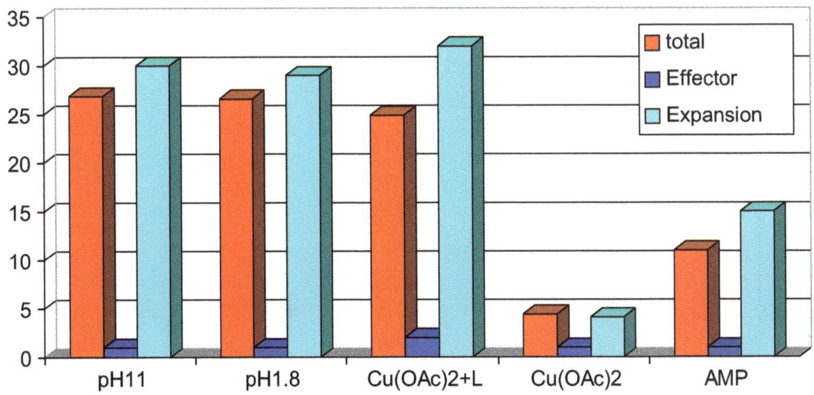

Figure 3.1 Weight increase (red, scaled per mg) of polymer gel **P1** compared to expansion V (green, in % volume); effector weight blue.[22a]
With permission from Wiley/VCH.

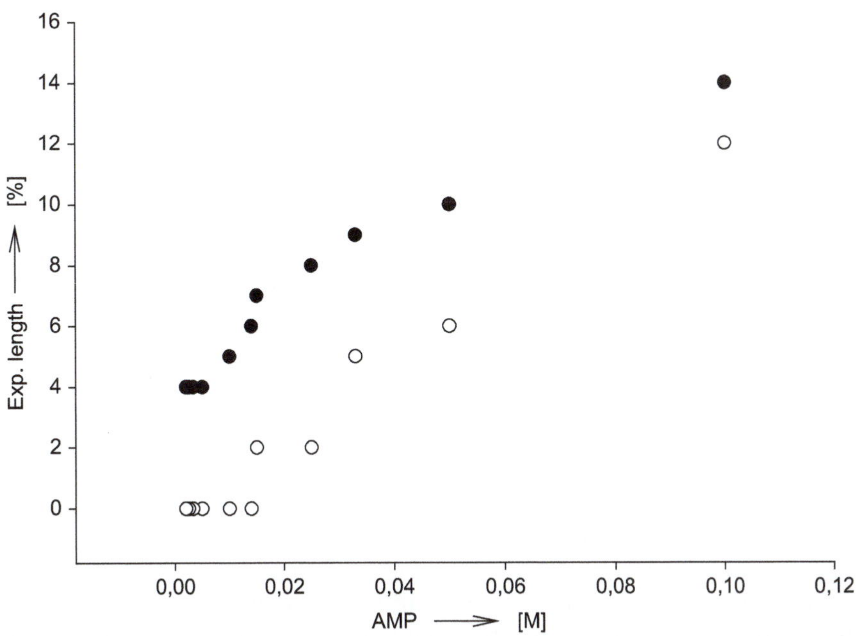

Figure 3.2 Expansion (length) as a function of AMP concentration; polymer **P1**, lower trace (○), in the absence of buffer; upper trace (●), in the presence of 0.02 M NaH_2PO_4 buffer;[22a] with permission from Wiley/VCH.

continuous size increase with solute concentration, depending also on the buffer used.[22] There is a small induction phase, as the solute first accumulates on the surface before it diffuses into the gel particle and starts swelling.

Similarly, the time dependence of size increase shows a continuous curve; as expected there is a faster swelling with increasing ratio *S*–*V* of

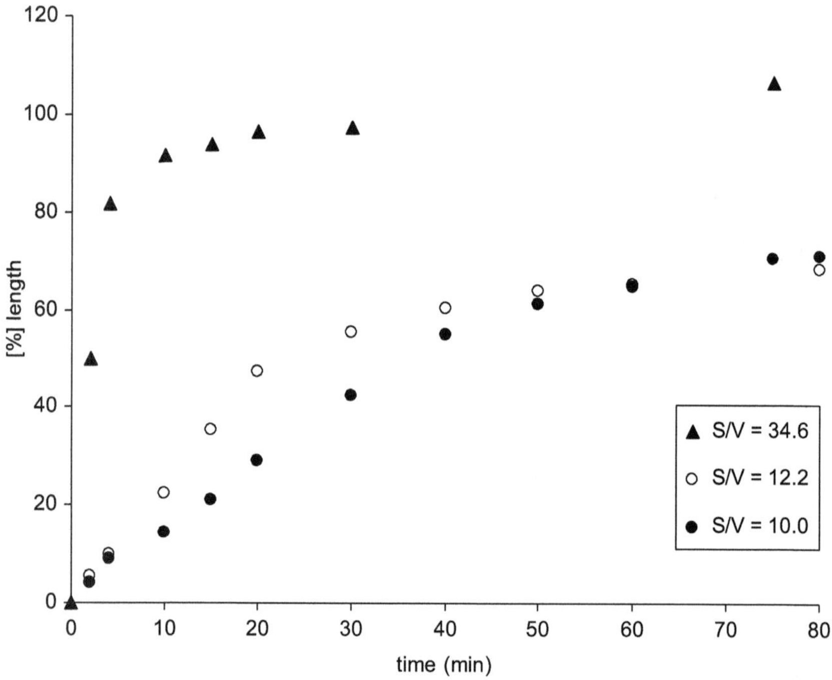

Figure 3.3 Velocity dependence on the surface to volume ratio *S–V*, measured with chitosan gel particle elongation induced by 50 mM L-histidine. Approximate half-life time $t_{1/2}$ for 50% of the maximum expansion: $t_{1/2} = 42$, 32 and 3 min for $S\text{–}V = 10.0$, 12.2 and 34.6, respectively.
From ref. 24, with permission from Wiley/VCH.

surface to volume of the gel particles (Figure 3.3).[24] Earlier it was shown that gel swelling is proportional to the inverse of the collective diffusion coefficient, and to the square of the gel particle radius.[25] The kinetics of size changes and the spectroscopically measured absorptions or desorptions of the effector (guest) molecules followed first order equations.[22a] Expansions with a chitosan gel film size of about $5 \times 2 \times 0.4$ mm showed a half-life time of $t_{1/2} = (3.5 \pm 2)$ min, rather independent of the effector nature, with the exception of slowly diffusing proteins. Typically, effector absorption and desorption are about twice as fast as swelling or contraction.

The observed sensitivity is a function of the gel particle size,[22a] as demonstrated in Figure 3.4.[26] This is due a general compartmentalization principle:[27] the smaller the particle volume the smaller the number of effector molecules needed to occupy all available binding sites inside the receptor particle; this very useful effect is limited only by a sufficient affinity between recognition sites and effector molecules, which must secure the binding of the maximum number of effector molecules to the available sites.

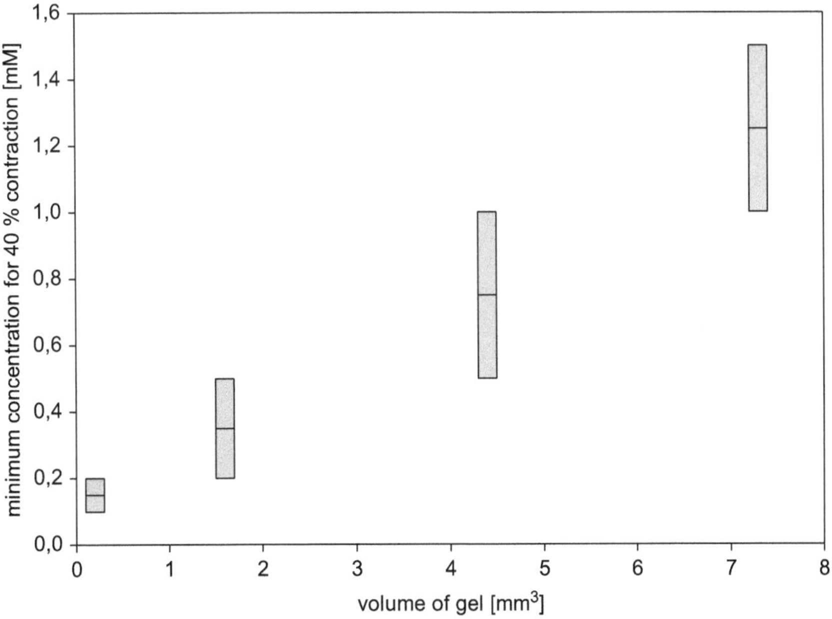

Figure 3.4 Sensitivity increase with decreasing particle size. Sensitivity of 40% contraction (in one-dimension) induced by D-O-dibenzoyl tartrate as a function of the gel particle size.[26]
From ref. 26. Copyright Wiley-VCH Verlag GmbH & Co. KGaA. Reproduced with permission.

3.3 Examples of Actuators

Chemically-stimulated hydrogels can be used, for example, in microfluidic systems,[28] as shown in Figures 3.5 and 3.6 with valves that open or close as a function of the fluid stream pH.[29] Miniaturization, as discussed earlier, allowed response times of around only 10 seconds.

In an intriguing experiment, oxidation (by Ce^{4+}) was used as a stimulus for a macroscopically measurable association between two gel particles, for example one being an anionic styrenesulfonate gel, the other one a ferrocene (FC) gel, which upon oxidation becomes anionic. In aqueous solution the two gel particles, in presence of Ce^{4+}, move together to form aggregates (Figure 3.7). In a different experiment, a cyclodextrin (CD)-containing gel was brought together with FC gel, exhibiting a strong association only in the absence of Ce^{4+}, because the oxidized cationic ferrocene gel does not bind to cyclodextrin. The strength of adhesion increased with the number binding partners in the particles.[30] If all three gels are brought together, they form trimeric aggregates, at one side held together by a hydrophobic/dispersive interaction between the ferrocene and cyclodextrin, at the other side by ion pairing.

A macroscopically visible metal-controlled adhesion was reported on the basis of a polyacrylamide hydrogel modified with both β-cyclodextrin (βCD)

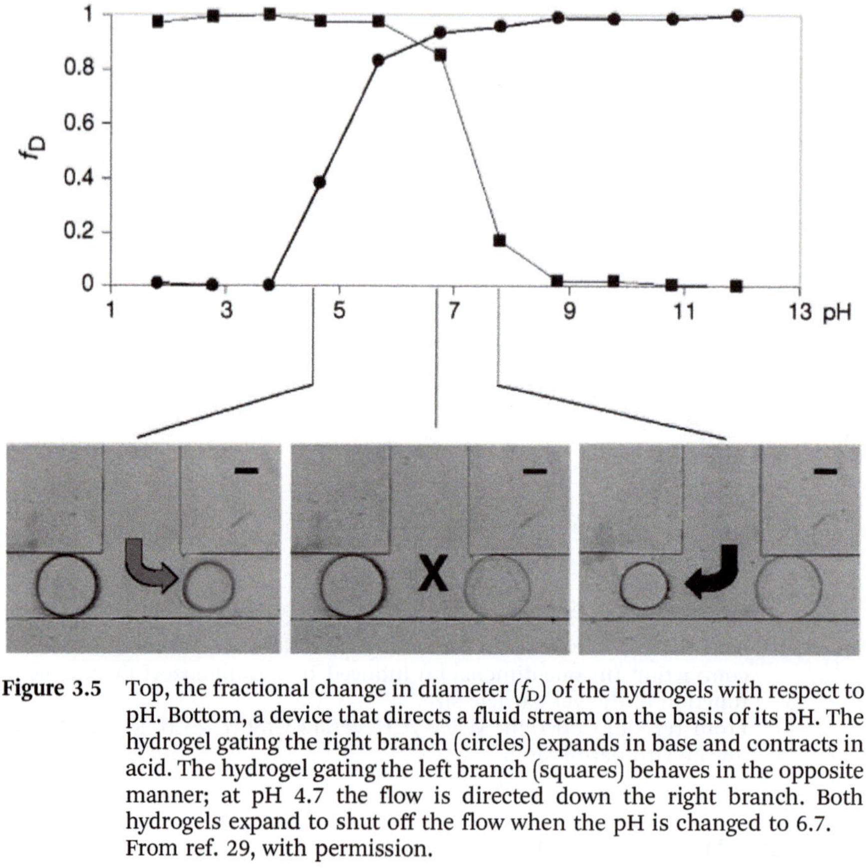

Figure 3.5 Top, the fractional change in diameter (f_D) of the hydrogels with respect to pH. Bottom, a device that directs a fluid stream on the basis of its pH. The hydrogel gating the right branch (circles) expands in base and contracts in acid. The hydrogel gating the left branch (squares) behaves in the opposite manner; at pH 4.7 the flow is directed down the right branch. Both hydrogels expand to shut off the flow when the pH is changed to 6.7. From ref. 29, with permission.

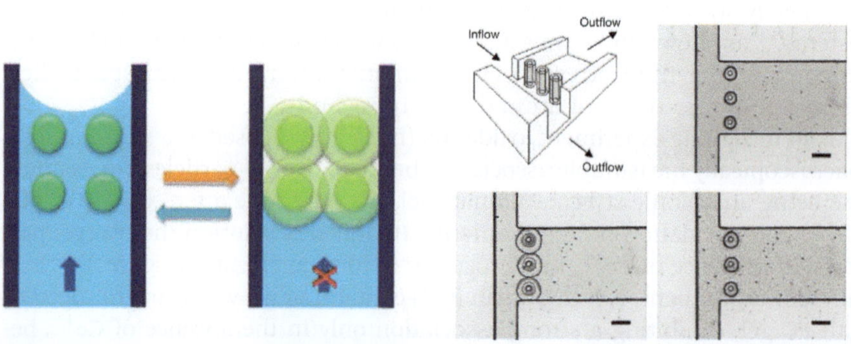

Figure 3.6 Volumetric control for microfluidics with an illustration of a typical set-up. From ref. 28 and 29, with permission.

moieties and 2,2′-bipyridyl moieties; the two moieties bind or crosslink together within the gel (Figure 3.8). In the presence of metal ions, such as Zn^{2+} or Cu^{2+}, the bipyridyl residue leaves the CD cavity and makes this unit free for binding to another gel, which contains *tert*-butyl groups as good binders to the CD group. In the presence of 0.1 M $CuCl_2$, the measured

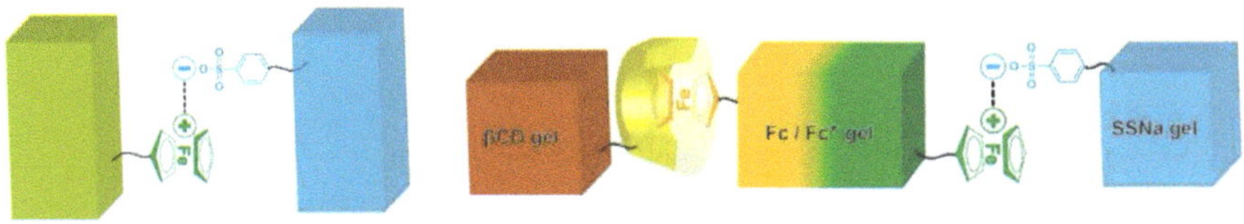

Figure 3.7 Macroscopic aggregates between gels containing either cyclodextrin, ferrocene or benzenesulfonate; for explanations, see text.
From ref. 30, with permission from Wiley/VCH.

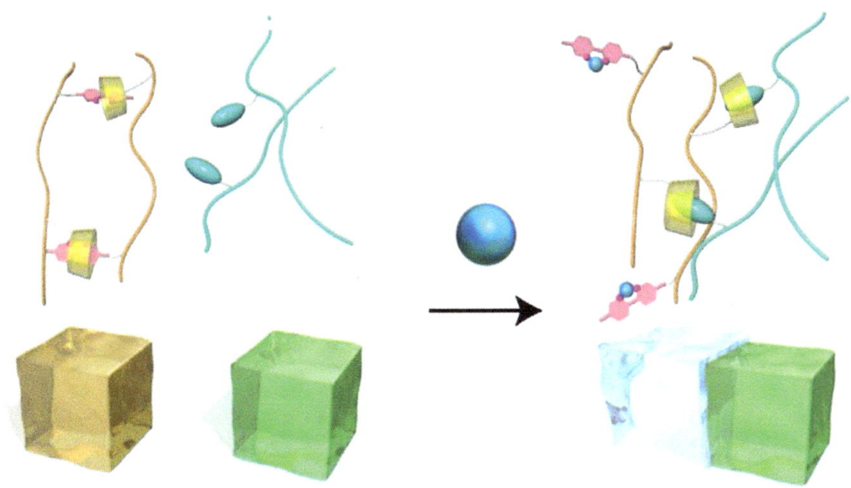

Figure 3.8 A gel particle bearing both β-cyclodextrin (CD) and bipyridyl moieties reacts with a particle bearing *tert*-butyl groups; in the presence of a metal ion, the CD is set free and associates with the *tert*-butyl gel.
Adapted from ref. 31.

adhesion strength between the particles was as high as 1000 ± 200 Pa, and was reversed by adding known strong chelators such as EDTA.[31]

A redox-triggered artificial muscle was described, in which a gel incorporating both CD and ferrocene (FC) units expands upon oxidation by Ce^{4+}, as the FC^{+} cation no longer binds into the CD cavity (Figure 3.9). The resulting movement is reversible; a weight attached to a gel strip was moved up and down in a redox circle.[32] A strong movement of gel particle could be demonstrated with a chitosan hydrogel filling a small tube with pores allowing contact with the surrounding solution.[33] Due to the water-like small compressibility of the hydrogel, a 200 g weight could be lifted five-fold by swelling induced by acetic acid; in pure water, chitosan contains only about 50% water, in 50 mM acetic acid, the water content increases to 96%.[34] The restriction within the tube also provides a simple practical solution to the problem of obtaining a unidirectional movement with molecular motors (Figure 3.10).

Hydrogels can provide a promising way to transform the chirality of molecules to a macroscopic entity,[35] including chiral sol–gel materials.[36] Chitosan is a natural source of chiral gels, which allows a direct way to translate enantiomeric recognition into macroscopic motions. Figure 3.11

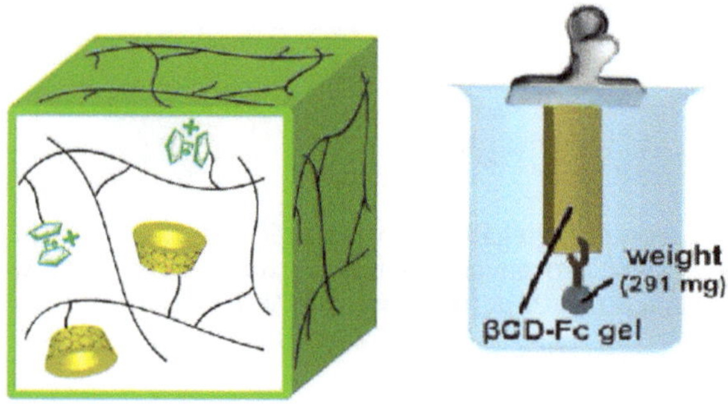

Figure 3.9 A redox-triggered artificial muscle with a gel incorporating both CD and ferrocene FC units; see text, adapted from ref. 32, with permission from Wiley/VCH.

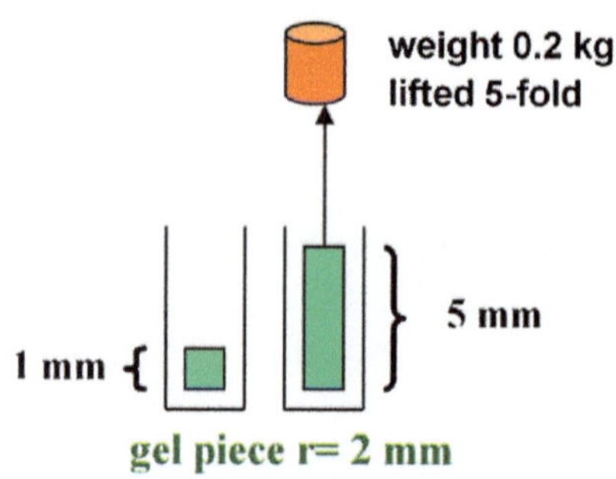

Figure 3.10 Mechanical motion of a chitosan gel; expansion triggered by acid. Adapted from ref. 33, with permission from RSC.

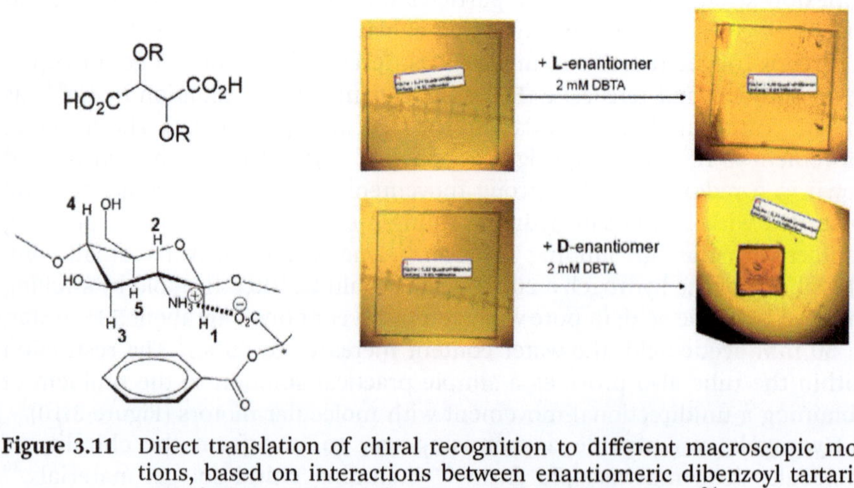

Figure 3.11 Direct translation of chiral recognition to different macroscopic motions, based on interactions between enantiomeric dibenzoyl tartaric acids (DBTA, R = COPhe) and chitosan gel film pieces.[34]

shows how the interaction of chiral dibenzoyl tartaric acids with a chitosan gel leads only with the D-enantiomer to a large contraction due to non-covalent crosslinking between the analyte carboxylate and the chitosan amino groups, whereas the L-enantiomer has a very small effect. Without benzoyl substituents, there is no significant discrimination; MAS-NMR spectra show a highly ordered supramolecular complex within the gel, with the phenyl ring undergoing a C–H–π interaction (with the axial protons 1, 3 and 5), and the carboxylate ion pairing between the analyte and the polymer backbone.[34]

3.4 pH-Effects on Gel Size Changes

Hydrogels responding to pH changes[17] are of interest for control of microfluidics (see Figure 3.6) and in particular for controlled drug release, which can occur as a result of squeezing out the drug upon shrinking of the gel particle.[37] pH-sensitive hydrogels can be effective for tissue-specific targeting, as pH values are different in each tissue and organs. Chapter 2 of this book by Peppas *et al.* contains many illustrative examples of such systems, which are of high interest also for oral drug administration. A typical example is a recent application of poly(methacrylic)-grafted ethyleneglycol nanoparticles to release doxorubicin locally in the colon for the treatment of colon cancers.[38] As observed with other gels, the pH of the surrounding medium has a distinct effect on the chemomechanical hydrogels if those bear any ionisable groups. If only one kind of ionisable function is present, such as in polyallylamine **P3**, one observes one deflection point in the pH-profile around the p*K* of the polymer.[39,40]

If both anionic and cationic groups are present, the pH profile becomes symmetric, as seen with the gel **P1**, which also contains free carboxylic groups (Figure 3.12).[22a] The increasing content of ionic groups

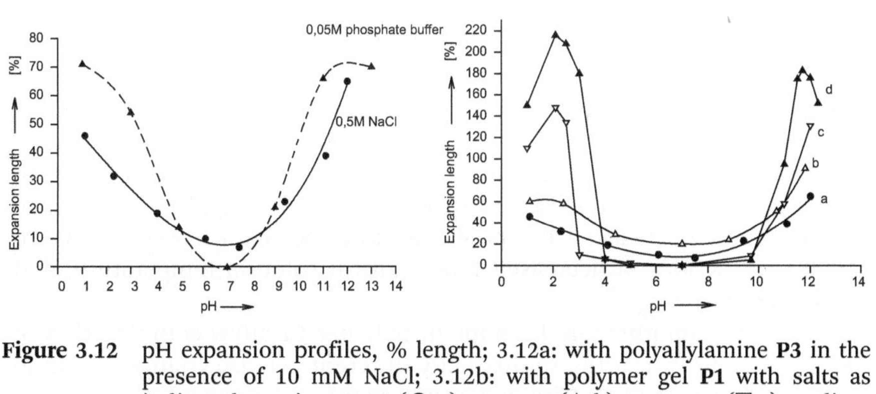

Figure 3.12 pH expansion profiles, % length; 3.12a: with polyallylamine **P3** in the presence of 10 mM NaCl; 3.12b: with polymer gel **P1** with salts as indicated; 5c: in 0.5 M (●-a), 0.05 M (△-b), 0.025 M (▽-c), sodium chloride, respectively, and in water with very dilute HCl or NaOH (▼-d). From ref. 22*a*. With permission from Wiley/VCH.

below and above pH 7 leads to the import of more water for solvation of the ions, as measured gravimetrically (see Figure 3.1), and thus the observed gel volume increases. However, measurements with different salt concentrations illustrate that a large part of the size changes are due to a change in the ionic strength that accompanies the addition of acids or bases (Figure 3.12b), in line with known electrolyte effects on swelling.[19]

Size changes in polyelectrolyte gels are not only a function of the applied pH, but are strongly moderated by the anion of the acid used. Anions, also those from added salts such as buffers, lead to crosslinking by ion pairing with protonated sites in gels. Related effects have been studied, in particular with polyallylamine **P5** gels.[41] The crosslinking effect becomes particularly visible with polydentate anions: if chloride inside the gel is replaced by added acetate, one observes 17% contraction at the same pH; with phosphate and with α,ω-dicarboxylates, one observes up to a 69% contraction (% in length).[42]

3.5 Metal Ion-triggered Size Changes

Different concentrations of alkali salts can lead to rather non-specific and smaller gel volume changes at constant pH, as discussed earlier, due to different degrees of non-covalent crosslinking. Generally, the uptake of ions increases the osmotic pressure inside the gel due to the formation of a Donnan potential, and thus leads to swelling. More selective responses are possible with polymers bearing crown ether moieties, which were obtained by polymerization of acryloylamidobenzo-18-crown-6.[43,44] The hydrogel films were incorporated in polymerized crystalline colloidal arrays (PCCA), which are crystalline colloidal arrays (CCA) of spherical polystyrene colloids. Optical detection is possible based on the volume increase, which changes the periodicity of the CCA, resulting in a shift in diffraction wavelength. The response of the PCCA particles towards Pb^{2+} was almost linear from concentrations as low as 10 μM up to 1 mM (Figure 3.13).[43] A new photonic crystal hydrogel with pendant crown ether groups also binds lead ions; the presence of ions in the initially electroneutral gel results in swelling, and shifting of diffraction wavelength as a function of the lead concentration.[44] Incorporation of crown ether 15C5 into a hydrogel was used for drug release stimulated by potassium ions—these were found to lead to shrinking of the gel and release of the loaded drug.[45] In another application, crown-containing microcapsules from poly *N*-isopropylacrylamide were shown to exhibit, with K ions, a decreased lower critical solution temperature, and swelling or shrinking.[46]

The ethylenediamine (ene) groups in polymer **P1** allow complexation of several metal ions, leading to distinctly different gel size changes of up to 390 vol% (Figure 3.14).[47] The maximum change is observed at a concentration where essentially all ene groups of a given gel piece are occupied, as established by spectroscopic metal content measurements. After the

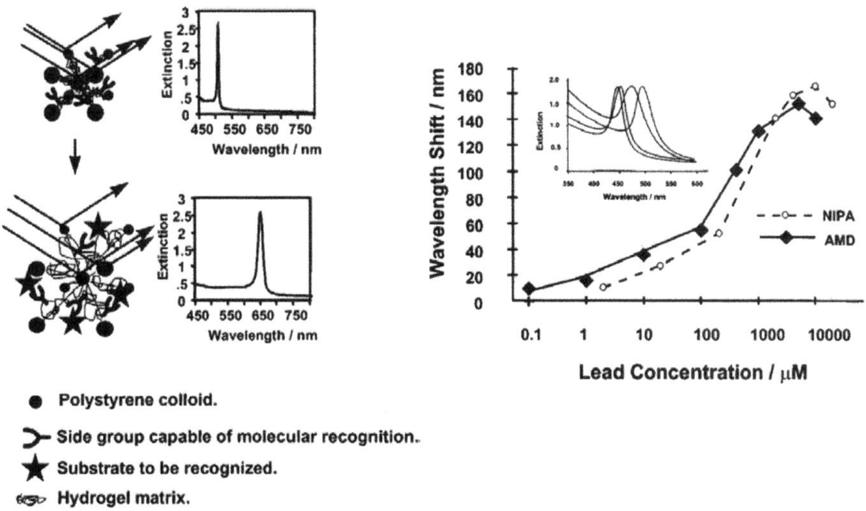

Figure 3.13 (a) Schematic description of a polymerized crystalline colloidal array (PCCA) sensor; the CCA Bragg diffraction monitors the hydrogel volume change induced by interaction of the polymer side group with a substrate; (b) shift in wavelength as a function of Pb^{2+} concentration with two crown ether-bearing polymers; the inset shows the extinction spectra of one sensor for 0.1, 10 and 100 μM Pb^{2+}.
From ref. 43.

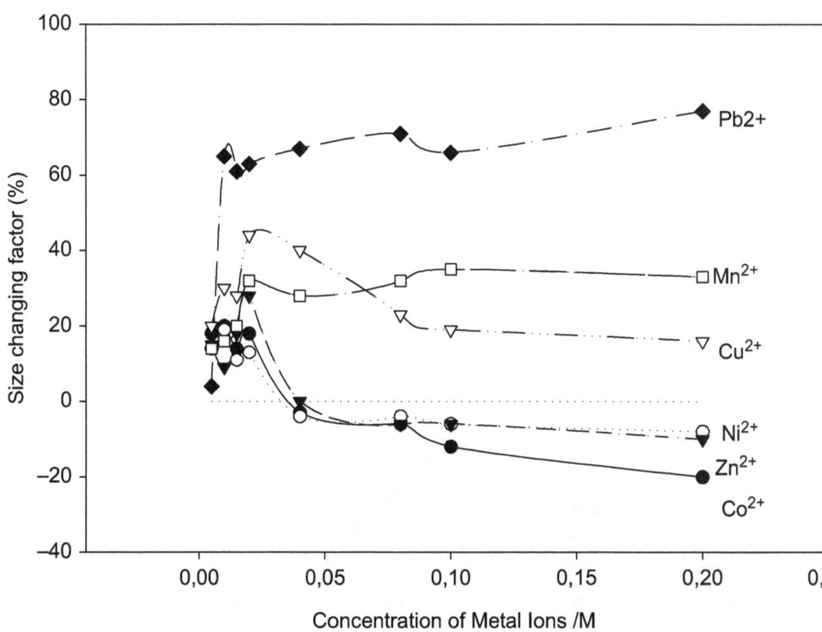

Figure 3.14 Size changes (% length increase or decrease) of the polymer film **P1** in the presence of various metal ions.
From ref. 47, with permission from RSC.

expansion reaches a maximum, one observes a partial decrease, which may be due to weaker bound additional ions, occupying other centers in the gel **P1**. Metal ions incorporated into hydrogels are of interest also in view of their capacity to bind peptides, as discussed in Section 3.7. They can also interact with proteins, which has been used for several purposes. Hydrogels containing glycidyl methacrylate–iminodiacetic acid as a metal-ion-chelating ligand were shown to be capable of decreasing protein release rates from hydrophilic tissue engineering scaffolds, which are often too fast.[48] Metal-binding hydrogels can be used as sorbents for protein separation.[49]

3.6 Gel Size Changes with Organic Solutes

The data in Scheme 3.2, taken from studies with gel polymer **P1**, indicate that swelling increases as expected with the size of the solutes. The absence of any measurable expansions by saturated frameworks shows that the cation–π effect of the charged nitrogen atoms of **P1** on aromatic residues of the solute is a dominating factor in binding. Nucleotides lead to similar gel expansions, with cooperativity between pH and different nucleobases.

Amino acids bearing different side chains can be distinguished by different gel size expansions, increasing with the size of the side group, if an anthryl substituent in the chosen chitosan derivative **P3** secures a lipophilic or stacking interaction (Scheme 3.3).[50]

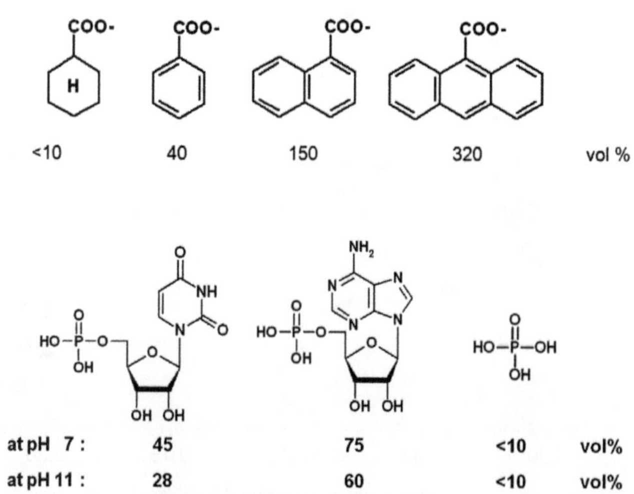

Scheme 3.2 Selectivity between different organic effector compounds with polymer **P1**; [0.02–0.05 M phosphate buffer; the effects at pH 11 are corrected for the difference between pH 7 and pH 11 alone (390 vol%)]. From ref. 22*a*.

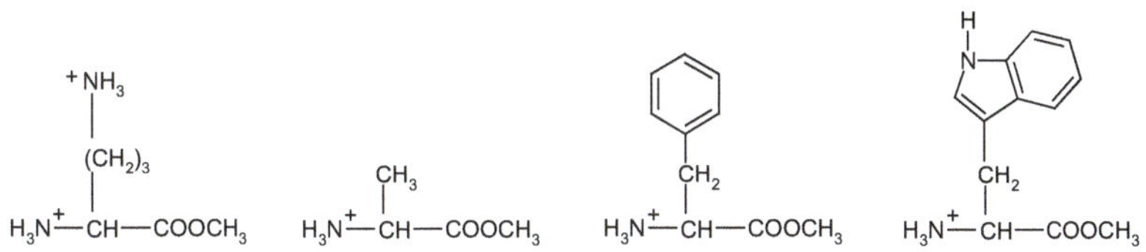

Scheme 3.3 Amino acid-selective size increase of chitosan gel **P3** bearing an anthryl substituent.[50]

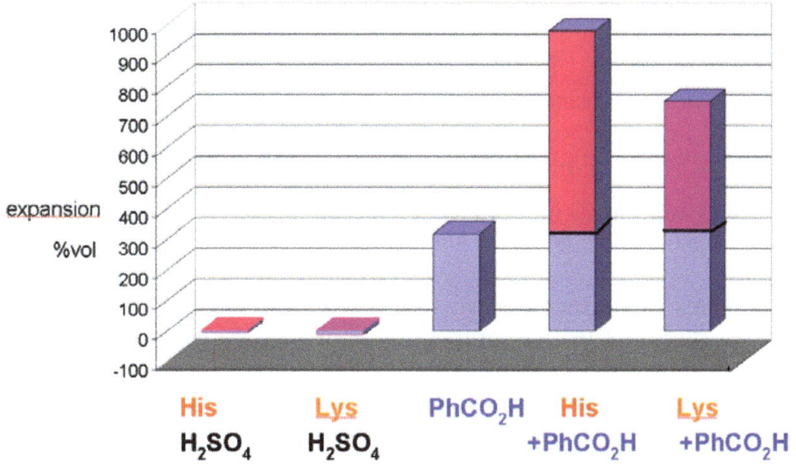

Figure 3.15 Cooperativity between benzoate and histidine or lysine in size expansions of chitosan gel; in the absence of benzoate (blue), no swelling is observed; from ref. 24, with permission from Wiley/VCH.

3.7 Cooperativity/Logical Gate Functions

Cooperativity between different solute (effector) compounds occurs more often with gels than in solution. Not only are most size changes, such as those induced by nucleotides, pH-dependent (Scheme 3.2), they also depend on the ionic strength or salts, thus representing simple logical AND gate functions, with either positive or negative cooperativity.[51] A very strong cooperativity was found in size expansions of chitosan gel between benzoate and basic amino acids such as histidine or lysine; in the absence of benzoate, no swelling was observed[24] (Figure 3.15). A logical gate system based on a crown ether-functionalized bis(urea) gelator with gel–sol transition controlled by three different molecular recognition events has been reported recently.[52]

Peptides as a target or effector for inducing gel motion are another example of strong cooperativity: with the gel **P1**, peptides alone show no measurable effect. The presence of metal ions such as Cu^{2+} or Zn^{2+},

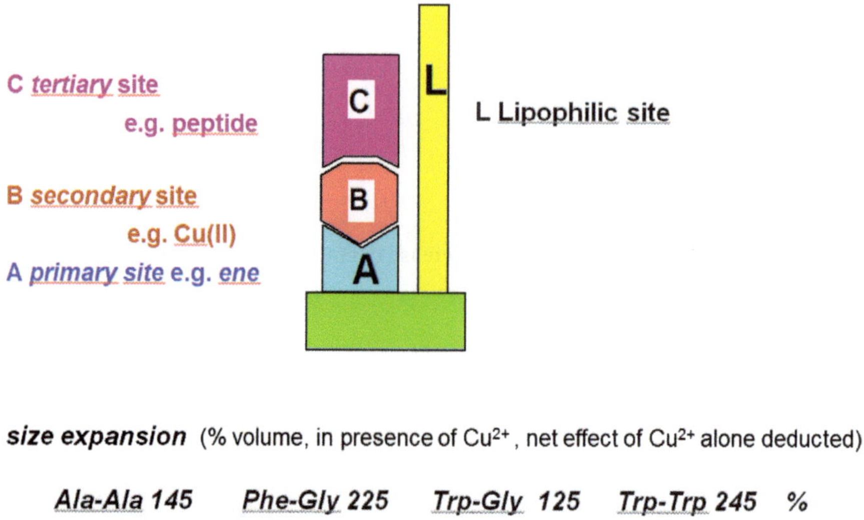

Figure 3.16 Cooperativity in gel **P1** size expansions; ternary complexes between the gel ene site, metal ion and peptide; see text.

however, leads to a sizeable expansion beyond the effect of the metal ions themselves (Figure 3.16). These metal ions, which are bound with the ene-units of polymer **P1** as the primary site, are known to complex peptides effectively; the lipophilic chain **L** in gel **P1** can interact differently with the side chains of the amino acids.[53]

3.8 Glucose as Effector/Gels as Sensor and Actuator

Carbohydrates are, in aqueous media, notoriously weak binders to artificial host compounds, due to their strong hydrogen bonding to bulk water. For that reason, the well-known[54] fast reaction of boronic acid with glucose has been widely used to develop chemomechanical hydrogels; these allow not only sensing, but eventually the release of drugs such as insulin. Obviously, such studies hold great promise for the development of minimally-invasive sensing and delivery devices for the treatment of diabetes, in addition to the biocompatibility, non-toxicity, non-immunogenicity, and chemical stability of such boronic acid-containing gels. Another intriguing way towards glucose-triggered insulin release relies on gels equipped with glucose oxidase and the ensuing enzyme-produced pH change.[55] pH-sensitive hydrogels have already been described for the oral administration of insulin.[56]

After the early paper by Kataoka *et al.*,[57] numerous studies appeared in the literature, of which we cite here only some recent papers and discuss principles involved in the reactions with gels containing boronic acids.[58–62] The essential mechanism is a strong and pH-dependent reversible covalent reaction, leading to enhanced accumulation of negative charge on the boron

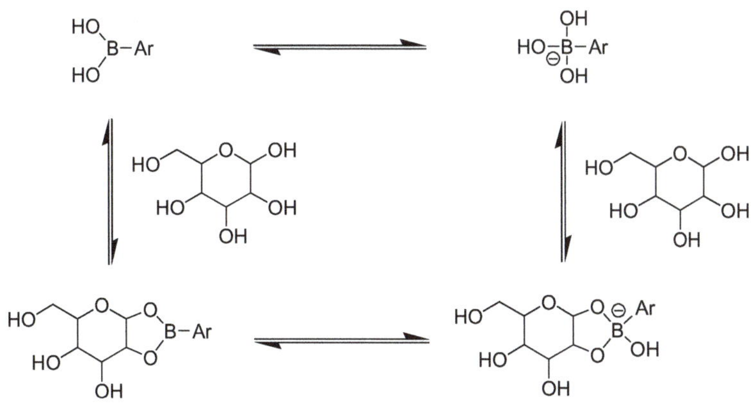

Scheme 3.4 Reaction pathways for boronic ester formation.

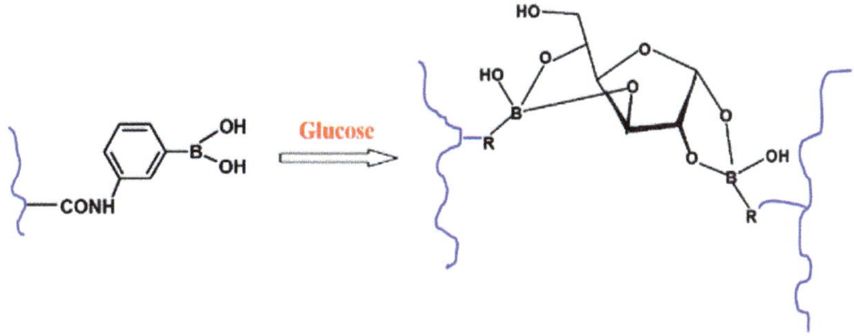

Scheme 3.5 Crosslinking between polymer chains (blue) by boronic ester formation with glucose in the furanose form.

(Scheme 3.4). The generation of that anion leads, as usual, with ions in hydrogels to swelling, which has been used early for the release of bound insulin.[57]

Recent studies[63] using polymerized crystalline colloidal (PCC) arrays as reporters of the volumetric change[64] have shown that electron-withdrawing substituents, which reduce the pK_a of the boronic acid, either increase or decrease the magnitude of response, and increase the response rate.

Contraction instead of swelling was observed with gels possessing a higher density of phenylboronic acid, which can lead to crosslinking according to Scheme 3.5.[65–67] Microgels based on *N*-alkylacrylamide derivatives, crosslinkers such as methylenebis(acrylamide), and phenylboronic acid show swelling with most biogenic saccharides, and only with glucose shrinking at a pH above the pK_a of the boronic acid derivative (Figure 3.17).[66]

Systems based on shrinking have the advantage of better preventing premature drug leakage, which can be a problem with particles relying on drug release by swelling.[68] Figure 3.18 shows how hydrogels might be used as chemical valves, incorporated in a silicon-based microdevice.[69]

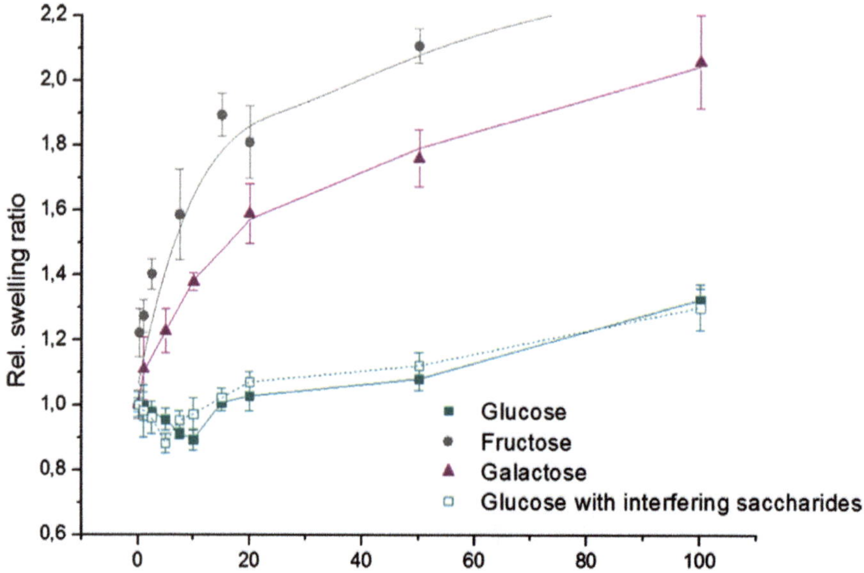

Figure 3.17 Relative swelling ratio with different sugars, and the response to glucose in the presence of fructose (0.05 mM) and galactose (0.05 mM).[66]
Reprinted with permission. Copyright 2001 American Chemical Society.

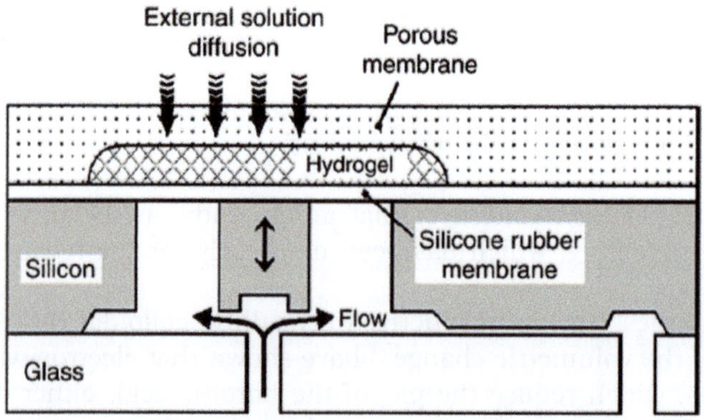

Figure 3.18 Hydrogel-based microvalve; when glucose diffuses through the membrane, the gel size changes and moves an embossment which opens or closes a microchannel.
From ref. 69 with permission.

In a different, and more simple, strategy, a hydrogel is used as a chemical corkscrew; interactions with saccharides and a gel obtained from the reaction of poly(methyl methacrylate), diethylenetriamine, alkyl amines, and 3-aminophenylboronic acid lead to a sizeable contraction.[70] Figure 3.19 illustrates how a capillary with the gel as a stopper is opened if immersed in 5 mM of glucose solution, releasing a coloured solution that serves as a

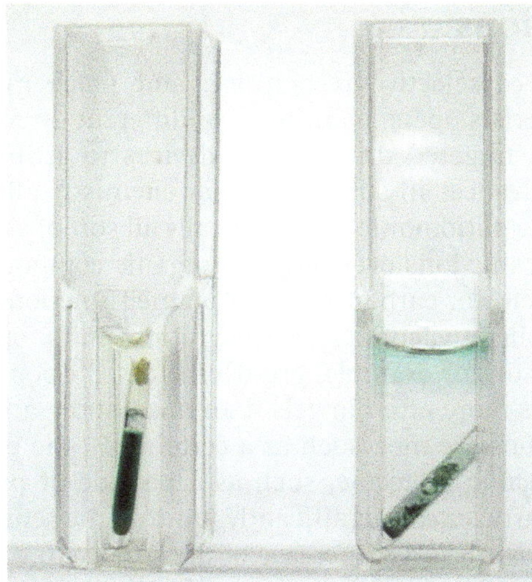

Figure 3.19 Illustration of possible drug release (blue coloured solution) from a capillary immersed in 5 mM of glucose solution; contraction of the hydrogel opens the vessel; from ref. 70 with permission from Wiley/VCH.

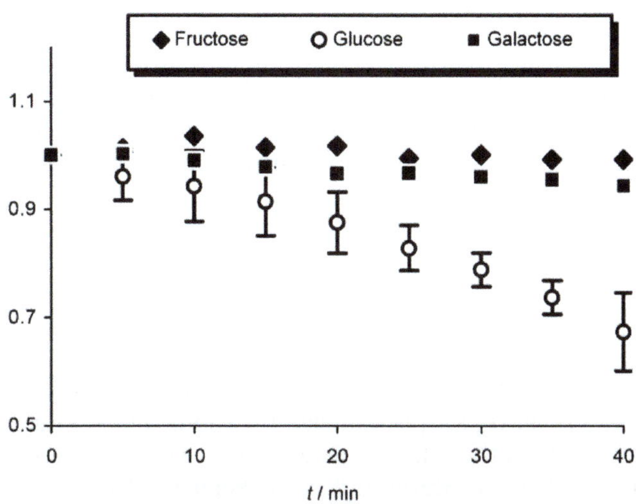

Figure 3.20 Reaction of a hydrogel with common blood sugars under physiological conditions (5 mM concentration); error bars indicate the reproducibility of the contraction with three different batches.
From ref. 70 with permission from Wiley/VCH.

model for insulin. The hydrogel exhibits a relatively fast reaction only with glucose, different from the two other common blood sugars fructose and galactose (Figure 3.20). In human blood plasma, 5 mM of added glucose leads to a sharp contraction within 5–10 minutes.

3.9 Conclusions

The combination of selective recognition and mechanical macroscopic motion in gel materials opens many ways in different directions. They span from new sensors, targeted drug release devices to artificial actuators or muscles. Recent advances in supramolecular chemistry allow the design of highly selective recognition units for practically all sort of analytes or targets, ranging from inorganic ions over biogenic organic compounds to peptides, proteins, nucleic acids or carbohydrates. Aptamer-functionalized hydrogels, containing short oligonucleotides as recognition units, open new ways to control the release of, for example, growth factors.[71] Cooperativity between two effectors occurs more often in gels than in solution, and leads to action only if two conditions are met, such as a certain pH, the presence of metal ions and of an organic molecule, such as a peptide or protein. Miniaturization of the gel particles can significantly improve the sensitivity, as long as the affinity is sufficiently large. Increase of the surface to volume ratio, such as with thin films, will speed up the response. The principles of smart systems for chemoselective sensors and actuators within one single unit have been largely explored, but their implementation for promising real life applications remains a great challenge.

Acknowledgements

I thank Nino Lomadze, Barbara Palm, Tianjun Liu and Kazuaki Kato for their contributions in my own group, the Alexander von Humboldt foundation and the Fonds der Chemischen Industrie for financial support, and Professor Alexander Sidorenko for careful reading of the manuscript and advice.

References

1. W. Kuhn, B. Hargitay, A. Katchalsky and H. Eisenberg, *Nature*, 1950, **165**, 514.
2. See e.g., Y. Osada, *Adv. Mater.*, 1991, **3**, 107–108; Y. Osada and J.-P. Gong, *Adv. Mater.*, 1998, **10**, 827; D. Kaneko, J. P. Gong and Y. Osada, *J. Mater. Chem.*, 2002, **12**, 2169; H. Brondsted and J. Kopecek, *ACS Symp. Ser.*, 1992, **480**, 285; C. S. Brazel and N. A. Peppas, *Macromolecules*, 1995, **28**, 8016; G. M. Eichenbaum, P. F. Kiser, D. Shah, S. A. Simon and D. Needham, *Macromolecules*, 1999, **32**, 8996; P. Gupta, K. Vermani and S. Garg, *Drug Discov. Today*, 2002, 7, 569.
3. N. A. Peppas, J. Z. Hilt, A. Khademhosseini and R. Langer, *Adv. Mater.*, 2006, **18**, 1345; L. A. Sharpe, A. M. Daily, S. D. Horava and N. A. Peppas, *Expert Opinion Drug Delivery*, 2014, **11**, 901; T. R. Hoare and D. S. Kohane, *Polymer*, 2008, **49**, 1993; C. J. Kearney and D. J. Mooney, *Nature. Mater.*, 2013, **12**, 1004; Y. Qiu and K. Park, *Adv. Drug Delivery Rev.*, 2012, **64**, 49.

4. A recent book with many examples: *Smart Materials for Drug Delivery: 2 Volumes*, ed. C. Alvarez-Lorenzo and A. Concheiro, RSC, Cambridge, UK, 2013.
5. Z. Qi and C. A. Schalley, *Acc. Chem. Res.*, 2014, **47**, 2222.
6. H. Yang, B. Yuan, X. Zhang and O. A. Scherman, *Acc. Chem. Res.*, 2014, **47**, 2106.
7. A. Richter, G. Paschew, S. Klatt, J. Lienig, K.-F. Arndt and H.-J. P. Adler, *Sensors*, 2008, **8**, 56; H. J. Schneider, K. Kato and R. M. Strongin, *Sensors*, 2007, 7, 1578.
8. H. Meng and J. L. Hu, *J. Intell. Mater. Syst. Struct.*, 2010, **21**, 859.
9. K. Osada, R. J. Christie and K. Kataoka, *J. R. Soc., Interface*, 2009, **6**, S325.
10. A. W. Orr, B. P. Helmke, B. R. Blackman and M. A. Schwartz, *Dev. Cell*, 2006, **10**, 11.
11. J. Bünsow, T. S. Kelby and W. T. S. Huck, *Acc. Chem. Res.*, 2010, **43**, 466.
12. M. V. Lee, M. T. Hoffman, K. Barnett, J. M. Geiss, V. S. Smentkowski, M. R. Linford and R. C. Davis, *J. Nanosci. Nanotechnol.*, 2006, **6**, 1639; S. A. Ruiz and C. S. Chen, *Soft Matter*, 2007, **3**, 168.
13. Z. Huang and R. Boulatov, *Pure Appl.Chem.*, 2010, **82**, 931; A. Noy, *Surface Interface Anal.*, 2006, **38**, 1429.
14. L. Y. Chu, A. S. Utada, R. K. Shah, J. W. Kim and D. A. Weitz, *Angew. Chem., Int. Ed.*, 2007, **46**, 8970.
15. A. K. Gaharwar, N. A. Peppas and A. Khademhosseini, *Biotechnol. Bioeng.*, 2014, **111**, 441, DOI: 10.1002/bit.25160; S. Mura, J. Nicolas and P. Couvreur, *Nat. Mater.*, 2013, **12**, 991.
16. Y. Jianga, J. Chena, C. Denga, E. J. Suuronenb and Z. Zhong, *Biomaterials*, 2014, **35**, 4969.
17. T. Tanaka, D. Fillmore, S.-T. Sun, I. Nishio, G. Swislow and A. Shah, *Phys. Rev. Lett.*, 1980, **45**, 1636; E. Oral and N. A. Peppas, *J. Biomed. Mater. Res. Part A*, 2004, **68A**, 439, (involving also glucose imprinting); K. S. Soppimath, A. R. Kulkarni and T. M. Aminabhavi, *J. Controlled Release*, 2001, **75**, 331; D. N. Robinson and N. A. Peppas, *Macromolecules*, 2002, **35**, 3668; M. Kaneko, M. Fukui, W. Takashima and K. Kaneto, *Synth. Met.*, 1997, **84**, 795; K. Suzuki, T. Yumura, Y. Tanaka and M. Akashi, *J. Bioact. Compat. Polym.*, 2001, **16**, 409; D. Kuckling, H. J. P. Adler, K. F. Arndt, L. Ling and W. D. Habicher, *Macromol. Symp.*, 1999, **145**, 65; B. A. Firestone and R. A. Siegel, *J. Biomater. Sci.-Polym. Ed.*, 1994, **5**, 433; E. Roux, M. Lafleur, E. Lataste, P. Moreau and J. C. Leroux, *Biomacromolecules*, 2003, **4**, 240, and references cited therein.
18. N. A. Peppas, *Hydrogels in Medicine and Pharmacy*, CRC, Boca Raton, 1987; *Polymeric and Self Assembled Hydrogels*, ed. X. J. Loh and O. Scherman, The Royal Society of Chemistry, Cambridge, UK, 2013; *Hydrogels: Synthesis, Characterization and Applications*, ed. V. F. Camara and L. J. Ferreira, Nova Science Pub. Inc., 2012; *Hydrogel Sensors and Actuators: Engineering and Technology*, ed. G. Gerlach, Springer, Berlin, 2012; *Reflexive Polymers and Hydrogels: Understanding and Designing Fast Responsive Polymeric Systems*, ed. N. Yui, R. J. Mrsny and J. Randall, CRC

Inc., Boca Raton, 2004; *Molecular Gels, Materials with Self-Assembled Fibrillar Networks*, ed. R. G. Weiss and T. Terech, Springer, Dordrecht, 2006; *Functional Molecular Gels*, ed. B. Escuder and J. F. Miravet, Royal Society of Chemistry, 2013.
19. S. Hirotsu, Y. Hirokawa and T. Tanaka, *J. Chem. Phys.*, 1987, **87**, 1392.
20. For recent theoretical work on hydrogels see M. A. Arndt and G. Sadowski, *J. Phys. Chem. B*, 2014, **118**, 10534.
21. H.-J. Schneider and R. M. Strongin, *Acc. Chem. Res.*, 2009, **42**, 1489.
22. (a) H. J. Schneider, T. J. Liu and N. Lomadze, *Eur. J. Org. Chem.*, 2006, 677; (b) H.-J. Schneider, T. Liu and N. Lomadze, *Angew. Chem., Int. Ed.*, 2003, **42**, 3544.
23. K. Ito, J. Chuang, C. Alvarez-Lorenzo, T. Watanabe, N. Ando and A. Y. Grosberg, *Prog. Polym. Sci.*, 2003, **28**, 1489; *Hydrogel Sensors Actuators: Engineering and Technology*, ed. G. Gerlach and K. F. Arndt, Springer, Berlin, 2008; A. Richter, G. Paschew, S. Klatt, J. Lienig, K. F. Arndt and H. J. P. Adler, *Sensors*, 2008, **8**, 561; S. Chaterji, I. K. Kwon and K. Park, *Progr. Polymer Sci.*, 2007, **32**, 1083, Reviews: (a) K. Ito, J. Chuang, C. Alvarez-Lorenzo, T. Watanabe, N. Ando and A. Y. Grosberg, *Prog. Polym. Sci.*, 2003, **28**, 1489–1515; (b) *Responsive Gels. Volume Transitions II*, ed. K. Dusek, Springer, Berlin, 1993.
24. K. Kato and H.-J. Schneider, *Eur. J. Org. Chem.*, 2009, 1042.
25. T. Tanaka and D. J. Fillmore, *J. Chem. Phys.*, 1979, **70**, 1214, see also T. Tanaka, E. Sato, Y. Hirokawa, S. Hirotsu and J. Peetermans, *Phys. Rev. Lett.*, 1985, **55**, 2455.
26. H. J. Schneider and K. Kato, *Angew. Chem., Int. Ed.*, 2007, **46**, 2694.
27. H.-J. Schneider, *Angew. Chem., Int. Ed.*, 2009, **48**, 3924.
28. D. H. Kang, S. M. Kim, B. Lee, H. Yoon and K.-J. Suh, *Analyst*, 2013, **138**, 6230.
29. D. J. Beebe, J. S. Moore, J. M. Bauer, Q. Yu, R. H. Liu, C. Devadoss and B.-H. Jo, *Nature*, 2000, **404**, 588.
30. M. Nakahata, Y. Takashima and A. Harada, *Angew. Chem., Int. Ed.*, 2014, **53**, 3617.
31. T. Nakamura, Y. Takashima, A. Hashidzume, H. Yamaguchi and A. Harada, *Nat. Commun.*, 2014, **5**, 4622.
32. M. Nakahata, Y. Takashima, A. Hashidzume and A. Harada, *Angew. Chem., Int. Ed.*, 2013, **52**, 5731.
33. K. Kato and H.-J. Schneider, *J. Mater. Chem.*, 2009, **19**, 569.
34. H.-J. Schneider and K. Kato, *Angew. Chem., Int. Ed.*, 2007, **46**, 2694.
35. D. K. Smith, *Chem. Soc. Rev.*, 2009, **38**, 684.
36. S. Marx and D. Avnir, *Acc. Chem. Res.*, 2007, **40**, 768.
37. Y. Qiu and K. Park, *Adv. Drug Delivery Rev.*, 2001, **53**, 321.
38. C. A. Schoener, H. N. Hutson and N. A. Peppas, *J. Biomed. Mater. Res., Part A*, 2013, **101A**, 2229.
39. K. Kato and H.-J. Schneider, *Eur. J. Org. Chem.*, 2008, 1378.
40. S. Yamada, H. Yamazaki and A. Kinoshita, *J. Appl. Polym. Sci.*, 1997, **9**, 79; see also M. S. Shin, S. J. Kim, S. J. Park, Y. H. Lee and S. I. Kim,

J. Appl. Polym. Sci., 2002, **86**, 498; K. Itano, J. Choi and M. F. Rubner, *Macromolecules*, 2005, **38**, 3450.
41. S. Yamada, H. Yamazaki and A. Kinoshita, *J. Appl. Polym. Sci.*, 1997, **9**, 79; M. S. Shin, S. J. Kim, S. J. Park, Y. H. Lee and S. I. Kim, *J. Appl. Polym. Sci.*, 2002, **86**, 498; H. Muta, M. Miwa and M. Satoh, *Polymer*, 2001, **42**, 6313; R. G. V. Rama, T. Konishi and N. Ise, *Macromolecules*, 1999, **32**, 7582.
42. K. Kato and H. J. Schneider, *Langmuir*, 2007, **23**, 10741.
43. J. H. Holtz, J. S. W. Holtz, C. H. Munro and S. A. Asher, *Anal. Chem.*, 1998, **70**, 780.
44. See also M. M. Ward Muscatello and S. A. Asher, *Adv. Funct. Mater.*, 2008, **18**, 1186.
45. P. Mi, X. J. Ju, R. Xie, H. G. Wu, J. Ma and L. Y. Chu, *Polymer*, 2010, **51**, 1648.
46. S. W. Pi, X. J. Ju, H. G. Wu, R. Xie and L. Y. Chu, *J. Colloid Interface Sci.*, 2010, **349**, 512, and references cited therein.
47. H. J. Schneider and T. J. Liu, *Chem. Commun.*, 2004, 100.
48. C. C. Lin and A. T. Metters, *J. Biomed. Mater. Res., Part A*, 2007, **83A**, 954.
49. N. Pekel, B. Salih and O. Guven, *Macromol. Symp.*, 2001, **169**, 329.
50. N. Lomadze and H. J. Schneider, *Tetrahedron*, 2005, **61**, 8694.
51. H. J. Schneider, N. Lomadze, L. Tianjun and B. Palm, *Adv. Mater.*, 2004, **16**, 613.
52. Z. Qi, N. L. Traulsen, P. Malo de Molina, C. Schlaich, M. Gradzielski and C. A. Schalley, *Chem. Sci.*, 2012, **3**, 2073.
53. N. Lomadze and H.-J. Schneider, *Tetrahedron Lett.*, 2005, **46**, 751.
54. J. Boeseken, *Adv. Carbohydr. Chem.*, 1949, **4**, 189.
55. S. R. Marek and N. A. Peppas, *AIChE J.*, 2013, **59**, 3578, and references cited therein.
56. A. M. Lowman, M. Morishita, M. Kajita, T. Nagai and N. A. Peppas, *J. Pharm. Sciences*, 1999, **88**, 933.
57. K. Kataoka, H. Miyazaki, M. Bunya, T. Okano and Y. Sakurai, *J. Am. Chem. Soc.*, 1998, **120**, 12694.
58. A. Kim, S. Mujumdar and R. Siegel, *Chemosensors*, 2014, **2**, 1.
59. A. Matsumoto, K. Yamamoto, M. Yoshida, K. Kataoka, T. Aoyagi and Y. Miyahara, *Chem. Commun.*, 2010, **46**, 2203.
60. S. A. Asher, V. L. Alexeev, A. V. Goponenko, A. C. Sharma, I. K. Lednev, C. S. Wilcox and D. N. Finegold, *J. Am. Chem. Soc.*, 2003, **125**, 3322; V. Alexeev, S. Das, D. Finegold and S. Asher, *Clin. Chem.*, 2004, **50**, 2353.
61. S.-H. Cho, P. Tathireddy, P. Rieth and J. Magda, *J. Appl. Polym. Sci.*, 2014, **131**, 40667.
62. F. Xue, Z. Meng, F. Wang, Q. Wang, M. Xue and Z. Xu, *J. Mater. Chem. A*, 2014, **2**, 9559.
63. C. Zhang, M. D. Losego and P. V. Braun, *Chem. Mater.*, 2013, **25**, 3239.
64. M. M. W. Muscatello, L. E. Stunja and S. A. Asher, *Anal. Chem.*, 2009, **81**, 4978.

65. A. Horgan, A. Marshall, S. Kew, K. Dean, C. Creasey and S. Kabilan, *Biosen. Bioelectron.*, 2006, **21**, 1838.
66. C. Ancla, V. Lapeyre, I. Gosse, B. Catargi and V. Ravaine, *Langmuir*, 2011, **27**, 12693.
67. X. Pan, X. Yang and C. R. Lowe, *J. Mol. Recognit.*, 2008, **21**, 205.
68. See e.g., W.-H. Chiang, W.-C. Huang, M.-Y. Shen, C.-H. Wang, Y. F. Huang, S.-C. Lin, C.-S. Chern and H.-C. Chiu, *PLoS One*, 2014, **9**, e92268.
69. R. A. Siegel, Y. Gu, M. Lei, A. Baldi, E. E. Nuxoll and B. Ziaie, *J. Controlled Release*, 2010, **141**, 303.
70. G. K. Samoei, W. Wang, J. O. Escobedo, X. Xu, H.-J. Schneider, R. L. Cook and R. M. Strongin, *Angew. Chem., Int. Ed.*, 2006, **45**, 5319.
71. M. R. Battig, Y. Huang, N. Chen and Y. Wang, *Biomaterials*, 2014, **35**, 8040; M. R. Battig, B. Soontornworajit and Y. Wang, *J. Am. Chem. Soc.*, 2012, **134**, 12410; B. Soontornworajit, J. Zhou, M. T. Shaw, T. H. Fan and Wang, *Chem. Commun.*, 2010, **46**, 1857.

CHAPTER 4

Chemosensitive Organogels

ARNAB DAWN,[a,†] BAPPADITYA ROY[a] AND SEIJI SHINKAI*[a,b]

[a] Institute for Advanced Study, Kyushu University, 744 Moto-oka, Nishi-ku, Fukuoka 819-0395, Japan; [b] Nanotechnology Laboratory, Institute of Systems, Information Technologies and Nanotechnologies (ISIT), 4-1 Kyudai-Shinmachi, Nishi-ku, Fukuoka 819-0388, Japan
*Email: shinkai_center@mail.cstm.kyushu-u.ac.jp

4.1 Introduction

Owing to their distinct appearance and physico–chemical properties that set them apart from conventional and well-defined solid, liquid and gas phase materials, gel systems have remained a topic of interest, wonder and scientific investigation since their discovery.[1] Their inherent softness comprising solid-like and liquid-like (in scientific terms 'visco-elastic') properties prompted gel systems to act as transformable functional materials. Invariably a gel system is composed of a smaller fraction of solid substrate (the gelator) and a higher fraction of liquid (the solvent). Although, defining a 'gel' still remains a matter of debate[2] the classification of gel systems is rather unambiguous and solely depends on the basis of differentiation as summarised in Figure 4.1.

Here, we will focus on some special properties of organogel systems involving low molecular-weight gelators (LMGs, with a molar mass ≤3000).[3,4] Different properties of organogels stem mainly from two parts: firstly, the chemical nature of the substrate (the LMG) and secondly, from their

†Present address: Department of Chemistry, Durham University, Science Site, South Road, Durham DH1 3LE, UK.

RSC Smart Materials No. 14
Chemoresponsive Materials: Stimulation by Chemical and Biological Signals
Edited by Hans-Jörg Schneider

Published by the Royal Society of Chemistry, www.rsc.org

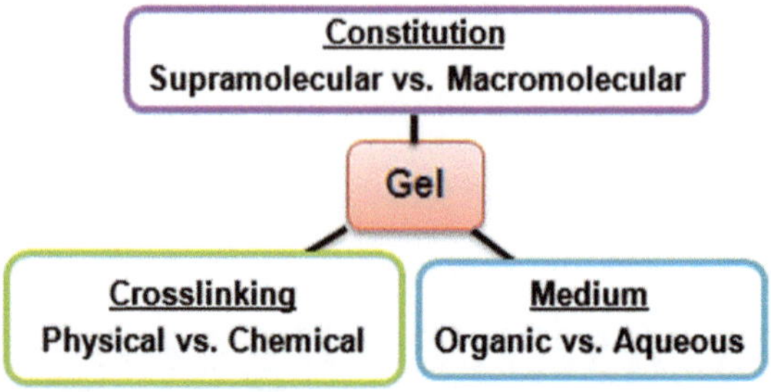

Figure 4.1 Classification of gels based on various factors.

non-covalent connectivity (supramolecular nature). Here, solvent molecules are immobilized by capillary action (specific interactions between gelator and solvent molecules can be found only in special cases) in the network of gelator molecules. The supramolecular nature of gelator–gelator connectivity induces unique reversibility in the system, which means that the solvent molecules can be freed upon breaking of the gel network and the system behaves like a fluid. A stimulus can act as a trigger to this phase transition and can initiate substantial alteration in physical (and chemical in some cases) properties of the system thereby 'responding' to the environment. As an adaptive soft material together with the 'smartness', a supramolecular gel system has the potential to become an indispensable part of several real-life applications, such as sensors, actuators, drug delivery, *etc*. In the forthcoming sections we will highlight a few very interesting and relatively new kinds of intelligent applications emerging from supramolecular organogel materials, which promise to maintain the relevance and applicability of this chemistry for the next generation of science.

4.2 Stimuli-sensitive Organogels in General

The supramolecular nature of organogels provides the systems with enough room to undergo various transitions in response to environmental stimuli. The transition could be anything from phase change to modifications of physical properties, such as optical, rheological, *etc*. In a more in-depth approach, we can divide the whole process into two parts, the 'cause' (*i.e.* the stimuli) and the 'effect' (*i.e.* the subsequent changes in the organogels). Either of these parts can be subdivided based on 'nature', into a physical or a chemical event (Figure 4.2). A physical stimulus can bring about a physical, a chemical or a combination of both physical and chemical, changes in the system. On the other hand, a chemical stimulus invariably causes a chemical change to a system, which in turn may initiate a subsequent physical change.

Among the different physical stimuli, a thermal stimulus is the most common and easy-to-generate, however, its destructive nature in some cases (application of heat can influence overall physical properties and chemical activity of a system and also a high temperature is not suitable for biological

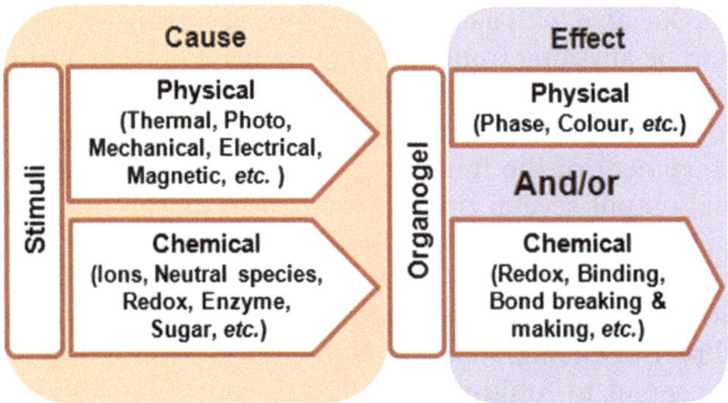

Figure 4.2 Different types of stimuli and their effect on organogels.

systems and processes) limits its application. Mechanical and photo stimuli constitute the second generation, whereas electrical and magnetic fields are more recent members in this category. Gel-to-sol phase transition is considered to be the most common and drastic physical change to an organogel system. Other noticeable physical changes include a visible colour change or a significant modification in optical property (absorption and emission). To date, several reviews have been published addressing the effect of physical stimuli on organogel systems, where inclusion of responsive functional groups in the gelator molecules was the key.[3b,e,5] Following a similar methodology, the design of organogels sensitive to chemical stimuli is a relatively new topic and is discussed only discretely in the literature so far.[6,7] Chemical sensing constitutes an integral and indispensable part of real-life applications. In one way, it can be used to detect pollutants, toxins, chemical weapons and explosives. At the same time, *in vivo* processes including drug delivery and the action of pharmaceuticals relies on the sensing of different chemicals present in the human body. Here, we will discuss the chemical sensing ability of organogel systems based on their molecular design and response to different classes of chemicals. Emphasis will be given to most recent reports based on the novelty in ideas, covering a wide range of chemicals.

4.3 Chemical Stimuli-sensitive Organogels

While acid–base, ion and redox are the most common chemical stimuli, any molecule by itself can act as a chemical stimulus as long as it is capable of initiating detectable changes. Based on this logic, chirality can also be considered as a chemical stimulus. The key in developing a chemical-sensitive organogel is the inclusion of one or more functional groups that can initiate detectable changes in terms of colour or appearance in response to the chemical stimuli. Considering ease of detection, visible changes are ideal; however, in some cases, minute differentiation requires a more in-depth study such as spectroscopic measurements. Response from a system is invariably associated with a chemical change occurring within the system. Changes could be anything from binding or release of a chemical species to bond formation, bond scission, electron transfer or more simply, modification of the aggregation mode. It would be worth mentioning here that the ordered arrangement of the functional groups in the gel phase intensifies the functional output several times compared to that in the randomly oriented solution state (*e.g.* aggregation-induced emission enhancement or oppositely, aggregation-assisted drastic quenching of emission, frequently take place in gel samples). Thus one could expect an amplified response from the gel phase sensing. Also, some molecules with multiple functional groups can respond to multiple stimuli and constitute a special class of multiresponsive materials. Thus, in some cases, it would be hard to classify the system in terms of a specific stimulus. As a more sensible approach, instead of discussing each stimulus individually, we will arrange the discussions by classes of stimuli. Thus acid–base (associated with exchange of protons) stimuli and ionic (cationic and anionic) stimuli will be discussed together because a proton and a metal ion frequently bind in a similar manner.

4.3.1 Ion-sensitive Organogels

Acid–base sensitive adaptive systems are particularly important because of their potential use in pH-triggered antitumour agent delivery.[8] Design of acid–base sensitive organogels essentially involves binding or release of proton to and from the organogelator. Proton-induced ring opening of spironaphthoxazine was explored in developing pH-sensitive organogels from LMGs **1a–1b**, which upon addition of *p*-toluenesulfonic acid (*p*-TSA), induced formation of a stable organogel in ethanol below the critical gelation concentration (in the absence of acid) with a concomitant change in colour from colourless to purple (Scheme 4.1).[9] Acetic acid and hydrochloric acid behaved similarly. Addition of an equimolar amount of triethylamine (TEA) destroyed the gel. Acid-induced transition of the molecule from a spiro

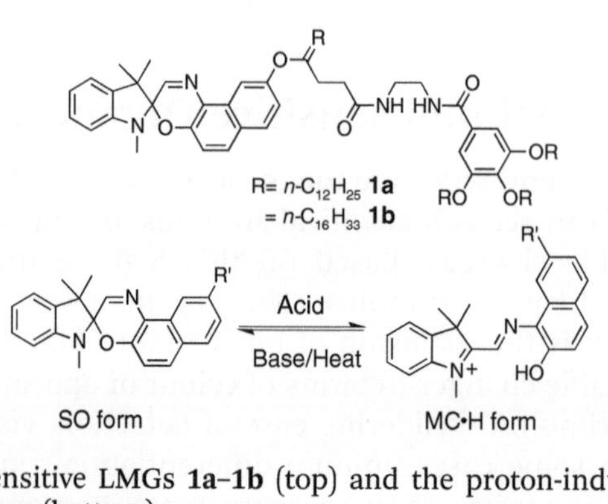

Scheme 4.1 pH-sensitive LMGs **1a–1b** (top) and the proton-induced ring opening reaction (bottom).

(SO) structure into a planar protonated merocyanine (MC · H) structure assisting π–π interactions has been attributed to this response mechanism. Acid-catalysed hydrolysis of imine was the key in developing an acid-responsive organogel system from a Schiff base molecule **2**, which is capable of undergoing self-assembly *via* three collective contributions of arene–perfluoroarene interactions, intermolecular hydrogen bonding and van der Waals interactions (Scheme 4.2).[10] Addition of a catalytic amount of *p*-TSA on top of the gel prepared from a *n*-hexane–ethyl acetate mixture caused a gel–sol phase transition as a result of cleavage of imine linkage. Decomposition of LMG **2** (into its two precursors) under acidic conditions left the system irreversible, which was the basic difference with LMG **1**.

Benzimidazole is a well-known functional group that has been integrated into various systems to act as a sensors for ions, where a significant alteration in its optical properties is the basis of detection. Compound **3** possesses four amide groups for anion binding and the benzimidazole moiety for anion and proton binding.[11] Addition of two equivalents of tetrabutylammonium hydroxide (TBAOH) to a DMSO:H_2O gel of **3** induced a transition from a yellow gel to an orange solution (with subsequent alterations in absorption and emission spectra), whereas the F^- counterpart of TBA could not make any significant difference. Interestingly, the orange solution turned into a yellow gel upon addition of one equivalent of trifluoroacetic acid (TFA) relative to OH^- ions with a simultaneous restoration of the optical properties. Further addition of TFA gradually shifted the emission band while keeping the gel phase intact (Figure 4.3). Gradual addition of TBAF (up to 10 equiv.) into a *o*-dichlorobenzene gel of **3** changed the state to a solution with subsequent changes in optical properties (large change in the case of emission and a small change in the case of absorption). While addition of TFA could restore the changes in the first instance, the presence of a large excess of TFA destroyed the gel. Based on these findings, complex logic gates with three inputs and three outputs have been constructed, where depending upon cases, H^+, OH^-, F^- and absorption maxima were the inputs while emission maxima, absorbance and physical appearance (gel or sol) were the outputs. Another combination of logic gate with

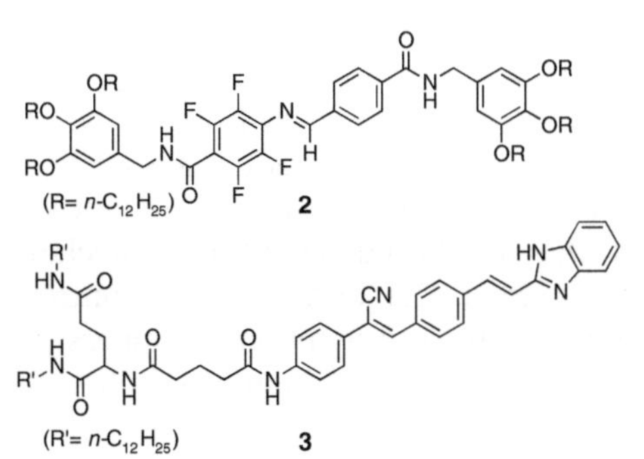

Scheme 4.2 Acid-sensitive LMGs.

Figure 4.3 Photos of the organogel of **3** in DMSO:H_2O (0.5 mM), showing responses to TBAOH and TFA (2 equiv.); the upper row is in natural light, the bottom row is under illumination at 365 nm (reproduced with permission).[11]

two outputs (emission intensity at 520 nm and naked eye detection of yellow and red fluorescence) and three inputs (TFA vapour, NH_3 vapour and irradiation of 400 nm light) was obtained using the xerogel films of **3** prepared from DMSO:H_2O. Interestingly, the xerogel film of **3** provided a fast response to the vapour of TFA (1.3 s).

Careful molecular design involving a tailored functional group can amplify the sensing action considerably (in terms of degree of changes in physical and/or chemical properties in response to a guest species), however, in some cases, an unexpected find may also provide a new strategy. A serendipitous discovery (as stated by the authors) of a urea-based molecular switch **4** based on [2]rotaxane is a beautiful example, where both gelation and switching properties could be controlled either by acid–base or by anions.[12] Gel–sol phase transition could be triggered either by translating the position of the interlocked macrocycle from the NH_2^+ centre to the urea station, or *vice versa*, or by changing the orientation of the carbonyl groups of the macrocycle. While sensing of acid–base pairs, $HClO_4$/*t*-BuOK, TFA/TEA and anions OAc^-/ClO_4^- was based on the first mechanism (Scheme 4.3), sensing of spherical anions, Cl^- and Br^- was controlled by the second mechanism. Terminal 3,5-di-*tert*-butylphenyl stoppers played an important role in gelation by retarding crystallisation of [2]rotaxanes.

An organogel made of a bolaform LMG **5** with L-lysine head groups and peripheral alkenes was found to be responsive to Ag^+ and Li^+ cations by executing gel-to-sol phase transition but was unresponsive to Na^+ and K^+

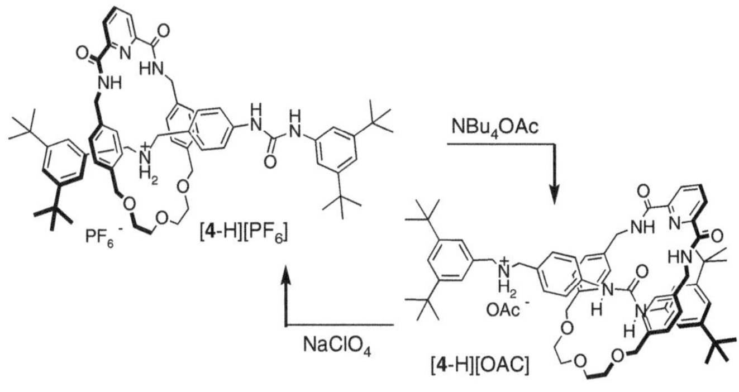

Scheme 4.3 Anion-induced switching of [2]rotaxane $[\mathbf{4}\text{-H}]^+$.

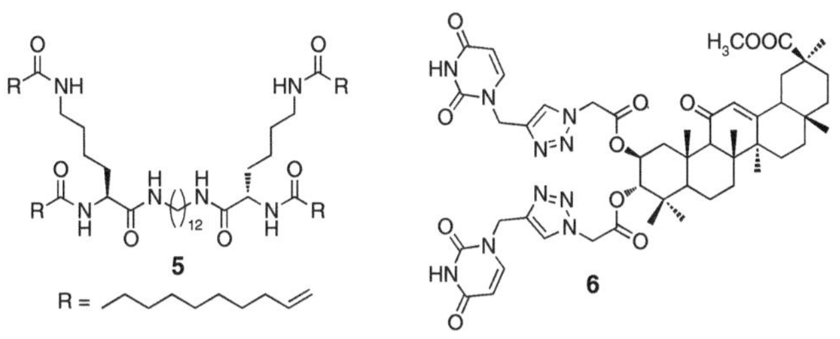

Scheme 4.4 Ag^+ and Hg^{2+} sensitive LMGs.

(all the gels were 3 mM, salt solutions were 90 mM, except for Ag^+ which was 30 mM) (Scheme 4.4).[13] While silver–alkene interactions played a key role in mediating a Ag^+ selective response, strong binding of Li^+ to the carbonyl group differentiated it from the Na^+ and K^+ ions with a relatively lower charge density. The organogel of LMG **6**, made from a functional tweezer based on uracil-appended glycyrrhetinic acid linked by 1,2,3-triazole, could show dual response to a cation Hg^{2+} and an anion F^-.[14] The 1,2-dichloroethane gel of **6** turned into a homogeneous solution on addition of 1 equiv. of F^-. On the other hand, with progressive addition of more than 2 equiv. of Hg^{2+}, Mg^{2+}, Cu^{2+}, Cd^{2+} and Zn^{2+} as perchlorate salts, only the gel treated with Hg^{2+} was transformed in to a sol. Disruption of intermolecular hydrogen bonding was the key in F^- selective response, whereas binding of Hg^{2+} with the 1,2,3-triazole unit destroyed the π–π interaction causing a gel-to-sol phase transition.

Anion binding by a hydrogen bond donor group present in a gelator is a well-studied phenomenon.[15] In this class, urea-based soft materials have received considerable interest and the competition between anion–urea binding and gelator–gelator self-assembly could be successfully fine-tuned in developing various anion-sensing soft materials.[16] In a beautiful example, carbamazepine (CBZ, a drug) single crystal grown in an anion responsive toluene gel of **7** was easily recovered by anion-induced destruction of the gel (Figure 4.4).[17]

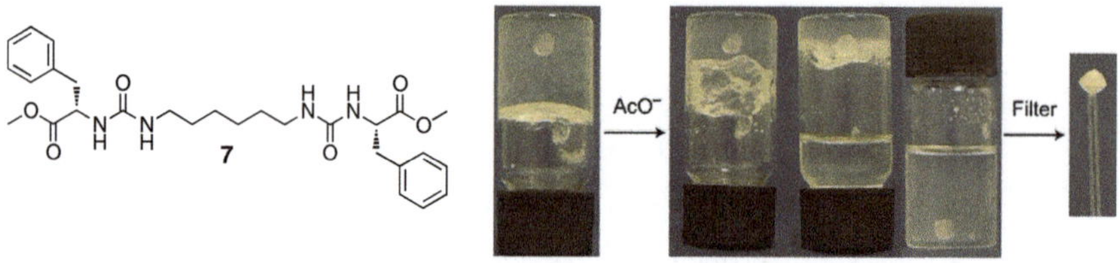

Figure 4.4 Anion responsive LMG **7** and recovery of CBZ crystal (form III) grown in the toluene:$CHCl_3$ (9 : 1) gel of **7** *via* gel dissolution upon addition of acetate ion (reproduced with permission).[17]

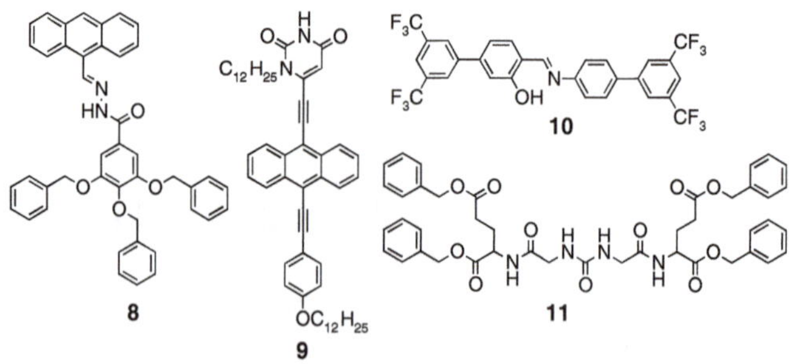

Scheme 4.5 Fluoride-sensitive LMGs.

The detection and sensing of fluoride ions has become a hot topic because of its important role in biological systems. Thus, apart from the multi-responsive systems, the design of a fluoride-specific detector is drawing considerable interest. An efficient 'naked-eye' detection of F^- has been performed utilising the LMG **8** based on a poly(aryl ether) dendron attached to an anthracene moiety (Scheme 4.5).[18] When a THF gel of **8** was exposed to a series of anions F^-, Cl^-, Br^-, I^-, ClO_4^-, CH_3COO^-, $H_2PO_4^-$, HSO_4^- as Bu_4N^+ salts, only F^- resulted in an intense variation in absorption spectra (colour changed from deep yellow to bright red) together with a subsequent gel-to-sol phase transition. Abstraction of a proton by a fluoride ion generated a more conjugated structure, thereby influencing the absorption behaviour and at the same time weakening of hydrogen bonding led to breaking of the gel. A noticeable detection can be made up to 0.1 equiv. of F^- with respect to **8**. Similarly, a DMSO gel of LMG **9** composed of an anthracene unit, a uracil unit and long alkyl chains could be used in highly selective fluoride detection (in contrast to Cl^-, Br^-, I^-, ClO_4^-, CH_3COO^-, HSO_4^-, $H_2PO_4^-$) through colorimetry and fluorimetry together with a phase transition (Figure 4.5).[19] Deprotonation of the uracil moiety by F^- was found to be responsible for the sensing mechanism. Interestingly, addition of methanol—a protic solvent—restored the system back to a gel. Such a system recovery can crucially enhance the applicability of a sensing system. Expanding the molecular design, a salicylidene aniline-based π-conjugated LMG **10** and glycine-glutamic acid based LMG **11** were also used for fluoride

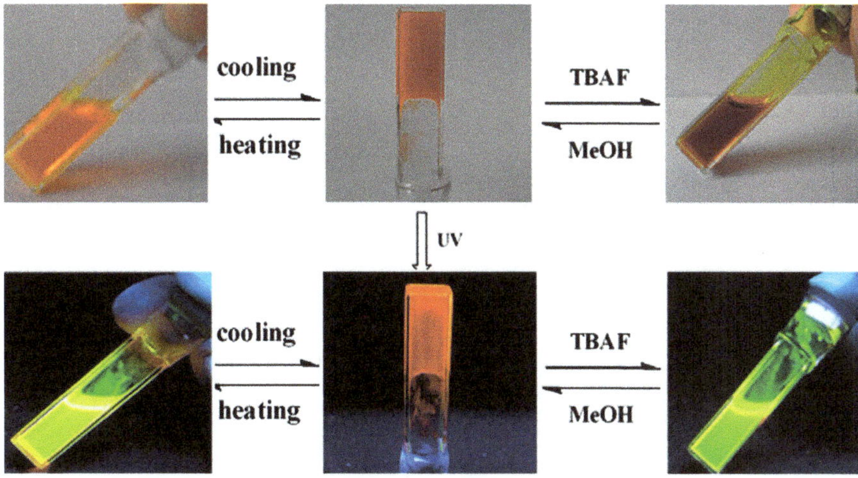

Figure 4.5 Illustration of the gel-to-sol transformation accompanied with colour and fluorescence changes of LMG **9** (10 mg mL^{-1}) in DMSO upon addition of fluoride ions (1 equiv.) under ambient conditions (reproduced with permission).[19]

detection. While the hydroxyl group played a key role in the former case exhibiting a triple-channel response (Figure 4.6),[20] the latter interacted with F^- *via* the urea and amide group (Figure 4.7).[21]

Detection of nitrite (NO_2^-) content in food preservatives could be a vital health issue, though its sensing *via* a supramolecular approach is rare. In a recent report, a two-component organogel system composed of a naphthalimide undecanoic acid **12** and diaminoanthraquinone (DAQ) was shown to provide a 'naked eye' detection of NO_2^-.[22] Addition of an aqueous $NaNO_2$ (100 equiv.) solution to the acetonitrile gel of **12** and DAQ caused a gel collapse with subsequent fading of the colour (Figure 4.8). The formation of a colourless triazole from the reaction of DAQ and NO_2^- was responsible for the sensing phenomenon. Collective contributions from colorimetric and fluorimetric detection were found to be capable of distinguishing NO_2^- from other common environmental ionic species (addition of 100 equiv. of NaF, $NaHCO_3$, Na_2SO_4, $NaSO_3$, Na_3PO_4 had little effect on the absorption behaviour of the gel).

Crown ethers are considered as the first generation of supramolecular macrocyclic hosts and offer an attractive platform to develop stimuli-responsive systems because of their excellent binding selectivity and binding strength. Recently, we reported an organogel system developed from crown-appended quaterthiophene **13** (Scheme 4.6), which could exhibit a unique allosterism-like sigmoidal response to alkali metal cations (Figure 4.9).[23] Addition of Li^+ (2 equiv.) and Na^+ (1.2 equiv.) both caused a gel precipitation, whereas an excess amount of Na^+ (up to 2 equiv.) turned the system into a solution. K^+ behaved differently and maintained the gel state initially at a low concentration (up to 1 equiv.), however, it turned the system directly

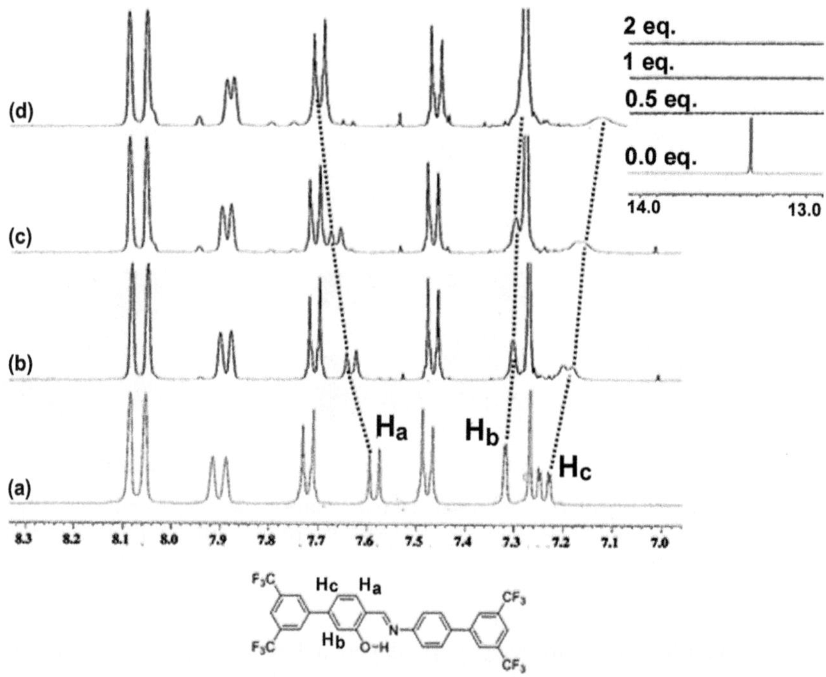

Figure 4.6 ^{1}H NMR titration spectra of **10**: (a) in the absence and in the presence of (b) 0.5, (c) 1.0, and (d) 2.0 equiv. of TBAF. The dotted lines indicate the directions of peak shifts and the corresponding protons are shown in the chemical structure at the bottom. The inset shows the disappearance of the hydroxyl peak of **10** after the addition of TBAF (reproduced with permission).[20]

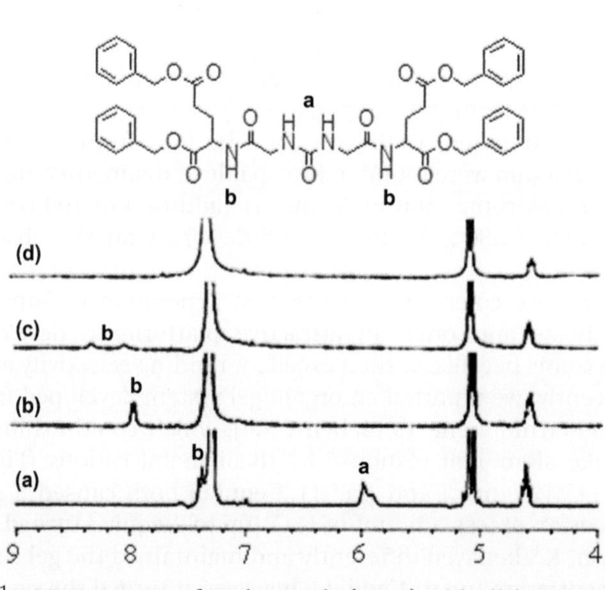

Figure 4.7 ^{1}H NMR spectra of **11** in $CDCl_3$ (40 M) with different amounts of F^- (a) free **11** (b) 0.5 equiv. (c) 1 equiv. (d) 5 equiv. F^- (reproduced).[21]

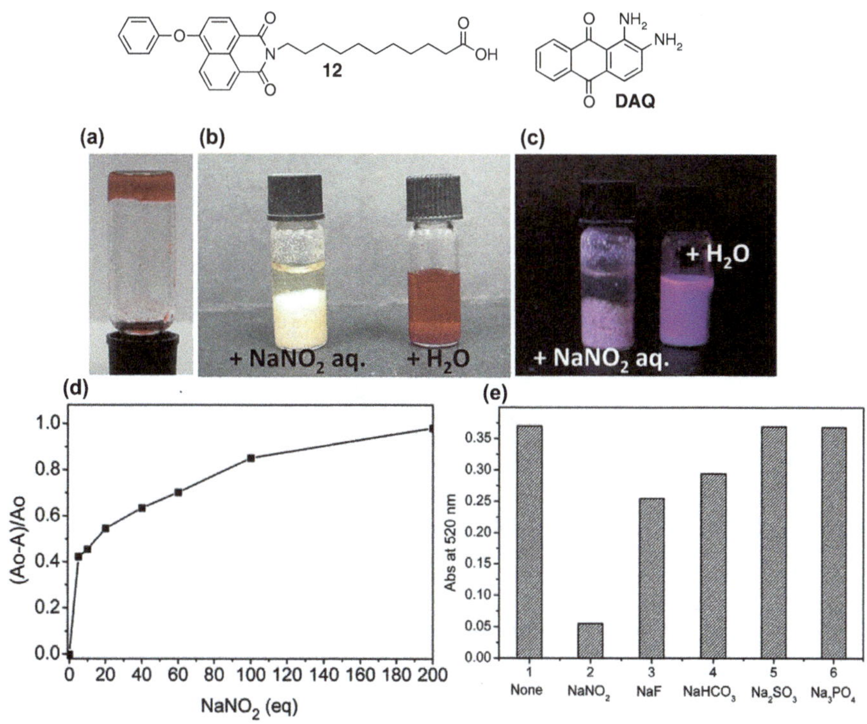

Figure 4.8 Nitrite responsive LMG **12** and **DAQ** (top); photo of (a) gel of **12** + **DAQ** in acetonitrile (25 mg mL^{-1}); gel with added $NaNO_2$ solution and pure water, in (b) natural light and (c) with 365 nm light irradiation; (d) absorbance change at 520 nm of the liquid part of gel from **12** + **DAQ** treated with water and different molar ratios of $NaNO_2$; (e) the absorbance at 520 nm with the addition of different types of anions (100 equiv. of DAQ) (reproduced).[22]

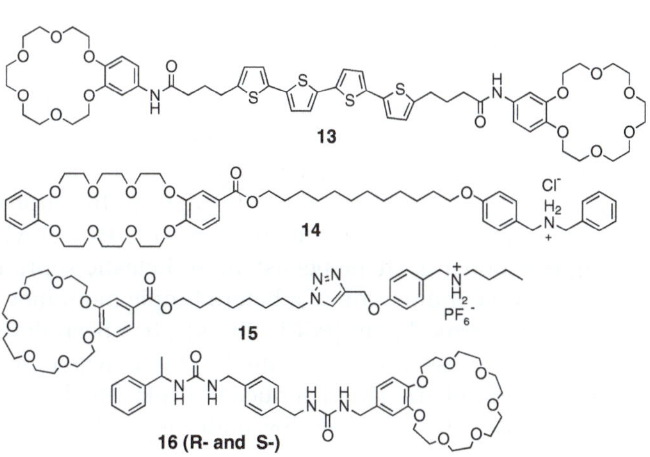

Scheme 4.6 Cation-responsive LMGs based on crown ethers.

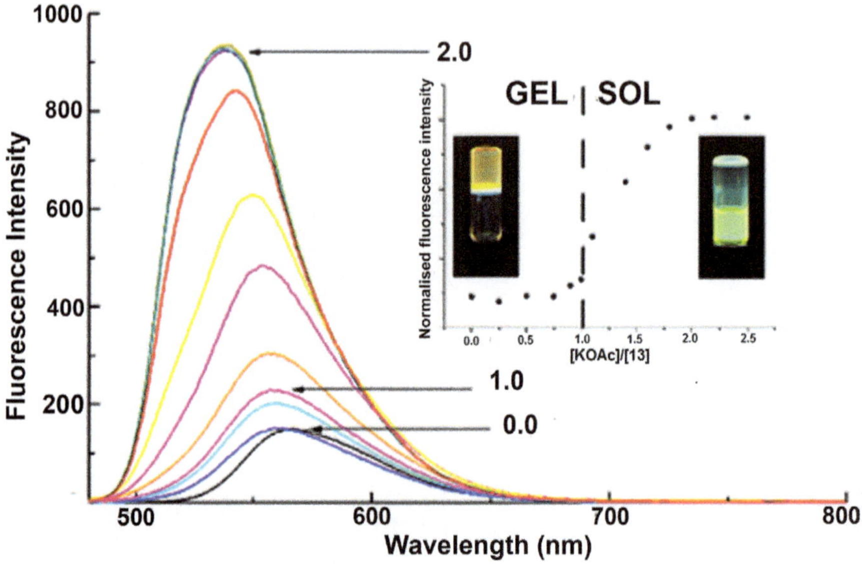

Figure 4.9 Fluorescence titration spectra of the gel and the collapsed gel of **13** (0.25 wt%) by KOAc at different equivalence; inset: plot of [KOAc] *vs.* normalized fluorescence intensity (reproduced).[23]

to a solution at a higher concentration. Cs^+ caused gel collapse even more efficiently (at 0.5 equiv.). Changes in the conformational and aggregation mode of a π-conjugated oligothiophene unit provided easy monitoring of cations *via* absorption and emission signals. Reported by another group, a crown ether appended supergelator **14**, which was a combination of cation binding dibenzo-24-crown-8 and acid–base sensitive organic ammonium, could exhibit multiple stimuli responsiveness.[24] A reversible gel–sol transition could be induced by addition and subsequent removal of equimolar amounts of organic cations, such as dibenzylammonium, 1,2-bis(pyridinium) ethane dication, paraquat and K^+. In contrast to Fe^{2+}, Fe^{3+} was bound efficiently with crown ether and the acetonitrile gel was turned into a sol in the presence of $FeCl_3$, followed by a gel recovery (by redox transition of Fe^{3+} to Fe^{2+}) upon addition of sodium ascorbate, a reducing agent. The acid–base pair TEA/HCl could induce a similar reversible transition. The organogel system also responded reversibly to the presence and the removal of Ag^+ (adding CF_3COOAg and removing AgCl by filtration) and *tert*-butoxycarbonyl (Boc) (adding Boc_2O and treating with HCl–ethyl acetate). Another quadruple-responsive, shape-persistent and elastic supramolecular polymer gel formed by a combination of a heteroditopic monomer **15** with a bridging ligand and the cross-linker $[PdCl_2(PhCN)_2]$ has been developed by the same group.[25] The cross-linked supramolecular polymer network gel showed a reversible gel–sol transition induced by pH (in the presence of acid–base pair TFA/TEA), heat, cation (by addition and removal of K^+)

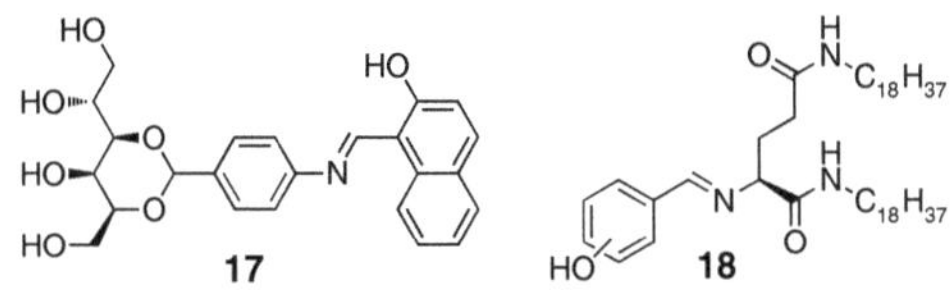

Scheme 4.7 LMGs sensitive to Cu^{2+}, Mg^{2+}, Ni^{2+}, Zn^{2+}.

and metal (by addition and removal of competition ligand to Pd). Very interestingly, this supramolecular polymeric material could be molded into shape-persistent, free-standing objects. Such a material is promising for exhibiting shape-memory functions. This phenomenon was controlled by the dynamically reversible complexation between benzo-21-crown-7 and dialkylammonium moieties and significant mechanical properties of the cross-linked supramolecular polymer. In an early approach towards *in vivo* application, the cation binding property of another crown ether based LMG **16** has been explored beautifully to immobilize and release an enzyme lipase B from *Candida antarctica* labelled with fluorescein isothiocyanate. At the same time it addressed a common problem related to the use of enzymes in a relatively nonpolar organic medium.[26]

A DMSO:H_2O gel of a salen-linked LMG **17**, composed of a naphthalene-based salicylideneaniline segment and a sorbitol group responded to Cu^{2+} (0.5 equiv.) through a reversible gel–sol transition accompanied with a colour change from yellow to colourless (Scheme 4.7).[27] The presence of other cations (Ca^{2+}, Mg^{2+}, Mn^{2+}, Zn^{2+}, Ni^{2+}, Cr^{2+}, Fe^{2+}, Co^{2+}) under the same conditions could not alter either the gel phase or the optical property of the gel. A chelate formation between Cu^{2+} and **17** facilitated by intramolecular charge transfer was found to be responsible for Cu^{2+} specific sensing. An organogel prepared from amphiphilic LMG **18** containing a Schiff base and L-glutamide exhibited interesting and diverse response behaviour to different chemical species.[28] While Zn^{2+} and Ni^{2+} destroyed the gel, it remained unaffected by Cu^{2+} and Mg^{2+}, although the presence of Cu^{2+} induced a chiral twisting of the gel fibre. The planarity of the gelator facilitated *via* the formation of intramolecular hydrogen bonding between the hydroxyl group and imine group and also the relative planarity of different metal–gelator complexes played a crucial role in perturbing overall aggregation mode. Finally, it was the inherent 'softness' of the system that became responsible for such adaptive behaviour.

4.3.2 Redox-sensitive Organogels

A redox-sensitive system is expected to respond to a redox process occurring within the system as a result of an external chemical trigger, which is in most cases either an oxidant or a reductant and in a few cases, a species capable of catalysing a redox process. Compared to other stimuli-sensitive systems, reports on redox-sensitive organogels are limited.[29] The key in developing a

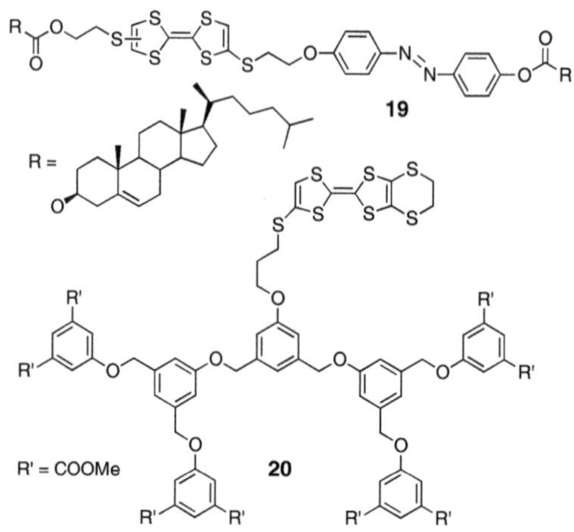

Scheme 4.8 Redox-responsive LMGs based on TTF.

redox-sensitive organogel would be to functionalise the LMG with a moiety that is capable of undergoing a redox transition and as a result, the system should exhibit detectable changes in its physico–chemical properties. Among one of the widely-used functional motifs in this category, tetrathiafulvalene (TTF) is a redox active group that can be reversibly transformed into radical cations and dications. An organogel of **19** composed of an electroactive TTF unit and a photoresponsive azobenzene unit coupled with gel-forming cholesterol units was capable of undergoing reversible gel–sol phase transition, fine-tuned by both redox and photostimuli (Scheme 4.8).[30] When 1.0 equiv. of $Fe(ClO_4)_3$ was placed above the gel of **19** in CH_2Cl_2, the gel was gradually destroyed leading to a dark green suspension. Formation of TTF^+ was the cause. Addition of an excess amount of ascorbic acid (reduced TTF^+ to TTF) turned the system to an orange solution which on subsequent heating and cooling regenerated the gel phase. Interestingly, the gel–sol transition could also be tuned by electrochemical oxidation and reduction. Alternatively, photoinduced isomerization of an azobenzene moiety could trigger a gel–sol transition. Reported by the same group, a similar gel–sol phase transition could be induced using another TTF-based, dendron-substituted LMG **20**.[31] Furthermore, a bicomponent gel of **20** with chloranil could be transformed into a solution only after the addition of Sc^{3+} or Pb^{2+} (Figure 4.10). Generation of TTF^+ *via* intermolecular electron transfer between TTF and chloranil, facilitated by Sc^{3+} or Pb^{2+}, was considered to be the responsible factor. Insertion of Mg strips causing reduction of TTF^+ followed by heating and cooling could restore the system into a gel phase again.

Organometallic compound ferrocene (Fc) is capable of undergoing a redox transformation to ferrocenium cation (Fc^+). This redox switching property of Fc/Fc^+ and the self-assembly property of peptides have been coupled to

Figure 4.10 Phase transition of the gel of **20** (30 mg mL^{-1}) and 1.0 equiv. of chloranil from a mixture of benzene and acetonitrile (1:1, v/v) into the solution after addition of $Sc(CF_3SO_3)_3$ and recovery of the gel phase after further stirring with magnesium strips, followed by heating and cooling (reproduced with permission).[31]

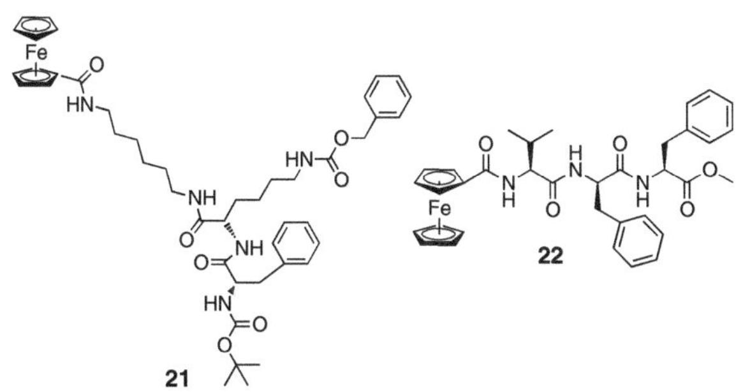

Scheme 4.9 Redox-responsive ferrocene-based LMGs.

develop redox-responsive ferrocene-peptide conjugates **21–22** (Scheme 4.9). An orange coloured acetone gel of **21** could be transformed into a green coloured solution upon addition of a small amount (less than an equimolar amount with respect to **21**) of $Fe(ClO_4)_3$ (Scheme 4.9). Addition of ascorbic acid followed by sonication recovered the gel state.[32] A sonication-induced gel of **22** in toluene underwent degradation upon treatment with an equimolar amount of oxidant $Fe(ClO_4)_3$ with a sharp colour change from light orange to deep blue.[33] The change in the redox state of the gelator caused a dramatic structural modification from a nanofibrillar network to a spherical micelle (Figure 4.11). Oxidation of **22** gave rise to Fc^+, which, together with the ClO_4^- counter ion, formed the core of the micelle with hydrophobic peptides facing up. Interestingly, reduction by ascorbic acid followed by sonication changed the morphology to the fibrillar network again.

A copper(II) complex of quinolinol-substituted L-glutamide **23** could collapse in the presence of ascorbic acid under a nitrogen atmosphere with a subsequent colour change from dark red to transparent yellow owing to the formation of Cu(I) (Scheme 4.10).[34] Oxidation by a flow of oxygen could restore the organogel again.

Coordination of Cu(II) with anionic quinolinol in square planar geometry played a key role in strengthening π–π stacking. Formation of Cu(I) caused

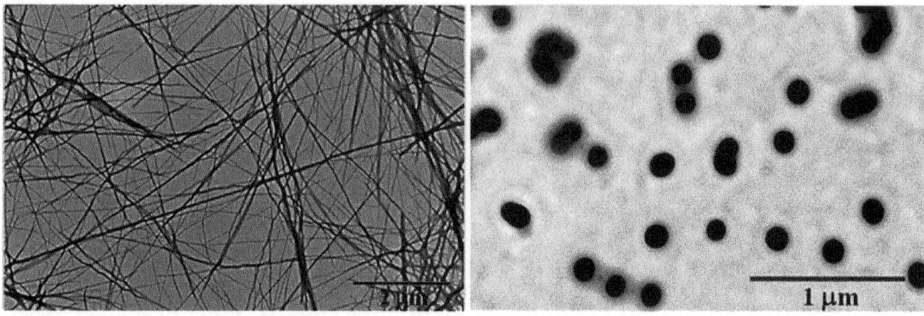

Figure 4.11 TEM images of self-assembled (left) and fully oxidized (right) **22** in the presence of $Fe(ClO_4)_3$ (reproduced).[33]

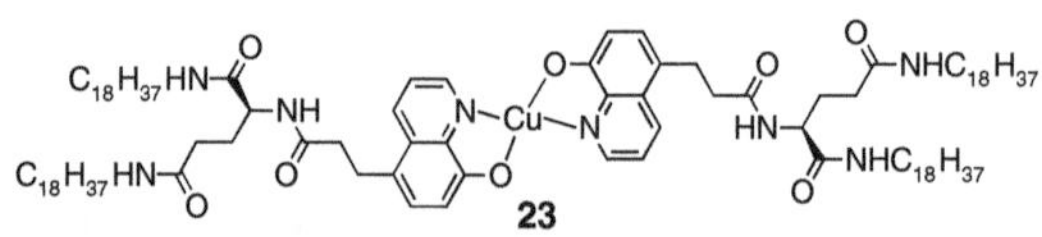

Scheme 4.10 Redox-responsive metallogelator based on copper(II).

deviation from the planarity which in turn resulted in a gel collapse. It would be worth mentioning here that in the first example of a redox responsive system based on a Cu(I)/Cu(II) redox couple reported by our group, the reduced form with Cu(I) constituted the gel phase whereas formation of Cu(II) was responsible for destruction of the assembled phase.[35a] This is just the opposite of what we received from LMG **23**. Such copper chelate complexes feature a structural interconversion between tetrahedral and planar geometry by redox treatments. The difference in their π–π stacking ability was further applied to a redox-controlled CNT solubilisation process.[35b] This concept would also be applicable for the redox control of gel formation. Therefore, depending on the ligand design, both forms of copper are equally useful for undergoing a self-assembly process.

4.3.3 Chirality-sensitive Organogels

For almost all biological polymers, homochirality is essential for their functions. For example, while L-amino acids are found naturally in proteins, most naturally-occurring sugars have a D-configuration. Similarly, chirality is a crucial issue for the function of drug molecules.[36] At the same time, chiral moieties play important roles in governing asymmetric reactions and in the formation of highly ordered nano-structures relevant to a natural or an artificial functioning system. Thus, a simple protocol to distinguish between right and left handed enantiomers is extremely important in the scientific sector as well as in everyday life. Integrating a chiral functional group into a LMG to develop a chiral self-assembled system is well-executed chemistry.[37] However, there are a relatively small number of reports on chiral recognition by organogel systems. The difficulty lies in the fact that the chemical and physical identities of the enantiomers of a chiral species are apparently the same except in their orientation around the chiral centre. The first example of chiral recognition based on a supramolecular gel system was reported by our group two decades ago by using gel-forming complexes of monosaccharides with cholesterylphenylboronic acid.[38] Here, the chirality (D- or L-) of the monosaccharides could be discriminated in terms of several parameters, such as the gel–sol phase transition temperature, the xerogel fiber structure and the gel stability. More recently, we developed an anthracene-based bicomponent component gelator that could differentiate between racemic and enantiomerically pure varieties of a solvent by exhibiting different gel melting temperatures.[39] After one year, the first example of visualized enantiomer recognition through enantioselective metallogel collapse was originated from the design of LMG **24**, which consists of terpyridine and chiral 1,1′-bi-2-naphthol (BINOL) functional groups (Scheme 4.11).[40] This compound formed an opaque green gel in chloroform upon sonication. When a solution of (*R*)-phenylglycinol (0.10 equiv.) in chloroform was added to the metallogel followed by sonication, the gel remained stable; on the other hand, when the same amount of (*S*)-phenylglycinol was applied, the metallogel collapsed under the same conditions (Figure 4.12). It was found that a two-fold higher amount of (*R*)-phenylglycinol compared to (*S*)-phenylglycinol was required to cause gel collapse. The enantiomer of **24** exhibited the opposite enantioselective gel collapse. The chiral recognition of the gel is compared with the fluorescence responses of **24** toward (*R*)- and (*S*)-phenylglycinol in solution. The enantioselectivity could be attributed to the enantioselective displacement of the Cu(II) ion from **24** by the amino alcohol. The more favorable interaction of **24** with (*S*)-phenylglycinol observed in solution might be greatly amplified in the supramolecular assembly of the gel network, leading to a visual response. In a similar way, the presence of (*S*)-1-amino-2-propanol led to gel collapse in contrast to the (*R*)-enantiomer. Another metallogel prepared from the LMG **23** described in the earlier section was capable of discriminating between L-and D-enantiomers of aromatic amino acids.[34] Addition of L-tyrosine enhanced

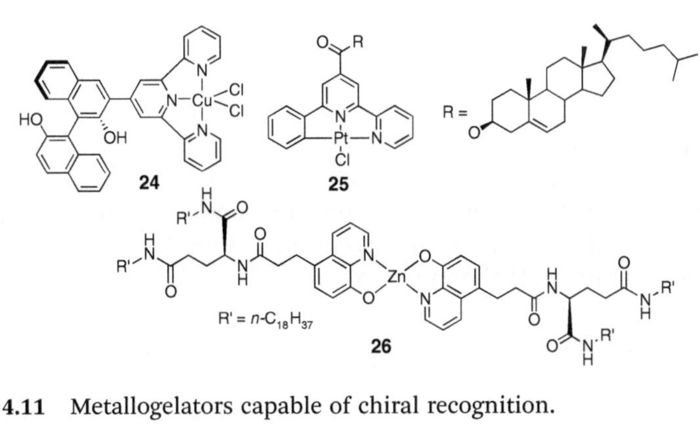

Scheme 4.11 Metallogelators capable of chiral recognition.

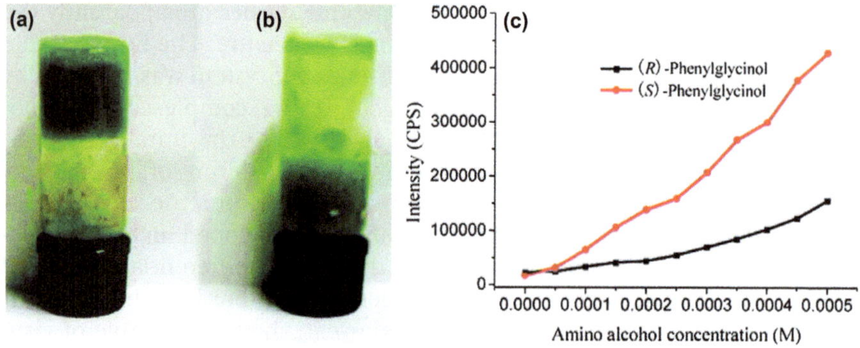

Figure 4.12 Enantioselective responses of the gel of **24** prepared in $CHCl_3$ toward (a) (*R*)-phenylglycinol (b) (*S*)-phenylglycinol and (c) fluorescence responses of **24** (5.0×10^{-7} M) in CH_2Cl_2–*n*-hexane (2 : 3) toward (*R*)- or (*S*)-phenylglycinol at λ_{em} = 396 nm (reproduced with permission).[40]

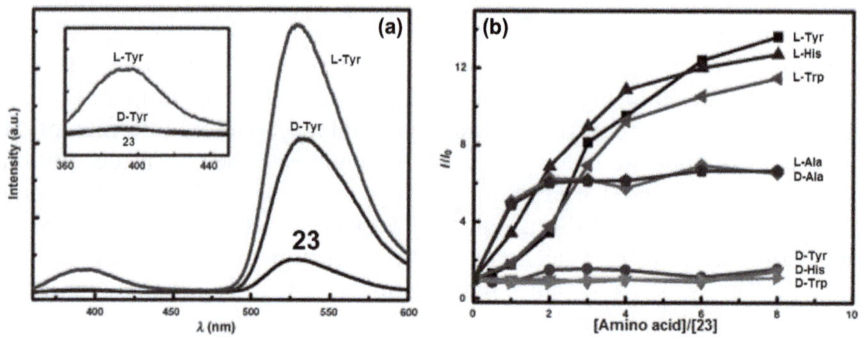

Figure 4.13 (a) Fluorescence spectra of gels prepared from **23** alone and in the presence of L- or D-tyrosine; the molar ratio of Tyr to **23** is 4. Inset: an enlargement in the range 360–450 nm; (b) dependence of the change in fluorescence on the molar ratio of the amino acids to **23** at 393 nm (λ_{exc} = 340 nm, slit = 5.0/5.0 nm) (reproduced with permission).[34]

the emission of the metal complex at 530 nm to a higher extent compared to D-tyrosine (Figure 4.13). Moreover, upon addition of only L-tyrosine, a new emission corresponding to a free ligand was observed. The coordination ability of amino acids was responsible for releasing the ligand partially from the system. On the other hand, the ligand-induced chirality in the Cu(II) site was responsible for enantioselective interaction with amino acids. Similarly, a Mg^{2+}-mediated organogel prepared from **18** showed a chiral recognition property by executing different fluorescent quenching behaviour to D- and L-tartaric acids (Figure 4.14).[28] Another example of visual chiral recognition by a metallogel system was based on the LMG **25** possessing a steroidal skeleton and a platinum pincer metallocycle.[41] This metal complex formed gels in various organic solvents *via* the collective contributions from a steroidal interaction, π–π stacking and Pt–Pt interaction. Binaphthyl derivative

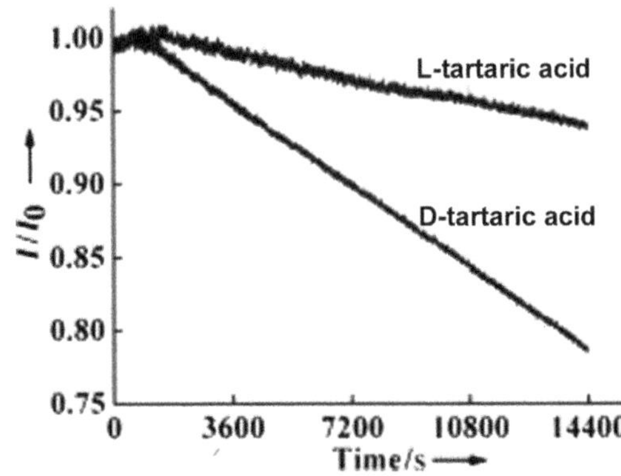

Figure 4.14 Fluorescence quenching of tartaric acid by Mg^{2+} mediated gel of **18** ('*o*-'variety) *versus* time (reproduced with permission).[28]

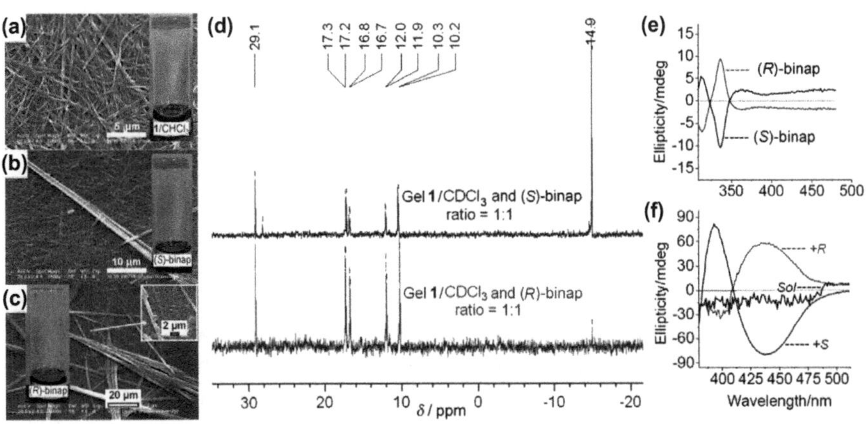

Figure 4.15 SEM images of (a) gel of **25** in $CHCl_3$, (b) gel of (**25** + 0.1 equiv. (*S*)-binap) in $CHCl_3$, (c) collapsed sol of (**25** + 0.1 equiv. (*R*)-binap) in $CHCl_3$; (d) ^{31}P NMR spectra of the solutions prepared with the gel of **25** in $CDCl_3$ and (*S*)-binap or (*R*)-binap and CD spectra of (e) (*R*)- or (*S*)-binap, (f) sol of **25** in $CHCl_3$ in the presence of (*R*)- or (*S*)-binap (reproduced with permission).[41]

2,2′-bis(diphenylphosphino)-1,1′-binaphthyl (binap) is a strong ligand and capable of ligand exchange with a chloride ligand coordinated to a metal centre. The addition of 0.1 equiv. of binap resulted in a striking difference in the behaviour of the respective gels. While the gel sample containing (*S*)-binap survived the heating and cooling sequence as a robust gel, the gel sample with the (*R*)-enantiomer collapsed (Figure 4.15). Similar phenomena were observed with other chiral di- and mono-phosphine ligands. A morphological investigation of the chloroform gels of **25** revealed that the presence of (*R*)-binap led to the formation of a dominating agglomerate of crystalline rods that were bulkier, denser and longer than those observed in the presence of (*S*)-binap. This difference in superstructure could be a responsible factor behind the enantioselective collapse of the gel network. The chiral compatibility of (*R*)-binap with the chiral cholesterol fragment was found to facilitate its coordination compared to the chirally unmatched (*S*)-binap.

A metal complex of Zn^{2+} with a quinolinol-functionalised L-glutamide **26** was found to form a fluorescent metallogel in several organic solvents.[42] While chiral complexes were CD (circular dichroism) silent in the chromophore absorption region, upon gelation, supramolecular chirality emerged which provided a tool for the chiral recognition of enantiomeric ligands. Interestingly, the gel formed in THF exhibited different emission maxima in the presence of (*R,R*)- or (*S,S*)-1,2-diaminocyclohexane accompanied by a visible colour change for the (*S,S*)-enantiomer. Using this system, enantiomeric diaminocyclohexanes could be detected by the naked eye up to a concentration of 0.15 mmol L^{-1}. Here too, the replacement of ligand **26** by a particular enantiomer which was (*S,S*)- in this case became favoured over the other because of the chiral selection of the supramolecular assembly. Different coordination geometry of the complexes from Li^{+} and Al^{3+} restricted these systems from showing any preference for ligand selection.

In all the examples discussed in this section, the coordination behaviour of the central metal ions led the spatial arrangement around the metal centre, which in turn perturbed the overall assembly in a constructive or a destructive manner. Apart from metallogelators, in a recent report, a naphthyl ring coupled with amphiphilic L-glutamide (compound **27**) was found to form emissive chiral nanotubes upon self-assembly.[43] Combined with enhanced fluorescence and supramolecular chirality, a film composed of the nanostructures by means of gelation in acetonitrile was used to sense organic amines. Interestingly, the nanotube film of **27** showed chiral recognition toward enantiomeric (*R*)- and (*S*)-1-(*p*-tolyl)ethylamine (*R*-TEA and *S*-TEA), in which the latter caused a higher fluorescence quenching (Figure 4.16).

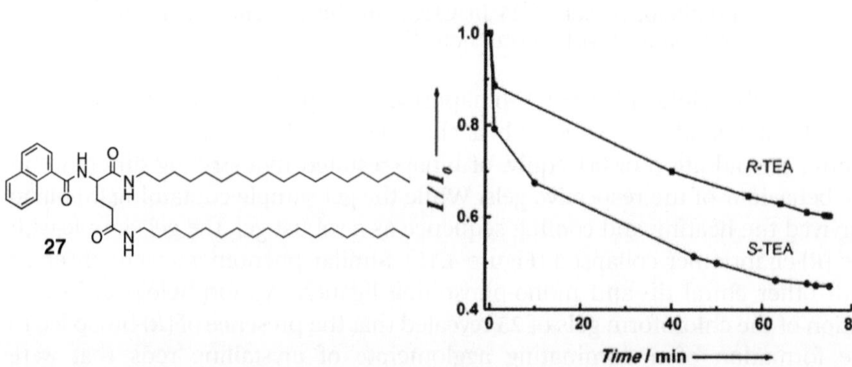

Figure 4.16 LMG 27 and fluorescence quenching efficiency of its film upon addition of chiral amines *vs.* time (reproduced with permission).[43]

4.3.4 Molecular Guest Sensitive Organogels

In this section, we will discuss organogel systems that respond to various kinds of non-ionic chemical species apart from those we have discussed in earlier sections. When such a species is introduced in a gel system, in addition to non-specific adsorption, a specific interaction may also take place depending on the complementarity of the respective systems. Such interactions may give rise to detectable signals in terms of drastic changes in the physico-chemical properties of any of the moieties. Since we reported one of the earliest examples of molecular recognition in an organogel system,[44] efforts have been made to design functional LMGs with interactive sites that can complement the guest molecules.[45] However, rationalising the molecular design remains a challenge. Recent reports have shown that a two-component hydrogen-bonded compound (**28**) self-assembled into nanofibers in cyclohexane upon gelation, and its fibrous film could be used as a sensor to detect and discriminate aromatic and aliphatic amines (Figure 4.17).[46] A physical adsorption of aniline (in the vapour phase) resulted in a drastic quenching of the fluorescence of the xerogel film, which could be recovered upon removal of the guest under gas blow. On the other hand, upon exposure to *n*-butylamine, the yellow xerogel films transformed into colourless films with a subsequent change in fluorescence

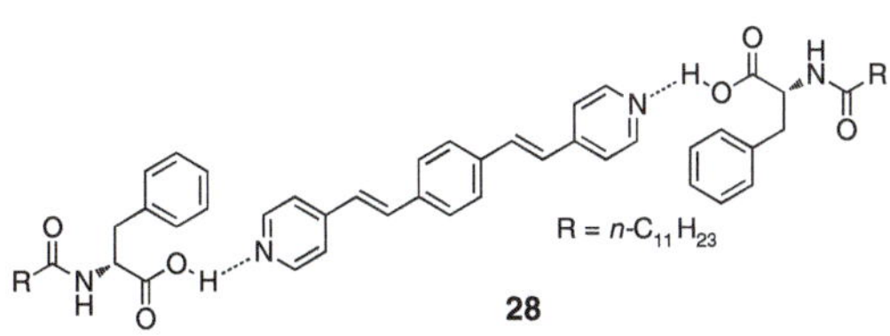

28

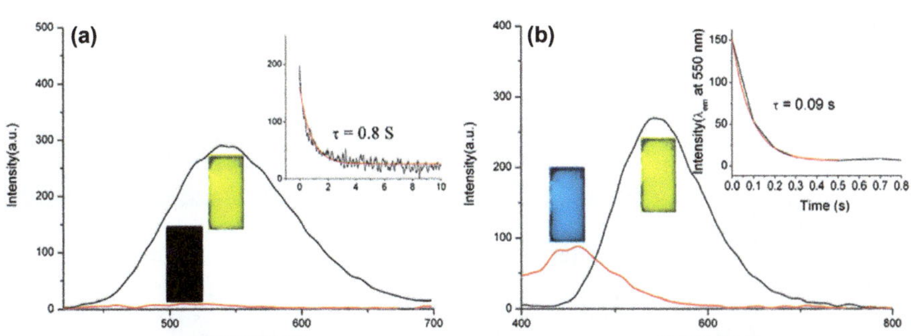

Figure 4.17 LMG **28** and fluorescence spectra of the cyclohexane xerogel film deposited on a glass slide before (black line) and after (red line) exposure to aniline vapour (a, 329 ppm) and *n*-butylamine (b, 80 000 ppm) for 30 s. The insets show the time-course of fluorescence quenching of the xerogel film after being added into a cell filled with the vapours of aniline and *n*-butylamine; the intensity was monitored at 550 nm (reproduced).[46]

from yellow-green to blue. In contrast to the aromatic amine, this change was irreversible as *n*-butylamine eventually destroyed the hydrogen bonding holding the two components in the LMG. Moreover, the sensing film exhibited a rapid response time and a low detection limit. In another remarkable work, a Co^{2+} coordinated thiacalix[4]arene (ligand **29**) supramolecular gel was used for the detection of volatile gases containing a chlorine atom.[47] When a filter paper coated with a red gel (prepared *in situ* from the mixtures of the metal and the ligand in aqueous/DMF mixture containing alkali) was exposed to toxic volatile compounds such as HCl, $SOCl_2$, $(COCl)_2$ and $COCl_2$, the colour changed to blue (Figure 4.18). A change in Co^{2+} coordination geometry from octahedral to tetrahedral was attributed to this colour change. This coordination alteration was initiated by the removal DMF molecules from the coordination site because of the strong intermolecular hydrogen bonding interaction between DMF and HCl, yielded by the volatile gases upon hydrolysis.

Nitroaromatic compounds such as 2,4,6-trinitrotoluene (TNT), 2,4-dinitrotoluene (DNT), and 2,4,6-trinitrophenol (picric acid) are well-known constituents in explosives and their detection is a crucial issue for homeland security as well as for environmental safety. In spite of the large number of reports available, a simple approach with an extremely low detection limit remains a challenge. It is particularly true for TNT due to its low vapour pressure compared to other explosives. Detection of TNT at the attogram level in a simple and low-cost method has been achieved by using a perfluoroarene-coupled oligo(*p*-phenylenevinylene) (OPVF) **30**, which showed a superior sensing capability in the gel form compared to that in the solution phase.[48] LMG **30** formed a stable gel in *n*-hexane with a subsequent change in fluorescence from green to yellow. The yellow emission changed further to an orange-red emission upon dip-coating the gel on a filter paper and the change could be reversed by heating the filter paper to 70 °C. Interestingly, TNT was found to be an efficient quencher for **30** but only in the gel

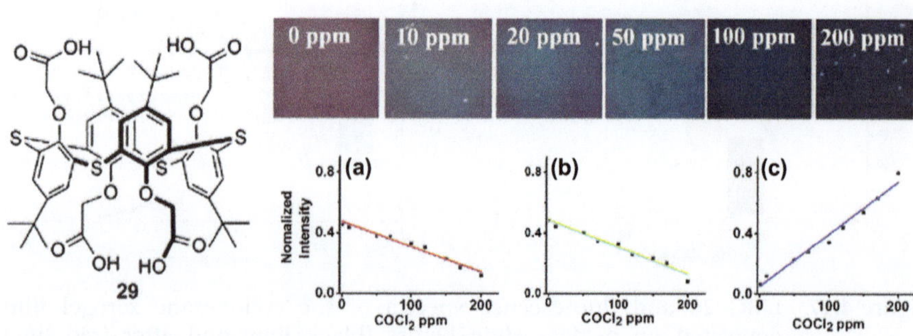

Figure 4.18 Ligand **29** (left); photographs of filter papers coated with the coordination polymer gel of Co^{2+} with **29** after diffusion of $COCl_2$ gas (0–200 ppm) and plots of $COCl_2$ against RGB intensities obtained by a smartphone: (a) red, (b) green, and (c) blue colours (reproduced with permission).[47]

phase. Test strips were prepared by dip-coating a Whatman filter paper with a gel sample. In the contact mode response, placing TNT crystals over a gel-coated strip for 5 s resulted in black spots under UV illumination as a result of fluorescence quenching. A similar experiment with 1,3,5-trinitroperhydro-1,3,5-triazine (RDX) crystal didn't leave any detectable mark (Figure 4.19). It was concluded that the TNT molecules were trapped inside the interstitial space of the gelator molecules without changing the molecular packing and the observed fluorescence quenching was due to the excited state processes. Such an interaction was assisted by hydrogen

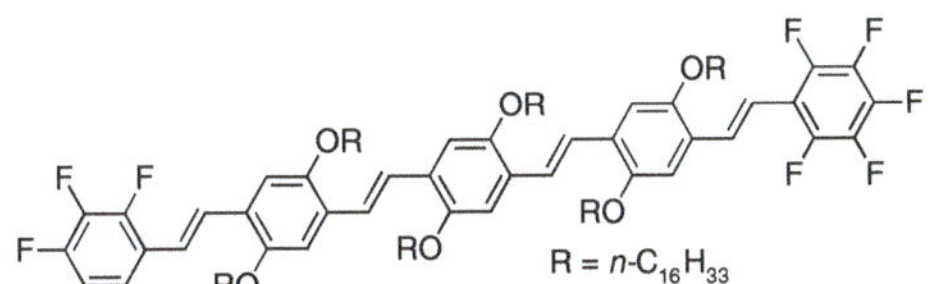

30

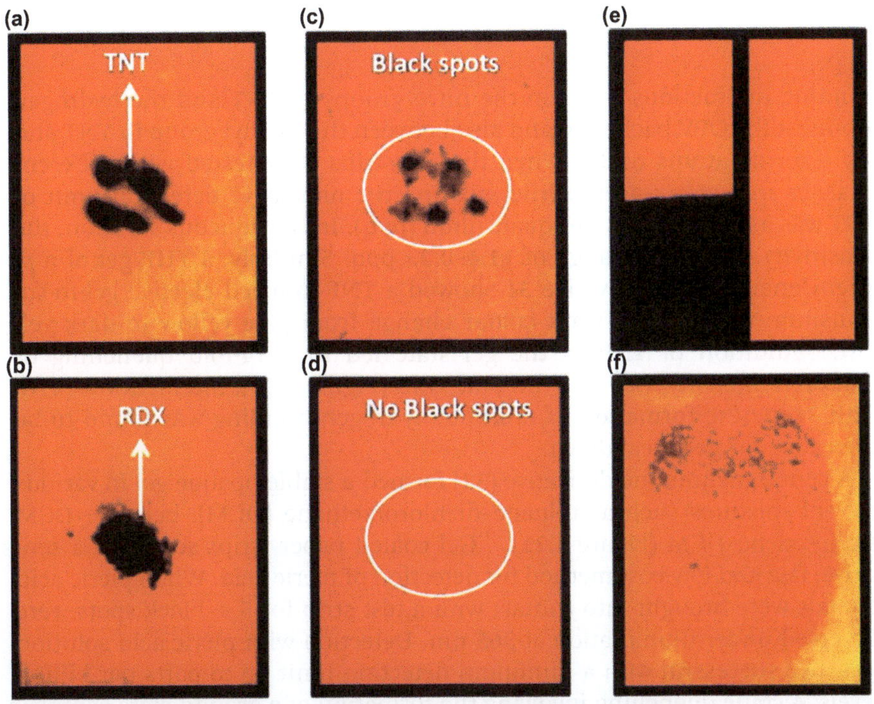

Figure 4.19 LMG **30** (left); photographs of test strips coated with the gel prepared from **30** under different experimental conditions: (a) TNT and (b) RDX crystals on top, (c and d) corresponding photographs upon removal of the crystals after 5 s (e) after dipping into solutions of TNT (left) and RDX (right) in acetonitrile (1×10^{-3} M), (f) thumb impression after rubbing with TNT crystals. All photographs were taken under 365 nm UV illumination (reproduced with permission).[48]

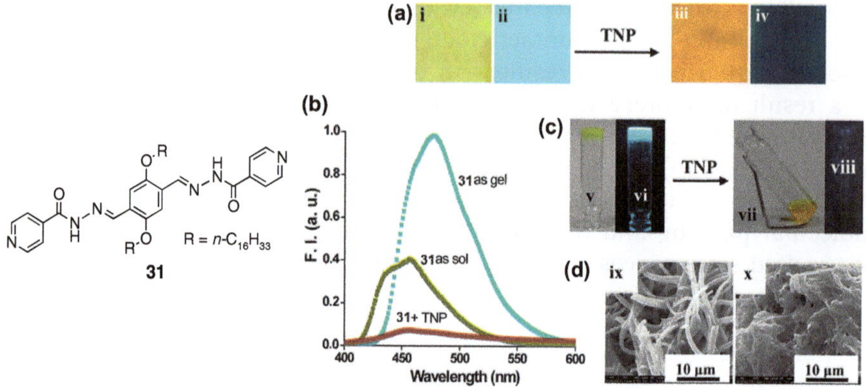

Figure 4.20 LMG **31** (left); (a) test strip detection of TNP under (i, iii) daylight and (ii, iv) UV light (365 nm); (b) fluorescence spectra of **31** in THF in gel and sol state in the absence and presence of 2 equiv. of TNP ($\lambda_{ex}=390$ nm); (c) gel of **31** in THF and in the presence of TNP under (v, vii) daylight and (vi, viii) UV light; (d) SEM images of (ix) **31** and (x) **31**+TNP in THF (reproduced with permission).[49]

bonding interactions between the nitro groups in TNT and the hydrogen atoms in the OPV backbone and also between the methyl group in TNT and the fluorine atoms of **30.** When the gel strips were tested with different analytes in a water–acetonitrile mixture, the naked-eye detection limit of TNT was as low as parts per quadrillion (ppq). In the aqueous solution, the sensitivity was in the range of 11.4–0.23 ppq. Similarly, a THF gel of a *p*-phenylenevinylene derivative **31** showed a TNP-induced elective gel-to-sol transition with a significant colour change from pale yellow to dark yellow.[49] Addition of TNP in the gel state led to a 14-fold quenching of emission intensity with a 24 nm blue shift (Figure 4.20). Interaction of TNP with **31** *via* the formation of multiple hydrogen bonding was found to be the cause for gel breakdown.

A pentacenequinone derivative (**32**) formed a stable opaque gel in various solvent mixtures such as toluene–dichloromethane (DCM), benzene–DCM and *o*-xylene–DCM (Figure 4.21).[50] Gel coated paper strips served as a sensitive, fast and low-cost method for detection of picric acid. When picric acid crystals were brought into contact with a test strip for 5 s, black spots were observed under illumination at 365 nm. Detection with picric acid solution was also successful with a minimum detection limit up to parts per billion levels. A static quenching involving the formation of a ground-state complex between **32** and picric acid was found to be responsible for fluorescence quenching. A highly porous gel network facilitated the penetration of picric acid and amplified the detection signal.

These findings highlight the unique capability of the self-assembled structures as more promising materials than the individual molecule for a specific function.

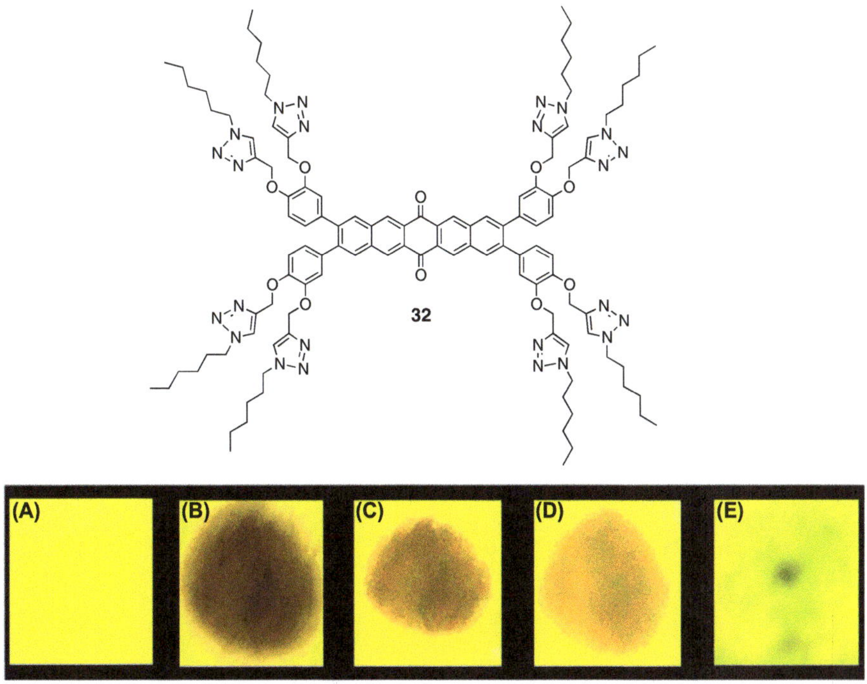

Figure 4.21 LMG **32** (top) and photographs (under 365 nm UV light) of fluorescence quenching of organogel of **32** on test strips for the visual detection of small amounts of picric acid: (a) test strip; picric acid of different concentrations (b) 1×10^{-3} M, (c) 1×10^{-5} M, (d) 1×10^{-7} M and (e) 1×10^{-9} M (reproduced with permission).[50]

4.3.5 Multifaceted Chemical Sensitivity and Extension to Biochemicals

In previous sections, we have attempted to categorise various organogel systems based on the nature of a stimulus. In every case, the application of a particular stimulus induced a significant and detectible alteration in the gel systems. However, there are cases where a highly organised self-assembly process involved in gel formation is capable of controlling the orientation of existing reactive or catalytic groups to bring them in to a critical proximity, which has been the basis of the use of organogels as reaction vessels and catalytic systems. Gel phase confinement of functional groups can even bring about amazing selectivity catalysing a physical or a chemical process. In the first example of supramolecular photochirogenesis carried out in an organogel medium, reported by our group, a cyclohexane gel of an anthracene-based bicomponent LMG **33** showed unprecedented stereoselectivity together with a significant enantioselectivity in the photocyaddition reaction of asymmetrically-substituted anthracene (Scheme 4.12).[51] A minute modification of LMG **33** to LMG **34**,

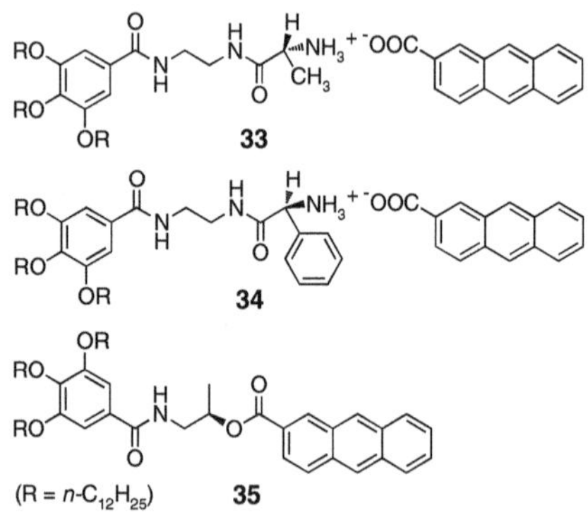

Scheme 4.12 Anthracene-based LMGs used for supramolecular photochirogenesis.

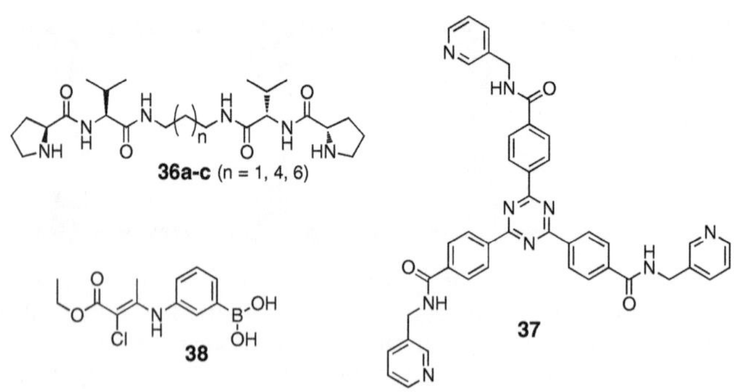

Scheme 4.13 LMGs used for catalysis and biosensing.

where only a methyl group was substituted with a bulky phenyl group, could inverse the stereoselectivity of the process.[39] Interestingly, LMG **35** with covalently-linked anthracene resulted in a high value of enantiomeric excess (ee).[52] In all cases, pre-orientation of the gelator molecules prior to the photoreaction, controlled *via* the supramolecular interactions, directed the selectivity of the process.

Although a drastic change in the gel itself might not always necessarily occur, its influence on the ongoing chemical event could be crucial. L-Proline derivatives **36**, which are able to form supramolecular gels in acetonitrile, showed a basicity increase in the gel state as compared to the solution (Scheme 4.13).[53] As a result, they behaved as enantioselective catalysts for the diffusion-controlled aldol reaction between acetone and 4-nitrobenzaldehyde in solution but produced a base-catalyzed aldol racemisation in the gel state. Further, nitromethane and nitroethane gels of **36c** acted as basic catalysts for the Henry nitroaldol reaction of these solvents with 4-nitrobenzaldehyde.[54] The catalytic gel from nitroethane was more active compared to that of nitromethane. In gel, the L-proline moiety acted as a basic catalyst promoting the reaction through an ion-pair type mechanism.

Further, in a recent study, a fibrous networked Pd(II) metallogel from a pyridine-based tripodal ligand (**37**) has shown higher catalytic activity in Suzuki–Miyaura coupling reactions than spherical gel.[55] The morphology of the gels was found to be largely dependent on the metal–ligand ratios.

Last but not the least, the use of organogels in biosensing and drug delivery[7a,56] is rare but not unprecedented. Unfortunately, their use in drug delivery is still quite limited by the toxicology information available on organogelators, as well as by the few pharmaceutically-accepted solvents used in gel systems.[57] A similar compatibility issue becomes relevant when designing a biosensitive organogel. Nevertheless, there are discrete examples in which innovative strategy coupled with thoughtful molecular design gave us some inspiring outputs. For example, the previously mentioned LMG **16** used for the first time for enzyme immobilisation in Pickering emulsions (a Pickering emulsion is an emulsion stabilized by solid particles)[58] and its release on demand could be initiated by a chemical stimuli.[26]

Glucose-responsive materials always draw considerable attention because of their potential in regulating insulin release for treating diabetes mellitus. Thus, there are several examples of glucose-responsive polymers functionalised with phenylboronic acid (PBA), which efficiently binds with sugar.[59] Extending this idea, a PBA derivative **38** was synthesised. As evaluated by UV-spectroscopy, a cyclohexane gel of **38** showed a prompt initial response to glucose at a low concentration of 0.012 mmol mL^{-1}, which is a critical concentration of venous plasma glucose for diabetes (Figure 4.22).[60] Also, this organogel showed an excellent sensitivity to glucose among the seven sugars (mannitol, galactose, lactose, maltose, sucrose, and fructose) tested.

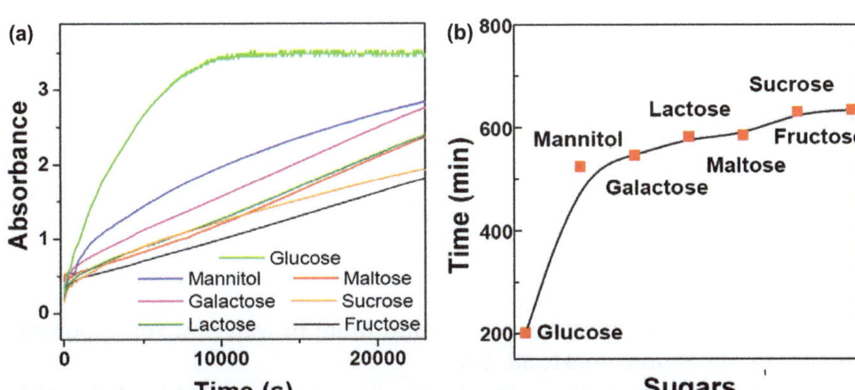

Figure 4.22 (a) Saccharide-response dynamic curve for the gels of **38** in saccharide solutions (3 mL, 0.18 mmol mL^{-1}), (b) sustained response time for total gel-to-sol transition in seven types of saccharide solutions. All the gels were made of 20 mg of **38** in 0.4 mL cyclohexane (reproduced with permission).[60]

The formation of the PBA–glucose complex eventually destroyed the intermolecular interactions among the gelator resulting in a gel breakdown.

4.4 Conclusions and Perspectives

Modern techniques in elucidating structural information, together with the involvement of crystal engineering, enhance the success rate in developing organogels as we could see a high influx of publications related to organogel systems during the last decade. However, rationalising the molecular design, especially in aiming for a particular gel function, still remains a challenge. The job becomes even more challenging in designing a stimuli-sensitive system in general, because of the involvement of a particular functional group that may crucially perturb the gel formation. Particularly, specific interactions with a foreign chemical species often take place in a chemoresponsive system. Thus, system restoration is much more difficult here. On the other hand, direct contact between a substrate and a host pave the way for chemical sensing. As we could see in the previous sections, there are several recent examples of organogel-based chemosensors, some of which have more practical appeal, while others fulfil academic interest mainly. Selectivity, specificity and robustness are the other issues that need to be addressed more carefully. Talking about the positive site, in most cases, self-assembled systems emerged as superior in translating the signals compared to their isotropic counterparts. Also, readily accessible synthetic routes, together with functional programming and fast response to external stimuli, are the advantages of these materials over polymeric gel systems. The complexity involved in the multiresponsive systems brings about the possibility of mimicking real-world complexity, which is still a fantasy. Finally, an early indication to bring this chemistry closer to biological systems would keep its relevance and promise intact for the next generation of smart materials.

Acknowledgements

Financial support (partly) from MEXT, Grant-in-Aid for Scientific Research B (No. 23350070).

References

1. First report of a molecular gel, lithium urate in water: A. von Lipowitz, *Ann. Chem. Pharm.*, 1841, **38**, 348.
2. (a) T. Graham, *Philos. Trans. R. Soc. London*, 1861, **151**, 183; (b) P. H. Hermans, *Colloid Science*, ed. H. R. Kruyt, Elsevier, Amsterdam, 1949, vol. II, p. 483; (c) P. Flory, *J. Discuss. Faraday Soc.*, 1974, **57**, 7.
3. Representative reviews on supramolecular gels: (a) P. Terech and R. G. Weiss, *Chem. Rev.*, 1997, **97**, 3133; (b) N. M. Sangeetha and U. Maitra, *Chem. Soc. Rev.*, 2005, **34**, 821; (c) P. Dastidar, *Chem. Soc. Rev.*,

2008, **37**, 2699; (d) A. R. Hirst, B. Escuder, J. F. Miravet and D. K. Smith, *Angew. Chem., Int. Ed.*, 2008, **47**, 8002; (e) A. Dawn, T. Shiraki, S. Haraguchi, S. Tamaru and S. Shinkai, *Chem. – Asian J.*, 2011, **6**, 266; (f) L. E. Buerklea and S. J. Rowan, *Chem. Soc. Rev.*, 2012, **41**, 6089; (g) G. Yu, X. Yan, C. Han and F. Huang, *Chem. Soc. Rev.*, 2013, **42**, 6697; (h) S. S. Babu, V. K. Praveen and A. Ajayaghosh, *Chem. Rev.*, 2014, **114**, 1973; (i) Y. Feng, Y.-M. He and Q.-H. Fan, *Chem. – Asian J.*, 2014, **9**, 1724.
4. Books and book chapters on supramolecular gels: (a) Low Molecular Mass Gelators: Design, Self-Assembly, Function, ed. F. Fages, *Topics in Current Chemistry*, Springer-Verlag, Berlin, 2005, p. 256; (b) *Molecular Gels, Materials with Self-Assembled Fibrillar Networks*, ed. R. G. Weiss and P. Terech, Springer, Dordrecht, 2006; (c) *Functional Molecular Gels*, ed. B. Escuder and J. F. Miravet, Royal Society of Chemistry, 2013.
5. X. Yang, G. Zhang and D. Zhang, *J. Mater. Chem.*, 2012, **22**, 38.
6. M. D. Segarra-Maset, V. J. Nebot, J. F. Miravet and B. Escuder, *Chem. Soc. Rev.*, 2013, **42**, 7086.
7. For chemosensitive hydrogels see: (a) N. A. Peppas, O. Fisher and M. Moore, *Chemoresponsive Materials: Stimulation by Chemical and Biological Signals*, ed. H.-J. Schneider, Royal Society of Chemistry, 2015, ch. II; (b) H.-J. Schneider, *Chemoresponsive Materials: Stimulation by Chemical and Biological Signals*, ed. H.-J. Schneider, Royal Society of Chemistry, 2015, ch. III.
8. (a) Y. Tao, L. Ai, H. Bai and X. Liu, *J. Polym. Sci., Part A: Polym. Chem.*, 2012, **50**, 3507; (b) A. S. Hoffman, *Adv. Drug Delivery Rev.*, 2013, **65**, 10; (c) R. K. Mishra, K. Ramasamy, N. A. Ahmad, Z. Eshak and A. B. A. Majeed, *J. Mater. Sci.: Mater. Med.*, 2014, **25**, 999; (d) Y. J. Wang, L. J. Chen, L. W. Tan, Q. Zhao, F. Luo, Y. Q. Wei and Z. Y. Qian, *Biomaterials*, 2014, **35**, 6972.
9. Y. Li, K. M.-C. Wong, A. Y.-Y. Tam, L. Wu and V. W.-W. Yam, *Chem. – Eur. J.*, 2010, **16**, 8690.
10. H. Wu, B.-B. Ni, C. Wang, F. Zhai and Y. Ma, *Soft Matter*, 2012, **8**, 5486.
11. P. Xue, R. Lu, J. Jia, M. Takafuji and H. Ihara, *Chem. – Eur. J.*, 2012, **18**, 3549.
12. S.-Y. Hsueh, C.-T. Kuo, T.-W. Lu, C.-C. Lai, Y.-H. Liu, H.-F. Hsu, S.-M. Peng, C-h. Chen and S.-H. Chiu, *Angew. Chem., Int. Ed.*, 2010, **49**, 9170.
13. W. Edwards and D. K. Smith, *Chem. Commun.*, 2012, **48**, 2767.
14. J. Lu, J. Hu, Y. Song and Y. Ju, *Org. Lett.*, 2011, **13**, 3372.
15. (a) H. Maeda, *Chem. – Eur. J.*, 2008, **14**, 11274; (b) M.-O. M. Piepenbrock, G. O. Lloyd, N. Clarke and J. W. Steed, *Chem. Rev.*, 2010, **110**, 1960; (c) M. Wenzel, J. R. Hiscock and P. A. Gale, *Chem. Soc. Rev.*, 2012, **41**, 480.
16. J. W. Steed, *Chem. Soc. Rev.*, 2010, **39**, 3686.
17. J. A. Foster, M.-O. M. Piepenbrock, G. O. Lloyd, N. Clarke, J. A. K. Howard and J. W. Steed, *Nat. Chem.*, 2010, **2**, 1037.
18. P. Rajamalli and E. Prasad, *Org. Lett.*, 2011, **13**, 3714.

19. L.-B. Xing, B. Yang, X.-J. Wang, J.-J. Wang, B. Chen, Q. Wu, H.-X. Peng, L.-P. Zhang, C.-H. Tung and L.-Z. Wu, *Langmuir*, 2013, **29**, 2843.
20. J. Lee, J. E. Kwon, Y. You and S. Y. Park, *Langmuir*, 2014, **30**, 2842.
21. M. Teng, G. Kuang, X. Jia, M. Gao, Y. Li and Y. Wei, *J. Mater. Chem.*, 2009, **19**, 5648.
22. Q. Xia, Y. Mao, J. Wu, T. Shu and T. Yi, *J. Mater. Chem. C*, 2014, **2**, 1854.
23. A. A. Sobczuk, S.-i. Tamaru and S. Shinkai, *Chem. Commun.*, 2011, **47**, 3093.
24. S. Dong, B. Zheng, D. Xu, X. Yan, M. Zhang and F. Huang, *Adv. Mater.*, 2012, **24**, 3191.
25. X. Yan, D. Xu, X. Chi, J. Chen, S. Dong, X. Ding, Y. Yu and F. Huang, *Adv. Mater.*, 2012, **24**, 362.
26. Z. Qi, C. Wu, P. Malo de Molina, H. Sun, A. Schulz, C. Griesinger, M. Gradzielski, R. Haag, M. B. Ansorge-Schumacher and C. A. Schalley, *Chem. – Eur. J.*, 2013, **19**, 10150.
27. K. Fan, J. Song, J. Li, X. Guan, N. Tao, C. Tong, H. Shen and L. Niu, *J. Mater. Chem. C*, 2013, **1**, 7479.
28. Q. Jin, L. Zhang, X. Zhu, P. Duan and M. Liu, *Chem. – Eur. J.*, 2012, **18**, 4916.
29. (a) S.-i. Kawano, N. Fujita and S. Shinkai, *Chem. – Eur. J.*, 2005, **11**, 4735; (b) T. Kitahara, M. Shirakawa, S.-i. Kawano, U. Beginn, N. Fujita and S. Shinkai, *J. Am. Chem. Soc.*, 2005, **127**, 14980; (c) J. Liu, P. He, J. Yan, X. Fang, J. Peng, K. Liu and Y. Fang, *Adv. Mater.*, 2008, **20**, 2508.
30. C. Wang, Q. Chen, F. Sun, D. Zhang, G. Zhang, Y. Huang, R. Zhao and D. Zhu, *J. Am. Chem. Soc.*, 2010, **132**, 3092.
31. X. Yang, G. Zhang, L. Li, D. Zhang, L. Chi and D. Zhu, *Small*, 2012, **8**, 578.
32. R. Afrasiabi and H.-B. Kraatz, *Chem. – Eur. J.*, 2013, **19**, 17296.
33. B. Adhikari and H.-B. Kraatz, *Chem. Commun.*, 2014, **50**, 5551.
34. W. Miao, L. Zhang, X. Wang, H. Cao, Q. Jin and M. Liu, *Chem. – Eur. J.*, 2013, **19**, 3029.
35. (a) S.-i. Kawano, N. Fujita and S. Shinkai, *J. Am. Chem. Soc.*, 2004, **126**, 8592; (b) K. Nobusawa, A. Ikeda, J.-i. Kikuchi, S.-i. Kawano, N. Fujita and S. Shinkai, *Angew. Chem., Int. Ed.*, 2008, **47**, 4577.
36. (a) A. J. Hutf and J. O'Grady, *J. Antimicrob. Chemother.*, 1996, **37**, 7; (b) S. Sangita and S. Roy, *J. Indian Chem. Soc.*, 2006, **83**, 315; (c) W. H. Brooks, W. C. Guida and K. G. Daniel, *Curr. Top. Med. Chem.*, 2011, **11**, 760.
37. (a) T. D. James, H. Kawabata, R. Ludwig, K. Murata and S. Shinkai, *Tetrahedron*, 1995, **51**, 555; (b) D. K. Smith, *Chem. Soc. Rev.*, 2009, **38**, 684; (c) P. Duan, H. Cao, L. Zhang and M. Liu, *Soft Matter*, 2014, **10**, 5428.
38. T. D. James, K. Murata, T. Harada, K. Ueda and S. Shinkai, *Chem. Lett.*, 1994, **23**, 273.
39. A. Dawn, N. Fujita, S. Haraguchi, K. Sada, S.-i. Tamaru and S. Shinkai, *Org. Biomol. Chem.*, 2009, 7, 4378.
40. X. Chen, Z. Huang, S.-Y. Chen, K. Li, X.-Q. Yu and L. Pu, *J. Am. Chem. Soc.*, 2010, **132**, 7297.

41. T. Tu, W. Fang, X. Bao, X. Li and K. H. Dotz, *Angew. Chem., Int. Ed.*, 2011, **50**, 6601.
42. W. Miao, L. Zhang, X. Wang, L. Qin and M. Liu, *Langmuir*, 2013, **29**, 5435.
43. X. Wang, P. Duan and M. Liu, *Chem. - Asian J.*, 2014, **9**, 770.
44. P. Mukhopadhyay, Y. Iwashita, M. Shirakawa, S-i. Kawano, N. Fujita and S. Shinkai, *Angew. Chem., Int. Ed.*, 2006, **45**, 1592.
45. (a) T. Tu, W. Fang and Z. Sun, *Adv. Mater.*, 2013, **25**, 5304; (b) S. Dong, B. Zheng, F. Wang and F. Huang, *Acc. Chem. Res.*, 2014, **47**, 1982.
46. P. Xue, Q. Xu, P. Gong, C. Qian, A. Ren, Y. Zhang and R. Lu, *Chem. Commun.*, 2013, **49**, 5838.
47. K. Y. Kim, S. Park, S. H. Jung, S. S. Lee, K.-M. Park, S. Shinkai and J. H. Jung, *Inorg. Chem.*, 2014, **53**, 3004.
48. K. K. Kartha, S. S. Babu, S. Srinivasan and A. Ajayaghosh, *J. Am. Chem. Soc.*, 2012, **134**, 4834.
49. N. Dey, S. K. Samanta and S. Bhattacharya, *ACS Appl. Mater. Interfaces*, 2013, **5**, 8394.
50. V. Bhalla, A. Gupta, M. Kumar, D. S. S. Rao and S. K. Prasad, *ACS Appl. Mater. Interfaces*, 2013, **5**, 672.
51. A. Dawn, N. Fujita, S. Haraguchi, K. Sada and S. Shinkai, *Chem. Commun.*, 2009, 2100.
52. A. Dawn, T. Shiraki, S. Haraguchi, H. Sato, K. Sada and S. Shinkai, *Chem. - Eur. J.*, 2010, **16**, 3676.
53. F. Rodriguez-Llansola, B. Escuder and J. F. Miravet, *Org. Biomol. Chem.*, 2009, 7, 3091.
54. F. Rodriguez-Llansola, B. Escuder and J. F. Miravet, *J. Am. Chem. Soc.*, 2009, **131**, 11478.
55. Y.-R. Liu, L. He, J. Zhang, X. Wang and C.-Y. Su, *Chem. Mater.*, 2009, **21**, 557.
56. For use of hydrogels in drug delivery see: (a) J. F. Miravet and B. Escuder, *Supramolecular Systems in Biomedical Fields*, ed. H.-J. Schneider, Royal Society of Chemistry, 2013, ch. XI, p. 331; (b) *Smart Materials for Drug Delivery*, ed. C. Alvarez-Lorenzo and A. Concheiro, Royal Society of Chemistry, 2013.
57. A. Vintiloiu and J.-C. Leroux, *J. Controlled Release*, 2008, **125**, 179.
58. Y. Chevalier and M.-A. Bolzinger, *Colloids Surf., A*, 2013, **439**, 23.
59. For glucose responsive hydrogels see: (a) H. J. van der Linden, S. Herber, W. Olthuis and P. Bergveld, *Analyst*, 2003, **128**, 325; (b) V. Ravaine, C. Ancla and B. Catargi, *J. Controlled Release*, 2008, **132**, 2; (c) J. D. Ehrick, M. R. Luckett, S. Khatwani, Y. Wei, S. K. Deo, L. G. Bachas and S. Daunert, *Macromol. Biosci.*, 2009, **9**, 864; (d) G. Ye and X. Wang, *Biosens. Bioelectron.*, 2010, **26**, 772; (e) R. A. Siegel, Y. Gu, M. Lei, A. Baldi, E. E. Nuxoll and B. Ziaie, *J. Controlled Release*, 2010, **141**, 303; (f) C. Zhang, M. D. Losego and P. V. Braun, *Chem. Mater.*, 2013, **25**, 3239; (g) S.-H. Cho, P. Tathireddy, L. Rieth and J. Magda, *J. Appl. Polym. Sci.*, 2014, 40667.
60. C. Zhou, W. Gao, K. Yang, L. Xu, J. Ding, J. Chen, M. Liu, X. Huang, S. Wang and H. Wu, *Langmuir*, 2013, **29**, 13568.

CHAPTER 5

Multi-stimuli Responsive Materials

ZHENHUI QI AND CHRISTOPH A. SCHALLEY*

Institut für Chemie und Biochemie, Freie Universität Berlin, Takustrasse 3, 14195 Berlin, Germany
*Email: c.schalley@fu-berlin.de

5.1 Introduction

Responsive materials adapting to their environment and responding to external stimuli are emerging as "smart materials".[1–11] In many real-world applications, different environmental changes occur. Single-stimulus responsiveness may therefore be of limited utility to accomplish a desired purpose. This is especially true when responsive materials are to be interfaced with biological systems with their incredible structural and dynamic complexity that expresses itself in the complex feedback-loop-governed spatiotemporal patterns created in the living cell. This generates a demand for materials that are responsive to multiple stimuli. For example, to achieve the best effects and to reduce side effects of drugs, it is highly desirable to precisely regulate drug carriers with respect to the released drug dosage and the release rate depending on a patient's individual health condition. A drug carrier system capable of sensing a number of different stimuli indicative of the patient's condition would not only create specificity for the location where the drug is needed, but could also regulate the amount of drug delivered automatically.

The aim of this chapter is to discuss the current state of the art in the design of multi-stimuli responsive materials. This area is diverse and does

RSC Smart Materials No. 14
Chemoresponsive Materials: Stimulation by Chemical and Biological Signals
Edited by Hans-Jörg Schneider

Published by the Royal Society of Chemistry, www.rsc.org

not only concern the vision of precise drug delivery used as just one of many examples above. Not only do the types of materials differ significantly, but also their area of application and of course the nature of the stimuli, which can roughly be categorized into physical signals such as temperature, pressure, electric and magnetic fields, and (bio)chemical stimuli ranging from simple ones such as the pH of the environment to specific non-covalent binders and even biocatalysts. We restrict our discussion on multi-stimuli responsive materials to those sensitive to the latter category and summarize the action of (bio)chemical signals on the materials. Materials sensitive to temperature,[12] electric[13] or magnetic fields,[14] or pressure changes[15] are nevertheless highly interesting, but will not be included here.

Therefore, in the first section of the chapter, recent achievements, mostly from the last five years, are summarized focusing on gels, surfaces and nanocarriers as a selection of representative materials. In particular, multi-stimuli responsive gels are chosen as a specific example to introduce the diverse strategies and elements that have been used to implement multiple responsiveness and fabricate multi-stimuli responsive materials. In the second section, recent attempts to program smart materials at a higher level of complexity are described including materials engineered to exert Boolean logic and their integration into signalling networks and cascade reactions.

5.2 Multi-stimuli Responsive Functional Materials

The design of materials that respond to multiple chemical stimuli is a process of integrating diverse responsive elements. The rapid development of multi-stimuli responsive gel materials provides us with a series of elegant examples to illustrate the design strategies and available elements that have been used for this purpose.

5.2.1 Dual-responsive Gels based on Dynamic Covalent Bonds

Gels represent a unique class of soft materials, because they combine the elastic behaviour of solids with the micro-viscous properties of fluids.[16,17] On the microscopic level, polymer gels and supramolecular gels usually consist of three-dimensional fibrous networks that span the volume of a liquid medium and entrap the solvent inside small cavities through the surface tension effect. They are ideally suited as responsive materials as it is often possible to convert the polymeric network structure into smaller, non-gelating units by external stimuli. This induces a gel–sol transition which liberates the contents of the small gel cavities. Responsive gels can therefore fulfil a number of different functions and have found applications as responsive reaction media, as devices for controlled drug delivery, for tissue engineering, and as self-healing materials.[18–21]

The development of dynamic combinatorial chemistry (DCC)[22,23] delivers a wide variety of reversible and thus dynamically forming covalent bonds

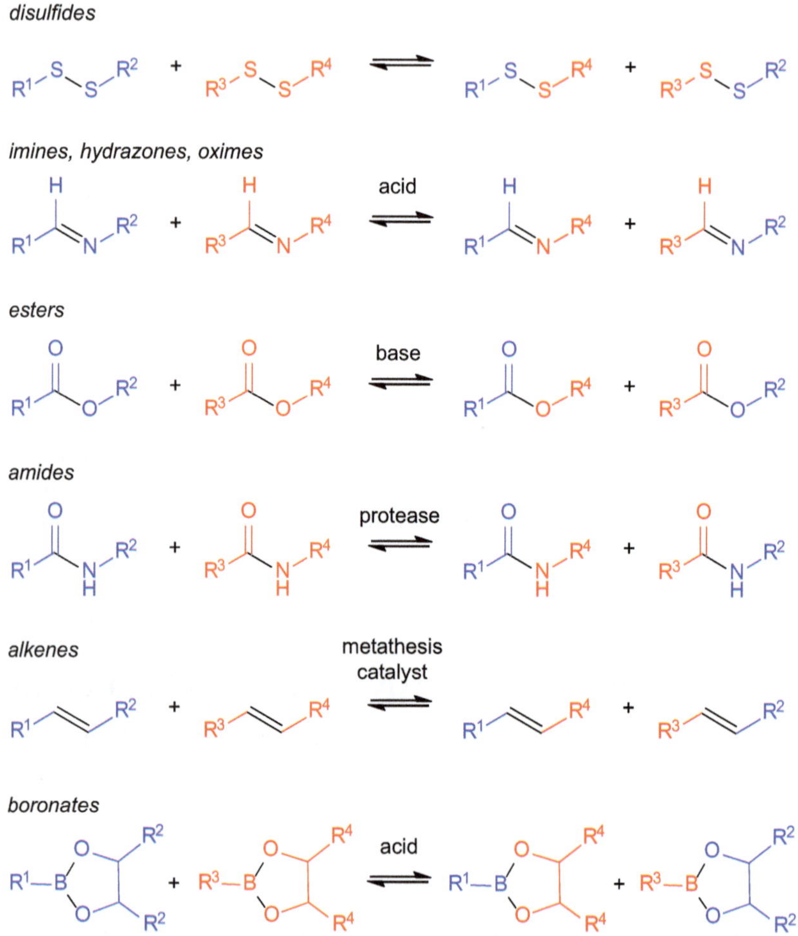

Figure 5.1 A brief overview of some frequently used dynamic covalent bonds.

(Figure 5.1) on which stimuli-responsiveness can be based.[24,25] One advantage is obvious: reversible bond formation occurs under thermodynamic control and changes in the environment caused by chemical stimuli will shift the equilibrium between different states. This sensitivity to the environmental conditions qualifies the dynamic covalent bond for accessing stimuli-responsive materials. A second advantage is that the reactivity of different dynamic covalent bonds is often orthogonal.[26] For example, imines and disulfides can each be made reversible under conditions that keep the other one irreversible.[27] Similarly, disulfide and thioester formation, imine and boronate condensation, or imine and olefin metathesis represent such orthogonal pairs. Even similar functional groups sometimes exhibit sufficiently pronounced reactivity differences. For example, acetals and ketals are both acid labile and are cleaved when the pH drops to below 5.0–5.5. Their half-lives are nevertheless very different: $t_{1/2}$(acetal) = 5.5 min,[28] while $t_{1/2}$(ketal) = 2 h.[29]

Multi-stimuli responsiveness has been implemented into polymer gels by using orthogonal dynamic covalent bonds in the network-forming polymeric chains.[30–32] One example is shown in Figure 5.2: Deng *et al.*[33] prepared environmentally adaptive self-healing hydrogels based on two types of dynamic covalent bonds by incorporating disulfides as well as acylhydrazones in the same cross-linked polymer through reaction of the three-armed aldehyde-terminated poly(ethylene oxide) **G1** with dithiodipropionic acid dihydrazide **G2**. In this reaction, a 3D dynamic polymeric network is formed. As the acylhydrazone forms and cleaves depending on the pH of the sample and the disulfide bond is redox-sensitive, the resulting hydrogels exhibit reversible gel–sol transitions in response to both triggers as the polymer network is disassembled either by the addition of HCl or dithiothreitol (DTT) and can be reconstructed by triethylamine or H_2O_2 addition, respectively. The hydrogel offers interesting self-healing capabilities at slightly acidic as well as slightly basic conditions, but becomes inert at pH 7, when both the acylhydrazone and the disulfide bond are kinetically inert.

Chen *et al.*[34] showed how to utilize boronate condensation to construct multi-responsive gels. They designed a group of dual-responsive hydrogels in which the stereochemistry of the gel plays an important role (Figure 5.3). They synthesized three glycopolymers, **PMan**, **PGal**, and **PGlc** with different pendent sugar stereoisomers (mannose, galactose and glucose). When these polymers are reacted with the benzoboroxole-containing polymer **PBOB**, hydrogels **Man-10**, **Gal-10**, and **Glc-10** form through the formation of the corresponding borates. These boronates are pH sensitive and thus pH is the first stimulus with which the hydrogels can be switched between the gel and the sol state.

Rheology revealed the gelation ability of the glycopolymers with **PBOB** to follow the sequence **PMan** > **PGal** > **PGlc**—inconsistent with their binding ability ranking. The authors attribute this finding to a glyco-stereoisomerism effect: when the sugars contain *cis*-diols as in **PMan** and **PGal**, the interaction between the two polymers is stronger as compared to **PGlc** (no *cis*-diol). This stereoisomerism effect only works with polymer-attached sugars, while free, no significant association constant differences were found for monosaccharide binding to benzoboroxole **BOB**. The addition of excess free glucose (1.34 equivalents per sugar moiety) as the second stimulus to hydrogels of **PGal** and **PGlc** induces a gel–sol transition, as glucose competes with the carbohydrate-substituted polymers, binds to **PBOB** and thus disrupts the gel network. In contrast, the strongest interaction with **PMan** is not affected by glucose, even when more than 4 equivalents of glucose per sugar moiety are added. As glucose does not bear any *cis*-diols, it is unable to compete with **PMan**. When fructose is added instead, the gel–sol transition is induced. This example thus not only shows how dual-response hydrogels can be designed based on dynamic covalent bonds, but also exhibits interesting stereoselectivity.

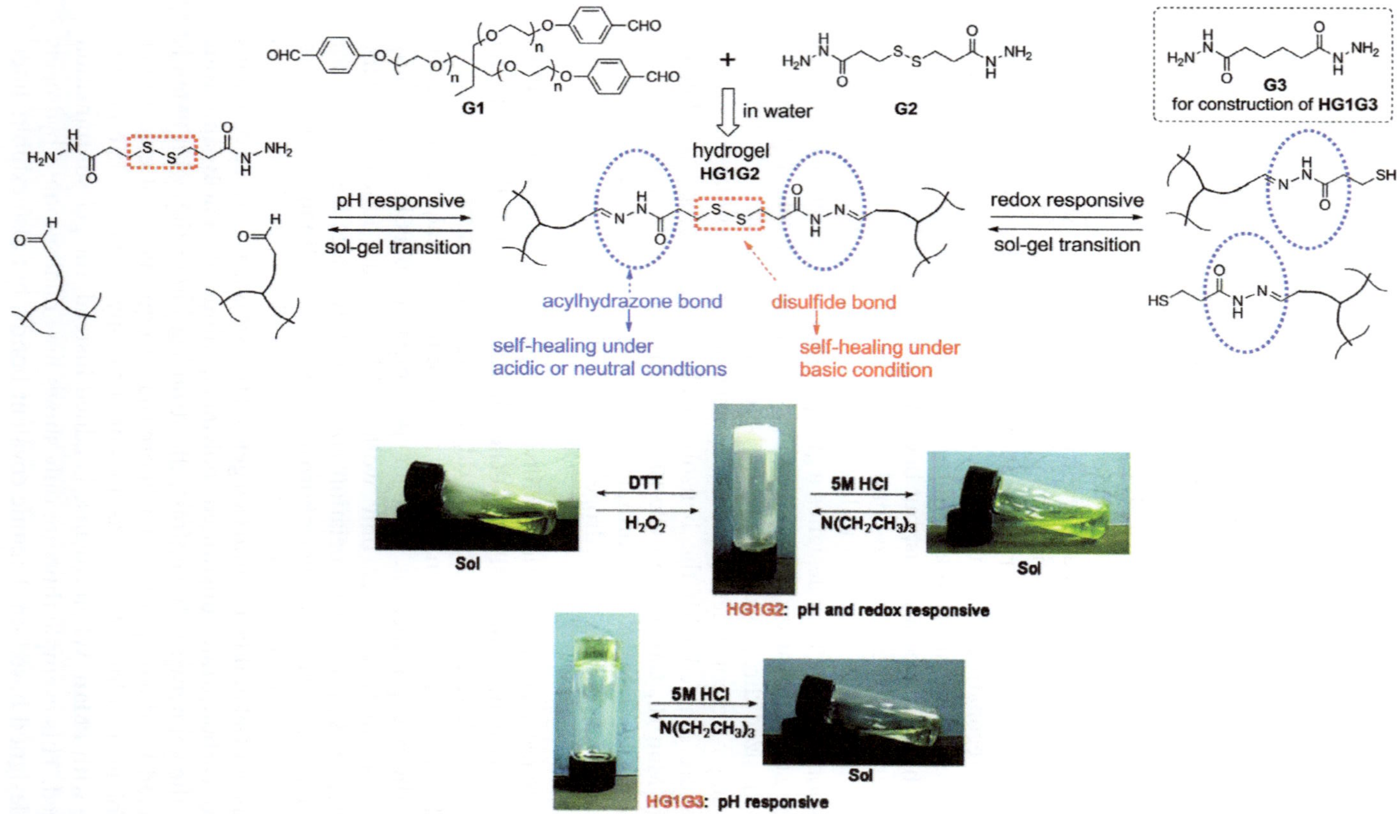

Figure 5.2 A dual-response hydrogel based on orthogonal dynamic covalent bonds. Sol–gel transitions can be triggered by pH changes that affect the acylhydrazone bond or by redox reactions that form and cleave the disulfide bond. The hydrogel exhibits self-healing ability.
Adapted from ref. 33 with kind permission from the American Chemical Society.

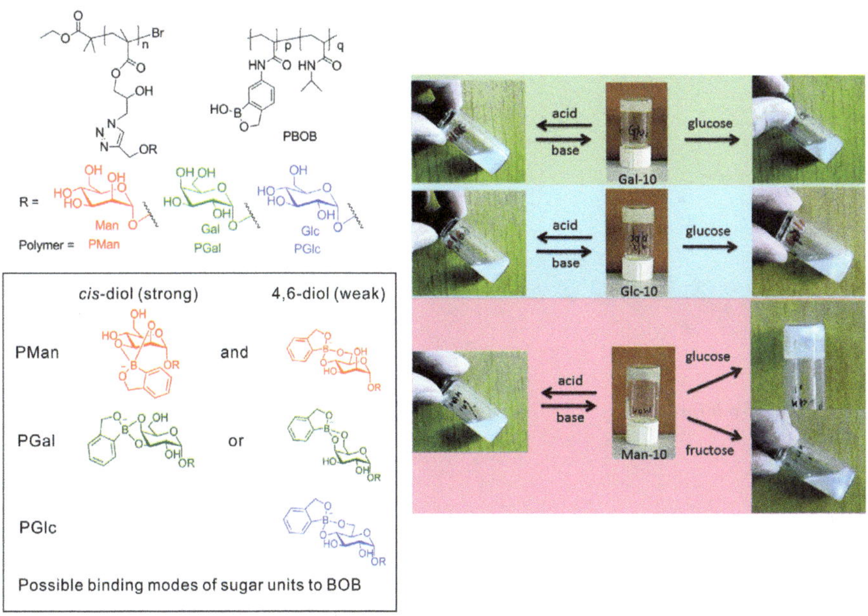

Figure 5.3 Dual-response hydrogels based on carbohydrate–boroxole interactions. The gels contain 10 wt% of the polymers and are denoted as **Man-10**, **Gal-10**, and **Glc-10**. The gelation ability and the capability to use glucose as a stimulus to induce gel–sol transitions depend on the sugar stereochemistry.
Adapted from ref. 34 with kind permission from the Royal Society of Chemistry.

5.2.2 Multi-stimuli Responsive Gels based on Non-covalent Bonds

Besides dynamic combinatorial chemistry, supramolecular chemistry offers a large number of molecular recognition motifs that are reversible and sensitive to the environment and thus versatile responsive elements for the design of multi-stimuli responsive materials (Figure 5.4).[35] In contrast to the dynamic covalent bond, it is usually difficult to make non-covalent binding motifs irreversible by environmental changes. Nevertheless, the broad variety of different interactions ranging from non-directional van der Waals forces through π–π stacking, cation–π and anion–π interactions to the many electrostatic interactions and directional forces, such as hydrogen bonding or metal coordination, offers many orthogonal combinations.[36]

Among the many binding motifs,[44] macrocycles and their host–guest chemistry[45] are particularly popular for the implementation of stimuli responsiveness,[46] as they are structurally well-defined host molecules (Figure 5.5). Until now, a large number of macrocyclic compounds with different constitutions, sizes, shapes and polarities have been developed.

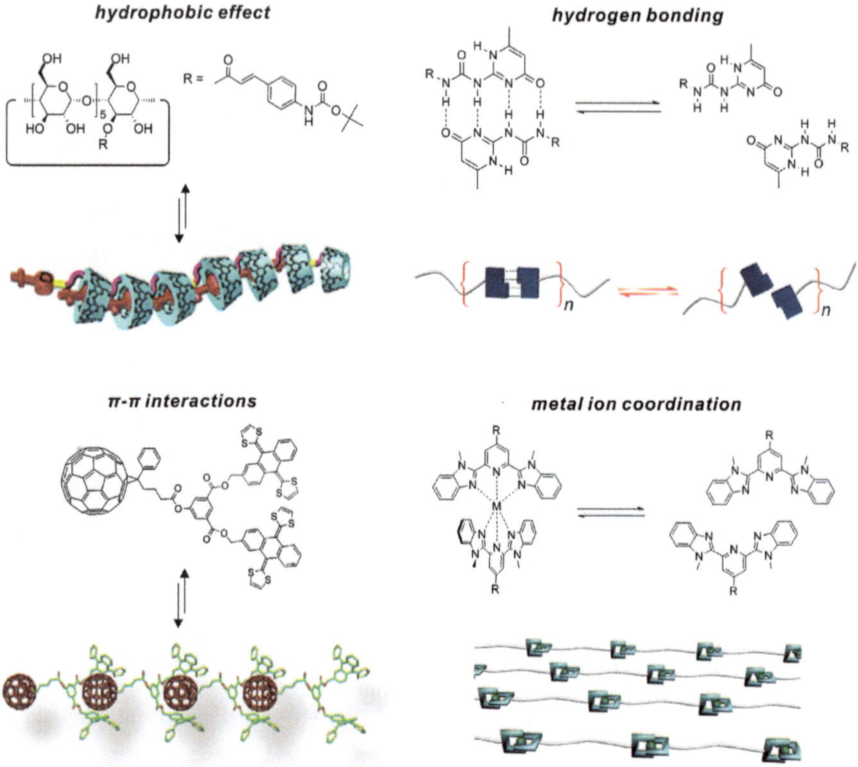

Figure 5.4 A selection of examples of supramolecular polymerization through different non-covalent recognition motifs.
Adapted from ref. 37–39 with kind permission from the American Chemical Society, Wiley-VCH and the Royal Society of Chemistry.

These macrocyclic hosts can be chemically modified to fine-tune their properties and can reversibly encapsulate guest molecules often with high selectivities. Furthermore, these host–guest complexes are sensitive to environmental variations. External stimuli—in particular the addition of competitive guests—are able to control their formation. These features have been successfully utilized to construct sophisticated supramolecular architecture including stimuli-responsive molecular machines.[47–50]

Examples of responsive supramolecular gels have recently been reviewed by Huang *et al.*[51] and provide ample evidence that multi-stimuli responsive gels can be based on non-covalent bonding between low-molecular weight gelator molecules. One example is shown in Figure 5.6. The crown-ether-substituted gelator **DB24-A** forms gels in acetonitrile.[52] It can even be regarded as a supergelator as it has a very low critical gelation concentration (cgc = 0.6 wt%). The chloride counterion is crucial for the high gelation ability and the critical gelation concentration increases significantly to 4.6 wt%, when the corresponding PF_6^- salt is used.

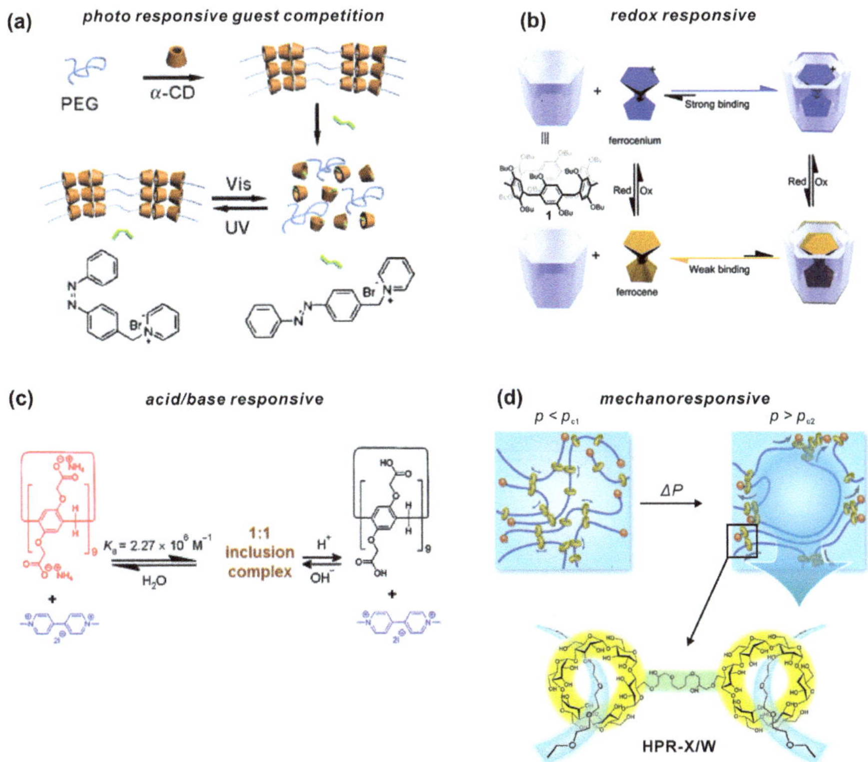

Figure 5.5 Selected examples of macrocycle-based responsive materials: (a) photoresponsive polymerization employing an azobenzene switch;[40] (b) redox-responsive ferrocenium–pillararene interactions;[41] (c) acid/base-switchable pillararene inclusion complexes;[42] (d) pressure-sensitive hydrogels.[43]
Adapted from these references with kind permission from Wiley-VCH and the Royal Society of Chemistry.

Overall, eight different stimuli have been shown to trigger gel–sol transitions that are reversible by a suitable counter-stimulus: organic salts such as dibenzyl ammonium (**DBA**) or paraquat hexafluorophosphate and alkali or transition metal cations such as Na^+, K^+ or Fe^{3+} compete with the secondary ammonium group incorporated in **DB24-A** for the crown ether and thus affect the gel network by reducing the chain length of the **DB24-A** supramolecular polymers. Similarly, deprotonation of the secondary ammonium ion with a base such as triethylamine or blocking it with a Boc group through di-*tert*-butyl dicarbonate addition also induce the collapse of gel. Furthermore, when the gel is formed in a suitable concentration range (between 0.6 and 4.6 wt%), counterion exchange can induce the gel–sol transition, for instance when $AgPF_6$ scavenges the chloride ion and leaves the gelator with PF_6^-.

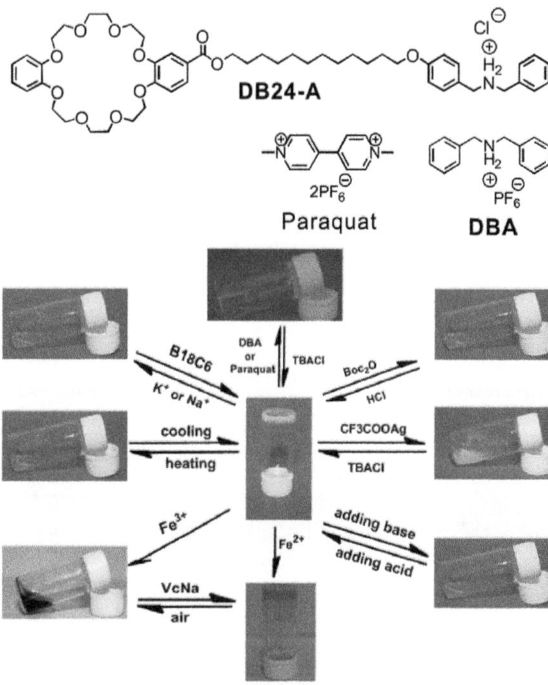

Figure 5.6 Top: chemical structures of the crown-ether-functionalized heteroditopic gelator **DB24-A** and two competing guest cations. Bottom: reversible gel–sol transitions of the **DB24-A** supramolecular gel in acetonitrile triggered by a total of eight different stimuli.
Adapted from ref. 52 with kind permission from Wiley-VCH.

This example elegantly demonstrates how the incorporation of macrocycles as an integral part of the gel structure makes their host–guest chemistry a rich source of stimuli responsiveness. Being directly involved in both gel formation and stimuli responsiveness is, however, only one role that the macrocycles can play.[46] They can also be used as a scaffold presenting conformationally more fixed binding sites. Or they can be appended as a side chain to a gel network.[53,54]

So far, the responsiveness of the gel has concerned gel–sol transitions. However, this is not the only way gels may respond to stimuli. Another option is to form macroscopic gel blocks that exhibit recognition sites on their exterior faces. If two different gel blocks have matching pairs of hosts and guests, the gel blocks can bind to each other and form assemblies on the macroscopic scale. Harada *et al.*[55,56] have followed this approach and were able to even provide evidence for self-sorting phenomena and photoswitchability by using, for example, azobenzenes and cyclodextrins exposed on the faces of small gel pieces.

Recently, Zhao and co-workers used a similar strategy, which was coined "Lego assembly" to implement multiple responsiveness in gel-assemblies based on host–guest interactions between cyclodextrin (CD) and ferrocene (Fc)

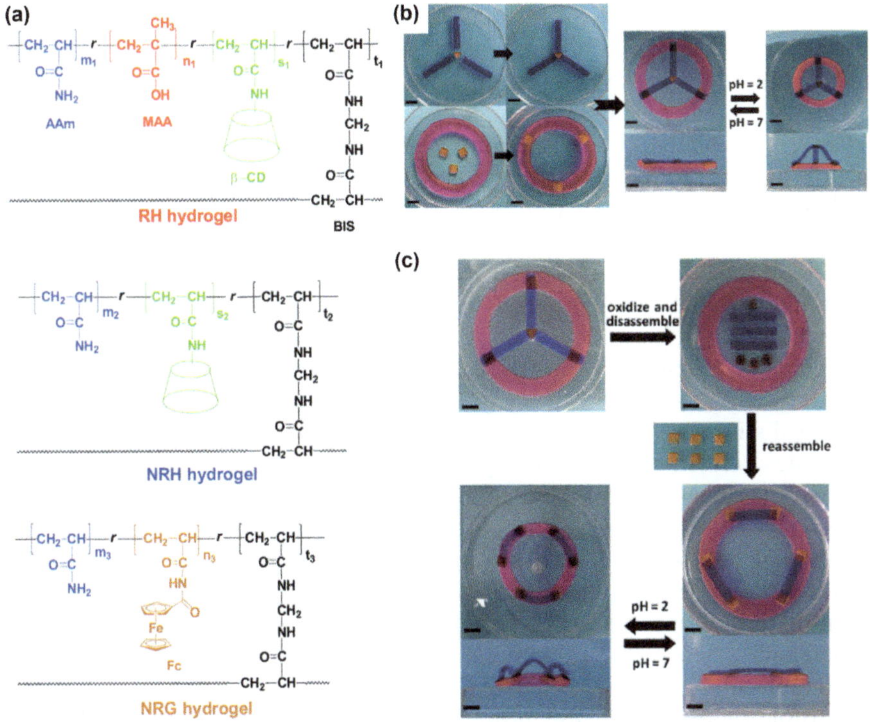

Figure 5.7 (a) The chemical structures of the three hydrogels, (b) the pH-regulated switching between "steering wheel" and "3D tower" assemblies, (c) supramolecular Lego: different stimuli lead to different 3D assemblies of the hydrogel subunits.
Adapted from ref. 57 with kind permission from Wiley-VCH.

units.[57] As shown in Figure 5.7, three types of acrylamide hydrogels were made: (i) a responsive host hydrogel (**RH**) contains both β-CD (the host) and carboxylic acids (the responsive unit). The **RH** hydrogel exhibits responsiveness towards pH and ionic strength owing to the carboxylic acid groups. The swelling ratio of the **RH** hydrogel increases from 7 to 22, when the pH is raised from 3 to 7. The swelling ratio instead decreases from nearly 100 to approximately 30 when the ionic strength changes from 1 mM to 20 mM. Therefore, the **RH** hydrogel exhibits a strong, reversible swelling–deswelling ability. (ii) In contrast, a non-responsive host hydrogel (**NRH**) is equipped only with the β-CD host, but does not have any carboxylic acid side chains. Thus, the swelling–deswelling capability is negligible. (iii) Likewise, the non-responsive guest hydrogel (**NRG**) functionalized with the ferrocene guest does not swell–deswell, but can be oxidized to the corresponding ferrocenium polymer. While ferrocene is a guest for β-CD, ferrocenium is not.

When each type of hydrogel is fabricated into a certain shape as shown in Figure 5.7(b) (**NRH**: blue strips, **NRG**: brownish triangle and square plates, **RH**: pink circle) and mixed, they assemble into a "steering wheel" assembly mediated by ferrocene–β-CD interactions at the faces. The interaction between the blue strips and the pink circle is mediated by the **NRG** subunits. Decreasing the pH causes deswelling and with it a size reduction of the **RH** gel circle, which pushes the non-affected blue **NRH** strips up into a "3D tower". This shape change is completely reversible as the 3D tower can switch back into the 2D steering wheel when the pH increases again. In a similar manner, the ionic strength can also trigger such a 3D shape change. By oxidation of the ferrocene units in the **NRG** hydrogel, the ferrocene–β-CD association is released and the "steering wheel" disassembles into its separate gel block components. The addition of fresh **NRG** units reconstitutes another circular assembly rather than the "steering wheel" as no functional **NRG** triangle is available. This work opens numerous possibilities for 3D assemblies of gels that can be programmed and reprogrammed by chemical stimuli even beyond simple gel–sol transitions.

Besides crown-ethers and cyclodextrins, an increasing variety of macrocycles has been utilized to construct multi-stimuli responsive gels: calixarenes,[58] calixpyrroles,[59] cucurbiturils,[60,61] pillararenes,[62] and macrocycles that cyclize and open through dynamic boronic acid–diol covalent bonds[63] have been employed. Recently, Liu *et al.* further demonstrated the combination of dynamic covalent chemistry and host–guest chemistry to be a practical strategy to implement multi-stimuli responsiveness.[64] Other supramolecular interactions, such as metal–ion coordination[65,66] and quadruple zwitterions,[67] also proved useful for an efficient construction of multiply-responsive gel materials.

5.2.3 Multi-stimuli Responsive Gels based on Biomolecules

Also, biomolecules have the potential to be incorporated as the responsive unit into stimuli-responsive materials. One of the benefits of using biohybrid materials is their potential to respond to natural biochemical signals including those associated with certain disease states. In addition, some suitable biomolecules, for example enzymes, enable the formation and cleavage of covalent bonds reversibly.[68] Figure 5.8 summarises three examples. In the first one (Figure 5.8(a)),[69] a C-terminally protected precursor peptide only assembles into micrometre-long fibres, when a suitable enzyme is present as the stimulus for the deprotection of the C-terminus. This example nicely illustrates that even (bio)catalysts can be used as the external stimuli to induce supramolecular order and make the assemblies emerge. In the second example,[70] a three-dimensional DNA nanostructure is switched by different stimuli depending on the sequence of the pink strand in Figure 5.8(b) between an open structure that does not show any Förster resonance energy transfer (FRET) between the two chromophores in the two corners of the tetrahedron and a closed structure which exhibits FRET. Finally, carbohydrates can be responsive units.[71] The rhodamine B–carbohydrate conjugates shown in Figure 5.8(c) quench the fluorescence of graphene oxide when attached to it. In the presence of Hep-G2 cancer cells

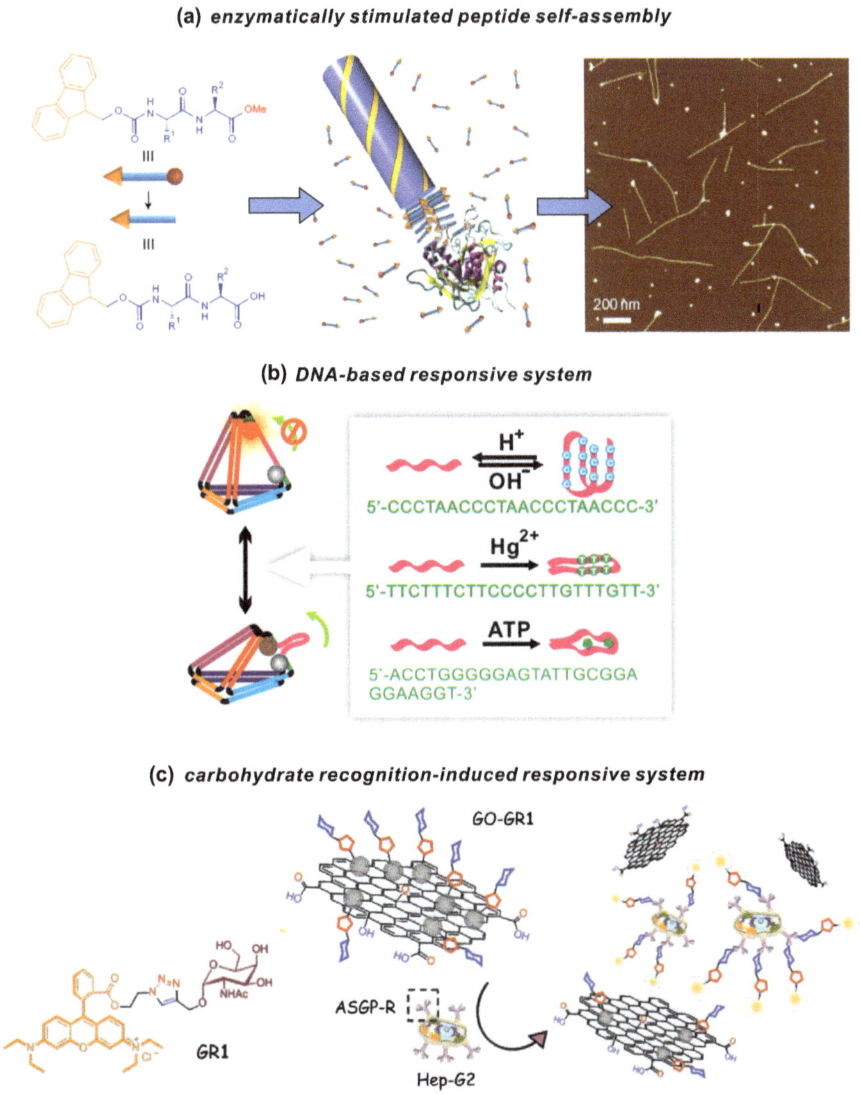

Figure 5.8 (a) Enzyme-catalysed peptide self-assembly. (b) Reconfigurable DNA tetrahedron that can be switched by different stimuli between an open and a closed structure (green arrow = FRET signal). (c) A carbohydrate recognition-based responsive system using a graphene oxide platform. Adapted from ref. 69–71 with kind permission from the Nature Publishing Group and Wiley-VCH.

expressing the corresponding asialoglycoprotein receptors, the sugar derivative dissociates from the graphene oxide and fluorescence appears.

In comparison to the large body of synthetic organic gels, examples of a successful incorporation of biomolecules in the construction of multi-stimuli responsive gels are relatively scarce.[72,73] Enzymes are not only

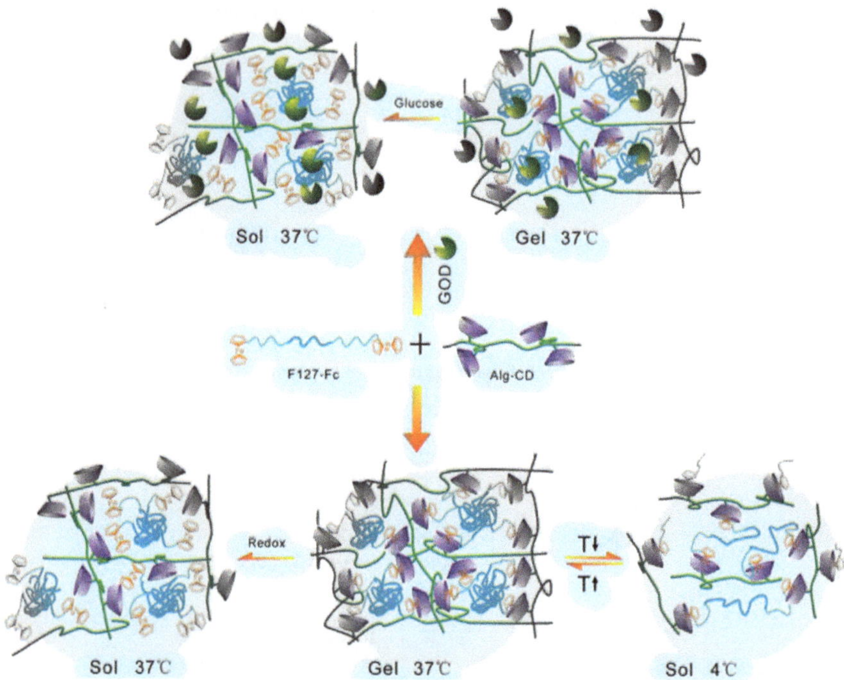

Figure 5.9 Schematic illustration of enzymes embedded in a multi-stimuli responsive hydrogel.
Reproduced from ref. 75 with kind permission from the Royal Society of Chemistry.

building blocks for hydrogels,[74] they can also be immobilized in the gels, which then respond to the enzyme's catalytic product (Figure 5.9).[75] A cyclodextrin-functionalized polymer forms hydrogels when the polymer strands are cross-linked by suitable divalent ferrocenes. Gel–sol transitions can be afforded by either temperature changes or by oxidation of ferrocene to ferrocenium, which is not a good guest for cyclodextrins and thus reduces the amount of cross-linking significantly. When glucose oxidase (GOD) is embedded in the gel, glucose addition also induces a gel–sol transition, because the hydrogen peroxide generated during glucose oxidation also leads to oxidation of the ferrocene.

Liu and his colleagues developed a versatile approach to hydrogels with designable responsive properties based on DNA self-assembly (Figure 5.10).[76] DNA hydrogels reversibly respond to a thermal stimulus because the assembly of the Y-scaffold and the linker loses its stability when the temperature is raised above the building block melting temperatures ($T_m = 63$ °C for the Y-scaffold, $T_m = 72$ °C for the linker). As the second stimulus, restriction endonucleases can be used that cut double-stranded DNA at specific restriction sites.

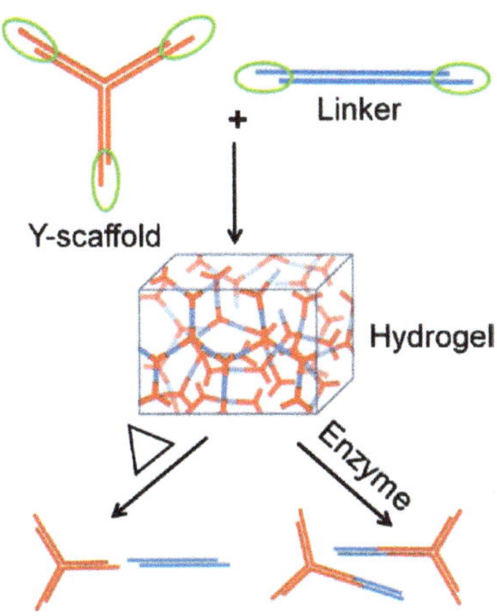

Figure 5.10 DNA hydrogel formation responding to temperature increases and the cleavage of the DNA linker double strands by the corresponding restriction enzyme.
Reproduced from ref. 76 with kind permission from Wiley-VCH.

5.2.4 Multi-stimuli Responsive Surfaces

Stimuli-responsive soft matter films deposited on surfaces have the potential to be applied as materials that can rapidly switch their transparency from opaque to transparent and back as well as their wetting properties from wettable to non-wettable. Also, the adhesion to, for example, biomolecules can be altered from sticky to non-sticky. Surfaces and coatings capable of a rapid release of a chemical load and self-healing coatings are also of great significance. Since we restrict the discussion to multi-stimuli responsive surface materials[77] here, the reader is referred to earlier reviews for surfaces that are responsive to a single stimulus.[78–81]

For the construction of multi-stimuli responsive surfaces, the responsive units need to be fixed to the surface. Certainly, the use of self-assembled monolayers (SAMs) as reliable and versatile platforms is the most widely used and is a quite easily-accomplished approach. The molecules constituting the SAM are anchored by covalent bonds (*e.g.* thiolates on gold or silyl ethers on silicon surfaces) and can be functionalized at the other end with suitable functional groups that allow addition of the subsequent layer covalently or non-covalently.

The first example to be discussed here is based on cucurbit[8]uril **CB8** as the host.[82] This pumpkin-shaped macrocycle can bind two guest molecules inside its cavity, when the guests come with matching electronic properties and form charge-transfer complexes. One such pair consists of an electron-poor and redox-active viologen dication and an electron rich, photochemically switchable azobenzene unit. Scherman *et al.* fixed a SAM of azobenzene-substituted alkylthiols on a micro-patterned gold surface.

Upon addition of **CB8** and a fluorescein-functionalized viologen dication, the complexes form on the surface indicated by a strong fluorescence observed exclusively at those patches, which are treated with the azobenzene-substituted thiol (Figure 5.11). When irradiated with light of a suitable wavelength, the *trans*-azobenzene switches to the *cis*-isomer leading to complex dissociation and the fluorescence vanishes completely. As a second stimulus orthogonal to the light-induced switching, an electrochemical reduction of the viologen dications with the gold surface as the working electrode gives rise to viologen cation radicals that tend to form dimers inside the **CB8** cavity rather than pairing up with azobenzene. The complex of two viologen cation radicals with **CB8** is not attached to the surface anymore and can be washed away. Again, fluorescence is lost. Both signals can be reversed and then the fluorescent complex is regenerated on

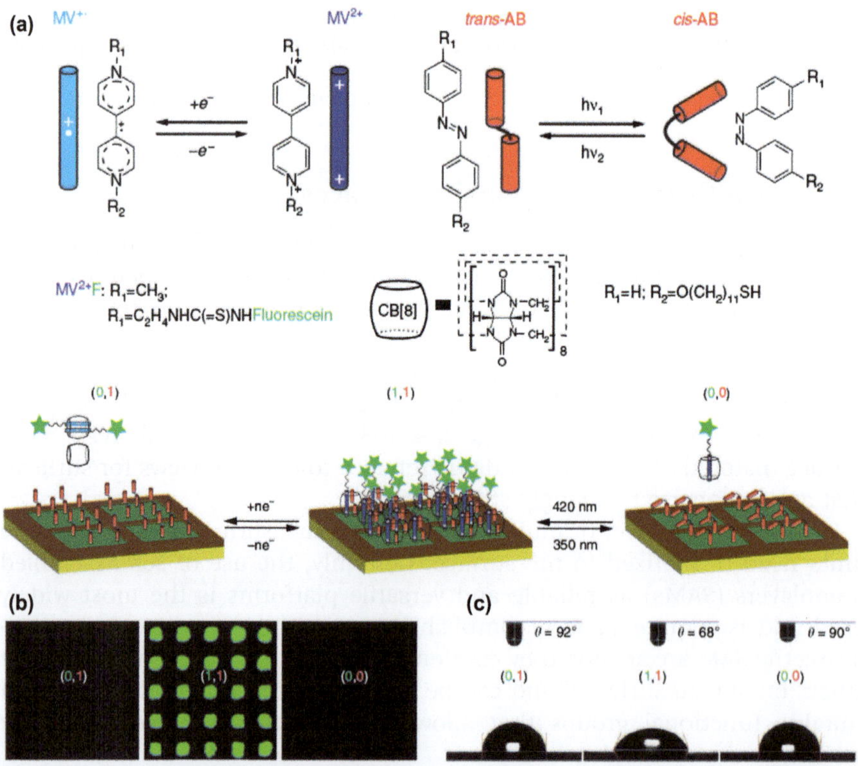

Figure 5.11 A cucurbit[8]uril-based dual-responsive surface. The **CB8** host–guest chemistry can be affected by light and/or redox stimuli when light-switchable azobenzenes and redox-active viologens are used as the guests. Fluorescein-labelled complexes only form on the surface when the azobenzene is in its *trans*-configuration and the viologen in its dicationic state.
Adapted from ref. 82 with kind permission from the Nature Publishing Group.

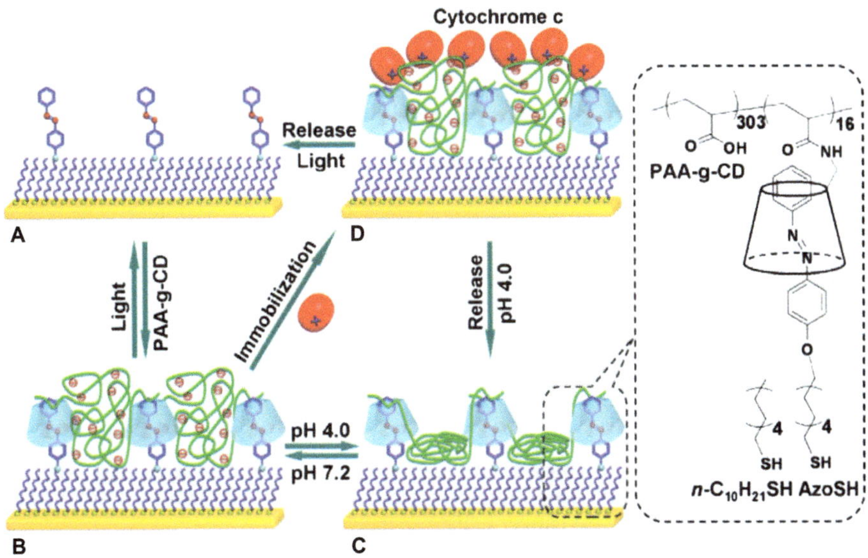

Figure 5.12 Light and pH as the two stimuli to control the adhesion of cytochrome c to a surface covered with an azobenzene switch.
Adapted from ref. 83 with kind permission from Wiley-VCH.

the surface. Even though the effect is not very pronounced, the construction and destruction of the complexes on the surface also changes the wetting abilities as determined by contact angle measurements.

Another example by Zhang *et al.* combines light and pH as the stimuli to make a surface dual responsive in controlling protein adhesion (Figure 5.12).[83] This surface is covered with a mixed azobenzene-functionalized SAM (**AzoSAM**). A pH-responsive poly(acrylic acid) polymer with β-cyclodextrin moieties (**PAA-g-CD**) is deposited and binds to the azobenzene through non-covalent interactions with the cyclodextrins. Under neutral conditions (pH 7.2), the negatively-charged **PAA** is an excellent adsorption matrix to efficiently capture positively-charged cytochrome c (**Cyt c**), a heme-containing metalloprotein with an isoelectric point at pH 10. When the pH value is reduced to 4.0, the neutral **PAA** loses its ability to capture **Cyt c** and thus releases the immobilized protein. When the surface is irradiated with light instead, the azobenzene switches from the *trans*- to the *cis*-isomer. The **PAA-g-CD** polymer thus does not bind to the *cis*-azobenzenes anymore and is removed from the surface. Consequently, the **Cyt c** is then desorbed as well.

The following two examples both make use of dynamic covalent bonds between phenylboronic acid and carbohydrates, but use these interactions for different purposes: while cell adhesion is controlled in the first example, the wettability is switched in the second.

In the first study,[84] Jiang *et al.* describe the fabrication of a dual-responsive surface that can reversibly capture and release targeted cancer cells by adjusting pH and glucose concentration. They deposited a SAM on

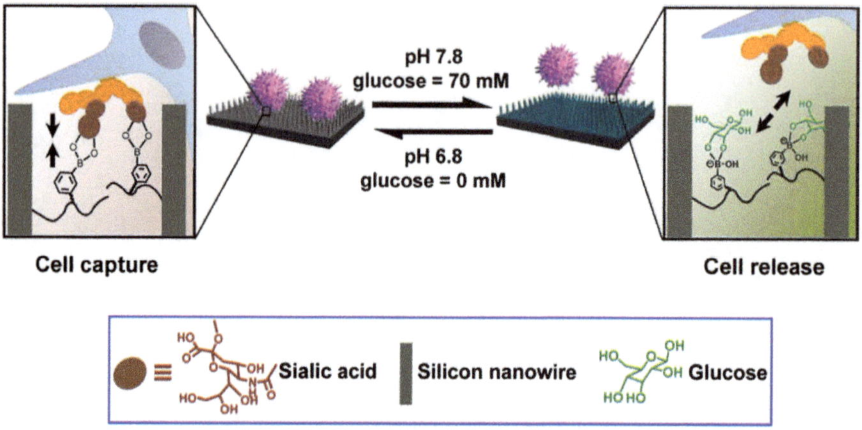

Figure 5.13 Schematic illustration of dual-responsive surfaces modified with a phenylboronic acid-containing polymer brush that can reversibly capture and release cancer cells.
Adapted from ref. 84 with kind permission from the American Chemical Society.

an array of aligned silicon nanowires, which was functionalized further with a phenylboronic acid-containing polymer brush (Figure 5.13).

As already outlined above, sugar binding to phenylboronic acid can be regulated by pH. In the absence of glucose, the resulting surface captures targeted cells due to the binding of sialic acid units exposed on the cell surface. A simultaneous increase of pH and glucose concentration to 7.8 and 70 mM, respectively, leads to competitive binding of glucose and a significantly reduced cell density on the surface. *Vice versa*, the cells are recaptured on the surface when the pH and glucose concentration are decreased to 6.8 and 0 mM, respectively. The dual-responsive switch between cell capture and release can be repeated for five cycles at least without negatively affecting cell viability (>95%).

The second study[85] reports a multi-stimuli responsive surface that reversibly switches between superhydrophilic and superhydrophobic states in response to glucose, temperature, and pH. Phenylboronic acid functionalized poly(*N*-isopropyl acrylamide) **pNIPAAm** copolymers were fixed onto a flat and a rough silicon substrate by surface-initiated atom transfer radical polymerization (Figure 5.14). It is well known that **pNIPAAm** polymers are sensitive to temperature changes. Furthermore, the neutral phenylboronic acid moiety is hydrophobic, while its charged form is hydrophilic—thus allowing its properties to be addressed by pH changes. When complexed to glucose, the hydrophilicity of charged phenylborates increases even further so that glucose addition is the third possible stimulus that will change wettability. As a result, the wetting of a pNIPAAm-*co*-PBA copolymer film indeed displayed responsiveness to these three stimuli. Obviously, the roughness of the substrate plays quite a significant role: for the flat

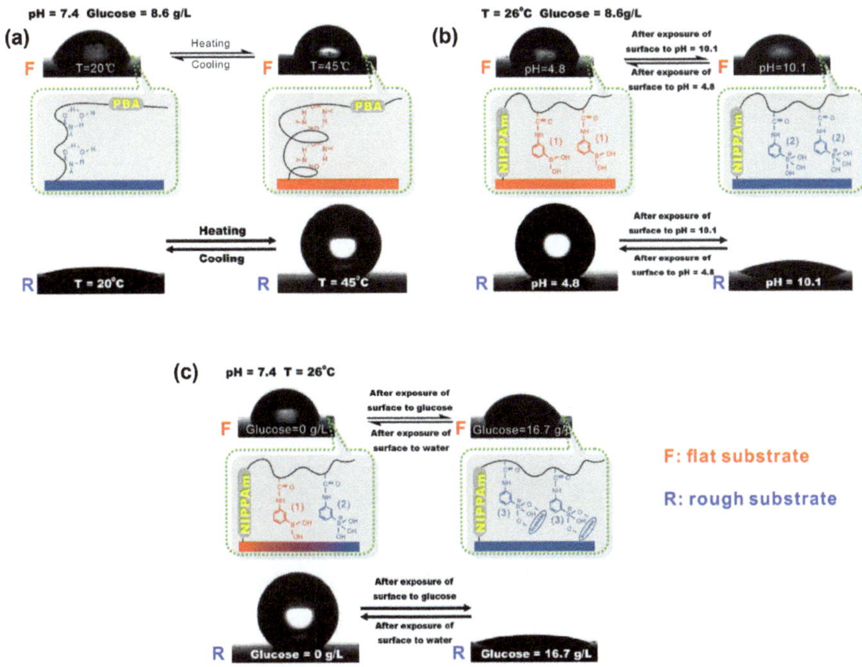

Figure 5.14 Schematic illustration of multi-stimuli responsive surfaces modified with a phenylboronic acid-containing **pNIPAAm** polymer, which exhibits switchable superhydrophilicity and superhydrophobicity: (a) variation of contact angles with temperature at pH 7.4 and $C_{\mathrm{glucose}} = 8.6\ \mathrm{g\ L^{-1}}$; (b) variation of contact angles with pH at $T = 26$ °C and $C_{\mathrm{glucose}} = 8.6\ \mathrm{g\ L^{-1}}$; (c) variation of contact angles with glucose concentration at pH 7.4 and $T = 26$ °C.
Adapted from ref. 85 with kind permission from Wiley-VCH.

substrate, changes in each of the three parameters only altered the contact angle in a narrow range from 63.1° to 80.2°. However, when the same experiments were carried out on rough substrates, the contact angle variation was much more pronounced and ranged from 8° (superhydrophilicity) to 152° (superhydrophobicity).

5.2.5 Multi-stimuli Responsive Surfaces with Gradients: Vectorial Devices

A lateral patterning of the surface-deposited soft matter films is also possible on a millimetre or centimetre scale. It is thus possible to tailor surfaces and the distribution of stimuli-responsive units on them. This adds a spatial dimension to multi-stimuli responsiveness and even extends the concept to surfaces that are covered by materials with programmed gradients.[86,87] They respond to stimuli in different sections of the surface so that vectorial devices can be constructed.

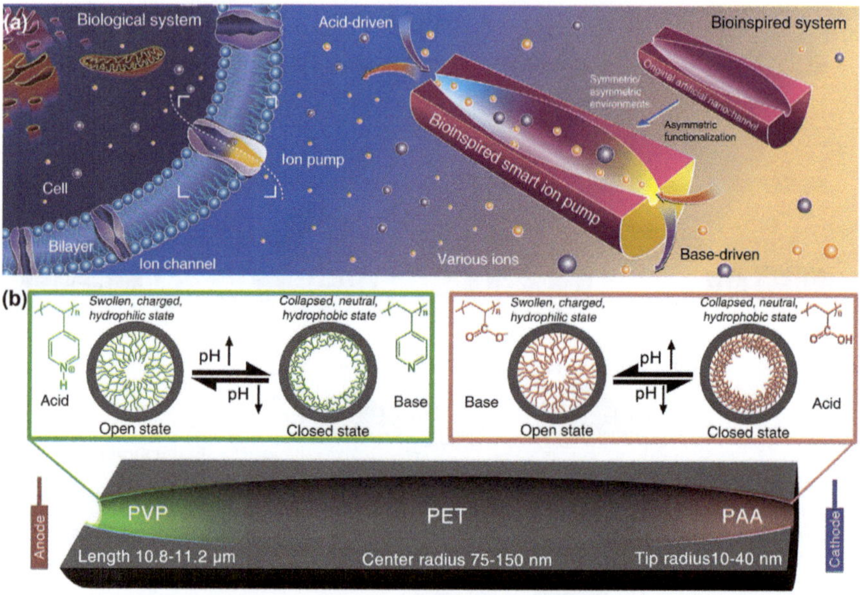

Figure 5.15 (a) Schematic illustration of a spatially resolved multi-stimuli responsive material that operates as an ion pump when pH changes are applied. (b) The dimensions of the channel and schematic representation of the locations of the three polymeric materials in it. Furthermore, the action principle is shown.
Adapted from ref. 88 with kind permission from the American Chemical Society.

Figure 5.15 shows an ion pump as an example, which was developed by Jiang *et al.*[88] A pH responsive dual-gate polyethylene terephthalate (**PET**) channel with radii in the upper nanometre range and a micrometre length that carries basic poly(4-vinylpyridine) **PVP** inside one gate and acidic poly(acrylic acid) **PAA** inside the other gate was manufactured. When the pH is raised, the protonated **PVP** is neutralized and changes into a collapsed hydrophobic state thus closing the gate at the **PVP** end of the channel for hydrophilic ions. The same conditions open the **PAA** gate, because the carboxylic acids are deprotonated and form a swollen and charged hydrophilic state. Upon decreasing the pH, these actions are reversed. The result is that ions can be pumped in a directional transport process driven by a sequence of pH changes.

5.2.6 Multi-stimuli Responsive Nanocarriers

Nanocarriers are soft or hard matter particles usually in a size range of 1–200 nm that carry a large number of peripheral functional groups and are used to deliver a bioactive agent to the target site in the body. They are aimed at improved therapeutic activity as compared to the free form of bioactive agent.[89]

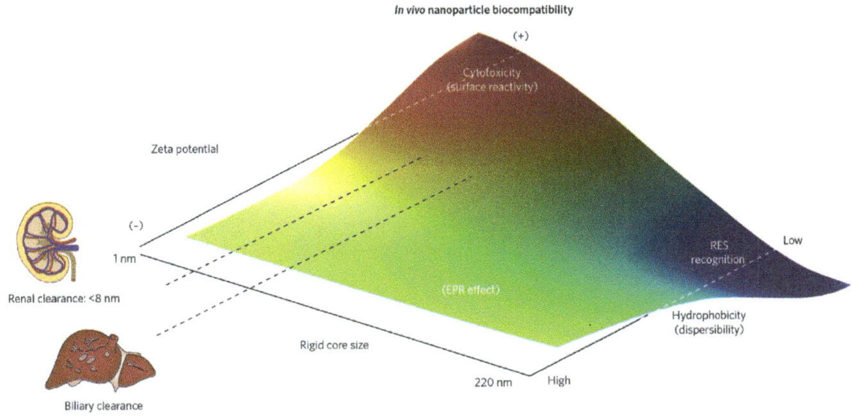

Figure 5.16 Physical characteristics of nanoparticles such as hydrophobicity, size and zeta potential determine their *in vivo* biocompatibility. Adapted from ref. 90 with kind permission from the Nature Publishing Group.

For specific drug delivery, biocompatibility and controlled release are two major issues to be addressed. For *in vivo* biocompatibility, a synergic interplay of parameters such as particle size, zeta potential and hydrophobicity are of importance (Figure 5.16).[90] By screening more than 130 different nanoparticle types including fullerenes, metal oxides, polymers, liposomes, dendrimers, quantum dots and gold colloids, the Nanotechnology Characterization Laboratory (NCL) at the National Cancer Institute in Maryland found that hydrophobicity, size and surface charge are the main parameters influencing nanoparticle biocompatibility.[91] Notably, nanocarriers with enhanced permeation and retention (EPR) in anticancer drug delivery systems are mostly those with an average size and a neutral surface charge. Nanocarriers with sizes over 10 nm need to be biodegradable, as an efficient clearance by the kidneys or biliary tract is only possible when the large structures are first decomposed into smaller pieces.

When the first requirement of biocompatibility is met, stimuli-responsiveness comes into play, in particular when the release of the drug is to occur at a specific target location in the body to ensure that side effects are minimized and the desired effects are maximized. The complexity of living systems is certainly a challenge here, but several mechanisms for targeted drug release[92] have been described (Figure 5.17). For example, in comparison with healthy tissue, pathological sites such as tumours typically exhibit properties somewhat different from healthy tissue, for example vascular abnormalities, weak acidity, elevated temperatures, over-expressed proteins and enzymes, hypoxia, high levels of certain metabolites or reactive small molecule species, and others.[93,94] Also, the properties change depending on the location inside the cell. Once the drug-loaded nanocarrier has entered the cell, it experiences diverse intracellular environments, for example intracellular pH gradients (pH 5.9–6.2 in early endosomes and

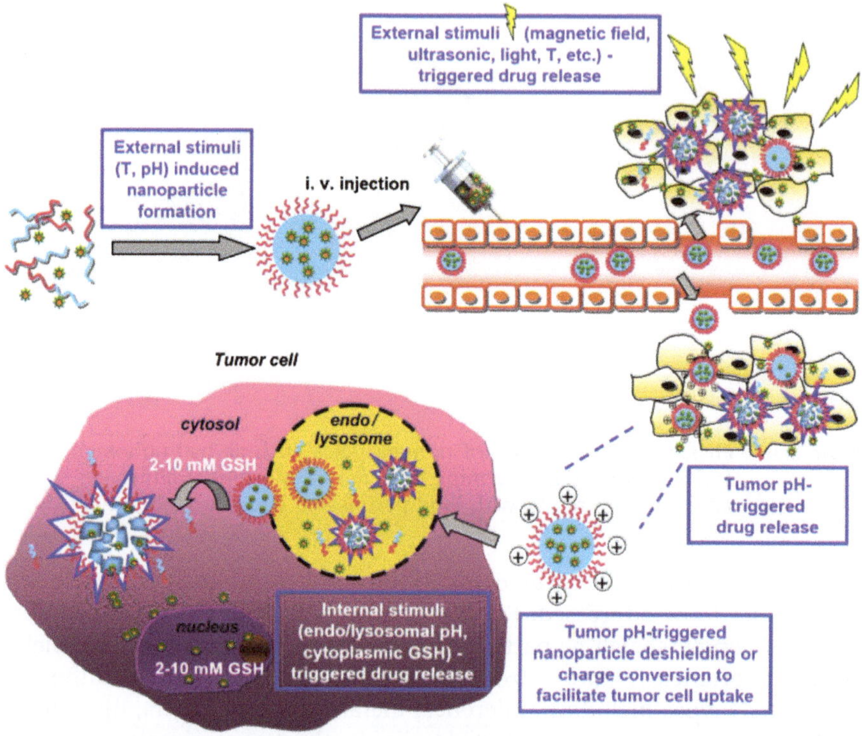

Figure 5.17 Different strategies to achieve a target-selective drug delivery and controlled release at the pathogenic site depending on differences in the environments.
Adapted from ref. 94 with kind permission from Elsevier.

pH 5.0–5.5 in late endosomes and lysosomes),[95] redox and oxidation gradients (high level cysteine or glutathione (GSH) in the cytoplasm and endolysosomes, oxidative microenvironment in mitochondria with a high concentration of H_2O_2).[96] Therefore, it is highly desirable to develop nanocarriers integrating multiple responsive modules that are adaptable to the characteristics of pathologic tissues and cells and triggering the drug release at the target disease site spontaneously and precisely. When more than one stimulus must be present to release the drug, a higher specificity for a certain pathogenic site can be achieved.

A broad variety of long-circulating liposomes,[97] polymeric micelles,[92] nano- or microgels,[98,99] and nanocomplexes[100,101] have been developed that are responsive to extracellular or intracellular stimuli innate to disease sites,[102] such as mildly acidic pH, temperature, enzymes (matrix metalloproteinase, β-glucuronidase, and phosphatase), oxidative or reductive microenvironments, and abnormal levels of bioactive molecules or metabolites. As this chapter cannot cover the huge body of literature available, we restrict the discussion to polymeric nanocarriers. Recent progress in silica nanoparticles is described elsewhere.[103–109]

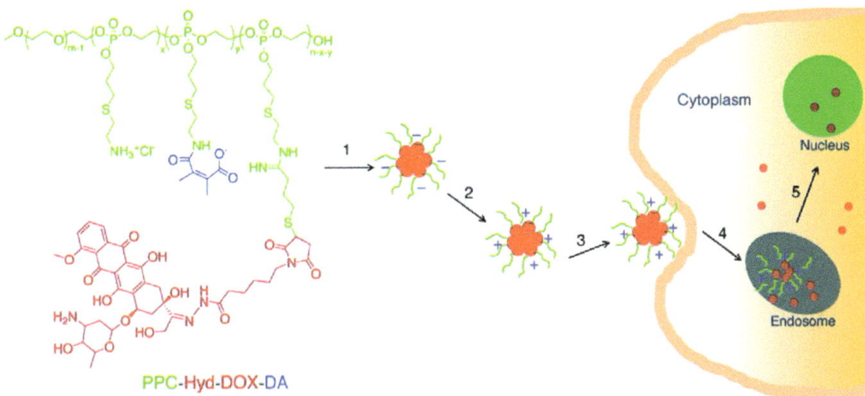

Figure 5.18 Schematic illustration of a dual pH-responsive nanocarrier. Numbers 1–5 refer to the following steps: (1) the conjugate self-assembles into negatively charged NPs in water; (2) the NPs become positively charged at the extracellular pH; (3, 4) the positively-charged NPs enter the cells by endocytosis; (5) **DOX** is released by hydrazone bond cleavage at the lower endo/lysosomal pH and diffuses into the nucleus.
Adapted from ref. 110 with kind permission from the American Chemical Society.

Many dual pH-responsive and dual redox-responsive nanocarriers have been designed and evaluated with respect to their capability to release anti-cancer drugs in a controlled way.[94] For instance, Wang *et al.* designed a dual pH-sensitive polymer–drug conjugate nanoparticle (**PPC-Hyd-DOX-DA** in Figure 5.18), in which a shell of PEG chains shields the nanoparticle surface.[110] At pH 7.4, the surface charge is negative, which quickly turns into positive when the 2,3-dimetyl maleic amide bond is cleaved upon incubation at pH 6.8. This pH corresponds to the extracellular pH of tumour tissue. The positive surface charge greatly facilitates internalization into the tumour cells. When the environmental pH is reduced to pH 5.0—the endo-/lysosomal pH—the release of the attached anticancer drug doxorubicin (**DOX**) is triggered owing to cleavage of the acid-labile acylhydrazone bond. This dual pH-sensitive nanoparticle exhibits enhanced inhibition to the progression of drug-resistant SK-3RD cancer cells.

Gu and Shen *et al.* developed a dual redox-sensitive polymer–drug conjugate nanocapsule (**OEG-2S-SN38** in Figure 5.19) which is able to cope with the tumour cells' redox heterogeneity.[111] Tumour cells are quite heterogeneous with respect to the levels of glutathione (**GSH**) and reactive oxygen species (**ROS**).[112] For an efficient drug release, it is therefore desirable to have a dual-stimuli responsive carrier that releases the drug at high **GSH** as well as at high **ROS** levels. To achieve this goal, the group attached the potent camptothecin-based topoisomerase I inhibitor 7-ethyl-10-hydroxyl-camptothecin (**SN38**) on a hydrophobic diester linker (**2S**). Upon oxidation of the linker to the more hydrophilic sulfoxide or sulfone, the phenol ester group, which conjugates **SN38** to the linker, is more rapidly hydrolysed and

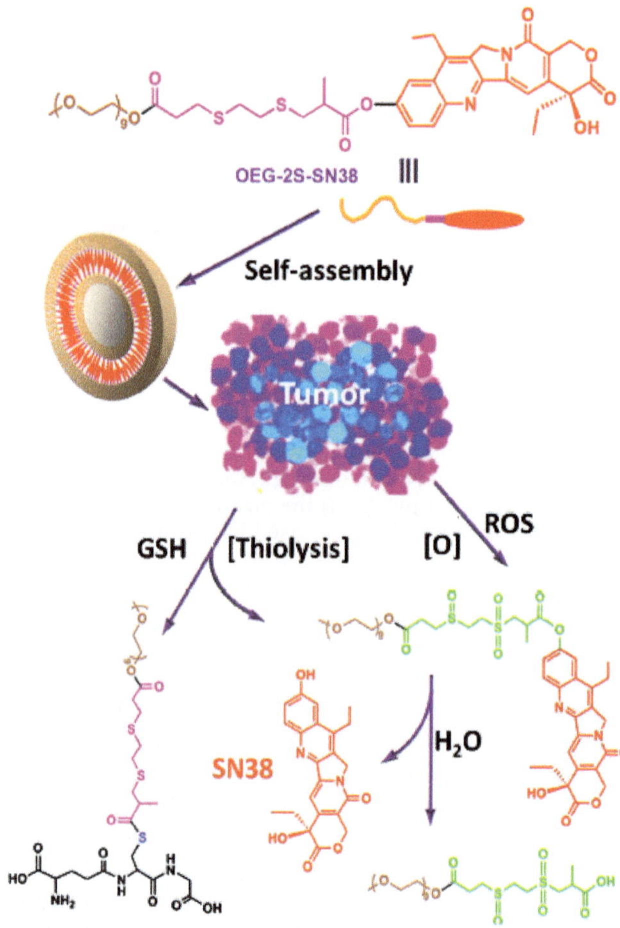

Figure 5.19 Schematic illustration of a dual redox-responsive nanocarrier. Adapted from ref. 111 with kind permission from Wiley-VCH.

the drug is released. On the other hand, the phenol ester can undergo **GSH**-triggered thiolysis and then also releases **SN38** under reductive conditions. Biological tests demonstrated that the nanocapsules decompose and quickly release the drug when triggered by either **GSH** or **ROS** and exhibit high *in vitro* cytotoxicity and *in vivo* anticancer activity. The high drug loading efficiency makes this nanocapsule system promising for translational nanomedicine.

By incorporating different responsive elements, dual- or multiple-responsive nanocarriers that not only react to different pH states or two redox stimuli, but to more diverse combinations, can be designed.[113,114] Zhong *et al.*'s redox and pH dual-responsive biodegradable micelles based on poly(ethylene glycol)-*SS*-poly(2,4,6-trimethoxybenzylidene-pentaerythritol carbonate) (**PEG-SS-PTMBPEC**) copolymers are a representative example

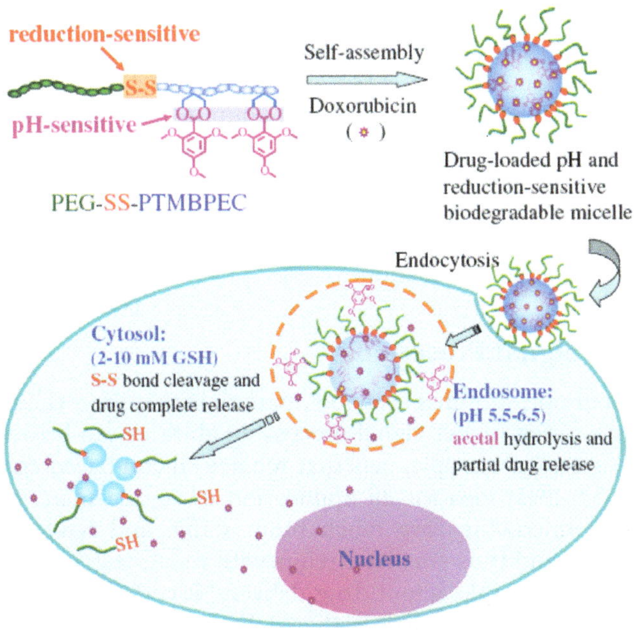

Figure 5.20 Illustration of the action principle of a redox- and pH-sensitive biodegradable nanocarrier for a two-step gradual intracellular release of the anticancer drug doxorubicin.
Reproduced from ref. 115 with kind permission from Elsevier.

(Figure 5.20). The copolymer has been examined for intracellular **DOX** release.[115] It bears an acetal group, which is sensitive to pH changes. Endocytosis into the mildly acidic endosomal compartments leads to cleavage of the acetals and to a partial release of the **DOX** that is trapped inside the micelles. However, the micelles as such survive acetal cleavage and finally leave the endosome and go into the cytosol. Here, reductive conditions (high **GSH** concentration) lead to the cleavage of the disulfide bonds connecting the parts of the copolymer. This destructs the micelle and the remaining **DOX** is released into the cytosol.

Similarly, Luo and Lam *et al.* prepared dual-responsive cross-linked micelles from cholic acid telodendrimers by applying boronic acid and catechol groups.[116] These micelles exhibit a rapid dual responsiveness to pH and the presence of competing diols.

5.3 Emerging Trends: Smart Multi-stimuli Responsive Materials

Even with the many exciting achievements in the development of multi-stimuli responsive materials in mind, the question remains: how could one mimic the much more sophisticated responses that regulate the complex processes in living organisms with respect to their sequences and spatio-temporal order? In this section, we focus on two main approaches to more complex stimuli: the first one is the fabrication of logic gate behaviour in responsive materials; the other one is the coupling of responsive materials to the permanently changing chemical environment as is, for example, encountered in oscillating reactions. As we will see, emergent properties can be found in such materials, for example gels that crawl forward or work as peristaltic transport devices driven by the chemical energy converted into heat during an oscillating reaction.

5.3.1 Logic Signal Processing

With more than one stimulus available, materials that exert the function of Boolean logic or even higher signal processing abilities can be designed. Of course, aiming at, for example, gels that replace silicon-based computers is probably not the best idea for an application of such a material, because current silicon microchips are much more stable and have much more highly integrated and thus smaller circuits with much faster reaction times. However, when one thinks about drug delivery, it certainly makes sense to develop, for example, nanocarriers that deliver their load only when two signals that characterize the pathogenic cells are present at the same time. This would represent an AND gate function and helps to increase the carrier's ability to distinguish between healthy and pathogenic tissue in the highly complex environment of a living body. Other logic gates, such as an XOR gate, might also be interesting, when the pathogenic site is identified by the presence of signal A, which, however, may also occur in healthy tissue, but then always together with signal B. Then, the carrier would need to sense A and the absence of B, while the simultaneous presence of both is not characteristic for the pathogenic site. Consequently, there are certainly interesting applications for materials that operate as logic gates.[117]

As a first example, we discuss here the low molecular weight bis-urea gelator **uBC7** which forms gels in acetonitrile that are responsive to three types of stimuli (Figure 5.21):[118] the macroscopic gel–sol transition can be triggered by complexation of K^+ to the crown ethers, the formation of pseudorotaxanes with secondary ammonium ions, or the disruption of the urea–urea hydrogen bonds by competition with chloride anions. All these stimuli can be reversed by scavenging the potassium ion with a cryptand, by deprotonation and subsequent dethreading of the secondary ammonium ion, or the precipitation of the chloride ions with Ag^+ or K^+, respectively. Depending on the choice of inputs as well as the additional components added to the gelator initially, a total of seven different logic gates were constructed. They include the NOT gate, which is controlled by a single stimulus, the OR, AND, XOR, NOR and XNOR gates, which connect two stimuli to determine the output, and finally the three-input INHIBIT gate. In addition, quite interesting behaviour is observed when one compares the OR with the AND gate. Both work based on the same pair of inputs. In fact, it is

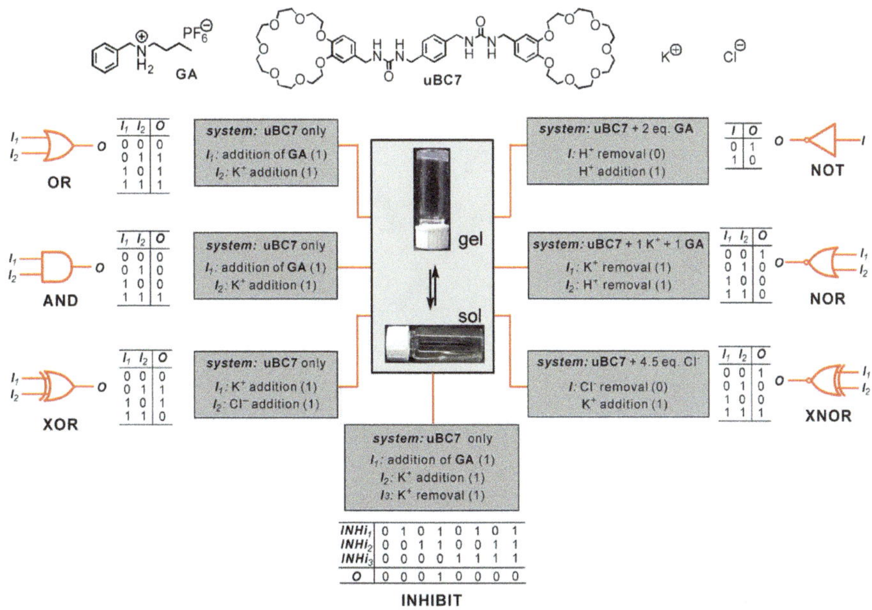

Figure 5.21 Seven logic gates constructed on the basis of multi-stimuli responsive bis-urea gelator **uBC7** and different combinations of chemical stimuli. Reproduced from ref. 118 with kind permission from the American Chemical Society.

the stoichiometry of the signal that decides which gate is realized. In the OR gate, the addition of 2 equivalents of **GA** or 2 equivalents of K^+ induces the gel–sol transition. If both inputs are active, a total of four equivalents of cations is present in the mixture and this, of course, also prevents gel formation. If one adds only 1 equivalent of **GA** or 1 equivalent of K^+, no gel–sol transition is observed. Adding both signals with 1 equivalent relative to **uBC7**, however, breaks the gel as at least one cation per crown is needed for the transition into a sol. Reducing the signal stoichiometry from two to one equivalents thus changes the OR gate into an AND gate. This behaviour can be considered a simple emergent property.

Hamachi's group also designed a supramolecular gel operating as AND, OR, NAND, and NOR gates.[119] Recently, they further developed a hydrogel–enzyme hybrid material displaying OR and processing ability with utility for medicine (Figure 5.22).[120] This system is based on redox-responsive peptide-based hydrogels that are capable of encapsulating different oxidase enzymes that do not lose their activity when immobilized within the gel's cavities. Gel–sol transitions are initiated when the boronic acid group is oxidatively cleaved by hydrogen peroxide. The source of the hydrogen peroxide is again an enzymatic oxidase reaction with enzymes specific for the presence of a certain biomarker.

Katz *et al.*[121] have developed responsive surface materials with logic processing capabilities.[122] Their design is based on networks of biocatalytic

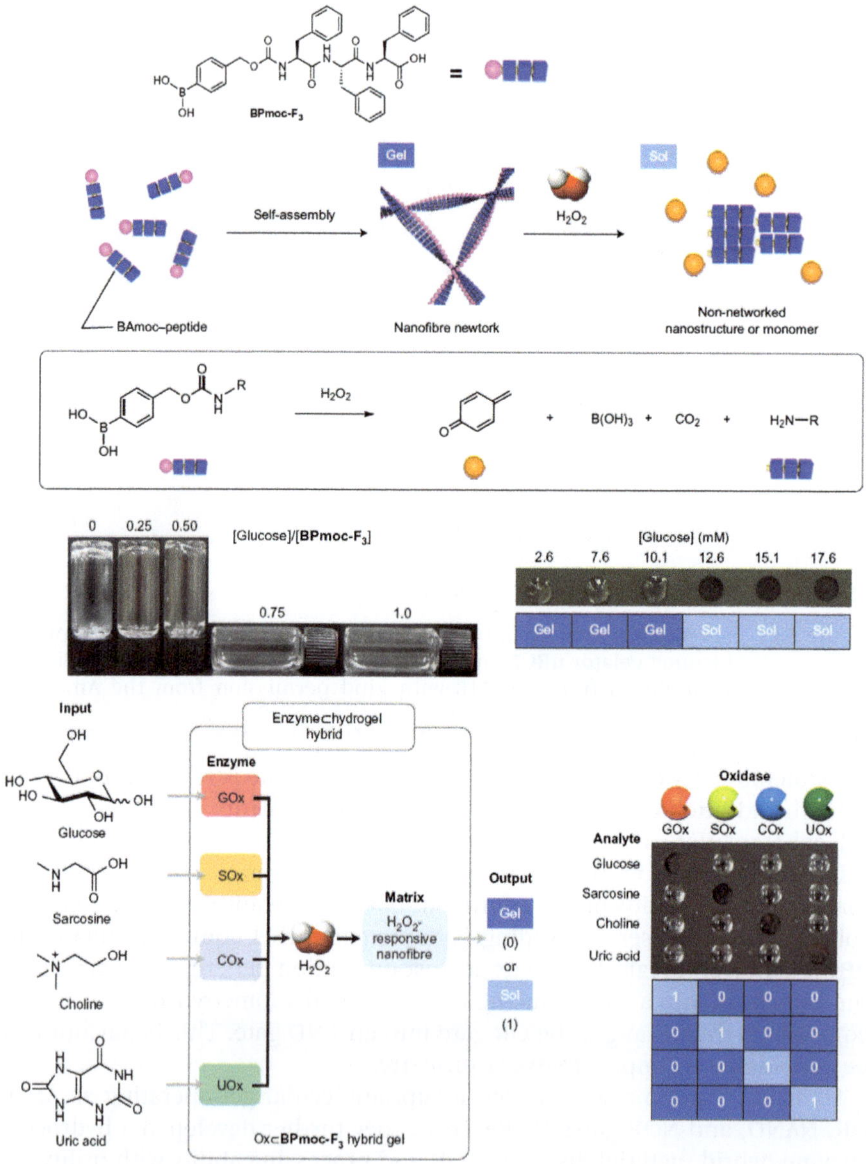

Figure 5.22 Hamachi's gel system, which not only detects the presence of various biomarkers, but also exhibits logic gate action regulating the release of therapeutic antibodies.
Reproduced from ref. 120 with kind permission from the Nature Publishing Group.

reactions proceeding simultaneously in multi-component ensembles.[123] For example (Figure 5.23), an electrode surface was covalently modified by polyelectrolyte brushes.[124] Under acidic conditions, the charged polymer brushes form a hydrophilic swollen state and allow the $[Fe(CN)_6]^{3-}$ redox

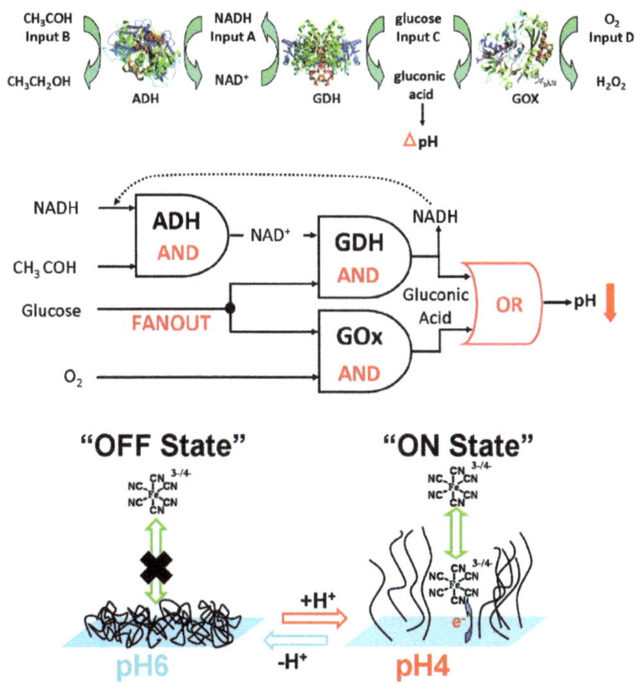

Figure 5.23 Schematic illustration of a biocatalytic cascade reaction used for introducing logic processing ability to reversibly switchable surface materials.
Adapted from ref. 124 with kind permission from the American Chemical Society.

probe permeating through the film to access the electrode, where it is reduced to $[Fe(CN)_6]^{4-}$. Under neutral or basic conditions, the uncharged polymer brushes exhibit very low permeability and thus inactivate the redox reaction at the electrode. The pH changes are caused by a system of concatenated enzyme catalysed processes. Three enzymes, *i.e.* alcohol dehydrogenase (**ADH**), glucose dehydrogenase (**GDH**) and glucose oxidase (**GOx**) represent three logic AND gates whose responses are finally connected with each other in an OR gate. A pH decrease is initiated when (i) NADH, acetaldehyde and glucose or (ii) glucose and oxygen or (iii) all four compounds are available to produce the final product gluconic acid. The acid reduces the initial pH from 6–7 to 4 and thus switches on the interface for the redox process. This enzymatic logic system can be reset through the addition of urease and urea. The urease catalyses the hydrolysis of urea into carbon dioxide and ammonia and thus increases the pH back to the initial value.

As stated above, nanocarriers that deliver drugs in an intelligent way by logically connecting several input signals are of great practical value as they allow more precise control of the drug delivery. Recently, proof-of-concept prototype nanocarriers were developed.[102]

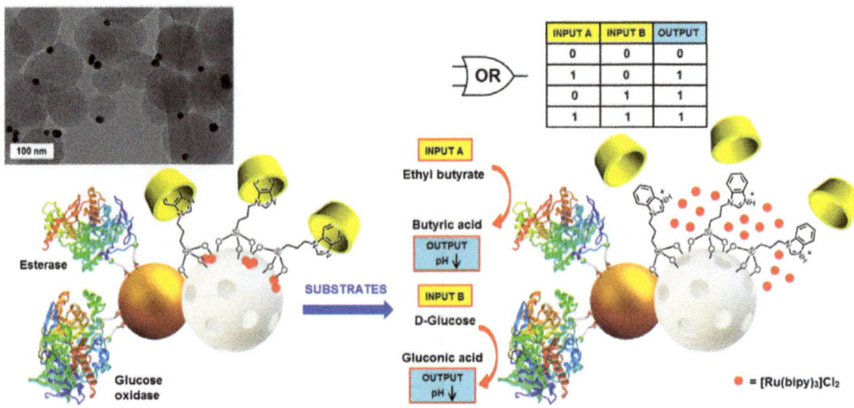

Figure 5.24 Illustration of the smart nanocarrier systems with an enzymatic logical OR gate function.
Reproduced from ref. 117 with kind permission from the American Chemical Society.

Pingarrón *et al.* reported a smart drug delivery system that consists of Janus-type nanoparticles with opposing gold and mesoporous silica nanoparticles (Figure 5.24).[117] The surface of the silica nanoparticles is functionalized with a pH-responsive β-cyclodextrin-based supramolecular nanovalve. The attached gold nanoparticle serves as a signal processing unit and carries two effectors, glucose oxidase and an esterase that are co-immobilized on the Au particle. In the presence of D-glucose or ethyl butyrate or both as the inputs, the immobilized enzymes catalyse the formation of either gluconic acid or butyric acid and thus reduce the local pH value surrounding the supramolecular nanovalve, which becomes protonated at the benzimidazole moiety. Protonation leads to the dissociation of the cyclodextrin and thus liberates the drug molecules entrapped within the pores of the silica nanoparticle.

Similar approaches have been taken by other groups.[125–127] Stoddart and Zink reported an AND gate drug delivery system,[128] again based on mesoporous silica particles. The cavities are functionalized inside with a light-responsive azobenzene and the surface is functionalized with pH-responsive cucurbit[6]uril–bisammonium stalk pairs. When the pores are loaded with any small molecules, the cargo remains trapped inside as long as the stalk is protonated and the cucurbituril is bound to it. But even when the cucurbituril dissociates under basic conditions, the azobenzenes still trap the cargo molecules until light induces their *cis–trans* isomerization.

5.3.2 Towards Emergent Properties: Materials Responding to Permanently Changing Environments

Most of the materials discussed so far have operated by assembling into a certain, thermodynamically favourable state. Applying a stimulus changes the potential energy surface so that another state is energetically more favourable in the presence of the stimulus. Again, the materials follow the stimulus and rearrange into the new thermodynamic minimum. This is the reason that reversible bonds are so useful for designing stimuli-responsive materials.

One of the most prominent features universal to living systems is that they operate far away from the thermodynamic minimum. This is driven by an energy flow through the system and self-organisation rather than self-assembly leads to the spontaneous formation of spatio-temporal patterns as expressed in self-sustaining oscillations such as the heartbeat, brain waves, the pulsatile secretion of hormones, cell cycles and biorhythms. Feedback loops, including positive ones that lead to self-accelerating processes, cause what are often called emergent properties—the unexpected and often surprising outcome of interesting behaviour far beyond simple predictability.

In the context of (multi)stimuli-responsive materials, it would be of great interest to mimic natural behaviour and create materials that respond to permanently changing environments. Oscillating reactions such as the Belousov–Zhabotinski reaction (BZR) provide exactly that and coupling the self-assembly processes in a material to the redox waves created by the BZR is thus one possibility to design materials that undergo property oscillations.

Overall, in the BZR, an organic substrate, for example malonic or citric acid, is oxidized by bromate as an oxidizing agent under acidic conditions. A metal catalyst, such as cerium ions, ferroin, or ruthenium tris(2,2′-bipyridine) ($Ru(bpy)_3{}^{2+}$), catalyses the redox waves. Waves are generated because the complex reactivity network of the BZR contains an autocatalytic step among the different reaction cycles. The concentration of bromide ions controls which of the processes predominates. The result is a spontaneous redox oscillation of the catalyst, which usually changes its colour or fluorescence so that the oscillations can be directly observed. When the BZR is performed in a thin layer without stirring, "redox waves" propagate through the medium.

The question is, of course, how to couple the BZR to materials. As early as 1995, Yoshida *et al.* reported a successful strategy (Figure 5.25).[129–131] The major idea was to use poly-NIPAAm-based hydrogels, which are known to exhibit temperature-dependent swelling–deswelling behaviour. At high temperature, the polymer network collapses and water is expelled. At lower temperature, the network swells and takes up significant amounts of water in its pores. When a $Ru(bpy)_3{}^{2+}$ redox catalyst is incorporated in the polymer, the swelling–deswelling occurs at different temperatures depending on the redox state of the complex. If the temperature is kept between the two transition temperatures of the two redox states, changes in the redox state will produce an oscillation of swelling and deswelling hydrogels when the medium trapped inside the gel is the BZR mixture—of course, without the catalyst that is incorporated in the polymer network.

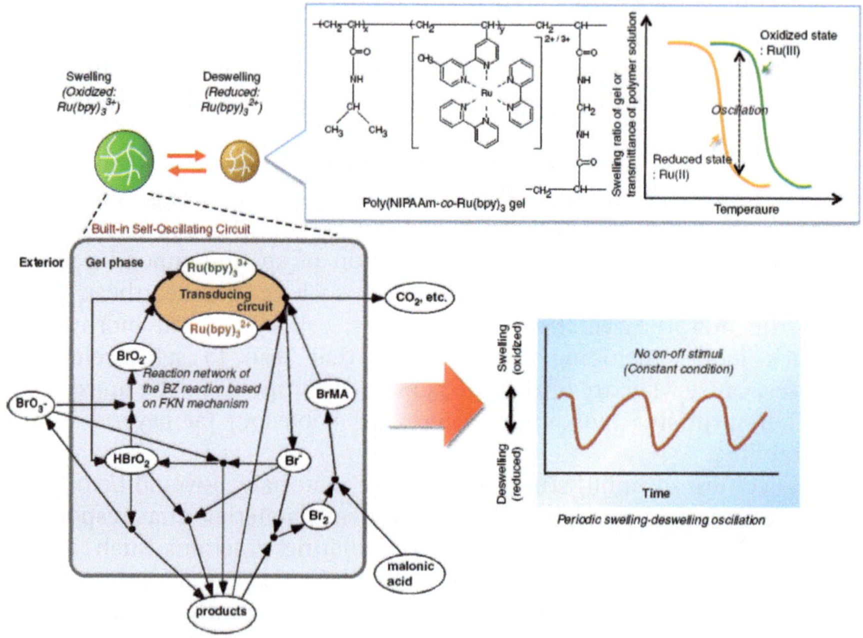

Figure 5.25 Design concept of a self-oscillating gel using the Belousov–Zhabotinsky reaction.
Reproduced from ref. 130 with kind permission from the Nature Publishing Group.

It is already intriguing to see how the redox oscillations can be converted into the macroscopically detectable property changes of the gel material. However, the basic design can be applied much beyond simple swelling–deswelling processes when the gels are designed properly. Figure 5.26 provides a brief overview.[130,132] The swelling–deswelling oscillations that propagate through the gel cause waves of thicker gel parts to move away from their origins. They, consequently, have a direction and can be used to transport particles.[133] If the gel is fabricated into a tube, peristaltic motion is observed running along the tube. Transport of, for example, CO_2 bubbles within the tube has been observed.[134] When gel stripes in which the gradient of the $Ru(bpy)_3{}^{2+}$ catalyst is perpendicular to the stripe plane are made, the stripes will bend differently depending on the oxidation state. If they are placed on a sawtooth-like surface, the gel stripes walk forward.[135]

Beyond these applications in gels and surface materials, the Yoshida group also developed self-oscillating micelles,[136] nanoparticles[137] and capsules.[138] These novel nanocarriers exhibit "temporal structure". Oscillations between unimer and micelle provide novel responsive scaffolds for drug delivery and controlled release. These chemical reaction networks are of great value to understand and mimic biological signalling networks[139,140] and have inspired a number of self-oscillation-based multiple responsive materials either on the nanometre or macroscopic scale.[11,141,142]

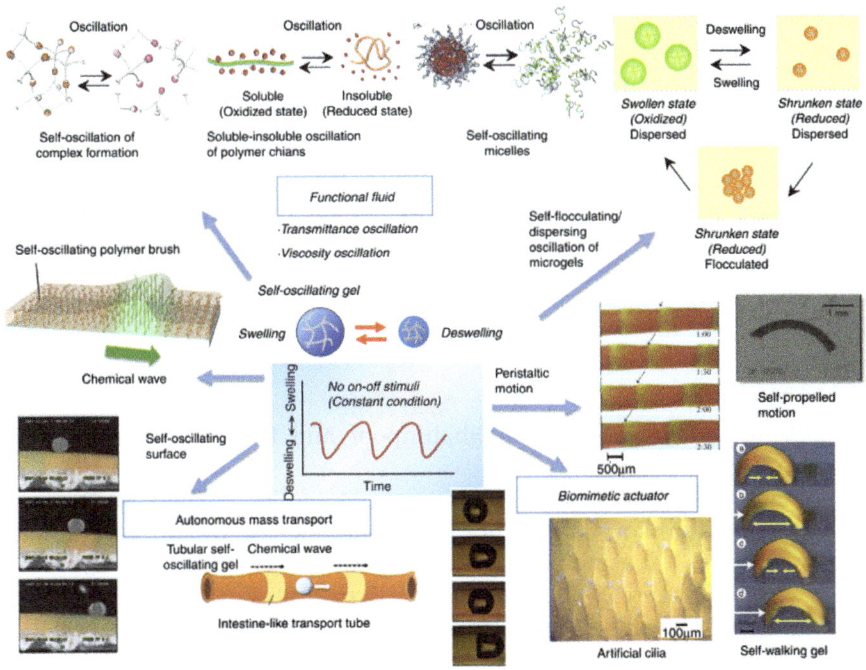

Figure 5.26 Evolution of the self-oscillating polymer gels as functional material systems.
Reproduced from ref. 130 with kind permission from the Nature Publishing Group.

5.4 Conclusions

Multi-stimuli responsive materials represent chemistry far beyond the paradigm of the pure compound that many chemists, in particular synthetic chemists, still often have in mind when thinking about chemistry. As the material itself and many stimuli are chemicals, most stimuli-responsive materials represent systems of molecules that are networked through reactivity and mutual interactions.

The field derives its origins mainly from two sources: supramolecular chemistry with non-covalent bonding and dynamic covalent chemistry. For both, the reversibility of the bonds formed is important and that is what makes the material adjustable to changing potential energy surfaces.

If one, however, follows the development of the field, one becomes immediately aware of the immense progress that has been made. Simple responsiveness turned into multiple responsiveness by incorporating more and diverse responsive units. Multiple stimuli have been connected with each other in logic gates. Finally, logic gates have been coupled to cascades of logic operations. Besides this, materials have been incorporated into vectorial devices. They have been used to control self-assembly on mesoscopic or even macroscopic dimensions and have been incorporated into functional devices that follow oscillating reactions and convert chemical energy into transport processes and the like. The responsiveness has thus started to become the basis for complex, but spontaneously-forming, dissipative structures in time and space.

Even if many of the systems discussed need optimization to be safely and reliably applied to solve everyday problems, the field of soft materials is approaching higher levels of complexity, which make them function as specific drug delivery devices, macroscopic transporters, sensors and in many other applications.

References

1. *Intelligent Materials*, ed. M. Shahinpoor and H.-J. Schneider, RSoC, Cambridge, 2008.
2. M. A. C. Stuart, W. T. S. Huck, J. Genzer, M. Müller, C. Ober, M. Stamm, G. B. Sukhorukov, I. Szleifer, V. V. Tsukruk, M. Urban, F. Winnik, S. Zauscher, I. Luzinov and S. Minko, *Nat. Mater.*, 2010, **9**, 101.
3. J. M. Spruell and C. J. Hawker, *Chem. Sci.*, 2011, **2**, 18.
4. *Handbook of Stimuli-Responsive Materials*, ed. M. W. Urban, Wiley-VCH, Weinheim, 2011.
5. *Smart Materials for Drug Delivery*, ed. C. Alvarez-Lorenzo and A. Concheiro, RSoC, Cambridge, 2013.
6. J. Zhuang, M. R. Gordon, J. Ventura, L. Li and S. Thayumanavan, *Chem. Soc. Rev.*, 2013, **42**, 7421.
7. E. G. Kelley, J. N. L. Albert, M. O. Sullivan and T. H. Epps III, *Chem. Soc. Rev.*, 2013, **42**, 7057.
8. K. Liu, Y. Kang, Z. Wang and X. Zhang, *Adv. Mater.*, 2013, **25**, 5530.
9. J. Huang and A. Heise, *Chem. Soc. Rev.*, 2013, **42**, 7373.
10. C. J. Kloxin and C. N. Bowman, *Chem. Soc. Rev.*, 2013, **42**, 7161.
11. A. Grinthal and J. Aizenberg, *Chem. Soc. Rev.*, 2013, **42**, 7072.
12. F. D. Jochum and P. Theato, *Chem. Soc. Rev.*, 2013, **42**, 7468.
13. C. Huang, S. J. Soenen, J. Rejman, B. Lucas, K. Braeckmans, J. Demeester and S. C. De Smedt, *Chem. Soc. Rev.*, 2011, **40**, 2417.
14. J. Thevenot, H. Oliveira, O. Sandre and S. Lecommandoux, *Chem. Soc. Rev.*, 2013, **42**, 7099.
15. L. D. Zarzar and J. Aizenberg, *Acc. Chem. Res.*, 2013, **47**, 530.
16. J. W. Steed, *Chem. Commun.*, 2011, **47**, 1379.
17. N. M. Sangeetha and U. Maitra, *Chem. Soc. Rev.*, 2005, **34**, 821.
18. M. D. Segarra-Maset, V. J. Nebot, J. F. Miravet and B. Escuder, *Chem. Soc. Rev.*, 2013, **42**, 7086.
19. A. R. Hirst, B. Escuder, J. F. Miravet and D. K. Smith, *Angew. Chem., Int. Ed.*, 2008, **47**, 8002.
20. Z. Qi, C. Wu, P. Malo de Molina, H. Sun, A. Schulz, C. Griesinger, M. Gradzielski, R. Haag, M. B. Ansorge-Schumacher and C. A. Schalley, *Chem. – Eur. J.*, 2013, **19**, 10150.

21. Z. Qi, N. L. Traulsen, P. Malo de Molina, C. Schlaich, M. Gradzielski and C. A. Schalley, *Org. Biomol. Chem.*, 2014, **12**, 503.
22. J.-M. Lehn, *Chem. Soc. Rev.*, 2007, **36**, 151.
23. S. J. Rowan, S. J. Cantrill, G. R. L. Cousins, J. K. M. Sanders and J. F. Stoddart, *Angew. Chem., Int. Ed.*, 2002, **41**, 898.
24. A. W. Jackson and D. A. Fulton, *Polym. Chem.*, 2013, **4**, 31.
25. A. Herrmann, *Chem. Soc. Rev.*, 2014, **43**, 1899.
26. A. Wilson, G. Gasparini and S. Matile, *Chem. Soc. Rev.*, 2014, **43**, 1948.
27. A. G. Orrillo, A. M. Escalante and R. L. E. Furlan, *Chem. Commun.*, 2008, 5298.
28. D. Steinhilber, T. Rossow, S. Wedepohl, F. Paulus, S. Seiffert and R. Haag, *Angew. Chem., Int. Ed.*, 2013, **52**, 13538.
29. R. A. Shenoi, B. F. L. Lai and J. N. Kizhakkedathu, *Biomacromolecules*, 2012, **13**, 3018.
30. D. E. Whitaker, C. S. Mahon and D. A. Fulton, *Angew. Chem., Int. Ed.*, 2013, **52**, 956.
31. C. B. Minkenberg, W. E. Hendriksen, F. Li, E. Mendes, R. Eelkema and J. H. van Esch, *Chem. Commun.*, 2012, **48**, 9837.
32. Y. Zhang, L. Tao, S. Li and Y. Wei, *Biomacromolecules*, 2011, **12**, 2894.
33. G. Deng, F. Li, H. Yu, F. Liu, C. Liu, W. Sun, H. Jiang and Y. Chen, *ACS Macro Lett.*, 2012, **1**, 275.
34. M. Lin, P. Sun, G. Chen and M. Jiang, *Chem. Commun.*, 2014, **50**, 9779.
35. F. Huang and L. Isaacs, *Acc. Chem. Res.*, 2014, **47**, 1923.
36. J. W. Steed and J. L. Atwood, *Supramolecular Chemistry*, Wiley, Chichester, 2nd edn, 2009.
37. M. Miyauchi, Y. Takashima, H. Yamaguchi and A. Harada, *J. Am. Chem. Soc.*, 2005, **127**, 2984.
38. X. Yan, F. Wang, B. Zheng and F. Huang, *Chem. Soc. Rev.*, 2012, **41**, 6042.
39. G. Fernández, E. M. Pérez, L. Sánchez and N. Martín, *Angew. Chem., Int. Ed.*, 2008, **47**, 1094.
40. X. Liao, G. Chen, X. Liu, W. Chen, F. Chen and M. Jiang, *Angew. Chem., Int. Ed.*, 2010, **49**, 4409.
41. L. Peng, A. Feng, M. Huo and J. Yuan, *Chem. Commun.*, 2014, **50**, 13005.
42. Z. Li, J. Yang, G. Yu, J. He, Z. Abliz and F. Huang, *Chem. Commun.*, 2014, **50**, 2841.
43. C. Katsuno, A. Konda, K. Urayama, T. Takigawa, M. Kidowaki and K. Ito, *Adv. Mater.*, 2013, **25**, 4636.
44. H.-J. Schneider and A. K. Yatsimirsky, *Chem. Soc. Rev.*, 2008, **37**, 263.
45. S. Dong, B. Zheng, F. Wang and F. Huang, *Acc. Chem. Res.*, 2014, **47**, 1982.
46. Z. Qi and C. A. Schalley, *Acc. Chem. Res.*, 2014, **47**, 2222.
47. V. Balzani, A. Credi, F. M. Raymo and J. F. Stoddart, *Angew. Chem., Int. Ed.*, 2000, **39**, 3348.
48. C. A. Schalley, K. Beizai and F. Vögtle, *Acc. Chem. Res.*, 2001, **34**, 465.

49. E. R. Kay, D. A. Leigh and F. Zerbetto, *Angew. Chem., Int. Ed.*, 2007, **46**, 72.
50. A. Coskun, M. Banaszak, R. D. Astumian, J. F. Stoddart and B. A. Grzybowski, *Chem. Soc. Rev.*, 2012, **41**, 19.
51. X. Yan, F. Wang, B. Zheng and F. Huang, *Chem. Soc. Rev.*, 2012, **41**, 6042.
52. S. Dong, B. Zheng, D. Xu, X. Yan, M. Zhang and F. Huang, *Adv. Mater.*, 2012, **24**, 3191.
53. J. A. Foster and J. W. Steed, *Angew. Chem., Int. Ed.*, 2010, **49**, 6718.
54. K. L. Liu, Z. X. Zhang and J. Li, *Soft Matter*, 2011, **7**, 11290.
55. A. Harada, R. Kobayashi, Y. Takashima, A. Hashidzume and H. Yamaguchi, *Nat. Chem.*, 2011, **3**, 34.
56. H. Yamaguchi, Y. Kobayashi, R. Kobayashi, Y. Takashima, A. Hashidzume and A. Harada, *Nat. Commun.*, 2012, **3**, 603.
57. C. Ma, T. Li, Q. Zhao, X. Yang, J. Wu, Y. Luo and T. Xie, *Adv. Mater.*, 2014, **26**, 5665.
58. H. Xu and D. M. Rudkevich, *Chem. - Eur. J.*, 2004, **10**, 5432.
59. B. Verdejo, F. Rodriguez-Llansola, B. Escuder, J. F. Miravet and P. Ballester, *Chem. Commun.*, 2011, **47**, 2017.
60. E. A. Appel, X. J. Loh, S. T. Jones, F. Biedermann, C. A. Dreiss and O. A. Scherman, *J. Am. Chem. Soc.*, 2012, **134**, 11767.
61. X. J. Loh, *Mater. Horiz.*, 2014, **1**, 185.
62. S. Y. Dong, J. Y. Yuan and F. H. Huang, *Chem. Sci.*, 2014, **5**, 247.
63. S. Ito, H. Takata, K. Ono and N. Iwasawa, *Angew. Chem., Int. Ed.*, 2013, **52**, 11045.
64. L. Zhou, J. Li, Q. Luo, J. Zhu, H. Zou, Y. Gao, L. Wang, J. Xu, Z. Dong and J. Liu, *Soft Matter*, 2013, **9**, 4635.
65. J. B. Beck and S. J. Rowan, *J. Am. Chem. Soc.*, 2003, **125**, 13922.
66. W. Weng, J. B. Beck, A. M. Jamieson and S. J. Rowan, *J. Am. Chem. Soc.*, 2006, **128**, 11663.
67. Y. Hisamatsu, S. Banerjee, M. B. Avinash, T. Govindaraju and C. Schmuck, *Angew. Chem., Int. Ed.*, 2013, **52**, 12550.
68. Z. Yang, G. Liang and B. Xu, *Acc. Chem. Res.*, 2008, **41**, 315.
69. A. R. Hirst, S. Roy, M. Arora, A. K. Das, N. Hodson, P. Murray, S. Marshall, N. Javid, J. Sefcik, J. Boekhoven, J. H. van Esch, S. Santabarbara, N. T. Hunt and R. V. Ulijn, *Nat. Chem.*, 2010, **2**, 1089.
70. H. Pei, L. Liang, G. Yao, J. Li, Q. Huang and C. Fan, *Angew. Chem., Int. Ed.*, 2012, **51**, 9020.
71. H.-L. Zhang, X.-L. Wei, Y. Zang, J.-Y. Cao, S. Liu, X.-P. He, Q. Chen, Y.-T. Long, J. Li, G.-R. Chen and K. Chen, *Adv. Mater.*, 2013, **25**, 4097.
72. Y. Gao, F. Zhao, Q. Wang, Y. Zhang and B. Xu, *Chem. Soc. Rev.*, 2010, **39**, 3425.
73. R. V. Ulijn, N. Bibi, V. Jayawarna, P. D. Thornton, S. J. Todd, R. J. Mart, A. M. Smith and J. E. Gough, *Mater. Today*, 2007, **10**, 40.
74. Y. Gao, Q. Luo, S. Qiao, L. Wang, Z. Dong, J. Xu and J. Liu, *Angew. Chem., Int. Ed.*, 2014, **53**, 9343.

75. L. Tan, Y. Liu, W. Ha, L.-S. Ding, S.-L. Peng, S. Zhang and B.-J. Li, *Soft Matter*, 2012, **8**, 5746.
76. Y. Xing, E. Cheng, Y. Yang, P. Chen, T. Zhang, Y. Sun, Z. Yang and D. Liu, *Adv. Mater.*, 2011, **23**, 1117.
77. H. Yang, B. Yuan, X. Zhang and O. A. Scherman, *Acc. Chem. Res.*, 2014, **47**, 2106.
78. H. Kuroki, I. Tokarev and S. Minko, *Annu. Rev. Mater. Res.*, 2012, **42**, 343.
79. J. Zhang and Y. Han, *Chem. Soc. Rev.*, 2010, **39**, 676.
80. B. Xin and J. Hao, *Chem. Soc. Rev.*, 2010, **39**, 769.
81. L. Zhai, *Chem. Soc. Rev.*, 2013, **42**, 7148.
82. F. Tian, D. Jiao, F. Biedermann and O. A. Scherman, *Nat. Commun.*, 2012, **3**, 1207.
83. P. Wan, Y. Wang, Y. Jiang, H. Xu and X. Zhang, *Adv. Mater.*, 2009, **21**, 4362.
84. H. Liu, Y. Li, K. Sun, J. Fan, P. Zhang, J. Meng, S. Wang and L. Jiang, *J. Am. Chem. Soc.*, 2013, **135**, 7603.
85. F. Xia, H. Ge, Y. Hou, T. Sun, L. Chen, G. Zhang and L. Jiang, *Adv. Mater.*, 2007, **19**, 2520.
86. J. Genzer and R. R. Bhat, *Langmuir*, 2008, **24**, 2294.
87. L. Tauk, A. P. Schröder, G. Decher and N. Giuseppone, *Nat. Chem.*, 2009, **1**, 649.
88. H. Zhang, X. Hou, L. Zeng, F. Yang, L. Li, D. Yan, Y. Tian and L. Jiang, *J. Am. Chem. Soc.*, 2013, **135**, 16102.
89. J. Khandare, M. Calderon, N. M. Dagia and R. Haag, *Chem. Soc. Rev.*, 2012, **41**, 2824.
90. A. E. Nel, L. Madler, D. Velegol, T. Xia, E. M. V. Hoek, P. Somasundaran, F. Klaessig, V. Castranova and M. Thompson, *Nat. Mater.*, 2009, **8**, 543.
91. S. E. McNeil, *WIREs Nanomed. Nanobiotechnol.*, 2009, **1**, 264.
92. L. M. Randolph, M.-P. Chien and N. C. Gianneschi, *Chem. Sci.*, 2012, **3**, 1363.
93. Z. Ge and S. Liu, *Chem. Soc. Rev.*, 2013, **42**, 7289.
94. R. Cheng, F. Meng, C. Deng, H.-A. Klok and Z. Zhong, *Biomaterials*, 2013, **34**, 3647.
95. R. F. Murphy, *J. Cell Biol.*, 1984, **98**, 1757.
96. Y.-M. Go and D. P. Jones, *Biochim. Biophys. Acta*, 2008, **1780**, 1273.
97. A. Rodríguez-Pulido, A. I. Kondrachuk, D. K. Prusty, J. Gao, M. A. Loi and A. Herrmann, *Angew. Chem., Int. Ed.*, 2013, **52**, 1008.
98. Y. Jiang, J. Chen, C. Deng, E. J. Suuronen and Z. Zhong, *Biomaterials*, 2014, **35**, 4969.
99. X. Zhang, K. Achazi, D. Steinhilber, F. Kratz, J. Dernedde and R. Haag, *J. Controlled Release*, 2014, **174**, 209.
100. Y.-C. Yeh, R. Tang, R. Mout, Y. Jeong and V. M. Rotello, *Angew. Chem., Int. Ed.*, 2014, **53**, 5137.
101. K. R. Raghupathi, J. Guo, O. Munkhbat, P. Rangadurai and S. Thayumanavan, *Acc. Chem. Res.*, 2014, **47**, 2200.

102. M. Motornov, Y. Roiter, I. Tokarev and S. Minko, *Prog. Polym. Sci.*, 2010, **35**, 174.
103. D. Tarn, C. E. Ashley, M. Xue, E. C. Carnes, J. I. Zink and C. J. Brinker, *Acc. Chem. Res.*, 2013, **46**, 792.
104. S. Wu, X. Huang and X. Du, *Angew. Chem., Int. Ed.*, 2013, **52**, 5580.
105. Z. Zhao, H. Meng, N. Wang, M. J. Donovan, T. Fu, M. You, Z. Chen, X. Zhang and W. Tan, *Angew. Chem., Int. Ed.*, 2013, **52**, 7487.
106. P. Yang, S. Gai and J. Lin, *Chem. Soc. Rev.*, 2012, **41**, 3679.
107. H. Yan, C. Teh, S. Sreejith, L. Zhu, A. Kwok, W. Fang, X. Ma, K. T. Nguyen, V. Korzh and Y. Zhao, *Angew. Chem., Int. Ed.*, 2012, **51**, 8373.
108. M. W. Ambrogio, C. R. Thomas, Y.-L. Zhao, J. I. Zink and J. F. Stoddart, *Acc. Chem. Res.*, 2011, **44**, 903.
109. Y.-W. Yang, Y.-L. Sun and N. Song, *Acc. Chem. Res.*, 2014, **47**, 1950.
110. J.-Z. Du, X.-J. Du, C.-Q. Mao and J. Wang, *J. Am. Chem. Soc.*, 2011, **133**, 17560.
111. J. Wang, X. Sun, W. Mao, W. Sun, J. Tang, M. Sui, Y. Shen and Z. Gu, *Adv. Mater.*, 2013, **25**, 3670.
112. A. Marusyk and K. Polyak, *Biochim. Biophys. Acta*, 2010, **1805**, 105.
113. F. Meng, Y. Zhong, R. Cheng, C. Deng and Z. Zhong, *Nanomedicine*, 2014, **9**, 487.
114. L. Wu, Y. Zou, C. Deng, R. Cheng, F. Meng and Z. Zhong, *Biomaterials*, 2013, **34**, 5262.
115. W. Chen, P. Zhong, F. Meng, R. Cheng, C. Deng, J. Feijen and Z. Zhong, *J. Controlled Release*, 2013, **169**, 171.
116. Y. Li, W. Xiao, K. Xiao, L. Berti, J. Luo, H. P. Tseng, G. Fung and K. S. Lam, *Angew. Chem., Int. Ed.*, 2012, **51**, 2864.
117. P. Díez, A. Sánchez, M. Gamella, P. Martínez-Ruíz, E. Aznar, C. de la Torre, J. R. Murguía, R. Martínez-Máñez, R. Villalonga and J. M. Pingarrón, *J. Am. Chem. Soc.*, 2014, **136**, 9116.
118. Z. Qi, P. Malo de Molina, W. Jiang, Q. Wang, K. Nowosinski, A. Schulz, M. Gradzielski and C. A. Schalley, *Chem. Sci.*, 2012, **3**, 2073.
119. H. Komatsu, S. Matsumoto, S.-i. Tamaru, K. Kaneko, M. Ikeda and I. Hamachi, *J. Am. Chem. Soc.*, 2009, **131**, 5580.
120. M. Ikeda, T. Tanida, T. Yoshii, K. Kurotani, S. Onogi, K. Urayama and I. Hamachi, *Nat. Chem.*, 2014, **6**, 511.
121. S. Minko, E. Katz, M. Motornov, I. Tokarev and M. Pita, *J. Comput. Theor. Nanosci.*, 2011, **8**, 356.
122. G. de Ruiter and M. E. van der Boom, *Acc. Chem. Res.*, 2011, **44**, 563.
123. M. Pita, S. Minko and E. Katz, *J. Mater. Sci.: Mater. Med.*, 2009, **20**, 457.
124. M. Privman, T. K. Tam, M. Pita and E. Katz, *J. Am. Chem. Soc.*, 2008, **131**, 1314.
125. M. Motornov, J. Zhou, M. Pita, V. Gopishetty, I. Tokarev, E. Katz and S. Minko, *Nano Lett.*, 2008, **8**, 2993.
126. M. Motornov, J. Zhou, M. Pita, I. Tokarev, V. Gopishetty, E. Katz and S. Minko, *Small*, 2009, **5**, 817.

127. G. Liu, X. Wang, J. Hu, G. Zhang and S. Liu, *J. Am. Chem. Soc.*, 2014, **136**, 7492.
128. S. Angelos, Y.-W. Yang, N. M. Khashab, J. F. Stoddart and J. I. Zink, *J. Am. Chem. Soc.*, 2009, **131**, 11344.
129. R. Yoshida, H. Ichijo, T. Hakuta and T. Yamaguchi, *Macromol. Rapid Commun.*, 1995, **16**, 305.
130. R. Yoshida and T. Ueki, *NPG Asia Mater.*, 2014, **6**, e107.
131. R. Yoshida, T. Takahashi, T. Yamaguchi and H. Ichijo, *J. Am. Chem. Soc.*, 1996, **118**, 5134.
132. R. Yoshida, *Adv. Mater.*, 2010, **22**, 3463.
133. Y. Murase, S. Maeda, S. Hashimoto and R. Yoshida, *Langmuir*, 2008, **25**, 483.
134. Y. Shiraki and R. Yoshida, *Angew. Chem., Int. Ed.*, 2012, **51**, 6112.
135. S. Maeda, Y. Hara, T. Sakai, R. Yoshida and S. Hashimoto, *Adv. Mater.*, 2007, **19**, 3480.
136. T. Ueki, M. Shibayama and R. Yoshida, *Chem. Commun.*, 2013, **49**, 6947.
137. H. Nabika, T. Oikawa, K. Iwasaki, K. Murakoshi and K. Unoura, *J. Phys. Chem. C*, 2012, **116**, 6153.
138. T. Narita, H. Takakura, N. Ogata, H. Kawakita and Y. Oishi, *Chem. Commun.*, 2013, **49**, 919.
139. X. Huang, M. Li, D. C. Green, D. S. Williams, A. J. Patil and S. Mann, *Nat. Commun.*, 2013, **4**, 2239.
140. R. J. R. W. Peters, M. Marguet, S. Marais, M. W. Fraaije, J. C. M. van Hest and S. Lecommandoux, *Angew. Chem., Int. Ed.*, 2014, **53**, 146.
141. O. Kuksenok, P. Dayal, A. Bhattacharya, V. V. Yashin, D. Deb, I. C. Chen, K. J. Van Vliet and A. C. Balazs, *Chem. Soc. Rev.*, 2013, **42**, 7257.
142. H. Zhou, X. Ding, Z. Zheng and Y. Peng, *Soft Matter*, 2013, **9**, 4956.

CHAPTER 6

Responsive Mesoporous Silica Nanoparticles for Targeted Drug Delivery

MONTSERRAT COLILLA[a,b] AND MARÍA VALLET-REGÍ*[a,b]

[a] Dpto. Química Inorgánica y Bioinorgánica, Facultad de Farmacia, Universidad Complutense de Madrid, Instituto de Investigación Sanitaria Hospital 12 de Octubre i+12, Pza. Ramón y Cajal s/n, 28040, Madrid, Spain; [b] Center on Bioengineering, Biomaterials and Nanomedicine (CIBER-BBN), Dpto. Química Inorgánica y Bioinorgánica, Facultad de Farmacia, Pza. Ramón y Cajal s/n, 28040, Madrid, Spain
*Email: vallet@ucm.es

6.1 Introduction

Since mesoporous materials were proposed as drug delivery systems (DDSs) in 2001,[1] mesoporous silica nanoparticles (MSNPs) have been widely explored for their potential applications in biotechnology and nanomedicine.[2–14] MSNPs own unique features such as uniform and tuneable pore sizes, high surface area, large pore volume and good biocompatibility, which make them suitable for effective encapsulation, carrying and release of therapeutic agents. Compared to organic DDSs, such as liposomes or polymeric nanoparticles (NPs), MSNPs exhibit the advantages of inorganic nanomaterials, such as thermal and chemical stability, mechanical strength, tuneable degradability and resistance to microbial attack under physiological conditions.[15,16] MSNPs can be easily synthetized on a large scale

RSC Smart Materials No. 14
Chemoresponsive Materials: Stimulation by Chemical and Biological Signals
Edited by Hans-Jörg Schneider

Published by the Royal Society of Chemistry, www.rsc.org

showing a great variety of morphologies and surface functionalities using diverse strategies that have been reviewed elsewhere.[17–20]

Much research effort is being devoted to developing DDSs owning targeting capability to achieve site-specific drug delivery to diseased tissues and cells. This is especially important in the oncology field, since most of the conventional antitumor drugs are highly cytotoxic and originate severe side effects in the patients which sometimes makes it necessary to suspend the treatment. NPs can be engineered to accumulate within solid tumours and once there to be selectively taken up by cancer cells.[21,22] This last property can be accomplished by locating targeting agents on the surface of the NPs. These targeting ligands are molecules or macromolecules capable of being selectively recognized by certain cell surface receptors present and/or over-expressed in tumour cells. Targeted drug delivery increases the selectivity and therefore efficacy of the therapy, which reduces the amount of administered drug and consequently decreases the risk of occurrence of side effects.

Targeted MSNPs can be loaded with different therapeutic drugs and the pore entrances can be blocked with sensitive gatekeepers. Upon exposure of internal or external release triggers, pore uncapping takes place and the payload is set free.[8,10,23] Among the different strategies to trigger drug release from DDSs, chemical and/or biological stimulation is receiving significant interest due to its non-invasive nature due to the absence of external mediation.[24] The performance of these smart DDSs relies on the diverse chemical and/or biological conditions (pH, enzymes, reducing agents, biochemicals, *etc.*) between diseased and normal tissues and also inside the cells, among the different compartments or organelles.

In this chapter we will provide an overview of the different synthetic strategies reported to date to develop chemical and/or biological stimuli-responsive MSNPs for targeted drug delivery.

6.2 Selective Targeting

One of the major limitations of cancer chemotherapy is the lack of selectivity of chemotherapeutics towards tumour cells, which leads to the administration of high drug doses to reach therapeutic levels.[25] MSNPs provide the opportunity to selectively carry the drugs to tumour cells. Targeted drug delivery to tumours can be achieved by either passive or active targeting or by a combination of both.[26–28]

6.2.1 Passive Targeting

Passive targeting is based on the "enhanced permeability and retention" (EPR) effect, which was discovered by Matsumura and Maeda in 1986.[29] They found that macromolecules larger than 50 kDa preferentially accumulated to the tumour interstitium and remained there for a long time. This effect, which is associated with the rapid growth of solid tumours, was attributed to the presence of fenestrations in the imperfect tumour blood vessels (permeability) and the poor lymphatic drainage (retention) in the tissue. When a solid tumour reaches a certain size, the normal vasculature present in its surroundings is not enough for nutrient and oxygen supply and the formation of new blood vessels from the surrounding capillaries (angiogenesis) is stimulated. Thus, new, irregular blood vessels are rapidly formed, which exhibit discontinuous epithelium and present fenestration of 200–2000 nm depending on the tumour type and location (Figure 6.1).[30] As the blood components reach tumour vessels, they extravasate through the fenestrations to the tumour interstitium. The defective lymphatic drainage in tumours permits small molecules to diffuse back to the blood circulation, but hinders the diffusion of NPs, which accumulate in the tumour interstitium (Figure 6.1).

The EPR effect is becoming the gold standard to design cancer-targeted nanomedicines. However, the successful accumulation of NPs in tumours requires long circulation lifetimes while avoiding renal clearance and uptake by the mononuclear phagocyte system (MPS). The three main features of MSNPs that influence their accumulation at tumour sites by passive targeting are size, shape and surface properties.

The optimal size of MSNPs to ensure longer circulation times, accumulation at tumour interstitium and cellular uptake is 50–300 nm.[31] Smaller MSNPs are difficult to prepare due to their inherent mesoporosity, whereas larger MSNPs cannot easily bypass biological barriers.

Few investigations have been reported regarding the effect of MSNPs' shape on *in vitro* and *in vivo* behaviour.[32,33] *In vitro* assays performed with

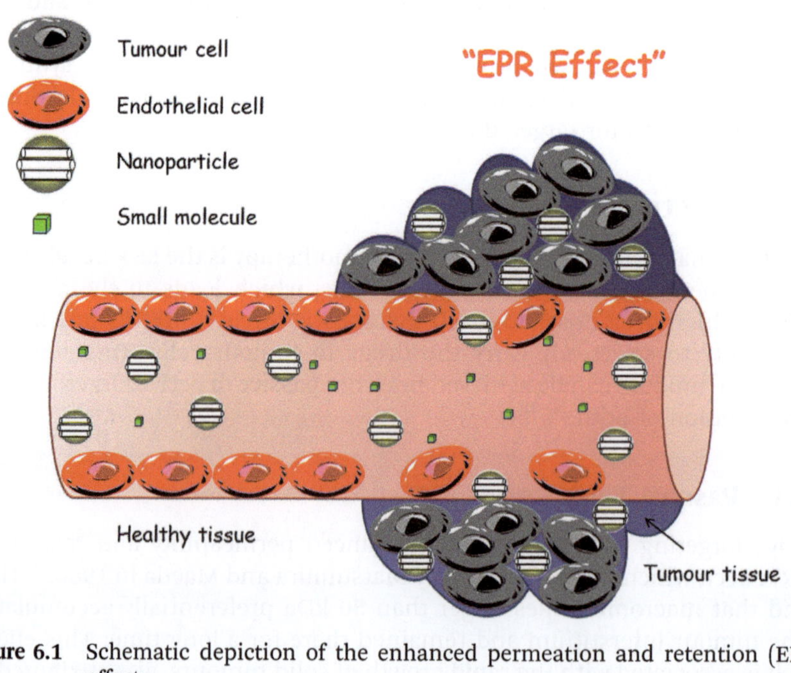

Figure 6.1 Schematic depiction of the enhanced permeation and retention (EPR) effect.

MSNPs exhibiting similar particle diameter, chemical composition and surface charge but different lengths, demonstrated that long rod-shaped particles were taken up in larger amounts and showed faster internalization rates than spherical ones.[32] The authors hypothesized that the larger contact area of the longitudinal axis of the rod-shaped MSNPs with the cell membrane, compared to spherical MSNPs, could promote cell internalization. However, these findings are the opposite of those found in worm-like elongated polymeric NPs, which exhibited negligible cellular uptake compared to spherical NPs of equal volume.[34] These ostensibly opposing observations, together with the difficulty of excluding the relationship of several chemical, electrostatic and morphological factors in control experiments, make the influence of the shape of MSNPs on cellular internalization and cellular fate an unsolved matter.

The surface properties of MSNPs can be tailored to control how the environment "sees" them. As soon as the NPs enter the bloodstream, they come into contact with a highly complex fluid that contains more than 3700 proteins and many other biomolecules, such as sugars, cholesterol, amino acids, *etc.*, which can interact with the NP. After a certain time, the surface of the NP will be completely covered by different proteins and molecules producing the "corona" of the nanocarriers.[35] Among the different blood proteins, there is a protein family called opsonins. These proteins are designed to stick to the surface of foreign bodies and allow the detection and rapid removal by the cells of the MPS system, such as macrophages. Grafting of hydrophilic polymers to the external surface of MSNPs has been found to be a good strategy to provide NPs with "stealth" properties. Polymers that are used extensively are polyethylene glycol (PEG) and its derivatives.[36–41] PEG reduces opsonisation and at the same time improves the stability of MSNPs in biological fluids.[37] The steric hindrance and repulsion effects of PEG chains against opsonins depend on their molecular weight, surface chain density and conformation. He *et al.* evaluated the effect of PEG molecular weight and packing density on the adsorption of serum proteins to MSNPs PEGylated using silane coupling chemistry.[38] The optimum PEG molecular weights for minimizing non-specific protein adsorption were in the range of 10 000–20 000, although this range could depend on the chemical approach used to PEGylate MSNPs. *In vivo* assays demonstrated that PEGylation of MSNPs decreased their arrest by liver, spleen and lung tissues, increased their blood-circulation lifetime and decreased biodegradation, which agrees with the finding of a lower level of degradation products compared to non-PEGylated MSNPs excreted in urine.[33,38]

The presence of a surface charge can alter the opsonisation profile of NPs, their recognition by cells in the organs of the MPS and their overall plasma circulation profile and intratumour processes.[22] Differing results have been found regarding the positive or negative effect of the surface charge of NPs on their accumulation at tumours. Parameters such as the ionization degree, the relative blood circulation times of the control NPs or the architecture of the construction might contribute to this controversy. Recently, the effects of spatial arrangement and accessibility of the charged molecules on the surface of MSNPs on their interaction with diverse cells and tissues *in vitro* and *in vivo* were investigated.[42] For this purpose, two types of MSNPs with almost identical size, shape, pore structure, colloid stability, PEGylation and ζ-potential, but differing in the exposure of amines on their surfaces were synthesized. MSNPs functionalized with distributed, obstructed amines provided both colloidal stability and protein corona neutrality, which correlated with minimal nonspecific binding *in vivo* and prolonged circulation times. On the contrary, MSNPs functionalized with exposed polyamines, although colloidally stable, immediately formed a protein corona and showed rapid nonspecific binding to endothelial and white blood cells (WBCs) and arrest within the chorioallantoic membrane. This report suggests that the combination of size and charge alone are poor predictors of *in vivo* behaviour, and charge exposure and its effect on protein corona formation and WBC scavenging should also be considered when designing MSNPs for *in vivo* applications.

Passive targeting of MSNPs to tumours has some limitations that cannot be disregarded.[22] The EPR effect is not universal for all types of solid tumours, as the degree of tumour vascularization and porosity of tumour vessels depend on the tumour type and state. PEGylation aimed at minimizing opsonisation also reduces the interactions between MSNPs and cell surfaces. Another major limitation of passive targeting concerns the interstitial fluid pressure (IFP) within solid tumours. The tumour interstitium comprises the bulk of tumour mass and consists of a crowded entanglement of collagen fibres, glycosaminoglycans and a highly viscous fluid. In contrast to healthy tissues, where interstitial flow is regulated to efficiently disseminate the fluid throughout the cell population, the disorganized structures in tumours strongly hinder the diffusion and fluid convection. Thus, the fluid within the tumour interstitium has a relatively high pressure that increases with tumour size and nearness to the centre of the tumour. IFP may hinder the penetration of the extravasated NPs into the tumour mass and, actually, in some cases it can even produce intravasation of NPs back to the bloodstream. This effect would produce the loss of the therapeutic effect and provoke systemic toxicity.[43] This lack of control may induce cancer cells to develop resistance towards different drugs (multiple drug resistance, MDR), which unavoidably would reduce any therapeutic effect.

A good alternative to overcome these limitations is providing NPs with active targeting capability by attaching targeting moieties to their surfaces. Active targeting would allow MSNPs to target cells through ligand–receptor interactions that favour receptor-mediated endocytosis and drug release inside the cell.

6.2.2 Active Targeting

Active targeting consists of using ligands attached to the surface of MSNPs. These ligands selectively bind molecules or receptors overexpressed in the

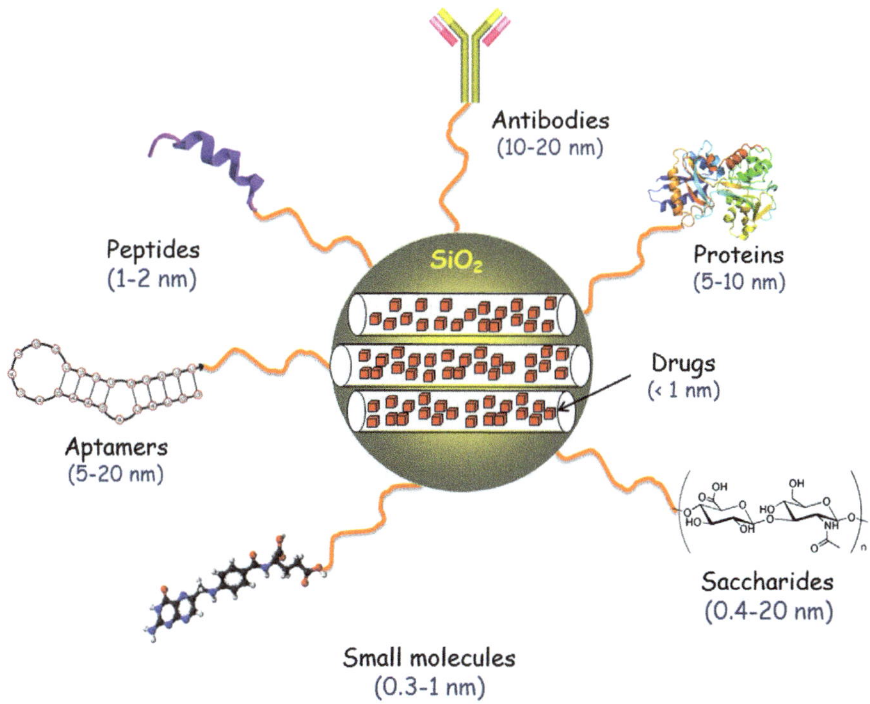

Figure 6.2 Active targeting possibilities on mesoporous silica nanoparticles (MSNPs).

surface of diseased organs, tissues, cells or organelles.[44] Characteristic ligands include antibodies, proteins, peptides, aptamers, saccharides and small molecules such as vitamins (Figure 6.2).

Active targeting is considered complementary to passive targeting to increase the efficiency of cancer nanotherapies. The targeting capability depends on both the abundance and function of the receptors overexpressed in the target and the ligand density in NPs.

Different strategies have been developed to graft ligands to the MSNPs' surfaces, such as carbodiimide-mediated COOH–NH_2 coupling, maleimide–SH coupling, *etc.* Let us summarize the most relevant reports regarding the conjugation of targeting ligands to MSNPs for specific recognition and enhanced cellular uptake by cancer cells.

Antibodies (Ab) have been conjugated to MSNPs *via* electrostatic interactions[45] or covalent bonds.[46–49] Usually, it is necessary to use PEG or other hydrophilic polymers as a linker between the Ab and the MSNPs to increase the dispersibility of the nanosystems, decrease the unspecific adsorption of other biomolecules in the environment and favour cellular uptake.[46–48] The Ab–MSNPs ratio also affects the specific cell recognition. For example, different amounts of anti-HER2/neu, a monoclonal Ab (mAb) that selectively recognizes HER2/neu glycoprotein, which is over-expressed in 25–30% of malignant human breast cancer, were conjugated to MSNPs through a PEG spacer. MSNPs with a high amount of the mAb showed greater selectivity and higher targeting efficiency towards HER2/neu over-expressing breast cancer cells (BT-474).[46] Within 6 h incubation, more than 90% of BT474 cells were targeted by these NPs. On the contrary, HER2/neu negative cell lines did not internalize significant amounts of these MSNPs. Very recently, the first example of antibody tumour targeting of MSNPs *in vivo* was reported.[48] The surface of PEGylated MSNPs was decorated with TRC105, a mAb that binds to CD105 receptors overexpressed in the tumour vasculature of many solid tumours. The specific targeting in this work was exclusively vascular CD105-based with no cancer cell targeting involved. Results indicated that vascular targeting led to a ~2-fold enhancement of tumour uptake when compared to that of passive targeting alone based on the EPR effect.

Proteins have been also conjugated to MSNPs for tumour targeting. For instance, transferrin (Tf) is a membrane glycoprotein that works together with its receptor (TfR) to aid cellular iron uptake. TfR can be overexpressed by up to 100-fold on tumour cells, which makes Tf an attractive targeting ligand for antitumour therapies using MSNPs. Different research groups have demonstrated the selective recognition and enhanced internalization of MSNPs conjugated to Tf by cancer cells overexpressing TfR, such as human pancreatic carcinoma (PANC-1),[50] human breast carcinoma (A549)[50] and human cervical adenocarcinoma (HeLa).[51]

The smaller sizes of peptides are linear or cyclic sequences of less than 50 amino acids. Their smaller size compared to proteins has several advantages, such as enhanced stability in the complex physiological environment and easier synthesis and conjugation. Also, peptides have several advantages compared to Ab. Peptides are cheaper and easier to manufacture than Ab and their use avoids the risk of an immune response by the host.

Among peptides, the RGD (Arg-Gly-Asp) family, which binds to endothelial $\alpha_v\beta_3$ integrins, has been widely investigated as a targeting ligand. The $\alpha_v\beta_3$ integrin is up-regulated in both tumour cells and angiogenic endothelial cells. Different cyclic and linear RGD peptides containing free thiol as part of cysteine[50–54] or free primary amine as part of lysine[55–57] have been used for conjugation to MSNPs and their targeting capability towards diverse tumour cells overexpressing $\alpha_v\beta_3$ integrins has been demonstrated. Recently, the influence of RGD conformation on cellular uptake was evaluated.[55] The results indicated that cyclic RGD showed higher internalization into tumour cells than linear sequences.

A cell-penetrating peptide derived from the human immunodeficiency virus (HIV) transactivator protein (TAT) has been conjugated to MSNPs to target the nuclei of cancer cells.[58–60] TAT peptide has demonstrated its efficiency in translocating MSNPs into cell nuclei by binding importin α and β receptors and then targeting the nuclear pore complexes of cancer cells and entering the nuclei. TAT-conjugated MSNPs have been used for direct intranuclear drug delivery of doxorubicin (DOX) in multidrug resistant (MCF-7/ADR) cancer cells.[59] TAT-MSNP nanocarriers were capable of increasing the intracellular, as well as intranuclear, drug concentrations much more effectively than free DOX or drugs delivered by MSNPs in the absence of TAT peptide.

Aptamers are single-stranded RNA or DNA or unnatural oligonucleotides that can fold by intramolecular interaction into specific three-dimensional conformations to recognize several types of targets with high affinity and specificity.[61] Aptamers have been conjugated to NPs to target specific receptors overexpressed in cancer cells.[62] They exhibit properties that make them attractive to use as targeting ligands, including their small size, non-immunogenic nature, ease of synthesis, characterization and modification, and high specificity and affinity for their target comparable to those of Ab. Aptamer AS141, which is the first aptamer entering clinical trials, has been attached to MSNPs.[63] AS141 interacts with nucleolin, which is overexpressed in certain tumour cells such as breast cancer (MCF-7). *In vitro* assays demonstrated that MCF-7 cells internalized higher amounts of DOX-loaded AS141-MSNPs than normal cells. Li *et al.* described the use of a lipid-coated MSNP conjugated with the thrombin-binding aptamer and loaded with the antitumour drug docetaxel (DXTL).[64] Thrombin is involved in several antitumour functions including proliferation and angiogenesis. *In vitro* assays using HeLa cells proved that the disruption of the PAR-1 receptor signaling caused by the aptamer provoked a significant increase in the cytotoxicity of DXTL.

Saccharides, such as lactobionic acid (LA),[65] mannose,[66] galactose[67] and hyaluronic acid (HA),[68,69] have also been used as targeting ligands since certain tumours overexpress carbohydrate receptors. Among polysaccharides, HA is attracting growing scientific interest as a targeting ligand because many types of tumour cells overexpress HA surface receptors such as CD44, RHAMM, CD168 and HARE. In recent work, HA has been conjugated to DOX-loaded MSNPs acting as both a capping and targeting agent.[69] The HA-MSNPs were internalized by cancer cells overexpressing the CD44 receptor. After endocytosis, hyaluronidase-1 enzyme in the lysosomal compartment degraded HA, allowing the release of DOX.

Small molecules with varied structures and properties are interesting targeting agents that offer many advantages compared to other ligands. They can be conjugated on the particle surface in higher amounts than big molecules such as antibodies or proteins. Besides, they are usually more stable than macromolecules, are less expensive and exhibit low immunogenicity. One of the most widely employed targeting molecules is folic acid (FA).[70] FA is a vitamin used by different enzymes during DNA replication. Many tumour cells overexpress folate receptors (FR) on their surface to capture as much as possible of this essential molecule. Zink *et al.* demonstrated the efficacy of FA conjugated MSNPs to reach tumours using *in vivo* murine models.[71] A small dose (0.5 mg per mouse) of MSNPs loaded with a potent cytotoxic drug (camptothecin, CPT) and decorated with FA was enough to induce strong tumour suppression. The particles were excreted through urine and faeces within a week of injection, which confirmed the biocompatibility of the treatment. FA has also been grafted onto poly(ethylene imine) (PEI)-coated MSNPs showing a significant enhancement in cellular uptake by cancer cells overexpressing high levels of FR.[72] An innovative approach was developed by Guo *et al.*,[73] who grafted FA to MSNPs acting as both redox-responsive capping agent and targeting ligand for efficient cancer cell targeted drug delivery. Another original research project consisted of conjugating methotrexate (MTX), an antitumour drug structurally similar to FA, to MSNPs acting both as targeting ligand and chemotherapeutic agent.[74] MTX-MSNPs induced apoptotic death in HeLa cancer cells, which overexpress FR, whereas no apoptosis was found in normal cells.

6.3 Stimuli-responsive Drug Delivery

In this section we will provide an overview of recent advances in the development of MSNPs as self-regulating drug delivery systems, *i.e.* those capable of releasing their payloads in response to chemical or biological stimulation taking place in certain sites inside the human body in the absence of external intervention. Such stimulation could be due to differences in the environmental conditions between diseased and normal cells, organs or tissues. Actually, the presence of diverse chemical and biological compositions can be found inside the cells, among the different compartments or organelles. Herein, we will focus on MSNP-based nanocarriers that release the entrapped cargoes upon exposure to pH variations, certain enzymes, small molecules or glutathione.

6.3.1 pH-Responsive Drug Delivery

pH is a stimulus broadly used to trigger drug release from MSNPs. Several pathologies exhibit pH values that differ from that found under normal conditions. For example, the extracellular pH of tumour tissues (6.5–7.0) and the pH of inflamed tissues and wounds (5.4–7.2) are more acidic than the pH of blood and normal tissues (7.4).[75] Non-healing wounds may present alkaline pH values up to 8.9.[76] Inside the cells, different pH values are found depending on the cell compartment or organelle.[77] The pH values in cytosol (7.4), the Golgi apparatus (6.4), endosomes (5.5–6.0) and lysosomes (5.0) are considerably different. Researchers have taken advantage of these pH variations to design smart drug delivery MSNPs. Depending on the components of the different pH-responsive nanosystems, they can be classified into four main groups: (1) MSNPs–polymer nanocomposites; (2) mechanically interlocked MSNPs; (3) MSNPs with inorganic capping; (4) MSNPs with grafted drugs.

6.3.1.1 Mesoporous Silica–Polymer Nanocomposites

These systems combine MSNPs with polymers bearing weak Brönsted acidic or basic functional groups.[24] pH modifications originate changes in the polymer ionization state, *i.e.* neutral or ionized, which provoke variations in its conformation and the affinity of its chains towards the aqueous medium. In the neutral state, polymer chains tend to minimize contact with water and adopt "globular" or collapsed conformations. In the ionized state, the polymer chains procure contact with water while minimizing the self-repulsive electrostatic forces and adopt "coil" or expanded conformations. The ionization degree of the polymers modulates their swelling/shrinkage, providing them with pH-responsive behaviour.

Diverse pH-sensitive polymers have been attached to MSNPs to design smart drug delivery nanocarriers. These nanosystems transport their payloads to the target without premature leakage. pH variations at the target produce conformational changes in the polymer that allow drug release from MSNPs. Liu *et al.*[78] reported the covalent grafting of poly(4-vinyl pyridine) (PVP) to MSNPs. At high pH values, pyridine groups in PVP are deprotonated and adopt a shrunken hydrophobic conformation that avoids the release of entrapped molecules. At acidic pH, the protonation of pyridine groups makes PVP adopt a swollen state that permits drug diffusion out of the pores. Polymers bearing tertiary amine groups, such as poly(2-(diethylamino)ethyl methacrylate) (PDEAEMA)[79] and poly[2-(dimethylamino)ethyl acrylate] (PDMAEA)[80] have also been used to design pH-responsive MSNPs. Figure 6.3 illustrates the performance of an "on" and "off" pH-switched nanosystem for intracellular doxorubicin (DOX) release.[80] MSNPs were coated with a nanoshell consisting of the pH-sensitive polymer PDMAEA to act as the pore valves and a PEG polymer for protection. At the neutral pH of 7.4, the tertiary amine

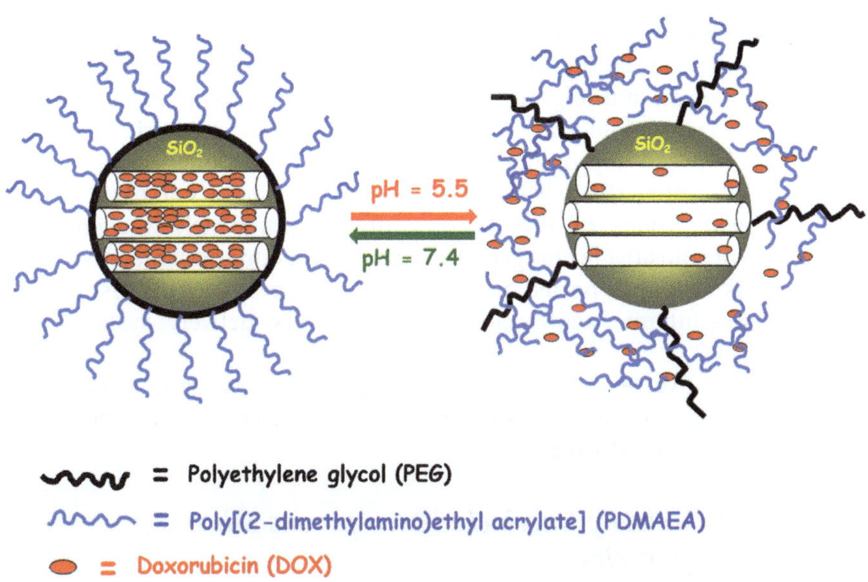

Figure 6.3 Schematic depiction for the reversible "on" and "off" pH-responsive release of doxorubicin from PEGylated MSNPs functionalized with poly[(2-dimethylamino)ethyl acrylate] (PDMAEA).[80]

groups ($-N(CH_3)_2$) of PDMAEA are deprotonated and the polymer carries no charge, causing the polymer to collapse on the MSNPs' surface and block the mesopores, which results in the trapping of DOX. By contrast, under mild acidic conditions (pH ~ 5), similar to those found in the endosome/lysosome intracellular compartments, the tertiary amine groups ($-N(CH_3)_2$) in the PDMAEA chains become protonated and positively charged, increasing the water solubility of the polymers. The solubilisation of the polymer results in the release of the cargo from the MSNPs. Aminopolysaccharides such as chitosan[81,82] and starch;[82] and poly(allylamine hydrochloride) (PAH) and poly(styrene sulfonate) (PSS);[83] have also been used to design MSNPs–polymer smart nanocomposites able to release drugs under acidic conditions.

Polymers bearing carboxylic acid groups able to protonate and deprotonate as a function of pH have also been used as pH-sensitive shells for MSNPs. The operating mechanism of these smart nanosystems can be tuned to permit drug release in response to either an increase or a decrease in pH depending on the chemical nature of the entrapped molecule. Drugs bearing very weak acidic or basic groups, such as salidroside ($pK_a \sim 10$), remain uncharged within the full range of physiological pH.[84–86] Since polymer–drug interactions are relatively weak, the driving force that governs molecule release is the conformational change in the polymer. Following this strategy, pH-responsive MSNP–poly(acrylic) (PAA) nanosystems have been engineered.[84,85] Drug delivery rates increased when the pH increased, since the polymer changed from a shrunken to a swollen state as carboxylic groups were deprotonated. This type of system would be suitable to treat certain pathologies that evolve with an increase in pH. It is also feasible to design MSNP–PAA nanocomposites able to release drugs under an acidic pH.[82,83,86] This approach requires the use of positively charged drugs, such as DOX, which possesses an amine group (pK_a of 8.2) that remains protonated over the full pH range of biological interest. At pH 7.4, the electrostatic attractive interactions between deprotonated carboxylic groups of PAA and positively charged DOX hinder the release of the drug. In contrast, at pH 5.5, the carboxylic acid groups of PAA become protonated, which weakens the interaction with positively charged DOX and permits drug release. A similar approach was recently reported by Zheng *et al.*, who prepared pH-responsive MSNPs–poly(L-glutamic acid) nanocarriers as DOX delivery systems for cancer therapy.[87]

Other innovative acid pH-switched drug delivery systems have been developed using coordination polymers,[88] natural gelatin[89] and a polymer-lipid[90] as coatings of MSNPs.

6.3.1.2 Mechanically Interlocked Mesoporous Silica Nanoparticles

The research groups of Zink and Stoddart were pioneers in using mechanically interlocked macromolecules to design innovative mechanized MSNPs with pH-responsive drug delivery capability.[6,91–93] Thus, curcurbit[6]uril

(CB[6]),[94,95] curcurbit[7]uril (CB[7])[96] and cyclodextrins (CDs)[97–101] have been widely explored because these polymacrocycles form inclusion complexes with a variety of guest molecules in aqueous media. The complexes reversibly dissociate upon exposure to changes in environmental pH.

As an illustrative example, Figure 6.4 schematically displays the controlled release operating-mechanism of MSNPs capped by pH-responsive PEI–CD polypseudorotaxane motifs.[97] The mesopores were filled with a guest molecule (calcein) and then blocked by threading of CD onto the surface-grafted pH chains at pH 11. At pH 5.5, the CDs were dethreaded from the PEI block owing to the weak interaction of the protonated PEI chain with the hydrophobic interior of CDs, allowing calcein release.

pH-Responsive CB[6]-based nanovalves operating under biologically relevant conditions have also been employed to design mechanically interlocked MSNPs. CB[6] is a pumpkin-shaped polymacrocycle consisting of six glycouril units strapped together by a pair of bridging methylene groups between nitrogen atoms.[94,102,103] CB[6] tends to form 1 : 1 inclusion complexes with different polymethylene derivatives such as diamino-alkanes, which have been widely investigated. The stability of these complexes depends highly on pH and therefore CB[6] is an attractive gatekeeper to design pH-responsive MSNP-based nanomachines.[95,104]

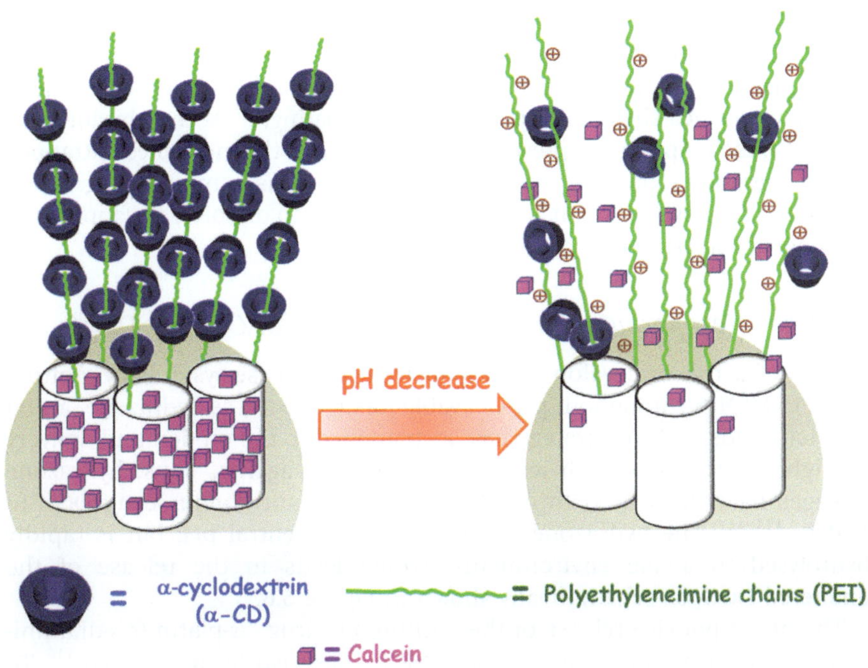

Figure 6.4 Schematic illustration of the pH-responsive release of calcein molecules from MSNPs modified with polyethyleneimine–α-cyclodextrin (PEI–α-CD) polypseudorotaxane motifs.[97]

Very recently, Zink and co-workers reported the combination of pH-operating nanovalves with polymer coatings on MSNPs to produce a new generation of NPs that exhibit characteristics of both components.[100] The nanovalves can trap molecules inside the pores of MSNPs without premature release, opening in response to pH changes. On the other hand, surface coatings with featured polymers can increase cellular uptake, avoid uptake by the MPS, provide protected space for storing siRNA and enhance the bio-distribution of MSNPs. This study demonstrates that the polymer coated on the MSNPs' surface produces enhanced biological properties without interfering with the on-demand release function of nanovalve-modified MSNPs.

6.3.1.3 Mesoporous Silica Nanoparticles with Inorganic Capping

pH-Responsive MSNPs have also been developed using inorganic compounds to block the pore openings. Two main strategies have been reported. The first one consists of using inorganic NPs, such as Au,[105,106] Fe_3O_4[107] and CeO_2[108] NPs, as pore blocking caps. These nanocaps are tethered to the pore outlets of MSNPs *via* acid cleavable chemical linkages such as boronate ester[109] or acetal bonds,[105,107] pH-sensitive host–guest interactions[100] or pH-dependent hybridization/dehybridization of complementary DNA strands.[106] The second approach relies on blocking the mesopores of MSNPs with pH-decomposable inorganic gatekeepers, such as ZnO quantum dots nanolids,[110] hydroxyapatite (HAp)[111] or layered double hydroxide (LDH)[112] nanosheets. Figure 6.5 schematically displays the pH-responsive operating drug release mechanism from MSNPs consisting of a magnetic Fe_3O_4 NP core and a mesoporous silica shell end-capped by a HAp coating.[111] At the physiological pH of 7.4, the HAp coating prevents the departure of the molecules entrapped in the mesopores, in this case ibuprofen, chosen as a model drug. Upon exposure of the system to an acidic pH, HAp starts to dissolve, uncapping the mesopores and triggering drug release.

6.3.1.4 Mesoporous Silica Nanoparticles with Grafted Drugs

It is also feasible to develop pH-responsive drug release systems by grafting drugs to MSNPs throughout pH-cleavable linkages such as hydrazone, acetal or ester bonds.[113] This strategy has been applied to the on-demand delivery of highly toxic drugs such as chemotherapeutic agents. Thus, hydrazone linkages have been used to covalently graft the antitumour drug DOX to MSNPs.[114–117] The hydrazone bond is stable at neutral pH, but is rapidly hydrolysed in acidic environments, which leads to the release of the unaltered DOX, as schematically shown in Figure 6.6.

The pH-responsive release of the antitumour drug *cis*-platin (*cis*-diamminedichloroplatinum(II), CDDP) from MSNPs has also been reported.[118] To design this smart system, carboxylate groups were anchored to the surface of MSNPs *via* pH-cleavable hydrazone bonds. At the physiological pH of 7.4, hydroxo-substituted CDDP was grafted onto the surface of MSNPs *via*

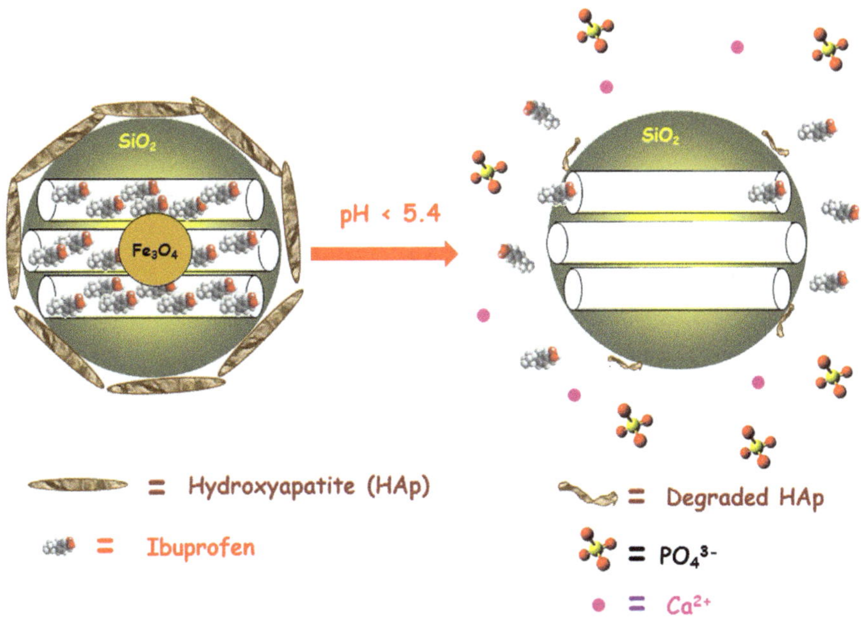

Figure 6.5 Schematic representation of pH-responsive ibuprofen release from MSNPs consisting of a Fe_3O_4 nuclei and mesoporous silica shells coated by a degradable hydroxyapatite (HAp) layer.[111]

coordination with carboxylate groups present at the pore surface. Upon cellular uptake by tumour cells, the acidic environment of endosomes (or lysosomes) provokes the cleaving of hydrazone bonds and triggers the release of the CDDP–carboxylate complex. Lastly, the released complex is converted to CDDP by the replacement of –OH groups by Cl^- ions present in the intracellular environment.

6.3.2 Enzyme-responsive Drug Delivery

Some diseased states involve enzymatic deregulation.[119,120] An anomalous increase of an enzymatic presence or activity in particular tissues or intracellular sites makes enzymes interesting stimuli to trigger drug release from MSNPs with high target specificity. Different enzyme-responsive MSNP drug delivery systems have been developed by using innovative approaches.[96,121–132] Among them, it has been possible to design enzyme-operating mechanically interlocked MSNPs by incorporating enzyme cleavable sites in the stalks of CD-based supramolecular nanovalves tethered to the pore openings.[121,122] In the presence of the specific enzyme, the sensitive bond is cleaved and the uncapping of pores and drug release take place. Another widely explored strategy relies on capping MSNPs with polymers bearing specific enzyme-cleavable sites. For instance different galacto-oligosaccharides[124] and galacto-polysaccharides[125] have been used

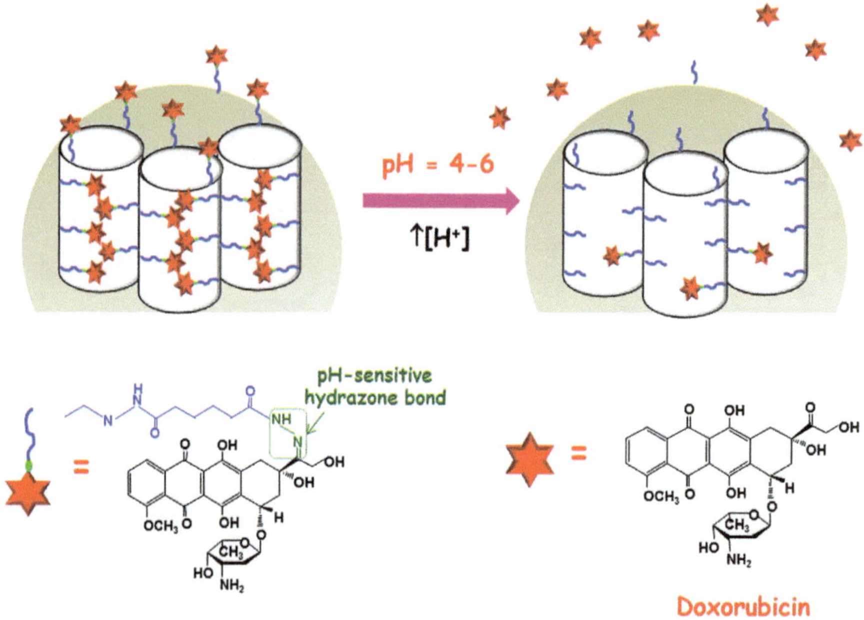

Figure 6.6 Schematic illustration of the pH-responsive drug release from MSNPs with doxorubicin covalently grafted to the mesoporous surface *via* hydrazone pH-sensitive linkers.[115]

as capping agents after grafting their silylated derivatives to the pore entrances of MSNPs. The presence of β-galactosidase triggers the rupture of β-glycosidic bonds, which reduces the size of the saccharide chain at the pore entrance and allows the release of the entrapped payload. Hyaluronic acid (HA) has also been used to design enzyme-responsive MSNPs.[96] As previously mentioned in the Selective Targeting section, HA plays a dual role, acting as both enzyme-sensitive capping agent and targeting ligand. HA degradation takes place in the presence of hyaluronidase-1, a lysosomal enzyme overexpressed in several tumour cells, which provokes pore uncapping and permits the release of the entrapped antitumour drugs. Gelatin and synthetic polypeptide,[126,130] gluconamide[128] and polyester[129] derivatives, have also been used as capping agents of MSNPs, which release antitumour drugs in response to metalloproteinases, amidases and esterases, respectively. Different *in vitro* and *in vivo* assays have demonstrated the capability of these smart nanosystems to treat diverse solid tumours.

Very recently, Mas *et al.* reported a new approach to design tailor-made scaffolds exhibiting enzyme-responsive drug release capability for potential application as biomaterials in bone tissue engineering.[132] These 3D smart systems consisted of macroporous scaffold supports incorporating enzyme-responsive gated MSNPs. The nanogates were designed to open in the presence of acid phosphatase (APase), an enzyme whose concentration is used to measure osteoclast activity in bone remodeling processes and as a

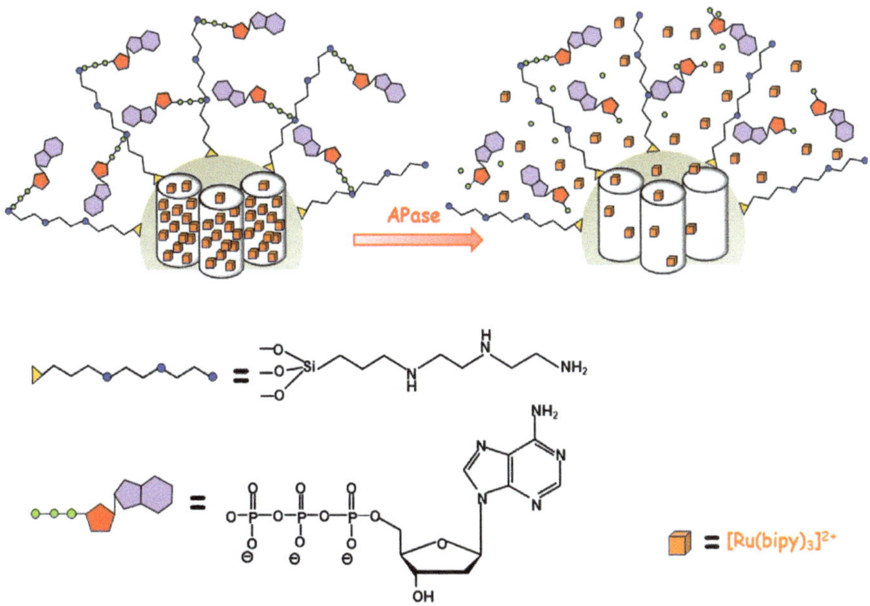

Figure 6.7 Schematic illustration of an enzyme-responsive drug delivery system consisting of MSNPs functionalized with 3-[2-(2-aminoethylamino) ethylamino]propyl-trimethoxysilane and capped with adenosine triphosphate (ATP). The release of the model molecule $[Ru(bipy)_3]^{2+}$ is triggered by acid phosphatase (APase).[132]

biochemical marker for the presence of bone metastases. The cargo release operating mechanism of APase-responsive MSNPs is shown in Figure 6.7. MSNPs were functionalized with 3-[2-(2-aminoethylamino)ethylamino]-propyl-trimethoxysilane, loaded with the model molecule $[Ru(bipy)_3]^{2+}$ and capped with adenosine triphosphate (ATP). The capping of the pores takes place by the formation of a dense network in the pore openings *via* supramolecular hydrogen-bonding interactions, and electrostatic forces between negatively charged ATP anions and positively charged ammonium groups on MSNPs. The presence of APase provokes the hydrolysis of phosphate–phosphate ATP bonds, which results in the disassembly of the polyamine–phosphate interactions and subsequent pore aperture and cargo release. This APase-driven release mechanism was maintained in the 3D scaffold, although slower release kinetics were found, most likely due to the different enzyme accessibility to the gated MSNPs embedded in the hybrid scaffold compared to free MSNPs.

6.3.3 Small Molecule-responsive Drug Delivery

Some diseases are characterized by the production or accumulation of unbalanced amounts of certain chemical species.[133] It is possible to design MSNPs capable of releasing their cargo upon exposure to these small molecules. These smart systems integrate both biomolecular recognition and responsive behaviour. Several MSNPs containing glucose-sensitive gatekeepers have been developed.[131,134,135] Aznar *et al.* described the use of glucose oxidase (GOx) conjugated with CDs as pore blockers of MSNPs functionalized with propylbenzimidazole groups to form an inclusion complex with the CDs.[131] In the presence of glucose, GOx decomposes this molecule into gluconic acid producing a pH drop which induces the protonation of benzimidazole and the dethreading of the inclusion complex and subsequent cargo release. Zhao *et al.* reported the synthesis of a glucose-responsive insulin release system by coating GOx and catalase (Cat) enzyme multilayer shells on the surface of MSNPs.[135] The decrease in pH caused by the decomposition of glucose by GOx activity results in the volume expansion of the enzyme multilayer shells and the subsequent release of insulin entrapped in the mesopores. H_2O_2 produced during the oxidation of glucose would reduce the activity of GOx and the sensitivity of the system to glucose. The presence of Cat in the enzymatic shell is aimed at overcoming this concern by decomposing H_2O_2 into H_2O and O_2. Furthermore, the O_2 produced is re-consumed during the oxidation of glucose. Figure 6.8 schematically displays the operating drug release mechanism of an innovative glucose-responsive MSNP based on the competitive combination between

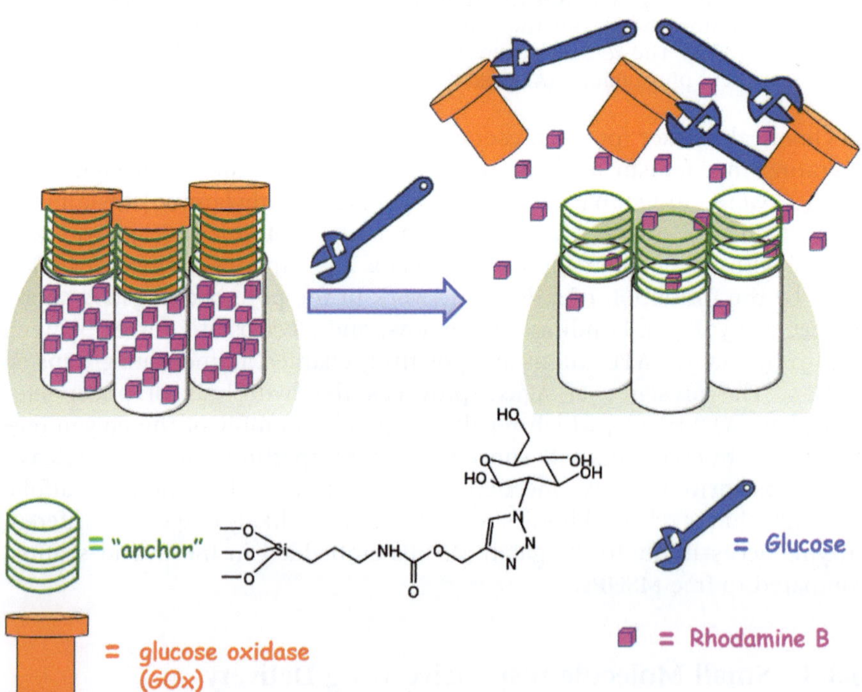

Figure 6.8 Schematic illustration of the glucose-responsive drug release behavior of a nanosystem consisting of MSNPs end-capped with glucose oxidase (GOx) enzyme.[136]

glucose oxidase, glucosamine and glucose.[136] The external surfaces of the MSNPs were first functionalized with D-(+)-glucosamine, an effective inhibitor of GOx, *via* a click-chemistry reaction with prop-2-yn-1-yl(3-(triethoxysilyl)-propyl)carbamate. Then, the MSNPs were loaded with rhodamine B, which was chosen as a model drug. Then GOx was used as a capping agent due to its capability to combine with D-(+)-glucosamine anchored outside the pores to form an enzyme-inhibitor (EI) complex that acts as a bio-gate. The opening of the nanogate takes place by a highly effective competitive combination of glucose (substrate) and GOx, which forms an enzyme–substrate complex that triggers pore uncapping and cargo release.

The development of sophisticated hydrogen peroxide (H_2O_2)-responsive MSNPs as release systems of therapeutic agents for Alzheimer's disease (AD) has also been described.[136] Elevated levels of trace metal ions contribute to neurodegeneration in AD, playing key roles in Aβ aggregate deposition and neurotoxicity that involves the formation of reactive oxygen species such as H_2O_2. MSNPs capable of releasing metal chelators in response to H_2O_2 have been developed. For this purpose, MSNPs were decorated with arylboronic acids, which form boronic esters with saccharides present in immunoglobulins, creating an organic shell that blocks cargo leakage. The presence of H_2O_2 provokes the complete breakage of arylboronic ester bonds, which results in the release of the immunoglobulin and the entrapped chelating agent molecules after that.

Antigen-responsive drug delivery from MSNPs has also been reported.[137] Thus, the pore outlets of MSNPs have been decorated with a certain hapten capable of being selectively recognized by an antibody, which acts as a nanocap. The presence of an antigen with a high affinity for the antibody produces a displacement reaction that triggers pore opening and cargo release.

Aptamer target-responsive drug delivery from MSNPs has also been reported.[138–140] A widely investigated aptamer target is adenosine triphosphate (ATP). Zhu *et al.* designed ATP-responsive nanosystems consisting of MSNPs whose pores were capped with Au NPs throughout an ATP-aptamer.[138] In the presence of ATP the nanocaps were removed by a competitive displacement reaction and the entrapped payload was released. Özalp *et al.* developed switchable ATP-responsive drug release MSNPs by using aptamers as nanovalves.[139] After drug loading, the pores were capped by grafting ATP aptamers with a hairpin form. In the presence of ATP the aptamer binds to ATP and its sequence changes to a molecular beacon-type hairpin structure that permits the release of the entrapped cargo. Figure 6.9 schematically shows the performance of innovative ATP-responsive MSNPs for drug delivery.[141] The mesopore outlets were functionalized with two single-strand DNA arms, *i.e.* ssDNA1 and ssDNA2 (*the complementary component*), which hybridized with ATP aptamer (*binding agent*) to form a sandwich-type DNA structure that blocked the nanochannels. In the absence of ATP (the *target biomolecule*), the mesopores are capped and the leakage of guest molecules is hindered. In the presence of ATP, a competitive reaction takes place owing

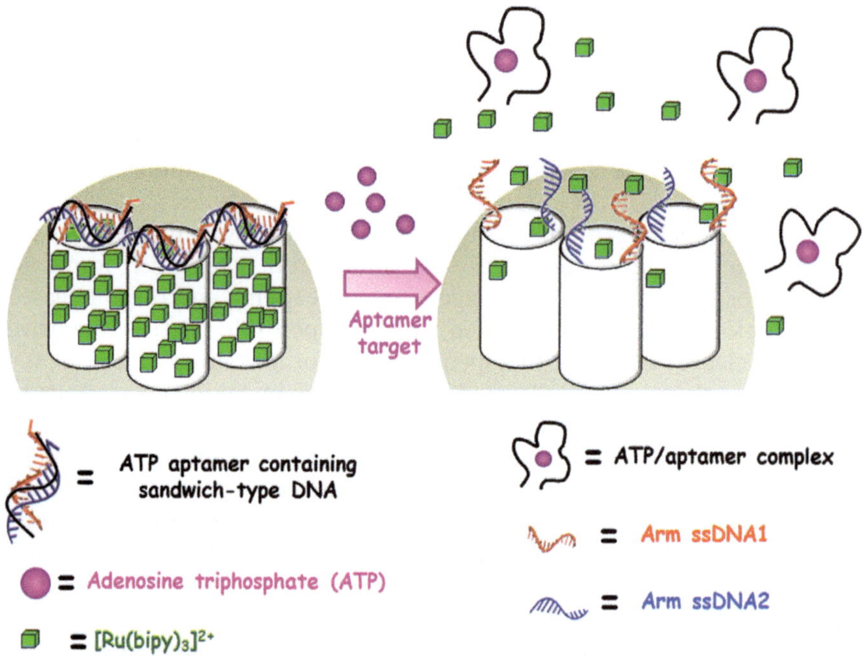

Figure 6.9 Schematic depiction of aptamer-based ATP-responsive MSNPs. MSNPs are functionalized with two single-stranded DNA arms capable of forming a sandwich-type DNA structure in the presence of an ATP aptamer.[141]

to the higher affinity and tighter binding of the ATP aptamer with ATP than with the ssDNA arms, which results in pore uncapping and release of guest molecules.

6.3.4 Redox-responsive Drug Delivery

Glutathione (GSH) is a reducing agent widely exploited as a release trigger due to an imbalance between GSH concentration outside and inside cells and also between normal and tumour tissues.[106,142] The intracellular compartments (cytosol, mitochondria and nucleus) contain GSH at a concentration of 2–10 μM, which is 1000 times higher than that in the extracellular media.[142] In addition, it has been observed than tumour tissues may reach GSH concentrations up to four-fold greater than normal tissues.[143] GSH is a tripeptide containing a thiol group of cysteine that is capable of breaking a disulfide (–S–S–) bond by reduction to two –SH groups. During this redox reaction, two GSH molecules convert into the oxidized state of glutathione (GSSG), which contains a disulfide bond. Advanced GSH-responsive MSNPs for drug delivery have been developed.[57,59,65,73,144–151] Sometimes, as a proof of concept prior to *in vitro* tests, *in vial* studies are carried out using other reducing agents, such as dithiothreitol (DTT) or mercaptoethanol (ME), as release triggers. In this chapter, GSH-responsive MSNPs have been categorized into four major groups depending on their components: (1) MSNPs–polymer nanocomposites; (2) inorganic capping of MSNPs; (3) capping of MSNPs with organic motifs; (4) grafting of drugs to MSNPs.

6.3.4.1 *Mesoporous Silica–Polymer Nanocomposites*

The combination of MSNPs and polymers to design GSH-responsive release systems has been carried out using two different approaches. The first one relies on using MSNPs as cores and covalently grafted polymers as shells. These polymers, such as poly(*N*-acryloxysuccinimide),[147] poly(oligo-(ethyleneglycol) acrylate)[59] or poly(acrylic acid),[152] are crosslinked by GSH-cleavable disulfide linkages. The presence of GSH provokes the rupture of disulfide bonds and switches on drug release from mesopores. The second approach consists of covalently grafting polymers to MSNPs throughout GSH-cleavable disulfide bonds. Different polymers, such collagen,[65] poly-(ethylene glycol)-*b*-poly(ε-caprolactone) (PEG–PCL)[153] and PEG,[149] have been used for this purpose. Figure 6.10 schematically displays this strategy. MSNPs are capped with PEG *via* disulfide bonds. GSH reducing molecules act as scissors and cut –S–S– bonds, which leads to the removal of the PEG gatekeepers and triggers the release of fluorescein, which was chosen as a model molecule, from mesopores.

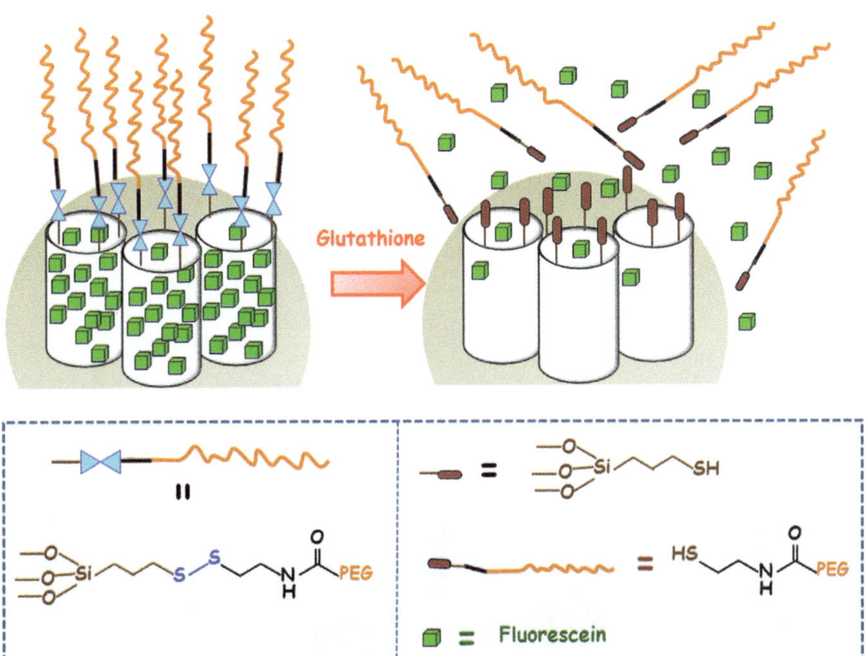

Figure 6.10 Schematic representation of the glutathione-responsive release behavior of a nanosystem consisting of MSNPs capped with polyethyleneglycol (PEG) *via* disulfide bonds.[149]

6.3.4.2 Inorganic Capping of Mesoporous Silica Nanoparticles

Another strategy consists of tethering inorganic blocking nanocaps, such as CdS,[144] Fe_3O_4[145] or Au[146] NPs, to the pore outlets of MSNPs *via* disulfide linkages. The presence of the reducing agent triggers pore opening and cargo release. Figure 6.11 illustrates the GSH-responsive operating release mechanism of MSNPs end-capped with CdS NPs *via* disulfide bonds.[144] For this purpose, MSNPs were functionalized with 2-(propyldisulfanyl)ethylamine. Then, cargo molecules (vancomycin or ATP) were loaded into the pores and *in situ* capping was achieved by allowing the pore surface-bound 2-(propyldisulfanyl)ethylamine functional groups to covalently capture the water-soluble mercaptoacetic acid-derivatized CdS NPs *via* amidation reaction (Figure 6.11). The resulting disulfide linkages between the MSNPs and the CdS NPs were cleaved by addition of reducing agents such as DTT and ME, which triggered the release of entrapped molecules.

6.3.4.3 Capping of Mesoporous Silica Nanoparticles with Organic Motifs

Blocking caps consisting of non-polymeric organic moieties have been linked to MSNPs *via* disulfide bonds to design GSH-responsive drug delivery

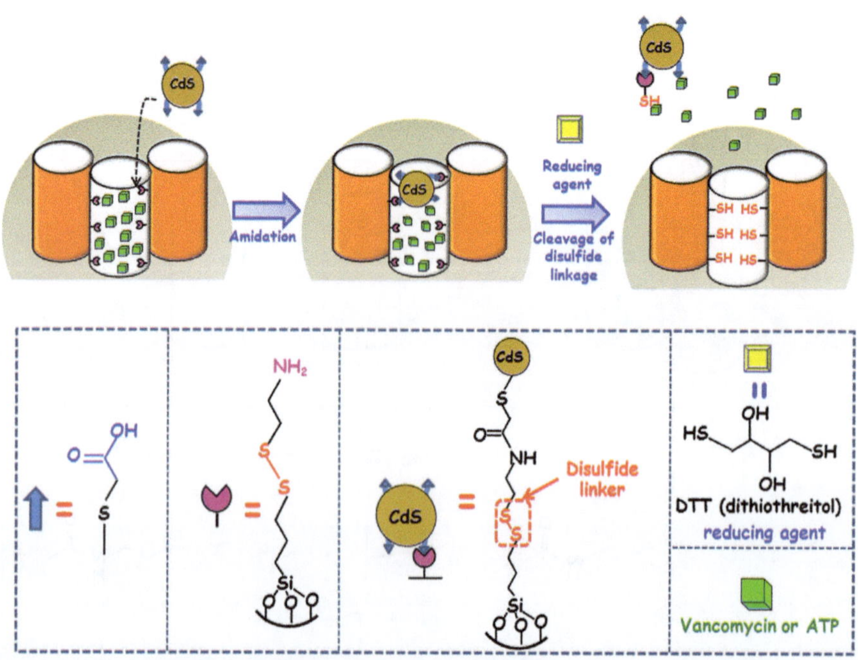

Figure 6.11 Schematic illustration of the glutathione-responsive drug delivery performance of MSNPs end-capped with CdS nanoparticles *via* disulfide linkages.[144]

systems. Small molecules such as folic acid (FA) have been grafted to MSNPs through disulfide linkages.[73] FA molecules act as efficient blocking caps because their molecular size is in the range of the pore diameter of MSNPs. The self-interaction between FA molecules *via* H-bonding also increases the pore capping efficiency. An added value of this system is the dual role of FA, since it acts not only as a capping agent but also as a targeting ligand. The efficient tumour-targeting and intracellular GSH-responsive drug release capabilities were assessed *in vitro*. Macromolecules, such as CDs[148,150] and poly(propylene imine) dendrimers (PPI) of generations I and II[154] grafted to the surface of MSNPs throughout reducible disulfide bonds are also capable of blocking the premature release of the trapped drugs until the carrier reaches the intracellular space.

Ma *et al.* have developed a GSH-responsive MSNP co-delivery system capable of simultaneously releasing drugs and ssDNA[155] (Figure 6.12). MSNPs were functionalized with amino-terminated alkyl chains containing disulfide bonds. Since the surface amino groups are protonated in aqueous solution at neutral or acidic pH, they can electrostatically interact with negatively charged ssDNA chains, which form a network that blocks the mesopores and prevents drug release. The reducing agent, DTT or GSH, effectively breaks the disulfide bond and thus collapses the gatekeepers provided by the gene network, leading to a simultaneous drug/ssDNA co-delivery.

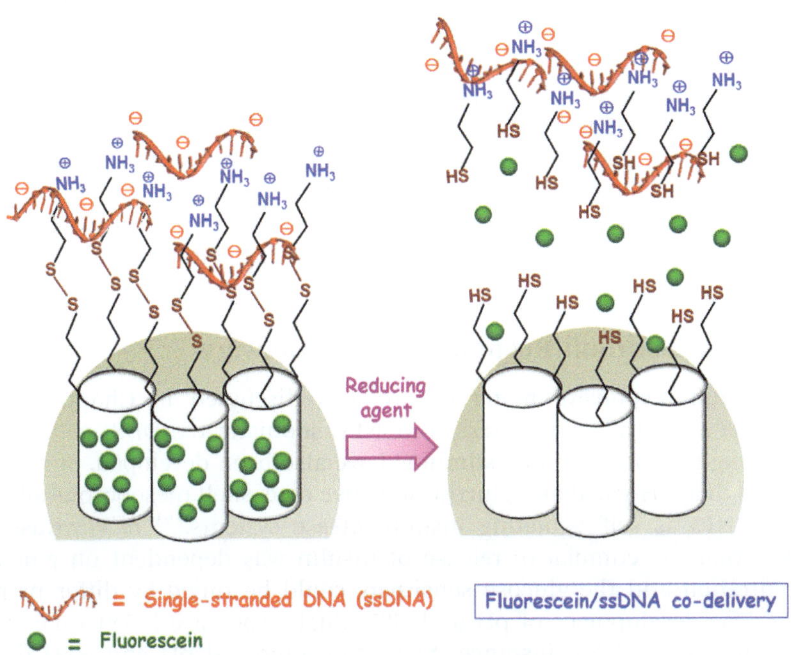

Figure 6.12 Schematic illustration of the glutathione-responsive co-delivery of drugs and single-stranded DNA (ssDNA) from MSNPs.[155]

Very recently, Zhang *et al.* reported the synthesis of cytochrome *c* (CytC) end-capped MSNPs as GSH-responsive drug delivery systems for liver tumour-targeted triplex therapy *in vitro* and *in vivo*.[156] CytC was immobilized onto the surface of MSNPs *via* intermediate linkers of disulfide bonds, acting as GSH-responsive gatekeepers. A G-rich oligonucleotide DNA aptamer, AS1411, was used as a specific targeting motif to tumour cells. Once MSNPs were endocytosed by tumour cells, the reducing microenvironment within cells, mainly due to the over-expression of GSH, led to the breakage of disulfide bonds for delivery of entrapped DOX. In this system, AS1411 and CytC not only acted as targeting ligand and end-capping agent, respectively, but also as therapeutic agents.

6.3.4.4 Redox-controlled Grafting of Drugs to Mesoporous Silica Nanoparticles

An interesting strategy aimed at delivering highly cytotoxic drugs or highly sensitive compounds consists of linking the cargo molecules to MSNPs *via* GSH-cleavable disulfide bonds. Using specific disulfide-containing linkers, potent antitumour drugs (*e.g.* paclitaxel, PTX)[157] or proteins (*e.g.* carbonic anhydrase, cytochrome c)[158,159] have been grafted to MSNPs for GSH-triggered intracellular release.[158] It is essential that despite being submitted to the grafting process, the molecules retain their activity once released.

An innovative strategy consists of using MSNPs as carriers of covalently-linked less toxic inactive prodrugs, which are later activated in the intracellular space. Thus, Ahn *et al.* have reported the grafting of a *cis*-platin (CDDP) prodrug to MSNPs.[160] The prodrug was the inactive oxidized Pt(IV) complex *c,c,t*-$[Pt(NH_3)_2Cl_2(OOCCH_2CH_2–COOH)]$. The prodrug grafting was accomplished by formation of amide bonds between terminal carboxylic groups of axial ligands in the complex amine groups of functionalized MSNPs. In the GSH-rich reductive environment of cancer cells, the active CDDP form was regenerated and released, inducing cellular death.

6.3.5 Multi-stimuli Responsive Drug Delivery

Multi-stimuli responsive materials are also discussed in Chapter 5 by Schalley *et al.* Within the context of MSNPs, sophisticated DDSs capable of responding to more than one stimulus have also been developed. Sun *et al.* designed dual pH-gated and glucose-sensitive core–shell mesoporous silica–polymer NPs as self-regulating insulin release systems.[161] *In vitro* assays showed that the cumulative release of insulin was dependent on glucose concentration and the glucose sensitivity could be tuned by different pH values. The development of pH and GSH dual-responsive MSNPs has also been reported.[53,162] For instance, Xiao *et al.* prepared pH and GSH dual-responsive tumour-triggered targeting MSNPs for delivering drugs into tumour cells.[53] A RGD-containing peptide was linked to the surface of

MSNPs through disulfide bonds. This peptide played a dual role: GSH-responsive gatekeeper and tumour-targeting ligand. Then, the nanosystem was provided with pH-responsive capability by coating it with a PEG-derivative *via* pH-sensitive benzoic-imine linkages. Under the acidic conditions of a tumour environment, the PEG coating became detached, switching on the targeting properties of the system by exposure of the RGD-containing peptide moieties. Upon tumour cell internalization, intracellular GSH provoked the removal of the peptide caps and triggered drug release.

6.4 Future Prospects

MSNPs have emerged as powerful candidates to develop novel nanomedicines for the treatment and diagnosis of diseases with a bad prognosis, such as cancer. Recent advances have permitted the engineering of MSNPs-based tumour-targeted stimuli-responsive DDSs. These nanosystems must fulfil several mandatory requirements, including biocompatibility and a non-toxic nature for all their constituents; efficacy to reach and accumulate in the target diseased tissue; and capability to release their payload in response to certain stimuli. Sophisticated MSNP-based tumour-targeted DDSs have been developed and their successful *in vitro* performance has been demonstrated. However, the therapeutic advantage of these systems for *in vivo* applications is a continuous matter of debate. This can be explained by the non-translatable conditions used during the *in vivo* tests and also by the differing conclusions derived from preclinical studies. Nonetheless, recent investigations suggest that favourable *in vitro* findings could be reproduced *in vivo*.

References

1. M. Vallet-Regí, A. Rámila, R. P. del Real and J. Pérez-Pariente, *Chem. Mater.*, 2001, **13**, 308.
2. M. Vallet-Regí, F. Balas and D. Arcos, *Angew. Chem., Int. Ed.*, 2007, **46**, 7548.
3. I. I. Slowing, J. L. Vivero-Escoto, C.-W. Wu and V. S. Y. Lin, *Adv. Drug Delivery Rev.*, 2008, **60**, 1278.
4. M. Manzano, M. Colilla and M. Vallet-Regí, *Expert Opin. Drug Delivery*, 2009, **6**, 1383.
5. Y. Zhao, J. L. Vivero-Escoto, I. I. Slowing, B. G. Trewyn and V. S. Y. Lin, *Expert Opin. Drug Delivery*, 2010, **7**, 1013.
6. Z. Li, J. C. Barnes, A. Bosoy, J. F. Stoddart and J. I. Zink, *Chem. Soc. Rev.*, 2012, **41**, 2590.
7. P. Yang, S. Gai and J. Lin, *Chem. Soc. Rev.*, 2012, **41**, 3679.
8. M. Colilla, B. González and M. Vallet-Regi, *Biomater. Sci.*, 2013, **1**, 114.
9. M. Vallet-Regí, M. Manzano and M. Colilla, *Biomedical Applications of Mesoporous Ceramics*, CRC Press, New York, 2013, p. 105.

10. M. Colilla and M. Vallet-Regí, in *Smart Materials for Drug Delivery*, ed. C. Álvarez-Lorenzo and Á. Concheiro, The Royal Society of Chemistry, London, UK, 2013, vol. 2, p. 63.
11. S. Mura, J. Nicolas and P. Couvreur, *Nat. Mater.*, 2013, **12**, 991.
12. V. Mamaeva, C. Sahlgren and M. Lindén, *Adv. Drug Delivery Rev.*, 2013, **65**, 689.
13. Y. Chen, H. Chen and J. Shi, *Expert Opin. Drug Delivery*, 2014, **11**, 917.
14. A. Baeza, E. Guisasola, A. Torres-Pardo, J. M. González-Calbet, G. J. Melen, M. Ramírez and M. Vallet-Regí, *Adv. Funct. Mater.*, 2014, **24**, 4625.
15. C. Barbé, J. Bartlett, L. Kong, K. Finnie, H. Q. Lin, M. Larkin, S. Calleja, A. Bush and G. Calleja, *Adv. Mater.*, 2004, **16**, 1959.
16. D. Avnir, T. Coradin, O. Lev and J. Livage, *J. Mater. Chem.*, 2006, **16**, 1013.
17. B. G. Trewyn, I. I. Slowing, S. Giri, H.-T. Chen and V. S. Y. Lin, *Acc. Chem. Res.*, 2007, **40**, 846.
18. F. Tang, L. Li and D. Chen, *Adv. Mater.*, 2012, **24**, 1504.
19. S.-H. Wu, C.-Y. Mou and H.-P. Lin, *Chem. Soc. Rev.*, 2013, **42**, 3862.
20. M. Vallet-Regí, *Bio-Ceramics with Clinical Applications*, John Wiley & Sons Ltd, Chichester, United Kindom, 2014.
21. I. Brigger, C. Dubernet and P. Couvreur, *Adv. Drug Delivery Rev.*, 2012, **64**, 24.
22. N. Bertrand, J. Wu, X. Xu, N. Kamaly and O. C. Farokhzad, *Adv. Drug Delivery Rev.*, 2014, **66**, 2.
23. A. Baeza, M. Colilla and M. Vallet-Regí, *Expert Opin. Drug Delivery*, 2015, **12**, 319.
24. C. Álvarez-Lorenzo and A. Concheiro, *Smart Materials for Drug Delivery*, The Royal Society of Chemistry, 2013, vol. 1, p. 1.
25. M. Ferrari, *Nat. Rev. Cancer*, 2005, **5**, 161.
26. R. Langer, *Nature*, 1998, **392**, 5.
27. O. C. Farokhzad and R. Langer, *ACS Nano*, 2009, **3**, 16.
28. F. Danhier, O. Feron and V. Préat, *J. Controlled Release*, 2010, **148**, 135.
29. Y. Matsumura and H. Maeda, *Cancer Res.*, 1986, **46**, 6387.
30. R. K. Jain and T. Stylianopoulos, *Nat. Rev. Clin. Oncol.*, 2010, 7, 653.
31. M. L. Etheridge, S. A. Campbell, A. G. Erdman, C. L. Haynes, S. M. Wolf and J. McCullough, *Nanomed. Nanotechnol. Biol. Med.*, 2013, **9**, 1.
32. X. Huang, X. Teng, D. Chen, F. Tang and J. He, *Biomaterials*, 2010, **31**, 438.
33. X. Huang, L. Li, T. Liu, N. Hao, H. Liu, D. Chen and F. Tang, *ACS Nano*, 2011, **5**, 5390.
34. J. Champion and S. Mitragotri, *Pharm. Res.*, 2009, **26**, 244.
35. P. d. Pino, B. Pelaz, Q. Zhang, P. Maffre, G. U. Nienhaus and W. J. Parak, *Mater. Horiz.*, 2014, **1**, 301.
36. H. Otsuka, Y. Nagasaki and K. Kataoka, *Adv. Drug Delivery Rev.*, 2003, **55**, 403.
37. V. Cauda, C. Argyo and T. Bein, *J. Mater. Chem.*, 2010, **20**, 8693.

38. Q. He, J. Zhang, J. Shi, Z. Zhu, L. Zhang, W. Bu, L. Guo and Y. Chen, *Biomaterials*, 2010, **31**, 1085.
39. V. Cauda, A. Schlossbauer and T. Bein, *Microporous Mesoporous Mater.*, 2010, **132**, 60.
40. A. S. Karakoti, S. Das, S. Thevuthasan and S. Seal, *Angew. Chem., Int. Ed.*, 2011, **50**, 1980.
41. H. Meng, M. Xue, T. Xia, Z. Ji, D. Y. Tarn, J. I. Zink and A. E. Nel, *ACS Nano*, 2011, **5**, 4131.
42. J. L. Townson, Y.-S. Lin, J. O. Agola, E. C. Carnes, H. S. Leong, J. D. Lewis, C. L. Haynes and C. J. Brinker, *J. Am. Chem. Soc.*, 2013, **135**, 16030.
43. T. Lammers, F. Kiessling, W. E. Hennink and G. Storm, *J. Controlled Release*, 2012, **161**, 175.
44. N. Bertrand, J. Wu, X. Xu, N. Kamaly and O. C. Farokhzad, *Adv. Drug Delivery Rev.*, 2014, **66**, 2.
45. Z. Deng, Z. Zhen, X. Hu, S. Wu, Z. Xu and P. K. Chu, *Biomaterials*, 2011, **32**, 4976.
46. C.-P. Tsai, C.-Y. Chen, Y. Hung, F.-H. Chang and C.-Y. Mou, *J. Mater. Chem.*, 2009, **19**, 5737.
47. K. Cheng, S. R. Blumen, M. B. MacPherson, J. L. Steinbacher, B. T. Mossman and C. C. Landry, *ACS Appl. Mater. Interfaces*, 2010, **2**, 2489.
48. F. Chen, H. Hong, Y. Zhang, H. F. Valdovinos, S. Shi, G. S. Kwon, C. P. Theuer, T. E. Barnhart and W. Cai, *ACS Nano*, 2013, **7**, 9027.
49. A. Milgroom, M. Intrator, K. Madhavan, L. Mazzaro, R. Shandas, B. Liu and D. Park, *Colloids Surf., B*, 2014, **116**, 652.
50. D. P. Ferris, J. Lu, C. Gothard, R. Yanes, C. R. Thomas, J.-C. Olsen, J. F. Stoddart, F. Tamanoi and J. I. Zink, *Small*, 2011, **7**, 1816.
51. W. Fang, Z. Wang, S. Zong, H. Chen, D. Zhu, Y. Zhong and Y. Cui, *Biosens. Bioelectron.*, 2014, **57**, 10.
52. K. Cheng, S. R. Blumen, M. B. MacPherson, J. L. Steinbacher, B. T. Mossman and C. C. Landry, *ACS Appl. Mater. Interfaces*, 2010, **2**, 2489.
53. D. Xiao, H.-Z. Jia, J. Zhang, C.-W. Liu, R.-X. Zhuo and X.-Z. Zhang, *Small*, 2014, **10**, 591.
54. L. He, Y. Huang, H. Zhu, G. Pang, W. Zheng, Y.-S. Wong and T. Chen, *Adv. Funct. Mater.*, 2014, **24**, 2754.
55. I. J. Fang, I. I. Slowing, K. C. W. Wu, V. S. Y. Lin and B. G. Trewyn, *Chem. – Eur. J.*, 2012, **18**, 7787.
56. G.-F. Luo, W.-H. Chen, Y. Liu, J. Zhang, S.-X. Cheng, R.-X. Zhuo and X.-Z. Zhang, *J. Mater. Chem. B*, 2013, **1**, 5723.
57. J. Zhang, Z.-F. Yuan, Y. Wang, W.-H. Chen, G.-F. Luo, S.-X. Cheng, R.-X. Zhuo and X.-Z. Zhang, *J. Am. Chem. Soc.*, 2013, **135**, 5068.
58. L. Pan, Q. He, J. Liu, Y. Chen, M. Ma, L. Zhang and J. Shi, *J. Am. Chem. Soc.*, 2012, **134**, 5722.

59. J.-T. Sun, J.-G. Piao, L.-H. Wang, M. Javed, C.-Y. Hong and C.-Y. Pan, *Macromol. Rapid Commun.*, 2013, **34**, 1387.
60. Z. Li, K. Dong, S. Huang, E. Ju, Z. Liu, M. Yin, J. Ren and X. Qu, *Adv. Funct. Mater.*, 2014, **18**, 3612.
61. J. Liu, Z. Cao and Y. Lu, *Chem. Rev.*, 2009, **109**, 1948.
62. O. C. Farokhzad, J. M. Karp and R. Langer, *Expert Opin. Drug Delivery*, 2006, **3**, 311.
63. L. Gao, Y. Cui, Q. He, Y. Yang, J. Fei and J. Li, *Chem. – Eur. J.*, 2011, **17**, 13170.
64. L.-L. Li, Q. Yin, J. Cheng and Y. Lu, *Adv. Healthcare Mater.*, 2012, **1**, 567.
65. Z. Luo, K. Cai, Y. Hu, L. Zhao, P. Liu, L. Duan and W. Yang, *Angew. Chem., Int. Ed.*, 2011, **50**, 640.
66. D. Brevet, M. Gary-Bobo, L. Raehm, S. Richeter, O. Hocine, K. Amro, B. Loock, P. Couleaud, C. Frochot, A. Morere, P. Maillard, M. Garcia and J.-O. Durand, *Chem. Commun.*, 2009, 1475.
67. M. Gary-Bobo, O. Hocine, D. Brevet, M. Maynadier, L. Raehm, S. Richeter, V. Charasson, B. Loock, A. Morère, P. Maillard, M. Garcia and J.-O. Durand, *Int. J. Pharm.*, 2012, **423**, 509.
68. M. Yu, S. Jambhrunkar, P. Thorn, J. Chen, W. Gu and C. Yu, *Nanoscale*, 2013, **5**, 178.
69. Z. Chen, Z. Li, Y. Lin, M. Yin, J. Ren and X. Qu, *Chem. – Eur. J.*, 2013, **19**, 1778.
70. H. Elnakat and M. Ratnam, *Adv. Drug Delivery Rev.*, 2004, **56**, 1067.
71. J. Lu, Z. Li, J. I. Zink and F. Tamanoi, *Nanomed. Nanotechnol. Biol .Med.*, 2012, **8**, 212.
72. J. M. Rosenholm, A. Meinander, E. Peuhu, R. Niemi, J. E. Eriksson, C. Sahlgren and M. Lindén, *ACS Nano*, 2008, **3**, 197.
73. R. Guo, L.-L. Li, W.-H. Zhao, Y.-X. Chen, X.-Z. Wang, C.-J. Fang, W. Feng, T.-L. Zhang, X. Ma, M. Lu, S.-Q. Peng and C.-H. Yan, *Nanoscale*, 2012, **4**, 3577.
74. J. M. Rosenholm, E. Peuhu, L. T. Bate-Eya, J. E. Eriksson, C. Sahlgren and M. Lindén, *Small*, 2010, **6**, 1234.
75. A. S. E. Ojugo, P. M. J. McSheehy, D. J. O. McIntyre, C. McCoy, M. Stubbs, M. O. Leach, I. R. Judson and J. R. Griffiths, *NMR Biomed.*, 1999, **12**, 495.
76. G. T. Gethin, S. Cowman and R. M. Conroy, *Int. Wound J.*, 2008, **5**, 185.
77. J. R. Casey, S. Grinstein and J. Orlowski, *Nat. Rev. Mol. Cell Biol.*, 2010, **11**, 50.
78. R. Liu, P. Liao, J. Liu and P. Feng, *Langmuir*, 2011, **27**, 3095.
79. J.-T. Sun, C.-Y. Hong and C.-Y. Pan, *J. Phys. Chem. C*, 2010, **114**, 12481.
80. Y. Tian, A. Glogowska, W. Zhong, T. Klonisch and M. Xing, *J. Mater. Chem. B*, 2013, **1**, 5264.
81. A. Popat, J. Liu, G. Q. Lu and S. Z. Qiao, *J. Mater. Chem.*, 2012, **22**, 11173.
82. J. Wang, H. Liu, F. Leng, L. Zheng, J. Yang, W. Wang and C. Z. Huang, *Microporous Mesoporous Mater.*, 2014, **186**, 187.

83. W. Feng, X. Zhou, C. He, K. Qiu, W. Nie, L. Chen, H. Wang, X. Mo and Y. Zhang, *J. Mater. Chem. B*, 2013, **1**, 5886.
84. C.-Y. Hong, X. Li and C.-Y. Pan, *J. Mater. Chem.*, 2009, **19**, 5155.
85. H. Peng, R. Dong, S. Wang, Z. Zhang, M. Luo, C. Bai, Q. Zhao, J. Li, L. Chen and H. Xiong, *Int. J. Pharm.*, 2013, **446**, 153.
86. L. Yuan, Q. Tang, D. Yang, J. Z. Zhang, F. Zhang and J. Hu, *J. Phys. Chem. C*, 2011, **115**, 9926.
87. J. Zheng, X. Tian, Y. Sun, D. Lu and W. Yang, *Int. J. Pharm.*, 2013, **450**, 296.
88. L. Xing, H. Zheng, Y. Cao and S. Che, *Adv. Mater.*, 2012, **24**, 6433.
89. Z. Zou, D. He, X. He, K. Wang, X. Yang, Z. Qing and Q. Zhou, *Langmuir*, 2013, **29**, 12804.
90. X. Zhang, F. Li, S. Guo, X. Chen, X. Wang, J. Li and Y. Gan, *Biomaterials*, 2014, **35**, 3650.
91. S. Saha, K. C. F. Leung, T. D. Nguyen, J. F. Stoddart and J. I. Zink, *Adv. Funct. Mater.*, 2007, **17**, 685.
92. K. K. Coti, M. E. Belowich, M. Liong, M. W. Ambrogio, Y. A. Lau, H. A. Khatib, J. I. Zink, N. M. Khashab and J. F. Stoddart, *Nanoscale*, 2009, **1**, 16.
93. M. W. Ambrogio, C. R. Thomas, Y.-L. Zhao, J. I. Zink and J. F. Stoddart, *Acc. Chem. Res.*, 2011, **44**, 903.
94. K. Kim, *Chem. Soc. Rev.*, 2002, **31**, 96.
95. S. Angelos, Y.-W. Yang, K. Patel, J. F. Stoddart and J. I. Zink, *Angew. Chem., Int. Ed.*, 2008, **47**, 2222.
96. T. Chen, N. Yang and J. Fu, *Chem. Commun.*, 2013, **49**, 6555.
97. C. Park, K. Oh, S. C. Lee and C. Kim, *Angew. Chem., Int. Ed.*, 2007, **46**, 1455.
98. L. Du, S. Liao, H. A. Khatib, J. F. Stoddart and J. I. Zink, *J. Am. Chem. Soc.*, 2009, **131**, 15136.
99. H. Meng, M. Xue, T. Xia, Y.-L. Zhao, F. Tamanoi, J. F. Stoddart, J. I. Zink and A. E. Nel, *J. Am. Chem. Soc.*, 2010, **132**, 12690.
100. J. Dong, M. Xue and J. I. Zink, *Nanoscale*, 2013, **5**, 10300.
101. C. Théron, A. Gallud, C. Carcel, M. Gary-Bobo, M. Maynadier, M. Garcia, J. Lu, F. Tamanoi, J. I. Zink and M. Wong Chi Man, *Chem. – Eur. J.*, 2014, **20**, 9372.
102. J. Lagona, P. Mukhopadhyay, S. Chakrabarti and L. Isaacs, *Angew. Chem., Int. Ed.*, 2005, **44**, 4844.
103. S. Liu, C. Ruspic, P. Mukhopadhyay, S. Chakrabarti, P. Y. Zavalij and L. Isaacs, *J. Am. Chem. Soc.*, 2005, **127**, 15959.
104. S. Angelos, N. M. Khashab, Y.-W. Yang, A. Trabolsi, H. A. Khatib, J. F. Stoddart and J. I. Zink, *J. Am. Chem. Soc.*, 2009, **131**, 12912.
105. R. Liu, Y. Zhang, X. Zhao, A. Agarwal, L. J. Mueller and P. Feng, *J. Am. Chem. Soc.*, 2010, **132**, 1500.
106. I.-P. Huang, S.-P. Sun, S.-H. Cheng, C.-H. Lee, C.-Y. Wu, C.-S. Yang, L.-W. Lo and Y.-K. Lai, *Mol. Cancer Ther.*, 2011, **10**, 761.

107. Q. Gan, X. Lu, Y. Yuan, J. Qian, H. Zhou, X. Lu, J. Shi and C. Liu, *Biomaterials*, 2011, **32**, 1932.
108. C. Xu, Y. Lin, J. Wang, L. Wu, W. Wei, J. Ren and X. Qu, *Adv. Healthcare Mater.*, 2013, **2**, 1591.
109. E. Aznar, C. Coll, M. D. Marcos, R. Martínez-Máñez, F. Sancenón, J. Soto, P. Amorós, J. Cano and E. Ruiz, *Chem. – Eur. J.*, 2009, **15**, 6877.
110. F. Muhammad, M. Guo, W. Qi, F. Sun, A. Wang, Y. Guo and G. Zhu, *J. Am. Chem. Soc.*, 2011, **133**, 8778.
111. C.-X. Zhao, L. Yu and A. P. J. Middelberg, *J. Mater. Chem. B*, 2013, **1**, 4828.
112. Q. Zheng, Y. Hao, P. Ye, L. Guo, H. Wu, Q. Guo, J. Jiang, F. Fu and G. Chen, *J. Mater. Chem. B*, 2013, **1**, 1644.
113. Y. Bae, S. Fukushima, A. Harada and K. Kataoka, *Angew. Chem., Int. Ed.*, 2003, **42**, 4640.
114. F. Jianquan, F. Gang, W. Xiaodan, Z. Fang, X. Yufei and W. Shuizhu, *Nanotechnology*, 2011, **22**, 455102.
115. C.-H. Lee, S.-H. Cheng, I. P. Huang, J. S. Souris, C.-S. Yang, C.-Y. Mou and L.-W. Lo, *Angew. Chem., Int. Ed.*, 2010, **49**, 8214.
116. J. E. Lee, D. J. Lee, N. Lee, B. H. Kim, S. H. Choi and T. Hyeon, *J. Mater. Chem.*, 2011, **21**, 16869.
117. Z.-Y. Li, Y. Liu, X.-Q. Wang, L.-H. Liu, J.-J. Hu, G.-F. Luo, W.-H. Chen, L. Rong and X.-Z. Zhang, *ACS Appl. Mater. Interfaces*, 2013, **5**, 7995.
118. C.-H. Lin, S.-H. Cheng, W.-N. Liao, P.-R. Wei, P.-J. Sung, C.-F. Weng and C.-H. Lee, *Int. J. Pharm.*, 2012, **429**, 138.
119. R. V. Ulijn, *J. Mater. Chem.*, 2006, **16**, 2217.
120. R. de la Rica, D. Aili and M. M. Stevens, *Adv. Drug Delivery Rev.*, 2012, **64**, 967.
121. K. Patel, S. Angelos, W. R. Dichtel, A. Coskun, Y.-W. Yang, J. I. Zink and J. F. Stoddart, *J. Am. Chem. Soc.*, 2008, **130**, 2382.
122. C. Park, H. Kim, S. Kim and C. Kim, *J. Am. Chem. Soc.*, 2009, **131**, 16614.
123. A. Schlossbauer, J. Kecht and T. Bein, *Angew. Chem., Int. Ed.*, 2009, **48**, 3092.
124. A. Bernardos, E. Aznar, M. D. Marcos, R. Martínez-Máñez, F. Sancenón, J. Soto, J. M. Barat and P. Amorós, *Angew. Chem., Int. Ed.*, 2009, **48**, 5884.
125. A. Bernardos, L. Mondragón, E. Aznar, M. D. Marcos, R. n. Martínez-Máñez, F. l. Sancenón, J. Soto, J. M. Barat, E. Pérez-Payá, C. Guillem and P. Amorós, *ACS Nano*, 2010, **4**, 6353.
126. N. Singh, A. Karambelkar, L. Gu, K. Lin, J. S. Miller, C. S. Chen, M. J. Sailor and S. N. Bhatia, *J. Am. Chem. Soc.*, 2011, **133**, 19582.
127. A. Agostini, L. Mondragón, C. Coll, E. Aznar, M. D. Marcos, R. Martínez-Máñez, F. Sancenón, J. Soto, E. Pérez-Payá and P. Amorós, *ChemistryOpen*, 2012, **1**, 17.
128. I. Candel, E. Aznar, L. Mondragon, C. d. l. Torre, R. Martinez-Manez, F. Sancenon, M. D. Marcos, P. Amoros, C. Guillem, E. Perez-Paya, A. Costero, S. Gil and M. Parra, *Nanoscale*, 2012, **4**, 7237.

129. A. Bernardos, L. Mondragón, I. Javakhishvili, N. Mas, C. de la Torre, R. Martínez-Máñez, F. Sancenón, J. M. Barat, S. Hvilsted, M. Orzaez, E. Pérez-Payá and P. Amorós, *Chem. - Eur. J.*, 2012, **18**, 13068.
130. J.-H. Xu, F.-P. Gao, L.-L. Li, H. L. Ma, Y.-S. Fan, W. Liu, S.-S. Guo, X.-Z. Zhao and H. Wang, *Microporous Mesoporous Mater.*, 2013, **182**, 165.
131. R. Villalonga, P. Díez, A. Sánchez, E. Aznar, R. Martínez-Máñez and J. M. Pingarrón, *Chem. - Eur. J.*, 2013, **19**, 7889.
132. N. Mas, D. Arcos, L. Polo, E. Aznar, S. Sánchez-Salcedo, F. Sancenón, A. García, M. D. Marcos, A. Baeza, M. Vallet-Regí and R. Martínez-Máñez, *Small*, 2014, **10**, 4859.
133. T. Miyata, T. Uragami and K. Nakamae, *Adv. Drug Delivery Rev.*, 2002, **54**, 79.
134. Y. Zhao, B. G. Trewyn, I. I. Slowing and V. S. Y. Lin, *J. Am. Chem. Soc.*, 2009, **131**, 8398.
135. W. Zhao, H. Zhang, Q. He, Y. Li, J. Gu, L. Li, H. Li and J. Shi, *Chem. Commun.*, 2011, **47**, 9459.
136. J. Geng, M. Li, L. Wu, C. Chen and X. Qu, *Adv. Healthcare Mater.*, 2012, **1**, 332.
137. E. Climent, A. Bernardos, R. n. Martínez-Máñez, A. Maquieira, M. D. Marcos, N. Pastor-Navarro, R. Puchades, F. l. Sancenón, J. Soto and P. Amorós, *J. Am. Chem. Soc.*, 2009, **131**, 14075.
138. C.-L. Zhu, C.-H. Lu, X.-Y. Song, H.-H. Yang and X.-R. Wang, *J. Am. Chem. Soc.*, 2011, **133**, 1278.
139. V. C. Özalp, F. Eyidogan and H. A. Oktem, *Pharmaceuticals*, 2011, **4**, 1137.
140. X. He, Y. Zhao, D. He, K. Wang, F. Xu and J. Tang, *Langmuir*, 2012, **28**, 12909.
141. J. Geng, M. Li, L. Wu, C. Chen and X. Qu, *Adv. Healthcare Mater.*, 2012, **1**, 332.
142. F. Q. Schafer and G. R. Buettner, *Free Radical Biol. Med.*, 2001, **30**, 1191.
143. P. Kuppusamy, H. Li, G. Ilangovan, A. J. Cardounel, J. L. Zweier, K. Yamada, M. C. Krishna and J. B. Mitchell, *Cancer Res.*, 2002, **62**, 307.
144. C.-Y. Lai, B. G. Trewyn, D. M. Jeftinija, K. Jeftinija, S. Xu, S. Jeftinija and V. S. Y. Lin, *J. Am. Chem. Soc.*, 2003, **125**, 4451.
145. S. Giri, B. G. Trewyn, M. P. Stellmaker and V. S. Y. Lin, *Angew. Chem., Int. Ed.*, 2005, **44**, 5038.
146. F. Torney, B. G. Trewyn, V. S. Y. Lin and K. Wang, *Nat. Nanotechnol.*, 2007, **2**, 295.
147. R. Liu, X. Zhao, T. Wu and P. Feng, *J. Am. Chem. Soc.*, 2008, **130**, 14418.
148. H. Kim, S. Kim, C. Park, H. Lee, H. J. Park and C. Kim, *Adv. Mater.*, 2010, **22**, 4280.
149. Y. Cui, H. Dong, X. Cai, D. Wang and Y. Li, *ACS Appl. Mater. Interfaces*, 2012, **4**, 3177.
150. P. Nadrah, U. Maver, A. Jemec, T. Tišler, M. Bele, G. Dražić, M. Benčina, A. Pintar, O. Planinšek and M. Gaberšček, *ACS Appl. Mater. Interfaces*, 2013, **5**, 3908.

151. B. Chang, D. Chen, Y. Wang, Y. Chen, Y. Jiao, X. Sha and W. Yang, *Chem. Mater.*, 2013, **25**, 574.
152. H. Li, J. Z. Zhang, Q. Tang, M. Du, J. Hu and D. Yang, *Mater. Sci. Eng., C*, 2013, **33**, 3426.
153. H. He, H. kuang, L. Yan, F. Meng, Z. Xie, X. Jing and Y. Huang, *Phys. Chem. Chem. Phys.*, 2013, **15**, 14210.
154. P. Nadrah, F. Porta, O. Planinsek, A. Kros and M. Gaberscek, *Phys. Chem. Chem. Phys.*, 2013, **15**, 10740.
155. X. Ma, K. T. Nguyen, P. Borah, C. Y. Ang and Y. Zhao, *Adv. Healthcare Mater.*, 2012, **1**, 690.
156. B. Zhang, Z. Luo, J. Liu, X. Ding, J. Li and K. Cai, *J. Controlled Release*, 2014, **192**, 192.
157. L. Yuan, W. Chen, J. Hu, J. Z. Zhang and D. Yang, *Langmuir*, 2012, **29**, 734.
158. J. Méndez, A. Monteagudo and K. Griebenow, *Bioconjugate Chem.*, 2012, **23**, 698.
159. J. Méndez, M. Morales Cruz, Y. Delgado, C. M. Figueroa, E. A. Orellano, M. Morales, A. Monteagudo and K. Griebenow, *Mol. Pharmaceutics*, 2013, **11**, 102.
160. B. Ahn, J. Park, K. Singha, H. Park and W. J. Kim, *J. Mater. Chem. B*, 2013, **1**, 2829.
161. L. Sun, X. Zhang, C. Zheng, Z. Wu and C. Li, *J. Phys. Chem. B*, 2013, **117**, 3852.
162. Z. Zhao, H. Meng, N. Wang, M. J. Donovan, T. Fu, M. You, Z. Chen, X. Zhang and W. Tan, *Angew. Chem., Int. Ed.*, 2013, **52**, 7487.

CHAPTER 7

Cyclodextrin-based Chemo- and pH-Responsive Polymer Systems for Pharmaceutical and Biomedical Applications

AKIRA HARADA* AND AKIHITO HASHIDZUME*

Department of Macromolecular Science, Graduate School of Science, Osaka University, 1-1 Machikaneyama-cho, Toyonaka, Osaka 560-0043, Japan
*Email: harada@chem.sci.osaka-u.ac.jp; akihito@chem.sci.osaka-u.ac.jp

7.1 Introduction

Stimuli-responsive polymer systems, of which physicochemical properties can be controlled by altering the external environments, have been of great importance because they are promising as smart soft materials applicable to various fields, including materials science, nanotechnology, pharmaceutics, and biomedical applications. Some pioneering works on stimuli-responsive polymer systems have been published so far.[1–17]

There are some types of stimuli that can be used for stimuli-responsive polymer systems, for example, temperature, chemicals, pH, electromagnetic fields, and redox. These stimuli can alter the environment of polymer systems, which causes small structural changes to the monomer units in the system. These small changes are accumulated over the whole polymer chain, leading to a drastic alteration of the system's properties.

RSC Smart Materials No. 14
Chemoresponsive Materials: Stimulation by Chemical and Biological Signals
Edited by Hans-Jörg Schneider

Published by the Royal Society of Chemistry, www.rsc.org

Table 7.1 Basic characteristics of cyclodextrins.[23]

CD	Number of glucose units	Molecular weight	Cavity diameter (Å)	Height (Å)
αCD	6	972	4.7–5.3	7.9 ± 1
βCD	7	1135	6.0–6.5	7.9 ± 1
γCD	8	1297	7.5–8.3	7.9 ± 1

Cyclodextrins (CDs) are a class of cyclic oligosaccharides composed of D-(+)-glucopyranose units linked *via* an α-1,4-glycoside bond.[18–22] α-Cyclodextrin (αCD), β-cyclodextrin (βCD), and γ-cyclodextrin (γCD), which are composed of 6, 7, and 8 glucopyranose units, respectively, are commercially available and generally used. CDs are toroidal molecules possessing a narrow rim of primary hydroxyl groups and a wider rim of secondary hydroxyl groups. The molecular dimension of a CD depends on the number of glucose (Glc) units, as listed in Table 7.1.[23] CDs are soluble in water whereas the cavities are relatively hydrophobic. Thus, CDs interact preferably with hydrophobic compounds, which fit the cavity to form inclusion complexes in aqueous solutions. Based on this phenomenon, CDs can be considered as simple biomimetic compounds with molecular recognition ability, and have been widely employed as molecular recognition motifs. Since CDs, *i.e.*, cyclic oligosaccharides, are nontoxic, CDs have been utilized in a variety of products, including sanitary and personal care items, foodstuffs, beverages, and pharmaceuticals.[24–28]

This chapter reviews briefly stimuli-responsive polymer systems based on CDs for pharmaceutical and biomedical applications. Here we focus on pH and chemicals, including chemical redox, as external stimuli. Readers who are interested in CD-based systems responsive to stimuli other than pH and chemicals can refer to other comprehensive review articles.[29–36]

7.2 Categories of Cyclodextrin-based Stimuli-responsive Polymer Systems

CD-based stimuli-responsive polymer systems consist of three components, *i.e.*, polymer, CD, and guest residues. Thus, CD-based stimuli-responsive polymer systems can be divided into six categories based on the molecular structure of the components: (1) CD/guest/polymer ternary systems, (2) CD/guest–polymer binary systems, (3) CD–polymer/guest binary systems, (4) CD–polymer/guest–polymer binary systems, (5) CD–guest–polymer systems, and (6) supramolecular polymer systems.

7.3 The Mechanism of Stimuli-responsiveness of Cyclodextrin-based Stimuli-responsive Polymer Systems

In CD-based stimuli-responsive polymer systems, polymer chains and inclusion complexes can be separately responsive to an alteration in the environment caused by external stimuli. When both respond at the same time, a remarkable change in the properties of the system can be expected.

In polymer solutions, the interaction of polymer chains with solvent molecules is changed depending on the external environment. A structural change of a monomer unit in the polymer chain may usually be small upon an alteration of the environment. However, because there are a number of monomer units on the polymer chain, these small changes are accumulated over the polymer chain, resulting in a considerable change in the global conformation of the polymer chain. When a polymer is very soluble in a solvent, the polymer chain takes an expanded conformation, resulting in a high solution viscosity. On the other hand, when a polymer is only slightly soluble in a solvent, the polymer chain adopts a compact conformation, leading to a decreased solution viscosity. Under non-solvent conditions, in which a polymer is insoluble, a number of polymer chains may aggregate at a polymer concentration higher than a certain level to cause macroscopic phase separation. When polymer chains are modified with associative residues (*e.g.*, hydrophobic residues in aqueous media) on the chain, a change in the polymer conformation depending on the solvent condition causes an alteration in association behavior of associative residues, for example, from intermolecular to intramolecular associations or *vice versa*, leading to a marked alteration in the solution viscosity, or gel-to-sol or sol-to-gel transition. In the case of amphiphilic block copolymers, aggregates, for example, micelles or vesicles, are formed or dissociated depending on a change in the environmental conditions. Gel-to-sol transition or decomposition of aggregates can be utilized for a variety of applications, including controlled release.

In the formation of inclusion complexes of CDs, the binding constant is strongly dependent on the shape and state of guest molecules and the medium. When CDs and guest residues are attached to polymer chains through covalent bonds, inclusion complexes between CDs and guest residues act as noncovalent crosslinks. In these systems, a shift of the equilibrium of the complexation means an alteration in the crosslinking density, resulting in a change in the size of polymer aggregates and the solution viscosity. Similarly, in the case of supramolecular polymer systems, which are composed of CD derivatives possessing a guest residue or of CD dimers and bifunctional guest compounds, the shift of equilibrium causes an alteration in the degree of polymerization (DP) of the supramolecular polymer, leading to a change in the solution viscosity.

7.4 Chemical and pH Stimuli for Cyclodextrin-based Polymer Systems

Chemicals can act as external stimuli for CD-based polymer systems. Chemical stimuli can be categorized into two classes, *i.e.*, modifiers of the medium quality and competitors for binding sites. The strength of noncovalent bonds, for example hydrogen bonding and hydrophobic interactions, is dependent on the medium. Since hydrogen bonding is stronger in less polar media, changing the solvent or adding polar or nonpolar additives may control hydrogen bonding. Since hydrophobic interactions are apparent attractive interactions due to the exclusion of hydrophobic solutes from the aqueous medium, the strength of hydrophobic interactions can be controlled by adding chemicals, for example, salts, urea, and surfactants. CDs include favorably hydrophobic guest compounds which fit the cavity in aqueous media. When a competitive guest or host is added to a solution of inclusion complexes, some of the initial inclusion complexes are dissociated and inclusion complexes with the competitor added are formed, depending on the concentration and binding constant. The shift of equilibrium can cause a change in the properties of the solution.

Since reductants and oxidants are chemicals, chemical redox is dealt with as a chemical stimulus in this chapter. A disulfide bond is often utilized as redox-responsive bonding because it is cleaved reductively with thiols, for example, dithiothreitol (DTT) and glutathione (GSH). It is known that metallocenes are redox-responsive guests for CDs.[37–41] For example, ferrocene (Fc), *i.e.*, the reduced state, is included rather strongly by βCD, whereas ferrocenium (Fc^+), *i.e.*, the oxidized state, is not included because of its positive charge.

pH can also act as an external stimulus for CD-based polymer systems. As the pH of the medium is increased, acids and bases are converted from their protonated form to the deprotonated one around their pK_a and pK_b, respectively. These conversions switch the acid and base moieties from charged to uncharged states or *vice versa*, resulting in structural or conformational changes of the polymer systems. It is known that CDs do not interact or interact more weakly with positively-charged species. Thus, CDs form inclusion complexes with weak bases, for example, amines and pyridines, at higher pH, whereas the inclusion complexes are dissociated at lower pH.

7.5 Examples of Cyclodextrin-based Chemo- and pH-Responsive Polymer Systems for Pharmaceutical and Biomedical Applications

7.5.1 Controlled Release

7.5.1.1 Cyclodextrin/Guest/Polymer Ternary Systems

A hydrogel system for controlled release was synthesized by crosslinking chitosan and poly(vinyl alcohol) (PVA) with glutaraldehyde in the presence of inclusion complexes of βCD with naproxen (NX) (Scheme 7.1).[42] For hydrogel preparation, the concentrations of chitosan and glutaraldehyde were fixed at 2 and 1 wt%, respectively, whereas the PVA concentration was varied in the range 0–10 wt%. The properties of hydrogel depended on

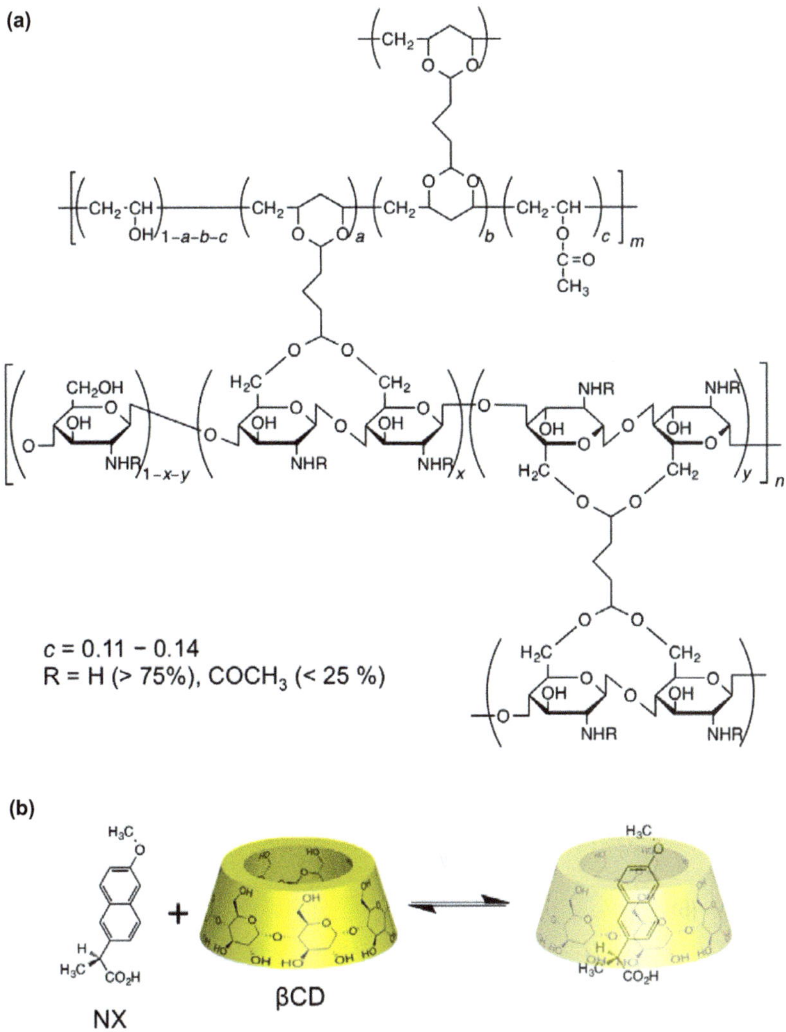

Scheme 7.1 Structures of a pH-responsive hydrogel composed of chitosan and PVA crosslinked with glutaraldehyde (a) and the equilibrium between βCD and NX (b).

the PVA content; the gel became stiffer with increasing the PVA content because of an increase in the crosslinking density. The swelling ratio of the hydrogels reached a maximum around pH 7 because of the stronger electrostatic repulsion and lower osmotic pressure, leading to faster release of NX, which was trapped in the hydrogel network, around pH 7. It is worth noting that release of NX was slower at a higher PVA content in the presence of βCD, indicating that the inclusion complexes were effective for the controlled release.

7.5.1.2 Cyclodextrin–Polymer Systems and Cyclodextrin–Polymer/Guest Binary Systems

pH-responsive microparticles for controlled release systems were prepared from hydroxypropyl-chitosan modified with carboxymethyl-βCD (Scheme 7.2a) and sodium tripolyphosphate (TPP).[43] Hydroxypropyl-chitosan, which was prepared from chitosan, was modified with carboxymethyl-βCD through amide coupling using 1-ethyl-3-(3-dimethylaminopropyl)carbodiimide as a coupling agent. The graft polymer was mixed with TPP at pH 4 to form microparticles. The microparticles captured a hydrophobic model drug, *i.e.*, ketoprofen (KTP), according to the Langmuir isotherm model. The amine moieties in the chitosan-based graft polymer were protonated at a lower pH (1.4), and the ammonium groups were deprotonated at a higher pH (7.4).

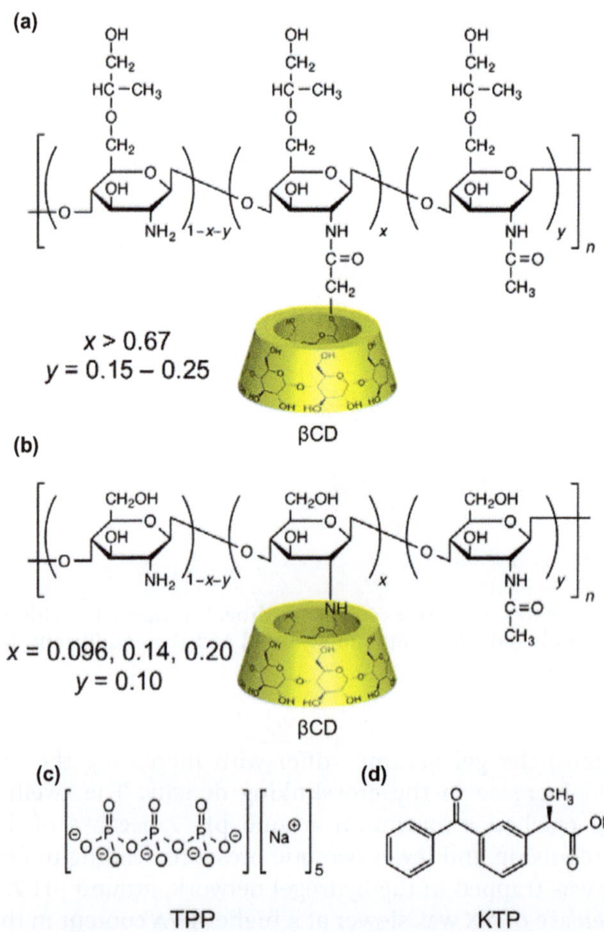

Scheme 7.2 Structures of a hydroxypropyl-chitosan modified with carboxymethyl-βCD(a), a chitosan modified with βCD (b), TPP (c) and KTP (d).

Since the chitosan-based graft polymer took a more compact conformation at pH 7.4 because of the more hydrophobic amine moieties, the encapsulated KTP molecules were released more slowly under these conditions. pH-responsive nanoparticles composed of chitosan possessing βCD side chains (Scheme 7.2b) were also reported as drug carriers for controlled release.[44] Chitosan was modified with mono-6-(*p*-toluenesulfonyl)-βCD. The degrees of modification for three CD-chitosan samples were determined to be 9.6, 14.0, and 20.0%, respectively. The chitosan samples were mixed with TPP to form nanoparticles of 200–590 nm in diameter through charge neutralization. The nanoparticles were characterized by electrophoretic light scattering, and ζ-potential values were evaluated to be +23 to +43 mV. Since the CD cavities in the nanoparticles were available for guest compounds, the nanoparticles encapsulated a hydrophobic model drug, KTP. The release kinetics of the nanoparticles was dependent on pH. At pH 4, amine residues in the chitosan were protonated and nanoparticles were then swollen. Thus, KTP was released faster under the acidic conditions. At pH 6.8, on the other hand, KTP was released slower. Since the inside of tumor cells is usually acidic, this system may be promising as a vehicle for cancer drugs.

A chemo-responsive βCD-carrying diblock copolymer was synthesized and the controlled release behavior of aggregates formed from the copolymer was investigated using several hydrophobic model drugs (Scheme 7.3).[45] Poly(ethylene glycol)-*b*-poly(β-benzyl-L-aspartate) (PEG-*b*-PBLA) was first prepared by anionic polymerization of β-benzyl-L-aspartate *N*-carboxyanhydride (NCA) using α-methoxy-ω-amino-PEG. PBLA was deprotected and then modified with ethylene diamine, followed by reaction with tosylated βCD to yield the diblock copolymer, PEG-*b*-PβCD. The degrees of PβCD block were determined to be 4, 7, and 14 by ^{1}H NMR. In the presence of hydrophobic compounds, the block copolymer samples formed micellar aggregates presumably because of the hydrophobicity of the inclusion complex. Hydrophobic model drugs, *i.e.*, coumarin 102, ibuprofen (IBU), indomethacin, rapamycin (RAP) and dexamethasone, loaded in the aggregates of PEG-*b*-PβCD of the longest PβCD block, were slowly released. Adding a competitive guest, for example, 1-adamantanecarboxylic acid (AdCA), accelerated the release of RAP.

A temperature- and pH-responsive poly(*N*-isopropylacrylamide) (PNIPAM)-based hydrogel containing crosslinked βCD was synthesized (Scheme 7.4a).[46] A βCD crosslinked with epichlorohydrin was modified with maleic anhydride (MAnh) to produce a maleate-carrying βCD-polymer. The multi-functional macromonomer was copolymerized with *N*-isopropylacrylamide (NIPAM) at NIPAM weight fractions of 0.3–0.7 wt% to form a hydrogel. The βCD cavities in the hydrogel included a guest model compound, *i.e.*, methyl orange. Since the hydrogel contained PNIPAM segments and carboxylic acid residues, the swelling ratio depended on temperature, pH, and NaCl concentration. As temperature was increased, the swelling ratio commenced to decrease remarkably at *ca.* 25, 30, and 35 °C at pH 3.0, 5.0, and 7.4, respectively. As pH was increased, the swelling ratio increased significantly

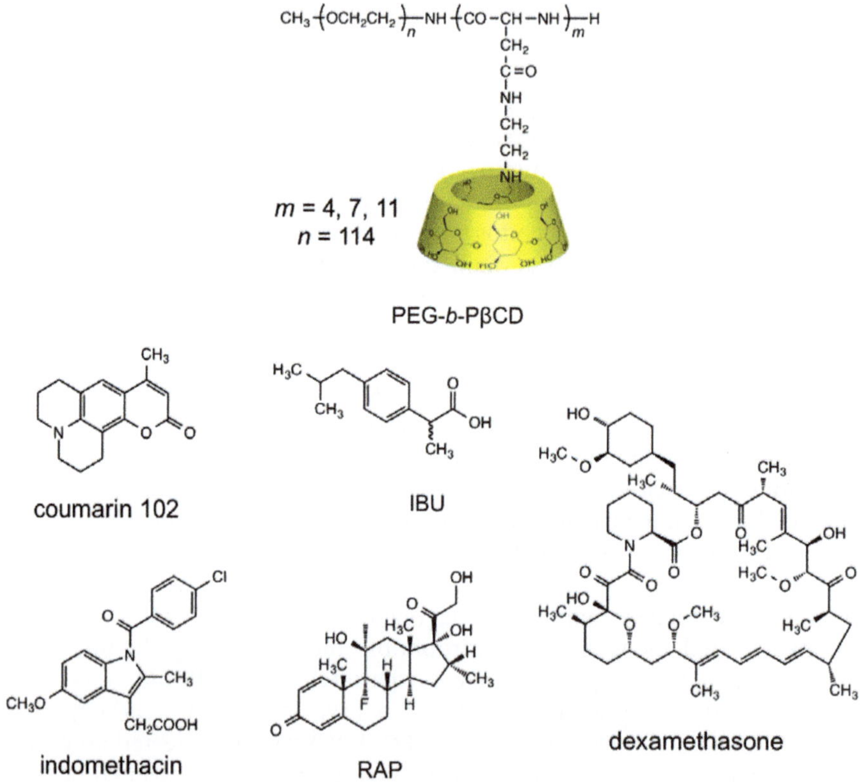

Scheme 7.3 Structures of PEG-*b*-PβCD and model drugs (coumarin 102, IBU, indomethacin, RAP, and dexamethasone).

in the pH region of 3–5. As NaCl concentration was increased, the swelling ratio decreased in two regimes of NaCl concentration, *i.e.*, $<10^{-2}$ and >1 M, presumably because of shielding of electrostatic repulsion and the salting-out effect. Temperature- and pH-responsive hydrogel samples possessing βCD residues were also synthesized from polysaccharides and NIPAM (Scheme 7.4b).[47] Polysaccharides, *i.e.*, xanthan and gellan gum (the degrees of substitution (DS) of pyruvate and acetate groups were 75 and 73%, respectively), were modified with MAnh to yield polysaccharides possessing maleate residues (DS of 10.8 and 15.4%, respectively). The maleate-polysaccharides were copolymerized with NIPAM in the presence of a crosslinker, *i.e.*, *N*,*N*′-methylenebis(acrylamide) (BIS) or a βCD acrylate crosslinker (DS ~ 3), to provide hydrogel samples. The concentration of the maleate-polysaccharides was fixed at 2.5% in hydrogel, and the crosslinker concentration was constant at 3% with respect to the mixture or comonomers. The swelling ratio of the hydrogel samples was also dependent on temperature, pH, and NaCl concentration; the swelling ratio was higher at lower temperatures and NaCl concentrations and at higher pH. The effects of pH and NaCl concentration on the swelling ratio were stronger for a

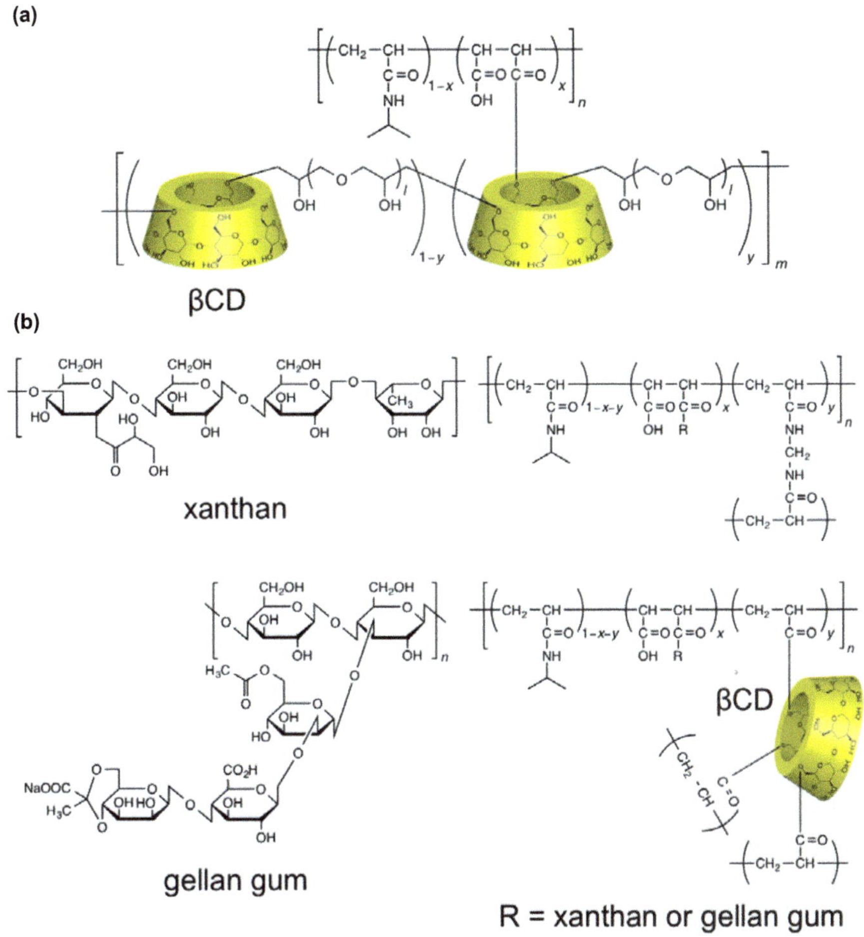

Scheme 7.4 Structures of temperature- and pH-responsive PNIPAM-based hydrogels containing βCD crosslinked with epichlorohydrin (a) and polysaccharides (b).

hydrogel sample of higher polysaccharide content because of a higher content of carboxylic acid residues.

An interpenetrating polymer network (IPN), which is defined as a combination of two or more polymers in network form that are synthesized in juxtaposition,[48] was prepared by crosslinking hydroxypropyl-βCD with ethylene glycol diglycidyl ether in the presence of an acrylic acid-based polymer, *i.e.*, Carbopol® 71G (Scheme 7.5a).[49] For IPN preparation, the concentrations of hydroxypropyl-βCD (20% in 0.2 M KOH) and glycol diglycidyl ether (0.6 equiv. of hydroxy groups in hydroxypropyl-βCD) were fixed, whereas that of Carbopol® 71G was varied in the region of 0–1%. The properties (*e.g.*, hardness, compressibility, and bioadhesion force) of the IPN were dependent on the content of Carbopol® 71G; the hardness,

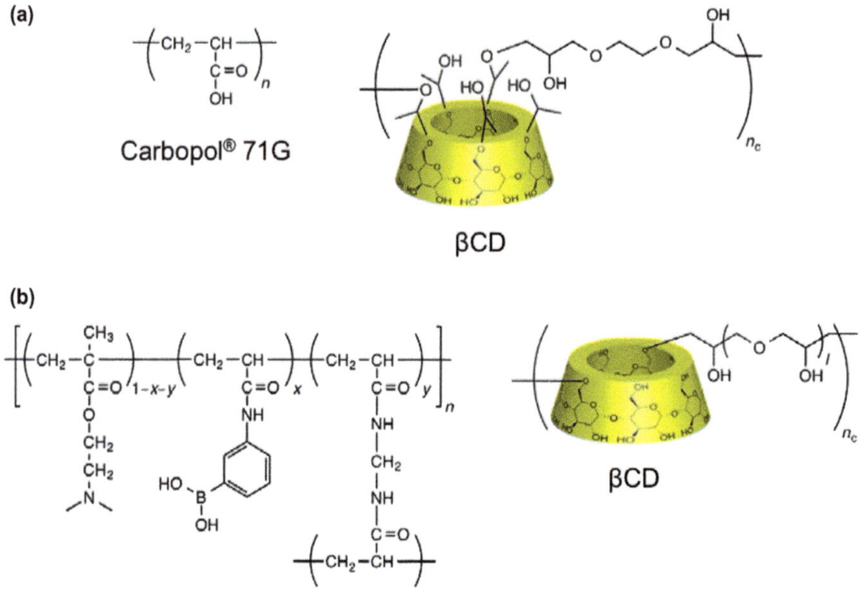

Scheme 7.5 Structures of network polymers for IPN; Carbopol® 71G and hydroxypropyl-βCD crosslinked with glycol diglycidyl ether (a) and crosslinked PDMEMA bearing phenylboronic acid residues and βCD crosslinked with epichlorohydrin (b).

compressibility, and bioadhesion force were higher at a higher content of Carbopol® 71G. Since βCD cavities in the IPN were available for guest compounds, the IPN encapsulated hydrophobic drugs, *i.e.*, estradiol and ketoconazole, through the formation of inclusion complexes. The rate of drug release was also dependent on the content of Carbopol® 71G and pH. At a higher pH (7.8), IPN was swollen because of electrostatic repulsion and the mesh size of the network was larger, resulting in faster release of the encapsulated hydrophobic drug molecules. A multi-stimuli responsive IPN was also synthesized from *N*,*N*-dimethylaminoethyl methacrylate (DMEMA), 3-(acrylamido)phenylboronic acid (AAPBA), and βCD crosslinked with epichlorohydrin (Scheme 7.5b).[50] The hydrogel was prepared by terpolymerization of DMEMA (0.36 mL), AAPBA (71.6 mg), and BIS (3.9 mg) in the presence of βCD crosslinked with epichlorohydrin (0–40 wt%). The swelling behavior of the hydrogel was dependent on temperature, pH, ionic strength, and Glc concentration. As temperature, pH, or ionic strength was increased, the swelling ratio decreased presumably because of the properties of DMEMA units. As Glc concentration was increased, the swelling ratio increased because Glc was substituted for a βCD residue in a boronic acid ester with AAPBA.[51] The βCD cavities in the IPN included hydrophobic drugs, *i.e.*, aminophylline and IBU. The release of IBU was much faster at pH 7.4 than that at pH 4 because of the swollen IPN.

A temperature- and pH-responsive miktoarm star polymer was synthesized (Scheme 7.6a).[52] A hyperbranched polymer ($M_n = 3.3 \times 10^4$, $M_w/M_n = 3.40$) was prepared from triallylammonium-carrying βCD and bis(dimethylsilyl)

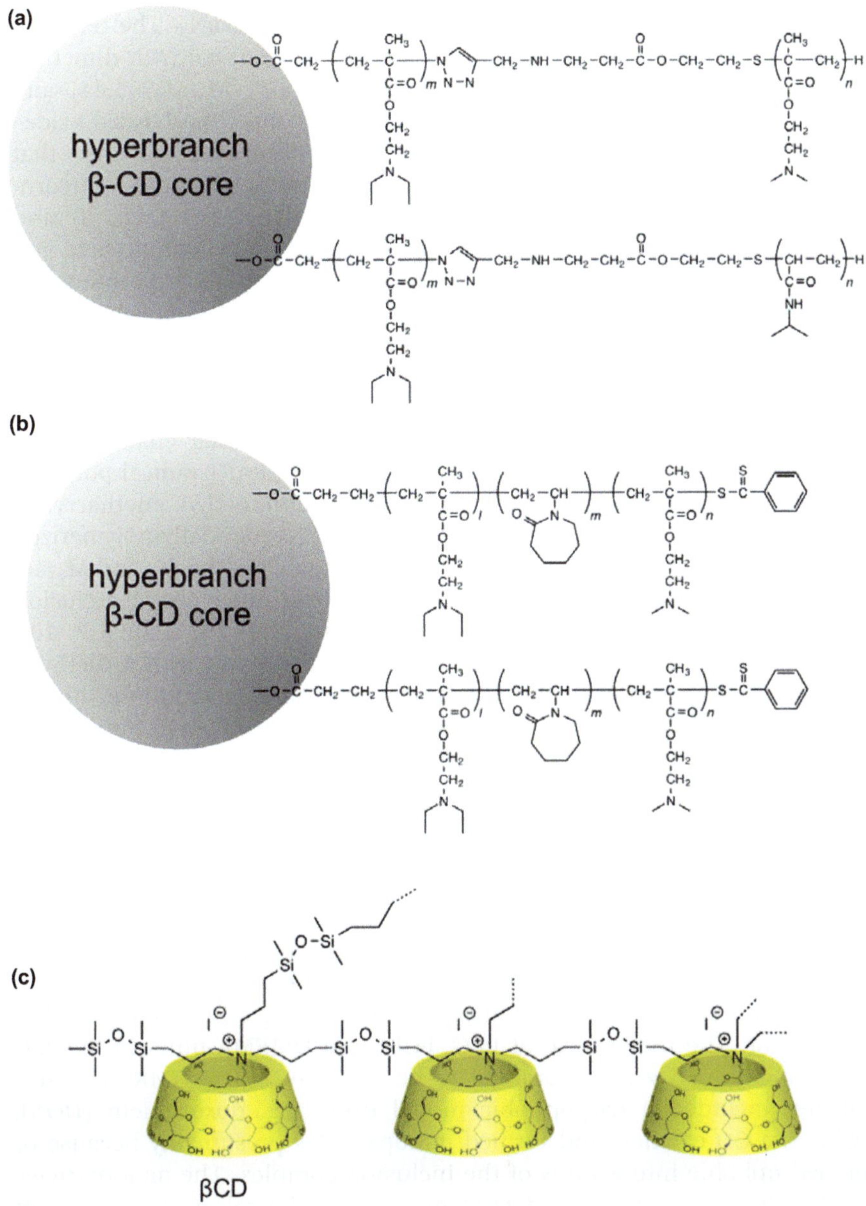

Scheme 7.6 Structures of a miktoarm star polymer (a) and star block polymers possessing a hyperbranched βCD-core (b), and a partial structure of the hyperbranched βCD-core (c).

ether by addition of Si–H to allyl groups, followed by coupling with allyl chloroacetate to form a macroinitiator for atom transfer radical polymerization (ATRP). Using the macroinitiator, DMEMA was polymerized by ATRP using CuCl and *N*,*N*,*N*′,*N*″,*N*″-pentamethyl diethylenetriamine to yield a star-shaped polymer ($M_n = 1.6 \times 10^5$, $M_w/M_n = 1.18$). The terminal chloride residues were converted into azide by treatment with NaN_3. The terminal azide residues were coupled with propargyl-carrying poly(*N*,*N*-dimethylaminoethyl methacrylate) (PDMEMA) ($M_n = 4.5 \times 10^3$, $M_w/M_n = 1.44$) and PNIPAM ($M_n = 4.7 \times 10^3$, $M_w/M_n = 1.41$) chains by copper(I)-catalyzed azide–alkyne cycloaddition (CuAAC) to form a miktoarm star polymer that possessed a hyperbranched βCD-core. Aqueous solutions of the miktoarm star polymer underwent phase separation at higher pH (≥*ca.* 8) and temperatures (≥*ca.* 35 °C), where DMEMA units were fully deprotonated and PNIPAM blocks were hydrophobic. Using the same type of hyperbranched βCD-polymer, temperature- and pH-responsive star block copolymers were also synthesized by reversible addition-fragmentation chain-transfer (RAFT) radical polymerization (Scheme 7.6b).[53] The hyperbranched polymer composed of triallylammonium-carrying βCD and bis(dimethylsilyl) ether was modified with dithiobenzoic acid to yield macromolecular chain transfer agent (macro-CTA) ($M_n = 3.0 \times 10^4$, $M_w/M_n = 1.90$) for RAFT radical polymerization. Utilizing the macro-CTA, *N*,*N*-diethylaminoethyl methacrylate (DEEMA), *N*-vinylcaprolactam, and DMEMA were sequentially polymerized by the RAFT mechanism to form star block copolymers. The M_n and M_w/M_n values of the star block polymers obtained were determined by size exclusion chromatography/multi-angle laser light scattering (9.4×10^4, 1.20; 1.7×10^5, 1.44; 1.3×10^5, 1.49, respectively). A solution of the star poly(*N*,*N*-diethylaminoethyl methacrylate) (PDEEMA) underwent phase separation at higher pH (≥*ca.* 5). Solutions of the star di- and triblock copolymers underwent phase separation at higher pH (∼12) and temperature (≥*ca.* 30 and 34 °C, respectively). The release of metronidazole (MET) was slower for the star triblock copolymer at pH 1.4 and 25 °C than those for the star PDEEMA polymer and the star diblock copolymer. Furthermore, the release of MET was slower at higher pH.

A pH-responsive carrier for hydrophobic drugs was realized using a star polymer that possessed a core containing ternary amine and βCD moieties (Scheme 7.7).[54] The star polymer was synthesized by ATRP of DMEMA and mono- and multi-methacrylate substituted βCD using a Br-carrying PEG as an initiator. The molar ratio of PEG, βCD, and DMEMA units in the star polymer obtained was determined to be 1 : 10 : 46 by ^{1}H NMR. The star polymer included a hydrophobic model drug, *i.e.*, doxorubicin (DOX), inside the CD cavities, and formed nanoparticles presumably because of the hydrophobic interactions of the inclusion complex. The nanoparticles slowly released the entrapped DOX molecules. The release was faster at lower pH (5.0) than that at pH 7.4 because the DMEMA units were protonated and the core was thus swollen under acidic conditions (Figure 7.1). The cytotoxicity study using HeLa cells demonstrated that DOX-loaded

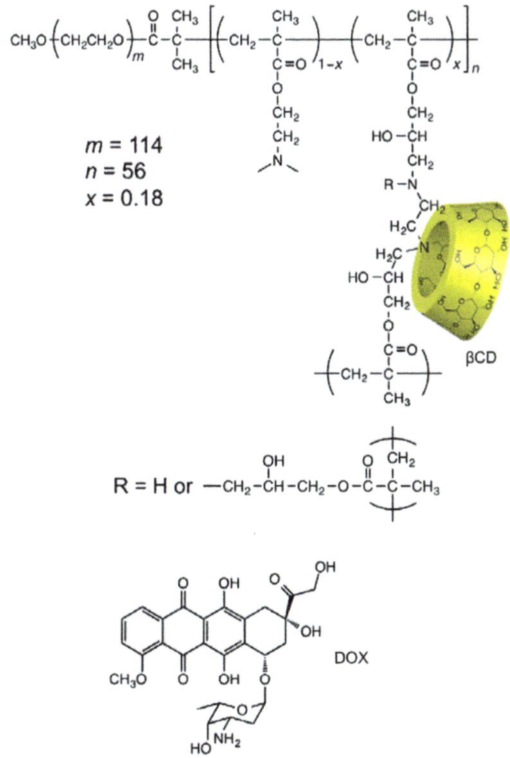

Scheme 7.7 Structures of a pH-responsive star block copolymer possessing a βCD-core and DOX.

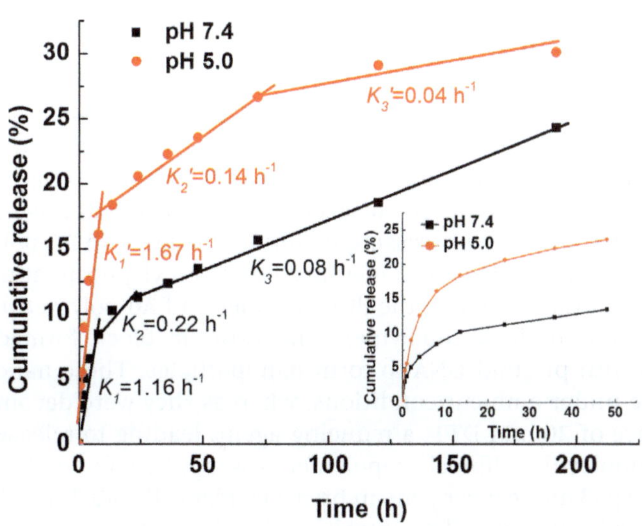

Figure 7.1 *In vitro* release profiles of DOX from nanoparticles at pH 5.0 and 7.4.

nanoparticles were more effective at higher DOX concentrations than free DOX molecules.

7.5.1.3 Cyclodextrin–Polymer/Guest–Polymer Binary Systems

pH-responsive microcapsules were prepared by the layer-by-layer (LbL) technique on $CaCO_3$ particles, followed by coring.[55] $CaCO_3$ particles were first prepared by mixing $CaCl_2$ and K_2CO_3 in the presence of poly(sodium 4-styrenesulfonate) (PSS) ($M_w = 7\times10^4$) and rhodamine B (RhB). After the $CaCO_3$ particles containing RhB were coated with poly(allylamine hydrochloride) ($M_w = 6\times10^4$) through electrostatic interactions, microparticles were prepared by the LbL technique utilizing self-assembly between dextran modified with βCD through an imine linkage and polyaspartic acid possessing adamantane (Ad) residues (Scheme 7.8a). The particles were cored by treatment with disodium ethylenediaminetetraacetate to form hollow particles containing RhB inside as a model drug. Since the imine bond was cleaved at acidic pH (5.5), the hollow particles were decomposed under acidic conditions, resulting in controlled release of RhB molecules. Microcapsules responsive to pH, which encapsulated DOX residues, were also prepared by the LbL technique.[56] $CaCO_3$ particles were first prepared by mixing $CaCl_2$ and K_2CO_3 in the presence of PSS ($M_w = 7\times10^4$) and fluorescein isothiocyanate (FITC)-grafted dextran ($M_w = 2\times10^4$). The LbL technique on the $CaCO_3$ particles using self-assembly between dextran possessing Ad groups through an imine linkage and carboxymethyl-dextran carrying βCD moieties (Scheme 7.8b) provided microparticles possessing an LbL outer layer and the $CaCO_3$ spherical core. The microparticles encapsulated Ad-modified DOX (Ad-DOX). At neutral pH (7.4), the particles were rather stable. At acidic pH (5.5), however, the $CaCO_3$ core was dissolved and the hydrazine bonds were easily cleaved, resulting in decomposition of particles and release of DOX molecules.

A gene delivery system was developed using a cationic star polymer linked to βCD through a disulfide bonding and a zwitterionic polymer bearing an Ad moiety at the chain end (Scheme 7.9).[57] A βCD-based cationic four-arm star PDMEMA was synthesized by ATRP of DMEMA using a multifunctional βCD initiator possessing disulfide linkages. An Ad-polymer was also synthesized by ATRP of 2-methacryloyloxyethyl phosphatidylcholine using an Ad initiator. Mixing the βCD-polymer and Ad-polymer provided a supramolecular diblock copolymer. The cationic block formed polyion complexes with plasmid DNA to form nanoparticles. These nanoparticles were stable under ambient conditions, whereas they were decomposed in the presence of 20 mM DTT, a reducing agent, leading to release of DNA. The supramolecular diblock copolymer was applied for delivery of the therapeutic p53 anti-cancer gene in breast cancer cells (MCF-7). The results demonstrated efficient cell apoptosis, indicative of the great potential of the supramolecular diblock copolymer for cancer gene therapy.

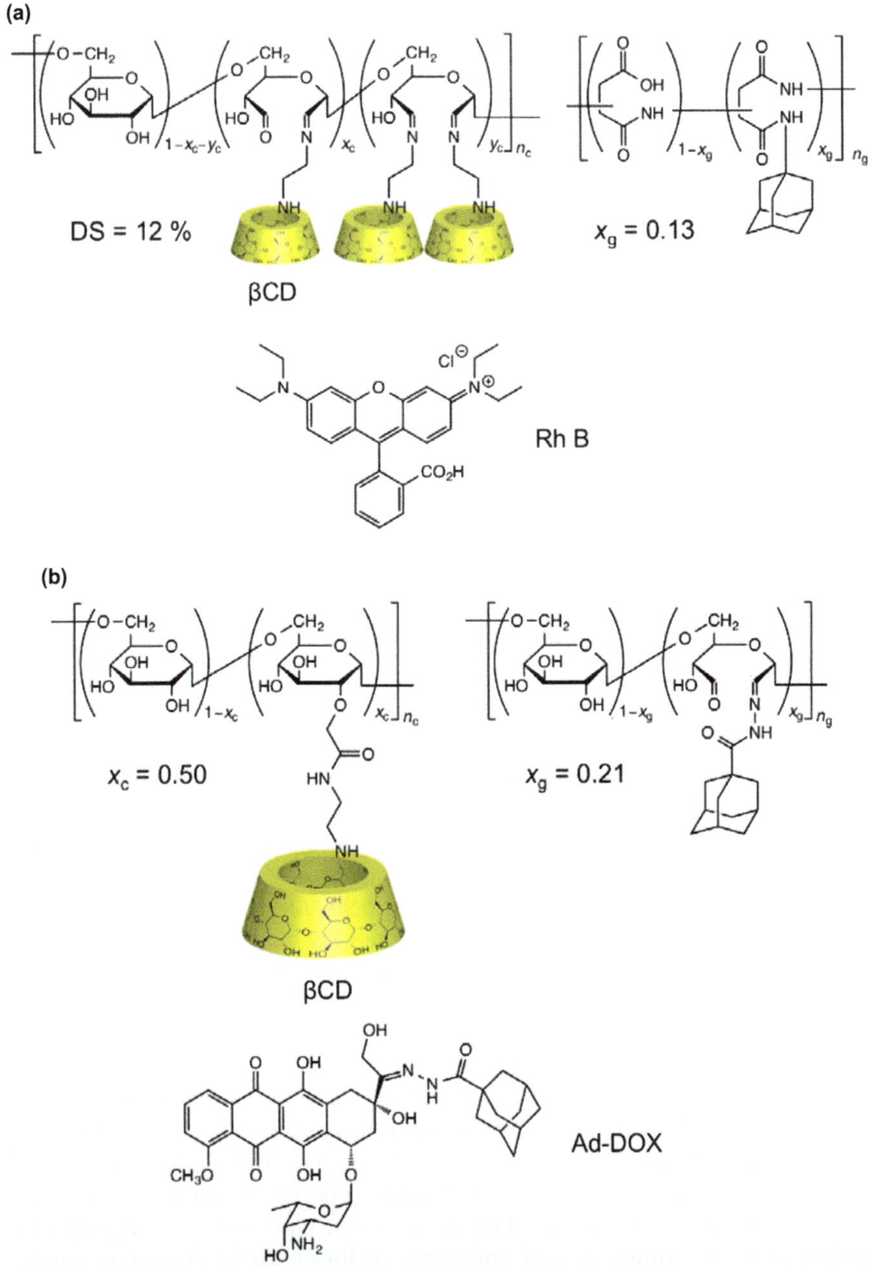

Scheme 7.8 Structures of βCD-polymers and Ad-polymers for the preparation of microparticles by the LbL technique (a dextran modified with βCD moieties and a polyaspartic acid modified with Ad moieties) and RhB (a), and dextrans modified with βCD and Ad moieties, and Ad-DOX (b).

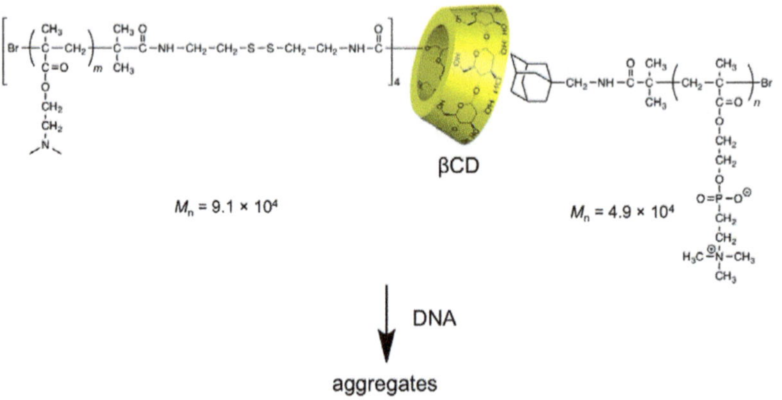

Scheme 7.9 A redox-responsive gene delivery system composed of four-arm star PDMEMA possessing a βCD-core and Ad-carrying poly(2-methoxyloyloxyethyl phosphatidylcholine).

7.5.2 Change in Aggregation Behavior

7.5.2.1 Cyclodextrin/Guest–Polymer Binary Systems

CD-based polypseudorotaxanes (PpRx)[34,58–69] have been widely utilized for biomedical applications. The pH-responsive formation of the polypseudorotaxane of αCD with a triblock copolymer, polyethyleneimine-*b*-PEG-*b*-polyethyleneimine (PEI-*b*-PEG-*b*-PEI), was reported (Scheme 7.10a).[70] The triblock copolymer, PEI-*b*-PEG-*b*-PEI, was prepared by polymerization of 2-methyl-2-oxazoline using PEG carrying two mesyl moieties at both ends, followed by hydrolysis in 10% hydrochloric acid. At basic pH (~11.0), αCD molecules included both the PEG and PEI blocks. At lower pH (<8.0), on the other hand, αCD molecules included only the PEG block because PEI blocks were protonated. The pH-dependent threading and dethreading of αCD on the PEI blocks was reversible. The pH-dependent aggregation behavior of a block copolymer, polyethylene oxide-*b*-polyacrylic acid (PEO-*b*-PAA) complexed with αCD was also investigated (Scheme 7.10b).[71] A block copolymer, PEO-*b*-poly(*t*-butyl acrylate), was prepared by ATRP using a Br-carrying PEO initiator, followed by deprotection of the *t*-butyl ester using trifluoroacetic acid. The diblock copolymer was mixed with αCD to form PpRx. Since PpRx is virtually insoluble in water, the diblock copolymer formed aggregates because of the interaction of the PpRx blocks. The type and size of aggregates formed were dependent on pH and time. At lower pH (2.5), square aggregates, which were considered as lamellae, were formed after 1 day, whereas spherical aggregates of 50–150 nm in diameter were formed after 7 days. At higher pH (6.5), spherical aggregates of *ca.* 100 nm in diameter were formed, and the aggregates were stable even after 7 days.

The L-phenylalanine (Phe)-responsive aggregation behavior of nanoparticles formed from αCD and ethylcellulose-based graft copolymers

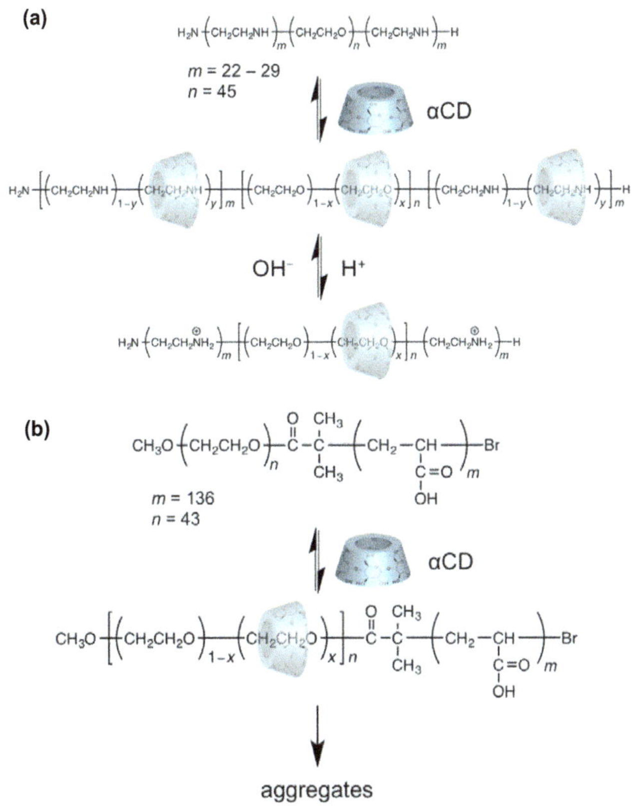

Scheme 7.10 Formation of pH-responsive PpRx from αCD and PEG-*b*-PEI-*b*-PEG (a) and from αCD and PEG-*b*-PAA (b).

possessing poly(ε-caprolactone) grafts (EC-*g*-PCL) was reported (Scheme 7.11).[72] The graft copolymers were prepared by ring opening polymerization of ε-caprolactone using ethylcellulose and stannous octanoate as a macroinitiator and a catalyst, respectively. The graft copolymers EC-*g*-PCL were mixed with αCD or αCD maleate in a mixed solvent of tetrahydrofuran and water at 60 °C. The mixture was dialyzed against water to form nanoparticles, in which αCD or its derivative formed PpRx with PCL grafts. When Phe was added as a competitive guest, the size of nanoparticles increased from *ca.* 200 nm to 5–6 μm for 24 h, indicative of disruption of the nanoparticles. The release of a model drug, *i.e.*, 5,10,15,20-tetrakis(4-hydroxyphenyl)-21*H*,23*H*-porphine (THPP), was accelerated by adding Phe. The Phe-triggered release of THPP was useful for *in vitro* photodynamic therapy for breast cancer cells (MCF-7).

Polyrotaxanes (PRx) of a triblock copolymer, PNIPAM-*b*-polypropylene glycol-*b*-PNIPAM (PNIPAM-*b*-PPG-*b*-PNIPAM), with βCD and γCD were synthesized by ATRP of NIPAM using bifunctional initiators of PPG/βCD and PPG/γCD PpRx possessing 2-bromopropionyl moieties at the ends

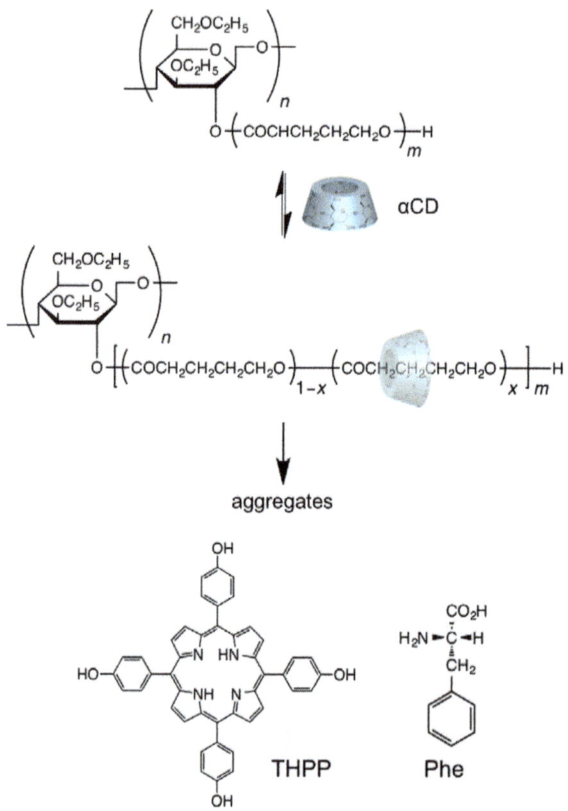

Scheme 7.11 Phe-responsive aggregation behaviour of PpRx formed from αCD and EC-*g*-PCL, and structures of THPP and Phe.

(Scheme 7.12a).[73] The PRx samples obtained were responsive to temperature because of the phase transition behavior of PNIPAM blocks in water. The response of PRx upon addition of *N*,*N*-dimethylformamide (DMF) was dependent on the type of CD. The βCD-based PRx exhibited good dispersion properties whereas the γCD-based PRx indicated a mosaic crystal structure after treatment with DMF. Chemical-redox responsive particles, possessing an αCD/PEG PRx core and PEG coronas, were reported (Scheme 7.12b).[74] PpRx was first formed from αCD and PEG carrying *o*-pyridyldisulfide groups at the ends. The PpRx was capped with 2-mercaptoethyl-3,4,5-tris(prop-2-yn-1-yloxy)benzamide through the formation of a disulfide bond. The propargyl groups were connected to azide-carrying PEG by CuAAC. Since PRx is practically insoluble in water, the block copolymer formed aggregates. The aggregates encapsulated hydrophobic compounds, *i.e.*, pyrene and calcein. In the presence of GSH, the disulfide bonding in the axis of the PRx was cleaved reductively, and αCD molecules were then dissociated, resulting in decomposition of the aggregates and release of the encapsulated hydrophobic compounds.

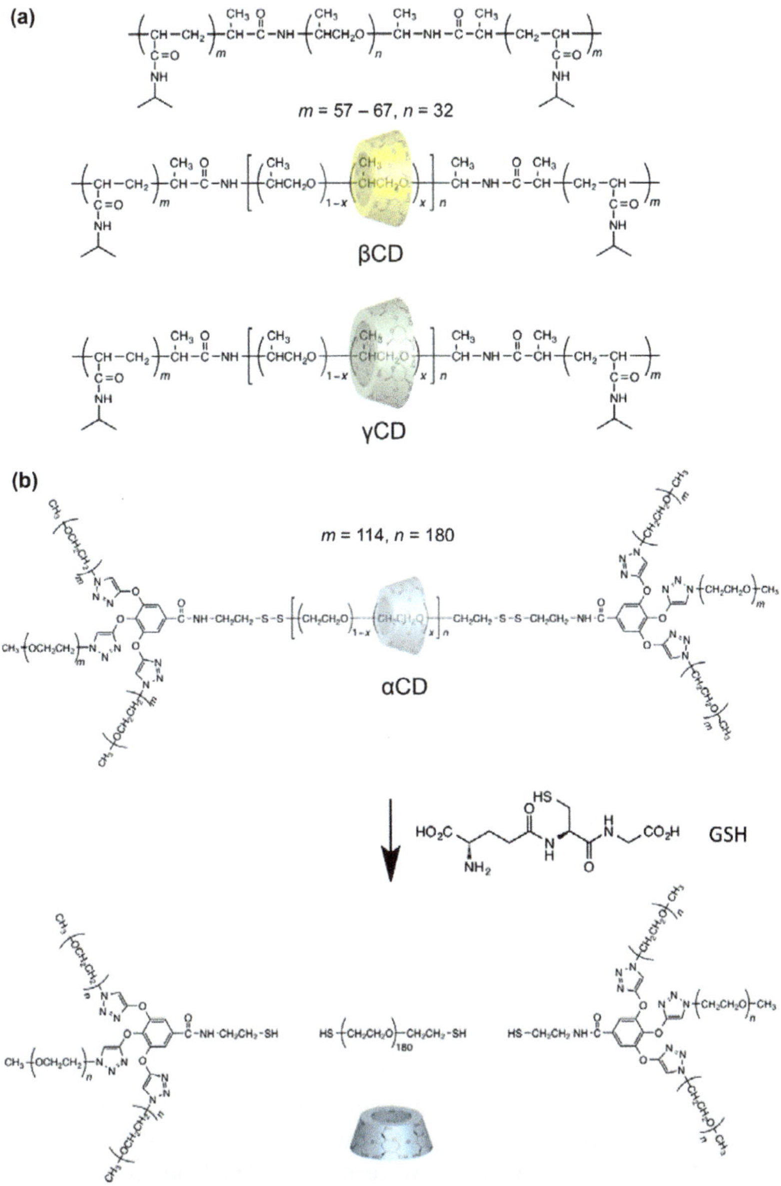

Scheme 7.12 Structures of DMF-responsive PRx of βCD/PNIPAM-*b*-PPO-*b*-PNIPAM and γCD/PNIPAM-*b*-PPO-*b*-PNIPAM (a) and a redox-responsive cleavage of PRx formed from αCD and branched PEG (b).

7.5.2.2 Cyclodextrin–Polymer Systems and Cyclodextrin–Polymer/Guest Binary Systems

A temperature- and pH-responsive miktoarm star polymer, which consists of different arms, possessing βCD at the center was synthesized by ATRP

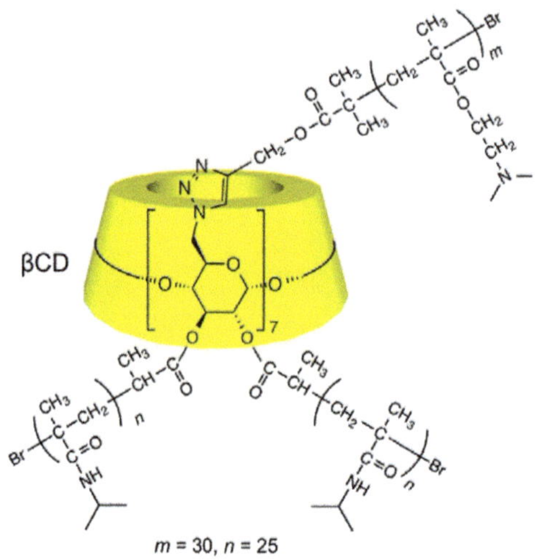

Scheme 7.13 Structure of a temperature- and pH-responsive miktoarm star polymer possessing a βCD-core.

and CuAAC (Scheme 7.13).[75] Heptakis(6-deoxy-6-azide)-βCD ($(N_3)_7$-βCD) was prepared with the corresponding iodide and NaN_3. The secondary hydroxyl groups in $(N_3)_7$-βCD were modified with 2-bromopropionic bromide to obtain a multi-functional initiator for ATRP ($(N_3)_7$-βCD-$(Br)_{14}$). Using $(N_3)_7$-βCD-$(Br)_{14}$ as an initiator, NIPAM was polymerized by ATRP to form a 14-arm star PNIPAM. The star polymer was coupled with alkyne-carrying PDMEMA, which was also prepared by ATRP, by CuAAC to yield a miktoarm star polymer possessing βCD at the center. Since PNIPAM and PDMEMA are responsive to temperature and pH, respectively, in aqueous solutions, the miktoarm polymer was also responsive to both stimuli. At higher pH (10), the miktoarm star polymer formed vesicles through hydrophobic association of the PDMEMA arms. At pH 4.5 and a higher temperature (45 °C), on the other hand, the miktoarm star polymer also formed vesicles through hydrophobic association of the PNIPAM arms.

pH-responsive multi-layered films were synthesized from a PDMEMA-based star polymer possessing βCD at the center (Scheme 7.14) and PSS ($M_w = 7\times10^4$).[76] The star polymers were synthesized from mono-, di-, tri-chloroacetylated βCD initiators by ATRP of DMEMA. Using a combination of the cationic star polymers and PSS, multi-layer films were prepared by the LbL technique at pH 6. The properties and morphology of the multi-layer films obtained were dependent on pH. At pH 2.75, the thickness of the multi-layer films increased because of the electrostatic repulsion of protonated PDMEMA blocks.

Dendrimers or hyperbranched polymers possessing CD moieties may act as ideal drug carriers because of their multi-functionality and narrow polydispersity of structure and size.[77–80] A pH-responsive βCD-carrying

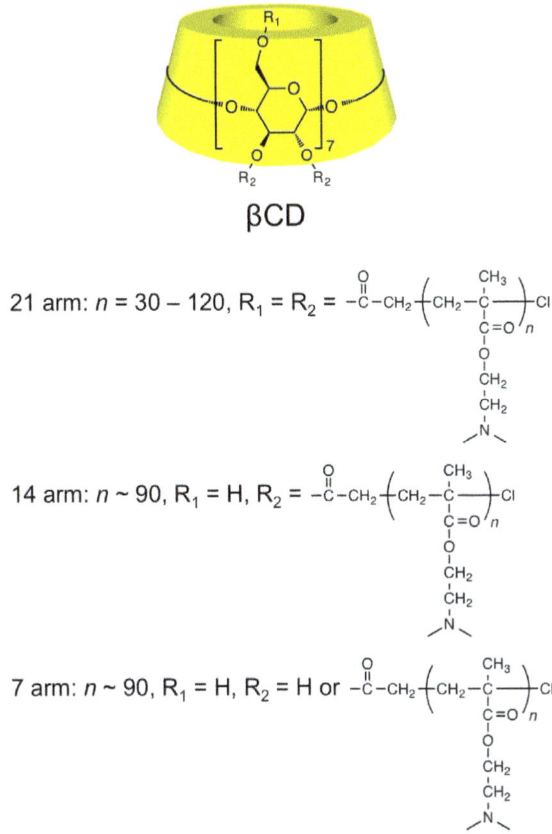

Scheme 7.14 Structure of 7-, 14-, and 21-arm star PDMEMA possessing a βCD-core.

hyperbranched polymer was synthesized (Scheme 7.15).[81] Amine moieties in a hyperbranched PEI of $M_w = 1 \times 10^4$ was partially modified with 5-hexynoic acid or 10-undecynoic acid under microwave irradiation to form PEI hyperbranched polymers carrying alkyne functionalities. The alkyne-modified PEI hyperbranched polymers were coupled with azide-βCD by CuAAC. The amount of βCD introduced was 0.6 mmol per 1 g PEI for both cases. The βCD hyperbranched polymers formed aggregates at pH 7; the hydrodynamic diameters (D_H) were determined to be 8 and 170 nm for hyperbranched polymers of 5-hexynoic acid and 10-undecynoic acid, respectively. When 1-adamantanecarboxylate (AdC) was added, the aggregates formed from the βCD hyperbranched polymer of 10-undecynoic acid were dissociated because of the formation of inclusion complexes and the negative charge of AdC. At higher pH (14), the aggregates were also dissociated, presumably because of the negative charge on βCD residues.

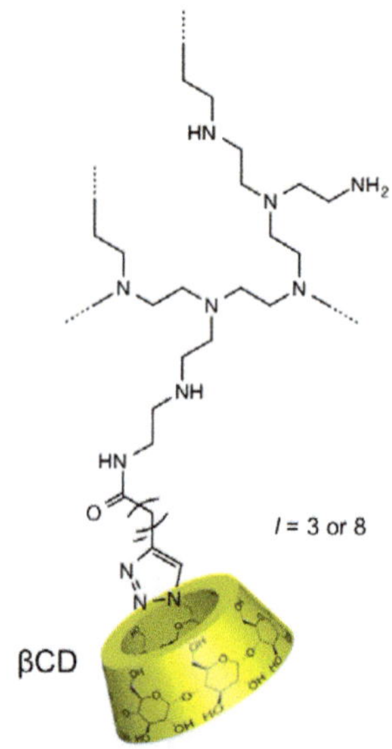

Scheme 7.15 Partial structure of a hyperbranched PEI possessing βCD moieties.

7.5.2.3 Cyclodextrin–Polymer/Guest–Polymer Binary Systems

A temperature- and pH-responsive supramolecular diblock copolymer was reported based on the formation of stable inclusion complexes of βCD with Ad residues (Scheme 7.16).[82] A βCD-polymer was synthesized by RAFT radical polymerization of 4-vinylpyridine (4VP) using a βCD-carrying chain transfer agent (CTA), and an Ad-carrying polymer was also synthesized by the RAFT technique of NIPAM using an Ad-carrying CTA. The βCD-polymer and Ad-polymer formed a supramolecular diblock copolymer. The aggregation behavior of the supramolecular diblock copolymer was investigated by static and dynamic light scattering at different temperatures and pH. At a higher temperature (60 °C), since the PNIPAM block is hydrophobic, the block copolymer formed aggregates, which were considered as spherical micelles based on the ratio of radius of gyration and hydrodynamic radius ($R_G/R_H = 0.81$) determined by light scattering measurements. At a higher pH (4.8), since the poly(4-vinylpyridine) (P4VP) block is rather hydrophobic, the supramolecular diblock copolymer formed aggregates through hydrophobic association of P4VP blocks, which might be vesicles based on the ratio of R_G/R_H (1.07).

The pH-responsive aggregation behavior of a βCD-polymer and dextran-based guest polymers was reported (Scheme 7.17a).[83] The βCD-polymer was prepared by crosslinking βCD with epichlorohydrin. Guest polymers were synthesized by modification of dextran with hydrocarbons (*i.e.*, undecyl or Ad) and cyclohexane-carboxylate (DS for hydrocarbon ~6%, DS for cyclohexane-carboxylate = 2–7%). The βCD-polymer and the guest polymers formed aggregates through the formation of inclusion complexes. At lower

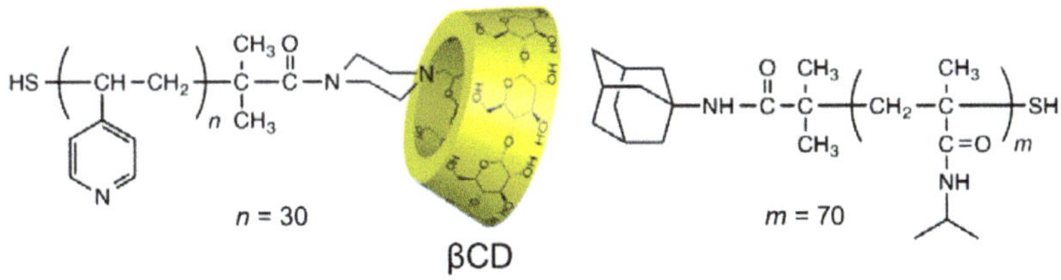

Scheme 7.16 Structures of βCD-carrying P4VP and Ad-carrying PNIPAM for the preparation of a supramolecular diblock copolymer, which exhibited temperature- and pH-responsive aggregation behaviour.

concentrations (0.5–2 g L^{-1}), the aggregation behavior was dependent on pH because the guest polymer possessed carboxylic acid residues. The polymer mixture formed negatively-charged stable aggregates at pH 7, where carboxylic acid residues in the modified dextran were fully deprotonated, and metastable larger aggregates at pH 4 (~apparent pK_a). On the other hand, the polymer mixture underwent precipitation at pH 2, where carboxylic acid residues were fully protonated. The aggregation behavior was also investigated for a mixture of βCD-carrying alternating copolymer of methyl vinyl ether (MVE) and maleic acid, and Ad-carrying three-arm PEO (Scheme 7.17b).[84] The βCD-polymers were prepared by coupling reaction of an alternating copolymer of MVE and MAnh with βCD in the presence of LiOH. The Ad-carrying three-arm PEO (PEO-Ad_3) was synthesized by the reaction of three-arm PEO ($M_n = 1.5\times10^4$) with adamantyl isocyanate. The βCD-polymer and PEO-Ad_3 formed aggregates through the formation of inclusion complexes. Since the βCD-polymer carried carboxylic acid residues, the aggregation behavior was pH-responsive. At neutral pH, the solution was transparent, whereas, at acidic pH (~2), phase separation occurred.

7.5.3 Gel-to-Sol and/or Sol-to-Gel Transition

7.5.3.1 Cyclodextrin/Guest–Polymer Binary Systems

A pH-responsive sol-to-gel transition was reported using complexes of αCD with a diblock copolymer, PEG-*b*-poly(L-lysine) (PEG-*b*-PLL) (Scheme 7.18a).[85] The diblock copolymer was prepared by the NCA polymerization method using a PEG-based macroinitiator. Complexation of αCD with PEG-*b*-PLL formed a diblock copolymer possessing an αCD/PEG PpRx block. At acidic pH (4.5), the diblock copolymer formed micellar aggregates through interaction of PpRx blocks. At basic pH (10), the PLL block was deprotonated and became more hydrophobic to form a hydrogel through the hydrophobic interaction of PLL blocks as well as the interaction of PpRx blocks. This micelle-to-gel transition was reversible. The same type of binary hydrogel system, responsive to temperature and pH, was prepared from αCD and a diblock copolymer, PEG-*b*-PDMEMA, in which αCD and the PEG block formed crystalline PpRx (Scheme 7.18b).[86] Two diblock copolymer samples

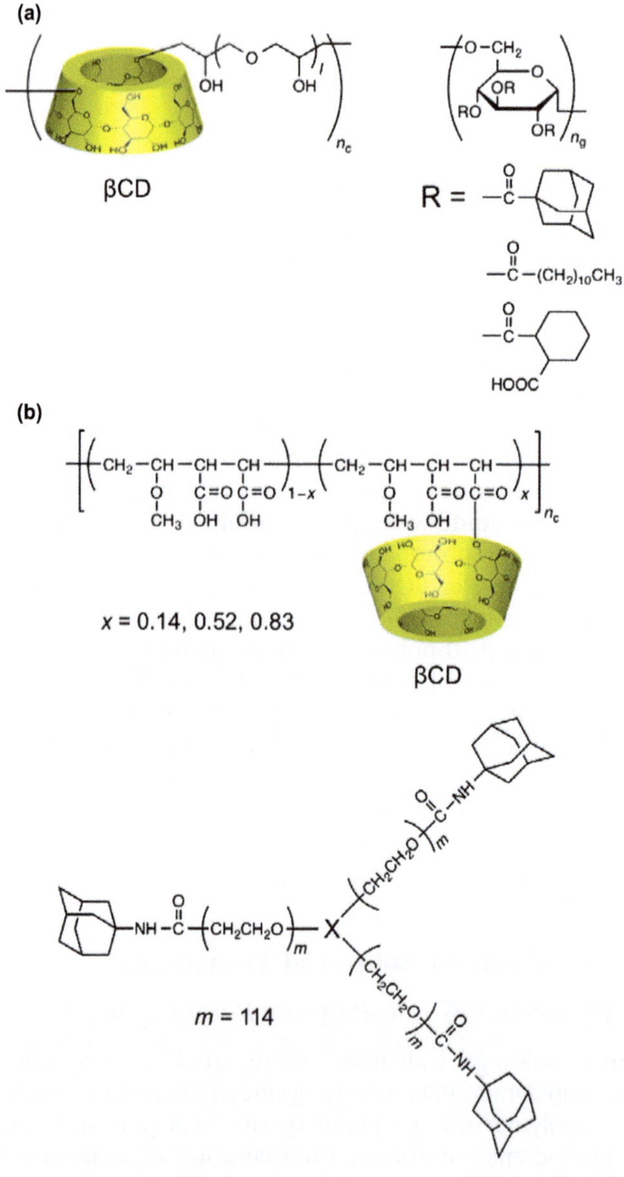

Scheme 7.17 Structures of βCD crosslinked with epichlorohydrin and a dextran modified with hydrocarbon (Ad or dodecyl) and cyclohexane-carboxylic acid moieties (a) and a βCD-carrying alternating copolymer and PEO-Ad_3 (b).

were prepared by ATRP of DMEMA using a PEG-based macroinitiator. The DP of PEG was fixed at 102, whereas the DP values of DMEMA were 96 and 46. The complexation of αCD with the PEG-*b*-PDMEMA of the longer PDMEMA block formed a precipitate at room temperature (25 °C), but a

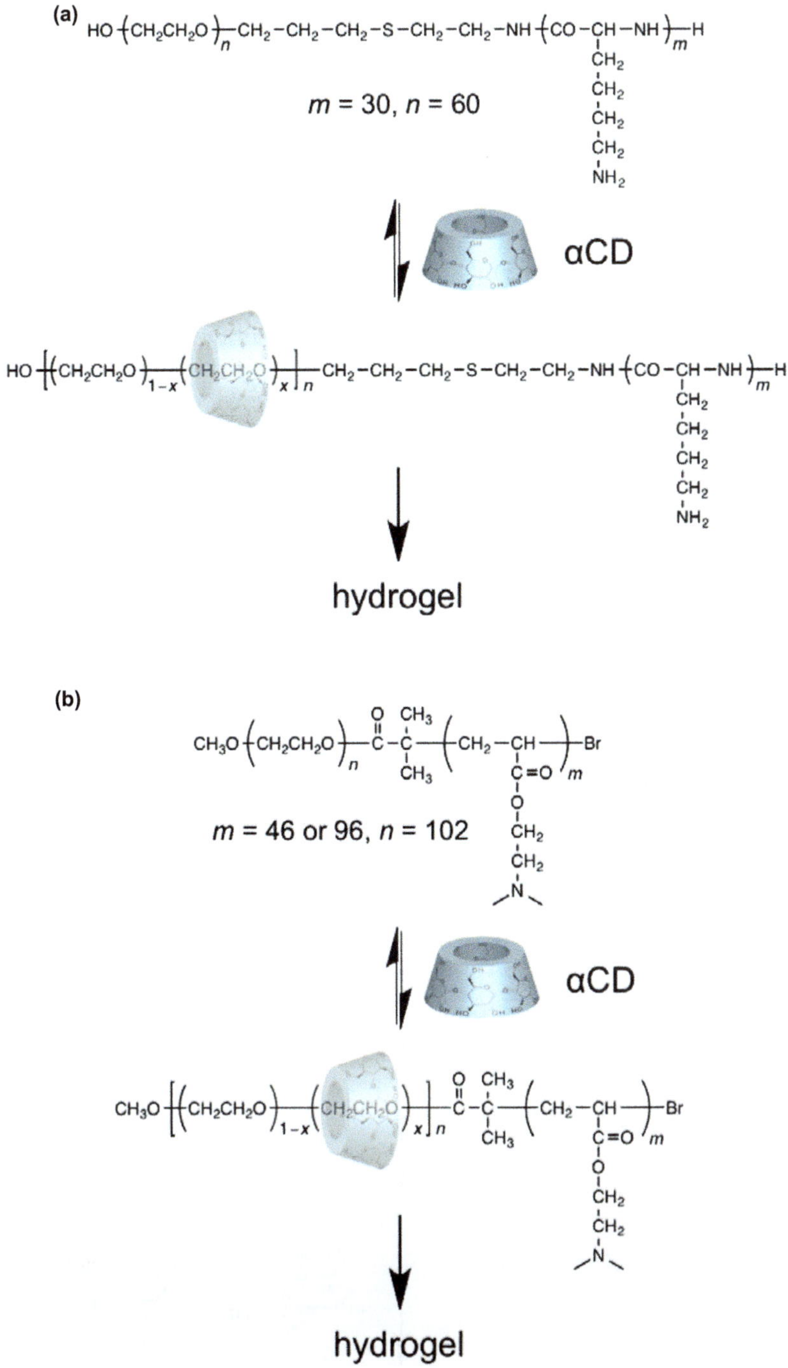

Scheme 7.18 Formation of a pH-responsive hydrogel from αCD/PEG-*b*-PLL PpRx (a) and from αCD/PEG-*b*-PDMEMA PpRx (b).

hydrogel at a higher temperature (60 °C). The hydrogel exhibited thixotropy. On the other hand, the complexation of αCD with the PEG-*b*-PDMEMA of the shorter PDMEMA block formed a hydrogel. At a lower concentration (1 g L^{-1}), the complexed block copolymers formed aggregates responsive to pH. As pH was increased from 1.0 to 8.0, the size of the aggregates increased from 10–20 nm to *ca.* 1 μm.

A ternary hydrogel system responsive to pH and Glc was prepared from αCD, a diblock copolymer, PEO-*b*-PVA, and PEO possessing phenylboronic acid residues at the chain ends (Scheme 7.19).[87] αCD and PEO form the PpRx, which is crystalline and practically insoluble in water, and a boronic acid residue forms rather stable esters with diols under basic conditions.[51] The ternary mixture formed a hydrogel at physiological pH (7.4) because of crosslinking by crystalline PpRx and esters of phenylboronic acid residues and alcohol resides in PVA. The release of FITC-tagged bovine serum albumin (BSA) loaded in the hydrogel was accelerated by adding a saccharide, *i.e.*, Glc or fructose (Fru). (Fru showed a more effective acceleration.) This is because some of the boronic acid ester residues with PVA were replaced with those with the Glc or Fru added, resulting in a decrease in the crosslinking density.

A temperature- and pH-responsive gel-to-sol transition was reported using a mixture of αCD and a PDMEMA-based copolymer possessing PEG grafts (Scheme 7.20).[88] The graft copolymer (PDMEMA-*g*-PEG) was prepared by radical copolymerization or ATRP of DMEMA and PEG methacrylate (PEGMA). The graft copolymer was mixed with αCD to form a hydrogel because αCD molecules included PEG grafts to form PpRx, which was practically insoluble in water. As pH was decreased, the mixture underwent

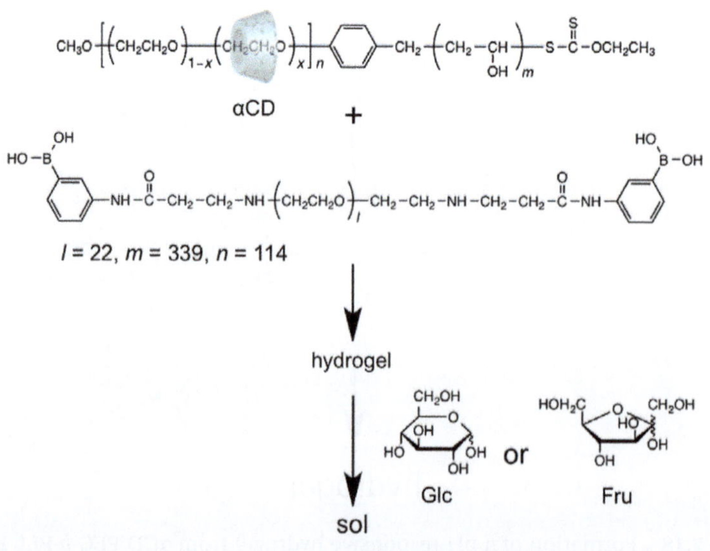

Scheme 7.19 Formation of a saccharide-responsive hydrogel from αCD/PEO-*b*-PVA PpRx and PEO possessing phenylboronic acid residues.

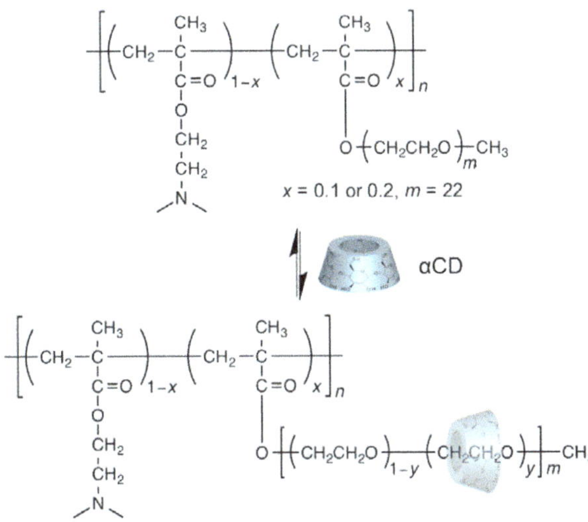

Scheme 7.20 Formation of PpRx from αCD and PDMEMA-*g*-PEG.

gel-to-sol transition. On the other hand, when the temperature was increased, the mixture exhibited gel-to-sol transition. This hydrogel system encapsulated and released 5-[(4,6-dichlorotriazin-2-yl)amino]fluorescein-BSA. The release was faster at a higher temperature and lower pH.

A pH-responsive hydrogel system was described using γCD and dextran possessing PEG-*b*-PEI grafts (Scheme 7.21).[89,90] A block copolymer PEG-*b*-PEI was prepared by ring opening polymerization of 2-ethyl-2-oxazoline using PEG monotosylate as a macroinitiator, followed by hydrolysis with hydrochloric acid. Chains of the PEG-*b*-PEI obtained were coupled with dextran to form a graft copolymer (dex-*g*-(PEG-*b*-PEI)). It is known that γCD molecules include two PEG chains to form a doubly-threaded PpRx.[62] The complexation behavior of mono-6-*O*-(2-sulfonato-6-naphthyl)-γCD (SN-γCD) with dex-*g*-(PEG-*b*-PEI) was investigated at different SN-γCD concentrations, temperatures, and pH by ^{1}H NMR.[89] The formation of PpRx of SN-γCD with the PEG-*b*-PEI grafts was dependent on pH because PEI blocks were protonated at lower pH (4) and cationic species are not included in the CD cavity. Thus, the complexation gave a network structure through the formation of doubly-threaded PpRx, and the network structure depended on pH. The pH-responsive rheological behavior of a binary mixture of γCD with dex-*g*-(PEG-*b*-PEI) was investigated.[90] Mixtures of γCD with dex-*g*-(PEG-*b*-PEI) exhibited viscoelastic properties because of the formation of doubly-threaded PpRx of γCD with PEG-*b*-PEI grafts. The viscoelastic behavior of the mixture was dependent on pH. At higher pH (10), the mixture exhibited higher viscosities and moduli because γCD molecules included both PEG and PEI blocks in the grafts. At lower pH (4), on the other hand, the mixture showed lower viscosities and moduli because γCD molecules included only PEG blocks.

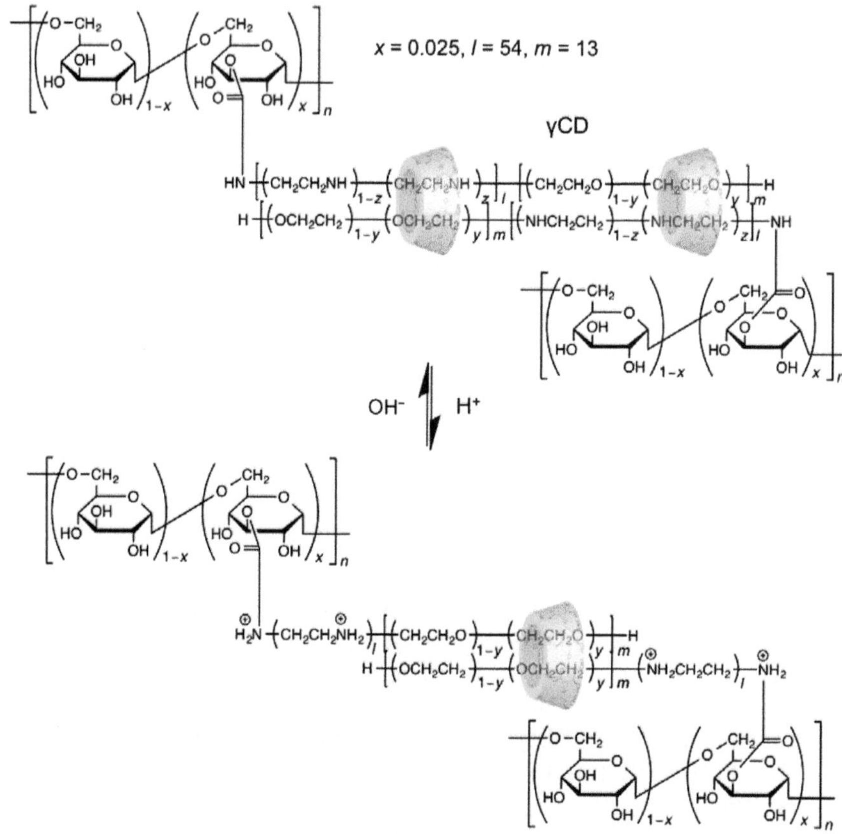

Scheme 7.21 A pH-responsive hydrogel system composed of γCD and dex-*g*-(PEG-*b*-PEI) on the basis of the doubly-threaded PpRx.

7.5.3.2 Cyclodextrin–Polymer/Guest Binary Systems

Poly(phenylacetylene)-based CD-polymers were synthesized, and two samples of them exhibited gelation properties in the presence of a chiral amine, (*S*)-1-phenylethylamine (Scheme 7.22).[91] Phenylacetylene monomers possessing CD moieties were synthesized and polymerized using [(norbornadiene)rhodium chloride]$_2$ as a catalyst. It is known that polymerization of phenylacetylenes using the rhodium catalyst yields selectively *cis*-transoid polymers, which take a dynamic helical conformation.[92–95] Poly(phenylacetylene)s modified with βCD residues through amide and ester linkages exhibited chiral-selective gelation properties. In the presence of (*S*)-1-phenylethylamine, the polymer mixtures formed a gel in a mixed solvent of water and dimethyl sulfoxide (DMSO) (6 : 4, v/v). In the presence of (*R*)-1-phenylethylamine, however, the polymer mixtures were sol.

A gel system responsive to multi-stimuli was prepared using an acrylate-based βCD-polymer and a stimuli-responsive bifunctional guest compound

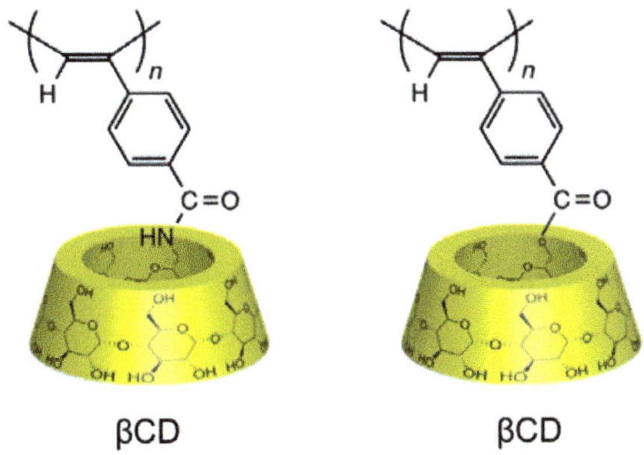

Scheme 7.22 Structures of βCD-carrying poly(phenylacetylene)s, which formed a gel in the presence of (*R*)-1-phenylethylamine.

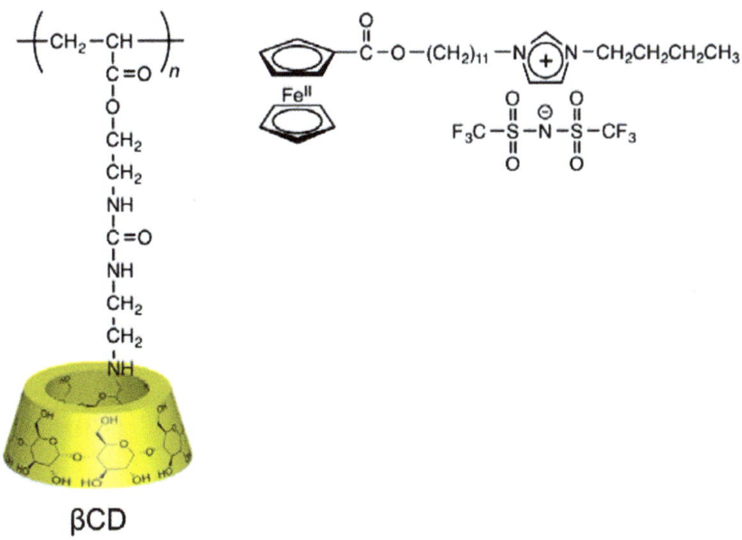

Scheme 7.23 Structures of an acrylate-based βCD-polymer and a bifunctional guest compound.

(Scheme 7.23).[96] The βCD-polymer was prepared by radical polymerization of an acrylate carrying βCD. The bifunctional guest was synthesized from ferrocene-carboxylic acid by coupling with a hydroxy-carrying imidazolium-type ionic liquid. Complexation of the βCD-polymer with the bifunctional guest in a mixed solvent of DMSO and chloroform (4 : 1, v/v) yielded a gel mixture because βCD residues included Fc moieties and bis(trifluoromethylsulfonyl)imide ($TFSI^-$) counter ions. The gel system was responsive to temperature, redox, and the counter ion. As temperature was increased up to *ca.* 40 °C, the gel was converted to a sol, and *vice versa*.

When NaClO, an oxidant, was added to the gel mixture, the Fc moiety in the guest was oxidized to Fc^+ and the inclusion complexes of βCD residues with the Fc^+ were dissociated, resulting in a gel-to-sol transition. When the counter ion of the ionic liquid was exchanged from $TFSI^-$ to Br^-, the gel mixture was also converted to a sol because βCD does not strongly include Br^-.

7.5.3.3 Cyclodextrin–Polymer/Guest–Polymer Binary Systems

A binary hydrogel system responsive to temperature and redox was prepared from a linear βCD-polymer crosslinked by epichlorohydrin and Pluronic® F127 (*i.e.*, a triblock copolymer, PEO-*b*-polypropylene oxide-*b*-PEO (PEO-*b*-PPO-*b*-PEO)) possessing Fc residues at both ends (Fc-F127-Fc) (Scheme 7.24).[97] It is known that Pluronic® F127 forms polymer micelles at temperatures higher than a certain level (≥*ca.* 25 °C) because of the hydrophobic interactions of PPO blocks.[98] Fc-F127-Fc thus formed micellar aggregates at higher temperatures. The micellar aggregates of Fc-F127-Fc were crosslinked by the linear βCD-polymer to form a binary hydrogel system. When the Fc residues were oxidized to Fc^+ with hydrogen peroxide, the inclusion complexes were dissociated, resulting in a gel-to-sol transition. Glucose oxidase (GOX) was mixed with the binary hydrogel system. When Glc was added to the mixture, Glc was oxidized by GOX to form hydrogen peroxide, resulting in a gel-to-sol transition. This system may be useful for sensing Glc. Hydrogel systems that sense Glc have the potential to deliver insulin.[99]

7.5.3.4 Supramolecular Polymer Systems

A chemoresponsive supramolecular hydrogel system was reported using cinnamoyl-carrying βCD derivatives (Scheme 7.25).[100,101] βCD derivatives

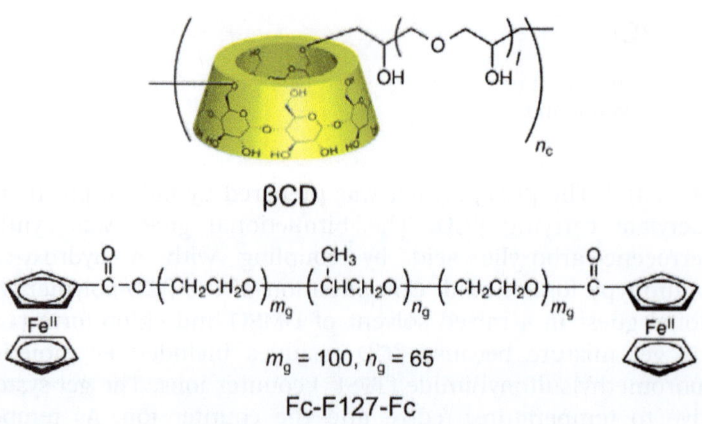

Scheme 7.24 Structures of βCD crosslinked with epichlorohydrin and Fc-F127-Fc.

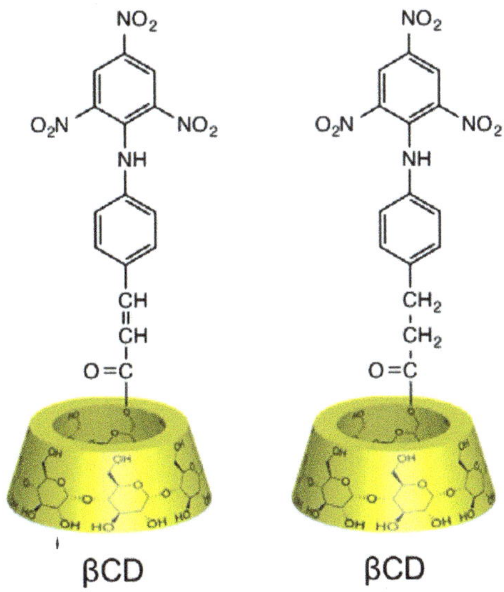

Scheme 7.25 Structures of cinnamoyl-carrying βCD derivatives that formed a chemo-responsive hydrogel.

modified with cinnamoyl moieties were synthesized. These βCD derivatives formed a hydrogel at higher concentrations. Atomic force microscopy indicated that these derivatives formed bundles of fibers. When a competitive guest and host, for example, AdCA and βCD, were added, the supramolecular polymer was decomposed, resulting in a gel-to-sol transition. When urea was added, the mixture underwent a gel-to-sol transition, indicating that hydrogen bonding was critical for the bundle formation.

7.5.4 Macroscopic Self-assembly

7.5.4.1 Cyclodextrin–Polymer/Guest–Polymer Binary Systems

Recently, the formation of macroscopic assemblies based on the molecular recognition of CDs was reported.[102–105] Polyacrylamide gel was modified with CD or guest (aliphatic or aromatic) residues to prepare CD-gels or guest-gels (Scheme 7.26). The interaction of CD-gels and guest-gels was investigated by agitating gel pieces of millimetre or centimetre scale in water. Gel pieces formed macroscopic assemblies depending on the binding constant of CD and guest residues.[102,103,105] It is worth noting that the strength of the interaction of the gel pieces increased with increasing the concentration of CDs and guest residues in the gels.[104] Since the macroscopic assemblies were formed from gel pieces through the formation of inclusion complexes of CD residues with guest residues on the gel interface, no assemblies were formed in the presence of competitive molecules, which masked the CDs or guest residues on the gel

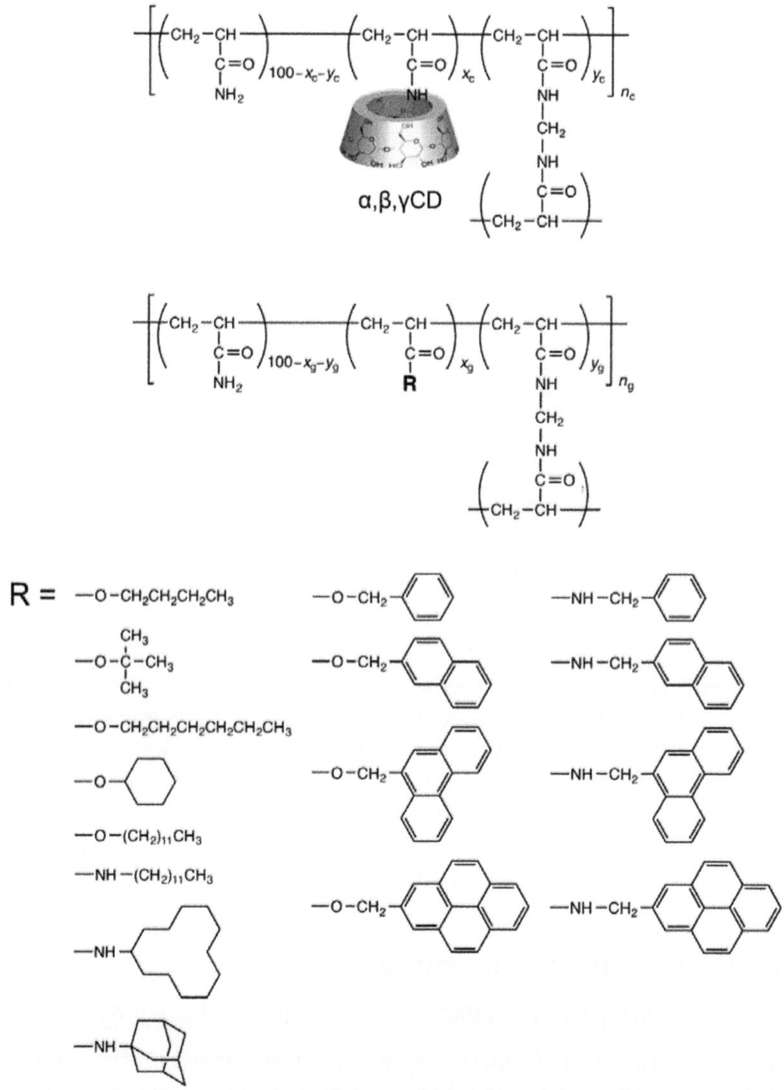

Scheme 7.26 Structures of CD-gels and guest-gels for macroscopic self-assembly.

surface. Thus, the formation of macroscopic gel assemblies was chemo-responsive.

The pH-responsive formation of macroscopic gel assemblies was reported utilizing the inclusion complex of βCD with dansyl (Dns) residues (Scheme 7.27).[106] βCD forms a stable inclusion complex with a dansyl residue with a binding constant of *ca.* 10^2 M^{-1} under neutral conditions. At acidic pH (≤3), a dansyl residue is protonated, and the inclusion complex of βCD with a dansyl residue is dissociated because of the positive charge on the Dns residue. The interaction of a gel possessing βCD residues (βCD-gel)

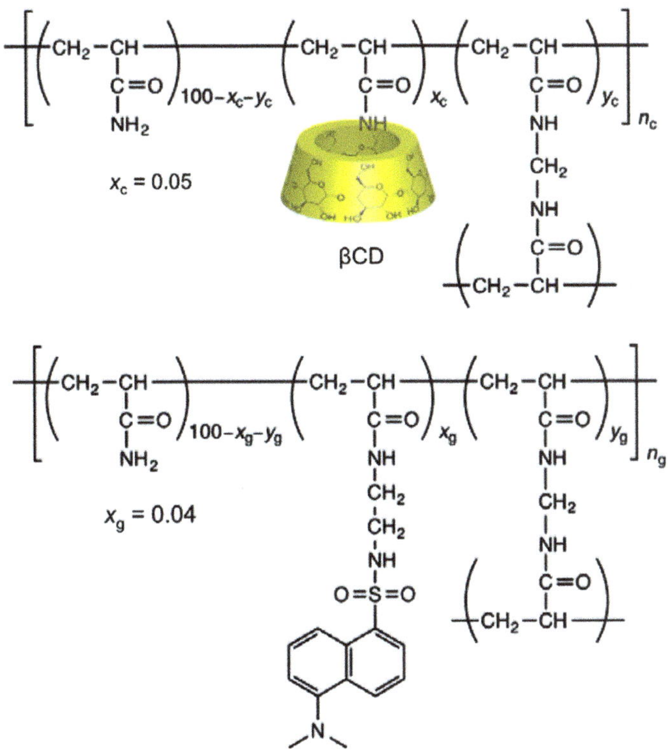

Scheme 7.27 Structures of βCD-gel and Dns-gel for pH-responsive self-assembly.

and a gel possessing Dns residues (Dns-gel) was investigated by agitating gel pieces in water at varying pH. At lower pH ($\leq$3), the gel pieces did not form any assemblies because of protonation of the Dns residues. At higher pH ($\geq$4), on the other hand, the gel pieces formed macroscopic assemblies.

The selectivity of the guest-gel was switched by changing the medium using a pyrenyl (Py) group as a guest for the formation of macroscopic gel assemblies (Scheme 7.28).[107] The interaction of a gel possessing Py residues (Py-gel) with αCD-gel, βCD-gel, and γCD-gel was investigated by agitating gel pieces in a mixed solvent of water and DMSO. In water, a considerable fraction of Py residues formed dimers on the surface of the Py-gel because of the hydrophobicity. When the dimeric Py groups on the gel surface were included in γCD residues on the γCD-gel, Py-gel pieces interacted favorably with γCD-gel pieces to form gel assemblies. In a mixed solvent of water and DMSO (1 : 1, v/v), on the other hand, Py residues existed in the monomeric form on the Py-gel surface. When the monomeric Py groups on the gel surface were included in βCD residues on the βCD-gel, Py-gel pieces interacted selectively with βCD-gel pieces to form gel assemblies. Furthermore, Py-gel pieces formed larger assemblies with the βCD-gel and γCD-gel pieces in a mixed solvent of water and DMSO (4 : 1, v/v).

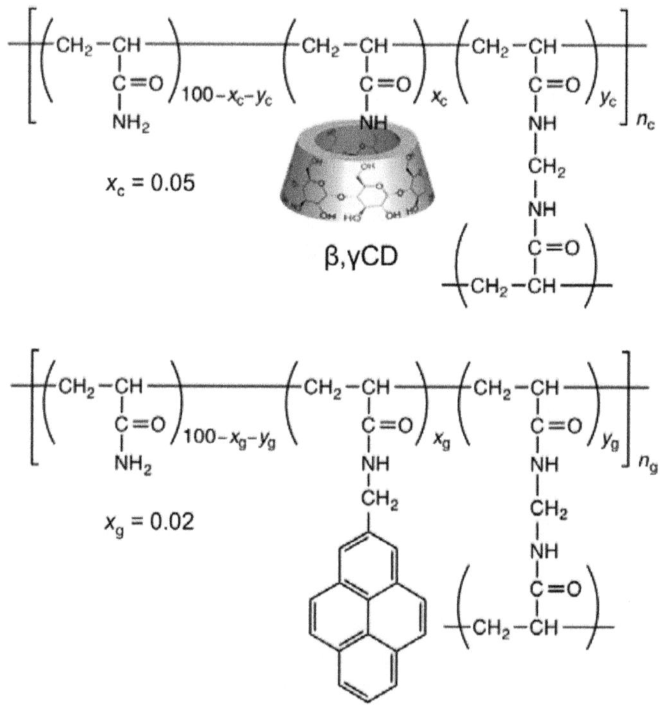

Scheme 7.28 Structures of CD-gels and Py-gel for medium-responsive self-assembly.

Temperature- and photo-responsive macroscopic gel assemblies have also been realized utilizing polyacrylamide-based gels possessing CDs and aromatic residues.[108,109]

7.5.5 Self-healing Materials

7.5.5.1 Cyclodextrin–Polymer/Guest–Polymer Binary Systems

A redox-responsive self-healing system was reported using the redox-responsive formation of inclusion complexes of βCD with Fc residues.[110] PAA ($M_w = 2.5\times10^5$) was modified with βCD or Fc residues through an amide linkage to prepare polymers possessing βCD and Fc residues (βCD-polymer and Fc-polymer) (Scheme 7.29). Mixing aqueous solutions of βCD-polymer and Fc-polymer gave a self-standing hydrogel at higher concentrations (>2 g L^{-1}). After a cube of the hydrogel was cut with a knife into two pieces, the gel pieces were brought into contact and then adhered to each other, resulting in the disappearance of the contact interface. However, when solutions of competitive molecules, *i.e.*, βCD and sodium AdC, were put on the cut surface of the gel pieces, the gel pieces did not adhere to each other, indicative of chemo-responsive self-healing. After Fc residues on the cut surface were oxidized to Fc^+ residues with NaClO, the gel pieces did not adhere to each other because βCD residues do not include practically

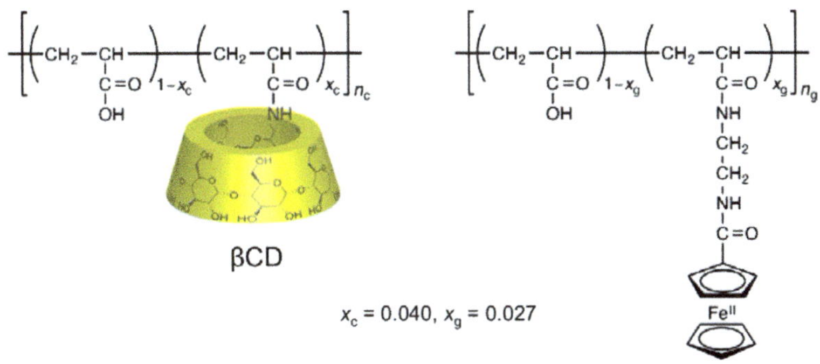

Scheme 7.29 Structures of the PAA-based βCD-polymer and Fc-polymer for a redox-responsive self-healing material.

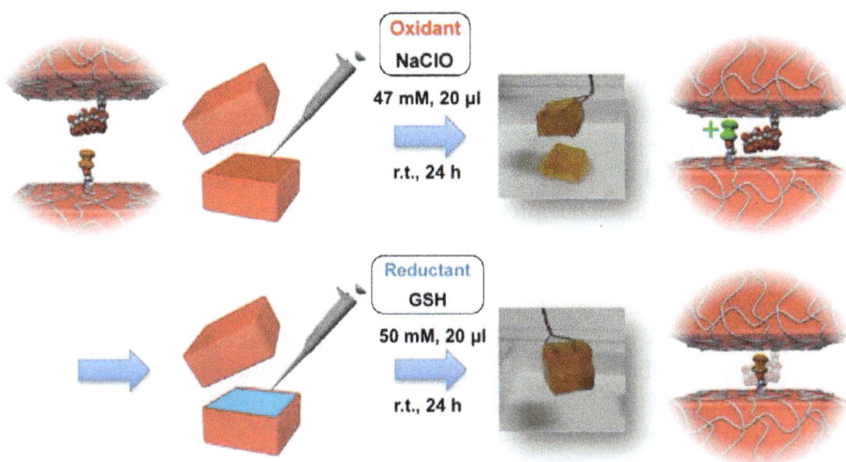

Figure 7.2 Redox-responsive healing experiment of a hydrogel formed from the βCD-polymer and Fc-polymer using oxidizing and reducing agents (NaClO and GSH, respectively).

Fc^+ residues. Whereas, when the Fc^+ residues were reduced back to Fc residues with GSH, the gel pieces recovered the self-healing ability (Figure 7.2).

7.5.5.2 Cyclodextrin–Guest–Polymer Systems

The inclusion complex of a βCD-monomer with an Ad-monomer was terpolymerized with acrylamide in an aqueous medium to give a hydrogel possessing crosslinks of the inclusion complexes (Scheme 7.30).[111,112] The hydrogel exhibited excellent self-healing ability. However, when a solution of a competitive guest, *i.e.*, sodium AdC, was put on the cut surface, the hydrogel lost its self-healing ability because the βCD or Ad residues were masked with the competitive molecules.

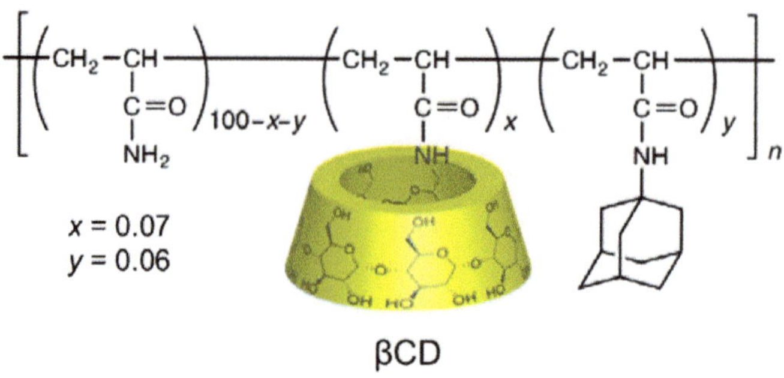

Scheme 7.30 Structure of a polyacrylamide modified with βCD and Ad moieties for a self-healing material.

7.6 Concluding Remarks

This chapter has reviewed CD-based chemo- and pH-responsive polymer systems for pharmaceutical and biomedical applications. Sections 7.2 and 7.3 described the categories and mechanism of cyclodextrin-based stimuli-responsive polymer systems. Section 7.4 dealt with chemical and pH stimuli for CD-based stimuli-responsive polymer systems. Then Section 7.5 provided an overview of a few examples of CD-based chemo- and pH-responsive polymer systems. Stimuli-responsive polymer systems for pharmaceutical and biomedical applications have attracted increasing interest from researchers over the last few decades. CD is an ideal material in these fields not only because of its nontoxicity but also because of its molecular recognition ability. We believe that CD-based polymer systems will be further developed utilizing the molecular recognition ability of CD and the effect of polymer chains to realize systems for practical use.

Abbreviations

Ad	adamantane
AdC	1-adamantanecarboxylate
AdCA	1-adamantanecarboxylic acid
ATRP	atom transfer radical polymerization
BIS	*N*,*N*′-methylenebis(acrylamide)
BSA	bovine serum albumin
CD	cyclodextrin(s)
αCD	α-cyclodextrin
βCD	β-cyclodextrin
γCD	γ-cyclodextrin
CuAAC	copper(I)-catalyzed azide-alkyne cycloaddition
DMEMA	*N*,*N*-dimethylaminoethyl methacrylate
DMSO	dimethyl sulfoxide
DOX	doxorubicin
DP	degree of polymerization

DS	degree of substitution
DTT	dithiothreitol
Fc	ferrocene
Fc^+	ferrocenium
FITC	fluorescein isothiocyanate
Glc	glucose
GSH	glutathione
IBU	ibuprofen
LbL	layer-by-layer
MAnh	maleic anhydride
NCA	*N*-carboxyanhydride
NIPAM	*N*-isopropylacrylamide
PAA	polyacrylic acid
PDMEMA	poly(*N*,*N*-dimethylaminoethyl methacrylate)
PEG	polyethylene glycol
PEI	polyethyleneimine
PEO	polyethylene oxide
PNIPAM	poly(*N*-isopropylacrylamide)
PpRx	polypseudorotaxane(s)
PRx	polyrotaxane(s)
PSS	poly(sodium 4-styrenesulfonate)
PVA	poly(vinyl alcohol)
RAFT	reversible addition-fragmentation chain-transfer

References

1. P. S. Stayton, T. Shimoboji, C. Long, A. Chilkoti, G. Ghen, J. M. Harris and A. S. Hoffman, *Nature*, 1995, **378**, 472–474.
2. B. Jeong and A. Gutowska, *Trends Biotechnol.*, 2002, **20**, 305–311.
3. E. S. Gil and S. M. Hudson, *Prog. Polym. Sci.*, 2004, **29**, 1173–1222.
4. C. d. l. H. Alarcon, S. Pennadam and C. Alexander, *Chem. Soc. Rev.*, 2005, **34**, 276–285.
5. D. Schmaljohann, *Adv. Drug Delivery Rev.*, 2006, **58**, 1655–1670.
6. N. Rapoport, *Prog. Polym. Sci.*, 2007, **32**, 962–990.
7. Y. Morishima, *Angew. Chem., Int. Ed.*, 2007, **46**, 1370–1372.
8. J. R. Capadona, K. Shanmuganathan, D. J. Tyler, S. J. Rowan and C. Weder, *Science*, 2008, **319**, 1370–1374.
9. C. He, S. W. Kim and D. S. Lee, *J. Controlled Release*, 2008, **127**, 189–207.
10. A. K. Bajpai, S. K. Shukla, S. Bhanu and S. Kankane, *Prog. Polym. Sci.*, 2008, **33**, 1088–1118.
11. S. Aoshima and S. Kanaoka, *Adv. Polym. Sci.*, 2008, **210**, 169–208.
12. M.-H. Li and P. Keller, *Soft Matter*, 2009, **5**, 927–937.
13. F. Liu and M. W. Urban, *Prog. Polym. Sci.*, 2010, **35**, 3–23.
14. M. Motornov, Y. Roiter, I. Tokarev and S. Minko, *Prog. Polym. Sci.*, 2010, **35**, 174–211.

15. D. Roy, J. N. Cambre and B. S. Sumerlin, *Prog. Polym. Sci.*, 2010, **35**, 278–301.
16. *Handbook of Stimuli-Responsive Materials*, ed. M. W. Urban, Wiley-VCH, Weinheim, Germany, 2011.
17. A. Hashidzume and A. Harada, in *Supramolecular Polymer Chemistry*, ed. A. Harada, Wiley-VCH, Weinheim, Germany, 2012, pp. 231–267.
18. M. L. Bender and M. Komiyama, *Cyclodextrin Chemistry*, Springer, Berlin, Germany, 1978.
19. J. Szejtli, *Cyclodextrins and Their Inclusion Complexes*, Akadémiai Kiadó, Budapest, Hungary, 1982.
20. *Cyclodextrins*, ed. J. Szejtli and T. Osa, Pergamon, Oxford, U.K., 1996.
21. J. Szejtli, *Cyclodextrin Technology*, Kluwer Academic Publishers, Dordrecht, 1988.
22. A. Harada, in *Large Ring Molecules*, ed. J. A. Semlyen, Wiley & Sons, Chichester, U.K., 1996, pp. 407–432.
23. W. Saenger, J. l. Jacob, K. Gessler, T. Steiner, D. Hoffmann, H. Sanbe, K. Koizumi, S. M. Smith and T. Takaha, *Chem. Rev.*, 1998, **98**, 1787–1802.
24. *Cyclodextrins in Pharmaceutics, Cosmetics, and Biomedicine: Current and Future Industrial Applications*, ed. E. Bilensoy, Wiley & Sons, Hoboken, New Jersey, 2011.
25. K. Uekama, F. Hirayama and T. Irie, *Chem. Rev.*, 1998, **98**, 2045–2076.
26. K. Uekama, F. Hirayama and H. Arima, in *Cyclodextrins and Their Complexes*, ed. H. Dodziuk, Wiley-VCH, Weinheim, 2006, pp. 381–422.
27. C. Ortiz Mellet, J. M. Garcia Fernandez and J. M. Benito, in *Supramolecular Systems in Biomedical Fields*, ed. H.-J. Schneider, The Royal Society of Chemistry, Cambridge, UK, 2013, pp. 94–139.
28. J. F. Miravet and B. Escuder, in *Supramolecular Systems in Biomedical Fields*, ed. H.-J. Schneider, The Royal Society of Chemistry, Cambridge, UK, 2013, pp. 331–354.
29. F. Yuen and K. C. Tam, *Soft Matter*, 2010, **6**, 4613–4630.
30. K. L. Liu, Z. Zhang and J. Li, *Soft Matter*, 2011, **7**, 11290–11297.
31. H. Tian and Q.-C. Wang, in *Molecular Switches*, ed. B. L. Feringa and W. R. Browne, Wiley-VCH, Weinheim, Germany, 2011, vol. 1, pp. 301–319.
32. G. Chen and M. Jiang, *Chem. Soc. Rev.*, 2011, **40**, 2254–2266.
33. A. Harada, *Acc. Chem. Res.*, 2001, **34**, 456–464.
34. A. Harada, A. Hashidzume, H. Yamaguchi and Y. Takashima, *Chem. Rev.*, 2009, **109**, 5974–6023.
35. A. Hashidzume and A. Harada, *Polym. Chem.*, 2011, **2**, 2146–2154.
36. A. Hashidzume, H. Yamaguchi and A. Harada, in *Molecular Machines and Motors*, ed. A. Credi, S. Silvi and M. Venturi, Springer, Berlin, Germany, 2014, vol. 354, pp. 71–110.
37. A. Harada and S. Takahashi, *J. Chem. Soc., Chem. Commun.*, 1984, 645–646.
38. A. Harada, Y. Hu, S. Yamamoto and S. Takahashi, *J. Chem. Soc., Dalton Trans.*, 1988, 729–732.

39. S. McCormack, N. R. Russell and J. F. Cassidy, *Electrochim. Acta*, 1992, **37**, 1939–1944.
40. D. Osella, A. Carretta, C. Nervi, M. Ravera and R. Gobetto, *Organometallics*, 2000, **19**, 2791–2797.
41. B. González, I. Cuadrado, B. Alonso, C. M. Casado, M. Morán and A. E. Kaifer, *Organometallics*, 2002, **21**, 3544–3551.
42. S. Das and U. Subuddhi, *Ind. Eng. Chem. Res.*, 2013, **52**, 14192–14200.
43. M. Prabaharan and J. F. Mano, *Macromol. Biosci.*, 2005, **5**, 965–973.
44. Z. Yuan, Y. Ye, F. Gao, H. Yuan, M. Lan, K. Lou and W. Wang, *Int. J. Pharm.*, 2013, **446**, 191–198.
45. J. Zhang, K. Ellsworth and P. X. Ma, *J. Controlled Release*, 2010, **145**, 116–123.
46. Y.-Y. Liu and X.-D. Fan, *J. Appl. Polym. Sci.*, 2003, **89**, 361–367.
47. M. Hamcerencu, J. Desbrieres, M. Popa and G. Riess, *Carbohydr. Polym.*, 2012, **89**, 438–447.
48. L. H. Sperling, *Interpenetrating Polymer Networks*, American Chemical Society, 1994, vol. 239, pp. 3–38.
49. C. Rodriguez-Tenreiro, L. Diez-Bueno, A. Concheiro, J. J. Torres-Labandeira and C. Alvarez-Lorenzo, *J. Controlled Release*, 2007, **123**, 56–66.
50. Y. Huang, M. Liu, L. Wang, C. Gao and S. Xi, *React. Funct. Polym.*, 2011, **71**, 666–673.
51. *Boronic Acids. Preparation, Application in Organic Synthesis and Medicine*, ed. D. G. Hall, Wiley-VCH, Weinheim, 1st edn, 2005.
52. W. Tian, X. Lv, C. Mu, W. Zhang, J. Kong, Y. Liu and X. Fan, *J. Polym. Sci., Part A: Polym. Chem.*, 2012, **50**, 759–771.
53. W. Tian, X. Lv, L. Huang, N. Ali and J. Kong, *Macromol. Chem. Phys.*, 2012, **213**, 2450–2463.
54. M. Zhang, Q. Xiong, J. Chen, Y. Wang and Q. Zhang, *Polym. Chem.*, 2013, **4**, 5086–5095.
55. C. Li, G.-F. Luo, H.-Y. Wang, J. Zhang, Y.-H. Gong, S.-X. Cheng, R.-X. Zhuo and X.-Z. Zhang, *J. Phys. Chem. C*, 2011, **115**, 17651–17659.
56. G.-F. Luo, X.-D. Xu, J. Zhang, J. Yang, Y.-H. Gong, Q. Lei, H.-Z. Jia, C. Li, R.-X. Zhuo and X.-Z. Zhang, *ACS Appl. Mater. Interfaces*, 2012, **4**, 5317–5324.
57. Y. Wen, Z. Zhang and J. Li, *Adv. Funct. Mater.*, 2014, **24**, 3874–3884.
58. A. Harada and M. Kamachi, *Macromolecules*, 1990, **23**, 2821–2823.
59. A. Harada and M. Kamachi, *J. Chem. Soc., Chem. Commun.*, 1990, 1322–1323.
60. A. Harada, J. Li and M. Kamachi, *Macromolecules*, 1993, **26**, 5698–5703.
61. A. Harada, J. Li and M. Kamachi, *Macromolecules*, 1994, **27**, 4538–4543.
62. A. Harada, J. Li and M. Kamachi, *Nature*, 1994, **370**, 126–128.
63. A. Harada, J. Li and M. Kamachi, *J. Am. Chem. Soc.*, 1994, **116**, 3192–3196.
64. A. Harada, *Coord. Chem. Rev.*, 1996, **148**, 115–133.
65. A. Harada, *Acta Polym.*, 1998, **49**, 3–17.

66. G. Wenz, B.-H. Han and A. Müller, *Chem. Rev.*, 2006, **106**, 782–817.
67. N. Yui and T. Ooya, *Chem. – Eur. J.*, 2006, **12**, 6730–6737.
68. J. Li and X. J. Loh, *Adv. Drug Delivery Rev.*, 2008, **60**, 1000–1017.
69. *Inclusion Polymers*, ed. G. Wenz, Springer, Berlin, Germany, 2009.
70. S. C. Lee, H. S. Choi, T. Ooya and N. Yui, *Macromolecules*, 2004, **37**, 7464–7468.
71. J. Huang, L. Ren and Y. Chen, *Polym. Int.*, 2008, **57**, 714–721.
72. H. Dong, Y. Li, H. Wen, M. Xu, L. Liu, Z. Li, F. Guo and D. Shi, *Macromol. Rapid Commun.*, 2011, **32**, 540–545.
73. J. Wang, L. Ye, A.-y. Zhang and Z.-g. Feng, *J. Mater. Chem.*, 2011, **21**, 3243–3250.
74. B. L. Tardy, H. H. Dam, M. M. J. Kamphuis, J. J. Richardson and F. Caruso, *Biomacromolecules*, 2014, **15**, 53–59.
75. Z. Ge, J. Xu, J. Hu, Y. Zhang and S. Liu, *Soft Matter*, 2009, **5**, 3932–3939.
76. Z. Guo, X. Chen, J. Xin, D. Wu, J. Li and C. Xu, *Macromolecules*, 2010, **43**, 9087–9093.
77. M. Osaki, Y. Takashima, H. Yamaguchi and A. Harada, *J. Org. Chem.*, 2009, **74**, 1858–1863.
78. S. Yoshida, D. Taura, Y. Takashima, A. Hashidzume, H. Yamaguchi and A. Harada, *Polym. Prep., Jpn.*, 2009, **58**, 5569–5570.
79. Y. Takashima, T. Oka, S. Yoshida, H. Yamaguchi and A. Harada, *Chem. Lett.*, 2011, **40**, 742–743.
80. H. Arima, K. Motoyama and T. Higashi, *Adv. Drug Delivery Rev.*, 2013, **65**, 1204–1214.
81. I. Böhm and H. Ritter, *Macromol. Chem. Phys.*, 2011, **212**, 1080–1085.
82. J. Zeng, K. Shi, Y. Zhang, X. Sun and B. Zhang, *Chem. Commun.*, 2008, 3753–3755.
83. A.-M. Layre, V. Wintgens, N.-M. Gosselet, F. Dalmas and C. Amiel, *Eur. Polym. J.*, 2009, **45**, 3016–3026.
84. G. Volet and C. Amiel, *Eur. Polym. J.*, 2009, **45**, 852–862.
85. R. Yuan and X. Shuai, *J. Polym. Sci., Part B: Polym. Phys.*, 2008, **46**, 782–790.
86. K. Sui, X. Shan, S. Gao, Y. Xia, Q. Zheng and D. Xie, *J. Polym. Sci., Part A: Polym. Chem.*, 2010, **48**, 2143–2153.
87. T. Yang, R. Ji, X.-X. Deng, F.-S. Du and Z.-C. Li, *Soft Matter*, 2014, **10**, 2671–2678.
88. L. Ren, L. He, T. Sun, X. Dong, Y. Chen, J. Huang and C. Wang, *Macromol. Biosci.*, 2009, **9**, 902–910.
89. Y.-K. Joung, H. S. Choi, T. Ooya and N. Yui, *J. Inclusion. Phenom. Macrocyclic Chem.*, 2007, **57**, 323–328.
90. Y.-K. Joung, T. Ooya, M. Yamaguchi and N. Yui, *Adv. Mater.*, 2007, **19**, 396–400.
91. K. Maeda, H. Mochizuki, K. Osato and E. Yashima, *Macromolecules*, 2011, **44**, 3217–3226.
92. T. Aoki, M. Kokai, K.-i. Shinohara and E. Oikawa, *Chem. Lett.*, 1993, **22**, 2009–2012.

93. E. Yashima, T. Matsushima and Y. Okamoto, *J. Am. Chem. Soc.*, 1995, **117**, 11596–11597.
94. E. Yashima, T. Nimura, T. Matsushima and Y. Okamoto, *J. Am. Chem. Soc.*, 1996, **118**, 9800–9801.
95. E. Yashima, K. Maeda and Y. Okamoto, *Nature*, 1999, **399**, 449–451.
96. C. Yuan, J. Guo, M. Tan, M. Guo, L. Qiu and F. Yan, *ACS Macro Lett.*, 2014, **3**, 271–275.
97. Y. Zhou, X. Fan, W. Zhang, D. Xue and J. Kong, *J. Polym. Res.*, 2014, **21**, 1–10.
98. P. Linse and M. Malmsten, *Macromolecules*, 1992, **25**, 5434–5439.
99. H.-J. Schneider, K. Kato and R. M. Strongin, *Sensors*, 2007, **7**, 1578–1611.
100. W. Deng, H. Yamaguchi, Y. Takashima and A. Harada, *Angew. Chem., Int. Ed.*, 2007, **46**, 5144–5147.
101. W. Deng, H. Yamaguchi, Y. Takashima and A. Harada, *Chem. - Asian J.*, 2008, **3**, 687–695.
102. A. Harada, R. Kobayashi, Y. Takashima, A. Hashidzume and H. Yamaguchi, *Nat. Chem.*, 2011, **3**, 34–37.
103. H. Yamaguchi, R. Kobayashi, Y. Takashima, A. Hashidzume and A. Harada, *Macromolecules*, 2011, **44**, 2395–2399.
104. Y. Zheng, A. Hashidzume, Y. Takashima, H. Yamaguchi and A. Harada, *Langmuir*, 2011, **27**, 13790–13795.
105. A. Hashidzume, Y. Zheng, Y. Takashima, H. Yamaguchi and A. Harada, *Macromolecules*, 2013, **46**, 1939–1947.
106. Y. Zheng, A. Hashidzume and A. Harada, *Macromol. Rapid Commun.*, 2013, **34**, 1062–1066.
107. Y. Zheng, A. Hashidzume, Y. Takashima, H. Yamaguchi and A. Harada, *Nat. Commun.*, 2012, **3**, 831.
108. Y. Zheng, A. Hashidzume, Y. Takashima, H. Yamaguchi and A. Harada, *ACS Macro Lett.*, 2012, **1**, 1083–1085.
109. H. Yamaguchi, Y. Kobayashi, R. Kobayashi, Y. Takashima, A. Hashidzume and A. Harada, *Nat. Commun.*, 2012, **3**, 603.
110. M. Nakahata, Y. Takashima, H. Yamaguchi and A. Harada, *Nat. Commun.*, 2011, **2**, 511.
111. T. Kakuta, Y. Takashima, M. Nakahata, M. Otsubo, H. Yamaguchi and A. Harada, *Adv. Mater.*, 2013, **25**, 2849–2853.
112. T. Kakuta, Y. Takashima and A. Harada, *Macromolecules*, 2013, **46**, 4575–4579.

CHAPTER 8

Polysaccharide-based Smart Materials

YUAN XIONG,[a] XUE QU,[b] CHANGSHENG LIU,[b]
WILLIAM E. BENTLEY,[c] XIAO-WEN SHI[a] AND
GREGORY F. PAYNE*[a,b,c]

[a] School of Resource and Environmental Science, Hubei Biomass-Resource Chemistry and Environmental Biotechnology Key Laboratory, Wuhan University, Wuhan 430079, China; [b] School of Material Science and Engineering, the State Key Laboratory of Bioreactor Engineering, Key Laboratory for Ultrafine Materials of Ministry of Education, East China University of Science and Technology, Shanghai 200237, China; [c] Institute for Bioscience and Biotechnology Research, and Fischell Department of Bioengineering, University of Maryland College Park, College Park, MD 20742, USA
*Email: gpayne@umd.edu

8.1 Introduction: Historical Perspective of Polysaccharides

For centuries, polysaccharides have been among the materials of choice for meeting mankind's needs for food, fuel and materials (*e.g.*, for paper, clothing and shelter), and they were the logical choice for a broad range of technical applications. For instance, polysaccharides have traditionally been added to foods and cosmetics to confer thickening, gelling and moisturizing properties (*e.g.*, pectin in jellies, xanthan gum in ice cream, and hyaluronic acid for personal care products). In addition, polysaccharides were obvious

RSC Smart Materials No. 14
Chemoresponsive Materials: Stimulation by Chemical and Biological Signals
Edited by Hans-Jörg Schneider

Published by the Royal Society of Chemistry, www.rsc.org

choices for applications in the life sciences where agar gels became the standard for the surface cultivation of microbes and the electrophoretic separation of biomacromolecules.

During the last century, however, polysaccharides have been overshadowed by synthetic polymers, proteins and nucleic acids since the remarkable advances in synthetic chemistry and molecular biology have enabled these latter polymers to be created with new and important material properties. For instance, the emergence of the petrochemical industry provided synthetic polymers that could often out-compete natural polymers for various applications ranging from those that required a low cost (*e.g.*, single use packaging) to those requiring high performance (*e.g.*, high temperature applications). Polymer scientists developed a range of capabilities to control structure and tailor properties, and these capabilities were enabled by advances in synthesis and characterization methods, along with the development of physical theories that related structure to properties.

At about the same time that polymer scientists were discovering ways to synthesize structure and function, life scientists were learning how biology created macromolecular structures and conferred function. Importantly, the protein and nucleic acid macromolecules are biosynthesized by templated mechanisms that yield unrivaled capabilities for controlling polymer sequence and size. For the case of nucleic acids, the link between structure and function is rather direct—three adjacent nucleotides possess the information that codes for a single amino acid in a protein sequence. For the case of proteins, the link between sequence, structure (*e.g.*, folding) and function (*e.g.*, enzymatic activity) is less well-understood. Yet there remains considerable effort to apply an ever-expanding array of experimental and computational tools for the *de novo* design of protein structure and function. Thus, biotechnology is poised to play a larger role in materials science, especially for nucleic acid and protein based materials.[1]

Polysaccharides lie somewhere between synthetic polymers and proteins in terms of their current materials science capabilities. Importantly, many of the advances in polymer science and biotechnology have not yet been fully adapted or applied to polysaccharides. Compared to proteins and synthetic polymers, the last half a century has witnessed only modest improvements in our capabilities to synthesize, sequence, characterize and model polysaccharides. Even without these advances, polysaccharides remain a unique and valuable resource for materials science.

Here, we review the capabilities and current limitations of polysaccharides. We start with a broad overview of polysaccharides and the unique features that confer "smartness" to polysaccharide-based materials. We conclude by highlighting the capabilities of two of the most-studied polysaccharide systems: chitosan and alginate. Our goal is to provide a broad sample rather than an exhaustive review and we only briefly discuss applications that have been extensively reviewed elsewhere (*e.g.*, controlled drug release[2] and regenerative medicine[3,4]). We also focus on the intrinsic properties of the polysaccharides and only briefly mention efforts to confer smart properties by chemical modification of the polysaccharide.

8.2 Why Polysaccharides? General Advantages

Table 8.1 lists the advantages common to polysaccharides. Polysaccharides are either recovered from plants and animals (*e.g.*, food or agricultural residues) or produced by fermentation processes, and thus they can be obtained from sustainable resources at reasonable volumes and costs.[5,6] Polysaccharides have a long history of safe use for a range of applications that involve direct contact or ingestion by humans.[7,8] Polysaccharides can also be degraded in the environment making them environmentally-friendly. In some cases, polysaccharides can also be digested or resorbed in the body and there appears to be increasing effort to explore polysaccharide-based materials for emerging medical applications (*e.g.*, regenerative medicine).[4,9] Since polysaccharides are generated by stereoselective biosynthetic reactions, the resulting polymer is chiral—a structural characteristic that may have significant but incompletely appreciated impact.[10] While these general characteristics can be advantageous, a polysaccharide must offer the necessary material properties to be appropriate for an intended application.

8.3 Common Features that Make Polysaccharides "Smart"

People were employing the smart properties of polysaccharides long before the term "smart" was ever applied to materials. Even today, some polysaccharides are so commonly used that their smart properties are taken for granted (*e.g.*, life scientists routinely cool agar solutions to induce gel formation). As illustrated by the following examples, some of the interaction mechanisms that confer smartness to polysaccharides are unique and warrant discussion.

Table 8.1 General characteristics of polysaccharides.

Advantages	Challenges
• Available and relatively inexpensive • Generally safe for food, cosmetic, pharmaceutical, and medical applications • Degradable for medical and environmental applications • Chiral • Stimuli-responsive (smart) properties • Bio-specific recognition by enzymes and lectins (sugar-binding proteins)	Structurally complex • May lack a precise sequence and size (biosynthesis is not templated) • Can have branched architecture • Can have modifications (acetylation and sulfation) that may vary between batches Properties may be inconsistent • Dependent on source, season and processing Properties may be difficult to tailor • Can't readily adjust composition, sequence and size

8.3.1 Stimuli-responsive Self-assembly

A characteristic feature of many "smart" polysaccharides is their ability to undergo reversible, stimuli-responsive sol–gel transitions to form three-dimensional hydrogel networks. Such a sol–gel transition is a form of self-assembly that is triggered by small changes in conditions and in many cases the transition occurs near ambient or physiological conditions. Figure 8.1a illustrates this behavior for the reversible, pH-responsive self-assembly of the aminopolysaccharide chitosan. Chitosan is obtained by the partial

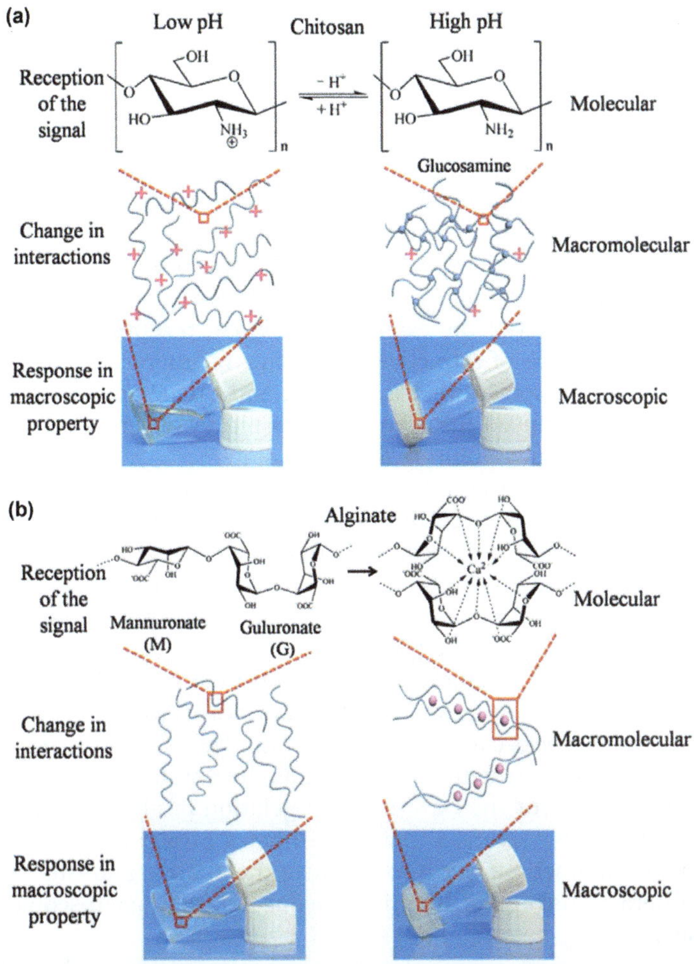

Figure 8.1 Smartness can result from stimuli-responsive self-assembly. (a) Chitosan's sol–gel transition is triggered by pH changes that deprotonate chitosan's primary amines converting this weak polyelectrolyte from a water soluble polycation (low pH) into a neutral hydrogel (neutral or basic pH). (b) Alginate's sol–gel transition is triggered by the addition of Ca^{2+} ions that form inter-chain associations (*i.e.*, physical crosslinks) with the guluronate (G) blocks.

de-acetylation of chitin which is the major structural polymer in the integument ("shells") of crustaceans and insects, and also in the cell walls of fungi. Chemical processes to de-acetylate chitin are incomplete and chitosan is formally a co-polymer composed of glucosamine and *N*-acetylglucosamine residues. In Figure 8.1a we show only glucosamine because these residues confer pH-responsiveness. At low pH, the primary amines of the glucosamine residues are protonated and positively charged making chitosan a cationic polyelectrolyte that is soluble in aqueous solutions. At higher pHs (greater than about 6.5), the amines become deprotonated, chitosan loses its charge and can undergo inter-chain self-associations that yield the hydrogel network.

As illustrated on the right in Figure 8.1a, chitosan's pH-responsive sol–gel transition is a self-assembly process that creates order over a hierarchy of length scales: de-protonation occurs at the level of the individual amine residues (molecular, sub-nanoscale) while the hydrogel network is macroscopic. It is important to emphasize that this pH-triggered self-assembly process is reversible and the macroscopic structure can be "erased" by re-dissolving the gel using mild acid to re-protonate the amines. On the left of Figure 8.1a, we adapt concepts and terminology proposed in the smart materials literature:[11,12] chitosan's amine protonation–deprotonation reaction serves to "receive" the external pH signal; reception of this signal "changes" the balance of attractive and repulsive interactions responsible for chitosan chain association–dissociation; and this change in inter-chain associations is transduced into the macroscopically-observable "response" of a transition between a solution and a gel state.

A second well-known example of stimuli-responsive polysaccharide self-assembly is the gelation of calcium-alginate illustrated in Figure 8.1b. Alginate is a weakly acidic polysaccharide that is either extracted from brown seaweed or generated by bacterial fermentation using *Pseudomonas* and *Azotobacter*. Alginate is a co-polymer of β-D-mannuronate (M) and its C-5 epimer α-L-guluronate residues (G) linked by 1–4 glycosidic bonds. Importantly, these monomers are not random but appear as homopolymeric blocks (MMMMM or GGGGG) or alternating blocks (MGMGMG). As illustrated in Figure 8.1b, the G blocks can associate with multivalent cations such as Ca^{2+} to generate "egg box" structures that serve as the physical crosslinks for the hydrogel network.[13] For the Ca^{2+}-induced gelation of alginate, Ca^{2+} serves as the signal that is "received" by the G blocks of alginate, the resulting Ca^{2+}–alginate interactions serve to organize the chains, and this "change" in chain interactions/organization is transduced into a macroscopic sol–gel "response".

8.3.2 Bio-specific Interactions

8.3.2.1 Reversible Bio-specific Interactions

There are many examples in which polysaccharides are specifically "recognized" and bound by proteins (or by protein domains). In some cases, protein-binding to the polysaccharide may be associated with catalysis (*e.g.*, a cellulose binding domain of a cellulase enzyme) and in other cases the binding may be associated with recognition (*e.g.*, immune cell receptors that recognize the polysaccharides of bacterial pathogens). The general term "lectin" refers to a carbohydrate-binding protein and the reversibility of lectin-binding suggests interesting possibilities for smart materials. For instance, the most widely studied plant lectin is Concanavalin A (ConA) which is a tetrameric protein with one binding site per monomeric subunit (four binding sites per tetramer). Binding is specific for α-D-mannosyl and α-D-glucosyl residues and since many glycoproteins have such residues, ConA has been used to selectively bind glycoproteins for biotechnology applications in enzyme immobilization and protein purification. For instance, ConA affinity chromatography employs ConA-functionalized solid supports to purify glycoproteins.[14] After glycoprotein binds to the ConA-functionalized support, the column can be washed to remove contaminating proteins, and then the glycoprotein can be eluted from the column by adding a sugar that out-competes the glycoprotein for ConA's binding site.

The fact that ConA has four binding sites and binds reversibly offers interesting opportunities for stimuli-responsive assembly and dis-assembly as illustrated in Figure 8.2a. In this example, a microcapsule is prepared using a layer-by-layer method with ConA and the glucose polymer glycogen. In this example, biospecific ConA–glycogen interactions are used for multilayer assembly which can be contrasted to more common layer-by-layer assembly methods that rely on electrostatic interactions between oppositely charged polyelectrolytes. These glycogen–ConA multilayers have been reported to be stable but can be triggered to disassemble in response to the addition of mannose and glucose.[15,16] These glycogen–ConA multilayers have been proposed as glucose-sensitive smart-coatings for the controlled disassembly and release of capsule contents (*e.g.*, drugs).

8.3.2.2 Irreversible Bio-specific Interactions

Polysaccharides, like other biopolymers, can be acted upon by enzymes and such reactions generally result in irreversible changes. Thus, a smart system involving an irreversible enzymatic reaction would be considerably different from the reversible stimuli-responsive approaches described earlier. For irreversible enzymatic systems, "smartness" would typically be based on whether the enzyme-catalyst and the polysaccharide-substrate become co-localized under appropriate conditions for the reaction to occur. Since enzymes that react with polysaccharides are specific, the examples of enzyme-based smart systems are also specific. For instance, Figure 8.2b illustrates a proposed tumor targeting drug delivery system. In this system, a drug-containing liposome is decorated with a cell-penetrating peptide and then an outer corona composed of the polysaccharide hyaluronic acid (HA). HA performs a protective function preventing premature cellular uptake until the liposome becomes localized in regions of cancer cells. Importantly, such tumor regions can also be enriched in the polysaccharide-degrading

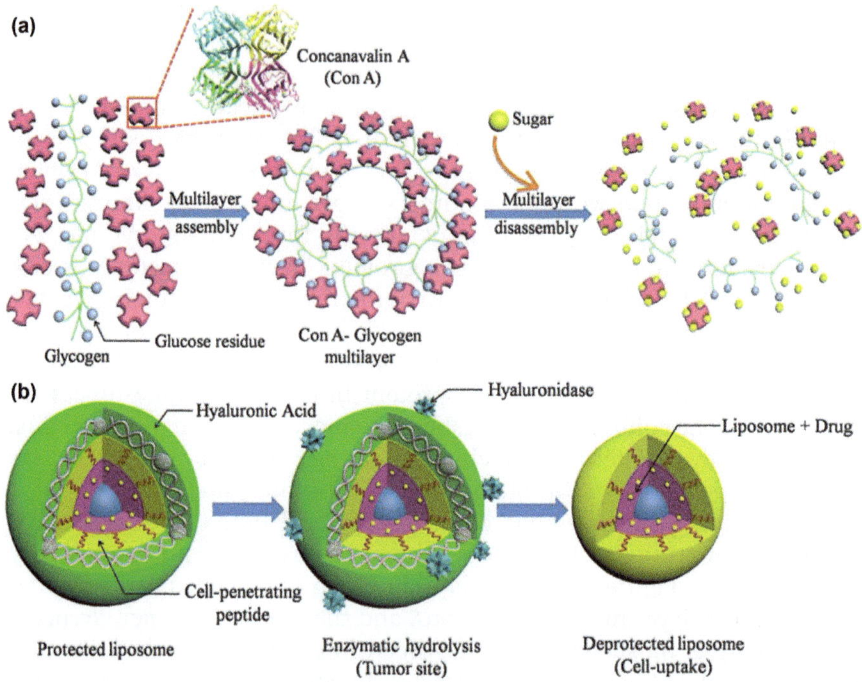

Figure 8.2 Smartness can be conferred by bio-specific interactions. (a) Lectins such as ConA are proteins that reversibly bind polysaccharides: interactions between polysaccharides and multi-subunit lectins enable the assembly of structures (*e.g.*, multilayers) that can be disassembled by competing sugars (adapted with permission from Gao and coworkers[15]). (b) Enzymes can selectively degrade polysaccharides: a drug-loaded liposome has a protective hyaluronic acid (HA) corona that facilitates targeted delivery to tumor regions that express high activities of HA-degrading enzymes (adapted with permission from Lv and Zhou[17]).

enzyme hyaluronidase which presumably serves a pathogenic role of degrading the extracellular matrix to facilitate the spreading of cancer cells.[18] Hyaluronidase in the tumor region hydrolyzes the outer HA corona of the liposome exposing the cell penetrating peptides that initiate cellular uptake and drug delivery.[17] While details of this drug-targeting example may be specific, the concept of using polysaccharide degrading systems to target delivery may be more general. For instance, the polysaccharide-degrading capabilities of the intestinal tract can be highly specific depending on the location, the individual, and the microbial population (*e.g.*, the microbiome),[19] and such variability may provide opportunities for targeting as well as for personalizing medicine and nutrition.[20]

8.3.3 Nanoscale and Hierarchical Supramolecular Assembly

Proteins are well-studied biopolymers and their organization is commonly classified into four levels of structure based on associations within an individual polypeptide chain or between chains; primary structure refers to the amino acid sequence, secondary structure refers to the localized ordering of a polypeptide chain into α-helices and β-sheets, tertiary structure refers to the folding of a polypeptide chain into a globular form, and quaternary structure refers to associations of multiple polypeptide chains. The organization of polysaccharides is less frequently examined, and probably the best known example is the anisotropic hierarchical structure of cellulose. Figure 8.3a illustrates that individual cellulose chains are organized into fibrils that are composed of highly ordered crystalline regions and disordered amorphous domains, and these fibrils are further organized into fibers. In some cases, smaller organized supramolecular polysaccharide structures are being examined for nanocomposites (*e.g.*, cellulose nanocrystals or whiskers) as well as for templates for growing nanoparticles.[5,21,22] Less well-known are the single, double and triple helical structures that some polysaccharides can form in aqueous solutions. Emerging research is showing that some helix-forming polysaccharides are capable of wrapping nanostructures (*e.g.*, carbon nanotubes) which may provide interesting opportunities in nanotechnology.[23] Polysaccharides may offer unique

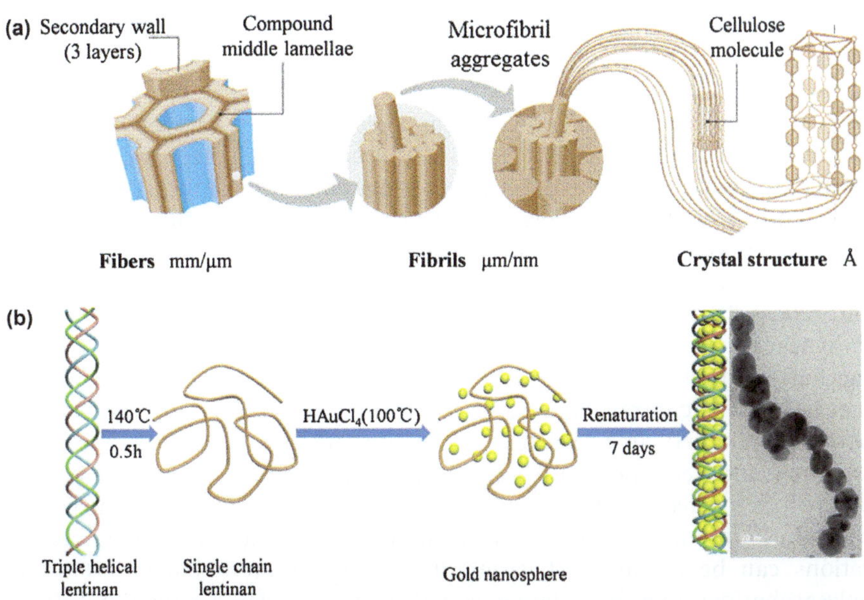

Figure 8.3 Smartness can emerge from polysaccharide's nanoscale and hierarchical supramolecular assembly. (a) The hierarchical structure of cellulose illustrates that many polysaccharides can undergo intra- and interchain associations to form crystalline domains, fibrils and fibers (reproduced with permission from Tingaut[21]). (b) Some polysaccharides have been observed to form helical structures in solution: the mushroom polysaccharide lentinan can form triple helices that serve to organize gold nanoparticles (adapted with permission from Xu[24]).

capabilities for these nano-applications because they can be biocompatible, chiral, selectively-degradable (by enzymes) and recognized by specific sugar-binding receptors (*e.g.*, for targeting).

In some cases, the supramolecular ordering of polysaccharides is readily reversible and can be triggered by externally imposed stimuli as illustrated by the example in Figure 8.3b. In this example, the branched neutral polysaccharide lentinan can exist as a triple helix in aqueous solution but can dissociate into individual chains at high temperature (>130 °C) or in dimethyl sulfoxide (DMSO). As illustrated in Figure 8.3b, these individual chains can (i) serve as a mild reducing agent for the reduction of gold ions into gold nanoparticles, (ii) stabilize the gold nanoparticles, and (iii) renature back into triple helices in water. Thus, lentinan may provide a simple and environmentally-friendly means to create gold nanoparticles and organize them over a hierarchy of length scales.[24]

8.4 Limitations and Challenges with Polysaccharides

While polysaccharides offer advantages (compared to synthetic polymers or protein/nucleic acid biopolymers), polysaccharides also have limitations. One major limitation is that there are only a finite number of possibilities. It is not guaranteed that a polysaccharide will be known (or even exist) that possesses the specific combination of properties required for an intended application. The representative structures in Scheme 8.1 illustrate that a diversity of polysaccharide structures and properties are known, yet this list is finite. Further, the current state of the art for polysaccharide "engineering" through either synthetic or biotechnological routes is not yet sufficiently advanced to allow the *de novo* design and synthesis of polysaccharides with arbitrarily defined properties.

Other characteristic difficulties of polysaccharides are listed in Table 8.1 and reflect their complex structures. Unlike proteins and nucleic acids, the biosynthesis of polysaccharides is not templated. As a result, a polysaccharide's molecular weight is not uniform and the sequence may not be precise. Also, in contrast to proteins and nucleic acids, some polysaccharides possess branched architectures and branching may also lack a precise structure and sequence. Further, polysaccharides that are obtained from nature often have modifications—hydroxyls can be sulfated, carboxylates can be esterified, and amines are commonly acylated—and these modifications can be a source of variability among different batches. Finally, polysaccharides may have larger substituents associated with lipids (*e.g.*, glycolipids) or proteins (*e.g.*, proteoglycans). Thus, while polysaccharides offer diverse structures and properties, these features may be inconsistent among different batches of materials. Subtle variations between batches are a major concern for medically important applications (*e.g.*, heparin anticoagulants) and this is especially true when the biological mechanism-of-action involves molecular recognition (*e.g.*, through protein or receptor binding). More general physicochemical properties, such as stimuli-responsive

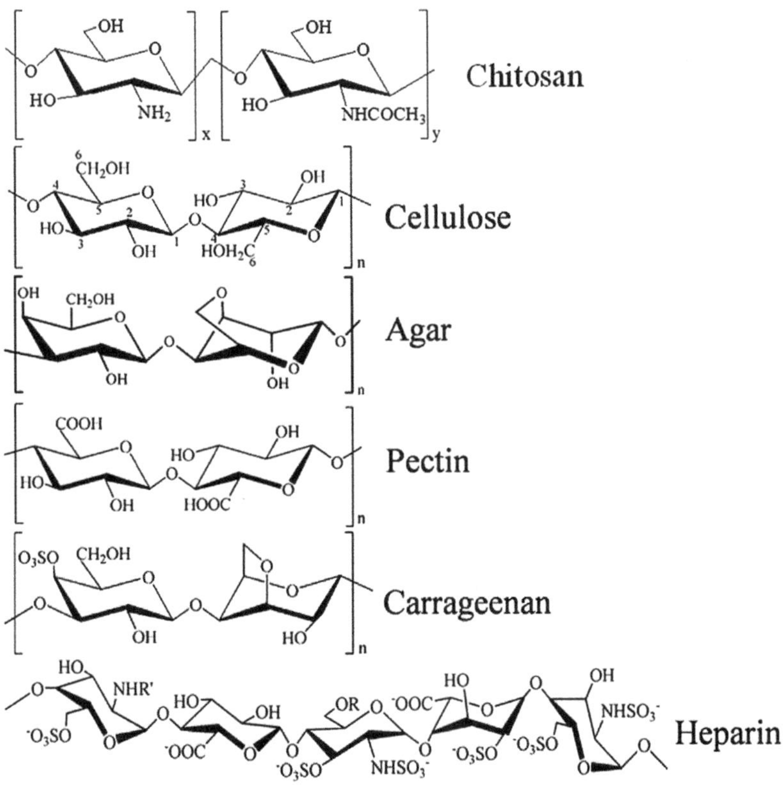

Scheme 8.1 Structures of common polysaccharides.

solubility, rely on rather dramatic changes and these smart properties may be less sensitive to small variations among batches.

8.5 Chemical Modification of Polysaccharides

In some cases, the limitations of polysaccharide can be addressed by chemically modifying these polymers. For instance, cellulose (Scheme 8.1) is insoluble in water yet a variety of chemical modifications are routinely used to create water-soluble derivatives.[25,26] Polysaccharide derivatization is facilitated by the abundance of moieties (*e.g.*, hydroxyl groups) that can readily undergo reaction. But this abundance also leads to difficulties in controlling which groups are modified (it is generally not possible to confine reactions to the 2, 3 or even 6 positions of cellulose). Polysaccharides are also commonly derivatized to confer stimuli-responsive functionality. For instance, acrylate modification may confer pH-responsive properties, poly(*N*-isopropylacrylamide) modification can confer thermal responsiveness, and modification with redox-active moieties (*e.g.*, thiols) can confer redox-responsiveness. Polysaccharide derivatives are not considered here but the interested reader is referred to recent reviews.[27,28]

8.6 Relevant Interaction Mechanisms for Polysaccharides

As indicated from the examples in Figures 8.1 through 8.3, polysaccharides can offer smart properties through various mechanisms. In the following, we summarize the interaction mechanisms that can be employed to confer "smartness".

8.6.1 General Interactions: Solvent and Electrostatic

It is well known that polymer solutions and gels are dramatically affected by the solvent and salt, and these effects can confer smart properties (see Chapters 2 and 3). The first entry in Table 8.2 indicates that polysaccharides also undergo significant interactions with the solvent which in most cases is an aqueous solution. If the polymer exists as individual chains, then favorable solvent–polymer interactions confer solubility. If the polymer is crosslinked into a network, then favorable polymer–solvent interactions result in a swelling of the network while the crosslinks constrain swelling and prevent

Table 8.2 Interaction mechanisms that allow a stimuli to be received and a response to be generated.

Interaction	Inducing stimuli	Response
General solvent effects	Chemical/solvent	Swell–collapse (crosslinked network) Dissolve–precipitate (non-crosslinked network)
General electrostatic effects	Salt, ions, other polyelectrolytes	Swell–collapse (salt) Dissolve–precipitate (salting out) Polyelectrolyte complexation (*e.g.*, layer-by-layer self-assembly)
Acid–base (pH-dependent electrostatic)	pH	Bind–release Swell–collapse Dissolve–precipitate Dissolve–gel
Specific ion/metal	Ion, metal	Gelation (*e.g.*, Ca^{2+} with pectin and alginate) Chelation (*e.g.*, chitosan)
Supramolecular structure (*e.g.*, helix formation)	*More investigation is required before generalizations are possible*	
Bio-specific (protein based; binding and catalysis)	Chemical and biological	*Future investigation will likely yield new interactions and applications*
Specialized chemical reactions (*e.g.*, Schiff-base and boronic acid complexes)	Chemical	Reversible complexation

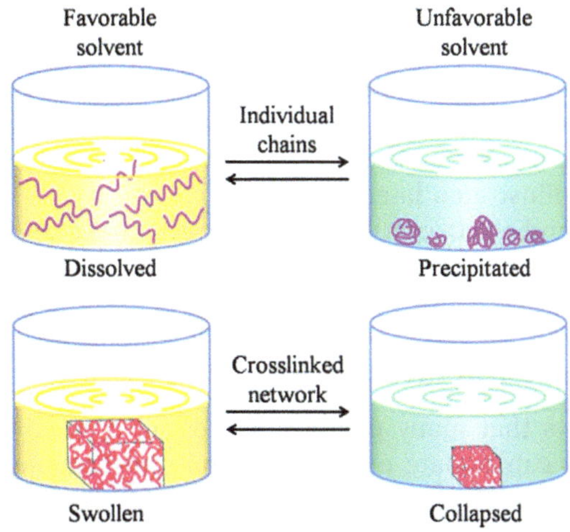

Figure 8.4 General solvent–polymer interactions can mediate polymer solubility–precipitation or the swelling–collapse of a polymer network. Favorable solvent–polymer interactions (*e.g.*, of a good solvent) promote polymer solubility and gel swelling. For charged polysaccharides (*i.e.*, polyelectrolytes), salts that screen electrostatic repulsions can promote precipitation (*e.g.*, salting out) and gel collapse.

the individual chains from dissolving.[29] Dissolution and swelling are illustrated in Figure 8.4 which also shows the reverse process—unfavorable solvent–polymer interactions can lead to polymer precipitation or network collapse. With respect to smart properties, crosslinked polysaccharide networks can often be induced to collapse or swell in response to changes in the solvent and these responses lead to changes in the network's permeability and size which can be used to control the release of the network's contents (*e.g.*, a drug) or to exert a force (*e.g.*, for chemomechanical transduction). As mentioned in the discussion of Figure 8.3b, solvent–polymer interactions may also promote the formation of a localized structure (*e.g.*, triple helices) for polysaccharides.

In some cases, salt can have generalized effects on a polysaccharide network by screening electrostatic interactions. As illustrated in Scheme 8.1, many polysaccharides have negatively charged residues—typically carboxylates and sulfates. The low pK_a of the sulfate residues makes sulfate-containing polysaccharides strong polyelectrolytes that possess a negative charge under most pH conditions. Repulsions of these "fixed" charges on the polymer backbone along with favorable solvent (water) interactions can confer solubility to such polysaccharides. For strong polyelectrolytes that are crosslinked, the electrostatic repulsions along with the osmotic pressure associated with the counterions can provide a considerable driving force for water uptake and swelling. Because added salt can screen electrostatic repulsions, polyelectrolytes can respond to salt additions by precipitating (*i.e.*, salting out) or collapsing as illustrated in Figure 8.4.[29]

Electrostatic interactions between oppositely charged polymers are often used to form polyelectrolyte complexes and to generate multilayers through sequential layer-by-layer (LbL) self-assembly. Polysaccharides are commonly used to create such complexes and multilayers. One unique feature that polysaccharides can offer is their ability to be degraded by hydrolytic enzymes as will be illustrated later.

These solvent and electrostatic effects are rather general characteristics of polymers[30] and are not specific to polysaccharides.

8.6.2 Acid–Base Interactions for Weak Polyelectrolytes

Scheme 8.1 shows that many polysaccharides have weakly acidic or basic substituents that can donate or accept protons to become charged or uncharged under different pH ranges. Typical weakly acidic substituents are carboxylates as illustrated by alginic acid in Scheme 8.2 and typical weakly basic substituents are primary amines as illustrated in Figure 8.1a for chitosan. Importantly, the physicochemical properties of an acidic or basic (macro)molecule are highly sensitive to its charge state and therefore to the pH conditions. Thus, the properties of a weak polyelectrolyte change dramatically when the pH is changed in a very narrow window around the polymer's pK_a. For instance, chitosan is soluble at pHs below its pK_a but undergoes a sol–gel transition in a narrow region near its pK_a. Four points to keep in mind when considering weak polyelectrolytes: first, protonation–deprotonation reactions are generally considered to be rapid and in local equilibrium, however, the local equilibrium states may vary with position if pH gradients exist (pH gradients are integral to an electrodeposition mechanism that will be discussed later). Second, polysaccharides with charged substituents generally have either anionic or cationic substituents but not both; this is in contrast to proteins that have both anionic and cationic residues and are characterized by an isoelectric point. Third, unlike small molecule acids and bases, it is not accurate to report a single pK_a value for a polyelectrolyte since the ability of one residue to gain or lose a proton depends on the charge on neighboring residues and electrostatic screening

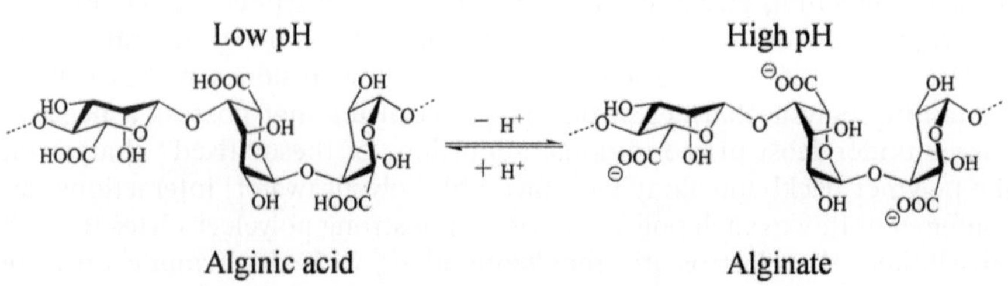

Scheme 8.2 Weakly acidic polysaccharides (*e.g.*, alginic acid) undergo dramatic changes in properties when the pH is changed near the pK_a. Similarly, weakly basic polysaccharides (*e.g.*, chitosan; Figure 8.1a) undergo dramatic changes in properties when the pH is changed near the pK_a.

(*e.g.*, by salt) of the fixed charges. Nevertheless, polyelectrolytes are commonly described by a pK_a, which serves as an operational measure for its ability to undergo protonation–deprotonation reactions. Finally, this generalization of weak polyelectrolytes considers only electrostatic interactions and may be oversimplified as it neglects chemical details (*e.g.*, of the counterion's structure) that can be subtle but have important consequences[31] (see Chapter 3).

Weak polyelectrolytes are commonly used to confer smart properties because solubility or swelling can be dramatically changed by changes in pH. Using the terminology above, information of the pH is "received" by the rapid and reversible protonation–deprotonation reaction, which "changes" the polymer's charge state and balance of attractive and repulsive interactions, and this change results in a macroscopically-observable "response" (*e.g.*, precipitation or gel collapse). Further discussion of pH-responsive smart properties of chitosan and alginate are provided later.

8.6.3 Specific Ion Interactions

Figure 8.1b shows that alginate can undergo specific interactions with multivalent cations (*e.g.*, Ca^{2+}) to generate the "eggbox" crosslinks responsible for gel formation (we further discuss alginate's metal ion induced smart properties later). In addition to alginate, other charged polysaccharides can undergo gelation in response to multivalent ions although the structural details of the interactions are less well characterized. For instance, pectin undergoes Ca^{2+}-ion induced gelation[32] and chitosan forms complexes with tripolyphosphate (TPP).[33] Chitosan can also chelate metals and these interactions have been used to remove metals from solutions,[34] bind metals as catalytic active sites[35] and anchor metals (*e.g.*, Ni^{2+}) to mediate the further assembly of proteins to chitosan (through the protein's histidine amino acid residues).[36]

8.6.4 Specific Supramolecular Interactions

In the previous discussion, we illustrated that polysaccharides can possess considerable supramolecular structure. Cellulose and chitin are highly ordered in nature (*e.g.*, Figure 8.3a), alginate can form macroscopic three-dimensional networks (*e.g.*, Figure 8.1b) and lentinan can form localized structural elements (*e.g.*, helices) in aqueous solutions (*e.g.*, Figure 8.3b). Our current understanding of such polysaccharide structures is limited (compared to our understanding of protein secondary and tertiary structures).[37] However, we anticipate that further study of the mechanisms and driving forces will reveal more examples of how supramolecular and hierarchical structure-formation "receive" stimuli that induce macromolecular "changes" to generate a macroscopic "response": thus, we anticipate that greater knowledge will reveal new ways that polysaccharides can perform smart functions.

8.6.5 Bio-specific Interactions

As discussed, polysaccharides can be acted upon by enzymes, and bound by receptors and lectins. As illustrated in Figure 8.2, such bio-specific interactions can confer smartness to polysaccharide-based systems. In the future, we anticipate further discoveries of catalytic/binding interactions (*i.e.*, the identification of new enzymes, receptors and lectins) and the development of new applications for such interactions. Such a prediction seems safe given the emergence of glycobiology which is providing the motivation for a broader study of carbohydrates and glycoconjugates due to their importance in various biological processes.

8.6.6 Specialized Chemical Reactions

A variety of somewhat standard chemical reaction methods have been applied to generate polysaccharide derivatives;[3] here we note two methods that are somewhat unique to polysaccharides. First, some polysaccharides with adjacent hydroxyl groups can be partially oxidized to convert these vicinal alcohols into dialdehydes. This partial oxidation is illustrated in Scheme 8.3a for the periodate oxidation of the C2 and C3 hydroxyls of the anhydroglucose unit of cellulose with cleavage of the C–C bonds and formation of the dialdehyde. The resulting aldehydes can readily react with amines to form Schiff bases. This partial oxidation and Schiff base formation provides a facile route to conjugate various substituents (including proteins) to polysaccharides.[38,39] Second, the diols of sugars can also form covalent complexes with phenylboronic acid as illustrated in Scheme 8.3b.[40] Networks containing phenylboronic acid moieties have been extensively investigated as sugar sensing smart systems as reviewed elsewhere.[41] In principle, both Schiff base and phenylboronic acid complexes are reversible and provide interesting opportunities for smart materials.

8.7 Example: Chitosan

As indicated in Figure 8.1a, chitosan is an aminopolysaccharide that possesses a unique combination of properties that have attracted considerable attention for a variety of applications. Since recent reviews consider the potential of chitosan as a biomedical material,[8,42–44] we highlight different features of chitosan's smart properties.

8.7.1 Functional Food

Potentially one of the most important but more controversial applications of chitosan's stimuli-responsive properties concerns its use as a dietary supplement to bind fats and cholesterol, thereby promoting weight loss and lowering cholesterol.[45,46] Chitosan can interact with membrane lipids through electrostatic and non-electrostatic interactions[47–49]

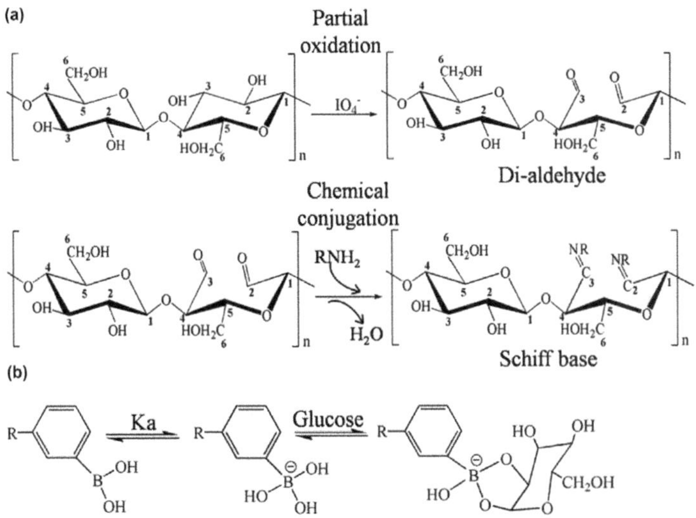

Scheme 8.3 Specialized chemical reactions common to polysaccharides. (a) The partial oxidation of polysaccharides can yield dialdehyde substituents that can further undergo Schiff base formation. (b) Reversible binding with boronic acids may enable the recognition of sugars as well as confer smart properties.

(*e.g.*, hydrophobic interactions and hydrogen bonding[50]), and chitosan's digestion in the intestinal tract appears to be limited. Based on the pH-responsive properties illustrated in Figure 8.1a, orally administered chitosan would be expected to dissolve or swell in the stomach as a result of the stomach's low pH. As chitosan transits through the intestine it would experience an increasing pH. At intermediate pHs, anionic fats would be expected to bind to cationic chitosan through electrostatic interactions. Non-electrostatic interactions have also been reported to be responsible for the binding of other non-ionic components (*e.g.*, cholesterol).[50] These binding interactions between chitosan and fats are not confined to individual fat molecules but may involve various structures (*e.g.*, micelles or vesicles)[47,48,50] to generate a diverse array of fat–chitosan complexes. As these fat–chitosan complexes pass through the digestive tract, the increasing pH should provide a tendency for chitosan to undergo gelation or aggregation which would serve to solidify these fat–chitosan complexes for subsequent excretion in feces.[51] While such physicochemical explanations are appealing, it is important to note that additional biological mechanisms may contribute to observed responses in such a complex system.[52] To date, there is not a general agreement on the efficacy of chitosan for weight loss or lowering cholesterol. However, it is important to note that while there may be no agreement on efficacy, there appear to be minimal concerns—we are unaware of reports of significant adverse side effects or drug interactions associated with the ingestion of chitosan.[46]

8.7.2 Biosensor and Microsystems Interface

In 2002, it was reported that chitosan could be electrodeposited at a cathode interface[53,54] and this has generated considerable interest as a means: to functionalize electrodes for biosensing applications;[55] to assemble biological components within microsystems;[56] to create biocompatible coatings;[57] and to generate matrices for controlled drug delivery.[58] Figure 8.5 illustrates the neutralization mechanism for the cathodic deposition of chitosan. Typically, a deposition solution is prepared at a low pH (<6) such that chitosan's amines are protonated and chitosan is soluble as illustrated in Figure 8.1a. Electrochemical reactions of water create the localized high pH at the cathode (>6.5) that deprotonates chitosan's amines, neutralizes the chains and induces the localized sol–gel transition to yield a hydrogel film at the electrode surface. While different labs use different deposition conditions, the chitosan film assembled on the electrode is stable after it is removed from the deposition bath: the film remains adhered to the electrode surface without the need for an applied potential although it can be re-dissolved by immersion in mild acid. Initial studies demonstrated that chitosan deposition is spatially selective in the lateral dimension (to the 1–10 μm)[59] while the thickness of the deposited film can be controlled by conditions.[60]

Interestingly, chitosan's electrodeposition can be considered to be a self-assembly process involving strong non-covalent inter-chain associations. As

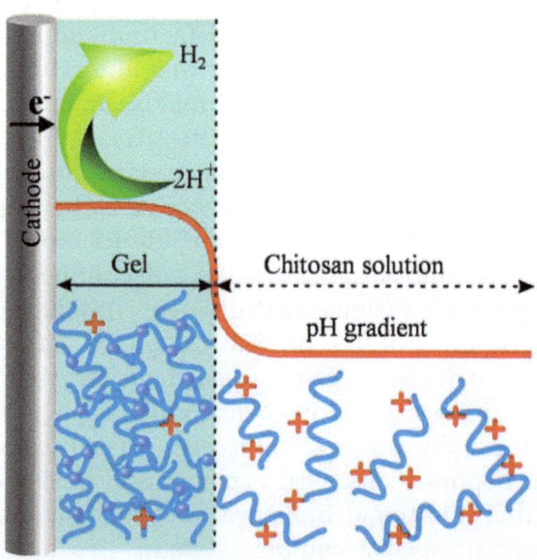

Figure 8.5 The neutralization mechanism for chitosan's cathodic electrodeposition. Electrolysis at the cathode leads to a net consumption of protons and a localized region of high pH adjacent to the cathode. This imposed stimulus is "received" by deprotonation of chitosan's amines which "changes" the balance of attractions and repulsions and induces a localized sol–gel transition (a macroscopic "response").

a result, the structure and properties of the deposited film are highly sensitive to the deposition conditions; once deposited, the films do not readily "anneal" to an equilibrium structure (although they can be re-dissolved in acid to erase the structure). The potential importance of this concept is illustrated by two recent studies. First, chitosan was electrodeposited from solutions containing different levels of salt and the results in Figure 8.6 show that the strength (*i.e.*, the elastic modulus) of the resulting wet films varied by several orders of magnitude depending on the salt concentration in the deposition solution.[61] In a second study, chitosan was deposited onto a wire using on–off pulses of current. Figure 8.7 shows that a complex multilayer structure was generated with layers growing during the on-pulse and interfaces generated during the off-pulse. These results demonstrate that complex structures and tailored properties can be generated by controlling the deposition conditions.[62] Remarkably, there are few fundamental theories currently available to explain these observations and to guide subsequent experimentation. Thus, while the results in Figures 8.6 and 8.7 demonstrate the broad design space available to control the deposited film's structure and properties, they also raise concerns of repeatability: small, seemingly insignificant changes in deposition conditions can yield large variations in the resulting structure and properties.

Chitosan electrodeposition has attracted much attention because a variety of mechanisms are available to confer functional properties to the electrodeposited films.[55] For instance, various components (*e.g.*, macromolecules or nanoparticles) can be dissolved or blended into the deposition solution, co-deposited with chitosan, and entrapped within the deposited films. Further, chitosan's amines are nucleophilic under neutral conditions and a variety of chemical, electrochemical[63] and biochemical[64] methods have been

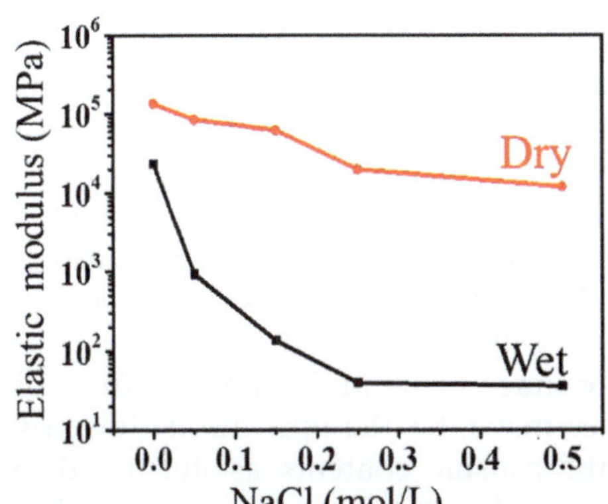

Figure 8.6 Chitosan's electrodeposition is a form of self-assembly involving strong non-covalent bonds. As a result, the properties of the deposited films (as measured by the elastic modulus) are sensitive to conditions experienced during the deposition process.
Adapted with permission.[61]

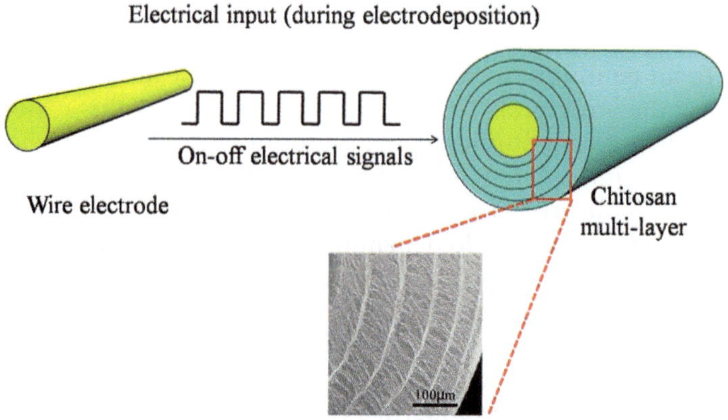

Figure 8.7 The sequence of electrical stimuli used to induce chitosan's cathodic electrodeposition can control the ultimate structure of the deposited film.
Adapted with permission.[62]

used to covalently attach nucleic acids, proteins and small molecules to confer functional properties to the films.

8.7.3 Capsules

Capsules are structures in which a membrane encloses (*i.e.*, encapsulates) the contents of an internal compartment. Capsules mimic the structure of cells which use lipid bilayer membranes to separate the extracellular and intracellular (*i.e.*, cytoplasmic) space, and to separate sub-cellular organelles to compartmentalize functions (*e.g.*, the mitochondria harvest energy). *In vitro*, capsules with polymer-based membranes can be readily generated to provide a versatile means of compartmentalization. Importantly, lipid membranes and polymer membranes are not equivalent but have significant differences. Lipid bilayer membranes tend to serve as better barriers for transport and thus sub-cellular organelles can exist at different pHs and redox potentials. Polymer membranes tend to be thicker (μm *vs.* nm) and mechanically more robust, but their selective permeability occurs at larger size cut-offs.[65] We illustrate the use of capsules to compartmentalize structure and function using two recent examples.

One potential advantage of building capsules from biologically derived polymers is that they may offer the opportunity for enzymatic cleavage to trigger release of the capsule contents as illustrated by the example in Figure 8.8. In this example, capsules were generated by adding a solution containing the gellan anionic polysaccharide into a solution containing chitosan dropwise. Complexation between the oppositely charged polysaccharides forms the capsule membrane (*i.e.*, interfacial shell). We should note that chitosan is a common component of polyelectrolyte complexes

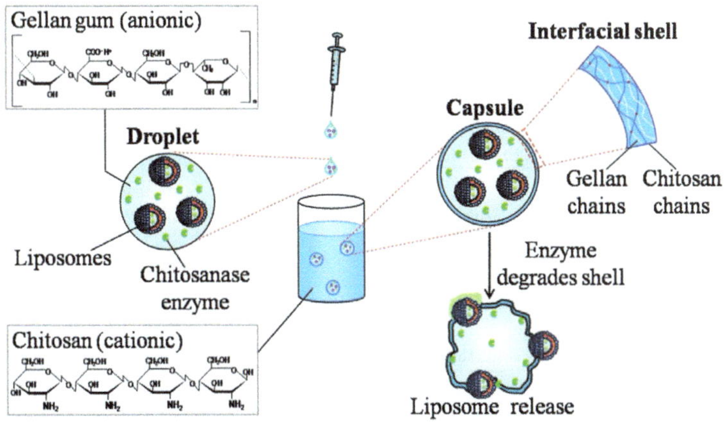

Figure 8.8 Polymer membrane capsules can be formed by complexation between cationic and anionic polysaccharides. Incorporation of a membrane-hydrolyzing enzyme serves as an "internal clock" that triggers self-destruction of the capsule and release of the encapsulated "cargo". Adapted with permission from Raghavan and co-workers.[66]

because it is one of the few positively charged polymers of natural origin. In the example in Figure 8.8, lipid vesicles (*i.e.*, liposomes) and a chitosan-hydrolyzing enzyme (chitosanase) were incorporated within a capsule by adding them to the gellan solution prior to adding the drops. The capsules could be freeze-dried and stored, but upon reconstitution with water, the enzyme became activated to begin its destruction of the capsule membrane. The time required to disrupt the capsule depended on how much enzyme was included, thus the enzyme level serves as an "internal clock" for release of the liposome cargo.[66]

In the second example, capsules were generated by dropping a chitosan solution into an alginate solution to form the alginate–chitosan polyelectrolyte membrane as illustrated in Figure 8.9a. Because of the semi-permeable properties of this membrane, chitosan chains from the initial solution were encapsulated within the capsule contents. Importantly, these encapsulated chitosan chains retain their pH-responsive self-assembling properties, while changes in pH in the external solution can be readily "transmitted" across the alginate–chitosan membrane. Thus it is possible to use changes in the external solution to induce a spatially-localized self-assembly of the encapsulated chitosan chains. Figure 8.9b illustrates an experiment in which the capsule was intermittently (in 2 minute intervals) transferred between a basic solution and water to generate a chitosan multilayer reminiscent of that shown in Figure 8.7. Figure 8.9c shows that these capsule multilayers could be used as a template for the generation of spherical iron oxide particles with complex internal multilayer structures.[67]

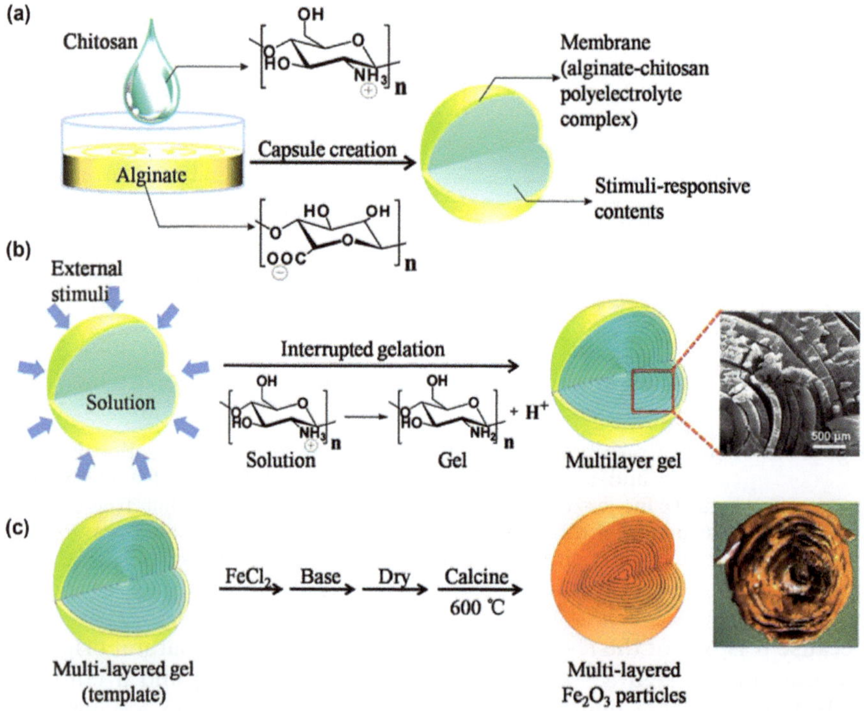

Figure 8.9 Encapsulation of the stimuli-responsive polymer allows external cues to trigger the localized responses within the capsule compartment. (a) Capsules are formed with stimuli-responsive contents. (b) Intermittent, externally imposed cues trigger multilayer gel formation. (c) Multilayer gel capsules serve as templates for the formation of iron oxide particles with complex internal structures. Reproduced with permission.[67]

8.7.4 Enzymes to Confer Smartness

There is considerable interest in enlisting the molecular recognition properties of enzymes to confer smartness to materials.[12,68,69] As illustrated by the examples in Figure 8.2b and 8.8, most enzymes that react directly with biopolymers tend to have hydrolytic activities that cleave structure. Efforts to enlist enzymes to build structure often use indirect approaches. One indirect approach involves the use of tyrosinase or phenol oxidase enzymes. These enzymes react with a broad range of phenolics (either natural or synthetic) and generate reactive *o*-quinones as illustrated in Figure 8.10a. These *o*-quinones can diffuse from the enzyme's active site and undergo covalent grafting reactions with chitosan's primary amine through mechanisms that likely include Schiff base or Michael-type adducts. Importantly, these reactions change the macroscopic properties of the chitosan.

Figure 8.10b illustrates one application of the tyrosinase-mediated grafting: as a smart colorimetric paper platform for the detection of troublesome

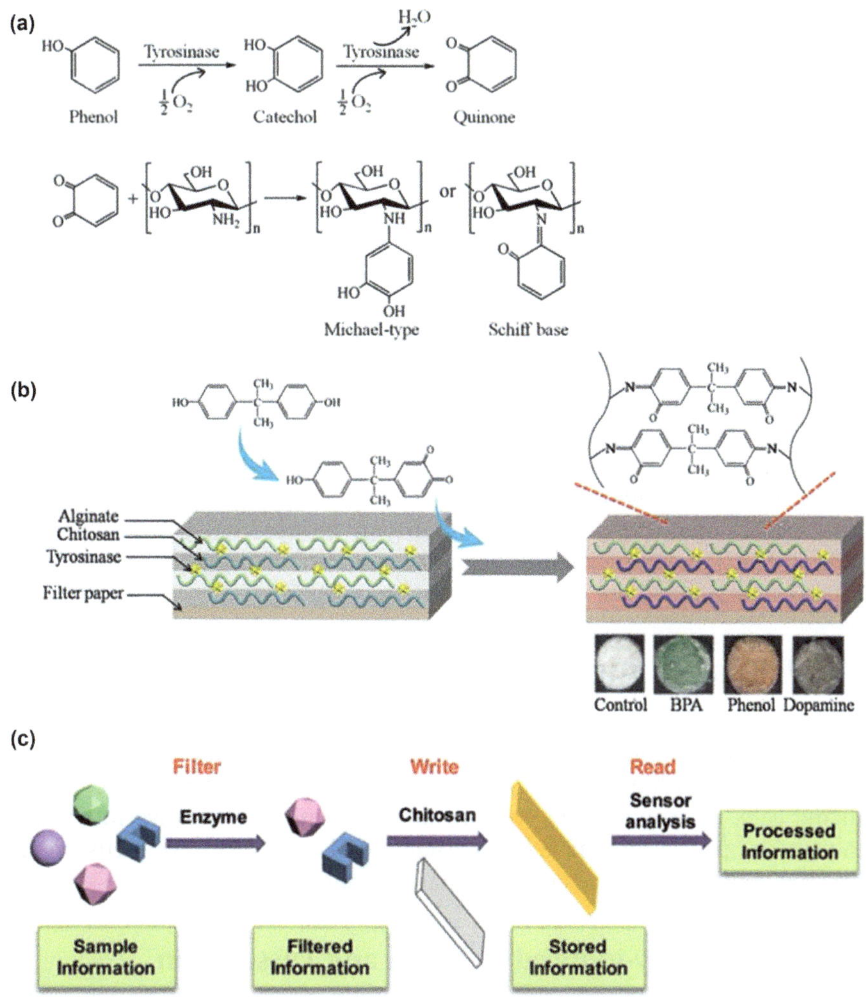

Figure 8.10 Enzymatic generation of an *o*-quinone signal is "received" through film grafting while "changes" in film properties are readily observable "responses". (a) Enzymatic *o*-quinone generation and the putative grafting reactions. (b) Chitosan-based "smart paper" can detect phenolic contaminants. Adapted with permission from Andreescu and co-workers[70] (BPA: bisphenol A). (c) Information processing analogy for enzymatic "writing".
Reproduced with permission.[71]

phenolic pollutants. In this example, tyrosinase was incorporated into an alginate–chitosan multilayer that was assembled onto filter paper. Reactions of the phenolic compounds resulted in discernable color changes in the paper. Visual and optical (*i.e.*, photographic) methods could then be used to detect and quantify the phenolic levels. This paper-based platform provides a simple and portable system that requires no additional reagents and should be useful for field applications. In addition, detection could be integrated with the imaging capabilities available with typical cell phones, which further enhances the convenience by limiting the need for dedicated or external hardware.[70]

Interestingly, phenolic compounds are common components of foods and represent one of the most abundant sources of antioxidants in the human diet. Thus, there has been considerable effort to detect and study dietary antioxidant phenols. Figure 8.10c illustrates an example in which thin chitosan films were prepared and placed in contact with solutions containing both tyrosinase and dietary antioxidants. Concepts from information processing were used: the enzyme serves as a "filter" to selectively identify (*i.e.*, oxidize) the phenolic antioxidants; the grafting reaction serves to "write" the *o*-quinone's "information" onto the initially transparent film; and this written information can be "read" using various sensor modalities. Specifically, the tyrosinase-mediated grafting to the chitosan film: (i) changes the film's color, which can be detected optically; (ii) changes the film's mass and mechanical properties, which can be detected by piezoelectric methods (*e.g.*, quartz crystal microbalance); and (iii) changes the film's redox properties, which can be detected electrochemically.[71]

8.8 Example: Alginate

Alginate, like chitosan, can form hydrogels under mild conditions and has attracted considerable attention for applications in medical and life sciences. Common examples include the use of alginate matrices for immobilizing cells for applications in biotechnology and regenerative medicine. Since these applications have been discussed elsewhere,[4,72] we focus on alternative applications that enlist alginate's smart properties.

8.8.1 Electrically-induced Gelation and Dissolution

As discussed earlier, chitosan undergoes cathodic electrodeposition by the neutralization mechanism shown in Figure 8.5. Shortly after this mechanism was reported, alginate was observed to electrodeposit at the anode by an analogous neutralization mechanism as illustrated in Figure 8.11a.[73] In this case, anodic electrolysis reactions generate the low pH that can locally neutralize alginate's carboxylate groups to induce its gelation as the neutral alginic acid polymer ($pK_a \approx 3.6$). Because various components can be co-deposited with alginic acid (*e.g.*, hydroxyapatite), this electrodeposition method was reported to provide a room temperature method to generate a bioactive composite coating for biomedical applications.[74]

A second mechanism is shown in Figure 8.11b and results in the electrodeposition of Ca^{2+}-alginate hydrogels. In this case, the deposition solution contains both sodium alginate and insoluble $CaCO_3$. In this mechanism, the anodic electrolysis reactions locally generate protons that are consumed by the solubilization of $CaCO_3$. This solubilization is

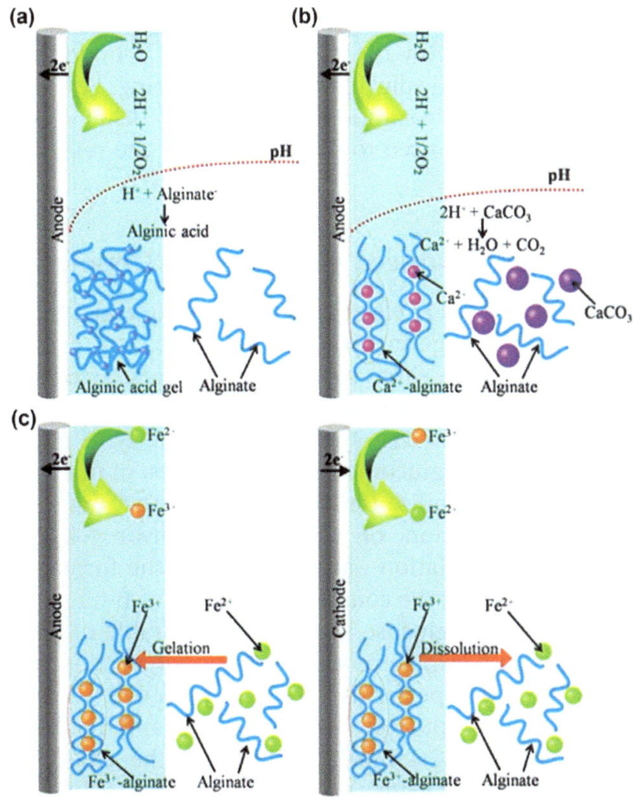

Figure 8.11 Three different mechanisms that enable alginate to "receive" an imposed electrical input, "change" the balance of interactions, and "respond" by undergoing gelation. (a) Anodically generated protons create the low pH conditions to neutralize alginate and trigger the formation of alginic acid hydrogels.[73] (b) Anodically generated protons react to solubilize $CaCO_3$, release Ca^{2+} and trigger the formation of Ca^{2+}-alginate hydrogels.[75] (c) Anodic oxidation of Fe^{2+} triggers the formation of Fe^{3+}-alginate hydrogels while cathodic reduction of Fe^{3+} to Fe^{2+} triggers dissolution of the Fe^{3+}-alginate hydrogels.[76]

accompanied by the release of Ca^{2+}, which induces the localized formation of the eggbox network junctions of Figure 8.1b that serve as the physical crosslinks for the Ca^{2+}-alginate hydrogel. Again, this Ca^{2+}-alginate deposition mechanism occurs under mild conditions thus allowing the co-deposition of viable populations of bacteria at specific electrode "addresses".[75] For instance, this Ca^{2+}-alginate deposition mechanism has been used to assemble a model biofilm composed of a complex bacterial population. This biofilm could be electrodeposited within an experimental microfluidic system that allows observation of the spatiotemporal dynamics of the interacting microorganisms.[77]

The third mechanism that allows alginate gels to be generated in response to electrical signals is shown in Figure 8.11c. In this electrodeposition mechanism, alginate's crosslinking is the result of Fe^{3+} ions that can be generated electrochemically by the oxidation of Fe^{2+}. Importantly, this reaction can be electrochemically reversed by reducing Fe^{3+} to Fe^{2+} and thereby initiating electro-dissolution of the gel.[76] This electrochemical deposition–dissolution was used to load and trigger the release of the antimicrobial enzyme lysozyme.[78]

8.8.2 Enzymatically-induced Gelation and Dissolution

As illustrated in Figure 8.11, electrical inputs induce alginate's response indirectly—through a change in pH, Ca^{2+} concentration, or $Fe^{2+/3+}$ redox state. In many cases, these same stimuli can be generated enzymatically. Figure 8.12a shows an enzyme-mediated alginate gelation mechanism analogous to that in Figure 8.11b. In this case, the enzyme glucose oxidase (GOx) converts glucose into gluconic acid, dissociation of this acidic product provides the proton to solubilize $CaCO_3$ and release Ca^{2+}, and the liberated Ca^{2+} induces the formation of Ca^{2+}-alginate hydrogel. The plot in Figure 8.12b shows that gelation—as measured by the formation of a self-supporting gel—occurs over the course of an hour depending on the glucose concentration (note: these glucose levels are comparable to levels found in blood and in common beverage products). Importantly, this example illustrates that the GOx enzyme performs a bio-specific conversion of a chemical input (glucose) into an input (proton generation and Ca^{2+} release) that is "received" by alginate's G blocks, "changes" the chain interactions/organization and ultimately leads to a macroscopic sol–gel "response".[79]

In a somewhat analogous manner, the dissolution mechanism in Figure 8.11c has also been extended from an electrically-imposed stimulus to an enzymatically-imposed stimulus as illustrated in Figure 8.13. In this case, dehydrogenase enzymes that generate NADH are used to provide the

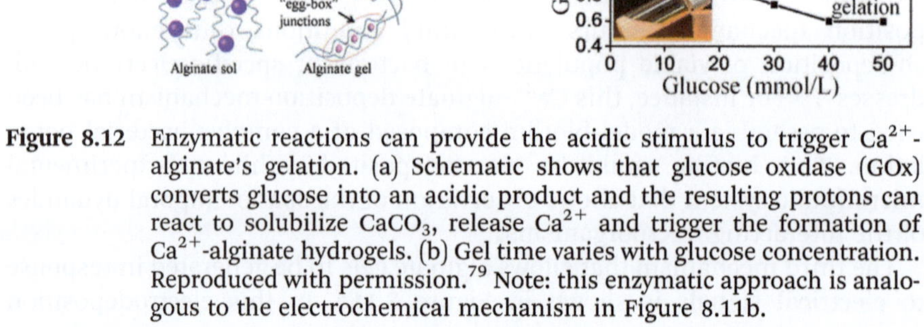

Figure 8.12 Enzymatic reactions can provide the acidic stimulus to trigger Ca^{2+}-alginate's gelation. (a) Schematic shows that glucose oxidase (GOx) converts glucose into an acidic product and the resulting protons can react to solubilize $CaCO_3$, release Ca^{2+} and trigger the formation of Ca^{2+}-alginate hydrogels. (b) Gel time varies with glucose concentration. Reproduced with permission.[79] Note: this enzymatic approach is analogous to the electrochemical mechanism in Figure 8.11b.

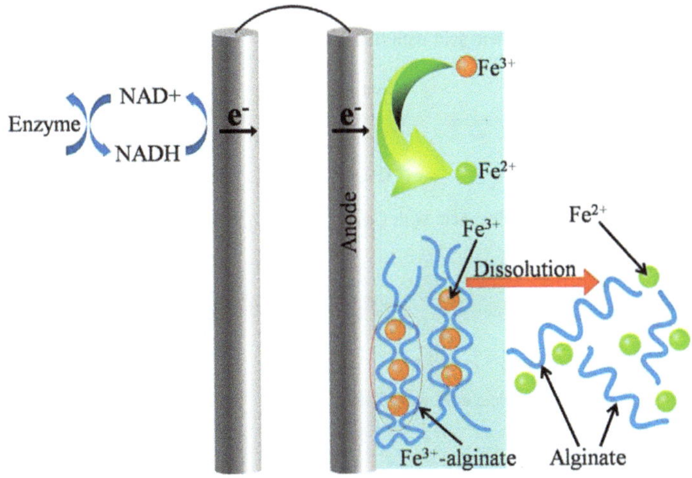

Figure 8.13 Enzymatic reactions can provide the redox stimulus to trigger dissolution of the Fe^{3+}-alginate hydrogel. Dehydrogenase enzymes generate NADH to provide the reducing equivalents that convert Fe^{3+} to Fe^{2+} and dissolve the Fe^{3+}-alginate hydrogel. Note: this enzymatic dissolution is analogous to the electrochemical dissolution mechanism in Figure 8.11c.
Adapted with permission from Katz and coworkers.[80]

reducing equivalents to convert Fe^{3+} to Fe^{2+} and thus dissolve the Fe^{3+}-alginate hydrogel. In this example, the enzyme converts a chemical signal (*i.e.*, the reactant), into the NADH input that is transmitted through a wire and "received" by the $Fe^{3+/2+}$ redox-reaction that "changes" interactions of the Fe^{3+}-alginate hydrogel and leads to the macroscopic "response" (dissolution). In this example, the sequence of information-processing events was further extended since the Fe^{3+}-alginate gel contained a conjugate of antibody and reporter. Enzyme-mediated dissolution of the gel released the conjugate enabling it to begin "searching" for its antigen. In essence, coupling enzymes with stimuli-responsive materials serves as a logic gate to process chemical and biochemical information.[80]

8.8.3 Microfluidic-induced Gelation

Recently, microfabrication technologies have been developed to make microfluidic devices that can controllably process small volumes of liquids. Several groups have enlisted such microfluidic devices to create alginate drops with the expected advantages that monodisperse drops can be generated with controlled sizes and properties, and microfluidic production should be scalable due to its parallel nature (production scales with the number of channels). Figure 8.14a illustrates that microfluidic contacting is used to generate alginate-containing drops dispersed within an insoluble continuous phase (*e.g.*, soybean oil).[81] In the top schematic, if all three inlet

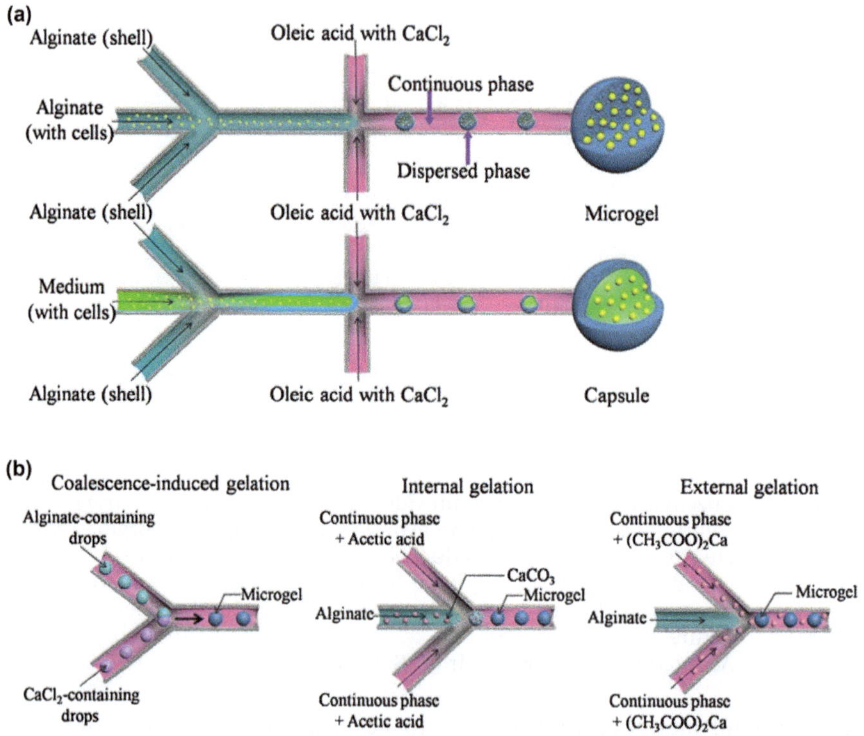

Figure 8.14 Microfluidic gelation of alginate provides a means to create microgels or capsules with controlled sizes and properties. (a) A typical approach employs a microfluidic system to controllably generate droplets (*e.g.*, of an alginate-containing aqueous phase) in an insoluble continuous phase (*e.g.*, soybean oil). Adapted with permission from Seiffert.[81] (b) Alginate's gelatin can be induced by: (i) coalescence (*e.g.*, of an aqueous alginate drop and an aqueous $CaCl_2$ drop), (ii) internal gelation (*e.g.*, by diffusion of acetic acid from the continuous phase into an alginate drop containing $CaCO_3$), or (iii) external gelation where Ca^{2+} (*e.g.*, from $(CH_3COO)_2Ca$) directly diffuses from the continuous phase into the alginate-containing drop.
Adapted with permission from Tumarkin and Kumacheva.[82]

channels on the left have alginate-containing solution, then subsequent gelation can generate microgel particles. The bottom schematic in Figure 8.14a illustrates that if the central channel on the left contains a separate solution (*e.g.*, cell-containing cultivation medium), then subsequent gelation in this case leads to a core–shell capsule structure.[81]

The gelation step can be performed using three general approaches as illustrated in Figure 8.14b.[82] *Coalescence-induced gelation* can occur if two separate aqueous phase drops are generated, one containing alginate and the other containing the stimulus for alginate's gelation (*e.g.*, $CaCl_2$). Coalescence of these drops initiates the formation of the Ca^{2+}-alginate gels. *Internal gelation* is typically performed by incorporating insoluble $CaCO_3$ in the alginate-containing solution and incorporating an acid (*e.g.*, acetic acid) that is miscible in both the aqueous dispersed phase and the non-aqueous continuous phase. Diffusion of the acid into the alginate-containing drop initiates the solubilization of $CaCO_3$ and release of Ca^{2+} that induces Ca^{2+}-alginate gelation. Finally, *external gelation* can be performed by adding the calcium to the continuous phase (*e.g.*, as $(CH_3COO)_2Ca$). The generation of Ca^{2+} (*e.g.*, by dissociation of $(CH_3COO)_2Ca$) at the oil–water interface releases Ca^{2+} and induces Ca^{2+}-alginate gelation.[82]

There has been considerable interest in enlisting microfluidics to create cell-containing alginate particles.[83,84] Here we consider a different application in which a microfluidically-generated Ca^{2+}-alginate microgel was prepared for the intestinal release of an oral formulation (the microgels contained a vitamin). The microgels remained intact under the low pH conditions of the stomach's gastric juice (possibly a transition from Ca^{2+}-alginate to alginic acid gels occurred). However, at higher pHs of the intestinal fluid, the alginate coating was observed to degrade.[85]

8.8.4 Controlling Structure of Ca^{2+}-alginate Gels by External Gelation

In preceding paragraphs, we indicated that Ca^{2+}-alginate's gelation can be initiated internally or externally, however, recent studies have demonstrated that the micro and macrostructures of these gels are not equivalent. When Ca^{2+}-alginate gels are formed by internal gelation, the sol–gel transition can sometimes occur homogeneously throughout the solution and the resulting gel can possess an isotropic structure. In contrast, the externally gelled samples rely on diffusion of Ca^{2+} ions and gelation is spatially localized to a gelation "front" that is normal to the interfacial surface.[86] As a result, Ca^{2+}-alginate gels formed by external gelation can have considerable anisotropic structure at both the micro[87] and macro scales[88,89] as illustrated in Figure 8.15. In the first case, external gelation was performed by adding alginate to a dialysis tube which was then immersed in a Ca^{2+}-containing solution. The outer regions of these Ca^{2+}-alginate gels were enriched in polymer (relative to gels in the center of the tube), and as illustrated in Figure 8.15a, the Ca^{2+}-alginate fibrils were aligned to be normal to the direction of Ca^{2+} diffusion.[87] Using a different experimental approach with a flat surface, it has been observed that gels can be formed that have pores oriented normal to the surface. This is illustrated in Figure 8.15b, which shows that parallel aligned capillary pores (white circles) are formed in the externally gelled alginate (alginate was stained to facilitate visualization). Importantly, this difference in structure between internally gelled (isotropic) and externally gelled Ca^{2+}-alginates was also accompanied by differences in properties (mechanical and mass transport).[89]

As discussed in these examples, anisotropic and "capillary" gels could have considerable technological significance (*e.g.*, to control the properties and porosity of tissue engineering scaffolds). Hopefully, observations such

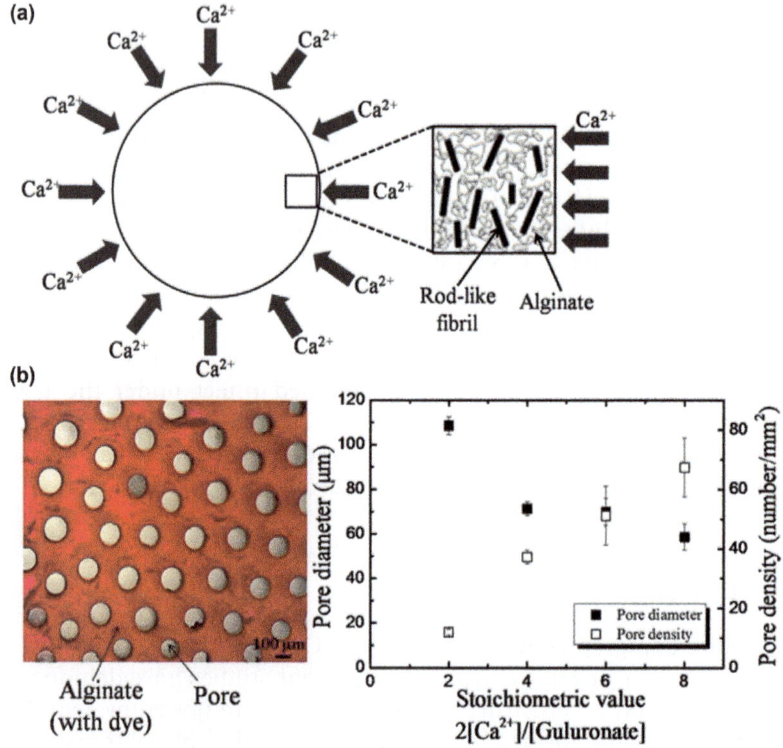

Figure 8.15 The formation of Ca^{2+}-alginate gels by external gelation can yield anisotropic structures and properties. (a) Diffusion of Ca^{2+} through a dialysis tube into an alginate-containing solution leads to gelation with the resulting fibrils having a preferred orientation. Adapted with permission from Maki *et al.*[87] (b) Control of external gelation from an upper surface can yield capillary pores (alginate gels were stained red to facilitate visualization while the pores appear white in the image). Adapted with permission from Strom and co-workers.[89]

as those in Figure 8.15 will induce more groups to investigate such gel-forming systems, which will both enhance our fundamental understanding of the underlying phenomena and accelerate the application of these phenomena.

8.9 Conclusions and Future Perspectives

During the last century there were large advances in understanding how polymers could be synthesized with controlled structures, properties and functions, and similarly large advances in characterizing the structure–property–function relationships for proteins and nucleic acids. During this time, there was comparatively little investment in the study of polysaccharides. Yet polysaccharides persist as important materials because they possess a unique combination of properties. Typically, they are biocompatible making them suitable for food, medical, and consumer applications. They are derived from renewable resources and are biodegradable, and thus may become important components of a sustainable materials portfolio. And they offer a unique combination of properties that are often especially relevant to the aqueous environments prevalent to many of the emerging materials applications in medical and life sciences. In this chapter we highlighted three major mechanisms by which polysaccharides confer smartness: stimuli-responsive self-assembly, biospecific molecular recognition, and nanoscale and hierarchical supramolecular assembly.

One of the most obvious smart characteristics of many polysaccharides is their ability to undergo reversible intra/inter-polymer associations in response to modest changes in conditions (temperature, pH and ions). These associations confer structure over a hierarchy of length scales from localized order (*e.g.*, triple helices) to macroscopic gels. Generally, these reversible intra/inter-polymer associations are poorly understood and the mechanistic details of how these associations form are often lacking, yet there is growing evidence that these details are important. Specifically, details of how the structure-inducing cues are imposed can dramatically affect the gel's structure and properties as evidenced by the templates in Figure 8.3b, the multilayers in Figure 8.7, and the "capillary" gels in Figure 8.15b. Currently, there are few theories that offer predictive abilities either for extrapolation to new experimental conditions or for control in manufacturing. This lack of understanding of the underlying phenomena makes it difficult to fully apply these remarkable observations yet it provides the motivation for further investigation. However, further investigation of these associations will likely require new tools and methods. In the future, we anticipate that the emergence of new theoretical and experimental tools will provide a greater understanding of polysaccharides that will enable their unique capabilities to be more broadly applied to materials science.

Acknowledgements

The authors gratefully acknowledge financial support from the Robert W. Deutsch Foundation, Defense Threat Reduction Agency (HDTRA1-13-0037). Also, we gratefully acknowledge financial support from the National Natural Science Foundation of China (Grants 21007049, 51373124, and 21334005), the High-end Foreign Experts Program (GDW20134200126), the National Basic Research Program of China (2012CB933600) and the 111 project (B14018).

References

1. G. F. Payne, E. Kim, Y. Cheng, H. C. Wu, R. Ghodssi, G. W. Rubloff, S. R. Raghavan, J. N. Culver and W. E. Bentley, *Soft Matter*, 2013, **9**, 6019–6032.

2. C. Alvarez-Lorenzo, B. Blanco-Fernandez, A. M. Puga and A. Concheiro, *Adv. Drug Delivery Rev.*, 2013, **65**, 1148–1171.
3. J. Thiele, Y. J. Ma, S. M. C. Bruekers, S. H. Ma and W. T. S. Huck, *Adv. Mater.*, 2014, **26**, 125–148.
4. K. Y. Lee and D. J. Mooney, *Prog. Polym. Sci.*, 2012, **37**, 106–126.
5. Z. Schnepp, *Angew. Chem., Int. Ed.*, 2013, **52**, 1096–1108.
6. V. Kumar and R. Fotedar, *Carbohydr. Polym.*, 2009, **78**, 813–819.
7. M. Rinaudo, *Prog. Polym. Sci.*, 2006, **31**, 603–632.
8. M. N. V. Ravi Kumar, *React. Funct. Polym.*, 2000, **46**, 1–27.
9. Y. Yamada, K. Hozumi, A. Aso, A. Hotta, K. Toma, F. Katagiri, Y. Kikkawa and M. Nomizu, *Biomaterials*, 2012, **33**, 4118–4125.
10. M. X. Zhang, G. Y. Qing and T. L. Sun, *Chem. Soc. Rev.*, 2012, **41**, 1972–1984.
11. M. Motornov, Y. Roiter, I. Tokarev and S. Minko, *Prog. Polym. Sci.*, 2010, **35**, 174–211.
12. D. Roy, J. N. Cambre and B. S. Sumerlin, *Prog. Polym. Sci.*, 2010, **35**, 278–301.
13. P. Sikorski, F. Mo, G. Skjak-Brk and B. T. Stokke, *Biomacromolecules*, 2007, **8**, 2098–2103.
14. B. Akkaya, H. Yavuz, F. Candan and A. Denizli, *J. Appl. Polym. Sci.*, 2012, **125**, 1867–1874.
15. Y. Zhu, W. J. Tong and C. Y. Gao, *Soft Matter*, 2011, 7, 5805–5815.
16. K. Sato, Y. Imoto, J. Sugama, S. Seki, H. Inoue, T. Odagiri, T. Hoshi and J. Anzai, *Langmuir*, 2005, **21**, 797–799.
17. T. Y. Jiang, Z. H. Zhang, Y. L. Zhang, H. X. Lv, J. P. Zhou, C. C. Li, L. L. Hou and Q. Zhang, *Biomaterials*, 2012, **33**, 9246–9258.
18. K. S. Girish, K. Kemparaju, S. Nagaraju and B. S. Vishwanath, *Curr. Med. Chem.*, 2009, **16**, 2261–2288.
19. A. El Kaoutari, F. Armougom, J. I. Gordon, D. Raoult and B. Henrissat, *Nat. Rev. Microbiol.*, 2013, **11**, 497–504.
20. X. He, M. L. Marco and C. M. Slupsky, *J. Agric. Food Chem.*, 2013, **61**, 9559–9574.
21. P. Tingaut, T. Zimmermann and G. Sebe, *J. Mater. Chem.*, 2012, **22**, 20105–20111.
22. E. Lam, K. B. Male, J. H. Chong, A. C. W. Leung and J. H. T. Luong, *Trends Biotechnol.*, 2012, **30**, 283–290.
23. M. Numata and S. Shinkai, *Chem. Commun.*, 2011, **47**, 1961–1975.
24. X. W. Jia, X. J. Xu and L. N. Zhang, *Biomacromolecules*, 2013, **14**, 1787–1794.
25. H. J. Prado and M. C. Matulewicz, *Eur. Polym. J.*, 2014, **52**, 53–75.
26. Y. Habibi, *Chem. Soc. Rev.*, 2014, **43**, 1519–1542.
27. X. Y. Qiu and S. W. Hu, *Materials*, 2013, **6**, 738–781.
28. M. Prabaharan and J. F. Mano, *Macromol. Biosci.*, 2006, **6**, 991–1008.
29. I. Tokarev and S. Minko, *Soft Matter*, 2009, **5**, 511–524.
30. E. M. White, J. Yatvin, J. B. Grubbs, J. A. Bilbrey and J. Locklin, *J. Polym. Sci., Part B: Polym. Phys.*, 2013, **51**, 1084–1099.

31. K. Kato and H. J. Schneider, *Eur. J. Org. Chem.*, 2009, 1042–1047.
32. I. Ventura, J. Jammal and H. Bianco-Peled, *Carbohydr. Polym.*, 2013, **97**, 650–658.
33. E. N. Koukaras, S. A. Papadimitriou, D. N. Bikiaris and G. E. Froudakis, *RSC Adv.*, 2014, **4**, 12653–12661.
34. J. Q. Albarelli, M. T. Luna, R. S. Vieira and M. M. Beppu, *Adsorpt. Sci. Technol.*, 2012, **30**, 227–240.
35. C. Jouannin, C. Vincent, I. Dez, A. C. Gaumont, T. Vincent and E. Guibal, *J. Appl. Polym. Sci.*, 2013, **128**, 3122–3130.
36. X. W. Shi, H. C. Wu, Y. Liu, C. Y. Tsao, K. Wang, E. Kobatake, W. E. Bentley and G. F. Payne, *Biomacromolecules*, 2008, **9**, 1417–1423.
37. M. Sletmoen and B. T. Stokke, *Biopolymers*, 2013, **99**, 757–771.
38. R. Gong, J. Zhang, J. Zhu, J. Wang, Q. Lai and B. Jiang, *Korean J. Chem. Eng.*, 2013, **30**, 1620–1625.
39. M. Rinaudo, *Polymers*, 2010, **2**, 505–521.
40. T. Ye, X. M. Jiang, W. T. Xu, M. M. Zhou, Y. M. Hu and W. T. Wu, *Polym. Chem.*, 2014, **5**, 2352–2362.
41. Q. Wu, L. Wang, H. J. Yu, J. J. Wang and Z. F. Chen, *Chem. Rev.*, 2011, **111**, 7855–7875.
42. M. Dash, F. Chiellini, R. M. Ottenbrite and E. Chiellini, *Prog. Polym. Sci.*, 2011, **36**, 981–1014.
43. R. Jayakumar, M. Prabaharan, P. T. S. Kumar, S. V. Nair and H. Tamura, *Biotechnol. Adv.*, 2011, **29**, 322–337.
44. A. Domard, *Carbohydr. Polym.*, 2011, **84**, 696–703.
45. J. Santas, J. Espadaler, R. Mancebo and M. Rafecas, *Food Chem.*, 2012, **134**, 940–947.
46. K. M. Shields, N. Smock, C. E. McQueen and P. J. Bryant, *Am. J. Health-Syst. Pharm.*, 2003, **60**, 1310–1313.
47. B. Krajewska, P. Wydro and A. Kyziol, *Colloids Surf., A*, 2013, **434**, 349–358.
48. O. Mertins and R. Dimova, *Langmuir*, 2013, **29**, 14545–14551.
49. D. Raafat, K. von Bargen, A. Haas and H. G. Sahl, *Appl. Environ. Microbiol.*, 2008, **74**, 3764–3773.
50. B. Krajewska, A. Kyziol and P. Wydro, *Colloids Surf., A*, 2013, **434**, 359–364.
51. P. Wydro, B. Krajewska and K. Hac-Wydro, *Biomacromolecules*, 2007, **8**, 2611–2617.
52. A. M. Walsh, T. Sweeney, B. Bahar and J. V. O'Doherty, *PLoS One*, 2013, **8**, 7.
53. L. Q. Wu, A. P. Gadre, H. M. Yi, M. J. Kastantin, G. W. Rubloff, W. E. Bentley, G. F. Payne and R. Ghodssi, *Langmuir*, 2002, **18**, 8620–8625.
54. J. Redepenning, G. Venkataraman, J. Chen and N. Stafford, *J. Biomed. Mater. Res. Part A*, 2003, **66**, 411–416.
55. W. Suginta, P. Khunkaewla and A. Schulte, *Chem. Rev.*, 2013, **113**, 5458–5479.

56. S. T. Koev, P. H. Dykstra, X. Luo, G. W. Rubloff, W. E. Bentley, G. F. Payne and R. Ghodssi, *Lab Chip*, 2010, **10**, 3026–3042.
57. A. R. Boccaccini, S. Keim, R. Ma, Y. Li and I. Zhitomirsky, *J. R. Soc., Interface*, 2010, 7, S581–S613.
58. S. Fusco, G. Chatzipirpiridis, K. M. Sivaraman, O. Ergeneman, B. J. Nelson and S. Pané, *Adv. Healthcare Mater.*, 2013, **2**, 1037–1044.
59. L. Q. Wu, H. M. Yi, S. Li, G. W. Rubloff, W. E. Bentley, R. Ghodssi and G. F. Payne, *Langmuir*, 2003, **19**, 519–524.
60. Y. Cheng, X. L. Luo, J. Betz, S. Buckhout-White, O. Bekdash, G. F. Payne, W. E. Bentley and G. W. Rubloff, *Soft Matter*, 2010, **6**, 3177–3183.
61. Y. Liu, B. Zhang, K. M. Gray, Y. Cheng, E. Kim, G. W. Rubloff, W. E. Bentley, Q. Wang and G. F. Payne, *Soft Matter*, 2013, **9**, 2703–2710.
62. K. Yan, F. Y. Ding, W. E. Bentley, H. B. Deng, Y. M. Du, G. F. Payne and X. W. Shi, *Soft Matter*, 2014, **10**, 465–469.
63. E. Kim, Y. Liu, X.-W. Shi, X. Yang, W. E. Bentley and G. F. Payne, *Adv. Funct. Mater.*, 2010, **20**, 2683–2694.
64. T. H. Chen, D. A. Small, L. Q. Wu, G. W. Rubloff, R. Ghodssi, R. Vazquez-Duhalt, W. E. Bentley and G. F. Payne, *Langmuir*, 2003, **19**, 9382–9386.
65. A. Gupta, J. L. Terrell, R. Fernandes, M. B. Dowling, G. F. Payne, S. R. Raghavan and W. E. Bentley, *Biotechnol. Bioeng.*, 2013, **110**, 552–562.
66. M. B. Dowling, A. S. Bagal and S. R. Raghavan, *Langmuir*, 2013, **29**, 7993–7998.
67. Y. Xiong, K. Yan, W. E. Bentley, H. Deng, Y. Du, G. F. Payne and X. W. Shi, *ACS Appl. Mater. Interfaces*, 2014, **6**, 2948–2957.
68. P. D. Thornton, R. J. Mart and R. V. Ulijn, *Adv. Mater.*, 2007, **19**, 1252–1256.
69. Z. Yang, G. Liang and B. Xu, *Acc. Chem. Res.*, 2008, **41**, 315–326.
70. R. S. Alkasir, M. Ornatska and S. Andreescu, *Anal. Chem.*, 2012, **84**, 9729–9737.
71. Y. Liu, E. Kim, M. E. Lee, B. Zhang, Y. A. Elabd, Q. Wang, I. M. White, W. E. Bentley and G. F. Payne, *Adv. Funct. Mater.*, 2014, **24**, 480–491.
72. M. Rinaudo, *Polym. Int.*, 2008, **57**, 397–430.
73. M. Cheong and I. Zhitomirsky, *Colloid Surf., A*, 2008, **328**, 73–78.
74. D. Zhitomirsky, J. A. Roether, A. R. Boccaccini and I. Zhitomirsky, *J. Mater. Process. Technol.*, 2009, **209**, 1853–1860.
75. X. W. Shi, C.-Y. Tsao, X. Yang, Y. Liu, P. Dykstra, G. W. Rubloff, R. Ghodssi, W. E. Bentley and G. F. Payne, *Adv. Funct. Mater.*, 2009, **19**, 2074–2080.
76. Z. Y. Jin, G. Guven, V. Bocharova, J. Halamek, I. Tokarev, S. Minko, A. Melman, D. Mandler and E. Katz, *ACS Appl. Mater. Interfaces*, 2012, **4**, 466–475.
77. Y. Cheng, C.-Y. Tsao, H.-C. Wu, X. Luo, J. L. Terrell, J. Betz, G. F. Payne, W. E. Bentley and G. W. Rubloff, *Adv. Funct. Mater.*, 2012, **22**, 519–528.
78. Z. Y. Jin, A. M. Harvey, S. Mailloux, J. Halamek, V. Bocharova, M. R. Twiss and E. Katz, *J. Mater. Chem.*, 2012, **22**, 19523–19528.

79. Y. Liu, V. Javvaji, S. R. Raghavan, W. E. Bentley and G. F. Payne, *J. Agric. Food Chem.*, 2012, **60**, 8963–8967.
80. S. Mailloux, J. Halamek and E. Katz, *Analyst*, 2014, **139**, 982–986.
81. S. Seiffert, *Angew. Chem., Int. Ed.*, 2013, **52**, 11462–11468.
82. E. Tumarkin and E. Kumacheva, *Chem. Soc. Rev.*, 2009, **38**, 2161–2168.
83. S. Sugiura, T. Oda, Y. Izumida, Y. Aoyagi, M. Satake, A. Ochiai, N. Ohkohchi and M. Nakajima, *Biomaterials*, 2005, **26**, 3327–3331.
84. W. H. Tan and S. Takeuchi, *Adv. Mater.*, 2007, **19**, 2696–2701.
85. K. S. Huang, C. H. Yang, C. P. Kung, A. M. Grumezescu, M. D. Ker, Y. S. Lin and C. Y. Wang, *Electrophoresis*, 2014, **35**, 330–336.
86. J. E. Maneval, D. Bernin, H. T. Fabich, J. D. Seymour and S. L. Codd, *Magn. Reson. Chem.*, 2011, **49**, 627–640.
87. Y. Maki, K. Ito, N. Hosoya, C. Yoneyama, K. Furusawa, T. Yamamoto, T. Dobashi, Y. Sugimoto and K. Wakabayashi, *Biomacromolecules*, 2011, **12**, 2145–2152.
88. R. Dittrich, G. Tomandl, F. Despang, A. Bernhardt, T. Hanke, W. Pompe and M. Gelinsky, *J. Am. Ceram. Soc.*, 2007, **90**, 1703–1708.
89. E. Schuster, J. Eckardt, A. M. Hermansson, A. Larsson, N. Loren, A. Altskar and A. Strom, *Soft Matter*, 2014, **10**, 357–366.

CHAPTER 9

Applications of Stimuli-responsive Functional Colloids

HUAI N. YOW,[a,c] SIMON BIGGS[b,c] AND OLIVIER J. CAYRE*[c]

[a] Briggs of Burton PLC, Briggs House, Derby Street, Burton-on-Trent DE14 2LH, UK; [b] Faculty of Engineering, Architecture & Information Technology, The University of Queensland, Brisbane QLD 4072, Australia; [c] Institute of Particle Science & Engineering, School of Chemical & Process Engineering, University of Leeds, Leeds LS2 9JT, UK

*Email: o.j.cayre@leeds.ac.uk

9.1 Introduction

The last 20 years have seen an ever increasing interest in opportunities for the development of functional colloidal systems using stimulus-responsive polymers for a range of different applications.[1] This has been driven in a large part by significant developments in living radical polymerisation techniques, which have facilitated access to a wide range of polymer and copolymer materials with excellent control over the molecular structure properties (*e.g.* molecular weight, block size ratios *etc.*). The ability to accurately control the properties of the polymers has amplified the opportunities available to exploit stimulus-responsive behaviour by, for example, changing the solvency of the polymers in solution. Tight control over the polymer properties generally results in much finer control over the transition from good to poor solvent behaviour. This, in turn, can be used to design a rapid behaviour switch in components of formulated products and chemical/biological sensors as we will discuss later.

RSC Smart Materials No. 14
Chemoresponsive Materials: Stimulation by Chemical and Biological Signals
Edited by Hans-Jörg Schneider

Published by the Royal Society of Chemistry, www.rsc.org

9.2 Stimuli-responsive Colloids

This section will briefly introduce the types of particles we have chosen to describe in this chapter. There are many examples of responsive particles in the academic literature, with a wide range of internal and surface structures, which cannot all be reviewed here. Therefore we concentrate on particles where the responsive behaviour is induced by the presence of an environmentally-responsive polymer either within the core or on the surface. This section mainly concentrates on introducing the types of systems discussed subsequently. In the next section, we review examples of work demonstrating specific applications for these particles.

9.2.1 Soft Responsive Particles

Here we describe the particles produced entirely from copolymers, specifically designed to posses a transition between different self-assembled structures or from unimers in solution in response to an environmental trigger.

9.2.1.1 Self-assembled Copolymer Aggregates

Copolymers that are able to self-assemble in solution have two or more monomers with a differing solubility in the solvent or a mixture of solvents acting as the bulk phase. The resulting copolymer aggregates can adopt a large range of morphologies as a function of their molecular weight, the arrangement of the monomers within the polymer, and their relative solubility towards the solvent. The most common aggregate structures for diblock copolymers are spherical and worm-like micellar structures[2] but a large range of other types is also possible and readers are referred to some seminal articles[3,4] and a review[5] for specific examples.

For brevity, we will restrict ourselves here to spherical aggregates, comprising the majority of the available literature. Some examples of such systems and their applications are given in Table 9.1.

It is worth noting that many of these copolymer materials contain reactive moieties that allow subsequent cross-linking reactions. These can, with an appropriate choice of block chemistry, be used to cross-link the core, the corona, or both parts of the copolymer aggregates. Indeed, by very careful choice of the copolymer microstructure, examples of selective cross-linking at only the chain ends, whether in the core or at the surface, have been demonstrated. The main advantage of cross-linking is, of course, to provide greater mechanical stability to the particulate phase. A thorough review of cross-linking reactions as they apply to copolymer aggregates, and some possible applications for the resultant particles is provided by O'Reilly *et al.*[6] It is worth noting that in many cases, the result of cross-linking the copolymer aggregates is to produce what is essentially a nanogel particle (see later); this is essentially true when the solubility of one or both blocks is responsive to some kind of stimulus.

Table 9.1 Stimuli-responsive copolymer aggregates.

Stimuli	Micelles/vesicles	Applications	References
• pH	• Poly(EO-lysine) or poly(EO-ester) with acetal-linked hydrophobic groups	• Encapsulation of active ingredients	54
	• Poly(NIPAM-MAA-ODA)	• Drug delivery	55
	• Poly(MPC-*b*-DPA)/Nile Blue-based label	• Sensor for live bio-imaging	18
	• Poly(EG)-*b*-poly(DEAEM-*s*-DMIBM) or poly(EG)-*b*-poly(DEAEM-*s*-DMIBM)	• Bio-nanoreactor • Size selective release	
• Gas	• Poly(EG-*b*-AD) or poly(EG)-*b*-poly(DEA-*co*-FMA)	• Encapsulation • Size selective separation	56, 57 56
• Biological reaction	• SC4A-myristoylcholine/cholinesterase	• Drug delivery	61
	• Poly(EG)-*b*-poly(LA)/proteinase K	• Wound dressing	60
• Electrochemical potential	• Poly(Sty-β-CD)/poly(EO-Fc)	• Encapsulation	59

9.2.1.2 *Micro- and Nanogels*

Micro- or nanogels consist of cross-linked polymer particles that can swell in a good solvent but that do not disintegrate (as would be the case for a simple latex particle). The swelling is driven by osmotic pressure as a result of enhanced polymer solvency and/or electrostatic repulsion from the presence of ionic monomer(s). The swelling–de-swelling mechanism can be triggered by a range of external stimuli that can affect the solvency, such as temperature, pH, ionic strength, solvent composition, and/or chemical/biological reactions. A multitude of excellent reviews exist highlighting the synthesis as well as the physical properties and applications of microgels.[7–12] In this book, Chapters 2, 3 and 12 also describe examples of gels, which correlate to the work presented in the following sections. Table 9.2 summarizes some examples of stimuli-responsive microgels, concentrating mainly on chemical and biological stimuli. For examples of temperature-responsive microgels, the following reviews will be of use: Guan and Zhang,[13] Lyon and Fernandez-Nieves.[14]

9.2.2 Solid Core-responsive Polymer Shell Particles

The particles in this section still rely on their polymer content to provide an environmental response but here the particle core is an organic, inorganic or metal solid and the responsive polymers are present on the particle surface (Figure 9.1). A multitude of examples have been reported in the literature and the examples given here will refer to the systems reviewed for their applications in the next section. Other examples with mesoporous silica particles as the core of such systems are also described in Chapter 6 of this book.

Table 9.2 Stimuli-responsive micro- and nanogels.

Stimuli	Micro-/nanogel	Applications	References
• pH	• Poly(AA) or poly(MAA)	• Rheological control	16, 17
	• Poly(NIPAM-AA-HEA)	• Micro-reactors	63, 64
		• Photonic crystal	67
	• Poly(DMAA-DMAEMA) or poly(DMAA-4VP)	• Micro-reactors	65
	• Poly(NIPAM-AA)	• Microgel lenses	68
		• Water purification	41
	• Poly(MAA-DMAEMA)	• Drug and protein delivery	44
	• OEI copolymer	• Drug delivery	45
	• PEG-chitosan	• Drug delivery	47
	• Poly(EA-MAA-BDDA) or poly(MMA-MAA-EGDMA)	• Regenerative medicine for IVD	8, 62
	• Poly(2-vinylpyridine) latex stabilized with monomethoxy-capped poly(ethylene glycol) steric shell.	• Foam stabilisation	25
• Solvent composition	• Cross-linked poly(MMA)	• Rheological control	15
	• Poly(VCL-AAEM)	• Micro-reactors	66
• Chemical reaction	• Poly(NIPAM) with cleavable DHEA	• Protein screening	40
	• PEG-poly(amino acid) with cleavable disulfide	• Drug delivery	48
	• PEG-P(HEMA-*co*-AC) with cleavable disulfide	• Drug delivery	49
• Biological reaction	• Poly(NIPAM-VI)/urease	• Membrane porosity control	39
	• Poly(NIPAM)/PBA	• Drug delivery	50

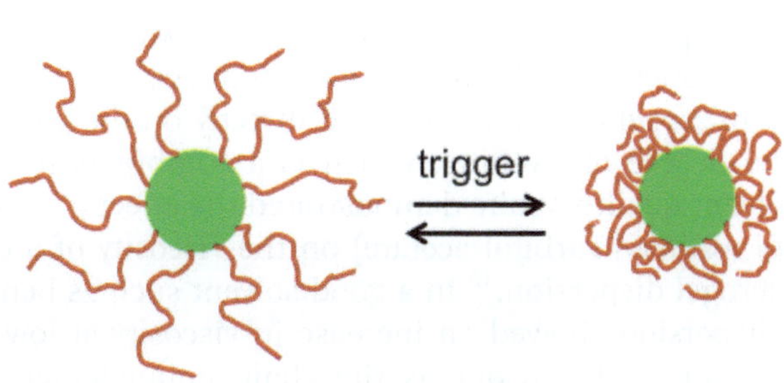

Figure 9.1 Schematic representation of a particle with a solid (*i.e.* non-responsive) core and a polymer shell that reversibly swells away from and contracts onto the particle core surface in response to a specific trigger.

Table 9.3 Stimuli-responsive solid core-responsive polymer shell particles.

Stimuli	Core/shell particles	Applications	References
• pH	• Silica/poly(2VP) with invertase/glucose oxidase/urease	• Sensor for molecular robotic systems	19
	• Silica/poly(Sty-*b*-4VP-EO) with invertase/glucose oxidase/urease	• Stimuli-responsive emulsifiers for biological sensor	21
	• Cadmium selenide/zinc sulfite quantum dots/poly(2VP-*b*-PyMMP)	• Ratiometric sensor	23
	• Poly(styrene) latex stabilized with poly(acrylic acid) steric shell	• Foam stabilisation	24
• Biological reaction	• Gold/single-strand DNA	• Sensor as colorimetric probe	22

The presence of the responsive polymer on the surface of the particles allows one to influence the colloidal stability of the particles in suspension as well as their behaviour towards interfaces. Indeed, several examples have been shown where particles can be flocculated upon changes in pH and electrolyte concentration, which can be used in sensing applications (Table 9.3). The adsorption of such particles at a liquid–liquid interface can also be controlled through changes in the external environment. This has been used to produce emulsions and foams of controlled stability.

9.3 Application-driven Examples

In this section, we provide recent examples of applications of the three different types of particles briefly reviewed earlier. This constitutes a non-exhaustive list of applications, which have been developed over the last decade.

9.3.1 Rheological Control

The first reported applications of microgel materials was in rheology control for coatings, cosmetics and pharmaceutical products.[15–17] The rheological properties of a microgel dispersion depend directly on the particle concentration and the degree of swelling, which is a combination of cross-link density and solvent quality. Wolfe demonstrated the effect of solvent quality (benzyl alcohol *vs.* butyl carbitol acetate) on the viscosity of a cross-linked poly(MMA) microgel dispersion.[15] In a good solvent such as benzyl alcohol, the microgel dispersion showed an increase in viscosity at low stresses as well as shear thinning behaviour as the chain entanglement and inter-particle interactions of overlapping microgels were broken during shearing (see Figure 9.2). Similar properties were also observed in pH-responsive poly(MAA) and poly(AA) microgels where, upon alkalisation, the increase in

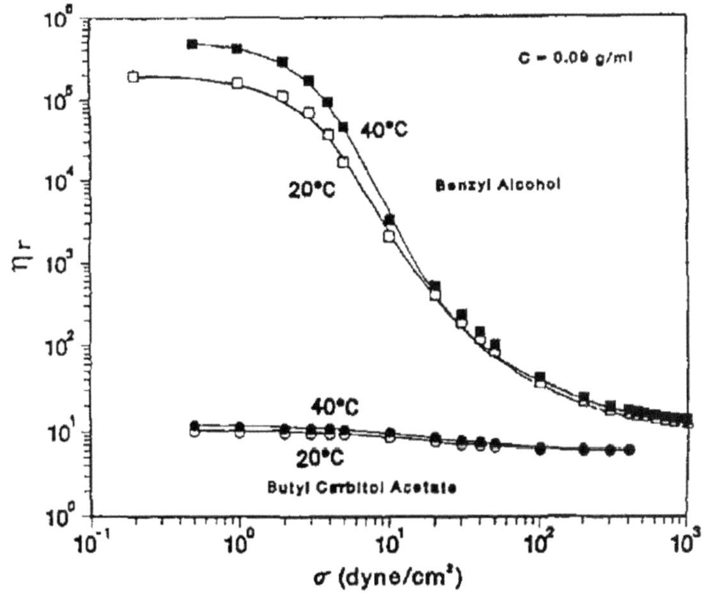

Figure 9.2 Effect of solvent quality on viscosity of microgel dispersions. Reproduced from ref. 15.

solubility and associated electrostatic repulsion between polymer chains caused swelling; this led to an overlapping of the microgels giving enhanced rheological properties.[16,17]

9.3.2 Chemical/Biological Sensors

Many diseases induce local pH changes as a result of an increase in lactic acid production as well as reduced buffering and perfusion. In most cases, the intracellular pH of the tumour cells remains at physiological levels. Most of the available diagnostic probes have several limitations, including a tendency to penetrate the cells and report intracellular pH, toxicity potential, targeting specificity issue and low signal-to-noise ratio. Madsen *et al.* designed a biocompatible pH-sensor, comprised of a pH-sensitive dye (Nile Blue-based) and pH-responsive diblock copolymer [poly(MPC-*b*-DPA)] that self-assembled into vesicles around pH 6.[18] These vesicles were successfully taken up by a multicell tumor spheroid and detected the growth of a diseased region by a colour change of the stained tissue from purple (at physiological pH) to blue (at low pH). These vesicles also exhibited varying degrees of absorption and fluorescence emissions depending on the local pH, allowing detection in far-red and infrared regions. This is more advantageous because of the lower background interference for deeper penetration into living tissues. These advantages enable the pH-sensitive dye-labelled poly(MPC-*b*-DPA) vesicles to be used as fluorescent sensors for live imaging in cell biology.

Another application of stimuli-responsive particles as chemical/biological sensors was demonstrated by Motornov *et al.* who synthesised pH-responsive hybrid systems with built-in Boolean logic.[19] These systems were comprised of polymeric-decorated nanoparticles with information-processing enzymes. An example was demonstrated by the use of invertase (which catalyses sucrose to glucose) and glucose oxidase (which catalyses glucose to gluconic acid) to yield an acidic environment that allows re-stabilisation and re-dispersion of the nanoparticles. The condition was reset by urease (which catalyses the hydrolysis of urea) to a neutral pH environment, inducing particle aggregation due to the isoelectric point of poly(2VP)-coated nanoparticles at pH 6.5. With these bio-computing hybrid systems, the output signal could be amplified by structural changes of the suspension, allowing easier analysis of complex biochemical analytes. Motornov *et al.* further applied these hybrid particle systems (comprising of silica particles with a poly(Sty-*b*-4VP-EO) shell) as emulsion stabilisers (or Pickering emulsifiers as here one refers to emulsions stabilised by particles only,[20] *i.e.* in the absence of traditional emulsifiers such as surfactants or proteins) and successfully encapsulated oil-soluble dyes (see Figure 9.3).[21] Motornov *et al.* investigated the inversion of these emulsions as a function of pH changes induced by the output signals catalysed by enzymatic reactions. A key advantage is the ability to use processed bio-information for simple regulation of chemical transport.

Wang *et al.* also synthesised bio-sensing stimuli-responsive hybrid particles, comprising gold nanoparticles in combination with target-free aptamers.[22] Aptamers possess high specificity and affinity; in the presence of the correct target, the aptamers bind to the target and release single-strand DNAs (ss-DNA). The highly charged ss-DNAs spontaneously bind to the gold nanoparticles, stabilising the particles against salt-induced aggregation (see Figure 9.4). The clear colour change in gold nanoparticle suspension (red for stabilised dispersion *vs.* blue for aggregates), even at a low stimuli/salt concentration, provides an effective visual detection that can be confirmed by UV-vis measurements. Wang *et al.* demonstrated the application of these bio-sensing stimuli-responsive systems with ATP-, potassium- and cocaine-based aptamers.

A similar colour-displaying, ratiometric pH sensor was also demonstrated by Paek *et al.*, who synthesised cadmium selenide/zinc sulfide (CdSe/ZnS) quantum dots stabilised with a poly(2VP-*b*-PyMMP) shell.[23] The colour change from blue to purple to red as a function of increasing pH was governed by the interspacing between the blue-emitting PyMMP block and the red-emitting quantum dots. As the poly(2VP) chains de-protonated and collapsed under high pH, the Förster resonance energy transfer from PyMMP to quantum dots was enhanced, allowing the apparent detection of a colour change from blue to red. This allows the development of a fast, effective sensor without expensive requirements, such as a photoluminescence detector (commonly used to monitor the intensity of quantum dots).

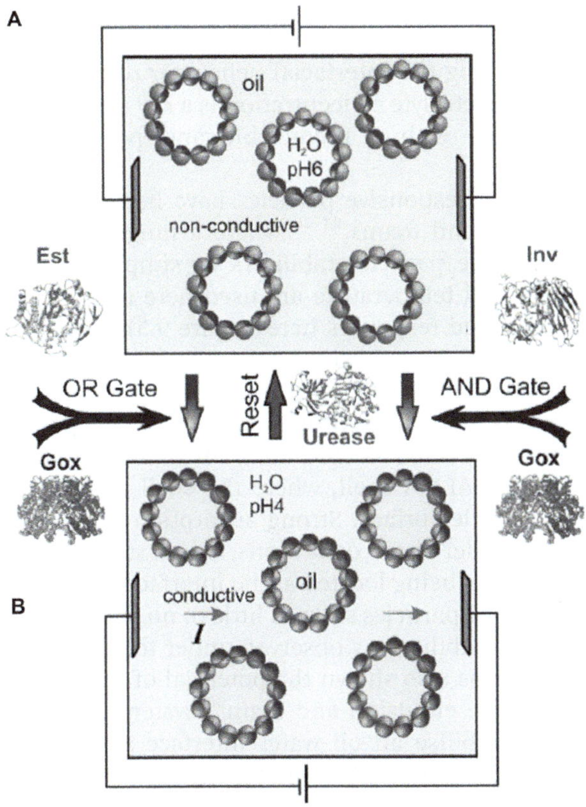

Figure 9.3 Schematic of pH-responsive hybrid system with built-in bio-information processing enzymes.
Reproduced from ref. 21.

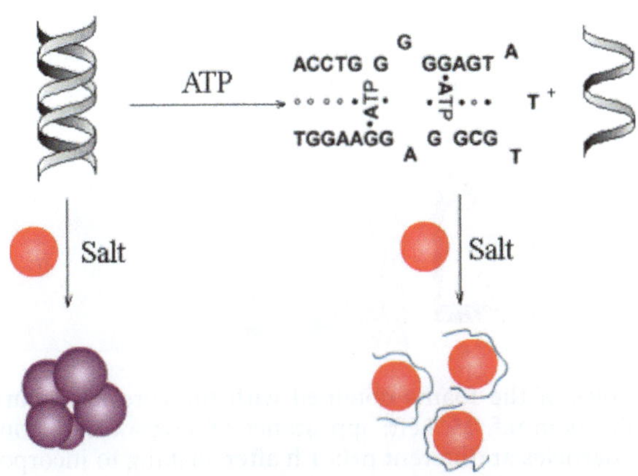

Figure 9.4 Schematic of gold nanoparticles with aptamer-target binding readout for ATP assay, with ss-DNA-stabilised gold nanoparticles possessing resistance against salt-induced aggregation.
Reproduced from ref. 22.

9.3.3 Smart Emulsions and Foams

The ability of controlling the interfacial behaviour of particles in response to changes in pH or electrolyte concentration is a key advantage in designing emulsions and foams with a predictable and potentially controllable stability.

Typically, core–shell responsive particles have been used in these cases to prepare emulsions and foams,[24–26] where a range of stabilities can be achieved with the same particle stabilisers by simply varying an external trigger. Mostly, pH and temperature are used here as triggers and we will concentrate on pH-based responses here (Figure 9.5).

For example, Fujii *et al.* synthesized poly(styrene) latex particles carrying a pH-responsive poly[2-(diethylamino) ethyl methacrylate] steric shell by dispersion polymerisation in isopropanol.[25] Once the particles were re-dispersed in water, they acted as efficient foam stabilisers at pHs above the pK_a ($\sim$7.5) of the PDEAEMA shell, where the shell was deprotonated and collapsed on the particle surface. Strong adsorption of the particles to the air–water interface under these conditions led to stable foams with clear evidence of the particles being located at the interface. At pHs below the pK_a of the polymer shell, the particles showed little to no affinity for the air–water interface and no foam ability was observed under these conditions.

A range of studies has also shown the potential of microgel[27–29] particles to controllably stabilise emulsion and foam[30] systems. Such particles also have the ability to stabilise an oil–water interface and can be controlled through changes in the trigger that the particle core responds to. In addition, microgel particles offer a different prospect altogether as it is now known that, once adsorbed, they can deform to maximise their area in contact with the interface.

Developments in the area of smart particle emulsifiers are expected to lead to emulsions and foams that will be destabilised on demand as is suggested by some academic studies investigating the efficiency of particle desorption from 2D interfaces[31] and from emulsion droplet surfaces.[32,33]

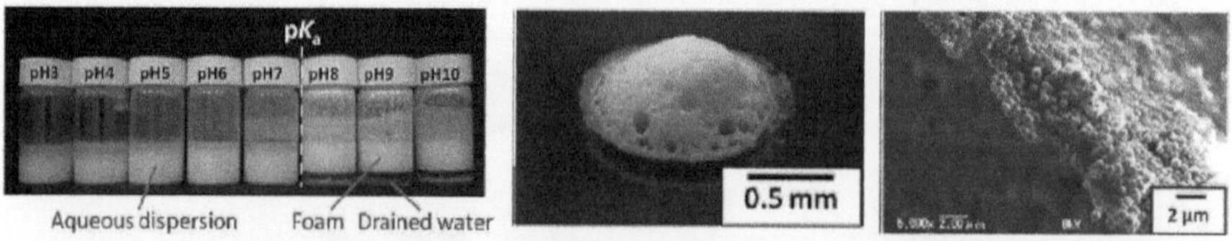

Figure 9.5 Examples of the foams obtained with the core–shell particles used as stabilisers in ref. 25. Left: appearance of 5 wt% dispersions of the core–shell particles at different pHs 1 h after shaking to incorporate air in the dispersion. Middle: a foam (dried on a glass substrate) created from a particle dispersion above the pK_a of the particle polymeric shell. Right: electron microscopy observation of the resulting particle layer between two gas bubbles from a foam created above the particle polymeric shell pK_a.

9.3.4 Particle Transfer Across Membranes and Separation Processes

The ability to change the affinity of a particle surface through the use of environmentally-responsive polymers as stabilisers also provides the opportunity to control the transport of these particles across interfaces, particularly for two immiscible (or poorly miscible) liquids. Through variation of certain physical[34,35] or chemical[36–38] characteristics of the continuous phase in which the particles are dispersed, one can dramatically decrease the solvency for the polymer stabiliser and comparatively favour the dispersion of the particles into a different phase. This is typically reported as having a significant potential for modelling transport across biological membranes, as the transport across these interfaces can be reversible.

A few examples of changes in pH, solvent types and electrolyte concentrations have been demonstrated to date. Edwards *et al.* grafted a PEGylated polymethacrylate cap on the surface of gold nanoparticles to facilitate reversible transport between aqueous and toluene phases.[38] By introducing an electrolyte to the water phase in which the particles are initially dispersed, the authors were able to induce the transfer of the particles into an upper toluene phase through the modification of the particle surface energy in the continuous phase. The transfer of the particles from salty water into the organic phase was reversed by promoting hydrogen bonding between the particles and the aqueous phase. This was achieved by replacing the water phase by an aqueous solution of citric acid, which led to the particles spontaneously transferring back into the aqueous phase. This reversible transfer was observed for five cycles (Figure 9.6).

Thermo- and pH-sensitive poly(NIPAM-VI) with encapsulated urease has been successfully synthesised and used to control the porosity of a cellulose membrane (see Figure 9.7).[39] These microgels catalyse the hydrolysis of

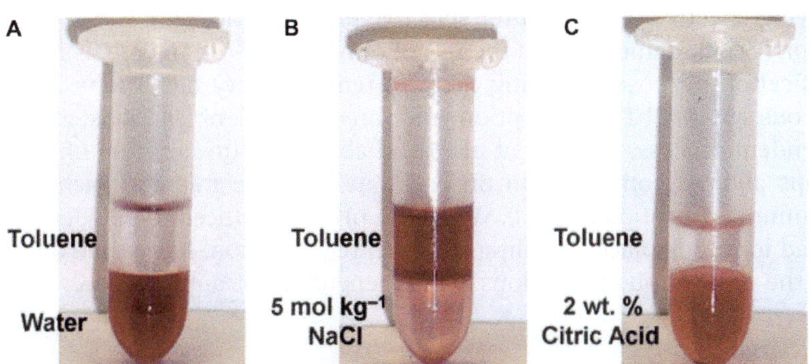

Figure 9.6 Gold nanoparticles capped with a poly(OEGMA) polymer driven to reversibly transport across a water–toluene interface through tuning the solvency of the aqueous phase for the stabilising polymer through addition of electrolyte and acid.
Reproduced from ref. 38.

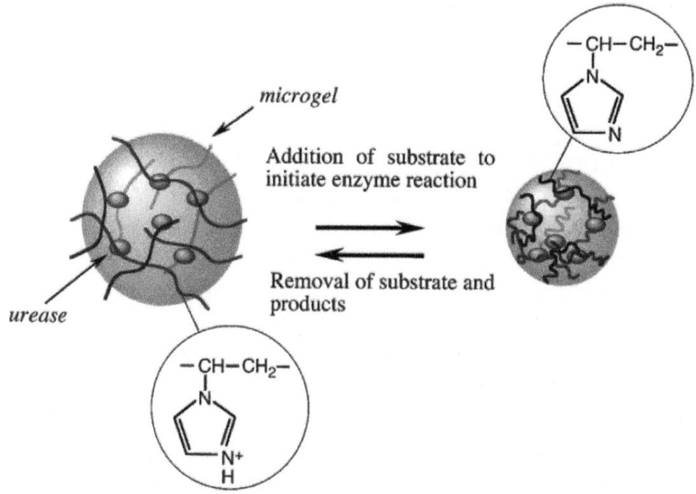

Figure 9.7 Schematic of enzymatically-induced swelling–de-swelling of poly(NIPAM-VI) microgel.
Reproduced from ref. 39.

urea into ammonia and carbon dioxide, resulting in an increase of pH surrounding the microgels. The collapse of microgels at high pH opens the gate pore between particles allowing transfer of liquid through the membrane.

Nayak and Lyon demonstrated an alternative mechanism to controlling the membrane porosity using microgels by designing a cleavable diol cross-linker, DHEA.[40] The complex microgels with protein binding ability manipulate the pore size of the microgel shell through cleavage of cross-links in the presence of sodium periodate. Thus, depending on the cross-link density, only specific protein structures will be allowed into the core, allowing effective separation for selective protein binding.

Morris *et al.* demonstrated the use of poly(NIPAM-AA) microgels for water purification purposes by using them to remove heavy metal ions from the aqueous system.[41] The adsorption, and thus removal, of lead ions, was a pH-dependent process, whereby at pH 5 and above, the dissociation of carboxyl groups allowed complexation of lead ions with the microgel, achieving a maximum adsorption at pH 8. When the pH was reduced to 3.5, desorption of lead ions took place, enabling microgel regeneration. It was also observed that the adsorption of lead ions was independent of temperature.

9.3.5 Active Ingredient Delivery Devices

Stimuli-responsive colloids, particularly microgels, have been used as active ingredient delivery devices. In the case of microgels, they are able to encapsulate various molecules within their large internal volume, as well as allow controlled and/or targeted release due to the dramatic swelling—and therefore porosity increase—in response to external stimuli. Various applications, including food,[42] personal care products, agrochemical formulations and biomedical uses[43,44] (*e.g.* drug delivery,[45–49] tissue engineering, imaging probes, *etc.*) have been investigated. Several examples of such applications are given later.

Recently, Urakami *et al.* synthesised disulphide-containing oligoethylenimine, OEI, nanogels.[45] These nanogels demonstrated pH-responsiveness, with an increase of amine protonation at low pH causing swelling under these conditions. These nanogels were also bio-degradable as the disulfide bonds are rapidly cleaved when exposed to a reductive intracellular environment. They also demonstrated bio-compatibility with minimal cytotoxicity against cells during incubation tests. These developments clearly show the potential of nanogels as carriers for delivery of drugs and genes.

Ding *et al.* also utilised similar disulfide bonds with PEG-poly(amino acid) nanogels to deliver an anticancer drug, DOX, *via* a reduction reaction in the presence of 10 mM glutathione, GSH.[48] These nanogels also demonstrated low *in vitro* cytotoxicity, indicating good compatibility of the nanogels as efficient drug delivery vehicles. Similar chemistry was also used by Chen *et al.* with PEG-poly(HEMA-*co*-AC) nanogels to deliver proteins *via* a reduction reaction with 10 mM dithiothreitol, DTT.[49]

Zhou *et al.* designed PEG-chitosan based nanogels by physically interpenetrating chitosan chains into non-linear PEG networks.[47] These nanogels are pH-responsive as well, but with additional control whereby the rate of pH-response can be regulated by external cooling–heating cycles. A model anti-cancer drug, 5-FU, capable of intermolecular complexing with the nanogel, allows higher encapsulation efficiency. When used in combination with chemo–cryo treatment, these nanogels exhibited much higher therapeutic efficacy compared to chemo- or cryo-treatment alone.

Polyampholyte microgel is an alternative system being explored for drug and protein delivery applications. Ho *et al.* recently synthesised biocompatible pH- and thermal-responsive poly(MAA-DMAEMA) microgels cross-linked with allyl methacrylate (AM).[44] Polyampholyte microgel has responsiveness in both low and high pH regions due to inclusion of both positive and negative charges on the same polymeric backbone. At low pH, the PDMAEMA segments protonate to give positive charges, while at high pH the PMAA segments acquire negative charges. The dissociation of NH^+ and COO^- ions in low and high pH respectively induce high osmotic pressure, causing the microgels to swell under these conditions (Figure 9.8). In contrast, at IEP, the polyampholyte microgel shrinks and retains the active ingredient until targeted delivery conditions are achieved. This new system allows the design of specific IEP characteristics, allowing precision tuning of encapsulation and release of the active ingredients.

The development of biologically-responsive microgels, demonstrated recently by Hoare and Pelton, has been mooted for applications in the drug delivery field. Hoare and Pelton synthesised amphoteric poly(NIPAM) microgels functionalized with aminophenyl boronic acid (PBA).[50] Depending

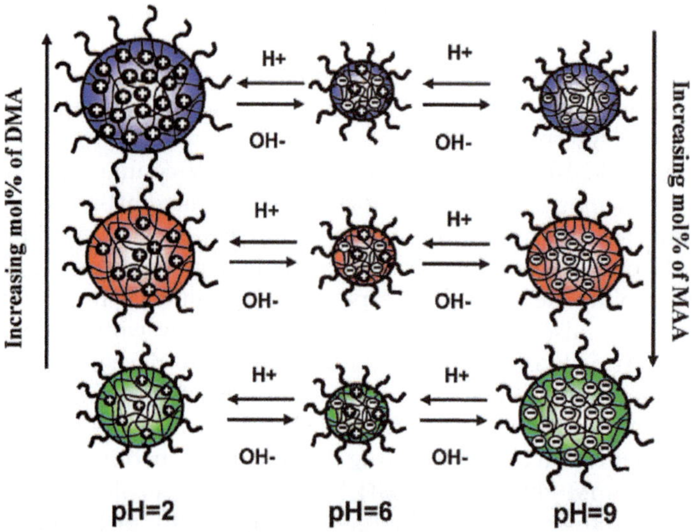

Figure 9.8 Schematic of swelling mechanisms for cross-linked polyampholyte microgels, poly(MAA-DMAEMA), at 25 °C.
Reproduced from ref. 44.

on the pH of the medium, in the presence of glucose, PBA can dissociate to give additional negative charges, which induce further swelling within the microgels. Preliminary investigations show that a high amount of insulin can be loaded into these microgels and the insulin can be selectively released *via* glucose-induced "on–off" switching under physiological conditions.

As an alternative to nano- or microgels, which are synthesised as cross-linked particles, a smartly-designed stimuli-responsive polymer can self-assemble to form micelles or vesicles to be used as active ingredient delivery devices.[51–53] Micelles are commonly used to encapsulate hydrophobic compounds, whereas vesicles are able to encapsulate both hydrophilic and hydrophobic ingredients within the aqueous internal and membrane portions respectively. An example of this was illustrated by Gillies *et al.* who designed pH-responsive linear-dendritic block copolymers with hydrophobic groups attached *via* acid-sensitive cyclic acetals.[54] These poly(EO) with either polylysine or polyester copolymers form stable micelles at neutral pH, but disintegrate in acidic conditions due to the loss of hydrophobic groups upon acetal hydrolysis. Gillies *et al.* demonstrated the controlled release of a model ingredient, Nile Red, and tracked the micelle disintegration into unimers over several hours upon introduction of a mild acidic environment (see Figure 9.9). Another example of pH-responsive polymeric micelles was shown by Taillefer *et al.* who successfully used poly(NIPAM-MAA-ODA) micelles to encapsulate and deliver a photoactive anticancer drug, AlClPc, during *in vivo* trials against mouse mammary cells.[55]

Over the last decade, novel external stimuli, such as gas,[56,57] electrochemical potential,[58,59] and enzymes,[60,61] have been investigated to expand

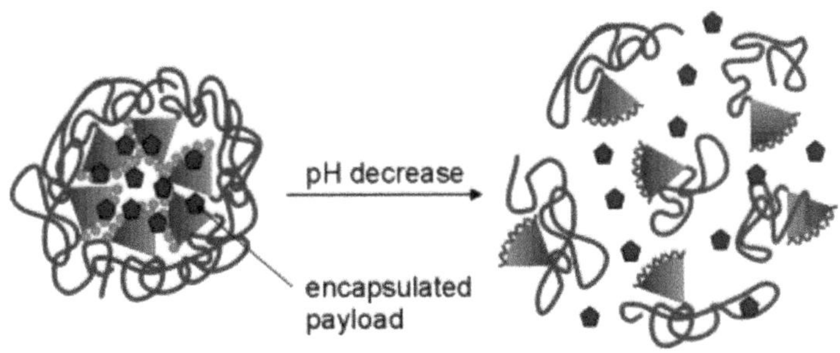

Figure 9.9 Schematic of active ingredient release from a pH-responsive micelle. Reproduced from ref. 54.

the applications of polymeric micelles and vesicles.[52] Gas-responsiveness, to CO_2, O_2 and N_2, is highly attractive as such gases are key metabolites in cells and typically present to levels that are high enough in the body to generate the required particle response. For example, Yan *et al.* synthesised CO_2-responsive poly(EG-*b*-AD) vesicles with tunable membrane permeation by regulating the amount of CO_2 introduced.[56] These vesicles were able to selectively separate and release their active ingredient (see Figure 9.10) as well as simulate enzymatic reactions using a model enzyme, myoglobin. An example of dual gas-responsive (CO_2 and O_2) vesicles was developed by Zhang and Zhu using poly(EG)-*b*-poly(DEA-*co*-FMA), where the DEA and FMA moieties provided the CO_2- and O_2-responsive behaviours, respectively.[57] Upon dispersion in water, the polymer self-assembled into vesicles that transformed into smaller micelles in a CO_2-rich environment and swelled by eight times as vesicles upon treatment with O_2.

Another example of a novel stimulus is enzyme-responsive vesicles as reported by Guo *et al.*, who designed a cholinesterase-responsive system for targeted delivery of an Alzheimer's disease drug. Cholinesterase is a key protein that is abundant in Alzheimer's disease.[61] Here, the authors employed host–guest complexes as superamphiphiles (with SC4A as the biocompatible host and natural myristoylcholine as the enzyme-cleavable guest) to self-assemble into vesicles. The myristoylcholine was cleaved upon exposure to the enzyme, resulting in complete disassembly of vesicles and fast release of the entrapped drugs.

9.3.6 Regenerative Medicine

An interesting example of responsive colloids providing an innovative solution to a health issue has been demonstrated by Saunders *et al.* These authors have explored the use of microgels based on poly(EA-MAA-BDDA) or poly(MMA-MAA-EGDMA) to improve the regeneration of damaged intervertebral discs (IVD).[8,62,63] The designed systems are pH-responsive microgel dispersions, which are injected unswollen at low pH into the damaged area. The associated pH change *in situ* swells the microgels and drives a transition of the dispersion from fluid-like to gel-like (Figure 9.11). This provides

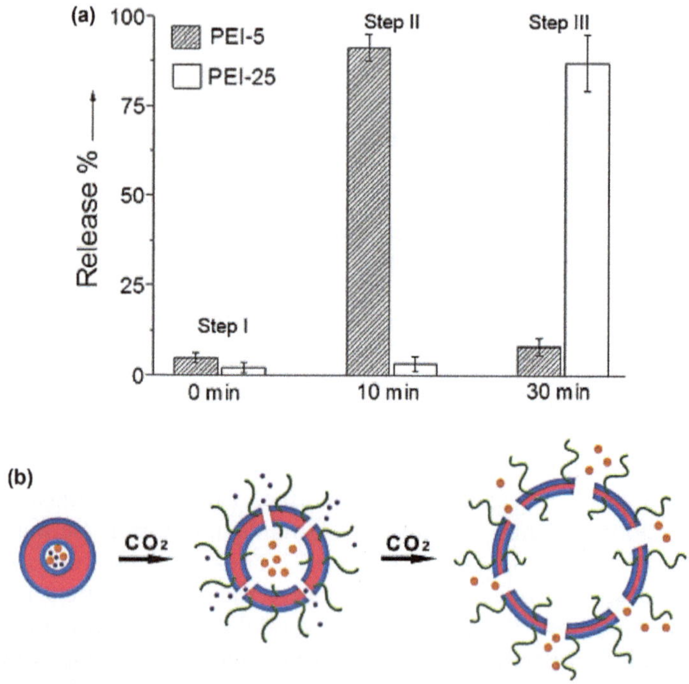

Figure 9.10 Schematic of CO_2-responsive poly(EG-*b*-AD) vesicles with tunable membrane permeation for size-selective release of nanoparticles (PEI-5 and PEI-25).
Reproduced from ref. 56.

additional support and raises the height of the degenerated IVDs, preventing further degeneration. The possibility to use encapsulates and achieve targeted release of biologic compounds thereby providing further avenues for tissue re-growth and eventual repair of the degenerated IVD has also been discussed by these authors. Preliminary compressive cycling tests indicate no differences between the mechanical properties of degenerated IVDs with injected microgels compared to healthy, non-degenerated IVDs.

9.3.7 Reactors

Soluble metal nanoparticles (NPs) or nanoclusters are crucial in various fields including catalysis, material science, *etc.* However, typical synthesis routes *via* metal atom precursors tend to be expensive and difficult to scale up. Hence, various researchers have investigated the use of microgels as micro-reactors to prepare NPs[64,65] (*e.g.* Fe_3O_4, CdS and Ag), as well as metal nanoclusters[66] (*e.g.* Pt and Pd). This is possible because microgels allow localization of reactive sites and controlled growth of the particles, leading to formation of NPs/nanoclusters with adjustable properties, including size, morphology and stabilization. Microgels also have high surface areas

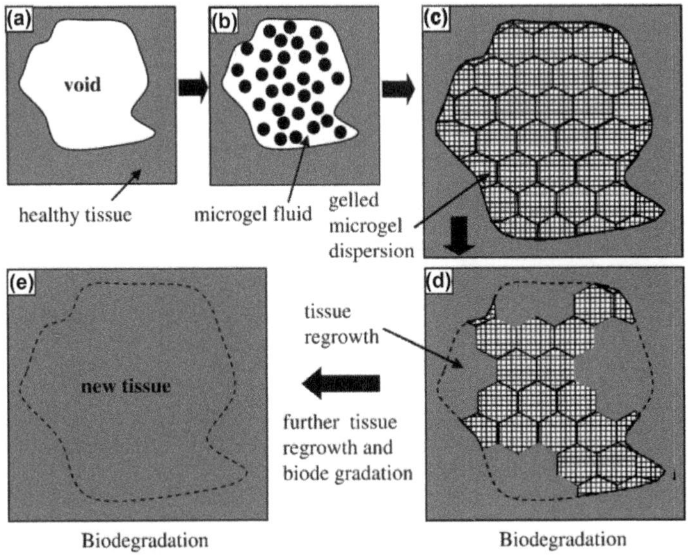

Figure 9.11 Schematic representation of regeneration of IVD using injectable pH-responsive microgels.
Reproduced from ref. 63.

allowing ease of accessibility to the particles as well as stimuli-responsiveness enabling control of the distance between the particles. The main drawback lies at high solid loadings within the microgel, whereby the presence and interactions of particles with microgel polymeric chains can affect the mobility of the chains and thus shift the transition temperature, as well as colloidal stability.

Microgels have also been used as reactors to synthesise conducting polymers, such as poly(Py).[67] This is due to the interest in the electrical properties of polymers with conjugated double bonds for use as advanced functional materials. The solvent-responsiveness of a VCL-rich microgel to alcohol additions induces greater swelling of the microgels leading to a larger deposition surface area for poly(Py). The higher loading of poly(Py)-containing microgel films confers a relatively higher conductivity.

9.3.7.1 *Optical Applications*

As mentioned in Section 9.3.7, microgels form a suitable template/reactor for the synthesis of metal nanoparticles *in situ* and by controlling the stimuli-responsiveness of the microgel, it is possible to control the distance between the NPs. These hybrid/composite microgels can be crystallised to fabricate photonic crystals (some of these examples are also given in Chapter 3). Xu *et al.* improved the crystallisation properties of these photonic crystals by encapsulating the composite microgels, poly(NIPAM-AA-HEA)/CdS/Ag, within a hydrophobic polymeric shell (Figure 9.12).[68] This gives dense,

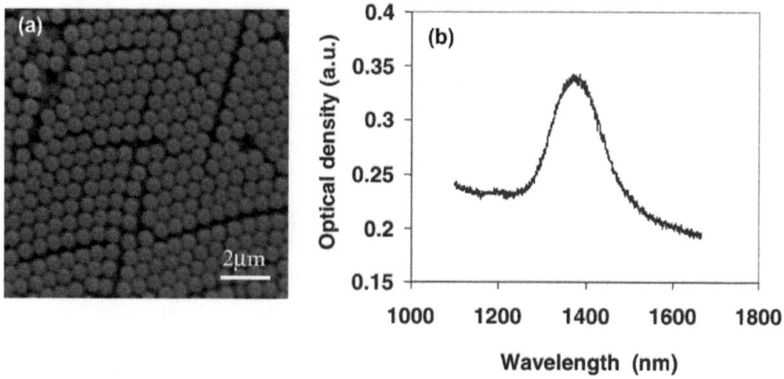

Figure 9.12 SEM image (a) and optical transmission (b) of photonic crystals, comprising a hybrid microgel, poly(NIPAM-AA-HEA), with CdS NPs. Reproduced from ref. 68.

negatively charged core–shell hybrid particles with narrow polydispersity and a smooth surface, allowing self-assembly into photonic crystals.

Microgels have also been used to synthesise a micro-lens array, commonly used in optical imaging systems, telecommunications and photolithography. Kim *et al.* utilised the pH-responsiveness of poly(NIPAM-AA) microgels to electrostatically control the assembly of these microgels on functionalised glass substrates.[69] The hemispherical shape of the microgels and refractive index contrast between the microgels and medium provide the lensing ability of the microgels. Also, the fast pH-responsiveness of the microgels allows easy tuning of the focal length of microlenses, enhancing the applicability of microgels in this optical application.

In another direction of study, photonic crystals (*i.e.* three-dimensional arrays of colloidal particles packed in regular patterns such as hexagonal or FCC packing), which can vary their optical properties (wavelength and width of band gap) in response to both changes in pH[70,71] and the presence of specific ions[72] or solvents,[73] were combined with various functional hydrogels. This led several authors to develop colour-based sensors, where a photonic crystal was embedded within a hydrogel matrix. In these systems, the responsive properties of the hydrogel are key to drive the changes in colour observed. Indeed, by forcing the hydrogel to swell or contract in response to changes in the chemical environment, the embedded colloidal array was forced to vary the inter-particle distance and thus its light diffraction properties.

An example from Tian *et al.* involved a system built from polymer latex particles and an acrylamide hydrogel, where the change in colour due to variation in the inter-particle distance in the colloidal array was driven by a response of the hydrogel to changes in humidity.[74] As seen in Figure 9.13, the authors were able to obtain drastic changes in band gap properties (and thus colour from grey to orange) of the developed device when the humidity varied from 20% to 100% over a significant number of cycles.

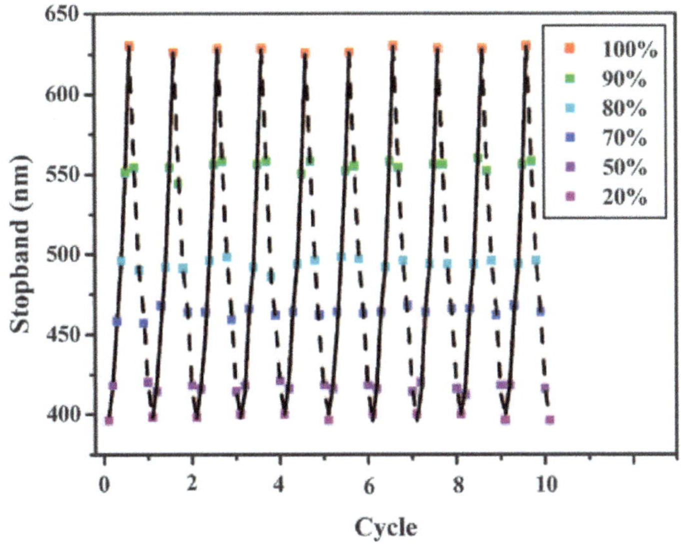

Figure 9.13 Variation in bandgap properties as a function of ambient humidity as a consequence of changes in inter-particle distance of a 3D array of tightly packed latex particles embedded within a humidity-sensitive hydrogel matrix.
Reproduced from ref. 74.

9.4 Conclusions

In this brief review, we have shown how advances in polymer synthesis capability over the last 10–20 years has allowed access to a much wider range of polymers with associated significant improvements in control over the molecular properties. This has opened up significant opportunities to design new materials for specific end use applications where the improved control over the polymer properties has allowed a significant amplification of the desired physical properties. The range of application examples given in this chapter highlight clearly this increased capability to design novel colloidal materials for highly-functional specific uses.

Abbreviations

2VP	2-Vinylpyridine
4VP	4-Vinylpyridine
5-FU	5-Fluorouracil
AA	Acrylic acid
AAEM	Acetoacetoxyethyl methacrylate
AC	Acryloyl carbonate
AD	(*N*-Amidino)dodecyl acrylamide
Ag	Silver
AlClPc	Aluminium chloride phthalocyanine
AM	Allyl methacrylate

BDDA	Butanediol diacrylate
CD	Cyclodextrin
CdS	Cadmium sulfite
DEA	*N,N*-Diethylaminoethyl methacrylate
DHEA	*N,N'*-(1,2-Dihydroxyethylene) bisacrylamide
DMAA	*N,N*-Dimethylacrylamide
DMAEMA	*N,N*-Dimethylamino-ethyl methacrylate
DOX	Doxorubicin
DTT	Dithiothreitol
EA	Ethyl acrylate
EG	Ethylene glycol
EGDMA	Ethylene glycol dimethacrylate
EI	Ethylene imine
EO	Ethylene oxide
Fc	Ferrocene
Fe_3O_4	Iron oxide
FMA	2,2,2-Trifluoroethyl methacrylate
GSH	Glutathione
HEA	2-Hydroxyethyl acrylate
HEMA	2-Hydroxyethyl methacrylate
IEP	Iso-electric point
IVD	Intervertebral disc
LA	Lactic acid
MAA	Methacrylic acid
MMA	Methyl methacrylate
NIPAM	*N*-Isopropylacrylamide
NP	Nanoparticle
ODA	Octadecyl acrylate
OEI	Oligoethylenimine
PBA	Aminophenyl boronic acid
Pd	Palladium
PEG	Poly(ethylene glycol)
Pt	Platinum
Py	Pyrrole
Sty	Styrene
SC4A	*p*-Sulfonatocalix[4]-arene
VCL	*N*-Vinylcaprolactam
VI	*N*-Vinylimidazole

References

1. O. J. Cayre, N. Chagneux and S. Biggs, Stimuli-responsive core-shell nanoparticles: synthesis and applications of polymer based aqueous systems, *Soft Matter*, 2011, 7, 2211–2234; S. Ganta, H. Devalapally, A. Shahiwala and M. Amiji, A review of stimuli-responsive nanocarriers for drug and gene delivery, *J. Controlled Release*, 2008, **126**, 187–204;

K. Letchford and H. A. Burt, Review of the formation and classification of amphiphilic block copolymer nanoparticulate structures: Micelles, nanospheres, nanocapsules and polymersomes, *Eur. J. Pharm. Biopharm.*, 2007, **65**, 259–269.
2. I. W. Hamley, *Block Copolymers in Solution: Fundamentals and Applications*, 2005, Wiley, Chichester.
3. L. F. Zhang, Y. Yu and A. Eisenberg, Ion-induced morphological changes in 'crew-cut' aggregates of amphiphilic diblock copolymers, *Science*, 1996, **272**, 1777–1779.
4. B. Li, M. A. Hillmyer and T. P. Lodge, Morphologies of multicompartment micelles formed by ABC miktoarm star terpolymers, *Langmuir*, 2006, **22**, 9409–9417.
5. J. Z. Du and R. K. O'Reilly, Advances and challenges in smart and functional polymer vesicles, *Soft Matter*, 2009, **5**, 3544–3561.
6. R. K. O'Reilly, C. J. Hawker and K. L. Wooley, Cross-linked block copolymer micelles: functional nanostructures of great potential and versatility, *Chem. Soc. Rev.*, 2006, **35**, 1068–1083.
7. B. R. Saunders and B. Vincent, Microgel particles as model colloids: Theory, properties and applications, *Adv. Colloid Interface Sci.*, 1999, **80**, 1–25.
8. B. R. Saunders *et al.*, Microgels: From responsive polymer colloids to biomaterials, *Adv. Colloid Interface Sci.*, 2009, **147–148**, 251–262.
9. B. H. Tan and K. C. Tam, Review on the dynamics and micro-structure of pH-responsive nano-colloidal systems, *Adv. Colloid Interface Sci.*, 2008, **136**, 25–44.
10. M. Das, H. Zhang and E. Kumacheva, Microgels: Old materials with new applications, *Annu. Rev. Mater. Res.*, 2006, **36**, 117–142.
11. A. Z. Pich and H.-J. Adler, Composite aqueous microgels: A overview of recent advances in synthesis, characterization and application, *Polym. Int.*, 2007, **56**, 291–307.
12. S. Dai, P. Ravi and K. C. Tam, pH-responsive polymers: Synthesis, properties and applications, *Soft Matter*, 2008, **4**, 435–449.
13. Y. Guan and Y. Zhang, PNIPAM microgels for biomedical applications: From dispersed particles to 3D assemblies, *Soft Matter*, 2011, **7**, 6375–6384.
14. L. A. Lyon and A. Fernandez-Nieves, The polymer/colloid duality of microgel suspensions, *Annu. Rev. Phys. Chem.*, 2012, **63**, 25–43.
15. M. Wolfe, Dispersion and solution rheology control with swellable microgels, *Prog. Org. Coat.*, 1992, **20**, 487–500.
16. B. Rodriguez, M. Wolfe and M. Fryd, Nonuniform swelling of alkali swellable microgels, *Macromolecules*, 1994, **27**, 6642–6647.
17. B. Barry and M. Meyer, The rheological properties of Carbopol gels I. Continuous shear and creep properties of Carbopol gels, *Int. J. Pharm.*, 1979, **2**, 1–25.
18. J. Madsen, I. Canton, N. Warren, E. Themistou, A. Blanaz, B. Ustbas, X. Tian, R. Pearsn, G. Battaglia, A. Lewis and S. Armes, Nile Blue-based

nanosized pH sensors for simultaneous far-red and near-infrared live bioimaging, *J. Am. Chem. Soc.*, 2013, **135**, 14863–14870.
19. M. Motornov, J. Zhou, M. Pita, V. Gopishetty, I. Tokarev, E. Katz and S. Minko, "Chemical transformers" from nanoparticle ensembles operated with logic, *Nano Lett.*, 2008, **8**, 2993–2997.
20. B. P. Binks, Particles as surfactants – similarities and differences, *Curr. Opin. Colloid Interface Sci.*, 2002, **7**, 21–41.
21. M. Motornov, J. Zhou, M. Pita, I. Tokarev, V. Gopishetty, E. Katz and S. Minko, An integrated multifunctional nanosystem from command nanoparticles and enzymes, *Small*, 2009, **5**, 817–820.
22. J. Wang, L. Wang, X. Liu, Z. Liang, S. Song, W. Li, G. Li and C. Fan, A gold nanoparticle-based aptamer target binding readout for ATP assay, *Adv. Mater.*, 2007, **19**, 3943–3946.
23. K. Paek, S. Chung, C.-H. Cho and B. J. Kim, Fluorescent and pH-responsive diblock copolymer-coated core-shell CdSe/ZnS particles for a colour-displaying, ratiometric pH sensor, *Chem. Commun.*, 2011, **47**, 10272–10274.
24. B. P. Binks, R. Murakami, S. P. Armes, S. Fujii and A. Schmid, pH-responsive aqueous foams stabilized by ionizable latex particles, *Langmuir*, 2007, **23**, 8691–8694.
25. S. Fujii, M. Mochizuki, K. Aono, S. Hamasaki, R. Murakami and Y. Nakamura, pH-responsive aqueous foams stabilized by hairy latex particles, *Langmuir*, 2011, **27**, 12902–12909.
26. J. Park, Z. Nie, A. Kumachev, A. I. Abdelrahman, B. P. Binks, H. A. Stone and E. Kumacheva, A microfluidic approach to chemically driven assembly of colloidal particles at gas-liquid interfaces, *Angew. Chem., Int. Ed.*, 2009, **48**, 5300–5304.
27. T. Ngai, S. H. Behrens and H. Auweter, Novel emulsions stabilized by pH and temperature sensitive microgels, *Chem. Commun.*, 2005, **3**, 331–333.
28. B. P. Binks, R. Murakami, S. P. Armes and S. Fujii, Effects of pH and salt concentration on oil-in-water emulsions stabilized solely by nanocomposite microgel particles, *Langmuir*, 2006, **22**, 2050–2057.
29. M. Destribats, M. Eyharts, V. Lapeyre, E. Sellier, I. Varga, V. Ravaine and V. Schmitt, Impact of pNIPAM microgel size on its ability to stabilize Pickering emulsions, *Langmuir*, 2014, **30**, 1768–1777.
30. D. Dupin, J. R. Howse, S. P. Armes and D. P. Randall, Preparation of stable foams using sterically stabilized pH-responsive latexes synthesized by emulsion polymerization, *J. Mater. Chem.*, 2008, **18**, 545–552.
31. M. D'Souza Mathew, M. S. Manga, T. N. Hunter, O. J. Cayre and S. Biggs, Behavior of pH-sensitive core-shell particles at the air-water interface, *Langmuir*, 2012, **28**, 5085–5092.
32. B. Brugger and W. Richtering, Magnetic, Thermosensitive microgels as stimuli-responsive emulsifiers allowing for remote control of separability and stability of oil in water-emulsions, *Adv. Mater.*, 2007, **19**, 2973–2978.

33. Y. Chen, Y. Bai, S. Chen, J. Ju, Y. Li, T. Wang and Q. Wang, Stimuli-responsive composite particles as solid-stabilizers for effective oil harvesting, *ACS Appl. Mater. Interfaces*, 2014, **6**, 13334–13338.
34. I. Robinson, C. Alexander, L. T. Lu, L. D. Tung, D. G. Fernig and N. T. K. Thanh, One-step synthesis of monodisperse water-soluble 'dual-responsive' magnetic nanoparticles, *Chem. Commun.*, 2007, 4602–4604.
35. D. Li and B. Zhao, Temperature-induced transport of thermosensitive hairy hybrid nanoparticles between aqueous and organic phases, *Langmuir*, 2007, **23**, 2208–2217.
36. D. Li and B. Sheng Zhao, Environmentally responsive "hairy" Nanoparticles: Mixed homopolymer brushes on silica nanoparticles synthesized by living radical polymerization techniques, *J. Am. Chem. Soc.*, 2005, **127**, 6248–6256.
37. Y. Guo and M. G. Moffitt, Semiconductor quantum dots with environmentally responsive mixed polystyrene/poly(methyl methacrylate) brush layers, *Macromolecules*, 2007, **40**, 5868–5878.
38. E. W. Edwards, M. Chanana, D. Wang and H. Mohwald, Stimuli-responsive reversible transport of nanoparticles across water/oil interfaces, *Angew. Chem., Int. Ed.*, 2008, **47**, 320–323.
39. K. Ogawa, B. Wang and E. Kokifuta, Enzyme-regulated microgel collapse for controlled membrane permeability, *Langmuir*, 2001, **17**, 4704–4707.
40. S. Nayak and L. A. Lyon, Ligand-functionalized core/shell microgels with permselective shells, *Angew. Chem., Int. Ed.*, 2004, **43**, 6706–6709.
41. G. Morris, B. Vincent and M. Snowden, Adsorption of lead ions onto *N*-isopropylacrylamide and acrylic acid copolymer microgels, *J. Colloid Interface Sci.*, 1997, **190**, 198–205.
42. Z. Wang, Y. Li, L. Chen, X. Xin and Q. Yuan, A study of controlled uptake and release of anthocyanins by oxidised starch microgels, *J. Agric. Food Chem.*, 2013, **61**, 5880–5887.
43. A. Kabanov and S. Vinogradov, Nanogels as pharmaceutical carriers: Finite networks of infinite capabilities, *Angew. Chem., Int. Ed.*, 2009, **48**, 5418–5429.
44. B. Ho, B. Tan, J. Tan and K. Tam, Inverse microemulsion polymerization of sterically stabilized polyampholyte microgels, *Langmuir*, 2008, **24**, 7698–7703.
45. H. Urakami, J. Hentschel, K. Seetho, H. Zeng, K. Chawla and Z. Guan, Surfactant-free synthesis of biodegradable, biocompatible and stimuli-responsive cationic nanogel particles, *Biomacromolecules*, 2013, **14**, 3682–3688.
46. N. Morinoto, S. Hirano, H. Takahashi, S. Loethen, D. Thompson and K. Akiyoshi, Self-assembled pH-sensitive cholesteryl pullulan nanogel as a protein delivery vehicle, *Biomacromolecules*, 2013, **14**, 56–63.
47. T. Zhou, C. Xiao, J. Fan, S. Chen, J. Shen, W. Wu and S. Zhou, A nanogel of on-site tunable pH-response for efficient anticancer drug delivery, *Acta Biomater.*, 2013, **9**, 4546–4557.

48. J. Ding, F. Shi, C. Xiao, L. Lin, L. Chen, C. He, X. Zhuang and X. Chen, One-step preparation of reduction-responsive poly(ethylene glycol)-poly(amino acid)s nanogels as efficient intracellular drug delivery platforms, *Polym. Chem.*, 2011, **2**, 2857–2864.
49. W. Chen, M. Zheng, F. Meng, R. Cheng, C. Deng, J. Feijen and Z. Zhong, In situ forming reduction-sensitive degradable nanogels for facile loading and triggered intracellular release of proteins, *Biomacromolecules*, 2013, **14**, 1214–1222.
50. T. Hoare and R. Pelton, Charge-switching, amphoteric glucose-responsive microgels with physiological swelling activity, *Biomacromolecules*, 2008, **9**, 733–740.
51. M. Stuart, W. Huck, J. Genzer, M. Muller, C. Ober, M. Stamm, G. Sukhorukv, I. Szleifer, V. Tsuruk, M. Urban, F. Winnik, S. Zauscher, I. Luzinov and S. Minko, Emerging applications of stimuli-responsive polymer materials, *Nat. Mater.*, 2010, **9**, 101–113.
52. A. Feng and J. Yuan, Smart nanocontainers: Progress on novel stimuli-responsive polymer vesicles, *Macromol. Rapid Commun.*, 2014, **35**, 767–779.
53. A. Hoffman, Stimuli-responsive polymers: Biomedical applications and challenges for clinical translation, *Adv. Drug Delivery Rev.*, 2013, **65**, 10–16.
54. E. Gillies, T. Jonsson and J. Frechet, Stimuli-responsive supramolecular assemblies of linear-dendritic copolymers, *J. Am. Chem. Soc.*, 2004, **126**, 11936–11943.
55. J. Taillefer, M.-C. Jones, N. Brasseur, J. Van Lier and J.-C. Leroux, Preparation and characterization of pH-responsive polymeric micelles for the delivery of photosensitizing anticancer drugs, *J. Pharm. Sci.*, 2000, **89**, 52–62.
56. W. J. Yan, Y. Yin and J. Yuan, Breathing polymersomes: CO_2-tuning membrane permeability for size-selective release, separation and reaction, *Angew. Chem., Int. Ed.*, 2013, **52**, 5070–5073.
57. Q. Zhang and S. Zhu, Oxygen and carbon dioxide dual responsive nanoaggregates of fluoro- and amino-containing copolymer, *Macro Lett.*, 2014, **3**, 743–746.
58. L. Peng, A. Feng, M. Huo and J. Yuan, Ferrocene-based supramolecular structures and their applications in electrochemical responsive systems, *Chem. Commun.*, 2014, **50**, 13005–13014.
59. Q. Yan, J. Yuan, Z. Cai, Y. Xin, Y. Kang and Y. Yin, Voltage-responsive vesicles based on orthogonal assembly of two homopolymers, *J. Am. Chem. Soc.*, 2010, **132**, 9268–9270.
60. K.-S. Tucking, S. Handschuh-Wang and H. Schonherr, Bacterial enzyme responsive polymersomes: A closer look at the degradation of PEG-*b*-PLA vesicles, *Aust. J. Chem.*, 2014, **67**, 578–584.
61. D.-S. Guo, K. Wang, Y.-X. Wang and Y. Liu, Cholinesterase-responsive supramolecular vesicle, *J. Am. Chem. Soc.*, 2012, **134**, 10244–10250.
62. J. Saunders, T. Tong, C. Le Maitre, T. Freemont and B. Saunders, A study of pH-responsive microgel dispersions: From fluid-to-gel transitions to

mechanical property restoration for load-bearing tissue, *Soft Matter*, 2007, **3**, 486–494.
63. T. J. Freemont and B. R. Saunders, pH-responsive microgel dispersions for repairing damaged load-bearing soft tissue, *Soft Matter*, 2008, **4**, 919–924.
64. J. Zhang, S. Xu and E. Kumacheva, Polymer microgels: Reactors for semiconductor, metal and magnetic nanoparticles, *J. Am. Chem. Soc.*, 2004, **126**, 7908–7914.
65. J. Zhang, S. Xu and E. Kumacheva, Photogeneration of fluorescent silver nanoclusters in polymer microgels, *Adv. Mater.*, 2005, **17**, 2336–2340.
66. A. Biffis and E. Sperotto, Microgel-stabilized metal nanoclusters: Improved solid-state stability and catalytic activity in Suzuki couplings, *Langmuir*, 2003, **19**, 9548–9550.
67. A. Pich, Y. Lu, V. Boyko, S. Richter, K.-F. Arndt and H.-J. P. Adler, Thermo-Sensitive poly(*N*-vinylcaprolactam-*co*-acetoacetoxyethyl methacrylate) microgels. 3. Incorporation of polypyrrole by selective microgel swelling in ethanol-water mixtures, *Polymer*, 2004, **45**, 1079–1087.
68. S. Xu, J. Zhang, C. Paquet, Y. Lin and E. Kumacheva, From hybrid microgels to photonic crystals, *Adv. Funct. Mater.*, 2003, **13**, 468–472.
69. J. Kim, M. Serpe and L. Lyon, Hydrogel microparticles as dynamically tunable microlenses, *J. Am. Chem. Soc.*, 2004, **126**, 9512–9513.
70. K. Lee and S. A. Asher, Photonic Crystal Chemical Sensors: pH and ionic strength, *J. Am. Chem. Soc.*, 2000, **122**, 9534–9537.
71. A. C. Sharma, T. Jana, R. Kesavamoorthy, L. Shi, M. A. Virji, D. N. Finewgold and S. A. Asher, A general photonic crystal sensing motif: Creatinine in bodily fluids, *J. Am. Chem. Soc.*, 2004, **126**, 2971–2977.
72. J. H. Holtz and S. A. Asher, Polymerized colloidal crystal hydrogel films as intelligent chemical sensing materials, *Nature*, 1997, **389**, 829–832.
73. H. Fudouzi and Y. Xia, Colloidal crystals with tunable colors and their use as photonic papers, *Langmuir*, 2003, **19**, 9653–9660.
74. E. Tian, J. Wang, Y. Zheng, Y. Song, L. Jiang and D. Zhu, Colorful humidity sensitive photonic crystal hydrogel, *J. Mater. Chem.*, 2008, **18**, 1116–1122.

CHAPTER 10

Responsive Membranes

SIAVASH DARVISHMANESH,[a] XIANGHONG QIAN*[b] AND S. RANIL WICKRAMASINGHE*[a]

[a] Ralph E. Martin Department of Chemical Engineering, University of Arkansas, Fayetteville, AR 72701, USA; [b] Department of Biomedical Engineering, University of Arkansas, Fayetteville, AR 72701, USA
*Email: xianghong.qian@uark.edu; ranil.wickramasinghe@uark.edu

10.1 Introduction

Membrane-based separation processes find numerous applications in many areas such as the chemical, pharmaceutical, food and biotechnological industries.[1] Membrane-based processes are attractive for a number of reasons. They often require lower energy input than competing processes, for example the desalination of sea water using reverse osmosis membranes,[2] they are easy to scale up, for example membrane adsorbers for bioseparations,[3] they are more environmentally benign, for example water treatment[4] or they are the only viable technology for niche applications, for example blood oxygenation.[5] Though successful membranes are manufactured for these applications, continued efforts are being made to produce more advanced membranes with tailored and improved properties.

Recently, there has been significant interest in developing stimuli-responsive membranes.[6,7] Inspired by nature, it is not difficult to imagine transport of chemicals or energy conversion processes at the molecular level, through synthetic materials that respond to an external stimulus. Membrane separations can be controlled through different mechanisms depending on the membrane structure (dense or porous) and the physicochemical properties of solution in contact with the membrane. Reflecting upon

RSC Smart Materials No. 14
Chemoresponsive Materials: Stimulation by Chemical and Biological Signals
Edited by Hans-Jörg Schneider

Published by the Royal Society of Chemistry, www.rsc.org

self-regulating biological membranes in nature, it is possible to imagine synthetic stimuli-responsive membranes where the performances of the membranes may be remarkably attuned in response to the changing environment.[8]

In general, for porous membranes, for example microfiltration membranes, where responsive groups located inside the pore surface of the membrane dominate, swelling or hydration of these responsive groups in the presence of an external cue will lead to the narrowing of the pores and hence reduced permeability. On the other hand, the opposite effect is often observed for dense membranes and membranes with very small pore sizes, for example nanofiltration (NF) membranes. If the responsive groups are predominately located on the membrane surface, swelling or hydration will lead to improved permeability.

Of particular interest is the considerable effort dedicated to the development of biologically inspired stimuli-responsive membranes. These smart separation barriers with switchable reversible physicochemical properties are attracting attention from various fields. There are a number of advantages arising from the use of stimuli-responsive membranes in the biomedical, pharmaceutical, fine chemical and other process industries. Currently stimuli-responsive membranes are key components in complex technological systems such as sensors, protein detection kits and drug delivery devices with potential application in tissue engineering.[7,9]

Synthetic stimuli-responsive membranes may be categorized into two groups. Numerous examples of external field responsive membranes have been described.[7] These include temperature, magnetic field, electric field and electromagnetic radiation (*e.g.* light). On the other hand, membranes responding to pH, salt ion, ionic strength and specific target/receptor interactions involve direct stimulation by chemical or biological cues. Our focus here is on the second direct stimulation of a responsive membrane by a specific target species.

This change in membrane properties in the presence of a target chemical or biological cue might happen through pore deformation across the membrane (internal effect) or/and to membrane surface properties (external effect), both of which affect the rate of mass transfer. Stimuli-responsive sites on the membrane at the molecular level must be activated and intensified to trigger membrane separation performance. Such changes have been used in many applications and devices to trigger the separation.[10] Here, an overview of stimuli-responsive membranes with their latest development is presented.

Preparation of membranes from stimuli-responsive materials has been achieved using only stimuli-responsive polymers and/or copolymers or by using these polymers as components of blends or as additives during membrane formation. Three methodologies have been developed for the development of stimuli-responsive membranes: (i) preparation of membranes from pre-synthesized stimuli-responsive polymers, copolymers, and polymer–additive mixtures. Membrane physical properties such as separation layer thickness, porosity, pore size, pore-size distribution and final barrier structure (symmetric *vs.* asymmetric) are adjusted in order to achieve the desired membrane performances. (ii) Localized polymerization of stimuli-responsive monomer and copolymer occurring in the presence of an initiator and a cross-linker. Responsive membranes with desirable mechanical strength can be obtained by varying the cross-linking degree. (iii) Surface modification of existing membranes by introducing responsive properties to the membrane surface leading to improved separation performance.[7,10] In the following sections, we describe three common types of direct stimuli-responsive membranes: pH-responsive, ionic strength-responsive and specific target/receptor based interactions.

10.2 pH-Responsive Membranes

Significant research studies have been devoted in recent years to the field of pH-responsive membranes. In response to a change in pH, these membranes show reversible structural and conformational changes leading to a reversible pore size change. Potential applications of these membranes include controlled drug (chemical) release, self-cleaning membranes, size and charge selective membranes, membranes as sensors, *etc.*

The performance of pH-responsive membranes is strongly dependent on their structures (porous or non-porous). Typically, for porous membranes, one or more layer(s) of pH-responsive polymer chains/brushes or cross-linked polymer gels (with pH-responsive groups) grafted on the pore walls can be used to trigger the change in membrane permeability and/or selectivity. The effective pore diameter of the membrane is controlled by the structural rearrangement of these grafted polymer chains. pH-responsive groups such as carboxyl, pyridine, imidazole, and dibutylamine, which are used to create or modify the membrane network (pore walls), can influence the adsorption (fouling) properties and function as a membrane adsorber. For non-porous membranes, the surface of the barrier layer is modified.

Since the hydration free energy is much larger for the charged group than the corresponding neutral one, a carboxyl group (*e.g.* acrylic acid) at a pH below its pK_a value, is protonated and charge neutral with limited hydration; accordingly, the volume of grafted polymer possessing the carboxyl group decreases. The polymer network swells when the pH exceeds its pK_a value as the carboxylic acid group becomes charged with enhanced hydration (Figure 10.1). Since protonation and deprotonation are strongly pH-dependent, particularly for weak acids, membrane pore size can be continuously adjusted by changing the pH, hence the rearrangement of the polymer chains, which then regulate the membrane separation performance. A pyridine group, on the other hand, at low pH, is protonated and positively charged. Polymers containing pyridine groups (*e.g.* polyvinyl pyridine) swell, due to increased hydration as well as the long-range electrostatic repulsion among these charged groups. As the pH increases, the polymer becomes less charged as the protonated pyridine groups release the protons. The degree of hydration decreases leading to polymer de-swelling.

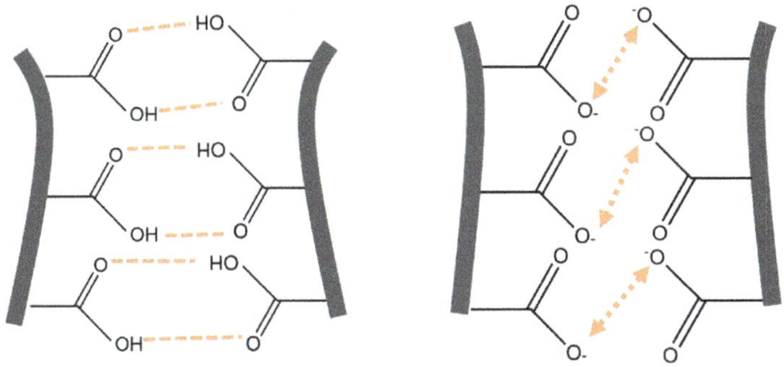

Figure 10.1 Effect of pH on the performance of a pH-responsive membrane functionalized by carboxyl groups.

Other functional groups, such as imidazole, dibutylamine and tertiary amine methacrylates have also been reported to be pH-responsive. These groups become positively charged at low pH and thus swell in acidic conditions.

It is worth mentioning that for non-porous membranes, the change in the degree of swelling of the polymeric network or grafted responsive layer affects the separation performances of the membrane. pH-responsive membranes may also be prepared by immersion precipitation in a pH-responsive graft copolymer solution on a support layer or even through layer-by-layer deposition of weak polyelectrolytes (PEL) on the membrane surface. In the latter approach, a pH stimulus occurs upon protonation/deprotonation of the ionizable groups leading to the structural change of PEL's top layer.

pH-Responsive membranes can be characterized using different experimental techniques such as:

Ion exchange capacity (IEC) measurement: In theory, the physical quantity of the anion or the cation groups in polymeric networks is referred to as the theoretical IEC, which is lower than measured IEC. IEC is measured by titration using a standard solution of NaOH or HCl. In general, the larger the IEC, the more tunable the membrane permeability becomes. The IEC of the reported pH-responsive membranes ranges from 0.01 to several mequiv. g^{-1}. A pH-responsive membrane with an IEC around 2 mequiv. g^{-1} can be used as an ion exchange membrane.

Membrane water flux, permeability and rejection of solutes as a function of pH: pH-responsive membranes spontaneously open and close their pores like a valve by means of swelling and shrinking in response to a change in solution pH. Normally, the pH of the feed solution is adjusted by adding HCl or NaOH. Water flux and solute rejection are then measured at a constant operating pressure at different pH values. Membranes containing carboxyl groups have a lower flux and higher rejection with increasing pH than membranes modified by pyridine. In the case of dense membranes, Donnan exclusion effects due to a charge on the membrane might also affect the rejection profile.

When producing membranes, the selection of the base membrane material is mainly driven by the necessity to achieve a high flux, high separation factor, high antifouling properties and high chemical and thermal stability. pH-responsive membranes are made from different polymers such as poly(ethersulfone) known as PES or poly(vinylidene fluoride) known as PVDF in combination with responsive copolymers as blend materials. Grafting of polymers on the surface of existing membranes has also been used to produce pH-responsive membranes. Recent progress in synthesizing pH-responsive membranes is summarized here.

pH-Responsive hollow fiber membranes were prepared using a blend of poly(methyl methylacrylate-*co*-4-vinyl pyridine) (PMMA–4VPy) and PES. Both polymers were dissolved in *N*,*N*-dimethylacetamide (DMAc) with the PES concentration at 20 wt% and the copolymer concentration ranging from 0 wt% to 3.6 wt%. The PES/PMMA–4VPy membranes are pH-responsive and adsorb Cu^{2+} ions.[11] PMMA–4VPy was prepared *via* free-radical solution polymerization. The effect of pH on water flux was studied. For PES membranes, no change in water flux was observed over the pH range 2.0 to 7.0. However, the water fluxes for the membranes prepared with poly (methyl methacrylate-*co*-vinyl pyridine) at concentrations of 0.6%, 1.2%, 1.8%, and 3.6% increased significantly.[11]

Biomimetic polydopamine (PD) coated PES ultrafiltration membranes were prepared and characterized for their hydrodynamic permeability, which is related to the membrane's intrinsic anti-fouling property, surface property by measuring platelet adhesion, and blood compatibility by measuring blood coagulation time (Figure 10.2). Modified PES membranes showed dramatic water flux decline, due to the coating of the PD layer. To avoid this, a pore-forming reagent was added to the dope solution and the membranes with tunable water flux were prepared. The blood compatibility of PD coated membranes was also enhanced. The surface hydrophobicity and water fluxes of the coated membranes can be adjusted by coating conditions such as coating time, pH and dopamine concentration. Platelet activation and adhesion decreased significantly. Fortunately, the PD layer did not interfere with the normal coagulation pathways, which enhanced blood compatibility.[12]

Hollow fiber membranes were prepared from blends of PES/poly(styrene-*alt*-maleic acid) (PSMA). PSMA was synthesized by free-radical polymerization of styrene (St) and maleic anhydride (MA). The membranes were incubated in NaOH solution at 50 °C and carboxylic hollow fiber membranes were then obtained. The carboxylic membranes displayed remarkable pH-responsive behavior due to the presence of the carboxyl groups. Under alkaline conditions, the carboxyl groups in the membranes deprotonate resulting in the swelling of the PSMA and a reduction in pore size. In the same way, when the feed solution is acidic, the carboxyl groups are protonated resulting in the de-swelling of PSMA and a higher water flux.[13]

pH-Responsive membranes were prepared from a blend of PES with poly(methyl methacrylate-*co*-maleic anhydride) (PMMA-MA). The copolymer was synthesized through free-radical polymerization. The membranes demonstrate significant pH-sensitivity and pH-reversibility due to the

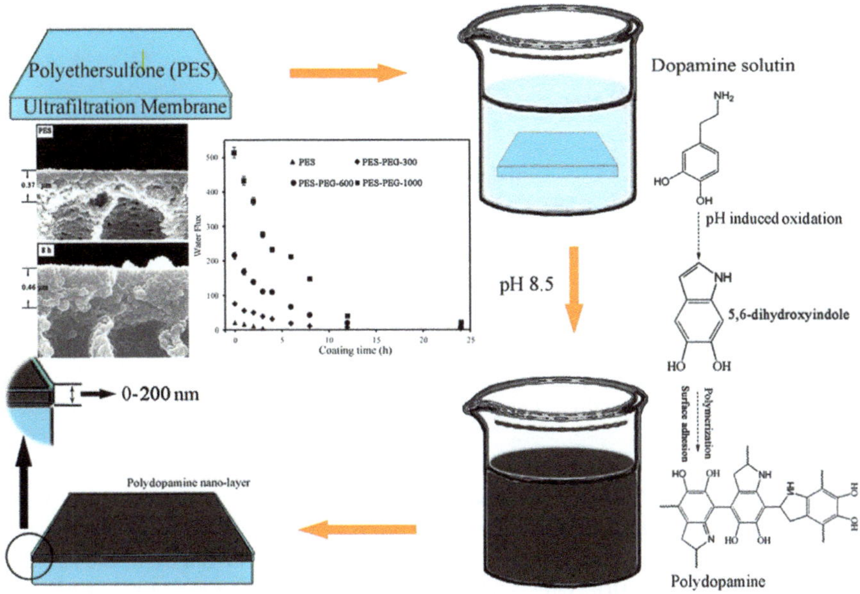

Figure 10.2 A graphic view of pH-prompted dopamine polymerization and adhesion onto a PES membrane.[12]

incorporation of carboxyl groups. The membranes also show hysteresis in water flux. The characteristic structure of the two adjacent COOH groups might lead to the hysteresis of ionization of these COOH groups. Through the permeability of the polyethylene glycol (PEG) solution, both pore size change and the electroviscous effects were studied. The calculated pore sizes for the membranes without copolymer showed little variation with pH. When the pH value increased from 3.0 to 11.0, the pore size decreased approximately three times for different concentrations of copolymer in the dope solution. At low pH, the carboxyl groups were protonated leading to polymer de-swelling and increased pore size; at high pH, the carboxyl groups became negatively charged leading to polymer swelling and decreased pore size. Furthermore, the modified hollow fiber membranes showed significant Cu^{2+} ion adsorption capacity. As a result, the modified hollow fiber membranes could be used for the removal of metal ions.[14]

In a similar study, poly(acrylonitrile-*co*-maleic anhydride) (PAN-MA) copolymer was synthesized by free-radical solution polymerization and then added to a PES polymer solution to prepare the blended dope solution. The hollow fibers displayed pH sensitivity and reversibility due to the presence of –COOH groups in the polymeric membrane matrix. Furthermore, the membranes were tested as membrane adsorbers for the removal of heavy metal ions. The membranes efficiently adsorbed Cu^{2+} from the aqueous solution. The membranes successfully adsorbed the other heavy metal ions tested including Ag^{+}, Fe^{2+}, and Fe^{3+} from their solutions. It was indicated that the membranes can be used to remove divalent heavy metal ions.[15]

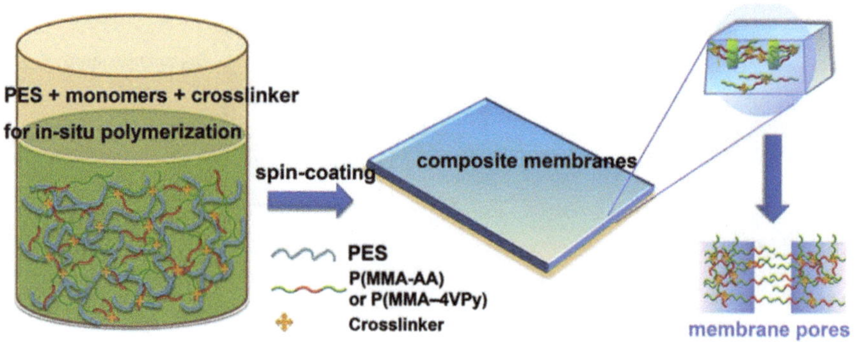

Figure 10.3 Assembly of PES/P(MMA-AA) and PES/P(MMA-4VPy) membranes *via in situ* cross-linked copolymerization.[16]

pH-Responsive membranes were prepared from blends of PES with copolymers *via in situ* cross-linked copolymerization of poly(methyl methacrylate-*co*-acrylic acid) (PMMA-AA) and PMMA–4VPy. The *in situ* cross-linking copolymerization was performed by adding three monomers as well as the initiator and the cross-linker to the PES/DMAc solution. The solution obtained was directly used to fabricate the membranes (Figure 10.3). It was shown that the PES/PMMA-AA and PES/PMMA–4VPy membranes displayed pH-responsive behavior over a wide range of pH. The anti-fouling property was enhanced due to the high-density chains at the membrane's pore surface. Moreover, the composite membranes showed good Cu^{2+} adsorption and the initial ion concentration affected the amount adsorbed.[16]

Thermo-pH responsive blend hollow fiber membranes were prepared by blending PES with a random terpolymer (a copolymer consisting of three distinct monomers) of poly(*N*-isopropylacrylamide-*co*-methacrylic acid-*co*-methyl methacrylate) P(NIPAM-MAA-MMA), which was synthesized *via* free-radical solution polymerization. Hollow fiber membranes were tested with PEG-8000 solutions at different pH values. The change in performance occurred at pH values between 7.0 and 10.0. The modified PES membranes were also reported to be thermo-responsive. The pore sizes could be altered indirectly by changing temperature. These membranes could be used for thermo-responsive separations as well.[17]

Nowadays, block copolymers are widely used in developing innovative membranes with uniform pore sizes due to their inherent ability to form highly ordered structures. Two amphiphilic block copolymers, polystyrene130-*block*-polyacrylic acid72 (PS130-*b*-PAA72) and PS130-*b*-PAA280, were blended with PES to fabricate pH-responsive composite membranes (Figure 10.4). Adding a relatively small amount of copolymer produced a membrane with a thinner skin layer than the base PES membrane. However, as the amount of blended copolymers increased, the thickness of the membrane skin layer increased, which decreased water flux. On the other hand, by increasing the fraction of blended PS130-*b*-PAA280 copolymers in

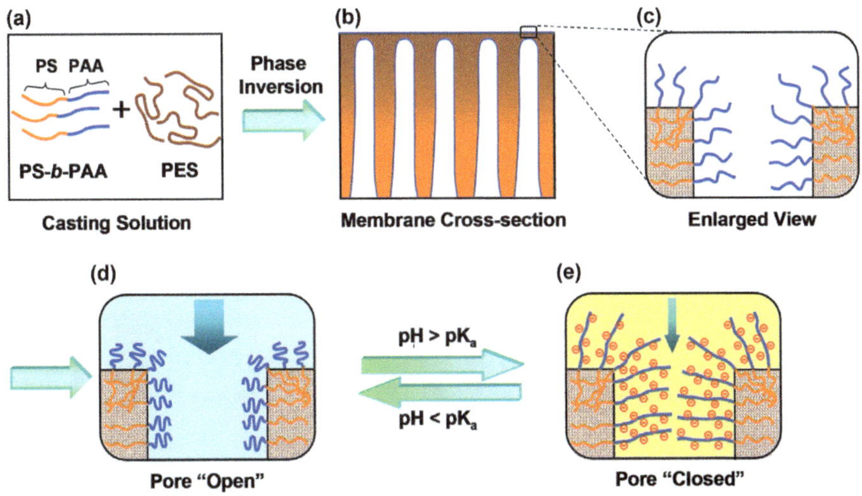

Figure 10.4 Graphic diagram of an assembly process and pH-responsive gating function of the stimuli-responsive membrane. Amphiphilic polystyrene-*block*-poly(acrylic acid) (PS-*b*-PAA) copolymers are mixed with poly(ether sulfone) (PES) (a) to fabricate an asymmetric PES composite membrane (b) with PAA blocks separated out on the membrane pore surface (c). The surface-segregated PAA chains on the membrane pore surfaces shrink at $pH < pK_a$ and thus result in the "open" state of the membrane pores (d), and swell broadly at $pH > pK_a$, resulting in the "closed" state of the membrane pores (e).[18]

the blend solution, the number of nano-pores increased across the membranes. These membranes display pH-responsive characteristics in the pH range from 3 to 8. As an example, the pore size of one of the composite membranes was reported to be 15.6 nm at pH 3 and increased to 7.0 nm at pH 8.[18]

Composite membranes with the mesoporous block copolymer PS-*block*-poly(2-vinyl pyridine) (P2VP) layer as the size-selective layer and the macroporous membrane as the robust supporting layer were generated by a selective swelling-induced pore-formation method. In this approach, block copolymer films were coated on a support layer, and then submerged in a bath of a swelling agent (polar solvent such as ethanol) to generate a mesoporous structure.

The generated composite membranes have high pore uniformity, robust mechanical properties, and a separating property that can be modified easily by changing the time of relaxation in the swelling agent bath. The active layer was reported to be between 4 to 7 μm thick and contained mesopores with diameters ranging from 8 nm to 30 nm, which affect the separation behavior of the membrane significantly (Figure 10.5). The membranes produced were able to completely separate 10 nm gold nanoparticles from a mixture containing 2 nm gold particles indicating the sharp membrane cut-off. Due to the presence of polyelectrolyte-natured PS-*block*-P2VP chains on the pore walls, the membranes possessed a gating effect demonstrating

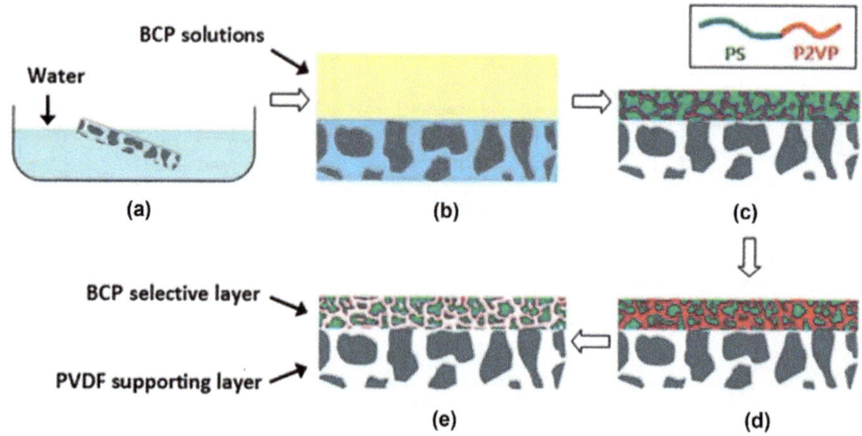

Figure 10.5 Graphic representation of the assembly processes of the composite membranes with mesoporous BCPs (PS-*b*-P2VP) as the selective layers and PVDF macroporous membranes as the supporting layers.[19]

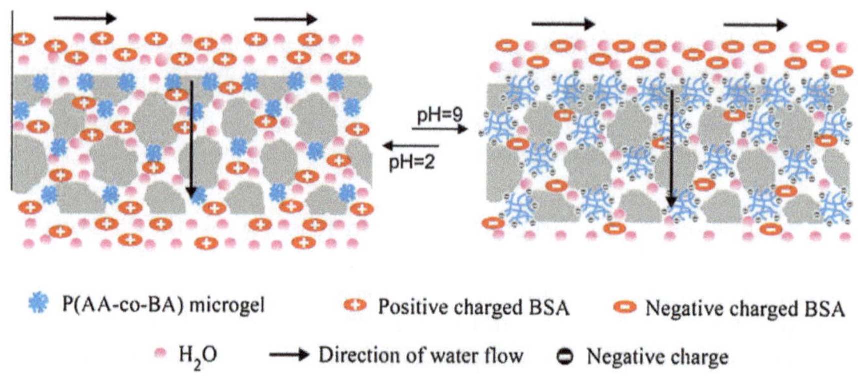

Figure 10.6 A schematic illustration of the response mechanism of pH-sensitive PVDF/P(AA-*co*-BA) membranes.[20]

reversible pH-responsive water flux. During membrane formation, PS-*block*-P2VP chains moved toward the membrane pore walls of the top layer thereby improving membrane hydrophilicity and fouling resistance.[19]

pH-Responsive membranes have been prepared by a phase inversion method using a blend solution containing poly (acrylic acid-*co*-butyl acrylate) (PAA-*co*-BA) microgels and PVDF (Figure 10.6). Both membrane surface and pore walls have been modified by PAA microgels. With the increase of the PAA-*co*-BA microgels in the blend solution, the membrane has larger pores, higher porosity, higher water flux, increased hydrophilicity and, consequently, enhanced anti-fouling property. Water flux displays a pH sensitivity between pH 3.7 and 6.3. When bovine serum albumin solution was filtrated, the flux reduced and membrane fouling increased gradually with increasing pH.[20]

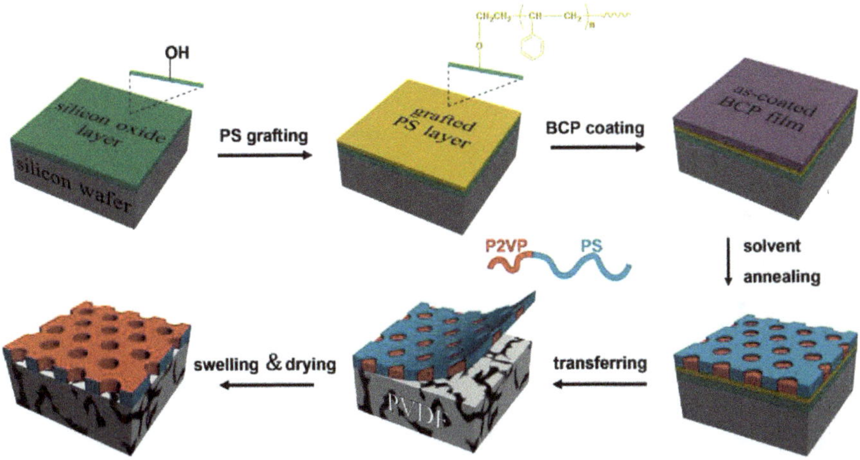

Figure 10.7 A graphic illustration of the fabrication process of isoporous BCP membranes supported on macroporous polyvinylidene fluoride (PVDF) membranes.[21]

Iso-structural syntheses of porous membranes from PS-*b*-P2VP, which have uniform and straight pores, hydrophilic properties and stimuli-responsive surfaces, have been developed on the surface of a PVDF membrane (Figure 10.7). Three critical steps were reported for membrane preparation: (1) perpendicular alignment of the PS-*b*-P2VP cylinders by solvent annealing, (2) relocating the block copolymer film onto the macroporous PVDF support, and (3) transformation of the solid PS-*b*-P2VP cylinders into straight pores by selective swelling in hot ethanol. The resulting membrane had a composite structure with the 85 nm thick iso-structural porous selective layer on top of the PVDF membrane. The membrane exhibits a very high water flux compared to the other isoporous membranes derived from block copolymers due to the high hydrophilicity of the highly permeable pores. Inherent pH-responsive functionality was observed through the presence of the polar PS-*b*-P2VP on the pore walls and membrane surface.[21]

Temperature-, pH-, and magnetically-responsive membranes were made by surface modification of PVDF. Poly(*N*-isopropylacrylamide) (PNIPAM), PAA, and Fe_3O_4 nanoparticles were attached to the PVDF membrane surface by ozone-mediated surface-initiated radical polymerization (Figure 10.8). PAA chains were attached to the PVDF membrane surface and amino-terminated poly(*N*-isopropylacrylamide) (PNIPAM-NH_2) was then joined onto the grafted PAA chains through an amidation reaction between the –COOH groups of the PAA chains and the amino groups of PNIPAM-NH_2. Magnetic Fe_3O_4 nanoparticles were then attached to the membrane surface through a reaction between the –COOH groups of the PAA chains and the –OH groups of the Fe_3O_4 nanoparticle surfaces. The stimuli-responsive property of the membrane was examined by measuring the water flux.

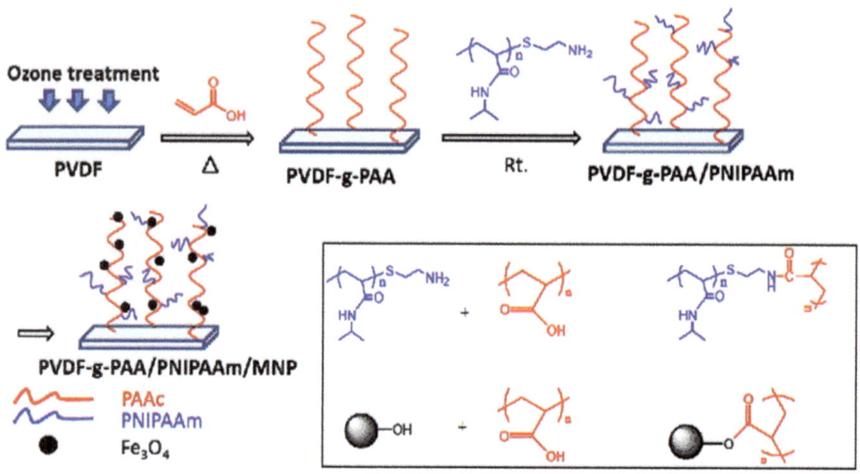

Figure 10.8 Synthesis route of a surface-modified PVDF membrane.[23]

A large change of water flux was observed between 30 to 42 °C under operation at pH 3.7 and pH 6.6, respectively. Under an alternating magnetic field, fouling was decreased significantly due to magnetically-induced micromixing as demonstrated earlier by Himstedt and coworkers.[22] The surface-modified PVDF membrane showed multiple stimuli-responsive characteristics with potential use in protein separation.[23]

Temperature- and pH-responsive porous membranes were developed through polymerization of 2-hydroxyethyl methacrylate with functional monomer methacrylic acid or NIPAM (Figure 10.9). The amount of functional poly(methacrylic acid) and PNIPAM chains on the membrane's surface was tunable by changing the amount of functional monomers in the polymerization solution. The poly(2-hydroxyethyl methacrylate/methacrylic acid) and poly(2-hydroxyethyl methacrylate/*N*-isopropylacrylamide) membranes showed responsive pH or temperature behavior. The flux of poly(2-hydroxyethyl methacrylate/methacrylic acid) membrane improved by 70% when the pH was reduced from 10.0 to 2.0. The flux of the poly(2-hydroxyethyl methacrylate/NIPAM) membrane increased by 150% when temperature was increased from 20 to 45 °C. The anti-fouling properties of the membrane were also strongly related to the pH or temperature of the filtration solution.[24]

Temperature- and pH-responsive cellulose membranes were produced by simultaneous grafting of PNIPAM and poly(2-(diethylamino) ethylmethacrylate) on a cross-linked cellulose membrane (Figure 10.10). The double-grafting process was achieved by using a diffusion device and initiated activators regenerated by electron transfer for atom-transfer radical polymerization. Surface-initiated atom-transfer radical polymerization (ATRP) is one of the favorable methods for grafting the membrane. ATRP initiators are first immobilized on the membrane surface, and responsive polymers are then grafted from the initiator sites. One of the benefits of ATRP is the ability to relatively independently vary chain length and density. Responsive

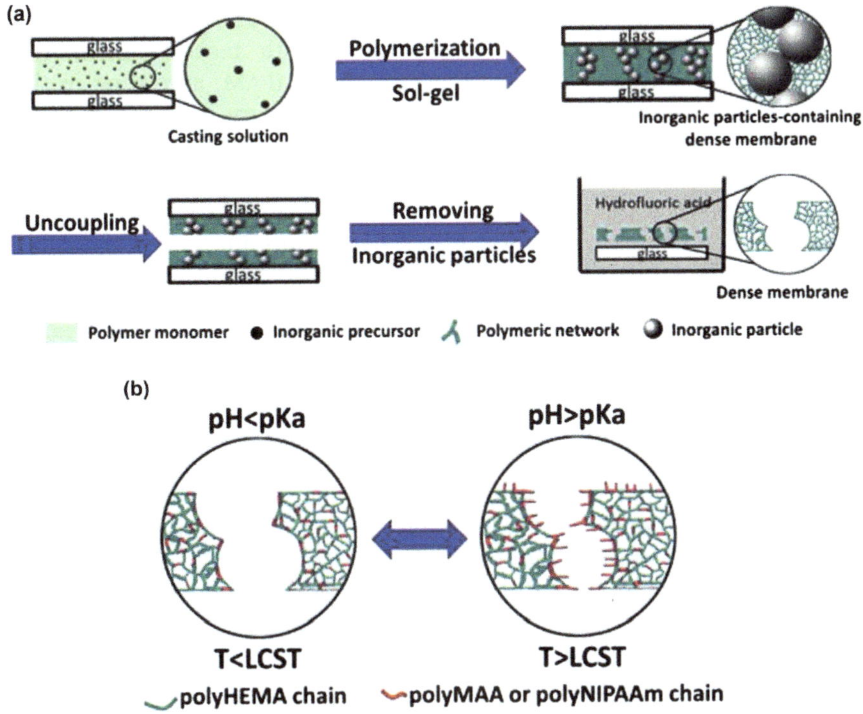

Figure 10.9 (a) Synthesis of stimuli-responsive polyhydroxyethylmethacrylate (polyHEMA) membranes. (b) A schematic representation of the pH- and temperature-responsive properties of poly(HEMA/MAA) and poly(HEMA/NIPAAm) membranes.[24]

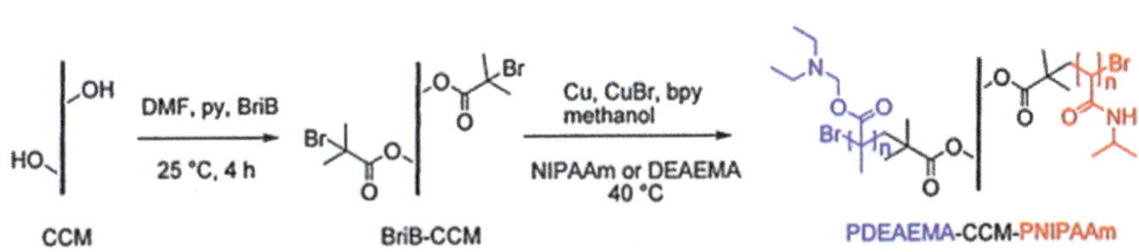

Figure 10.10 Surface-initiated activators regenerated by electron transfer (ARGET)-ATRP of PDEAEMA–CCM–PNIPAAm from a cross-linked cellulose membrane.[25]

membranes by surface-initiated ATRP are normally made through multiple steps after immobilization of ATRP initiators on the membranes. ATRP is a redox-initiated polymerization reaction; the transition metal ion is vulnerable to reaction with oxygen, and the activity of the metal ion is reduced in contact with oxygen. Thus ATRP polymerization needs an oxygen-free atmosphere during the whole synthesis route.

Surface-initiated activators regenerated by electron transfer were established to allow reactions to be performed in the presence of a limited amount of oxygen by adding reducing agents. After the reaction, a membrane with high grafting thickness of the stimuli-responsive polymers was obtained. Thicknesses increased linearly with reaction time. The membrane was hydrophilic below the LCST of PNIPAM, according to the results of water contact angle measurements, whereas this surface was hydrophobic above the LCST. The poly(2-(diethylamino)ethyl methacrylate) (PDEAEMA)-grafted surface showed wettability changes in response to pH changes between acidic and basic conditions. The hydrophilicity of the (PDEAEMA–cross-linked cellulose–PNIPAM) membrane was reversible on changing the temperature of the solution between 20 °C and 50 °C and between pH 1.0 and pH 7.0. This dual-responsive cellulose membrane was reported as a potential membrane for specific applications in water treatment, targeted separations and drug delivery *etc.*[25]

In a recent study, Himstedt *et al.* grafted PAA chains from the surface of commercial nanofiltration membranes using UV initiated free-radical polymerization. It was shown that at higher pH values the grafted chains change their conformation and swell because of deprotonation of carboxylic groups. The switch between more swollen and collapsed conformations is reversible by changing the pH of the solution.[26,27] The investigators considered the use of these responsive nanofiltration membranes for sugar separations.

10.3 Ionic Strength Responsive Membranes

Solution ionic strength affects the degree of swelling that is observed for pH-responsive membranes. For example, in a high ionic strength solution, deprotonated PAA chains swell much less due to charge shielding by small ionic species present in the solution. Thus all pH-responsive membranes also respond to ionic strength when acid groups are deprotonated and base groups are protonated. However, the polymers used to prepare these responsive membranes do not need to be neutral. Polyelectrolytes are polymers containing multiple ionizable groups. Strong acid or base groups (*e.g.* quaternary amines) are generally insensitive to pH. However, at high salt concentration, when the ionic strength of the solution approaches that inside the polyelectrolyte, electrostatic screening results in conformational changes. These membranes respond to changes in ionic strength but not pH.[10]

In a recent publication, de Grooth *et al.*[28] describe the formation of poly(zwitterion) based multilayer membranes that respond to ionic strength. They modified 10 kDa hollow fiber ultrafiltration membranes using poly(zwitterion) poly(*N*-(3-sulfopropyl)-*N*-(methacryloxyethyl)-*N*,*N*-dimethylammonium betaine) (PSBMA), combined with the cationic poly(diallyldimethylammonium chloride) (PDADMAC). Zwitterionic multilayers were formed *via* the layer-by-layer (LbL) deposition technique. The ionic strength during film deposition was found to influence the layer growth.

The zwitterionic systems display a number of interesting effects. An 'antipolyelectrolyte effect' can be observed. Unlike weak polyelectrolytes (see Section 10.2: pH-responsive membranes) where increasing ionic strength leads to reduced swelling, increasing ionic strength for strong polyelectrolytes leads to an increase in swelling by the disruption of the strong electrostatic interactions between the oppositely charged groups in the zwitterionic layers. de Grooth *et al.* observed an increase in membrane permeability of more than 100% at 1.5 M NaCl compared to the permeability in the absence of NaCl. This is due to the increased swelling of the zwitterionic layers at higher salt concentrations. The ability to adjust membrane permeability based on solution ionic strength independent of changes in pH could lead to a new class of ionic strength responsive membranes.

Recently, a number of investigators have proposed the use of stimuli-responsive membranes for hydrophobic interaction chromatography (HIC).[29,30] HIC depends on reversible interactions between the hydrophobic patches on proteins and hydrophobic ligands. Proteins are loaded in a high ionic strength buffer, usually containing ammonium sulfate and eluted at low ionic strength. Recent studies have highlighted the advantages of membrane-based HIC compared to traditional resin-based HIC.[29,30]

The effect of solution ionic strength is very different to the 'anti-polyelectrolyte effect' noted for poly(zwitterion)-based multilayer membranes. Previous investigators[29,30] attached poly (*N*-vinylcaprolactam) (PVCL) at the surface of PVDF or regenerated cellulose membranes. PVCL exhibits a lower critical solution temperature (LCST) at around 30 °C. When grafted to a support membrane, the grafted PVCL nanolayer will swell and collapse below and above its LCST[31] respectively. At temperatures above the LCST, the collapsed and more dehydrated conformation of PVCL promotes interactions with the hydrophobic patches on the proteins in solution. Below the LCST, the swollen, more hydrated, conformation of PVCL promotes desorption of adsorbed proteins.

Solution ionic strength affects the LCST of polymeric species.[32] As the ionic strength increases, the LCST decreases. The actual decrease in LCST depends on the ionic species present.[33,34] In the case of PVCL, the LCST reduces from about 30 °C in DI water to 20 °C in a 1 M KCl solution.[32] Consequently, when conducting HIC at 25 °C at high ionic strength, PVCL adopts a dehydrated conformation promoting protein interactions. In low salt concentrations, the LCST rises above room temperature, causing the PVCL to swell and promote protein desorption. Using a responsive ligand that changes its conformation during loading and elution, higher protein recoveries were obtained.[29] Development of stimuli-responsive ligands that promote protein adsorption during loading and desorption during elution could lead to the development of membrane adsorbers that display higher capacities and recoveries than conventional membrane adsorbers that make use of non-responsive ligands.

10.4 Biologically-responsive Polymer Systems

Biologically-inspired membranes are becoming more and more important in the framework of controlled drug delivery, bio-sensing, smart films/matrices for biomedical applications, *etc.* Contemporary research has focused on stimuli-responsive polymers and smart membranes that respond to biological molecules (biomolecules). Membrane polymers can be mixed by biological components (prepared *via* phase inversion) or the chosen membrane can be functionalized with specific functional groups to interact with biologically-related species (grafting to/from the surface).[35] Common biologically inspired responsive polymers and membranes with high potential applications are listed below:

- Antigen-responsive membranes
- Redox/thiol-responsive membranes
- Enzyme-responsive membranes
- Glucose-responsive polymers

Each category is briefly described here.

10.4.1 Antigen-responsive Membranes

Antigen–antibody interactions, or antigen–antibody reactions, are very specific interactions throughout the immune response between antibodies produced by white blood cells and antigens.

Interactions between antigens and antibodies occur through a variety of non-covalent interactions, such as *van der Waals* interactions, electrostatic interactions, hydrogen bonding, and hydrophobic interactions. Antibodies are employed in a number of immunological tests and, due to their specific interactions with antigens, have been employed in a variety of responsive polymeric structures.[36–39] Generally, antigen–antibody interactions have been used in polymeric structures in the following way:

- To initiate and magnify responses in hydrogels containing physically entrapped antibodies.
- In the polymerization of monomers with antibodies or antigens in networks.
- Employment of antigen–antibody pairs as reversible cross-linkers within polymeric networks.

Antigen-responsive membranes were prepared with antibody/antigen moieties within the network on a supported layer (Figure 10.11). A cross-linked dextran was grafted with both an antigen and an antibody onto the surface of a nylon support. The permeability of large solutes through these membranes is dependent on the presence of the soluble antigen. The permeability of these membranes was tested by blue-dextran diffusion. Addition and removal of the antigen in sequence showed that these permeability changes are reversible. When free antigen was added to the membrane, antibody–antigen cross-links were broken by competitive binding of the free antigen. The disruption of the cross-links resulted in reversible hydrogel swelling. A potential application is membrane gates for selective diffusion in response to the presence of an antigen.

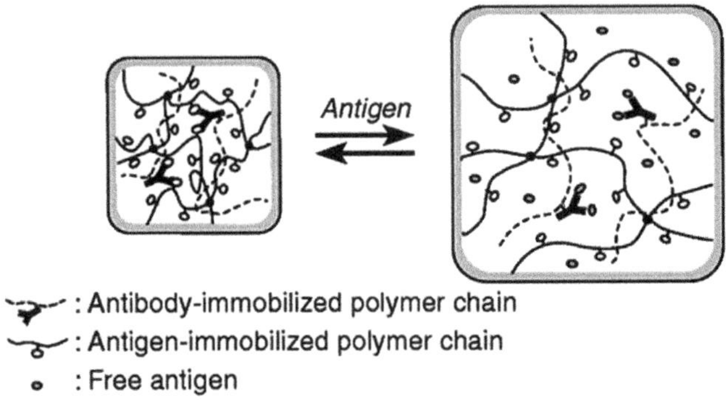

Figure 10.11 Illustration of a suggested mechanism for the swelling of an antigen-antibody semi-inter penetrating hydrogel in response to a free antigen.[36]

$$2\,RSH \rightleftarrows RS - SR + 2\,H^+ + 2e^-$$

Scheme 10.1

10.4.2 Redox/Thiol-responsive Membranes

Redox/thiol-responsive membranes have been explored mainly in the field of controlled drug delivery. Disulfide bonds are usually formed from the oxidation of sulfhydryl (–SH) groups, especially in biological environments.[40] The chemical conversion is presented in Scheme 10.1.

Disulfide bonds can be reversibly converted to thiols by exposure to various reducing agents and/or in the presence of other thiols. Polymers containing disulfide linkages can be considered both redox- and thiol-responsive.

The successful application of gene therapy is largely dependent on the development of a membrane system that can selectively and efficiently deliver the gene or therapeutic agents to the target cells with minimal toxicity.[41–43] A successful drug carrier was synthesized by copolymerization of a (1) pyridyl disulfide containing acryloyl monomer with (2) methacrylic acid and (3) butyl acrylate. The resulting terpolymers were pH- and thiol-sensitive and demonstrated membrane-disruptive properties necessary for effective gene delivery.

10.4.3 Enzyme-responsive Membranes

Enzyme-responsive polymeric systems undergo macroscopic property changes when the systems are activated by the effective catalytic actions of enzymes. The sensitivity of enzyme-responsive systems is very high, since enzymes are extremely selective. Enzyme-responsive materials are typically composed of two parts: (1) an enzyme-sensitive substrate and (2) a component that controls interactions that lead to macroscopic changes. The macroscopic structure of the polymeric substrate can change in response to the catalytic action of the enzyme, for example swelling/shrinking of the structure, or change of surface properties. This approach was used to synthesize a nanostructured signal-responsive thin hydrogel membrane coupled with enzyme-based systems. Membrane structure changes occurred by biochemical input signals through the enzyme system. The enzyme systems transduced and amplified the biochemical signals and their transformation into gated transport of molecules through the membrane. This membrane was suggested for drug delivery systems that control the release of drugs in response to chemical signals received *in situ* from tissues or blood.[44–47]

10.4.4 Glucose-responsive Polymers

Polymers that respond to glucose have received considerable attention because of their potential applications in both glucose sensing and insulin delivery applications. In such a smart delivery system, insulin release is automatically triggered by a rise in blood glucose levels (see Chapter 2 for more details). In this context, glucose-responsive polymers can be used for developing a self-regulated insulin delivery system.[48]

Glucose-responsive polymeric systems are typically based on:

10.4.4.1 Enzymatic Oxidation of Glucose by Glucose Oxidase

An example is a glucose-responsive membrane that was prepared for insulin release by grafting pH-responsive PAA chains onto a porous membrane support. Glucose oxidase was then immobilized on the pH-responsive PAA chains (Figure 10.12). The response time of the membrane to glucose was reported to be very short.[49]

10.4.4.2 Binding of Glucose with Concanavalin-A

Another insulin delivery system was based on the competitive binding of insulin by glucose from concanavalin-A implemented in a polymeric membrane device that is permeable to glucose and insulin and non-permeable to concanavalin-A, which can be implanted into the peritoneal cavity.[50]

10.4.4.3 Reversible Covalent Bond Formation between Glucose and Boronic Acids

Generally, boronic acid-based responsive polymers have been hydrogels that swell or collapse when exposed to glucose. Boronic acids can be made to reversibly complex with sugars. They are generally employed as glucose sensors and as ligand moieties for chromatography purposes.[51] The water solubility of boronic acids can be tuned by changes in diol concentration due to the presence of an anionic boronate group (see Chapter 3).

Polymers containing neutral boronic acid groups are generally hydrophobic, however, the presence of anionic boronate groups increases water

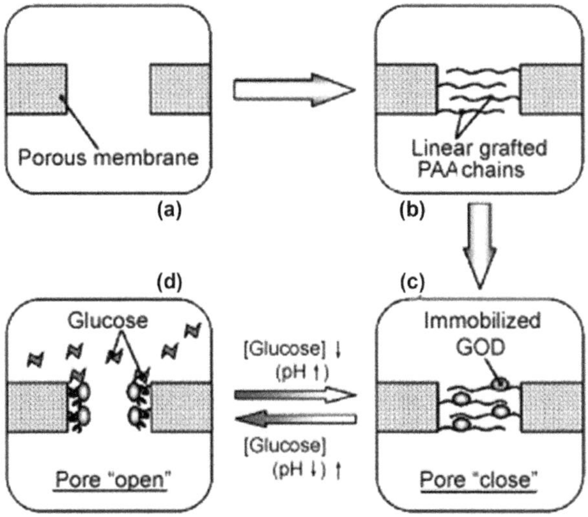

Figure 10.12 Schematic illustration of the preparation process route and the principle of glucose-responsive control of the permeation through the gating membrane: (a) porous membrane substrate; (b) pH-responsive gating membrane with poly(acrylic acid) (PAA) gates prepared using a plasma-graft pore-filling polymerization to graft linear PAA chains into the pores of the membrane substrate; (c) glucose-responsive gating membrane prepared by immobilizing glucose oxidase (GOD) onto the PAA-grafted membrane. At neutral pH in the absence of glucose, the carboxyl groups of the grafted PAA chains are dissociated and negatively charged, therefore the membrane gates "close" because the repulsion between negative charges make the PAA chains extend; and (d) when glucose concentration increases, GOD catalyzes the oxidation of glucose into gluconic acid, thereby lowering the local pH in the microenvironment, protonating the carboxylate groups of the grafted PAA chains, therefore the gates "open" because of the reduced electrostatic repulsion between the grafted PAA chains in the pores.[49]

solubility. As the concentration of glucose is increased, the ratio of the anionic form to the neutral form increases, and the hydrophilicity of the system increases and the polymer swells. Therefore the solubility of boronic acid-containing polymers is dependent on the concentration of compatible diols in the surrounding medium.

10.5 Insights from Quantum Mechanical and Molecular Mechanical Simulations

Membranes that respond to pH and salt ions or ionic strength in solutions have been investigated extensively experimentally as already described. However, simulations of these responsive membranes at the molecular level remain limited. Our understanding of these systems is mostly based on experimental observations. Several recent studies[33,34,52–56] using quantum mechanical and molecular mechanical calculations to gain insights into the pH and salt ion responsiveness of membrane properties have been conducted and are briefly described here. Since the pH responsiveness critically depends on the pK_a value of the polymeric acid grafted or deposited on the membrane's surface, accurate determination of the pK_a value is necessary in order to understand membrane behavior. It is known that the pK_a values of acids or conjugate bases are strongly dependent on their local environments. The pK_a value of a polymeric acid is usually different from its corresponding monomer value due to the change in its chemical and electrostatic environment. Combined quantum mechanical and molecular mechanical calculations show that the pK_a values are found to increase by 2 log units for both strong and weak polymeric acids.[53] First, pK_a relates to the dissociation of the first proton in a compound with multiple acid groups. Figure 10.13 shows the first pK_a value of the acid sites except the two end groups (mid) as well as the average first pK_a value for all the sites on the methacrylic acid oligomers as a function of degree of polymerization (DP). It can be seen that the first pK_a value increases as DP increases up to about 8–10. The pK_a value remains more or less the same for longer polymer chains at around 7.4 in excellent agreement with experimental results.

Besides pH, salt or ionic strength responsive polymers have also been investigated using classical molecular dynamics simulations.[33,34,55,56] PNIPAM is a thermo-responsive polymer exhibiting a coil-to-globule phase transition at its lower critical solution temperature (LCST) in aqueous solution.[57] Below its LCST, PNIPAM adopts a strongly hydrated extended coil conformation. Above its LCST, the polymer becomes partially dehydrated and forms a folded globule conformation. The LCST for PNIPAM in pure water is around 305 K. Besides temperature, PNIPAM is also responsive to salt ions and their concentrations. Salt ions have been shown to suppress the

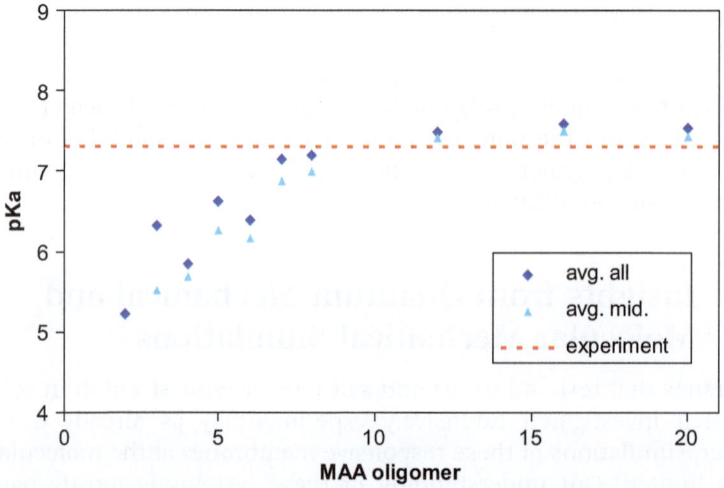

Figure 10.13 Average pK_a values for methacrylic acid oligomers with DP = 2–20.[53]

LCST and significantly affect its coil-to-globule transition. Most salts are found to shift the transition temperature. Earlier studies show that anions tend to decrease the transition temperature.[57] The effect of salt anions on LCST reduction follows the order $CO_3^{2-} > SO_4^{2-} > S_2O_3^{2-} > H_2PO_4^{-} > F^{-} > Cl^{-} > ClO_4^{-} > Br^{-} > NO_3^{-} > I^{-} > SCN^{-}$, similar to the Hofmeister series. Recent studies for PNIPAM in various sulfate salt solutions show that cations exert significant impact on PNIPAM LCST transition as well as on its wettability and swellability.[58] In particular, it has been found that cationic effects depend strongly on the charges of the ions. It appears that trivalent cations, such as the Al^{3+} ion, have the strongest effect on reducing LCST followed by the divalent ions, and finally the monovalent ions with the order $Al^{3+} > Zn^{2+} > Mg^{2+} > NH_4^{+} > Na^{+} > Li^{+}$ when the salt solution activity rather than salt concentration is taken into account. Some researchers attribute this ion specificity to the effects of salt ions on bulk water,[59,60] whereas others ascribe these ion-specific phenomena to the direct and specific ion–ion–polymer interactions.[61] A fundamental understanding of salt effects on chemical and biological processes is critical to many chemical and biological applications.

In order to understand the ion specificity of the LCST transitions of the thermo-responsive PNIPAM in various salt solutions, classical molecular dynamics (MD) simulations were conducted for PNIPAM in salt solutions. Here the specific cation–PNIPAM interactions in chloride salt solutions with monovalent Li^{+}, Na^{+}, K^{+}, Rb^{+} and Cs^{+} ions and divalent Ca^{2+} and Mg^{2+} ions were investigated. Experimentally, it was found that the effects of cations on LCST are sensitive to the anions present in the salt solution.[62] For chloride salt solutions involving alkali ions, the LCST of PNIPAM is reduced to 298 K in 1 M LiCl and to 293–294 K in 1 M NaCl, KCl, RbCl and CsCl salt solutions,[63,64] a reduction of 7–12 K compared to its value in pure water. For the divalent Mg^{2+} and Ca^{2+} ions, so far, the LCSTs of PNIPAM in their chloride salt solutions have not been measured experimentally, but it is expected to be influenced even more strongly by the salt ions. The details of the simulations conducted have been included in our earlier publications.[33,34,65]

For monovalent salts, MD simulations were performed for a 50 DP PNIPAM in 1 M LiCl, NaCl, KCl, RbCl and CsCl solutions above and below their respective LCSTs. The radius of gyration, the number of water molecules associated with the first hydration shell of the polymer and the pair distribution functions between the salt ions and various atoms on PNIPAM were evaluated. The water number in the first hydration shell includes all the water molecules less than 3.5 Å away from the heavy atoms on the polymer surface as was defined earlier.[34,55] All of the simulations start with a coil-like initial structure for PNIPAM. At simulation temperatures above LCST, a coil-to-globule transition was observed before the first 200 ns simulations. However, when starting with a folded collapsed initial structure, a globule-to-coil transition at simulation temperature below LCST was not observed in the course of several hundred ns simulations for all the salt systems. The

Table 10.1 The estimated radius of gyration (R_g) and first hydration water number (N_w) of PNIPAM at temperatures above (T_H) and below (T_L) its estimated LCST in various salt solutions.[56]

	LiCl	NaCl	KCl	RbCl	CsCl	$MgCl_2$	$CaCl_2$
R_g@T_H (Å)	10	10	10	11	11	10	10
R_g@T_L (Å)	18	17	22	20	14	21	21
N_w@T_H	340	360	360	370	370	350	360
N_w@T_L	500	510	570	550	490	590	600

time scale required for PNIPAM unfolding appears to be much longer than that of the folding process.

The radius of gyration in the salt solutions was calculated during the entire course of MD simulations. The calculated average radius of gyration at lower temperatures is estimated to be at around 18, 22, 20 and 14 Å for PNIPAM in Li^+, K^+, Rb^+ and Cs^+ chloride solutions, respectively, as shown in Table 10.1. The radius of gyration at higher temperatures appears to be rather similar to each other in the range between 9 and 11 Å for PNIPAM in all these salt solutions. It is clear that at a lower temperature, when the polymer is hydrated, the size of the polymer or degree of hydration is salt dependent. Further, the dynamics of the PNIPAM's coil-to-globule transition at a higher temperature also depends on the specific cations present in the salt solutions. The number of first hydration water molecules associated with the polymer determined within the radius of 3.5 Å from the PNIPAM surface is plotted in Figure 10.14. When the polymer is collapsed in a partially dehydrated globular state, the first hydration water numbers are close to 350 in all cases. However, in the extended coil conformation, the first hydration water number is strongly cationic-dependent, with the highest water number observed at around 600 in KCl and the lowest one at around 500 in CsCl. This difference in the degree of hydration agrees with previous experimental results on the water contact angle and wettability for PNIPAM in different salt solutions.[58] However, the differences in radius of gyration and degree of hydration do not appear to be correlated with the LCST transition temperature in these salt solutions. The estimated radius of gyration and the first hydration water number of PNIPAM in different salt solutions above and below its LCST are listed in Table 10.1.

MD simulations for PNIPAM in the 0.5 M divalent $MgCl_2$ and $CaCl_2$ solutions were conducted at 320 K and 280 K, respectively, at temperatures above and below their expected LCSTs. The coil-to-globule transition appears to occur rather rapidly in less than 20 ns in both salt solutions as shown in Figure 10.15. The radii of gyration for both coiled conformations at 280 K are in the range between 21 and 24 Å. PNIPAM at the lower temperature appears to be highly hydrated in $MgCl_2$ and $CaCl_2$ solutions with the first hydration water numbers varying slightly in the 580–600 range as shown in Figure 10.15 and listed in Table 10.1. Similarly, the radii of gyration and first hydration water numbers for the polymer at a higher temperature after the coil-to-globule transition are only slightly different for the two salt solutions.

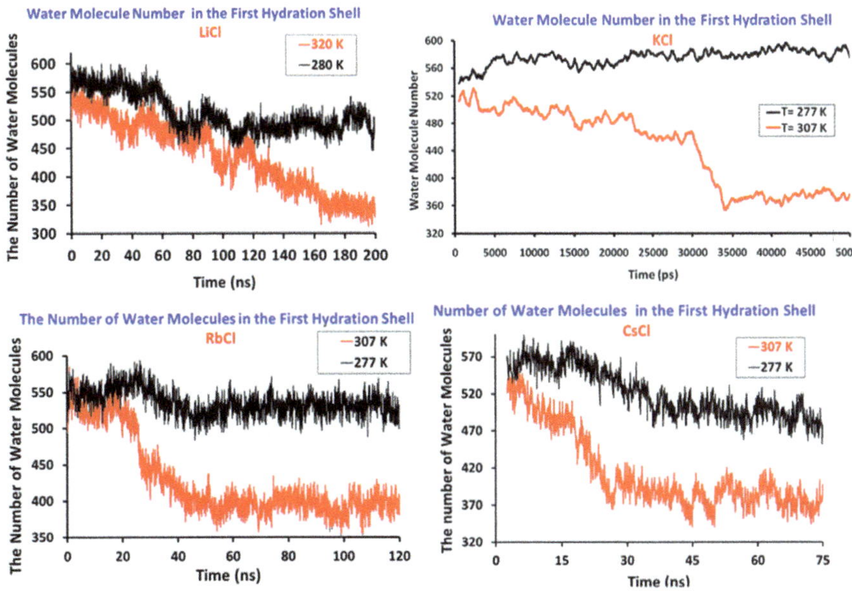

Figure 10.14 The number of water molecules in the first hydration shell of PNIPAM in 1 M alkali chloride (LiCl, KCl, RbCl and CsCl) salt solutions during MD simulations at temperatures above and below their respective LCSTs.[55]

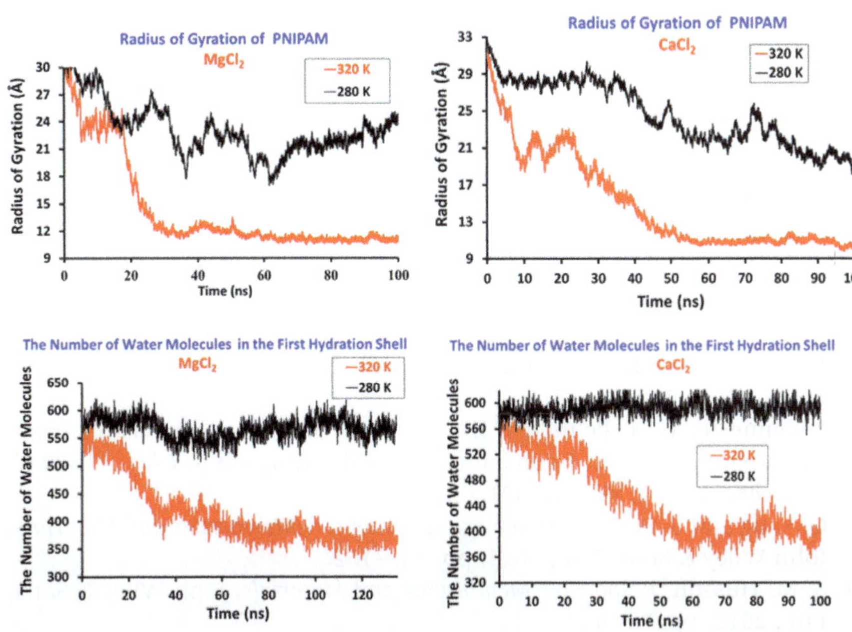

Figure 10.15 Radius of gyration and the number of water molecules in the first hydration water shell associated with PNIPAM in divalent 0.5 M $CaCl_2$ and $MgCl_2$ salt solutions during MD simulations at temperatures above and below their respective expected LCSTs.[56]

10.6 Outlook

Membrane-based separation processes continue to find growing applications. Tremendous advances have been made in incorporating pH, ionic strength and specific target/receptor based interactions into membranes such that the membrane properties can be modulated. However, when one considers biological systems, the level of complexity of the responsive behavior that has been imparted into synthetic membranes is much less. Recently, fundamental molecular dynamic simulations are starting to elucidate the molecular level interactions that lead to the macroscopic responsive behavior we are trying to impart in synthetic membranes. As these studies increase our understanding of the molecular level interactions that occur, it is likely that far more sophisticated responses can be imparted perhaps even involving a cascade of responses analogous to biological systems.

Acknowledgements

Funding for this work was obtained from the National Science Foundation CBET1066505 and the Arkansas Research Alliance.

References

1. A. K. Pabby, S. S. H. Rizvi and A. M. S. Requena, *Handbook of Membrane Separations: Chemical, Pharmaceutical, Food, and Biotechnological Applications*, CRC press, 2008.
2. N. N. Li, A. G. Fane, W. S. W. Ho and T. Matsuura, *Advanced Membrane Technology and Applications*, John Wiley & Sons, 2011.
3. R. van Reis and A. Zydney, *J. Membr. Sci.*, 2007, **297**, 16–50.
4. W. S. W. Ho and K. K. Sirkar, *Membrane Handbook*, Springer, 1992.
5. J. D. Bronzino, *The Biomedical Engineering Handbook*, CRC Press, 2000, vol. 1.
6. D. Bhattacharyya, T. Schäfer, S. R. Wickramasinghe and S. Daunert, *Responsive Membranes and Materials*, John Wiley and Sons, 2012.
7. D. Wandera, S. R. Wickramasinghe and S. M. Husson, *J. Membr. Sci.*, 2010, **357**, 6–35.
8. J. Vanneste, W. J. Peumans, E. J. M. Van Damme, S. Darvishmanesh, K. Bernaerts, J. M. C. Geuns and B. Van der Bruggen, *J. Chem. Technol. Biotechnol.*, 2014, **89**, 354–371.
9. Q. Yang and S. R. Wickramasinghe, *Responsive Membranes and Materials*, John Wiley & Sons, Ltd., 2012, pp. 143–162.
10. S. M. Husson, *Responsive Membranes and Materials*, John Wiley & Sons, Ltd., 2012, pp. 73–96.
11. R. Wang, T. Xiang, W. Yue, H. Li, S. Liang, S. Sun and C. Zhao, *J. Membr. Sci.*, 2012, **423–424**, 275–283.
12. C. Cheng, S. Li, W. Zhao, Q. Wei, S. Nie, S. Sun and C. Zhao, *J. Membr. Sci.*, 2012, **417–418**, 228–236.

13. T. Xiang, M. Tang, Y. Liu, H. Li, L. Li, W. Cao, S. Sun and C. Zhao, *Desalination*, 2012, **295**, 26–34.
14. W.-W. Yue, T. Xiang, W.-F. Zhao, S.-D. Sun and C.-S. Zhao, *Sep. Sci. Technol.*, 2013, **48**, 1941–1953.
15. T. Xiang, H. Fu, W.-W. Yue, S.-D. Sun and C.-S. Zhao, *Sep. Sci. Technol.*, 2013, **48**, 1627–1635.
16. Z. Han, C. Cheng, L. Zhang, C. Luo, C. Nie, J. Deng, T. Xiang and C. Zhao, *Desalination*, 2014, **349**, 80–93.
17. H. Li, J. Liao, T. Xiang, R. Wang, D. Wang, S. Sun and C. Zhao, *Desalination*, 2013, **309**, 1–10.
18. T. Luo, S. Lin, R. Xie, X.-J. Ju, Z. Liu, W. Wang, C.-L. Mou, C. Zhao, Q. Chen and L.-Y. Chu, *J. Membr. Sci.*, 2014, **450**, 162–173.
19. Z. Wang, X. Yao and Y. Wang, *J. Mater. Chem.*, 2012, **22**, 20542.
20. Y. He, X. Chen, S. Bi, W. Fu, C. Shi and L. Chen, *React. Funct. Polym.*, 2014, **74**, 58–66.
21. W. Sun, Z. Wang, X. Yao, L. Guo, X. Chen and Y. Wang, *J. Membr. Sci.*, 2014, **466**, 229–237.
22. H. H. Himstedt, Q. Yang, L. P. Dasi, X. Qian, S. R. Wickramasinghe and M. Ulbricht, *Langmuir*, 2011, **27**, 5574–5581.
23. C.-C. Hsu, C.-S. Wu and Y.-L. Liu, *J. Membr. Sci.*, 2014, **450**, 257–264.
24. R. Zhang, Y. Su, J. Peng, X. Fan, Z. Jiang, X. Zhao, J. Liu, Y. Li and J. Zhao, *Polymer*, 2014, **55**, 1347–1357.
25. X. Qiu, X. Ren and S. Hu, *Carbohydr. Polym.*, 2013, **92**, 1887–1895.
26. H. H. Himstedt, H. Du, K. M. Marshall, S. R. Wickramasinghe and X. Qian, *Ind. Eng. Chem. Res.*, 2013, **52**, 9259–9269.
27. H. H. Himstedt, K. M. Marshall and S. R. Wickramasinghe, *J. Membr. Sci.*, 2011, **366**, 373–381.
28. J. de Grooth, M. Dong, W. M. de Vos and K. Nijmeijer, *Langmuir*, 2014, **30**, 5152–5161.
29. M. Kuczewski, N. Fraud, R. Faber and G. Zarbis-Papastoitsis, *Biotechnol. Bioeng.*, 2010, **105**, 296–305.
30. J. H. Vogel, H. Nguyen, R. Giovannini, J. Ignowski, S. Garger, A. Salgotra and J. Tom, *Biotechnol. Bioeng.*, 2012, **109**, 3049–3058.
31. M. Prabaharan, J. J. Grailer, D. A. Steeber and S. Gong, *Macromol. Biosci.*, 2008, **8**, 843–851.
32. Y. Maeda, T. Nakamura and I. Ikeda, *Macromolecules*, 2001, **35**, 217–222.
33. H. Du and X. Qian, *J. Polym. Sci., Part B: Polym. Phys.*, 2011, **49**, 1112–1122.
34. H. Du, R. Wickramasinghe and X. Qian, *J. Phys. Chem. B*, 2010, **114**, 16594–16604.
35. E. Gil and S. Hudson, *Prog. Polym. Sci.*, 2004, **29**, 1173–1222.
36. T. Miyata, N. Asami and T. Uragami, *Macromolecules*, 1999, **32**, 2082–2084.
37. T. Miyata, N. Asami and T. Uragami, *Nature*, 1999, **399**, 766–769.
38. Z.-R. Lu, P. Kopečková and J. Kopeček, *Macromol. Biosci.*, 2003, **3**, 296–300.

39. R. Zhang, A. Bowyer, R. Eisenthal and J. Hubble, *Biotechnol. Bioeng.*, 2007, **97**, 976–984.
40. P. C. Jocelyn, *Sulfur and Sulfur Amino Acids*, Elsevier, 1987, vol. 143.
41. S. Cerritelli, D. Velluto and J. A. Hubbell, *Biomacromolecules*, 2007, **8**, 1966–1972.
42. J. Meng, Z. Cao, L. Ni, Y. Zhang, X. Wang, X. Zhang and E. Liu, *J. Membr. Sci.*, 2014, **461**, 123–129.
43. L. Wong, C. Boyer, Z. Jia, H. M. Zareie, T. P. Davis and V. Bulmus, *Biomacromolecules*, 2008, **9**, 1934–1944.
44. S. Toledano, R. J. Williams, V. Jayawarna and R. V. Ulijn, *J. Am. Chem. Soc.*, 2006, **128**, 1070–1071.
45. R. V. Ulijn, *J. Mater. Chem.*, 2006, **16**, 2217.
46. S. Guo, A. Sugawara-Narutaki, T. Okubo and A. Shimojima, *J. Mater. Chem. C*, 2013, **1**, 6989.
47. Z. Yang, H. Gu, D. Fu, P. Gao, J. K. Lam and B. Xu, *Adv. Mater.*, 2004, **16**, 1440–1444.
48. T. Miyata, T. Uragami and K. Nakamae, *Adv. Drug Delivery Rev.*, 2002, **54**, 79–98.
49. L.-Y. Chu, Y. Li, J.-H. Zhu, H.-D. Wang and Y.-J. Liang, *J. Controlled Release*, 2004, **97**, 43–53.
50. C. Sugnaux, L. Lavanant and H.-A. Klok, *Langmuir*, 2013, **29**, 7325–7333.
51. K. Kataoka, H. Miyazaki, T. Okano and Y. Sakurai, *Macromolecules*, 1994, **27**, 1061–1062.
52. H. Dong, H. Du and X. Qian, *J. Phys. Chem. A*, 2008, **112**, 12687–12694.
53. H. Dong, H. Du and X. Qian, *J. Phys. Chem. B*, 2009, **113**, 12857–12859.
54. H. Dong, H. Du, S. R. Wickramasinghe and X. Qian, *J. Phys. Chem. B*, 2009, **113**, 14094–14101.
55. H. Du and X. Qian, *Carbohydr. Res.*, 2011, **346**, 1985–1990.
56. H. Du, S. R. Wickramasinghe and X. Qian, *J. Phys. Chem. B*, 2013, **117**, 5090–5101.
57. Y. Zhang, S. Furyk, D. E. Bergbreiter and P. S. Cremer, *J. Am. Chem. Soc.*, 2005, **127**, 14505–14510.
58. H. Fu, X. Hong, A. Wan, J. D. Batteas and D. E. Bergbreiter, *ACS Appl. Mater. Interfaces*, 2010, **2**, 452–458.
59. F. Vanzi, B. Madan and K. Sharp, *J. Am. Chem. Soc.*, 1998, **120**, 10748–10753.
60. Q. Zou, B. J. Bennion, V. Daggett and K. P. Murphy, *J. Am. Chem. Soc.*, 2002, **124**, 1192–1202.
61. A. W. Omta, *Science*, 2003, **301**, 347–349.
62. P. Lo Nostro and B. W. Ninham, *Chem. Rev.*, 2012, **112**, 2286–2322.
63. M. Annaka, Y. Amo, S. Sasaki, Y. Tominaga, K. Motokawa and T. Nakahira, *Phys. Rev. E: Stat., Nonlinear, Soft Matter Phys.*, 2002, **65**, 031805.
64. R. Freitag and F. Garret-Flaudy, *Langmuir*, 2002, **18**, 3434–3440.
65. H. B. Du, S. R. Wickramasinghe and X. H. Qian, *Procedia Eng.*, 2012, **44**, 485–488.

CHAPTER 11

Responsive Layer-by-layer Films

SRY D. HUJAYA,[a] BENJAMIN M. WOHL,[b] JOHAN F. J. ENGBERSEN*[a] AND JOS M. J. PAULUSSE*[a]

[a] Department of Controlled Drug Delivery, MIRA Institute for Biomedical Technology and Technical Medicine, Faculty of Science and Technology, University of Twente, P.O. Box 217, 7500 AE, Enschede, The Netherlands; [b] Department of Chemistry and iNano Interdisciplinary Nanoscience Centre, Aarhus University, Aarhus 8000, Denmark
*Email: j.f.j.engbersen@utwente.nl; j.m.j.paulusse@utwente.nl

11.1 Introduction

Thin multilayers, often referred to as polyelectrolyte multilayers, multilayer films, multilayered thin films, multilayered capsules (free-standing nano- or microsized), or simply multilayers, are a class of smart materials distinguished by the layer-by-layer (LbL) fabrication technique, *i.e.* they are built up one layer at a time. As illustrated in Figure 11.1, fabrication involves incubating a substrate alternatingly in two or more aqueous deposition solutions of complementary macromolecules. This results in the formation of self-assembled layers with intricate architectures on the molecular level. The properties of each layer can be tuned by the deposition solution, and/or deposition technique. Factors such as pH, ionic strength, and temperature affect how macromolecules in solution interact with macromolecules on the surface and hence determine the layer properties and architecture. In addition to conventional dip-coating techniques, multilayers have been fabricated in various other ways such as alternate[1,2]

RSC Smart Materials No. 14
Chemoresponsive Materials: Stimulation by Chemical and Biological Signals
Edited by Hans-Jörg Schneider

Published by the Royal Society of Chemistry, www.rsc.org

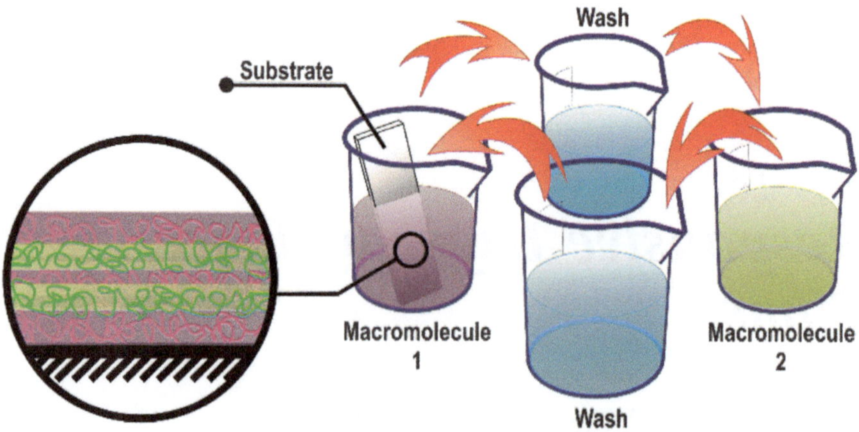

Figure 11.1 Schematic illustration of layer-by-layer assembly *via* alternate dipping in two complementary macromolecule solutions and an illustration of the resulting multilayers.

or simultaneous[3] spraying, spin coating,[4,5] agitation,[6] de-wetting,[7] inkjet printing,[8] microfluidic,[9] and flexible role coating,[10] endowing different properties to the multilayer products. The resulting multilayers can also be treated post-assembly in various ways to further alter their properties and functionalities.

The versatility of the LbL fabrication technique allows for the incorporation of a wide choice of materials to be used as both multilayer components and substrates on which layer build-up occur. Multilayers have been built not only on inorganic solids (*e.g.* glass,[11] gold electrodes,[12] metal stents,[13] *etc.*), but also on organic templates (ionic liquids,[14] polymeric membranes,[15] plastics,[16,17] 3D tissue engineering scaffolds,[18] polymeric microneedles[19–21]) and even biological objects (living cells,[22,23] or as sacrificial templates[24]), in all shapes and sizes. Likewise, for multilayer components, it is possible to combine polymers with inorganic materials and render the multilayers magnetically responsive,[25,26] electroconductive[27,28] or optically active.[29] Smaller molecules, such as therapeutic compounds, may be incorporated through conjugation to one of the polymeric components or through physical entrapment within the multilayer,[30] while the all-aqueous deposition conditions are ideally suited to fragile biomacromolecules such as enzymes. Multilayers may also be constructed based on different driving forces. In addition to conventional electrostatic interactions, multilayers have been built through hydrogen bonding interactions,[31,32] van der Waals interactions,[33] hydrophobic interactions,[34,35] stereocomplexation of chiral polymers,[36–38] charge transfer,[39–41] host–guest,[42,43] specific biorecognition[44,45] (including DNA hybridization[46,47]), and covalent bonding.[48–50] Due to these versatilities, multilayers have found applications in a wide range of fields such as nanoelectronics and sensors,[51–54] separation membranes,[55–57] mechanical enhancing,[58] superhydrophobic,[59,60] scratch resistant,[61] antireflection,[62] anticorrosion,[63] and anti-fouling[64] coatings, bioreactors,[65,66] and tissue engineering and biomedical applications.[31,67–70]

The intricate architecture of a multilayered system sets it apart from other smart systems such as hydrogels and other coatings, which can be considered as homogeneous media. A multilayer is divided into regions with distinct differences in material composition. These unique physicochemical characteristics of multilayers will be briefly explained in the first part of this chapter, followed by examples of their chemical and biological responsiveness. In the following sections, responsiveness attributed to the use of specifically tailored polymers is described. The final section discuses conclusions, challenges and prospects of the future development of chemically and biologically responsive multilayered systems.

11.2 LbL Fabrication-specific Responsiveness

The physicochemical aspects of multilayers have been mostly investigated for those built through electrostatic interactions as classic examples dating back to the early reports on LbL assembly first reported on poly(styrenesulfonate) (PSS) and poly(allylamine hydrochloride) (PAH) by Decher *et al.* in the early 1990s.[71] Stratification and the molecular growth mechanism have been matters of continuous debate. As indicated previously, the multilayer architecture is a unique property that can be potentially exploited to induce responsiveness. In this respect, the build-up profile is an exclusive characteristic of multilayers, which is known to have direct correlation to its architectural aspects. Most multilayers can be classified into two distinct build-up profiles: linear and supralinear or exponential, with some exceptions described as combinations of two linear functions due to the influence of underlying substrates and electrostatic interactions.[72–74] The different build-up profiles, experimental findings, and proposed molecular mechanisms reported to date are briefly described here in relation to the possible responsiveness.

The multilayer growth mechanism is generally described as originating from overcompensation of surface functionality (*i.e.* the driving force for layer deposition) thereby facilitating the formation of the next layer. The process is driven by entropy increases through the liberation of polymer-associated water and ions. More specific molecular phenomena, however, determine whether build-up follows a linear or exponential profile. This build-up is usually followed by incremental increases in either mass or thickness as a function of deposition cycle or layer number. For example, while a 10-bilayered linearly growing multilayer may reach ∼100 nm in thickness, exponentially growing multilayers easily reach micro- or even millimetre range thicknesses at the same number of layers.[75]

Multilayers with linear build-up profiles make up most of the early reports on LbL assembly together with electrostatic driving force. Architecturally, these multilayers possess stratification with interpenetration between neighboring layers.[76] The polymer chains within the multilayer are described to be frozen and immobilized, and the overall layer is "fuzzy" and highly disordered despite observable stratification.[77] The proposed molecular mechanism for these systems involves the assumption that incoming polymer chains form complexes only with the topmost layer[78] and thus 1 : 1 deposition and overcompensation are maintained in every cycle of deposition. In 2000, Decher and co-workers proposed a three-zone model consisting of a charged zone closest to the substrate whose nature is influenced by the substrate, a zone at the interface with air or solution which is charged to facilitate new layer deposition, and a neutral "bulk" zone in-between, which grows along with the addition of new layers (Figure 11.2).[79]

Exponentially growing multilayers were first discovered by Hubbell and co-workers in 1999 on ((poly(L-lysine) (PLL)#alginate (Alg)) multilayers.[81] (The symbol # indicates alternating multilayers from the given materials.) Laugel *et al.* found through isothermal titration microcalorimetry that higher exothermic electrostatic interactions tend to lead to linearly growing multilayers while weaker exothermic or endothermic interactions tend to lead to exponentially growing multilayers.[82] Thus, it was found that linearly growing multilayers can be rendered into exponentially growing layers by increasing the deposition temperature and/or increasing ionic strength of the deposition solutions to mask the strong electrostatic interactions between the polyelectrolytes.[82] Two models have been proposed for the molecular mechanism of exponentially growing multilayers. In the first model it was proposed that exponential profiles are a result of accumulation of surface roughness with increasing layer number.[83–85] The increase in surface roughness is attributed to the more globular conformation of polyelectrolytes when no significant charge repulsion takes place. This higher roughness leads to increases in surface area, resulting in the increased deposition of incoming polymers. This model is supported by the finding that multilayers that grow exponentially through dip-coating may be turned into linearly growing layers through spin coating, as was reported for hydrogen

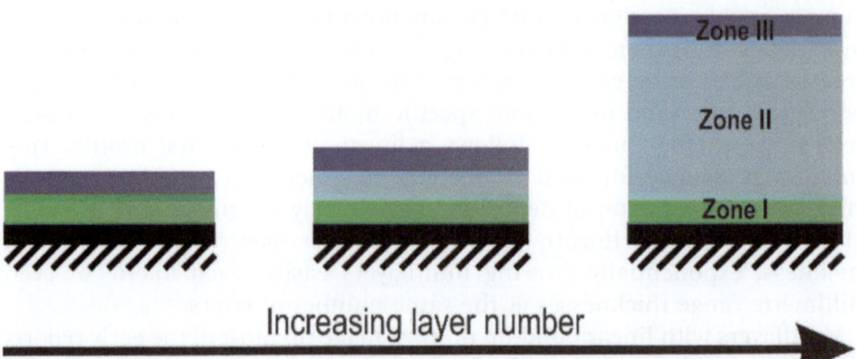

Figure 11.2 A schematic illustration of the three-zone model as proposed by Decher and co-workers (adapted from ref. 79 and 80).

bonded (poly(ethylene glycol) (PEG)#poly(acrylic acid) (PAAc)) multilayers,[16] and more recently for (PAH#PAAc) through application of high gravity fields,[86] two deposition techniques that reduce potential surface roughness.

In an alternative model, Lavalle and co-workers proposed that incoming polyelectrolytes not only interact with the outermost layer, but also diffuse into the core of the multilayer, through liberation of small counter ions from the multilayers.[87] The diffusion of the polyelectrolyte into the entire multilayer is accompanied by diffusion of smaller extents of the same polyelectrolyte out of the multilayer, which is limited by the presence of an electrostatic barrier at the film–solution interface. Diffusion halts when overcompensation is achieved by the polyelectrolytes complexating with the outermost layer, and the electrostatic barrier becoming too high for additional diffusion. As such, the amount of deposited polyelectrolyte depends not only on the fixed amount of outer polymer layers, but also on the relative thickness of the multilayer, resulting in an exponential increase with increasing layer number. This model is supported most notably by a confocal laser scanning microscopy (CLSM) observation that fluorescein isothiocyanate (FITC)-labeled PLL diffuses through the entire multilayer consisting of unlabeled (PLL#hyaluronic acid (HA)). On the other hand, Texas red (TR)-labeled HA deposited into a distinct layer without diffusion, identifies PLL as the diffusing species (Figure 11.3).[88]

Very interestingly, these findings indicate that compared to linearly growing multilayers, much higher mobility of polymers is observed for the exponentially growing layers, not only within the multilayer but also in the deposition solution where the two actively exchange during layer deposition.[89–91] In direct relation to chain mobility, it was found that faster growth was observed with polyelectrolytes of lower molecular weight,[92] and higher polydispersity.[93]

To maintain electroneutrality, multilayers also incorporate counter ions from solutions to compensate for excess charge of the polyelectrolytes. These multilayers are said to be extrinsically compensated and such compensation

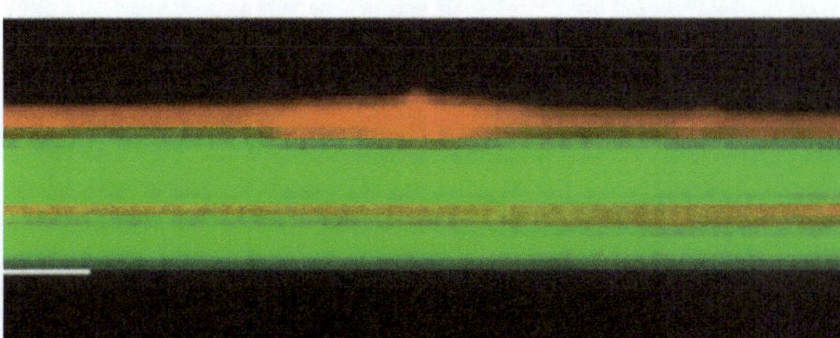

Figure 11.3 Vertical CLSM image of exponentially growing $(PLL\#HA)_{13.5}\#HA^{TR}\#(PLL\#HA)_4\#(PLL^{FITC}\#HA^{TR})$ with the glass substrate indicated by the white line. The green fluorescent layers emerged following the last deposition step of FITC-conjugated PLL.
Reproduced with permission from ref. 88.

gives rise to Donnan potential.[94,95] However, some multilayers have also been reported to display selectivity in ion incorporation. For example, Schlenoff *et al.* reported that no salt ions were present within a linearly growing poly(diallyldimethylammonium chloride) (PDADMAC)#PSS multilayer when the amount of positive and negative charges were approximately equal, indicating intrinsic charge compensation (*i.e.* charges of one polyelectrolyte are completely compensated by charges of the counter polyelectrolyte).[96] The same research group further indicated the possibility to "dope" the intrinsically compensated multilayers with selected ions by increasing the ionic strength of the incubation solution.[97] These features are highly beneficial for applications where high ion selectivity and permeability are required such as for separation and electronics applications.

For multilayer capsules, similar to liposomal or polymersome systems, multilayer capsules with weaker interactions undergo fusion.[98,99] Sukhorukov and co-workers reported salt-induced fusion of (PDADMAC#PSS) upon slow increase in salt concentration,[99] while Volodkin and co-workers reported fusion of (PAH#PSS) induced not by salt, but by elevated acidity ($[H^+] > 0.1$ M) (Figure 11.4).[98] The stronger interactions in the latter system were proposed to be the reason for its apparent non-responsiveness to salt. Fusion of two or more capsules upon contact is entropically driven by liberated water molecules, due to decreases in surface area. The kinetics of the fusion were found to be a function of layer number in relation to layer density, and the H^+ concentration. Thinner multilayers and elevated H^+ concentration increased the onset of the fusion. Volodkin and co-workers also identified that polymer exchange took place during fusion through the use of fluorescently-labeled PAH-containing multilayer capsules, which were mixed with non-fluorescently labeled capsules. The authors found that upon contact, non-fluorescently labeled capsules gradually became more fluorescent, depending on whether the final layers of the fusing capsules were of similar (enhanced) or opposing charges.

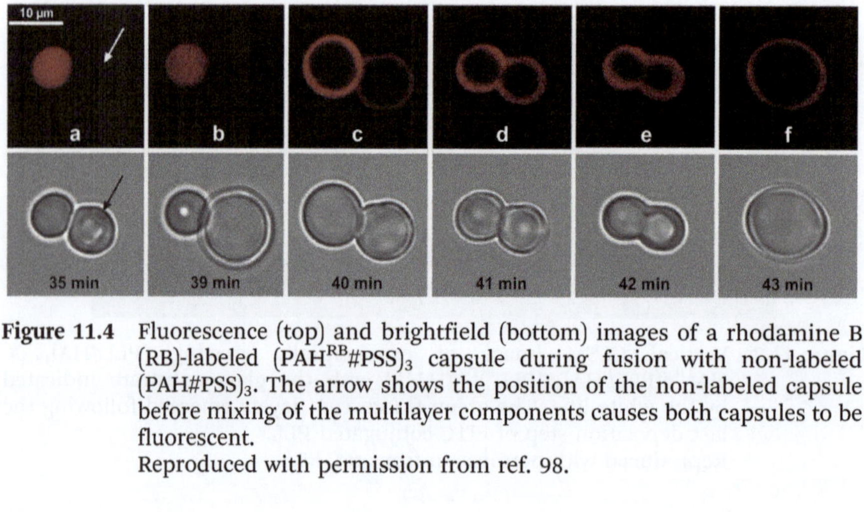

Figure 11.4 Fluorescence (top) and brightfield (bottom) images of a rhodamine B (RB)-labeled $(PAH^{RB}\#PSS)_3$ capsule during fusion with non-labeled $(PAH\#PSS)_3$. The arrow shows the position of the non-labeled capsule before mixing of the multilayer components causes both capsules to be fluorescent.
Reproduced with permission from ref. 98.

These findings indicate that multilayer properties, such as responsiveness, can be introduced not only through the choice of the main multilayer components, but also through the physical deposition aspects.

11.2.1 Physical Deposition Aspects

During LbL assembly, the presence of supporting components such as different types and concentrations of salt in the deposition solution, may exert detrimental features onto the resulting multilayers. In general, higher salt concentrations result in increased polymer entanglement due to screening of charges in the polyelectrolyte, which leads to increased material deposition, and increased surface roughness (Figure 11.5).[100] Salt also increases the mobility of polyelectrolytes within the multilayer and therefore enhances multilayer stability by introducing higher degrees of entanglement through rearrangement. Multilayers built in the absence of salts are therefore much less stable and readily deconstruct upon introduction of salt or changes in pH.[101] However, when the salt concentration in the deposition solution is too high, polyelectrolytes tend to remain in solution and no multilayer build-up is observed.

The type of salt influences multilayer thickness,[102] porosity,[103,104] stiffness,[105] as well as swelling properties.[106] Stiffness and porosity are two surface characteristics known to play a role in cellular behavior.[107,108] For example, by simply changing the pore size of (PAH#PAAc) multilayers to the nanometre range to mimic natural basement membranes, human corneal epithelial cells (HCEC) were found to proliferate and migrate at twice the speed, as compared to cells cultured on (PAH#PAAc) multilayers with submicron porosity.[104]

In contrast to the usual trend of increasing thickness with increasing ionic strength, multilayers containing strong polyzwitterions (polymers are

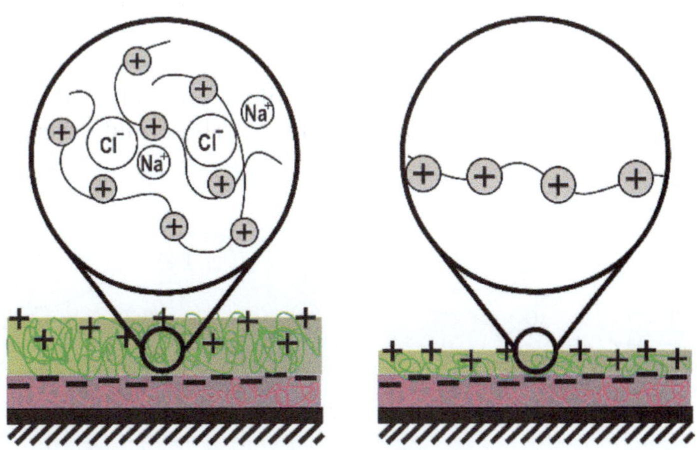

Figure 11.5 Schematic illustration of the effect of salt on polyelectrolyte multilayer build-up.

neutral over a wide range of pH) as one of the multilayer components display a reversed trend of decreasing deposition with increasing ionic strength.[15] For example, de Vos and co-workers studied the multilayer build-up characteristics of polysulfobetaine (a strong polyzwitterion) with PDADMAC and found that deposition was optimal at low salt concentration. This observation was explained as an effect of antipolyelectrolyte properties of polyzwitterions, where a more globular conformation of polyzwitterions is more readily attainable at low ionic strength as the internal charges in polyzwitterions are not screened (Figure 11.6). Thus, deposition onto a polycation layer is accompanied by relatively low entropic gain, leading to thicker layers. At higher ionic strength, screening of internal charges of polyzwitterions leads to higher entropic gain upon layer deposition, resulting in thinner layers.[96] The authors further demonstrated that the multilayers can be fabricated on hollow fiber membranes to provide ionic strength responsiveness, where increased permeability is observed upon increases in NaCl concentration.[15] The reversed trend of decreased deposition with increasing salt concentration was also observed when NaBr was used instead of NaCl; this is attributed to the stronger interaction of the bromide ion with the polycation in the multilayer.[109]

The role of secondary substituents is not only limited to salts. Combining additional polymer in a blending fashion may help induce enhanced responsiveness. It was recently reported that alginate (Alg) incorporation into poly(ethylene imine) (PEI)#DNA enhanced DNA release from the multilayer.[110] The effect was not only observed with alginate as an additional layer (*i.e.* as (PEI#DNA#PEI#Alg)), but also when alginate was added separately into the incubation medium to provide an alginate-triggered release.

For weaker polyelectrolytes (*e.g.* those which contain carboxylic acid groups and amines), in addition to ionic strength, the pH of deposition solutions also determines the growth behavior of multilayers, as well as their responsiveness. A classic example was reported by Shiratori and Rubner on (PAH#PAAc) multilayers.[111] In general, a deposition pH that results in increased proportions of charged groups will lead to decreased deposition.[84,101,112] Notably, the thickest films are obtained when pH is closest to the pK_a of the ionizable groups.[113] These multilayers possess

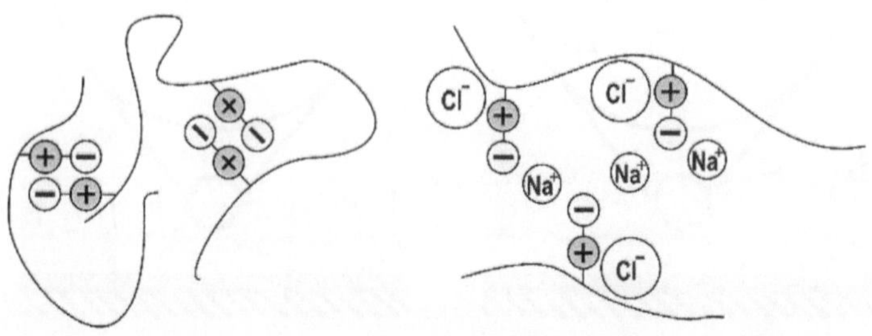

Figure 11.6 Schematic illustration of the antipolyelectrolyte properties of polyzwitterions in the absence (left) and presence (right) of salt.

pH-responsiveness, which depends strongly on the deposition pH.[101] Multilayers fabricated at a pH that facilitates strong electrostatic interactions will be less likely to disassemble due to the stronger multivalent interactions, whereas multilayers that are fabricated at a pH that facilitates deposition of macromolecules of more globular conformation are significantly more sensitive to changes in pH. Likewise, multilayers that are fabricated at a pH at which most of the ionizable groups are not charged, will disassemble rapidly at a pH that ionizes the groups and causes charge repulsion.

Unlike multilayers based on electrostatic interactions, hydrogen bonded multilayers are characterized by a more limited pH range for deposition. Owing to this, these multilayers are able to readily disassemble at physiologically relevant pH, making them interesting for biomedical applications where pH responsiveness is desired. Some of the most commonly used hydrogen bond acceptor components are PEG, poly(*N*-vinylpyrrolidone) (PVPON), and the temperature-responsive poly(*N*-isopropylacrylamide) (PNIPAM), while the hydrogen bond donor components include poly(vinyl alcohol) (PVA), carboxylic acid-containing PAAc, and poly(methacrylic acid) (PMAAc). To obtain hydrogen bonded multilayers, assembly is carried out below the pK_a of the carboxylic acid containing component. Subsequently, upon increase of pH above the pK_a, a critical disintegration pH exists at which the hydrogen bonds are disrupted, leading to rapid multilayer disassembly.[32,114] Higher salt concentrations may decrease the critical disintegration pH, as this increases the degree of ionization of the carboxylic acid groups.[115] By simply choosing the appropriate component, the stability of the multilayers can be maintained over a wider pH range, as was reported by Erel-Unal and Sukhishvili through the use of tannic acid (TA) with a pK_a of 8.5 and a branched structure for enhanced multivalency.[116]

To obtain prolonged and sustained release, Sakamoto and co-workers reported the use of plotted agarose scaffolds as substrates for the hydrogen bonded multilayer fabrication of PEG, PAAc, and lysozyme.[117] They demonstrated sustained lysozyme release for up to four weeks with release kinetics dependent not only on the main multilayer components, but also on agarose concentration (*i.e.* porosity), and the type of precursor layer. This system also demonstrates the influence of the multilayer substrate on the multilayer properties.

The underlying substrates often only have a minor influence on the properties of the multilayer itself, and often persist only in the first few layers.[118] The substrate effect can be more pronounced when the type of interaction between polymer pairs is different from that between the polymers and the substrate, or when the substrate possesses less functionality (*e.g.* lower charge density in the case of electrostatic interactions).[119] To reduce the effect of the substrate, precursor layers from PEI, PSS, and PAH are sometimes deposited before the main multilayer components are deposited.[120–123] Especially when used for biomedical applications, the identity of the precursor layer has to be carefully considered, as most of the well-studied polyelectrolyte pairs are not biodegradable and may be cytotoxic. For example, high molecular weight PEI has been found to be potentially cytotoxic.[124]

For multilayer capsules, sacrificial substrates based on $CaCO_3$,[125,126] SiO_2,[99] or melamine formaldehyde (MF)[127] were found to lead to matrix-type, instead of shell-type, multilayers, due to the incorporation of the multilayer components into the pores of the substrate during build-up. By choosing the appropriate core dissolution method, these matrix-type capsules can be forced into a shell-type multilayer, which retains gaseous hollow lumen.[98] Another example illustrating the influence of the substrate on multilayers was reported by Tsukruk and co-workers, who utilized cubic $CdCO_3$ as templates to be sacrificed for multilayer capsule preparation and found that in comparison to capsules prepared using a spherical SiO_2 template, the multilayer was softer and more permeable.[128]

Drug crystals have also been used as substrates to automatically entrap the drug within the multilayer to achieve very high drug loading, and controlled release.[129,130] Möhwald and co-workers encapsulated ibuprofen within multilayers of chitosan (CHI) and dextran sulfate (DS) by utilizing the limited solubility of ibuprofen at $pH < 7$.[129] At pH 7.4, the ibuprofen crystals readily dissolved and diffused through the pores of the multilayers, leaving the empty multilayer shell intact (Figure 11.7). The release kinetics were found to depend on the size of the crystals (smaller crystals dissolved faster), number of layers (faster with thinner multilayers), and pH (much slower at pH 1.4 due to the low solubility of ibuprofen), but release kinetics were in all cases significantly slower than the dissolution rate of the uncoated drug crystals.

Almodóvar *et al.* found from their studies on carbohydrate-based multilayers that hydrophilic multilayer surfaces are usually obtained from pairs of carbohydrates that are either both weak polyelectrolytes, or both strong polyelectrolytes, while multilayers from a combination of weak and

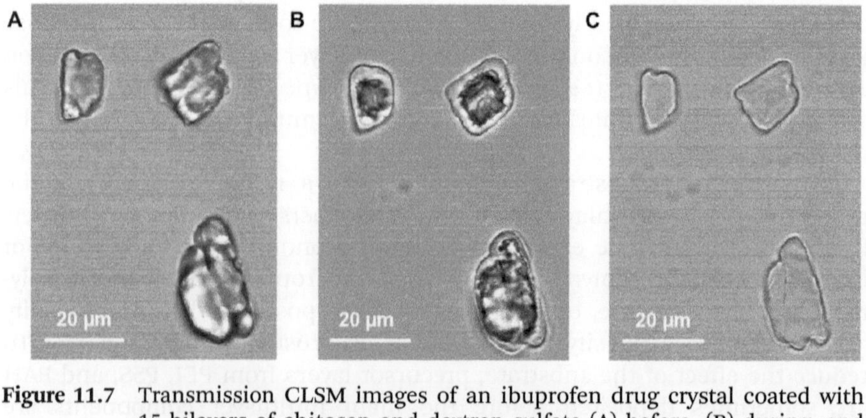

Figure 11.7 Transmission CLSM images of an ibuprofen drug crystal coated with 15 bilayers of chitosan and dextran sulfate (A) before, (B) during, and (C) after complete ibuprofen core dissolution.
Reproduced with permission from ref. 129.

strong polyelectrolytes tend to be hydrophobic.[131] This phenomenon was discussed based on the different predisposition of strong *versus* weak polyelectrolytes to ion pairing. This eventually leads to the ability of multilayers to swell and change the hydrophilicity or hydrophobicity of the resulting surface.

11.2.2 Post-assembly Treatment

Post-assembly treatments are aimed at addressing various needs. For tissue engineering applications, formation of additional fibronectin layers[132] or the tripeptide RGD[133,134] is a straightforward way to promote protein adsorption and cell attachment. Treating (PDADMAC#PSS) multilayers with various NaCl concentrations post-assembly has also been reported to result in different cell migration behavior.[135] Diffusion of multilayer components can be prevented by fabricating barrier layers that are impermeable to these components.[136] For example, for exponentially growing multilayers, the addition of several bilayers of linearly growing multilayers may result in the prevention of diffusion and offer a way to compartmentalize a multilayered construct. Examples of this approach are given in Section 11.2.3.

Depending on chain mobility within a multilayer, annealing steps can be carried out to smoothen the multilayer surface. In the case of weaker interactions, such as hydrogen bonding, this may be induced through simple incubation in deionized water,[137] while in the case of stronger electrostatic interactions, exposure to high salt concentration is used to first weaken the interactions and induce rearrangement.[138,139] However, some exponentially growing electrostatically-based multilayers were found to self-heal simply upon exposure to water.[140] This swelling behavior may lead to disintegration as a function of time, salt concentration, and polymer molecular weight (slower with higher MW). For (PVPON#PAAc), it has been reported that such disintegration proceeds gradually from top to bottom,[137] in contrast to the much more rapid responsiveness to pH where hydrogen bonding is directly affected. Quite recently, Voegel and co-workers reported that the restructuring of exponentially growing (PLL#HA) multilayers through an increase or decrease in ionic strength may result in the formation of holes within the multilayers.[141]

Free-standing multilayers (both as macroscopic films or microscopic capsules) have also been prepared through post-assembly treatment. To obtain microscopic capsules, SiO_2, polystyrene, or $CaCO_3$ substrates that are already coated with multilayers are subjected to dissolution. These processes may often cause changes in the properties of the obtained capsules. For example, MF core removal has been associated with increased osmotic pressure, which may cause rupture of the multilayer capsule.[142] The use of an organic solvent for PS core removal has been shown to affect the stability of the resulting capsule.[143] SiO_2 and MF removal is carried out at low pH with HF and HCl respectively and is therefore limited to multilayers with high stability at acidic pH.[144]

For macroscopic free-standing multilayers, hydrogen bonded multilayers serve as convenient sacrificial layers that are easily disassembled at mild pH. For example, (PEG#PAAc) disassembles at pH 5.6–6.3.[72] In addition to the hydrogen bonded multilayers, polyzwitterion-based multilayers have been utilized as sacrificial layers to obtain free-standing multilayers upon disassembly at pH $\geq$ 12.[145] Robust, defect-free mechanically detachable free-standing multilayers have also been reported for (CHI#HA) by choosing hydrophobic polypropylene as the multilayer substrate.[17] Similar attempts were reported for (PEG#PAAc) multilayers through the use of Teflon as a substrate.[16] Interestingly, the free-standing (PEG#PAAc) multilayers differ depending on whether dip coating (linearly growing) or spin coating (exponentially growing) was used for their fabrication. The exponentially growing multilayers were found to result in opalescent films, while the linearly growing multilayers were transparent (Figure 11.8). Recently, Rubner and co-workers reported the fabrication of sacrificial multilayers from bovine submaxillary mucin (BSM) and lectin jacalin (JAC) which can be conveniently disassembled in the presence of melibiose (a sugar), which competes for binding with mucin.[146] The authors further demonstrated that such an on-demand disassembly may be used to release the free-standing multilayer to further perform as a drug-releasing "backpack" for cell monolayers *in vitro* (Figure 11.9). The multilayer segment containing magnetic nanoparticles serves as the model "backpack" in this study.

The majority of post-fabrication treatments are aimed at increasing mechanical strength and stability, especially for hydrogen bonded multilayers. Multivalent ions have, for example, been introduced to increase the stability of hydrogen bonded multilayers at pH 7.1.[147] Heating is a popular way of forming amide crosslinks in (PAH#PAAc) multilayers.[148–151] Such crosslinks are therefore often introduced to act as barrier layers within a multilayered construct,[148] to control release rate,[150] and reduce permeability and increase ion-transport selectivity for separation membrane applications.[151] Other popular methods of forming covalent interlayer crosslinks include carbodiimide chemistry,[152–154] click chemistry,[2,155,156] thiol oxidation,[157,158] and glutaraldehyde crosslinking.[17,159] Liu recently reviewed

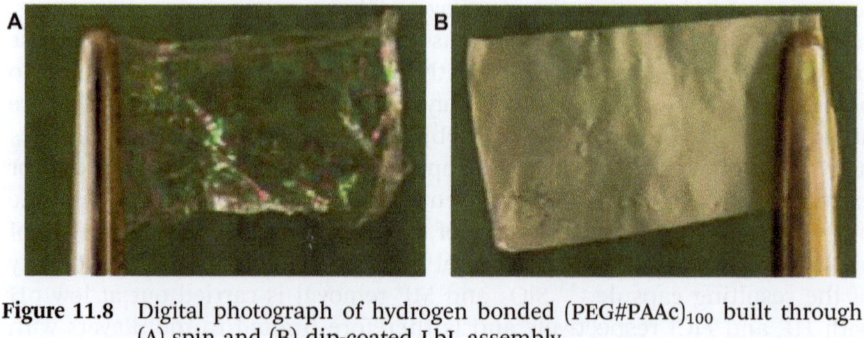

Figure 11.8 Digital photograph of hydrogen bonded $(PEG\#PAAc)_{100}$ built through (A) spin and (B) dip-coated LbL assembly.
Reproduced with permission from ref. 16.

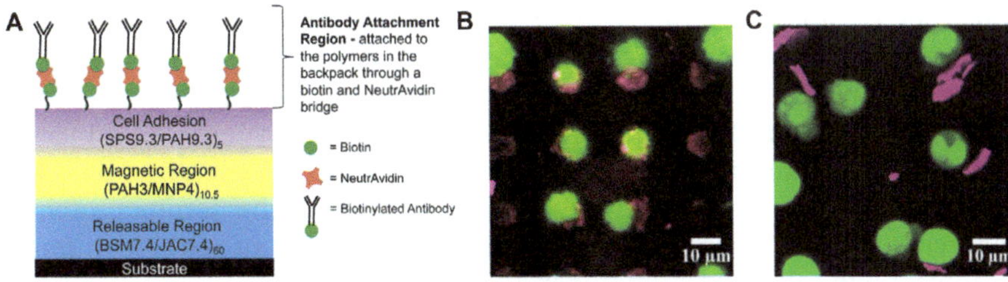

Figure 11.9 (A) Multilayer structure consisting of a (BSM#JAC) base layer as a sacrificial layer that can be disassembled upon introduction of melibiose. The magnetic region consists of (PAH#magnetic nanoparticle (MNP)) multilayers as a backpack. The cell adhesion top layer consists of (PSS#PAH) multilayers for enhanced cell adhesion, further covalently coupled with an IgG antibody for cell targeting. (B) Monocytes attaching to the backpack *via* the IgG antibody prior to the addition of melibiose. (C) Monocytes with the attached backpack following sacrificial layer disassembly through the addition of melibiose.
Reproduced with permission from ref. 146.

various stabilization techniques to enhance the stability of multilayer capsules.[160] Covalent crosslinking has been reported to increase multilayer stiffness. In a study by Picart and co-workers, increasing EDC concentration for crosslinking (PLL#HA) multilayers resulted in increased myoblast cell adhesion.[161] Additionally, crosslinks may also be introduced to add additional functionality and responsiveness. For example, Shu *et al.* prepared disulfide crosslinked (CHI#DS) multilayer capsules.[158] Compared to the non-crosslinked capsules, the disulfide-crosslinked capsules are more stable at the low pH encountered in the stomach, while being responsive to reducing enzymes and intracellular glutathione.

Controlled loading and release of small molecules, inorganic particles, and macromolecules has been achieved by utilizing pH as a trigger, when one of the multilayer components is a weak polyelectrolyte.[162–165] For example, the charge density within (PAH#HA) multilayers depends on pH, governing the ionization state of both amine groups of PAH, as well as the carboxylic acid groups of HA. Thus, loading of chromotrope 2R, a dye containing two sulfonate groups, is optimal when the pH is low due to the presence of excess positive charges in the multilayer,[162] while its release is optimal at high pH where the decrease in PAH charge density weakens the electrostatic interactions with the dye. (PAH#PSS) multilayer capsules can also be "open" for incorporation of FITC-dextran at $pH < 6$ and "closed" from further incorporation at $pH > 8$.[163]

11.2.3 Multilayer Architecture-specific Responsiveness

Increasing the bilayer number of multilayers is associated with decreasing Young's modulus or stiffness due to more pronounced hydration.[152,166,167] As a consequence of the decrease in stiffness, substantially increased spreading of smooth muscle cells (SMC) cultured on top of (PLL#HA) was observed when the number of bilayers was 20 instead of 60.[152] The same research group also reported that in addition to decreased chondrosarcoma cell adhesion on (PLL#poly(L-glutamic acid) (PGA)) with increasing bilayer number, cellular adherence may also be diminished entirely when the PGA layer is the topmost layer, due to the prevention of the serum proteins' adherence.[168] Interestingly, the phenotypic properties of SaOS-2 (human osteoblast-like cells) and human periodontal ligament cells were reported to be maintained only when the multilayer topmost layer is negatively charged.[169]

For the exponentially growing multilayers, the high mobility of polyelectrolytes in and out of the multilayer serves as a very convenient way to load additional materials homogeneously into the entire multilayered construct without the need for specific interactions between the materials to be loaded and the layer component. Lavalle and co-workers reported the loading of Oregon Green 488-labelled paclitaxel into exponentially growing (PLL#HA) multilayers.[136] By means of CLSM it was observed that paclitaxel was evenly distributed throughout the multilayer at an extent that was proportional to the concentration of paclitaxel in the loading solution and achieving up to 50 times higher concentration in the multilayer as compared to the concentration in the loading solution. The authors deposited (PAH#PSS) layers on top of the loaded multilayer and showed that this capping layer prevented the release of paclitaxel from the multilayer, while facilitating good HT29 cell adhesion. As long as the capping layer was not thick enough for complete surface coverage, cells could still internalize paclitaxel. The authors further showed through another publication that when a single poly(lactic-*co*-glycolic acid) (PLGA) layer was used instead of the non-degradable (PAH#PSS) layer, cells could degrade the PLGA layer and internalize the fluorescently-labeled PLL (Figure 11.10).[170]

Using a hydrolytically degradable polymer (poly(β-amino ester), here: polymer 1), Hammond and co-workers reported that different top-down film degradation profiles were observed depending on whether heparin (HEP) or dextran sulfate was used as the counter polyelectrolyte.[148] Build-up of (polymer 1#HEP) proceeded in an exponential manner with HEP as the diffusive species, while build-up of (polymer 1#DS) proceeded linearly. Erosion of the exponentially growing (polymer 1#HEP) proceeded in a linear manner, while that of (polymer 1#DS) proceeded through an early linear phase followed by leveling off of the erosion speed. The authors further demonstrated that by introducing one bilayer of thermally crosslinked (PAH#PAAc) in between (polymer 1#DS) on the bottom and (polymer 1#HEP) on top, sequential erosion of the two hydrolyzable compartments could be achieved with a delay that depends on the crosslinking degree of the (PAH#PAAc) barrier layer (*i.e.* the length of thermal crosslinking duration). Crosslinking for over 1.5 h fully halted hydrolysis of the bottom region.

Another example of the influence of multilayer architecture on its responsiveness was given by Erel-Unal and Sukhishvili with their hydrogen

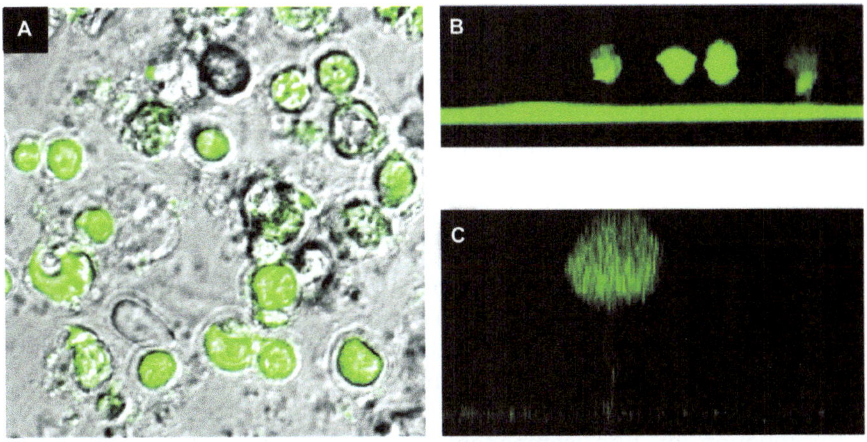

Figure 11.10 (A) Overlaid image of brightfield and green channel of bone marrow cells after 17 h of culture on exponentially growing (PLL#HA)$_{20}$#PLL^{FITC}#PLGA#PLL multilayers. The cells are green due to internalization of PLL^{FITC} from the multilayer. (B) CLSM images of the cross section of the multilayer with cells on top. (C) Higher magnification of the CLSM image showing the formation of a pseudopod through the non-fluorescent, biodegradable PLGA layer. Reproduced with permission from ref. 170.

bonded multilayers.[115] They showed that by combining (poly(*N*-vinylcaprolactam) (PVCL)#poly(L-aspartic acid) (PLAA)) with a critical disintegration pH of 3.3 with (PVCL#TA) with a critical disintegration pH of 9.5, it was possible to not only shift the disintegration profiles, but also introduce a two-step pH response.[115]

11.3 Polymer-specific Responsiveness

A different approach to obtain responsive multilayers is to introduce polymer-specific responsiveness. This means that the responsiveness is not a consequence of the employed assembly methodology, but is inherent to the film components or due to the interaction between them. One approach that has been studied thoroughly is the incorporation of polymers with inherent responsiveness into the multilayers. When engineered in the right manner, these polymers will respond to specific stimuli by changing their properties (*e.g.* charge, hydrophobicity), leading to swelling or disassembly of the multilayer, which can be used to facilitate, for example, the release of a therapeutic cargo. A similar approach involves the incorporation of active components (*e.g.* enzymes, micelles, and liposomes) to achieve a responsive system. By retaining their activity within the multilayer, they can endow responsiveness on the entire assembly. In contrast, the incorporation of (bio)degradable polymers may result in films susceptible to hydrolytic and enzymatic degradation. Finally, multilayers may be assembled in a way that they require chemical stabilization through inter-layer bonds to remain stable. By engineering specific degradable or responsive linkages into the constructs, these multilayers can be rendered responsive. In the following sections these different approaches will be reviewed and prominent and novel examples will be discussed to give the reader an overview of the available approaches found in the literature.

11.3.1 Disruption of Multilayers with Responsive Polymers

The first reported multilayers were built up of alternating layers of polyanions and polycations. These systems are, especially in the case of weak polyelectrolytes, considered inherently responsive. Changes in pH can change the charge ratio of these polymers drastically around their pK_a and thus lead to uncompensated charges that destabilize the film. Similarly, (de)protonation of film components in multilayers stabilized through hydrogen bonding are susceptible to changes in pH. Increases in salt concentrations have similar effects by shielding the polymer charges. In both cases these effects can lead to enhanced permeability, but also complete film disassembly, depending on the precise circumstances (*i.e.* pH, salt concentration, employed film components) and importantly the presence of secondary chemical crosslinks between the layers. Assembly as well as disassembly of these films have been thoroughly studied by several authors and have been reviewed in greater detail elsewhere.[114,171–175]

A different approach to endow multilayers with responsiveness through charge-shifting polymers was first reported by De Geest *et al.*[176] By developing polymers with a side chain that flipped charge upon hydrolytic cleavage they were able to construct multilayers that slowly degraded over time depending on the pH (Figure 11.11). Other authors have subsequently reported similar systems with both positive and negative initial charges.[177–179] Degradation kinetics can be fine-tuned by small variations in the side chain chemistry as well as the degree of functionalization of the polymer.[177,178,180]

Ma *et al.* reported a different stimulus for the charge-shifting of a polymer within a multilayer.[181] The authors demonstrated that multilayers constructed from polycationic and polyanionic poly(ferrocenyl silane)s

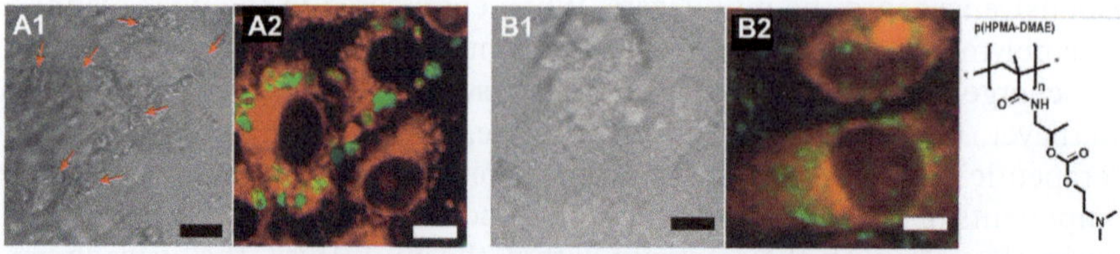

Figure 11.11 Brightfield (1) and overlaid fluorescence (2) images of VERO-1 cells after 60 h of culture with DS^{FITC}-filled nondegradable (PAH#PSS) (A) and charge shifting (p(HPMA-DMAE)#PSS) (B) microcapsules. Scale bar = 10 μm. The arrows indicate the presence of intact capsules. Reproduced with permission from ref. 176.

responded to oxidation by $FeCl_3$ with swelling, enhanced permeability and ultimately disassembly. These effects were caused by the excessive positive charges induced on the ferrocene during oxidation. Interestingly, by applying capping layers of (PSS#PAH) they were able to retain the enhanced permeability, while preventing destabilization of the multilayer, thus allowing for control over multilayer permeability. Several other authors reported on redox responsive multilayers responding to electric potentials.[182,183]

An important aspect of the degradation behavior is the time frame in which this process takes place. Liang *et al.* have reported in a recent publication the assembly of multilayer capsules based on the pH-responsive polymer poly(2-diisopropylaminoethyl methacrylate) (PDPA).[184] Through copolymerization with lauryl methacrylate they obtained a polymer that allowed for single component multilayer capsules based on the hydrophobic interactions between C_{12} chains. This resulted in capsules that swell rapidly in response to endosomal pH (*i.e.* pH 6) due to protonation of PDPA. The authors demonstrated this process to be reversible and showed controlled release of different model drugs.[184]

Yet another trigger for changes in the charge ratio of polymers has been reported to be carbohydrates.[185] This effect was achieved by synthesizing a copolymer of the polycation poly(2-(dimethylamino)ethyl methacrylate) (PDMAEMA) and phenylboronic acid (PBA) groups. Phenylboronic acid exists in equilibrium between its neutral trigonal form and its anionic tetrahedral form. PBAs are further known to form a reversible cyclic boronic ester with diols (Figure 11.12). This reaction lowers the pK_a of the boronic acids by 2–4 units, thus shifting the equilibrium towards the negatively charged form. This has triggered extensive research into PBAs as carbohydrate sensors.[186–190] De Geest *et al.* demonstrated that multilayers composed of the PDMAEMA–PBA copolymer and PSS were responsive to glucose close to physiologically relevant concentrations (5–10 mM), though regrettably not at physiological pH.[185] Further optimization of the initial pK_a of the boronic acid group[191] employed might lead to systems sensitive at physiological pH. In addition to multilayered systems, PBA has also been utilized extensively in hydrogel systems. Notably, some PBA-equipped hydrogels have shown selectivity against various sugars in blood plasma (discussed in more detail in Chapter 3).

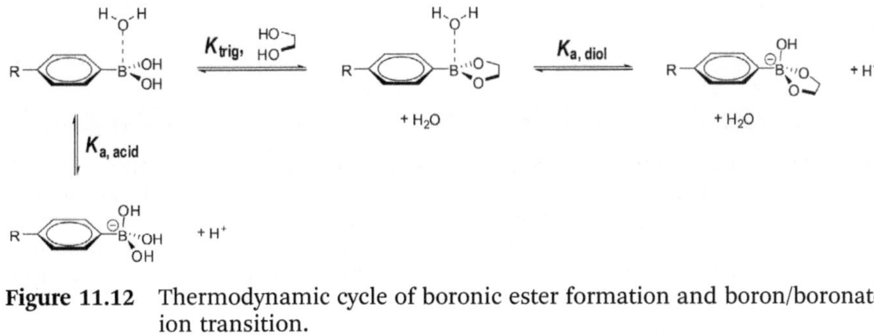

Figure 11.12 Thermodynamic cycle of boronic ester formation and boron/boronate ion transition.

11.3.2 Multilayers Composed of Inherently Active, Responsive, or Degradable Polymers

Multilayers owe their stability to the multiplicity of interactions between consecutive layers. Therefore, upon degradation of one of the multilayer components and resulting decreases in polymer molar mass, multilayer disassembly is observed. Several research groups have investigated polymers with (bio)degradable links in their backbone. A well-studied example is poly(β-amino ester)s, which degrade through hydrolysis.[192] By monitoring the release of complementary PSS it was observed that release periods could be controlled by varying the hydrophobicity of the polymer. Further control was achieved by combining different polymers into a single multilayered construct.[192] Applicability of this system in gene therapy was investigated by replacing the polyanion PSS with plasmid DNA (pDNA). The authors observed the release of transcriptionally active pDNA over a range of three days.[193] The authors further investigated the application of the multilayers for transcutaneous delivery of pDNA into the skin of mice *in vivo*, showing potential applicability for vaccine delivery (Figure 11.13).[21] Similarly, Blacklock *et al.* investigated the assembly and dissolution of multilayers of pDNA and an artificial peptide with reducible disulfide linkages in the backbone.[194] Multilayers disassembled selectively in the presence of a reducing agent, releasing the co-assembled pDNA. The same research group further demonstrated through the use of reducible hyperbranched poly(amido amine) (RHB) that higher and longer lasting transfection could be achieved as compared to (DNA#PEI) multilayers.[195] The authors speculate that highly localized release in the reducing microenvironment of the cellular membrane could lead to enhanced gene delivery efficiency.

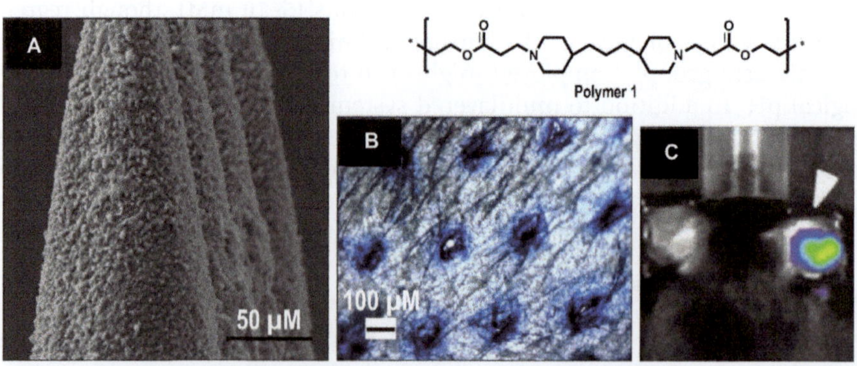

Figure 11.13 (A) Scanning electron microscope image of (polymer 1#pDNA)-coated PLGA microneedle arrays. (B) Optical micrograph of mouse ear following penetration of multilayer-coated PLGA microneedle arrays and staining with trypan blue. (C) Representative bioluminescence signal of treated mouse ear with efficacy that depends on the layer number and treatment duration.
Reproduced with permission from ref. 21.

Polymers are an essential part of biological systems and as such enzymes exist to modify and process these natural polymers. Natural polymers (*i.e.* polypeptides, nucleic acids, polysaccharides) have been demonstrated to retain their availability to these enzymes within multilayers. Following addition of an enzyme that is able to degrade at least one of the layer components, film disassembly was achieved. Reported enzymes consist of proteases,[176,196–200] nucleases,[201] lysozyme,[202] chitosanase,[203] and hyaluronidase.[204]

Similarly, enzymes can be incorporated into multilayers as active components themselves. This allows for the construction of biocatalytic thin films, a concept well-established in the literature.[205] A large number of enzymes have been incorporated into multilayers and shown to retain enzymatic activity.[206,207] A recent example highlights their potential in the biomedical field—Andreasen *et al.* thoroughly studied the incorporation of β-galactosidase into multilayers.[208] These multilayers have potential in drug delivery through surface-mediated enzyme prodrug therapy, where a systemically administered prodrug is locally activated by β-galactosidase, thus allowing for external control in dosing as well as highly localized delivery.[208]

Alternative active components have been incorporated into multilayers as well.[209–211] For example, Addison and Biggs *et al.* reported the incorporation of pH-responsive block copolymer micelles (BCM) into multilayers. These can be used as drug reservoirs within the multilayer that retain their pH responsiveness and thus allow for a controlled release from the multilayer. Both planar multilayers as well as capsules were obtained from positively charged BCMs of poly(2-(dimethylamino)ethyl methacrylate)-*block*-poly(2-(diethylamino)ethyl methacrylate) (PDMAEMA-*b*-PDEAEMA) and PSS. It was further demonstrated that PSS could be exchanged for anionic BCMs to obtain alternating layers of micelles.[212–214] Similarly, films stabilized by hydrogen bonding were reported between PNIPAm-*b*-PDEA micelles and tannic acid.[215] These films remained stable at slightly acidic pH, but released hydrophobic cargo from the BCM core. Both thermoresponsive polymers and BCMs have been used to construct multilayers and endow them with responsiveness.[216,217] Liposomes have been incorporated into multilayers as well, either as responsive reservoirs[218] or for the encapsulation of active components (*i.e.* enzymes) to protect them from environmental influences, while maintaining their activity.[219]

A quickly growing application of nucleic acid technology is aptamers.[220,221] Aptamers are short sequences of single stranded DNA or RNA that are able to bind a large variety of analytes with very high selectivity. The specific sequence for a specific analyte is usually selected from a large pool of random sequences in an iterative process. The binding ability of aptamers originates from their three-dimensional structure, which folds around the analyte of interest.[220,221] Sultan *et al.* co-assembled aptamers into (PAH#PSS) multilayer capsules.[222] Interestingly, the authors found that binding the analyte to the incorporated aptamers resulted in an enhanced permeability of the multilayer. It is speculated that the analyte triggers conformational

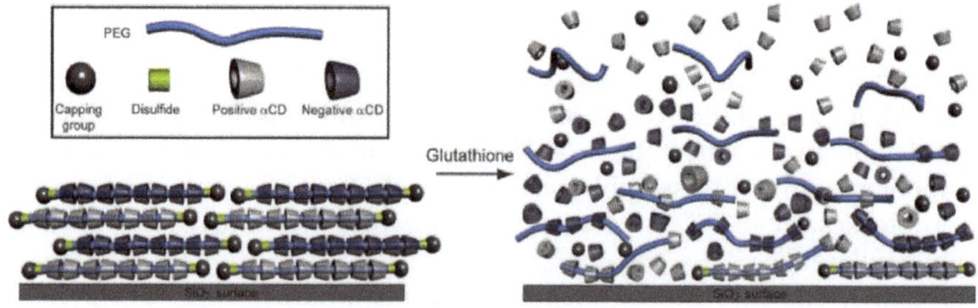

Figure 11.14 Schematic illustration of multilayers from dimethylethylenediamine- (positively charged) and glycine- (negatively charged) functionalized PEG-threaded cyclodextrins with disulfide capping. The presence of a reducing agent such as glutathione results in multilayer disassembly. Reproduced with permission from ref. 225.

changes of the aptamers inside the film, thus disturbing the multilayer enough to increase its permeability. The same group later demonstrated multilayer capsules with aptamers as structural components in the capsule core. These capsules ruptured on addition of the analyte.[223] Both modes (*i.e.* enhanced permeability, capsule rupture) have applications in drug delivery and sensor designs by incorporating aptamers specific for therapeutically relevant targets. The potential of this system was recently demonstrated by another group that employed an aptamer-containing multilayer as a switchable barrier between an etchant and a plasmonic nanoparticle sensor.[224] Upon binding of the analyte, the enhanced permeability of the film allowed access of the etchant to the surface-immobilized nanoparticles, resulting in a visible color change. The authors argue that this general approach may be extrapolated to a wide range of aptamers to construct facile dip-and-read sensors.[224]

Dam *et al.* recently explored a novel approach by threading cyclodextrins onto PEG and locking them in place with a disulfide-stabilized cap (Figure 11.14).[225] Post-modification of cyclodextrins with charged moieties allowed for LbL assembly of these rod-shaped stiff supramolecular structures. Physiological concentrations of glutathione, a reducing agent present in the cytoplasm of cells, resulted in release of the capping group through disulfide cleavage and rapid film disassembly. The biocompatibility of both PEG and cyclodextrins is a major advantage, while the latter may also have potential for use as drug carriers, making these systems interesting candidates for drug delivery.[225]

11.3.3 Responsive Multilayers Through Specific Disruption of Inter-layer Bonds

An alternative approach to building up layers of interacting components is by forming bonds that are susceptible to the presence of external chemical moieties. As discussed in the previous sections, in the case of crosslinks introduced subsequent to assembly, these linkages can be engineered to be susceptible to commonly employed degradation pathways (*i.e.* hydrolytic, enzymatic, reducing). An advantage of introducing these inter-layer cross-links as part of the assembly process is the reduced number of total assembly steps, as well as avoiding potentially toxic and difficult to remove crosslinking agents.[226] Build-up of multilayers through host–guest interactions between subsequent layers is another effective approach. These interactions can later be disrupted by introducing moieties that competitively bind to these bonds, replacing the inter-layer interactions and thus disrupting the multilayer.[227–229] Interesting in this approach is the fact that these systems have the potential to be reversible, as removal of the stimuli allows reformation of the interlayer bonds and thus re-stabilization of the multilayer.

Sato *et al.* capitalized on the well-studied interactions between lectins and carbohydrates,[227,228] as well as between streptavidin and biotin.[230] In the case of multilayers composed of lectins (concanavalin A) and glycopolymers, addition of carbohydrates with higher affinity to the lectin resulted in film destabilization. Similarly, multilayers of respectively streptavidin- and biotin-labeled polymers were destabilized by the presence of biotin.[227,228,230] A glucose-triggered release of insulin was proposed from multilayer capsules constructed from lectins and glycopolymers. However, this system needs to be further improved in its glucose sensitivity to become practical for *in vivo* applications.

As discussed earlier, phenylboronic acids are able to bind polyols in a dynamic manner through the formation of cyclic boronic esters. As such, formation of multilayers between phenylboronic acid (PBA)-functionalized polymers and polyols has received considerable attention. Addition of sugars with high binding constants for PBA effects disruption of these multilayers. Films of PBA-functionalized PAAc and mannan responded to glucose only at high pH,[229] whereas multilayers of PVA and PBA-functionalized poly(acrylamide) responded at near physiological pH.[231] As is the case with boronic acids in general, including the examples presented throughout this chapter, these multilayers demonstrate much higher sensitivity towards other sugars or carbohydrates besides mere glucose. For example, sensitivity to fructose has been reported to be one order of magnitude higher than for glucose.[229] Regrettably, side-by-side comparisons of sensitivity to different sugars are often missing. As such the challenge lies in developing systems sensitive selectively to glucose at a physiologically relevant concentration range (5–10 mM), while being able to function at physiological pH (see also Chapter 3).

Multilayers that disassemble in the presence of chelating agents were reported by Krass *et al.*[232] by the construction of multilayers from a pyridine functionalized polymer and transition metal ions. Addition of chelating agents disrupts the metal ion coordination bonds and thus leads to film disruption. Similarly, multilayers fabricated from alternating layers of cyclodextrin- and adamantane-grafted chitosan were shown to disassemble upon addition of molecules that compete with this host–guest interaction.[233]

Multilayers assembled through click-chemistry between azide- and alkyne-functionalized dextran were reported.[226] Conjugation of azide and alkyne groups was achieved through a carbonate linkage. Hydrolytic degradation of this linkage resulted in multilayer disassembly under physiological conditions.

Nucleic acids hold a special place within natural polymers as they allow for extremely specific inter- and intra-polymer interactions due to base pair hybridization. The potential of this is well illustrated by the impressive array of 2D and 3D structures that have been assembled in this field in what has been coined DNA origami.[234–236] Johnston *et al.* have made use of specific base pair hybridization to assemble multilayers composed purely of DNA.[237] By introducing a binding sequence specific to a certain restriction enzyme, they were able to demonstrate specific degradation of the multilayers (Figure 11.15). A system reported later using a peptide analogue of DNA showed similar results, but allowed for greater stability in biological media.[238] DNA multilayers can be programmed to degrade in the presence

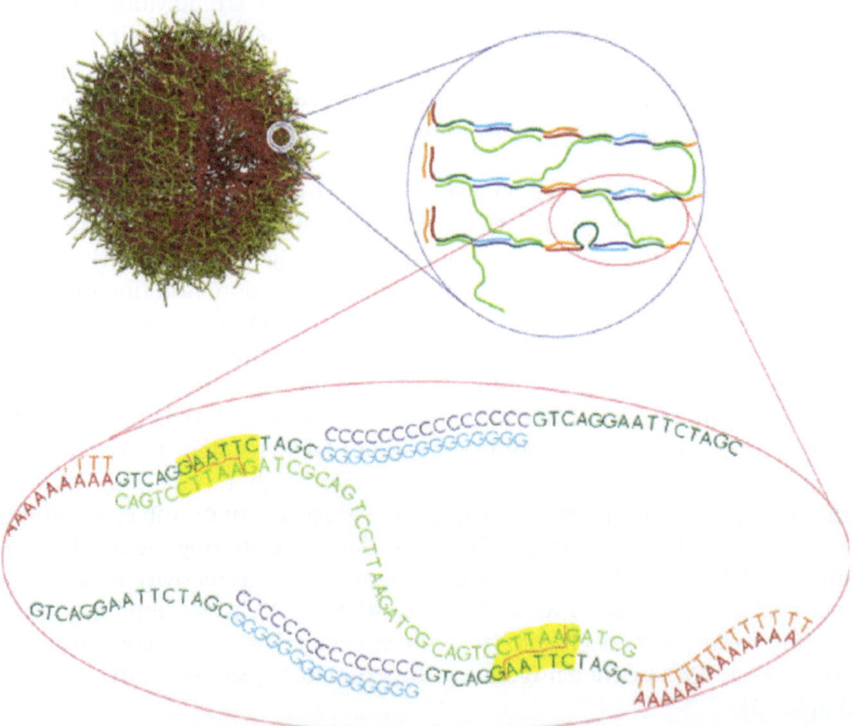

Figure 11.15 Schematic illustration of a DNA-hybridization driven multilayer of (A_{15}-X_{15}-G_{15}#T_{15}-X_{15}-C_{15}). The unhybridized X strands (dark green) in the multilayers were treated with a triblock complementary X strand (light green) to introduce interlayer crosslinking for improved stability. The yellow highlights indicate the specific sequence to be cleaved by the *Eco*RI restriction enzyme to induce multilayer dissolution. Reproduced with permission from ref. 237.

of complementary DNA sequences that specifically target the sequences responsible for film stabilization.[239]

11.4 Conclusion and Prospects

Layer-by-layer assembly offers a very versatile and particularly economical method, not only for the functionalization of practically any existing surface, but also as a complete delivery system by itself. The ability to tune the thin film both in thickness and amount of incorporated materials, in combination with the more intricate details of multilayer architectures, is a major advantage of the LbL assembly technique. Among others, multilayers can be tuned to load specific active compounds to achieve very high loadings within the multilayer, and hence on the surface of the coated materials. The multilayered architecture and the interplay of various materials with different properties as multilayer components offer a direct approach to achieve sequential release of cooperative active agents, in addition to providing a simple means to tune release or degradation profiles.

Drawn by these promises, a lot of research has been invested in the fundamental molecular understanding of the LbL assembly process and the resulting multilayer products, as well as in the translation of existing and novel biomaterials and functional polymers to be exploited for LbL assembly. The extensive research in the development of new smart biomaterials is anticipated to continue, which may give rise to novel multilayer systems. A deeper understanding into specific multilayer systems will not only help in their development towards a perceived goal, but also provide additional information on the physicochemical aspects of LbL assemblies in general. This in turn may further help to establish new opportunities for applications.

Despite substantial research in the LbL assembly field, several major challenges remain to be addressed. These include the need to increase the mechanical stability[160] of these thin film systems to lengthen shelf life, and the need to find faster and more reproducible methods for large-scale production. The 2004 report by Shiratori and co-workers on roll-to-roll processes for the fabrication of large-area thin multilayers is one such example.[10] Although many new LbL fabrication techniques have been reported over the past years, a lot of them still require long processing times, or complex devices, especially in the case of fabricating smaller multilayer capsules. Attempts to decrease the size of multilayer constructs from micro- to nanocapsules have so far received only limited attention, though these nanostructures may have tremendous application potential in nanomedicine. For therapeutic applications, detailed *in vivo* studies need to be carried out to assess the true potential of these multilayers. The work of Hammond and co-workers referred to in this chapter provides the first report on *in situ* DNA delivery from multilayer-coated microneedles *in vivo*,[21] while Lynn and co-workers recently reported an *in vivo* study on the same multilayer coatings on stents.[13] Other authors have also reported *in vivo* studies of multilayer capsules.[240,241] With a lot of opportunities yet to be discovered, the development of novel multilayers that provide orchestrated responses to various biological and chemical stimuli is expected to continue well into the future.

List of Abbreviations

Alg	alginate
BCM	block copolymer micelles
BSM	bovine submaxillary mucin
CHI	chitosan
CLSM	confocal laser scanning microscopy
DS	dextran sulfate
EDC	1-ethyl-3-(3-dimethylaminopropyl)carbodiimide
FITC	fluorescein isothiocyanate
HA	hyaluronic acid
HCEC	human corneal epithelial cells
HEP	heparin
JAC	lectin jacalin
LbL	Layer-by-layer
MF	melamine formaldehyde
MNP	magnetic nanoparticle
PAAc	poly(acrylic acid)
PAH	poly(allylamine hydrochloride)
PBA	phenylboronic acid
PDADMAC	poly(diallyldimethylammonium chloride)
PDMAEMA	poly(2-(dimethylamino)ethyl methacrylate)
PDMAEMA-*b*-PDEAEMA	poly(2-(dimethylamino)ethyl methacrylate)-*block*-poly(2-(diethylamino)ethyl methacrylate)
pDNA	plasmid DNA
PDPA	poly(2-diisopropylaminoethyl methacrylate)
PEG	poly(ethylene glycol)
PEI	poly(ethylene imine)
PGA	poly(L-glutamic acid)
PLAA	poly(L-aspartic acid)
PLGA	poly(lactic-*co*-glycolic acid)
PLL	poly(L-lysine)
PMAAc	poly(methacrylic acid)
PNIPAM	poly(*N*-isopropylacrylamide)
polymer 1	poly(β-amino ester)
PSS	poly(styrenesulfonate)
PVA	poly(vinyl alcohol)
PVCL	poly(*N*-vinylcaprolactam)
PVPON	poly(*N*-vinylpyrrolidone)
RB	rhodamine B
RHB	reducible hyperbranched poly(amido amine)
SMC	smooth muscle cells
TA	tannic acid

References

1. J. B. Schlenoff, S. T. Dubas and T. Farhat, Sprayed Polyelectrolyte Multilayers, *Langmuir*, 2000, **16**, 9968–9969.
2. K. L. Cho, H. Lomas, A. J. Hill, F. Caruso and S. E. Kentish, Spray Assembled, Cross-Linked Polyelectrolyte Multilayer Membranes for Salt Removal, *Langmuir*, 2014, **30**, 8784–8790.
3. C. H. Porcel, A. Izquierdo, V. Ball, G. Decher, J. C. Voegel and P. Schaaf, Ultrathin coatings and (poly(glutamic acid)/polyallylamine) films deposited by continuous and simultaneous spraying, *Langmuir*, 2005, **21**, 800–802.
4. S. S. Lee, J. D. Hong, C. H. Kim, K. Kim, J. P. Koo and K. B. Lee, Layer-by-layer deposited multilayer assemblies of ionene-type polyelectrolytes based on the spin-coating method, *Macromolecules*, 2001, **34**, 5358–5360.
5. H. E. Karahan, L. Eyuboglu, D. Kiyilar and A. L. Demirel, pH-stability and pH-annealing of H-bonded multilayer films prepared by layer-by-layer spin-assembly, *Eur. Polym. J.*, 2014, **56**, 159–167.
6. Y. Fu, S.-J. Li, J. Xu, M. Yang, J.-D. Zhang, Y.-H. Jiao, J.-C. Zhang, K. Zhang and Y.-G. Jia, Facile and Efficient Approach to Speed up Layer-by-Layer Assembly: Dipping in Agitated Solutions, *Langmuir*, 2010, **27**, 672–677.
7. B. S. Shim, P. Podsiadlo, D. G. Lilly, A. Agarwal, J. Leet, Z. Tang, S. Ho, P. Ingle, D. Paterson, W. Lu and N. A. Kotov, Nanostructured thin films made by diewetting method of layer-by-layer assembly, *Nano Lett.*, 2007, **7**, 3266–3273.
8. M. Toda, Y. Chen, S. K. Nett, A. N. Itakura, J. Gutmann and R. Berger, Thin Polyelectrolyte Multilayers Made by Inkjet Printing and Their Characterization by Nanomechanical Cantilever Sensors, *J. Phys. Chem. C*, 2014, **118**, 8071–8078.
9. C. Priest, A. Quinn, A. Postma, A. N. Zelikin, J. Ralston and F. Caruso, Microfluidic polymer multilayer adsorption on liquid crystal droplets for microcapsule synthesis, *Lab Chip*, 2008, **8**, 2182–2187.
10. K. Fujimoto, S. Fujita, B. Ding and S. Shiratori, Fabrication of Layer-by-Layer Self-Assembly Films Using Roll-to-Roll Process, *Jpn. J. Appl. Phys.*, 2005, **44**, L126.
11. K. Ren, L. Fourel, C. G. Rouvière, C. Albiges-Rizo and C. Picart, Manipulation of the adhesive behaviour of skeletal muscle cells on soft and stiff polyelectrolyte multilayers, *Acta Biomater.*, 2010, **6**, 4238–4248.
12. E. Lojou and P. Bianco, Buildup of polyelectrolyte-protein multilayer assemblies on gold electrodes. Role of the hydrophobic effect, *Langmuir*, 2004, **20**, 748–755.
13. E. M. Saurer, C. M. Jewell, D. A. Roenneburg, S. L. Bechler, J. R. Torrealba, T. A. Hacker and D. M. Lynn, Polyelectrolyte Multilayers Promote Stent-Mediated Delivery of DNA to Vascular Tissue, *Biomacromolecules*, 2013, **14**, 1696–1704.

14. L. C. Bradley and M. Gupta, Formation of Heterogeneous Polymer Films via Simultaneous or Sequential Depositions of Soluble and Insoluble Monomers onto Ionic Liquids, *Langmuir*, 2013, **29**, 10448–10454.
15. J. de Grooth, M. Dong, W. M. de Vos and K. Nijmeijer, Building Polyzwitterion-Based Multilayers for Responsive Membranes, *Langmuir*, 2014, **30**, 5152–5161.
16. J. Seo, J. L. Lutkenhaus, J. Kim, P. T. Hammond and K. Char, Effect of the layer-by-layer (LbL) deposition method on the surface morphology and wetting behavior of hydrophobically modified PEO and PAA LbL films, *Langmuir*, 2008, **24**, 7995–8000.
17. A. L. Larkin, R. M. Davis and P. Rajagopalan, Biocompatible, Detachable, and Free-Standing Polyelectrolyte Multilayer Films, *Biomacromolecules*, 2010, **11**, 2788–2796.
18. C. Holmes, J. Daoud, P. O. Bagnaninchi and M. Tabrizian, Polyelectrolyte Multilayer Coating of 3D Scaffolds Enhances Tissue Growth and Gene Delivery: Non-Invasive and Label-Free Assessment, *Adv. Healthcare Mater.*, 2014, **3**, 572–580.
19. P. C. DeMuth, A. V. Li, P. Abbink, J. Liu, H. Li, K. A. Stanley, K. M. Smith, C. L. Lavine, M. S. Seaman, J. A. Kramer, A. D. Miller, W. Abraham, H. Suh, J. Elkhader, P. T. Hammond, D. H. Barouch and D. J. Irvine, Vaccine delivery with microneedle skin patches in nonhuman primates, *Nat. Biotechnol.*, 2013, **31**, 1082–1085.
20. E. M. Saurer, R. M. Flessner, S. P. Sullivan, M. R. Prausnitz and D. M. Lynn, Layer-by-Layer Assembly of DNA- and Protein-Containing Films on Microneedles for Drug Delivery to the Skin, *Biomacromolecules*, 2010, **11**, 3136–3143.
21. P. C. DeMuth, X. Su, R. E. Samuel, P. T. Hammond and D. J. Irvine, Nano-Layered Microneedles for Transcutaneous Delivery of Polymer Nanoparticles and Plasmid DNA, *Adv. Mater.*, 2010, **22**, 4851–4856.
22. Y. Kim, A. L. Larkin, R. M. Davis and P. Rajagopalan, The Design of In Vitro Liver Sinusoid Mimics Using Chitosan-Hyaluronic Acid Polyelectrolyte Multilayers, *Tissue Eng., Part A*, 2010, **16**, 2731–2741.
23. B. Wang, Functional Multilayered Polyelectrolyte Assemblies on Biological Cells, *Cell Surface Engineering: Fabrication of Functional Nanoshells*, The Royal Society of Chemistry, 2014, pp. 4–27.
24. R. Georgieva, S. Moya, E. Donath and H. Baumler, Permeability and conductivity of red blood cell templated polyelectrolyte capsules coated with supplementary layers, *Langmuir*, 2004, **20**, 1895–1900.
25. H. Ai, Layer-by-layer capsules for magnetic resonance imaging and drug delivery, *Adv. Drug Delivery Rev.*, 2011, **63**, 772–788.
26. F. Ridi, M. Bonini and P. Baglioni, Magneto-responsive nanocomposites: Preparation and integration of magnetic nanoparticles into films, capsules, and gels, *Adv. Colloid Interface Sci.*, 2014, **207**, 3–13.
27. R. L. Muhlbauer, T. L. Pruyn, W. T. Puckett and R. A. Gerhardt, Effect of graphitic filler size and shape on the microstructure, electrical

percolation behavior and thermal properties of nanostructured multilayered carbon films deposited onto paper substrates, *J. Mater. Res.*, 2014, **29**, 472–484.
28. C. Q. Peng, Y. S. Thio and R. A. Gerhardt, Conductive paper fabricated by layer-by-layer assembly of polyelectrolytes and ITO nanoparticles, *Nanotechnology*, 2008, **19**, 505603.
29. P. Yang and N. Murase, Preparation and photoluminescence of SiO 2 -CdTe multilayer film by layer-by-layer self-assembly, *J. Phys.: Conf. Ser.*, 2009, **152**, 012009.
30. R. C. Smith, M. Riollano, A. Leung and P. T. Hammond, Layer-by-Layer Platform Technology for Small-Molecule Delivery, *Angew. Chem., Int. Ed.*, 2009, **48**, 8974–8977.
31. G. K. Such, A. P. R. Johnston and F. Caruso, Engineered hydrogen-bonded polymer multilayers: from assembly to biomedical applications, *Chem. Soc. Rev.*, 2011, **40**, 19–29.
32. E. Kharlampieva, V. Kozlovskaya and S. A. Sukhishvili, Layer-by-Layer Hydrogen-Bonded Polymer Films: From Fundamentals to Applications, *Adv. Mater.*, 2009, **21**, 3053–3065.
33. M. Matsusaki, H. Ajiro, T. Kida, T. Serizawa and M. Akashi, Layer-by-Layer Assembly Through Weak Interactions and Their Biomedical Applications, *Adv. Mater.*, 2012, **24**, 454–474.
34. J. Zhao, F. S. Pan, P. Li, C. H. Zhao, Z. Y. Jiang, P. Zhang and X. Z. Cao, Fabrication of Ultrathin Membrane via Layer-by-Layer Self-assembly Driven by Hydrophobic Interaction Towards High Separation Performance, *ACS Appl. Mater. Interfaces*, 2013, **5**, 13275–13283.
35. T. Serizawa, S. Hashiguchi and M. Akashi, Stepwise Assembly of Ultrathin Poly(vinyl alcohol) Films on a Gold Substrate by Repetitive Adsorption/Drying Processes, *Langmuir*, 1999, **15**, 5363–5368.
36. T. Akagi, T. Fujiwara and M. Akashi, Inkjet Printing of Layer-by-Layer Assembled Poly(lactide) Stereocomplex with Encapsulated Proteins, *Langmuir*, 2014, **30**, 1669–1676.
37. T. Serizawa, K. Hamada and M. Akashi, Polymerization within a molecular-scale stereoregular template, *Nature*, 2004, **429**, 52–55.
38. D. Kamei, H. Ajiro, C. Hongo and M. Akashi, Solvent Effects on Isotactic Poly(methyl methacrylate) Crystallization and Syndiotactic Poly(methacrylic acid) Incorporation in Porous Thin Films Prepared by Stepwise Stereocomplex Assembly, *Langmuir*, 2008, **25**, 280–285.
39. Y. Shimazaki, S. Ito and N. Tsutsumi, Adsorption-Induced Second Harmonic Generation from the Layer-by-Layer Deposited Ultrathin Film Based on the Charge-Transfer Interaction, *Langmuir*, 2000, **16**, 9478–9482.
40. Z. Liu, Y. Yi, J. Gauczinski, H. Xu, M. Schönhoff and X. Zhang, Surface Molecular Imprinted Layer-by-Layer Film Attached to a Porous Membrane for Selective Filtration, *Langmuir*, 2011, **27**, 11806–11812.
41. O. Crespo-Biel, B. Dordi, D. N. Reinhoudt and J. Huskens, Supramolecular layer-by-layer assembly: Alternating adsorptions of guest- and

host-functionalized molecules and particles using multivalent supramolecular interactions, *J. Am. Chem. Soc.*, 2005, **127**, 7594–7600.
42. C. Li, G.-F. Luo, H.-Y. Wang, J. Zhang, Y.-H. Gong, S.-X. Cheng, R.-X. Zhuo and X.-Z. Zhang, Host–Guest Assembly of pH-Responsive Degradable Microcapsules with Controlled Drug Release Behavior, *J. Phys. Chem. C*, 2011, **115**, 17651–17659.
43. A. Van der Heyden, M. Wilczewski, P. Labbe and R. Auzely, Multilayer films based on host-guest interactions between biocompatible polymers, *Chem. Commun.*, 2006, 3220–3222.
44. S. Takahashi, K. Sato and J.-i. Anzai, Layer-by-layer construction of protein architectures through avidin-biotin and lectin-sugar interactions for biosensor applications, *Anal. Bioanal. Chem.*, 2012, **402**, 1749–1758.
45. T. Crouzier, C. H. Beckwitt and K. Ribbeck, Mucin Multilayers Assembled through Sugar–Lectin Interactions, *Biomacromolecules*, 2012, **13**, 3401–3408.
46. K. Müller, J. F. Quinn, A. P. R. Johnston, M. Becker, A. Greiner and F. Caruso, Polyelectrolyte Functionalization of Electrospun Fibers, *Chem. Mater.*, 2006, **18**, 2397–2403.
47. A. P. R. Johnston, E. S. Read and F. Caruso, Multilayer Films on Planar and Colloidal Supports: Sequential Assembly of Like-Charged Polyelectrolytes, *Nano Lett.*, 2005, **5**, 953–956.
48. D. E. Bergbreiter and K.-S. Liao, Covalent layer-by-layer assembly-an effective, forgiving way to construct functional robust ultrathin films and nanocomposites, *Soft Matter*, 2009, **5**, 23–28.
49. C. R. Kinnane, K. Wark, G. K. Such, A. P. R. Johnston and F. Caruso, Peptide-Functionalized, Low-Biofouling Click Multilayers for Promoting Cell Adhesion and Growth, *Small*, 2009, **5**, 444–448.
50. Y.-h. Li, D. Wang and J. M. Buriak, Molecular Layer Deposition of Thiol-Ene Multilayers on Semiconductor Surfaces, *Langmuir*, 2010, **26**, 1232–1238.
51. Z. Q. Tou, C. C. Chan and S. Leong, A fiber-optic pH sensor based on polyelectrolyte multilayers embedded with gold nanoparticles, *Meas. Sci. Technol.*, 2014, **25**, 7.
52. D. Sarauli, C. G. Xu, B. Dietzel, B. Schulz and F. Lisdat, A multilayered sulfonated polyaniline network with entrapped pyrroloquinoline quinone-dependent glucose dehydrogenase: tunable direct bioelectrocatalysis, *J. Mater. Chem. B*, 2014, **2**, 3196–3203.
53. J. Luo, Y. Z. Chen, Q. Ma, R. Liu and X. Y. Liu, Layer-by-layer assembled ionic-liquid functionalized graphene-polyaniline nanocomposite with enhanced electrochemical sensing properties, *J. Mater. Chem. C*, 2014, **2**, 4818–4827.
54. G. A. Evtugyn, V. B. Stepanova, A. V. Porfireva, A. I. Zamaleeva and R. R. Fakhrullin, Electrochemical DNA Sensors Based on Nanostructured Organic Dyes/DNA/Polyelectrolyte Complexes, *J. Nanosci. Nanotechnol.*, 2014, **14**, 6738–6747.

55. W. X. Xu, X. F. Li, J. Y. Cao, H. Z. Zhang and H. M. Zhang, Membranes with well-defined ions transport channels fabricated via solvent-responsive layer-by-layer assembly method for vanadium flow battery, *Sci. Rep.*, 2014, **4**, 9.
56. N. Joseph, P. Ahmadiannamini, R. Hoogenboom and I. F. J. Vankelecom, Layer-by-layer preparation of polyelectrolyte multilayer membranes for separation, *Polym. Chem.*, 2014, **5**, 1817–1831.
57. X. Wang, N. Xia and L. Liu, Boronic Acid-Based Approach for Separation and Immobilization of Glycoproteins and Its Application in Sensing, *Int. J. Mol. Sci.*, 2013, **14**, 20890–20912.
58. J. H. Park, B. S. Kim, Y. C. Yoo, M. S. Khil and H. Y. Kim, Enhanced mechanical properties of multilayer nano-coated electrospun nylon 6 fibers via a layer-by-layer self-assembly, *J. Appl. Polym. Sci.*, 2008, **107**, 2211–2216.
59. L. Zhai, F. Ç. Cebeci, R. E. Cohen and M. F. Rubner, Stable Superhydrophobic Coatings from Polyelectrolyte Multilayers, *Nano Lett.*, 2004, **4**, 1349–1353.
60. U. Manna, M. J. Kratochvil and D. M. Lynn, Superhydrophobic Polymer Multilayers that Promote the Extended, Long-Term Release of Embedded Water-Soluble Agents, *Adv. Mater.*, 2013, **25**, 6405–6409.
61. X. Wang, Y. Wang, S. Bi, Y. G. Wang, X. G. Chen, L. Y. Qiu and J. Q. Sun, Optically Transparent Antibacterial Films Capable of Healing Multiple Scratches, *Adv. Funct. Mater.*, 2014, **24**, 403–411.
62. J. Hiller, J. D. Mendelsohn and M. F. Rubner, Reversibly erasable nanoporous anti-reflection coatings from polyelectrolyte multilayers, *Nat. Mater.*, 2002, **1**, 59–63.
63. D. V. Andreeva, E. V. Skorb and D. G. Shchukin, Layer-by-Layer Polyelectrolyte/Inhibitor Nanostructures for Metal Corrosion Protection, *ACS Appl. Mater. Interfaces*, 2010, **2**, 1954–1962.
64. X. Zhu, D. Janczewski, S. S. C. Lee, S. L.-M. Teo and G. J. Vancso, Cross-Linked Polyelectrolyte Multi layers for Marine Antifouling Applications, *ACS Appl. Mater. Interfaces*, 2013, **5**, 5961–5968.
65. Z. F. Dai, H. Mohwald, B. Tiersch and L. Dahne, Nanoengineering of polymeric capsules with a shell-in-shell structure, *Langmuir*, 2002, **18**, 9533–9538.
66. A. M. Yu, I. Gentle, G. Q. Lu and F. Caruso, Nanoassembly of biocompatible microcapsules for urease encapsulation and their use as biomimetic reactors, *Chem. Commun.*, 2006, 2150–2152.
67. P. T. Hammond, Building biomedical materials layer-by-layer, *Mater. Today*, 2012, **15**, 196–206.
68. A. Shukla and B. Almeida, Advances in cellular and tissue engineering using layer-by-layer assembly, *Wiley Interdiscip. Rev.: Nanomed. Nanobiotechnol.*, 2014, **6**, 411–421.
69. L. L. del Mercato, M. M. Ferraro, F. Baldassarre, S. Mancarella, V. Greco, R. Rinaldi and S. Leporatti, Biological applications of LbL

multilayer capsules: From drug delivery to sensing, *Adv. Colloid Interface Sci.*, 2014, **207**, 139–154.
70. B. M. Wohl and J. F. J. Engbersen, Responsive layer-by-layer materials for drug delivery, *J. Controlled Release*, 2012, **158**, 2–14.
71. G. Decher, J. D. Hong and J. Schmitt, Buildup ultrathin multilayer films by a self-assembly process .3. Consecutively alternating adsorption of anionic and cationic polyelectrolytes on charged surfaces, *Thin Solid Films*, 1992, **210**, 831–835.
72. S. S. Ono and G. Decher, Preparation of Ultrathin Self-Standing Polyelectrolyte Multilayer Membranes at Physiological Conditions Using pH-Responsive Film Segments as Sacrificial Layers, *Nano Lett.*, 2006, **6**, 592–598.
73. J. Jeon, V. Panchagnula, J. Pan and A. V. Dobrynin, Molecular Dynamics Simulations of Mutilayer Films of Polyelectrolytes and Nanoparticles, *Langmuir*, 2006, **22**, 4629–4637.
74. C. Peng, Y. S. Thio and R. A. Gerhardt, Effect of Precursor-Layer Surface Charge on the Layer-by-Layer Assembly of Polyelectrolyte/Nanoparticle Multilayers, *Langmuir*, 2011, **28**, 84–91.
75. Y. Li, X. Wang and J. Sun, Layer-by-layer assembly for rapid fabrication of thick polymeric films, *Chem. Soc. Rev.*, 2012, **41**, 5998–6009.
76. J. Schmitt, T. Grunewald, G. Decher, P. S. Pershan, K. Kjaer and M. Losche, Internal structure of layer-by-layer adsorbed polyelectrolyte films – A neutron and X-ray reflectivity study, *Macromolecules*, 1993, **26**, 7058–7063.
77. B. F. Abu-Sharkh, Structure and mechanism of formation of polyelectrolyte multilayers, *Polymer*, 2006, **47**, 3674–3680.
78. M. Castelnovo and J. F. Joanny, Formation of polyelectrolyte multilayers, *Langmuir*, 2000, **16**, 7524–7532.
79. G. Ladam, P. Schaad, J. C. Voegel, P. Schaaf, G. Decher and F. Cuisinier, In Situ Determination of the Structural Properties of Initially Deposited Polyelectrolyte Multilayers, *Langmuir*, 1999, **16**, 1249–1255.
80. G. Decher, Fuzzy nanoassemblies: Toward layered polymeric multicomposites, *Science*, 1997, **277**, 1232–1237.
81. D. L. Elbert, C. B. Herbert and J. A. Hubbell, Thin Polymer Layers Formed by Polyelectrolyte Multilayer Techniques on Biological Surfaces, *Langmuir*, 1999, **15**, 5355–5362.
82. N. Laugel, C. Betscha, M. Winterhalter, J.-C. Voegel, P. Schaaf and V. Ball, Relationship between the Growth Regime of Polyelectrolyte Multilayers and the Polyanion/Polycation Complexation Enthalpy, *J. Phys. Chem. B*, 2006, **110**, 19443–19449.
83. R. A. McAloney, M. Sinyor, V. Dudnik and M. C. Goh, Atomic Force Microscopy Studies of Salt Effects on Polyelectrolyte Multilayer Film Morphology, *Langmuir*, 2001, **17**, 6655–6663.
84. B. Schoeler, E. Poptoshev and F. Caruso, Growth of Multilayer Films of Fixed and Variable Charge Density Polyelectrolytes: Effect of Mutual Charge and Secondary Interactions, *Macromolecules*, 2003, **36**, 5258–5264.

85. D. M. DeLongchamp, M. Kastantin and P. T. Hammond, High-contrast electrochromism from layer-by-layer polymer films, *Chem. Mater.*, 2003, **15**, 1575–1586.
86. C. Jiang, X. Liu, C. Luo, Y. Zhang, L. Shao and F. Shi, Controlled exponential growth in layer-by-layer multilayers using high gravity fields, *J. Mater. Chem. A*, 2014, **2**, 14048–14053.
87. P. Lavalle, C. Picart, J. Mutterer, C. Gergely, H. Reiss, J. C. Voegel, B. Senger and P. Schaaf, Modeling the buildup of polyelectrolyte multilayer films having exponential growth, *J. Phys. Chem. B*, 2004, **108**, 635–648.
88. C. Picart, J. Mutterer, L. Richert, Y. Luo, G. D. Prestwich, P. Schaaf, J. C. Voegel and P. Lavalle, Molecular basis for the explanation of the exponential growth of polyelectrolyte multilayers, *Proc. Natl. Acad. Sci. U. S. A.*, 2002, **99**, 12531–12535.
89. H. W. Jomaa and J. B. Schlenoff, Accelerated exchange in polyelectrolyte multilayers by "catalytic" polyvalent ion pairing, *Langmuir*, 2005, **21**, 8081–8084.
90. P. Lavalle, C. Gergely, F. J. G. Cuisinier, G. Decher, P. Schaaf, J. C. Voegel and C. Picart, Comparison of the Structure of Polyelectrolyte Multilayer Films Exhibiting a Linear and an Exponential Growth Regime: An in Situ Atomic Force Microscopy Study, *Macromolecules*, 2002, **35**, 4458–4465.
91. N. S. Zacharia, D. M. DeLongchamp, M. Modestino and P. T. Hammond, Controlling Diffusion and Exchange in Layer-by-Layer Assemblies, *Macromolecules*, 2007, **40**, 1598–1603.
92. L. Richert, P. Lavalle, E. Payan, X. Z. Shu, G. D. Prestwich, J.-F. Stoltz, P. Schaaf, J.-C. Voegel and C. Picart, Layer by Layer Buildup of Polysaccharide Films: Physical Chemistry and Cellular Adhesion Aspects, *Langmuir*, 2003, **20**, 448–458.
93. B. Sun, C. M. Jewell, N. J. Fredin and D. M. Lynn, Assembly of Multilayered Films Using Well-Defined, End-Labeled Poly(acrylic acid): Influence of Molecular Weight on Exponential Growth in a Synthetic Weak Polyelectrolyte System, *Langmuir*, 2007, **23**, 8452–8459.
94. E. J. Calvo and A. Wolosiuk, Donnan Permselectivity in Layer-by-Layer Self-Assembled Redox Polyelectrolye Thin Films, *J. Am. Chem. Soc.*, 2002, **124**, 8490–8497.
95. N. Laugel, F. Boulmedais, A. E. El Haitami, P. Rabu, G. Rogez, J.-C. Voegel, P. Schaaf and V. Ball, Tunable Synthesis of Prussian Blue in Exponentially Growing Polyelectrolyte Multilayer Films†, *Langmuir*, 2009, **25**, 14030–14036.
96. J. B. Schlenoff, H. Ly and M. Li, Charge and Mass Balance in Polyelectrolyte Multilayers, *J. Am. Chem. Soc.*, 1998, **120**, 7626–7634.
97. T. R. Farhat and J. B. Schlenoff, Doping-Controlled Ion Diffusion in Polyelectrolyte Multilayers: Mass Transport in Reluctant Exchangers, *J. Am. Chem. Soc.*, 2003, **125**, 4627–4636.

98. M. A. Pechenkin, H. Mohwald and D. V. Volodkin, pH- and salt-mediated response of layer-by-layer assembled PSS/PAH microcapsules: fusion and polymer exchange, *Soft Matter*, 2012, **8**, 8659–8665.
99. R. Zhang, K. Koehler, O. Kreft, A. Skirtach, H. Moehwald and G. Sukhorukov, Salt-induced fusion of microcapsules of polyelectrolytes, *Soft Matter*, 2010, **6**, 4742–4747.
100. R. v. Klitzing, J. E. Wong, W. Jaeger and R. Steitz, Short range interactions in polyelectrolyte multilayers, *Curr. Opin. Colloid Interface Sci.*, 2004, **9**, 158–162.
101. E. Tjipto, J. F. Quinn and F. Caruso, Assembly of Multilayer Films from Polyelectrolytes Containing Weak and Strong Acid Moieties, *Langmuir*, 2005, **21**, 8785–8792.
102. M. Salomäki, P. Tervasmäki, S. Areva and J. Kankare, The Hofmeister Anion Effect and the Growth of Polyelectrolyte Multilayers, *Langmuir*, 2004, **20**, 3679–3683.
103. V. Ball, J.-C. Voegel and P. Schaaf, Effect of Thiocyanate Counterion Condensation on Poly(allylamine hydrochloride) Chains on the Buildup and Permeability of Polystyrenesulfonate/Polyallylamine Polyelectrolyte Multilayers, *Langmuir*, 2005, **21**, 4129–4137.
104. C. S. Hajicharalambous, J. Lichter, W. T. Hix, M. Swierczewska, M. F. Rubner and P. Rajagopalan, Nano- and sub-micron porous polyelectrolyte multilayer assemblies: Biomimetic surfaces for human corneal epithelial cells, *Biomaterials*, 2009, **30**, 4029–4036.
105. M. Salomäki, T. Laiho and J. Kankare, Counteranion-Controlled Properties of Polyelectrolyte Multilayers, *Macromolecules*, 2004, **37**, 9585–9590.
106. M. Salomäki and J. Kankare, Specific Anion Effect in Swelling of Polyelectrolyte Multilayers, *Macromolecules*, 2008, **41**, 4423–4428.
107. M. M. Stevens and J. H. George, Exploring and engineering the cell surface interface, *Science*, 2005, **310**, 1135–1138.
108. D. E. Discher, P. Janmey and Y. L. Wang, Tissue cells feel and respond to the stiffness of their substrate, *Science*, 2005, **310**, 1139–1143.
109. X. Gong and C. Gao, Influence of salt on assembly and compression of PDADMAC/PSSMA polyelectrolyte multilayers, *Phys. Chem. Chem. Phys.*, 2009, **11**, 11577–11586.
110. W. W. Hu and S. L. Tsou, The effect of alginate on DNA delivery from layer-by-layer assembled films, *Carbohydr. Polym.*, 2014, **101**, 240–248.
111. S. S. Shiratori and M. F. Rubner, pH-Dependent Thickness Behavior of Sequentially Adsorbed Layers of Weak Polyelectrolytes, *Macromolecules*, 2000, **33**, 4213–4219.
112. S. A. Sukhishvili and S. Granick, Layered, Erasable Polymer Multilayers Formed by Hydrogen-Bonded Sequential Self-Assembly, *Macromolecules*, 2001, **35**, 301–310.
113. O. Mermut and C. J. Barrett, Effects of Charge Density and Counterions on the Assembly of Polyelectrolyte Multilayers, *J. Phys. Chem. B*, 2003, **107**, 2525–2530.

114. S. A. Sukhishvili and S. Granick, Layered, erasable, ultrathin polymer films, *J. Am. Chem. Soc.*, 2000, **122**, 9550–9551.
115. I. Erel-Unal and S. A. Sukhishvili, Hydrogen-Bonded Hybrid Multilayers: Film Architecture Controls Release of Macromolecules, *Macromolecules*, 2008, **41**, 8737–8744.
116. I. Erel-Unal and S. A. Sukhishvili, Hydrogen-bonded multilayers of a neutral polymer and a polyphenol, *Macromolecules*, 2008, **41**, 3962–3970.
117. S. Mehrotra, D. Lynam, R. Maloney, K. M. Pawelec, M. H. Tuszynski, I. Lee, C. Chan and J. Sakamoto, Time Controlled Protein Release from Layer-by-Layer Assembled Multilayer Functionalized Agarose Hydrogels, *Adv. Funct. Mater.*, 2010, **20**, 247–258.
118. C. C. Buron, C. Filiâtre, F. Membrey, C. Bainier, D. Charraut and A. Foissy, Early steps in layer-by-layer construction of polyelectrolyte films: The transition from surface/polymer to polymer/polymer determining interactions, *J. Colloid Interface Sci.*, 2007, **314**, 358–366.
119. R. Advincula, E. Aust, W. Meyer and W. Knoll, In Situ Investigations of Polymer Self-Assembly Solution Adsorption by Surface Plasmon Spectroscopy, *Langmuir*, 1996, **12**, 3536–3540.
120. C. Peng, Y. S. Thio and R. A. Gerhardt, Effect of Precursor-Layer Surface Charge on the Layer-by-Layer Assembly of Polyelectrolyte/Nanoparticle Multilayers, *Langmuir*, 2012, **28**, 84–91.
121. J. T. Zhang, L. S. Chua and D. M. Lynn, Multilayered thin films that sustain the release of functional DNA under physiological conditions, *Langmuir*, 2004, **20**, 8015–8021.
122. C. Brunot, B. Grosgogeat, C. Picart, C. Lagneau, N. Jaffrezic-Renault and L. Ponsonnet, Response of fibroblast activity and polyelectrolyte multilayer films coating titanium, *Dent. Mater.*, 2008, **24**, 1025–1035.
123. M. Onda, Y. Lvov, K. Ariga and T. Kunitake, Sequential reaction and product separation on molecular films of glucoamylase and glucose oxidase assembled on an ultrafilter, *J. Ferment. Bioeng.*, 1996, **82**, 502–506.
124. C. Brunot, L. Ponsonnet, C. Lagneau, P. Farge, C. Picart and B. Grosgogeat, Cytotoxicity of polyethyleneimine (PEI), precursor base layer of polyelectrolyte multilayer films, *Biomaterials*, 2007, **28**, 632–640.
125. D. V. Volodkin, A. I. Petrov, M. Prevot and G. B. Sukhorukov, Matrix Polyelectrolyte Microcapsules: New System for Macromolecule Encapsulation, *Langmuir*, 2004, **20**, 3398–3406.
126. D. V. Volodkin, N. I. Larionova and G. B. Sukhorukov, Protein Encapsulation via Porous CaCO3 Microparticles Templating, *Biomacromolecules*, 2004, **5**, 1962–1972.
127. N. G. Balabushevich, O. P. Tiourina, D. V. Volodkin, N. I. Larionova and G. B. Sukhorukov, Loading the Multilayer Dextran Sulfate/Protamine Microsized Capsules with Peroxidase, *Biomacromolecules*, 2003, **4**, 1191–1197.

128. O. Shchepelina, M. O. Lisunova, I. Drachuk and V. V. Tsukruk, Morphology and Properties of Microcapsules with Different Core Releases, *Chem. Mater.*, 2012, **24**, 1245–1254.
129. X. Qiu, S. Leporatti, E. Donath and H. Möhwald, Studies on the Drug Release Properties of Polysaccharide Multilayers Encapsulated Ibuprofen Microparticles, *Langmuir*, 2001, **17**, 5375–5380.
130. H. Ai, S. A. Jones, M. M. de Villiers and Y. M. Lvov, Nano-encapsulation of furosemide microcrystals for controlled drug release, *J. Controlled Release*, 2003, **86**, 59–68.
131. J. Almodóvar, L. W. Place, J. Gogolski, K. Erickson and M. J. Kipper, Layer-by-Layer Assembly of Polysaccharide-Based Polyelectrolyte Multilayers: A Spectroscopic Study of Hydrophilicity, Composition, and Ion Pairing, *Biomacromolecules*, 2011, **12**, 2755–2765.
132. C. R. Wittmer, J. A. Phelps, W. M. Saltzman and P. R. Van Tassel, Fibronectin terminated multilayer films: Protein adsorption and cell attachment studies, *Biomaterials*, 2007, **28**, 851–860.
133. C. Picart, R. Elkaim, L. Richert, T. Audoin, Y. Arntz, M. D. Cardoso, P. Schaaf, J. C. Voegel and B. Frisch, Primary cell adhesion on RGD-functionalized and covalently crosslinked thin polyelectrolyte multilayer films, *Adv. Funct. Mater.*, 2005, **15**, 83–94.
134. P.-H. Chua, K.-G. Neoh, E.-T. Kang and W. Wang, Surface functionalization of titanium with hyaluronic acid/chitosan polyelectrolyte multilayers and RGD for promoting osteoblast functions and inhibiting bacterial adhesion, *Biomaterials*, 2008, **29**, 1412–1421.
135. L. L. Han, Z. W. Mao, J. D. Wu, Y. Y. Zhang and C. Y. Gao, Influences of surface chemistry and swelling of salt-treated polyelectrolyte multilayers on migration of smooth muscle cells, *J. R. Soc., Interface*, 2012, **9**, 3455–3468.
136. J. M. Garza, P. Schaaf, S. Muller, V. Ball, J.-F. Stoltz, J.-C. Voegel and P. Lavalle, Multicompartment Films Made of Alternate Polyelectrolyte Multilayers of Exponential and Linear Growth, *Langmuir*, 2004, **20**, 7298–7302.
137. Y. Guan, S. Yang, Y. Zhang, J. Xu, C. C. Han and N. A. Kotov, Fabry – Perot Fringes and Their Application To Study the Film Growth, Chain Rearrangement, and Erosion of Hydrogen-Bonded PVPON/PAA Films, *J. Phys. Chem. B*, 2006, **110**, 13484–13490.
138. S. T. Dubas and J. B. Schlenoff, Swelling and Smoothing of Polyelectrolyte Multilayers by Salt, *Langmuir*, 2001, **17**, 7725–7727.
139. R. A. McAloney, V. Dudnik and M. C. Goh, Kinetics of Salt-Induced Annealing of a Polyelectrolyte Multilayer Film Morphology, *Langmuir*, 2003, **19**, 3947–3952.
140. X. Wang, F. Liu, X. Zheng and J. Sun, Water-Enabled Self-Healing of Polyelectrolyte Multilayer Coatings, *Angew. Chem., Int. Ed.*, 2011, **50**, 11378–11381.
141. H. Mjahed, G. Cado, F. Boulmedais, B. Senger, P. Schaaf, V. Ball and J.-C. Voegel, Restructuring of exponentially growing polyelectrolyte

multilayer films induced by salt concentration variations after film deposition, *J. Mater. Chem.*, 2011, **21**, 8416–8421.
142. C. Y. Gao, S. Moya, H. Lichtenfeld, A. Casoli, H. Fiedler, E. Donath and H. Mohwald, The decomposition process of melamine formaldehyde cores: The key step in the fabrication of ultrathin polyelectrolyte multilayer capsules, *Macromol. Mater. Eng.*, 2001, **286**, 355–361.
143. C. Déjugnat and G. B. Sukhorukov, pH-Responsive Properties of Hollow Polyelectrolyte Microcapsules Templated on Various Cores, *Langmuir*, 2004, **20**, 7265–7269.
144. A. A. Antipov and G. B. Sukhorukov, Polyelectrolyte multilayer capsules as vehicles with tunable permeability, *Adv. Colloid Interface Sci.*, 2004, **111**, 49–61.
145. Z. Gui, J. Qian, B. Du, M. Yin and Q. An, Fabrication of free-standing polyelectrolyte multilayer films: A method using polysulfobetaine-containing films as sacrificial layers, *J. Colloid Interface Sci.*, 2009, **340**, 35–41.
146. R. Polak, T. Crouzier, R. M. Lim, K. Ribbeck, M. M. Beppu, R. N. M. Pitombo, R. E. Cohen and M. F. Rubner, Sugar-Mediated Disassembly of Mucin/Lectin Multilayers and Their Use as pH-Tolerant, On-Demand Sacrificial Layers, *Biomacromolecules*, 2014, **15**, 3093–3098.
147. J. F. Quinn and F. Caruso, Multivalent-Ion-Mediated Stabilization of Hydrogen-Bonded Multilayers, *Adv. Funct. Mater.*, 2006, **16**, 1179–1186.
148. K. C. Wood, H. F. Chuang, R. D. Batten, D. M. Lynn and P. T. Hammond, Controlling interlayer diffusion to achieve sustained, multiagent delivery from layer-by-layer thin films, *Proc. Natl. Acad. Sci. U. S. A.*, 2006, **103**, 10207–10212.
149. W. Tong and C. Gao, Stable microcapsules assembled stepwise from weak polyelectrolytes followed by thermal crosslinking, *Polym. Adv. Technol.*, 2005, **16**, 827–833.
150. M. C. Kittredge, T. S. Durst and K. W. Kittredge, Heat-induced crosslinking of nylon-like weak polyelectrolyte multilayer films: UV-visible studies of methylene blue loading and rates of release, *Thin Solid Films*, 2010, **518**, 3949–3953.
151. M. L. Bruening and D. M. Sullivan, Enhancing the Ion-Transport Selectivity of Multilayer Polyelectrolyte Membranes, *Chem. – Eur. J.*, 2002, **8**, 3832–3837.
152. L. Richert, A. J. Engler, D. E. Discher and C. Picart, Elasticity of native and cross-linked polyelectrolyte multilayer films, *Biomacromolecules*, 2004, **5**, 1908–1916.
153. S. Y. Yang, D. Lee, R. E. Cohen and M. F. Rubner, Bioinert solution-cross-linked hydrogen-bonded multilayers on colloidal particles, *Langmuir*, 2004, **20**, 5978–5981.
154. F. Gilde, O. Maniti, R. Guillot, J. F. Mano, D. Logeart-Avramoglou, F. Sailhan and C. Picart, Secondary Structure of rhBMP-2 in a Protective Biopolymeric Carrier Material, *Biomacromolecules*, 2012, **13**, 3620–3626.

155. G. K. Such, J. F. Quinn, A. Quinn, E. Tjipto and F. Caruso, Assembly of Ultrathin Polymer Multilayer Films by Click Chemistry, *J. Am. Chem. Soc.*, 2006, **128**, 9318–9319.
156. B. G. De Geest, W. Van Camp, F. E. Du Prez, S. C. De Smedt, J. Demeester and W. E. Hennink, Degradable Multilayer Films and Hollow Capsules via a 'Click' Strategy, *Macromol. Rapid Commun.*, 2008, **29**, 1111–1118.
157. S.-F. Chong, R. Chandrawati, B. Städler, J. Park, J. Cho, Y. Wang, Z. Jia, V. Bulmus, T. P. Davis, A. N. Zelikin and F. Caruso, Stabilization of Polymer-Hydrogel Capsules via Thiol–Disulfide Exchange, *Small*, 2009, **5**, 2601–2610.
158. S. Shu, X. Zhang, Z. Wu, Z. Wang and C. Li, Gradient cross-linked biodegradable polyelectrolyte nanocapsules for intracellular protein drug delivery, *Biomaterials*, 2010, **31**, 6039–6049.
159. W. Tong, C. Gao and H. Möhwald, Manipulating the Properties of Polyelectrolyte Microcapsules by Glutaraldehyde Cross-Linking, *Chem. Mater.*, 2005, **17**, 4610–4616.
160. P. Liu, Stabilization of layer-by-layer engineered multilayered hollow microspheres, *Adv. Colloid Interface Sci.*, 2014, **207**, 178–188.
161. K. F. Ren, T. Crouzier, C. Roy and C. Picart, Polyelectrolyte multilayer films of controlled stiffness modulate myoblast cell differentiation, *Adv. Funct. Mater.*, 2008, **18**, 1378–1389.
162. S. E. Burke and C. J. Barrett, pH-Dependent Loading and Release Behavior of Small Hydrophilic Molecules in Weak Polyelectrolyte Multilayer Films, *Macromolecules*, 2004, **37**, 5375–5384.
163. G. B. Sukhorukov, A. A. Antipov, A. Voigt, E. Donath and H. Möhwald, pH-Controlled Macromolecule Encapsulation in and Release from Polyelectrolyte Multilayer Nanocapsules, *Macromol. Rapid Commun.*, 2001, **22**, 44–46.
164. J. A. Hiller and M. F. Rubner, Reversible Molecular Memory and pH-Switchable Swelling Transitions in Polyelectrolyte Multilayers, *Macromolecules*, 2003, **36**, 4078–4083.
165. S. Srivastava, V. Ball, P. Podsiadlo, J. Lee, P. Ho and N. A. Kotov, Reversible Loading and Unloading of Nanoparticles in "Exponentially" Growing Polyelectrolyte LBL Films, *J. Am. Chem. Soc.*, 2008, **130**, 3748–3749.
166. K. Ren, T. Crouzier, C. Roy and C. Picart, Polyelectrolyte multilayer films of controlled stiffness modulate myoblast cell differentiation, *Adv. Funct. Mater.*, 2008, **18**, 1378–1389.
167. A. Schneider, G. Francius, R. Obeid, P. Schwinté, J. Hemmerlé, B. Frisch, P. Schaaf, J.-C. Voegel, B. Senger and C. Picart, Polyelectrolyte Multilayers with a Tunable Young's Modulus: Influence of Film Stiffness on Cell Adhesion, *Langmuir*, 2005, **22**, 1193–1200.
168. L. Richert, P. Lavalle, D. Vautier, B. Senger, J. F. Stoltz, P. Schaaf, J. C. Voegel and C. Picart, Cell interactions with polyelectrolyte multilayer films, *Biomacromolecules*, 2002, **3**, 1170–1178.

169. P. Tryoen-Toth, D. Vautier, Y. Haikel, J. C. Voegel, P. Schaaf, J. Chluba and J. Ogier, Viability, adhesion, and bone phenotype of osteoblast-like cells on polyelectrolyte multilayer films, *J. Biomed. Mater. Res.*, 2002, **60**, 657–667.
170. J. M. Garza, N. Jessel, G. Ladam, V. Dupray, S. Muller, J.-F. Stoltz, P. Schaaf, J.-C. Voegel and P. Lavalle, Polyelectrolyte Multilayers and Degradable Polymer Layers as Multicompartment Films, *Langmuir*, 2005, **21**, 12372–12377.
171. A. A. Antipov, G. B. Sukhorukov, S. Leporatti, I. L. Radtchenko, E. Donath and H. Möhwald, Polyelectrolyte multilayer capsule permeability control, *Colloids Surf., A*, 2002, **198–200**, 535–541.
172. A. I. Petrov, A. A. Antipov and G. B. Sukhorukov, Base-acid equilibria in polyelectrolyte systems: From weak polyelectrolytes to interpolyelectrolyte complexes and multilayered polyelectrolyte shells, *Macromolecules*, 2003, **36**, 10079–10086.
173. A. A. Antipov and G. B. Sukhorukov, Polyelectrolyte multilayer capsules as vehicles with tunable permeability, *Adv. Colloid Interface Sci.*, 2004, **111**, 49–61.
174. S. A. Sukhishvili and S. Granick, Layered, erasable polymer multilayers formed by hydrogen-bonded sequential self-assembly, *Macromolecules*, 2002, **35**, 301–310.
175. K. Sato, K. Yoshida, S. Takahashi and J.-i. Anzai, pH- and sugar-sensitive layer-by-layer films and microcapsules for drug delivery, *Adv. Drug Delivery Rev.*, 2011, **63**, 809–821.
176. B. G. De Geest, R. E. Vandenbroucke, A. M. Guenther, G. B. Sukhorukov, W. E. Hennink, N. N. Sanders, J. Demeester and S. C. De Smedt, Intracellularly degradable polyelectrolyte microcapsules, *Adv. Mater.*, 2006, **18**, 1005–1009.
177. J. Zhang and D. M. Lynn, Ultrathin multilayered films assembled from "charge-shifting" cationic polymers: Extended, long-term release of plasmid DNA from surfaces, *Adv. Mater.*, 2007, **19**, 4218–4223.
178. X. Liu, J. Zhang and D. M. Lynn, Ultrathin multilayered films that promote the release of two DNA constructs with separate and distinct release profiles, *Adv. Mater.*, 2008, **20**, 4148–4153.
179. X. Liu, J. Zhang and D. M. Lynn, Polyelectrolyte Multilayers Fabricated from 'Charge-Shifting' Anionic Polymers: A New Approach to Controlled Film Disruption and the Release of Cationic Agents from Surfaces, *Soft Matter*, 2008, **4**, 1688–1695.
180. B. Sun and D. M. Lynn, Release of DNA from polyelectrolyte multilayers fabricated using 'charge-shifting' cationic polymers: Tunable temporal control and sequential, multi-agent release, *J. Controlled Release*, 2010, **148**, 91–100.
181. Y. Ma, W. F. Dong, M. A. Hempenius, H. Möhwald and G. Julius Vancso, Redox-controlled molecular permeability of composite-wall microcapsules, *Nat. Mater.*, 2006, **5**, 724–729.

182. Y. Ma, W.-F. Dong, M. A. Hempenius, H. Mohwald and G. J. Vancso, Redox-controlled molecular permeability of composite-wall microcapsules, *Nat. Mater.*, 2006, **5**, 724–729.
183. K. C. Wood, N. S. Zacharia, D. J. Schmidt, S. N. Wrightman, B. J. Andaya and P. T. Hammond, Electroactive controlled release thin films, *Proc. Natl. Acad. Sci. U. S. A.*, 2008, **105**, 2280–2285.
184. K. Liang, G. K. Such, A. P. R. Johnston, Z. Zhu, H. Ejima, J. J. Richardson, J. Cui and F. Caruso, Endocytic pH-triggered degradation of nanoengineered multilayer capsules, *Adv. Mater.*, 2014, **26**, 1901–1905.
185. B. G. De Geest, A. M. Jonas, J. Demeester and S. C. De Smedt, Glucose-responsive polyelectrolyte capsules, *Langmuir*, 2006, **22**, 5070–5074.
186. H. Fang, G. Kaur and B. H. Wang, Progress in boronic acid-based fluorescent glucose sensors, *J. Fluoresc.*, 2004, **14**, 481–489.
187. H. Shibata, Y. J. Heo, T. Okitsu, Y. Matsunaga, T. Kawanishi and S. Takeuchi, Injectable hydrogel microbeads for fluorescence-based in vivo continuous glucose monitoring, *Proc. Natl. Acad. Sci. U. S. A.*, 2010, **107**, 17894–17898.
188. M. Ikeda, K. Fukuda, T. Tanida, T. Yoshii and I. Hamachi, A supramolecular hydrogel containing boronic acid-appended receptor for fluorocolorimetric sensing of polyols with a paper platform, *Chem. Commun.*, 2012, **48**, 2716–2718.
189. M. B. Lerner, N. Kybert, R. Mendoza, R. Villechenon, M. A. B. Lopez and A. T. C. Johnson, Scalable, non-invasive glucose sensor based on boronic acid functionalized carbon nanotube transistors, *Appl. Phys. Lett.*, 2013, **102**.
190. C. Zhang, M. D. Losego and P. V. Braun, Hydrogel-Based Glucose Sensors: Effects of Phenylboronic Acid Chemical Structure on Response, *Chem. Mater.*, 2013, **25**, 3239–3250.
191. D. G. Hall, *Boronic Acids – Preparation and Applications in Organic Synthesis and Medicine*, John Wiley & Sons, 2005.
192. J. Zhang, N. J. Fredin, J. F. Janz, B. Sun and D. M. Lynn, Structure/property relationships in erodible multilayered films: Influence of polycation structure on erosion profiles and the release of anionic polyelectrolytes, *Langmuir*, 2006, **22**, 239–245.
193. E. M. Saurer, C. M. Jewell, J. M. Kuchenreuther and D. M. Lynn, Assembly of erodible, DNA-containing thin films on the surfaces of polymer microparticles: Toward a layer-by-layer approach to the delivery of DNA to antigen-presenting cells, *Acta Biomater.*, 2009, **5**, 913–924.
194. J. Blacklock, H. Handa, D. Soundara Manickam, G. Mao, A. Mukhopadhyay and D. Oupicky, Disassembly of layer-by-layer films of plasmid DNA and reducible TAT polypeptide, *Biomaterials*, 2007, **28**, 117–124.
195. J. Blacklock, Y. Z. You, Q. H. Zhou, G. Mao and D. Oupicky, Gene delivery in vitro and in vivo from bioreducible multilayered polyelectrolyte films of plasmid DNA, *Biomaterials*, 2009, **30**, 939–950.

196. K. Ren, J. Ji and J. Shen, Tunable DNA release from cross-linked ultrathin DNA/PLL multilayered films, *Bioconjugate Chem.*, 2006, **17**, 77–83.
197. C. J. Ochs, G. K. Such, B. Städler and F. Caruso, Low-fouling, biofunctionalized, and biodegradable click capsules, *Biomacromolecules*, 2008, **9**, 3389–3396.
198. C. J. Ochs, G. K. Such, Y. Yan, M. P. Van Koeverden and F. Caruso, Biodegradable click capsules with engineered drug-loaded multilayers, *ACS Nano*, 2010, **4**, 1653–1663.
199. C. J. Ochs, G. K. Such and F. Caruso, Modular assembly of layer-by-layer capsules with tailored degradation profiles, *Langmuir*, 2011, **27**, 1275–1280.
200. T. Borodina, E. Markvicheva, S. Kunizhev, H. Möhwald, G. B. Sukhorukov and O. Kreft, Controlled release of DNA from self-degrading microcapsules, *Macromol. Rapid Commun.*, 2007, **28**, 1894–1899.
201. T. Serizawa, M. Yamaguchi and M. Akashi, Time-controlled desorption of ultrathin polymer films triggered by enzymatic degradation, *Angew. Chem., Int. Ed.*, 2003, **42**, 1115–1118.
202. C. Picart, A. Schneider, O. Etienne, J. Mutterer, P. Schaaf, C. Egles, N. Jessel and J. C. Voegel, Controlled degradability of polysaccharide multilayer films in vitro and in vivo, *Adv. Funct. Mater.*, 2005, **15**, 1771–1780.
203. Y. Itoh, M. Matsusaki, T. Kida and M. Akashi, Enzyme-responsive release of encapsulated proteins from biodegradable hollow capsules, *Biomacromolecules*, 2006, **7**, 2715–2718.
204. A. Szarpak, D. Cui, F. Dubreuil, B. G. De Geest, L. J. De Cock, C. Picart and R. Auzély-Velty, Designing hyaluronic acid-based layer-by-layer capsules as a carrier for intracellular drug delivery, *Biomacromolecules*, 2010, **11**, 713–720.
205. K. Ariga, Q. Ji, T. Mori, M. Naito, Y. Yamauchi, H. Abe and J. P. Hill, Enzyme nanoarchitectonics: organization and device application, *Chem. Soc. Rev.*, 2013, **42**, 6322–6345.
206. G. Decher, M. Eckle, J. Schmitt and B. Struth, Layer-by-layer assembled multicomposite films, *Curr. Opin. Colloid Interface Sci.*, 1998, **3**, 32–39.
207. J. F. Rusling, E. G. Hvastkovs, D. O. Hull and J. B. Schenkman, Biochemical applications of ultrathin films of enzymes, polyions and DNA, *Chem. Commun.*, 2008, 141–154.
208. S. O. Andreasen, B. Fejerskov and A. N. Zelikin, Biocatalytic polymer thin films: optimization of the multilayered architecture towards in situ synthesis of anti-proliferative drugs, *Nanoscale*, 2014, **6**, 4131–4140.
209. L. Hosta-Rigau, P. Schattling, B. M. Teo, M. E. Lynge and B. Stadler, Recent progress of liposomes in nanomedicine, *J. Mat. Chem. B*, 2014.
210. B. M. Teo, L. Hosta-Rigau, M. E. Lynge and B. Städler, Liposome-containing polymer films and colloidal assemblies towards biomedical applications, *Nanoscale*, 2014, **6**, 6426–6433.
211. Z. Zhu and S. A. Sukhishvili, Layer-by-layer films of stimuli-responsive block copolymer micelles, *J. Mater. Chem.*, 2012, **22**, 7667–7671.

212. S. Biggs, K. Sakai, T. Addison, A. Schmid, S. P. Armes, M. Vamvakaki, V. Bütün and G. Webber, Layer-by-layer formation of smart particle coatings using oppositely charged block copolymer micelles, *Adv. Mater.*, 2007, **19**, 247–250.
213. T. Addison, O. J. Cayre, S. Biggs, S. P. Armes and D. York, Incorporation of block copolymer micelles into multilayer films for use as nano-delivery systems, *Langmuir*, 2008, **24**, 13328–13333.
214. T. Addison, O. J. Cayre, S. Biggs, S. P. Armes and D. York, Polymeric microcapsules assembled from a cationic/zwitterionic pair of responsive block copolymer micelles, *Langmuir*, 2010, **26**, 6281–6286.
215. I. Erel, Z. Zhu, A. Zhuk and S. A. Sukhishvili, Hydrogen-bonded layer-by-layer films of block copolymer micelles with pH-responsive cores, *J. Colloid Interface Sci.*, 2011, **355**, 61–69.
216. J. Zhou, M. V. Pishko and J. L. Lutkenhaus, Thermoresponsive Layer-by-Layer Assemblies for Nanoparticle-Based Drug Delivery, *Langmuir*, 2014, **30**, 5903–5910.
217. T. Liao, M. D. Moussallem, J. Kim, J. B. Schlenoff and T. Ma, N-isopropylacrylamide-based thermoresponsive polyelectrolyte multilayer films for human mesenchymal stem cell expansion, *Biotechnol. Progr.*, 2010, **26**, 1705–1713.
218. M. Ciobanu, B. Heurtault, P. Schultz, C. Ruhlmann, C. D. Muller and B. Frisch, Layersome: Development and optimization of stable liposomes as drug delivery system, *Int. J. Pharm.*, 2007, **344**, 154–157.
219. M. Michel, Y. Arntz, G. Fleith, J. Toquant, Y. Haikel, J.-C. Voegel, P. Schaaf and V. Ball, Layer-by-Layer Self-Assembled Polyelectrolyte Multilayers with Embedded Liposomes: Immobilized Submicronic Reactors for Mineralization, *Langmuir*, 2006, **22**, 2358–2364.
220. A. D. Keefe, S. Pai and A. Ellington, Aptamers as therapeutics, *Nat. Rev. Drug Discovery*, 2010, **9**, 537–550.
221. K.-M. Song, S. Lee and C. Ban, Aptamers and Their Biological Applications, *Sensors*, 2012, **12**, 612–631.
222. Y. Sultan and M. C. Derosa, Target binding influences permeability in aptamer-polyelectrolyte microcapsules, *Small*, 2011, **7**, 1219–1226.
223. X. Zhang, D. Chabot, Y. Sultan, C. Monreal and M. C. Derosa, Target-molecule-triggered rupture of aptamer-encapsulated polyelectrolyte microcapsules, *ACS Appl. Mater. Interfaces*, 2013, **5**, 5500–5507.
224. B. Malile and J. I. L. Chen, Morphology-based plasmonic nanoparticle sensors: Controlling etching kinetics with target-responsive permeability gate, *J. Am. Chem. Soc.*, 2013, **135**, 16042–16045.
225. H. H. Dam and F. Caruso, Construction and Degradation of Polyrotaxane Multilayers, *Adv. Mater.*, 2011, **23**, 3026–3029.
226. B. G. De Geest, W. Van Camp, F. E. Du Prez, S. C. De Smedt, J. Demeester and W. E. Hennink, Degradable multilayer films and hollow capsules via a 'click' strategy, *Macromol. Rapid Commun.*, 2008, **29**, 1111–1118.

227. K. Sato, Y. Imoto, J. Sugama, S. Seki, H. Inoue, T. Odagiri, T. Hoshi and J. I. Anzai, Sugar-induced disintegration of layer-by-layer assemblies composed of concanavalin A and glycogen, *Langmuir*, 2005, **21**, 797–799.
228. K. Sato, D. Kodama and J. I. Anzai, Sugar-sensitive thin films composed of concanavalin A and sugar-bearing polymers, *Anal. Sci.*, 2005, **21**, 1375–1378.
229. T. Levy, C. Déjugnat and G. B. Sukhorukov, Polymer microcapsules with carbohydrate-sensitive properties, *Adv. Funct. Mater.*, 2008, **18**, 1586–1594.
230. H. Inoue, K. Sato and J. I. Anzai, Disintegration of layer-by-layer assemblies composed of 2-iminobiotin-labeled poly(ethyleneimine) and avidin, *Biomacromolecules*, 2005, **6**, 27–29.
231. Z. Ding, Y. Guan, Y. Zhang and X. X. Zhu, Layer-by-layer multilayer films linked with reversible boronate ester bonds with glucose-sensitivity under physiological conditions, *Soft Matter*, 2009, **5**, 2302–2309.
232. H. Krass, G. Papastavrou and D. G. Kurth, Layer-by-layer self-assembly of a polyelectrolyte bearing metal ion coordination and electrostatic functionality, *Chem. Mater.*, 2003, **15**, 196–203.
233. A. Van Der Heyden, M. Wilczewski, P. Labbé and R. Auzély, Multilayer films based on host-guest interactions between biocompatible polymers, *Chem. Commun.*, 2006, 3220–3222.
234. T. Torring, N. V. Voigt, J. Nangreave, H. Yan and K. V. Gothelf, DNA origami: a quantum leap for self-assembly of complex structures, *Chem. Soc. Rev.*, 2011, **40**, 5636–5646.
235. I. Saaem and T. H. LaBean, Overview of DNA origami for molecular self-assembly, *Wiley Interdiscip. Rev.: Nanomed. Nanobiotechnol.*, 2013, **5**, 150–162.
236. B. Sacca and C. M. Niemeyer, DNA Origami: The Art of Folding DNA, *Angew. Chem., Int. Ed.*, 2012, **51**, 58–66.
237. A. P. R. Johnston, L. Lee, Y. Wang and F. Caruso, Controlled degradation of dna capsules with engineered restriction-enzyme cut sites, *Small*, 2009, **5**, 1418–1421.
238. A. L. Becker, A. P. R. Johnston and F. Caruso, Peptide nucleic acid films and capsules: assembly and enzymatic degradation, *Macromol. Biosci.*, 2010, **10**, 488–495.
239. L. Lee, A. P. R. Johnston and F. Caruso, Programmed degradation of DNA multilayer films, *Small*, 2014, **10**, 2902–2909.
240. Q. Zhao, B. Han, Z. Wang, C. Gao, C. Peng and J. Shen, Hollow chitosan-alginate multilayer microcapsules as drug delivery vehicle: doxorubicin loading and in vitro and in vivo studies, *Nanomed.: Nanotechnol., Biol. Med.*, 2007, **3**, 63–74.
241. L. Yu, Y. Gao, X. Yue, S. Liu and Z. Dai, Novel Hollow Microcapsules Based on Iron – Heparin Complex Multilayers, *Langmuir*, 2008, **24**, 13723–13729.

CHAPTER 12

Responsive Materials by the Self-assembly of Low Molecular Weight Gelators

EMILY R. DRAPER AND DAVE J. ADAMS*

Department of Chemistry, University of Liverpool, Crown Street, Liverpool L69 7ZD, UK
*Email: d.j.adams@liverpool.ac.uk

12.1 Introduction

Hydrogels are a class of soft materials that have a wide range of applications.[1,2] These include use in cosmetics,[3] medicine,[4] food[5] and electronics.[6] Gels consist of a dilute cross-linked network that consists mainly of liquid. Some examples are made up of up to 99.9% water, however, they still exhibit many properties of a solid. This composition gives the gels interesting physical properties. Gels are formed from long fibres that are dissolved in solution. These then entangle and cross link, trapping the solution and forming a self-supporting gel matrix (Figure 12.1). When the solvent is water, these gels are called hydrogels. The high water content of these materials leads to increased biocompatibility and, as a result, hydrogels can be used in biological applications such as drug delivery, therapeutics and cell culture.[7–11] Biologically relevant hydrogels will be the main focus of this chapter.

Hydrogels can be made from high molecular weight polymers that occur naturally, such as pectin and collagen, and also from synthetic polymers.[12,13] In the latter case, the polymer fibres are cross-linked irreversibly *via* covalent

RSC Smart Materials No. 14
Chemoresponsive Materials: Stimulation by Chemical and Biological Signals
Edited by Hans-Jörg Schneider

Published by the Royal Society of Chemistry, www.rsc.org

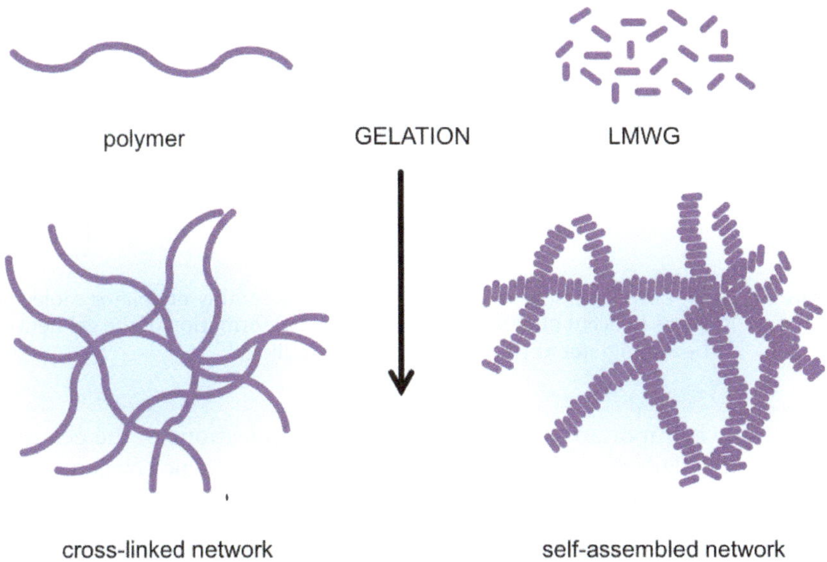

Figure 12.1 Scheme showing a polymer gel network and a LMWG gel.

bonds; this makes gelation irreversible and results in stiff gels as bonds are less easily broken by force or other stimuli. There are many good reviews that discuss the formation and use of these gels in detail as well as Chapters 2, 3 and 7.[12,14–17] Another class of non-polymeric hydrogels is formed by low molecular weight gelators (LMWGs), which are molecules that are typically less than 1000 Daltons.[1,2] These LMWGs self-assemble through non-covalent interactions such as hydrogen bonding, π–π stacking and ion pairing to form fibres, which then entangle.[18] These will be the focus of this chapter.

LMWGs can be peptides, quaternary ammonium salts, functionalised sugars, metallic soaps and aliphatic hydrocarbons.[1,10,19,20] These gelators can have a wide variety of structures. Examples are shown in Figure 12.3. Peptides, bile salts and functionalised sugars are naturally occurring in the body and are ideal for use in biomedical applications.[9,21] To form hydrogels, typically, the LMWGs are initially dissolved or suspended in water. These molecules can only then start to self-assemble in water when a suitable trigger is applied, which results in the LMWGs becoming significantly less soluble. As the molecules become less soluble, they assemble into long fibres in order to minimise their interactions with the surrounding water. These fibres then interact further with each other by cross-linking and entanglement resulting in the gel network (Figure 12.2). This self-assembly of fibres is still poorly understood, but the use of circular dichroism and X-ray scattering gives us some insight into the fibre structure,[22,23] while rheology can be used to gain some information about the entanglement of the network.[24]

An advantage of gels formed from LMWGs compared to cross-linked polymeric hydrogelators is that gelation can be reversible. This is due to the gel structure being held together by non-covalent interactions. This

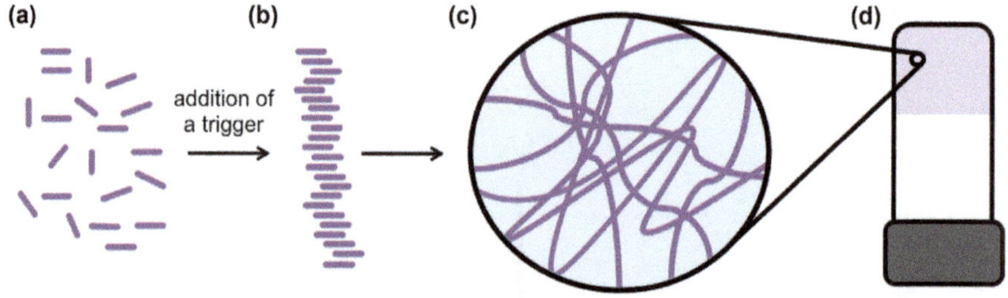

Figure 12.2 (a) LMWG dissolved in solution (b) self-assembly of gelator molecules starts to occur after addition of a trigger (c) formation of the gel network by entanglement (d) a self-supporting bulk gel.

reversibility is important when thinking about applications of the gels such as the controlled release of guest molecules.[25] The weak covalent forces holding fibres together can also allow the gels to quickly recover back to their original strength upon breaking. This so-called thixotropic behaviour means that the gels are often suitable for injection as the gel may be broken and injected into tissue.[26] Once the hydrogel has reached the desired place, it could then return to its original gelled state. As LMWGs can be made from sugars, peptides and other naturally occurring molecules, they can be easily broken down by the body with potentially little or no toxicity and removed once they have served their purpose. Also, as LMWGs are made from naturally occurring molecules, their production is cheap and relatively easy to scale up.

As noted, gelation arises from a solubilisation change from highly soluble to less soluble. For most molecules, such a change in solubility results in precipitation or crystallisation; crystallisation tends to be the most thermodynamically stable outcome. In the event of gelation, the implication is that crystallisation is inhibited, and hence fibre formation is favoured. Gelation is highly likely to be a kinetically controlled process and may well not represent the thermodynamic minimum. Nonetheless, there are a significant number of studies that attempt to link a crystal structure to the potential packing in the fibre phase.[27,28] Evidence that gel formation is kinetically controlled comes from observations that crystallisation can occur over time from some gel phases, implying that there is a more thermodynamically stable phase.[19,29] However, this is a rare observation and so may not hold for all gel phases.

Another important point that is rarely discussed is evolution of the gel network. This links with the comments regarding the most thermodynamically stable structure earlier, but it is clear from a small number of reports that the gel network can change over time. Hence, the time at which the gel properties are measured is important and these properties may not be constant. Donald and co-workers looked at the dynamic properties of gelation using microrheology of Fmoc-derivative hydrogels.[30] Microrheology is the use of a collection of techniques that use Brownian motion to monitor gel formation *via* particle tracking of trapped probe molecules. They showed how G′ changed during gelation and that longer incubation periods of the gel led to different bulk rheological properties. Smith's group also observed a change in rheological properties and morphology upon aging of gels.[31] They used a mixture of a protected sorbitol gelator and a modified cholesterol gelator for this study in which the rheological strength increased upon aging. They contributed this change in gel strength to be from a change in nanoscale morphology. This again illustrates that gel properties may change over time, but more investigation into this is needed.

12.2 Gelator Design

Design of the LMWG is important when considering the method of gelation and the desired application, as environment is crucial for the formation and stability of the gel. Many gelators are discovered by accident and the structure then altered slightly to form new gelators.[32] As the structures of known gelators can be quite diverse, it is often difficult to predict which molecules will self-assemble and form a gel.[33] Even small changes in the gelator molecule or reverse in sequence of moiety can prevent gelation from occurring. Not only does the structure of the molecule need to be well understood, but also how these molecules interact with each other to form fibres or aggregates, and how these in turn elongate and interact with other fibres to form the gel matrix. As previously mentioned, the self-assembly of these fibres into supramolecular assemblies is still poorly understood, however, the structure of individual fibres is becoming better known and easier to predict.[32] Although LMWG structures are hard to design, they often have a few common features. They usually require hydrophobic interactions, π–π stacking, hydrogen bonding and electrostatic interactions to drive assembly. Amino acids provide excellent candidates for these materials as they have a variety of functional groups that can be added in sequence to provide different interactions and therefore give different properties.[34]

The first amino acid based hydrogelator was discovered over 100 years ago by Gortner and Hoffman.[35] However, it was not until 80 years later that Menger *et al.* used dibenzoyl cysteine to show the importance of hydrogen bonding and π–π interactions between molecules in forming gels using rheology.[36] Now, there are lots of examples of short peptide materials that form hydrogels. Short peptides often need a simple aromatic protecting group on the *N*-terminus of the molecule in order to form a gel. Examples of these protecting groups are anthracene,[37] naphthalene,[38] fluorenylmethoxycarbonyl (Fmoc),[39] cinnamoyl[40] and azobenzene,[41] to name but a few. The aromatic groups provide hydrophobic characteristics to the otherwise more hydrophilic short peptide sequences. These groups play an important role in self-assembly as they provide π–π interactions, hydrogen bonding and Van der Waals forces due to their hydrophobicity. The aromatic groups at the *N*-terminus not only help with assembly but also can be used for oxidative, reductive, and photoresponsive applications.[42,43]

Another class of peptide hydrogelators is peptide amphiphiles (PA) (**1**, Figure 12.3). These PAs consist of longer peptide sequences with a long alkyl chain at the *N*-terminus. PAs were first reported by Tirrell and coworkers using collagen-based materials.[44] The Stupp group have used PA hydrogelators for a large variety of applications.[45,46] The self-assembly of these materials depends on the collapse of this long hydrophobic alkyl chain to form cylindrical fibres.[45]

Longer peptides do not need these extra groups on them in order to gel. There are many examples of oligopeptides that form β-sheets.[47,48] These differ from fibril-like α-helices as they have lateral attractions between different strands. These attractions include H-bonding, hydrophobic and electrostatic interactions. β-sheets then form higher order structures such as ribbons and tapes that can themselves further interact to form more complex structures depending on concentration. These oligopeptides are fully

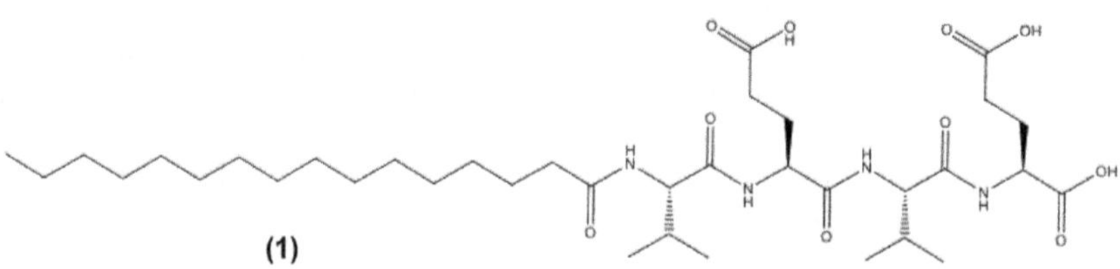
(1)

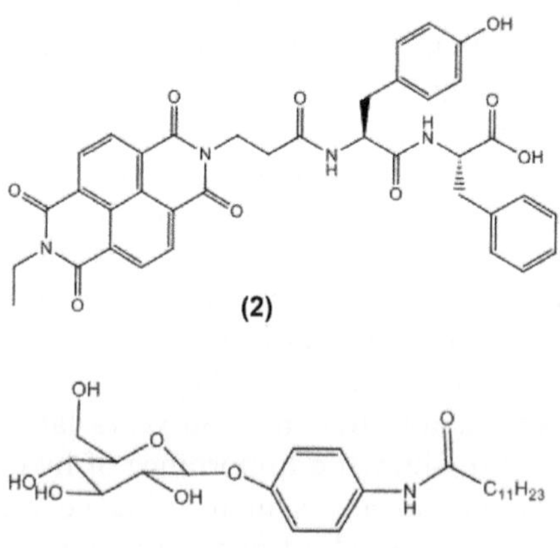
(2)

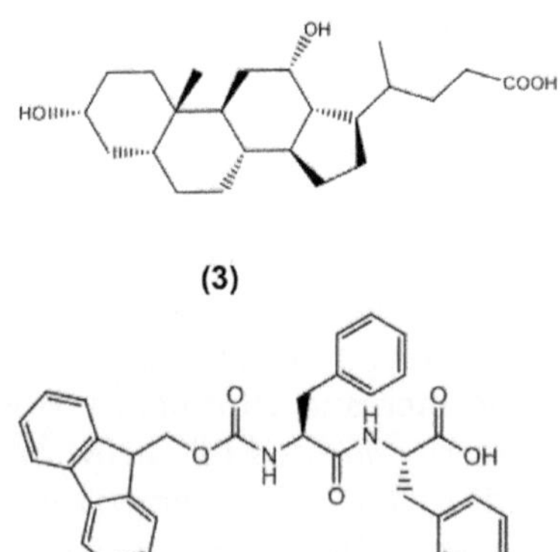
(3)

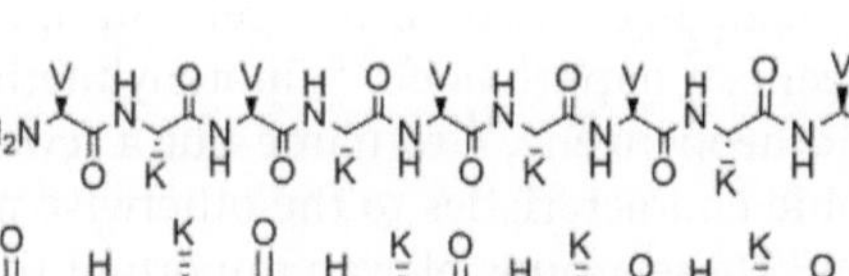
(4)

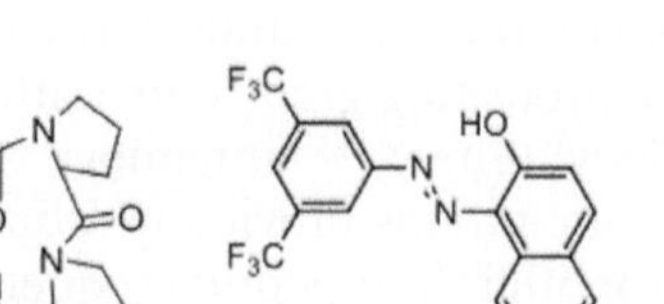
(5)

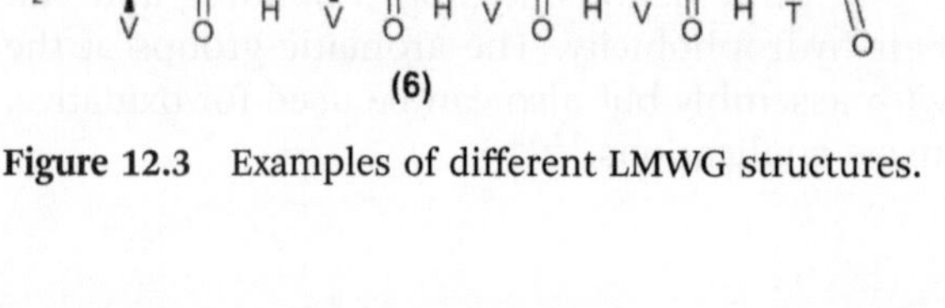
(6)

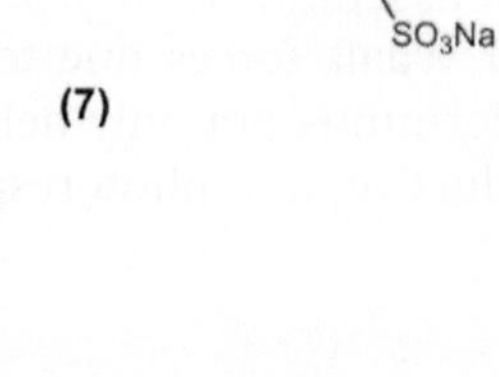
(7)

Figure 12.3 Examples of different LMWG structures.

soluble in solution and exist as random coiled structures, which then form gels upon addition of a trigger.

The groups of Pochan and Schneider have made a series of LMWGs that form on the basis of a β-hairpin structure, where the peptide folds on itself like a hairpin and is stabilised by hydrogen bonds. These gelator molecules have a central tetrapeptide with a high type II' β-turn propensity causing the hairpin.[47,49] To either side of this are chains of alternating hydrophobic and hydrophilic amino acids. In the case of MAX1 (**6**, Figure 12.3), these alternating amino acids are valine and lysine.[49] The self-assembly occurs *via* hydrogen-bonding and the hydrophobic association of the hairpins. The hydrophobic faces of the molecules interact to minimise their interaction with the surrounding water molecules. These types of LMWGs can be triggered by many external stimuli, such as pH and addition of salt, resulting in gels with different rheological properties.

As noted, many LMWGs are found by serendipity. These gelators are then often modified slightly to give different properties, for example by changing an amino acid or changing the protecting group. This approach can give gels with different mechanical properties and different responsive behaviours. However, this method does not always work. For example, the LMWG dipeptide naphthalene-ValGly is able to form gels, but changing the order of the amino acids to naphthalene-GlyVal resulted in a non-gelator.[50] Similarly naphthalene-AlaVal is a non-gelator, but the addition of a bromine atom on to the naphthalene resulted in an effective LMWG.[38] These examples show that there is a delicate balance of intermolecular forces needed for molecular packing and then further interactions for gelation to occur which are not fully understood. This makes designing a new LMWG extremely difficult.

Although there are lots of hydrogelators based on amino acids, other small molecules can be used such as bile salts and functionalised sugars. These are perhaps not as widely researched but offer potentially good biocompatibility for use in medicinal applications. Other LMWGs have been synthesised using thiophenes[51] and dye molecules such as perylene diimides[52,53] and naphthalene diimides[54] (for example **2**, Figure 12.3), but these need to be functionalised in order to improve solubility in water. These types of LMWGs are promising materials for use in organic electronics[44] and in organic solar cells.[55]

When designing or choosing an existing gelator for a specific application, it is important to consider the method of gelation. For example, if an enzymatic trigger is to be used, then functional groups that can be cleaved or produced by the specific enzyme need to be present. This is also true when using a pH switch; groups that can be protonated or deprotonated need to be included in the molecule in order for the triggered gelation to be possible. Different methods of gelation can result in gels with different rheological properties, so this also needs to be considered.[56] For example, a gel with a specific G' is needed for applications such as cell growth and cell differentiation as it mimics the extracellular matrix (ECM) and physically supports cell growth.[24] For gels used in cell imaging, the absolute G' may be less important.

12.3 Gelation Triggers

As mentioned previously, LMWGs require a trigger in order to self-assemble. There are several potential triggers that have been described such as a solvent switch,[57] pH switch,[58] temperature change,[59] shear force,[60] addition of a salt[61] and addition of enzymes.[62] All of these methods work by making the LMWG less soluble and so inducing self-assembly. One of the most commonly used methods is a heat-cool method in which a LMWG suspension or solution is heated to increase the gelator solubility. The solution is then cooled, decreasing the solubility of the LMWG and triggering self-assembly. Another common assembly method is a solvent switch approach, where molecules are first dissolved in a water-miscible organic solvent, followed by the rapid addition of water as an anti-solvent, again reducing solubility and inducing self-assembly. These two methods have been reviewed extensively,[56,63–65] and, although they are good methods for forming gels, they are not necessarily the most suitable for use in biological applications and so will not be discussed further.

Methods that are more suitable for biocompatible applications include a pH switch, addition of enzymes, the addition of salts and metal ions and the addition of glucose. These will be discussed further in this chapter. As mentioned previously, the trigger used depends on the structure of the preferred gelator.

12.3.1 Change in pH

A pH trigger can be used when the gelator molecule has a functional group that can be deprotonated and protonated, for example carboxylic acids or amines. In the case of carboxylic acids, the LMWG can be freely dissolved in water above the pK_a. However, when acid is added and the pH is adjusted to below the pK_a, the carboxylate is re-protonated, decreasing the solubility of the LMWG and inducing self-assembly leading to gelation. In the case of an amine, when the pH is raised above the pK_a of the LMWG, deprotonation of the amine reduces the solubility, again resulting in self-assembly of the molecules. Using a pH switch to induce gelation can also be affected by salts present solution as mentioned in Chapter 3.

There are many examples of gelators that use an acid trigger for gelation. Of course, the final pH of the gel is important when considering its application. For example, if used in the body, the gel will need to be stable at biological pH. A LWMG that is pH-triggered *in vivo* would need to be quite sensitive, as the biological pH does not change significantly. If used *in vitro* for applications such as cell culture, the final pH gel again needs to be considered, so gels with a low pK_a would not be suitable. In addition, the final pH often affects the strength of the gel, again determining what the gel could be used for, as shown by Nanda *et al.*[66] Many researchers use hydrochloric acid to lower the pH of gelator solutions. Yang *et al.* prepared a small library of naphthalene-based dipeptide gelators and dissolved them at high pH. They then used a small amount of 1 M HCl to lower the pH of the solution below the gelator's pK_a. Transparent gels were formed at a final pH ∼ 2.[50] A pH trigger has also used been used to form gels using Fmoc-PhePhe (5, Figure 12.3). As Fmoc-PhePhe has a pK_a of 9.9, gelation can occur at physiological pH and so has a possible application for cell culture, although Fmoc-PhePhe has a variety of reported mechanical properties,[56] and gelation is not very reproducible using this method.[67–69] This gelation can be reversible by raising the pH again; for example van Bommel *et al.* demonstrated that when using a pH-responsive cyclohexane trisamide-based hydrogelator, the gel-to-sol-to-gel transition could occur many times without loss in gelation ability by cycling the pH.[70]

In some cases, how the pH is adjusted can affect the final gel properties.[71] For example, with a very hydrophobic LMWG, the addition of hydrochloric acid to change the pH can result in inhomogeneous turbid gels with irreproducible rheological data. This is caused by the rate of the pH change. On adding a mineral acid to the solution, the solution immediately in contact with the acid gels very quickly compared to the rest of the solution. In addition to this, the acid now has to diffuse through the already formed gel in order to lower the pH throughout the rest of the solution (Figure 12.4b, left). Adams *et al.* showed that by slowing the pH drop by utilising the hydrolysis of glucono-δ-lactone (GdL, Figure 12.4a), more homogeneous transparent hydrogels with highly reproducible rheological data were prepared.[72] This is a result of the fast dissolution of the GdL in the solution, followed by slow hydrolysis and acidification, resulting in a uniform pH drop (Figure 12.4b, right).

Another method of pH-triggered gelation includes the use of a UV light generated photoacid, which breaks down to release protons upon exposure to the UV light.[73] The same principle is used in electrochemically-triggered gels.[74,75] Here, a molecule that is able to undergo electrochemical reduction, such as hydroquinone, is used. When current is generated, hydroquinone is reduced to quinone and the protons released during this reaction lower the pH and cause gelation to occur. These two methods are useful as they allow gelation to happen only where the stimulus is applied rather than giving a

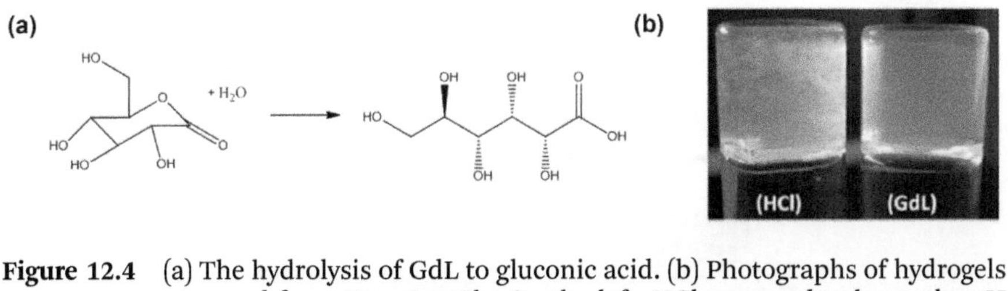

Figure 12.4 (a) The hydrolysis of GdL to gluconic acid. (b) Photographs of hydrogels prepared from FmocLeuGly. On the left, HCl was used to lower the pH to 3.9; turbid inhomogeneities are visible in this gel. On the right, GdL was used to lower the solution pH to 3.9; a transparent homogeneous gel is formed.
Adapted from ref. 77.

bulk gelation of the material, *i.e.* when using HCl or GdL. Hence, these methods allow gel patterning, with potential applications in cell culture and differentiation.

12.3.2 Use of an Enzyme

Enzymes work in two ways, by making or breaking bonds. Enzymatic triggers for LMWGs work by synthesising the gelator *in situ*, either by cleaving a solubilising group from a precursor molecule or by forming the LMWG *via* reaction between two soluble precursors (Figure 12.5). Enzymatic triggers are useful as they are highly specific and can be used in biological systems.[43] They can be used in both intracellular and extracellular environments and in complex fluids such as blood.[76] Other conditions need to be considered when using enzymes as triggers, such as pH and temperature. The enzyme's normal function can be slowed or the enzyme even denatured if not in a suitable environment.

Different enzymes have specific functions, making enzymatically-triggered gelation highly selective.[77] For example, using tyrosine phosphatase, an enzyme that breaks phosphorus–oxygen bonds in tyrosine phosphate containing gelator precursors, the hydrophilic charged phosphate group is removed, and replaced with an alcohol group. The molecule therefore becomes less soluble. With the right molecules, fibres are then formed leading to gelation. This concept has been used by the groups of Xu,[78] Ulijn,[79] Yang[80] and others[81] using tyrosine phosphates.

Similarly, hydrolysis at the *C*-terminus of ester terminated peptide pregelators to the corresponding carboxylic acid can be achieved by using esterases. Unlike phosphatases and esterases where a bond is broken, enzymes such as thermolysin can be used to make covalent bonds by reverse

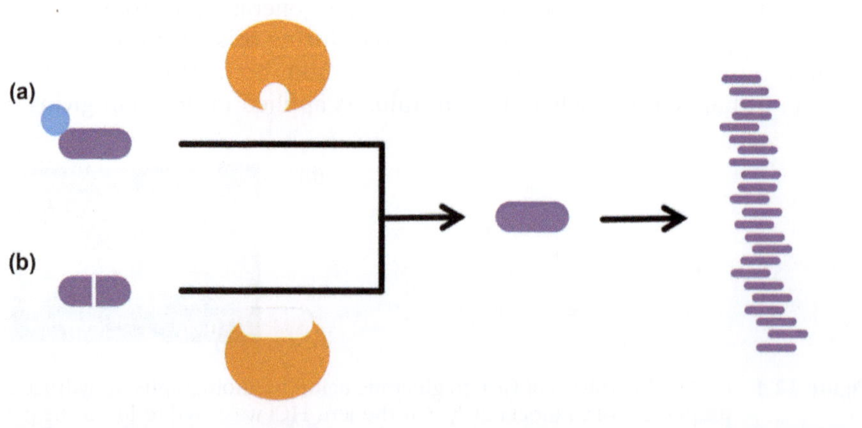

Figure 12.5 Schematic showing enzymatic self-assembly by (a) bond breaking and (b) bond making.

hydrolysis.[82,83] This protocol can be used to form a LMWG from two non-gelator precursors, promoting molecular self-assembly and forming the gel matrix. For example, Hughes *et al.* have used this method to couple two amino acid derivatives to form an Fmoc protected dipeptide that forms a hydrogel.[84] Szkolar *et al.* also used this reverse hydrolysis method to couple two tetrapeptide sequences together to form an octapeptide gelator that was used to encapsulate cells for 3D cell culture.[83] Amide bond formation is thought to be favoured rather than hydrolysis as the amine group was stabilised by the formation of the gelation and the subsequent self-assembly.[85] This method again allows gelation to happen *in situ* as gelation only occurs at the site of the enzymes. The use of enzymes as triggers has a potential application for studying the enzymes themselves and also for potential screening of enzyme inhibitors.[86]

As described for the pH-triggered hydrogels, the rate at which the gel is formed seems to control the homogeneity of the gel and the morphology of the fibres within it.[87] This is also true of the rate at which the enzymes produce the gelators. It has been shown that when less enzyme is present, the assembly affords more uniform fibres and reproducible gels due to slower gelator formation.[79] Another way of achieving more uniform fibres is to use two different enzymes as an enzyme switch. Here, the two enzymes work counter-actively, one making and the other breaking the same covalent bond. This is quite common in nature to regulate biological functions. An example where this has been used is by Xu and coworkers.[88] They use kinase (an enzyme that catalyses the formation of P–O bonds) and phosphatases (an enzyme that breaks P–O bonds). This method utilises the LMWGs' ability to reversibly switch between the solution and gel states. The phosphatase removes the phosphate group from the pentaphosphate derivative gelator causing it to self-assemble. This dephosphorylation is interrupted by kinase phosphorylating the tyrosine moiety. This slows down the rate of self-assembly resulting in more uniform fibres.

12.3.3 Addition of Metal Ions and Salts

Metal ion triggers for gelation, such as calcium and zinc ions, work by enhancing ionic strength. Metal ions can bind to specific groups on gelator molecules and encourage self-assembly.[89] Metal ion triggered gels are advantageous as they are able to form in biological pH and can be tuned using temperature and strengthened by the use of salts. Zinc ions have been used by Schneider and Pochan to trigger gelation using peptide-based LMWGs. The peptides chosen bind specifically to Zn^{2+} and form β-hairpin gels.[47] The amount of zinc present in the gels can be controlled by the amount of peptide used in the gel network. This means the gel could be tuned to release different amounts of zinc ions, which are often used in topical wound healing as it is has anti-bacterial character as well as stimulating epithelialisation of the wound.[90] Other metals such as Hg^{2+}, Cd^{2+} and Pb^{2+} can be used in a similar way, promoting β-sheet formation.[91] This mechanism

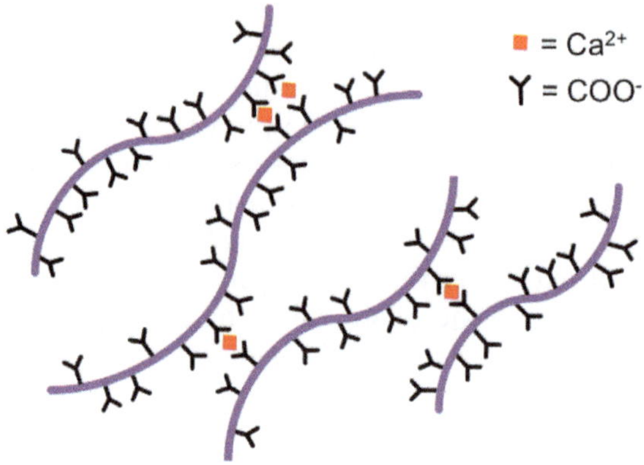

Figure 12.6 Gelation by the use of Ca^{2+} ions.

could be utilised for heavy metal detection in water. The identity of the metal can also be used to control the final gel properties as different ions result in different rates of gelation. It was also shown through circular dichroism that the metal ions formed part of the network by cross-linking fibres by a chelating effect.[92]

Elsewhere, Stupp's group showed that Ca^{2+} can be used as a trigger when using peptide amphiphiles that contain phosphoric acid groups.[93] Shi *et al.* then showed that gelators that contain several carboxylic acid groups could do the same.[94] Ca^{2+} gelation works by bridging fibres together through these acid groups *via* the metal ions (Figure 12.6). Adams' group also showed that naphthalene-containing dipeptides could be triggered by the addition of calcium chloride, as the viscous solutions contain worm-like micelles, which can be bridged by the ions, even at high pH.[61,95]

12.3.4 Specific Protein–Peptide Interactions

The use of specific peptide–protein interaction has only recently been reported by Zhang *et al.* for the formation of hydrogels using LMWGs.[96] Although enzymatic triggers are a type of peptide–protein driven gelation which involves the making or breaking of covalent bonds, proteins can also have non-covalent interactions with a gelator molecule. Using specific non-covalent interactions to cause gelation has already been shown with polymers but Yang's group showed the first example using LMWGs.[97] They used a TIP-1 protein that is expressed in the bacteria *E. coli* with a naphthalene based gelator containing a TIP-1 protein binding motif. When this protein is added to the gelator solution, a gel is formed almost instantly. This is due the protein having multiple binding positions and immobilising the LMWG through hydrogen-bonding and other non-covalent forces, causing gelation to occur.

By using different LMWGs with different binding affinities for the TIP-1 protein, gels with different rheological properties could be produced and therefore tailored for different applications. These gels also exhibit thixotropic behaviour and fast recovery so could be used for the encapsulation of drug molecules or cells. Also, due to the protein having multiple binding sites, other molecules could be bound. This could lead to multifunctional hydrogels for the delivery of biologically active molecules.

12.3.5 Use of More Than One Trigger

There are several examples where gelation has induced or been enhanced by the use of more than one trigger; this has also been shown in Chapters 3 and 5. This often allows gelation to be tailored for specific applications or environments. For example, glucose-induced gels use the presence of an enzyme and the resulting pH drop to trigger gelation.[98] Glucose is enzymatically converted into gluconic acid, which lowers the pH triggering gelation, similar to GdL as mentioned previously. This form of gelation is particularly useful for the potential treatment of diabetes. The use of both enzymes and a pH trigger enables specificity.

Phosphatase enzymes have been shown to form gels by cleaving phosphate groups as previously discussed. When Ca^{2+} was added to these preformed gels, Mann and coworkers showed that the gels were strengthened.[99] This was attributed to calcium ions being mopped up by the cleaved phosphate groups producing calcium phosphate.

In some cases, naphthalene dipeptides can form gels triggered both by $CaCl_2$ and by GdL.[38,95] As the addition of $CaCl_2$ results in a gel at high pH and GdL causes gelation at low pH, addition of GdL to a $CaCl_2$-triggered gel results in a gel-to-sol-to-gel transition as the salt-induced gel changes to a pH-induced gel. This has the potential to create single gelator systems with interesting and dramatic pH gradients.

12.4 How Triggers Affect Rheological Properties

Since gelation appears to be a kinetically controlled process, the method of gelation is very important as this controls the rate of nucleation and fibre growth. These in turn control the microstructure of the resulting gel and hence the strength of the bulk gel matrix.[100] As different triggers work at different rates, there is an opportunity to access different microstructures and hence rheological properties of gels and hence tune the properties for specific applications.[65] This approach is a relatively recent observation. Stupp investigated how the molecular structure of gelators affects the 1D microstructure formed.[101] They used 4 PAs with identical composition but with the sequence of the 4 amino acids changed. This change on the molecular level led to different self-assembled microstructures which were then studied by cyro-SEM and small angle X-ray scattering (SAXS). This work demonstrated the importance of the amino acid side positioning and interactions on the supramolecular architecture of the gel network.

Several LMWGs can be triggered by multiple triggers, such as Fmoc-Phe-Phe,[56] MAX1, and Fmoc-Tyr. For example, MAX1, developed by Schneider and Pochan's groups, can form gels at specific pH, temperature and salt concentrations. Different triggers can be used, all of which give gels with different rheological properties.[100] When the pH is above that of the pK_a of the gelator, the lysines are no longer charged and so MAX1 can self-assemble. In a pH 9 buffer at 25 °C with MAX1 at 2 wt%, the storage modulus (G′) was reported as ~1200 Pa. When prepared using a NaCl again with MAX1 at 2 wt% at a pH of 7.4, 20 °C with 20 mM of NaCl, the resulting gel had a G′ of ~100 Pa. Higher concentrations of salt was found to lead to stronger gels, with 400 mM of NaCl giving a gel with a G′ of 3000 Pa. The final rheological properties of this gel also depended on the temperature, pH and concentration of gelator present. As the self-assembly on the molecular level did not change (circular dichroism demonstrated that hairpin folding occurs in all cases), the difference in rheological properties must be due to different microstructure arrangements and different levels of fibre entanglement.

This difference in mechanical properties from different triggers is useful as gels can therefore be tailored for specific functions.[63] For example, the gels could be formed at physiological pH and temperature just by changing the trigger used. The G′ of the gels is important, for example gels with different mechanical properties are needed for the growth of different cells in cell culture and differentiation. These gels need to be stiffer to support the growth of cells, whereas weaker gels are needed, for example, if a sample is injected as it needs to be able to flow through a needle.[102]

12.5 Applications

The use of biopolymer and polymer hydrogels is already in place in many industries, including, for example, the pharmaceutical industry for cell culture and in drug delivery.[12] These materials are suitable for such applications because they are rheologically strong and stable materials that are biocompatible. LMWGs also offer a great potential for many uses. They can form a variety of different strength hydrogels and also often show little or no toxicity. LMWGs offer unique potential for new materials due to their reversibility, thixotropic behaviour and multi-stimuli responsiveness. The next section will discuss the use of these materials as drug delivery vehicles, cell culture media, biological sensors, therapeutics and for other applications.[11,103]

12.5.1 Sensors in Biological Systems

Biological sensors are useful materials as they can be used to screen for specific molecules. They could screen for specific indicators for diseases such as cancer and diabetes, where early indications show up in complex fluids such as blood and urine. These materials could also be used to detect the presence of certain enzymes or bacteria. Cross-linked polymer hydrogels as sensors often work by the swelling of the gel network.[104,105] Being only held together by non-covalent forces, LMWG hydrogels offer different responses and so can be used for more applications. Biological sensing could also be used in targeted drug delivery by sensing certain environments or cells and then responding. These responses include gelation, dissolution of the gel, a change in rheological strength, and fluorescence. The potential here is very large and we only give here specific examples.

Hydrogels have been used by Kiyonaka *et al.* as a rapid and high-throughput assay for biological events of proteins.[106] LMWGs are suitable for this application as the aqueous pores formed in the gel network are a suitable environment for enzymes. A further advantage of using these semi-wet arrays as opposed to dry ones is that the enzymes are less likely to become denatured and they show a greater sensitivity towards enzyme activity. The hydrophobic gel fibres act as the site for monitoring the reactions. A glycosylated amino acid LMWG was used at 0.1 wt% to entrap lysyl-endopeptidase (LEP). By using a substrate that was made from a penta-peptide bearing a lysine and *C*-terminated with a hydrophobic fluorescent molecule (in this case 5-dimethylaminonapthalene-1-(*N*-2-aminoethyl) sulfonamide (DANSen)), they were able to monitor the enzymes' activity. The LEP cleaves the DANSen from the pentapeptide. As DANSen is hydrophobic, it migrates from the aqueous environment of the gel cavity to the gel fibres; this changes the emission wavelength of the DANSen and there was an increase in fluorescence intensity. This was visualised by a change in colour from a pinky yellow before hydrolysis of the peptide to light green after. The change in colour can be monitored and rates of reaction determined. Different peptides were also used with specific enzymes such as glutamyl endopeptidase to illustrate further this colour change as a means of monitoring the enzyme activity. Adding a LEP inhibitor (N^{α}-tosyl-lysine chloromethylketone) to the DANSen-bearing pentapeptide resulted in the fluorescent spots becoming smeared depending on the concentration of the inhibitor. In this way, these arrays could be used to monitor the activity of the enzymes and also to screen for enzyme inhibitors. The use of LMWGs allows the potential to tailor for the monitoring of different reactions and screening of lots of different enzyme inhibitors.

Fmoc-PhePhe hydrogels have been used as a biosensor for detecting analytes such as glucose and toxic phenolic compounds. The detection of glucose is important for diagnosing diabetes and eventually targeted insulin delivery (see Chapters 2 and 3). Kim *et al.* encapsulated enzymes and quantum dots (CdTe and CdSe) within the Fmoc-PhePhe hydrogel.[107] For the detection of glucose, the enzyme glucose oxidase (GOx) was used. For detection of phenol and hydroquinone, the enzyme horseradish peroxidase (HRP) was used. Both enzymes and quantum dots were added to a solution of Fmoc-PhePhe before gelation so when gelation was triggered *via* a solvent switch they became trapped in the gel matrix. The quantum dots adhered to the gel fibres making them highly photoluminescent and the enzymes were immobilised in the gel pores. Detection of the analyte occurred by the conversion of the molecules into photoluminescent quenching agents by the enzymes. The photoluminescent quenchers work by acting as electron acceptors that decrease the intensity of the quantum dots. Hence, the gels go from red to colourless in the presence of an analyte. The use of the strong Fmoc-PhePhe gel is advantageous as it enables the simple incorporation of the enzymes and quantum dots with very little leakage whilst maintaining their properties. The porous nature of the gel network also allows facile movement of the analytes throughout the system. This system has the potential to become a versatile platform for the detection of analytes that could be used for the detection of diseases such as diabetes.

Another way of using hydrogels as a way of detecting the presence of glucose was used by Zhuo's group.[98] They used two different gelators, naphthalene-PhePheAsp and naphthalene-PhePheLysLys-$CONHCH_3$, with glucose oxidase. The oxidation of glucose in the presence of glucose oxidase forms gluconic acid and so causes a drop in pH. In the case of naphthalene-PhePheAsp solution, this drop in pH acts as a trigger and self-assembly occurs forming a gel (Figure 12.7a and b). On the other hand, the naphthalene-PhePheLysLys-$CONHCH_3$ gels go through a gel-to-sol transition when the pH is decreased by the formation of gluconic acid (Figure 12.7c and d). This is due to the lysine residues becoming protonated, causing the gelator to be more soluble resulting in the gel breaking down. The rate of gelation or dissolution could be used to determine how much glucose was present, as with more glucose present the response happened faster. This kind of system provides a very simple, visual means of determining the presence of glucose *via* its metabolism by glucose oxidase.

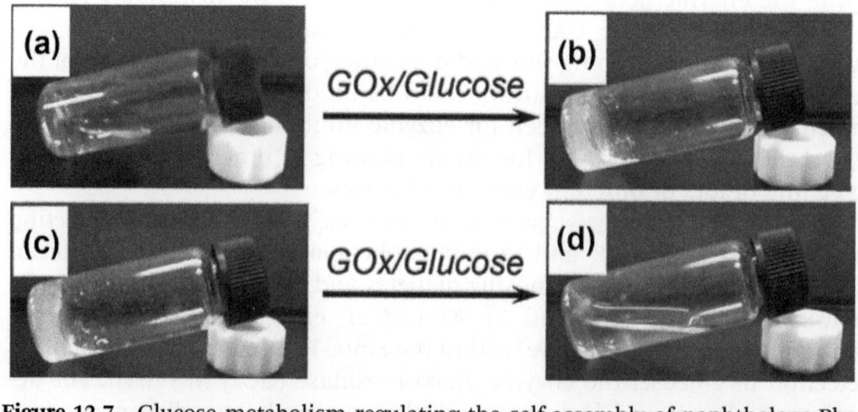

Figure 12.7 Glucose metabolism regulating the self-assembly of naphthalene-PhePheAsp from aqueous solution (a) to hydrogel (b); glucose metabolism regulating the self-assembly of gelator naphthalene-PhePheLysLys-$CONHCH_3$ from hydrogel (c) to aqueous solution (d).
Adapted from ref. 98. Copyright (2012) with permission from John Wiley and Sons.

Karp's group developed a way of sensing and monitoring pH within cell environments so that pH change during different cell processes could be determined.[108] This is important for targeted drug delivery as it reports what happens inside the cell in real time and so the fate of a potential drug can be determined. They used tri(hydroxymethyl)aminomethane (TRIS) and dodecanoic acid to make an LMWG which self-assembled into gel fibres. By incorporating a pH-sensitive dye (pHrodo) during self-assembly of the TRIS amphiphile LMWG, the dye was encapsulated into the gel. As the gel had very little buffering effect, this offered an ideal environment for pH sensing which could be monitored with fluorescence microscopy. Over time and within a pH range of 2 to 8.5, the dye did not leak out showing stability within the gel. When placed into a biological environment, the LMWG was internalised into the cell by macrophages through phagocytosis. This was shown by the discrete intense fluorescence of the pHrodo sensing the acidified environment of the phagosomes. Fluorescence was also observed in the cytoplasm and a small amount in the nucleus of the cells. This differs from pHrodo-*E. coli* which did not internalise into the cytoplasm or nucleus. Thus, pHrodo-TRIS gels are capable of reporting multiple cell states. These too could be used for the delivery of a wide variety of other sensors and even drug molecules.

Song's group has constructed LMWGs that can be used to make supramolecular logic gates that respond to multiple stimuli.[109] Logic gates are devices that use one or more inputs (in this case stimuli) to produce a single response. These can be used for intelligent soft materials to perform functions in response to various stimuli, and so could be used in sensors, drug delivery and organic electronics. These are also discussed in more depth in Chapters 3 and 5. The LMWGs were based on sorbitol and naphthalene-based salicylideneanilines. These gels were found to respond to copper by a reversible sol-to-gel transition with a colour change from yellow to colourless. The gels also showed separate responses to UV light, addition of anions and change in temperature, all of which resulted in a gel-to-sol transition. The photoresponse was caused by an isomerisation upon irradiation with UV light which destroyed the gel state. When the sample was returned to visible light, the gel re-formed. The response towards anions was due to the deprotonation of the phenolic group leading to dissolution of the gel. Again, this is reversible upon re-protonation. From these responses, an OR logical gate was constructed where only one of the stimuli is needed to cause a response. The presence of the stimulus was represented as 1 and if not present 0. This showed that one or more of these stimuli caused a response in which the gel dissolved, which is represented by 1. A more complex Inhibit (INH) logic gate could also be prepared (Figure 12.8b), where one of the stimuli caused dissolution, for example an anion, and the other stimulus caused gelation, for example a cation. They found that the presence of both stimuli cancelled each other out and so the gel remained intact. By using Cu(II) and EDTA, two responses or outputs could be reported—the dissolution of the gel and the change in colour of the gel. This could then be

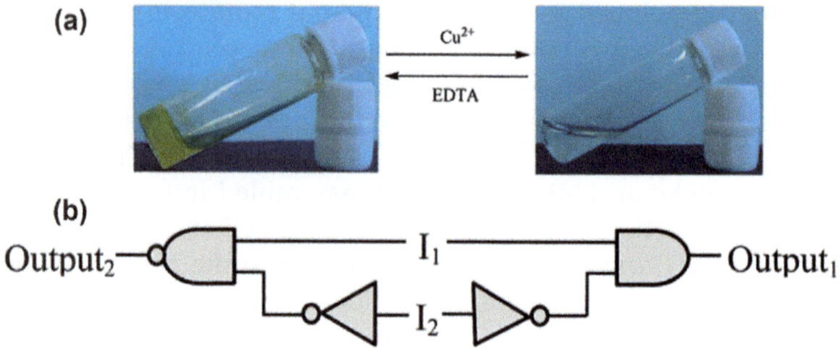

Input		Output1		Output2	
Cu^{2+}	EDTA	INH gel state		INH naked eyes	
0	0	Gel	0	yellow	1
1	0	Sol	1	colorless	0
0	1	Gel	0	yellow	1
1	1	Gel	0	yellow	1

Figure 12.8 (a) Photographs displaying the responses of the gel in the presence of chemical stimuli: Cu^{2+} and EDTA. (b) Representation of a combination logic gate and truth table of the logic gate.
Reproduced from ref. 109.

used to form a two input, two output response from the hydrogel logic gate (Figure 12.8a). This system is a promising candidate for use as a molecular logic gate and for use in environment sensitive actuators and drug-delivery systems.

12.5.2 Drug Delivery and Therapeutics

Drug delivery and therapeutics are highly researched and widely publicised areas as well as a multi-million pound industry.[110,111] Although drugs have been developed that are effective at treating illnesses or alleviating symptoms, the targeted and sustained delivery of these drugs is still an issue. Drugs may need to able to target specific areas of the body where they are needed, this lowers the amount of drug needed in administration and the potential of toxic effects to other parts of the body. Sustained delivery allows the drugs to be released continuously at a controlled rate to allow the maximum activity of the drug to be maintained. This also means that the patient would potentially have to take the drug less often. Many ways of making drug delivery specific and enabling the sustained release of drug molecules have been investigated and trialled. Hydrogels are a promising way of doing this due to their biocompatibility and low toxicity. The encapsulated molecules are able to flow through the pores of the hydrogel to allow long term drug delivery. Significant work has been done here using polymeric hydrogels. The multi-stimulus response of LMWG hydrogels means that there are potential advantages as hydrogels could to be tailored to respond to certain environments in the body, therefore allowing targeted delivery. This topic is also thoroughly discussed in Chapter 2 using polymer hydrogels.

Normally, when creating a gelator to be used *in vivo*, biocompatibility is a major factor. Turning this on its head, Xu *et al.* have used the cytotoxicity of a gelator as a way of controlling cell fate.[112] This is useful in the treatment of cancerous cells such as HeLa cells, which are cervical cancer cells. Naphthalene-PhePhe-$NHCH_2CH_2OH$ was used in this study, which has a minimum gelation concentration of 0.08 wt%. They then attached a butyric diacid group *via* the OH group to make a gelator precursor. This butyric diacid is susceptible to cleavage and so, upon addition of an esterase enzyme, the hydrophilic precursor is converted into the hydrophobic LMWG and self-assembly occurs. Gelation was shown to occur at 37 °C over a pH ranging from 0–14. When the precursor was placed in a culture of HeLa cells, the amount taken into the HeLa cells was monitored by UV-vis spectroscopy of the naphthalene group in solution and in the cells. After three days, the cells had taken up 32% of the precursor molecules. The UV-vis spectra also showed the naphthalene peak to be identical to that of the LMWG, indicating that gelation had occurred inside the cells due to esterases in the cells. The dead HeLa cells were found to contain the hydrogel, but the cells that remained alive did not show any hydrogel fibre formation, demonstrating that the presence of the hydrogel must be causing cell death. Tests showed that a concentration of 0.04 wt% of the precursor was needed to cause cell death. When the precursor was placed in a culture of fibroblast cells, little cell death occurred. This specificity towards HeLa cells was explained by the higher expression of esterases in the HeLa cells compared to the fibroblast cells. This specific necrosis of the HeLa cells *via* intracellular hydrogelation offers great potential in the treatment of cervical cancer and other cancers.

Patil *et al.* have also investigated the treatment of cancer using hydrogels to allow therapeutic delivery of small interfering RNA (siRNA) to cells to silence the cancer cell gene expression.[113] They used B-vitamin based bolaamphiphiles (VBs) as LMWGs to facilitate the delivery of the siRNA. The use of vitamin B makes the gelators biocompatible. The siRNA complexes with the VB gel fibres that form lamellar type structures, this aids with the delivery of the siRNA. The structure of the VB gels could also have the potential for wound healing and tissue culture. The group used fluorescein-tagged vascular endothelial growth factor siRNA to monitor the uptake and internalisation of siRNA into various different cells. The results showed that uptake of siRNA in cancer cells (HeLa MCF7, Huh7 and MDA-MB-231) using VB gel based on vitamin B7 was three times more likely to occur than in normal cells. The ability of the different VBs to deliver siRNA was linked to their hydrogelation properties as the gels that form at lower concentrations have greater siRNA delivery capabilities. These VB gelators were used in mice that had been given cancer. VB gelators were then injected into the blood stream and were shown to be able to deliver the siRNA specifically to the tumours. This research shows promising results in the application of chemotherapeutic drug delivery due to its specificity to cancerous cells and its uptake of the molecules.

Yang's group incorporated the anticancer drug taxol into a gelator precursor that they used to inhibit the growth of cancer cells.[114] The gelator precursor was synthesised by covalently attaching the hydrophobic drug molecule taxol onto succinic acid (SA) and oxidised glutathione (GSSG). These are both biocompatible and GSSG makes the gelator precursor more hydrophilic. Gelation was triggered by the addition of 2.2 molar equivalents of Na_2CO_3 in a PBS buffer. This gel was then injected into subcutaneous lymphoma tumours and breast tumours in mice. The gel then reformed after being injected. Self-hydrolysis of taxol-SA-GSSG resulted in the release of taxol from the molecule and into the tumour after about 4 days. The injection of taxol hydrogels inhibited the growth of both lymphoma and breast cancer in the mice more than the same dose of free taxol used intravenously. Also, the use of taxol hydrogels was shown to be less toxic to the mice. 40 mg kg^{-1} of free taxol was fatal to mice as compared to 300 mg kg^{-1} of the taxol hydrogel needed for a fatal dose. The taxol-hydrogel showed less toxicity than free taxol and so could be used at higher concentrations although concentrations of the taxol hydrogel higher than 40 mg kg^{-1} caused skin wounds on the mice after 4–7 days. The development of this taxol-based LMWG showed high anti-cancer efficacy by growth inhibition and prevention of cancer cell metathesis. The local injection of the taxol gelator into solid tumours lowers the concentration of taxol in the blood, so side effects to other tissues is reduced implying that this approach may have excellent potential for use in chemotherapy.

Hydrogels have been used in regenerative medicine for the stimulation of new bone growth. Calcium phosphate minerals have been studied in detail in a few forms such as hydroxyapatite and calcium phosphate, with hydroxyapatite being the most stable and what is present in bone and teeth and calcium phosphate being the least stable. The less stable forms of calcium phosphate are able to dissolve and so could be a source of calcium ions for use in bone regeneration.[115] Rapaport's group used an acidic β-sheet forming peptide hydrogel as an extracellular matrix (ECM) loaded with β-tricalcium phosphate (β-TCP) as a method of delivery of calcium ions to bone defects to promote new bone growth.[116] The gelator used was ProAsp(PheAsp)$_5$Pro (P_{FD}-5). This is an acidic peptide that is able to form β-sheets in the presence of calcium ions, which neutralise the six carboxyl groups on the peptide. Amphiphilic acidic peptides were shown to promote calcium phosphate mineralisation and hydroxyapatite formation. Gels were prepared in a saline solution with $NaHCO_3$ and various amount of β-TCP were added. This solution could be placed into a syringe and gelation

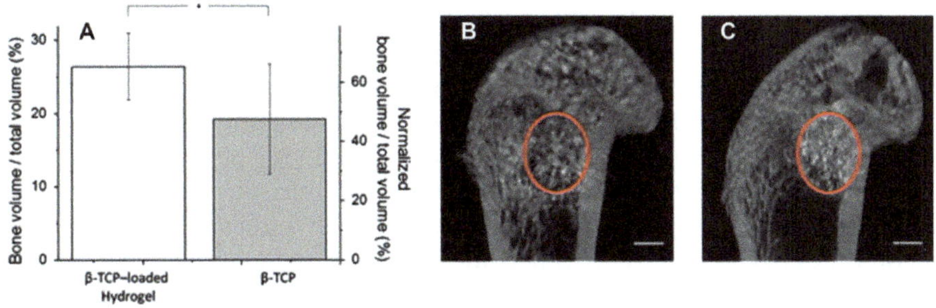

Figure 12.9 (a) Absolute (white) and normalized (grey) bone volume/total volume values according to micro-CT analysis for the porous β-TCP-loaded hydrogel compared to the value of healthy bone volume/total volume. The asterisk denotes significant difference (according to the Mann–Whitney U-test): $*p < 0.05$. (b) and (c) 360 1 mm thick section coronal images of femora treated with porous β-TCP-loaded hydrogel (b) and porous β-TCP particles only. The regions of the hole are marked in red circles. Note that the residual β-TCP particles appear brighter than the bone tissue. Scale bars represent 0.5 mm.
Adapted from ref. 116. Copyright (2012) with permission from Elsevier.

triggered by the addition of $CaCl_2$, so that the gel could be injected into rat bone defects. The calcium-loaded hydrogel improved bone regeneration *in vivo* using the two different clinically approved forms of β-TCP (Figure 12.9).

Mineral reabsorption occurred faster than with use of just β-TCP. The group postulated that the calcium-loaded hydrogel helped dissolve the calcium ions used for bone regrowth and also provided a reservoir of calcium and phosphate ions for bone regeneration. The delivery of calcium ions is also known to promote osteoblast differentiation, so new bone cells could also be made. This research is encouraging for new regenerative medicine for the repair of broken bones, or the treatment of osteoporosis.

Pochan and Schneider developed the LMWG MAX1 that showed antimicrobial properties.[49] This is important for implantation of biomaterials for use in regenerative applications and the treatment of chronic and traumatic wounds so it is very important to reduce the potential infection. Hydrogels have good potential for treating wounds as they keep the wound hydrated, encouraging the wound to heal from the inside out. The gels can be used to deliver cells or drugs to the wound, aiding healing. However, a hydrogel environment is also potentially an ideal environment for bacteria that cause infection and so would prevent healing and in some cases the removal of the implant. Gels that encapsulate antimicrobial agents have been made, but MAX1 itself has been shown to be antibacterial. As mentioned previously, MAX1 is a 20-residue peptide that forms β-hairpin bends to form gels using a variety of triggers. Here, the addition of a cell culture medium at pH 7.4 triggered self-assembly. The β-sheet gel has a polycationic lysine-rich surface due to the valine-rich hydrophobic faces collapsing during gelation. This leads to the surface being antibacterial. The charges on the surface interact with negatively charged bacteria resulting in disruption of the cell membrane. MAX1 showed antibacterial activity to a wide variety of both Gram negative bacteria (*Escherichia coli*) and Gram positive bacteria (*Klebsiella pneumonia*), both of which are common in hospital environments. This was measured using live-dead assays and by measuring the amount of β-galactosidase leakage from the bacteria, which indicates that the cell membranes of the bacteria had become compromised leading to cell death. In co-culture experiments, it was shown that MAX1 gels could be used for fibroblast cell adhesion and proliferation yet inhibited the growth of bacteria. MAX1 shows excellent potential for use in tissue regeneration in non-sterile environments.

12.5.3 Cell Culture and Differentiation

Cell culture and differentiation is important for use in tissue engineering. Tissue engineering is the development of biological substitutes that can restore, maintain or improve biological tissue function. The need for tissue engineering has arisen as medical implants often cannot replicate the full functions of the tissue they have replaced. There is also a shortage of donor tissue to be transplanted as well as an increase in chronic diseases due to life expectancy increasing. Tissue engineering requires a material that can provide a scaffold for cell growth. This scaffold mimics the ECM. Cells can be grown on the surface or imbedded in the scaffold. Hydrogels can provide this scaffold for both surface cell growth and 3D-cell growth due to their fibrous structure. Hydrogels are able to provide cell-stimulating factors in the water, as they are biocompatible and able to gel in biological media such as plasma and blood. Biocompatibility allows these hydrogels to be used both extracellularly and intracellularly and allows complete degradation in the body without any toxicity. LMWGs also allow the material to be tailored for unique biological properties due to the wide variety of molecules that can be used, as well as the multi-stimuli response behaviour.

The ability of the hydrogel to mimic the ECM is vitally important when using a gel as a cell culture medium. The ECM is a very complex material with dozens of different molecules contained within it, thereby making mimicking this environment difficult. The ECM is also used in nature to support the function of its corresponding tissues. By tailoring the ECM, biological outcomes such as cell differentiation can be tuned. This can be quite difficult due to the complexity and number of variants in the ECM. Collier and coworkers used factorial optimisation to look at the effect of modifying an 11-residue peptide (Q11) with various ligands, such as Arg-GlyAspSer (RGDS).[117] These peptides were then co-assembled with the Q11 derivatives to examine the effect on endothelial cell growth. A high-throughput method was used to vary different concentrations of each of the gelators. This investigation showed that when the normal cell supportive RGDS modified Q11 gelator was co-assembled with a TyrIleGlySerArg

(YIGSR) modified Q11 gelator, which again normally encouraged cell growth, the co-assembly had an inhibitory effect on the endothelial cells. This effect was greater when a higher concentration of RGDS modified Q11 was used, although the reason for this was unclear. The use of high-throughput techniques to study the co-assembly of gelators for their use as cell culture media is particularly useful when more than one biologically active component needs to be present in the gel scaffold. This method allows optimisation of gel matrices for specific gel growth.

Fmoc-PhePhe (Fmoc-FF) and Fmoc-ArgGlyAsp (Fmoc-RGD) (Figure 12.10A) have been used as a 3D-scaffold that enables the attachment of anchorage dependent cells.[68] The mixture of the two gelators forms a stiff gel matrix that has bioactive motifs on the surface of the fibres. As well as providing a fibrous structure to mimic that of the ECM, the Fmoc-RGD provides cell adhesion as RGD acts as an attachment site *via* the cell's surface receptors. Fmoc-FF has been shown to form hydrogels at neutral pH. The switch used for the co-assembly of these two gelators was a pH switch. Fmoc-FF was dissolved in water at pH 10 using NaOH to adjust the pH and Fmoc-RGD was dissolved in water at pH 3 using HCl to adjust the pH. These two solutions were mixed together in various ratios and the pH was then neutralised to pH 7 and kept at 37 °C where self-assembly occurred. The addition of Dulbecco's Modified Eagle Medium (DMEM) was used to accelerate the self-assembly and gels were formed after 1 minute. The addition of DMEM aids self-assembly as it contains metal ions that reduce molecular repulsion of gelator molecules. The use of DMEM also proved a simple way to encapsulate cells for 3D cell culture. In this study, human adult dermal fibroblasts (HDFa) cells were encapsulated within the hydrogel. The hydrogel did not seem to affect the cell viability as no dead cells were detected upon gelation. After 24 hours, the cells had spread throughout the gel, suggesting direct adhesion to

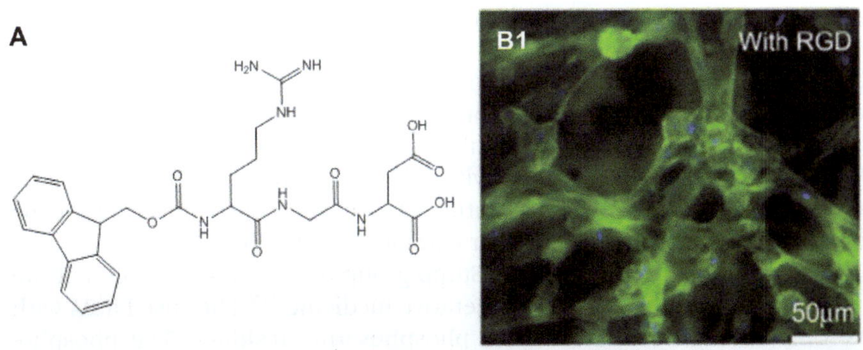

Figure 12.10 (A) The structure of Fmoc-RGD. (B1) Cell adhesion and morphology in the Fmoc-FF/RGD hydrogel, human adult dermal fibroblasts (HDFa) are well-spread in the Fmoc-FF/RGD hydrogels, and form a three-dimensional cell network 48 h post culture (B1).
Adapted from ref. 68. Copyright (2009) with permission from Elsevier.

the gel fibres. After 48 hours, the cells had formed a 3D-network with highly elongated cells showing that they were aligning along the gel fibres *via* the attachment to the RDG moiety (Figure 12.10B1). This was proven by the addition of a RGD-blocking antibody. In the presence of this antibody, the cells were rounded and disorganised in the gel and did not spread throughout the gel network, implying attachment to RGD sequences had not occurred. The use of these simple LMWGs as bioactive hydrogels has potential as a low cost system for cell culture. They also have the potential for the integration of other bioactive ligands to supply other functions to the hydrogel.

Stupp's group have used gel 'noodles' to promote the growth of aligned human mesenchymal stem cells (hMSCs) to form a 'cellular wire'.[118] These could serve as cell bridges for migration from one site to another. They could also be used for the growth of cardiomyocytes that show electrical activity and so, by being highly aligned, would propagate electrical signals. The gel was made from a peptide amphiphile containing a C_{16} alkyl tail and the peptide sequence $V_3A_3E_3COOH$ triggered by the addition of ions that screen the amino acid's charges. The viscosity of the solution was increased by heating the PA solutions to 80 °C then cooling to 25 °C. The resulting gel was then fourfold stronger than the unheated gel. Polarised optical microscopy showed that these heated gels had areas of birefringence suggesting that self-assembly had already started to occur, whereas the unheated solution showed no birefringence. Gel noodles were formed by taking the heated solution and manually drawing it onto a salty medium *via* pipette or needle. These gels now showed birefringence along the whole length of the noodle and also alignment in SEM (Figure 12.11a and c). This suggested that there was macroscopic alignment of already formed fibres. The unheated solution also formed gel noodles but under SEM, showed more randomly entangled fibres (Figure 12.11b and d).

These observations were further confirmed by small-angle X-ray scattering. It was possible to then add hMSCs into the PA solutions and draw out gel noodles. The cells remained viable during the process and optical, fluorescence and electron microscopy showed that the cell bodies and filopodia were aligned with the PA fibres. This process was repeated with cardiomyocytes, which again showed alignment along the fibres. Furthermore, these cells showed electrical activity. This method could be used in regenerative medicine for the restoration of electrical communication for the treatment of spinal cord injuries or cardiac arrhythmia.

Further work carried out by the Stupp group used PAs for the cell culture and differentiation for use in regenerative medicine.[119] They used a PA with the cell binding motif RGDS and phosphoserine residues. The phosphoserine residues were used to promote the formation of hydroxyapatite in the cell culture medium. The fibres were able to nucleate the growth of crystalline hydroxyapatite due to the presence of phosphoserine amino acids. These mineralised hydrogel matrices promoted the differentiation of hMSCs into osteogenic cells. PA gels were formed by adding the sterilised PAs to an

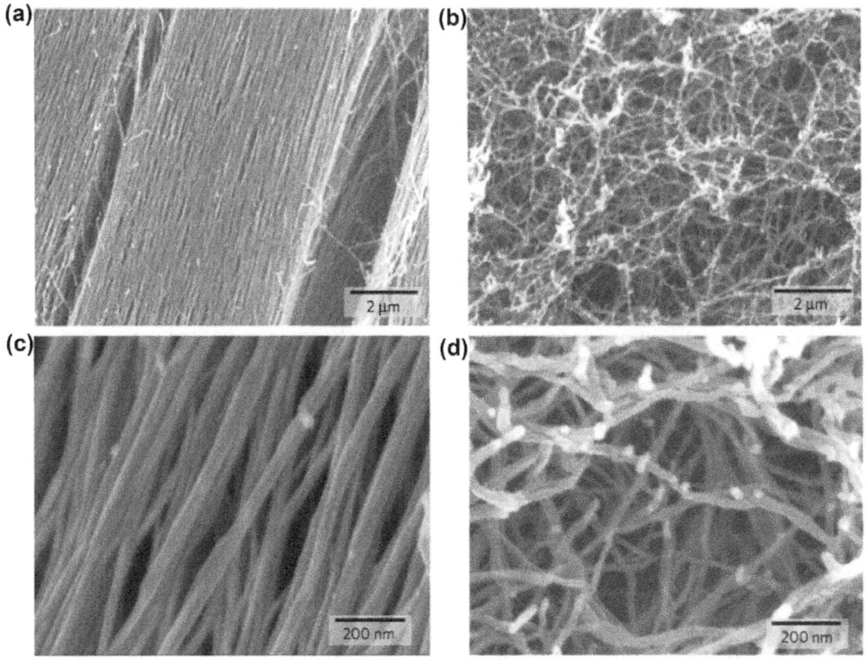

Figure 12.11 SEM evidence of massive alignment *versus* isotropy of nanofibre bundles. (a) and (c) show aligned nanofibre bundles in macroscopic strings formed by dragging thermally treated amphiphile solutions onto a $CaCl_2$ solution. (b) and (d) show an isotropic network of nanofibre bundles formed by adding $CaCl_2$ to unheated amphiphile solutions. Adapted from ref. 118. Copyright (2010) with permission from the Nature Publishing Group.

osteogenic medium containing $CaCl_2$ (mineralisation medium); this triggered self-assembly. When hMSCs were incorporated into or on top of the hydrogel, they were able to adhere to the gel fibres *via* the RDGS and were inhibited by the mineralisation of the fibres. The gels containing the hMSCs were incubated for four weeks and then the cells were analysed for any differentiation that may have occurred. This was done by monitoring the amount of osteopontin (OP) and alkaline phosphatase (ALP)—these are both osteogenic markers. ALP is a glycoprotein that is involved in bone mineralisation and OP is also a glycoprotein and is expressed in osteoblastic maturation. The cells cultured using the mineralised PA gels showed significant levels of both ALP and OP indicating that cell differentiation had occurred. This again could be used in regenerative medicine for the treatment of osteoporosis or for promoting new bone growth in broken bones.

12.5.4 Other Applications

Xu has used the gelation of incorporated naphthalene Phe-Phe-Lys-Tyrosine phosphate functionalised with different fluorophores for imaging inside living cells.[120] This cell imaging can be used to determine what happens to small molecules when they enter cells. By being able to track the fate of these small molecules and see where they assemble, gelators can be designed for use in applications such as drug delivery or chemotherapy. The fluorophores used in this study were nitrobenzoxadiazole (NBD), dansyl (DNS), 4-(*N*,*N*-dimethylsulfonamoyl)-2,1,3-benzoxadiazole (DBD) and rhodamine. When free in solution, the gelator precursors fluoresce weakly, but upon self-assembly (triggered by the enzymatic cleavage of the phosphate group on the tyrosine), the fluorescence intensity increased. LMWGs were able to diffuse into a cell and gel in the presence of the phosphatases within the cell. After gelation, the cells could be imaged using fluorescence microscopy. The LMWGs showed spatiotemporal behaviour (shown in Figure 12.12). The NBD-functionalised LMWG assembled near the endoplasmic reticulum of the cell, whereas the DNS-functionalised LMWG localised in the cell membrane. DBD-functionalised LMWGs were found to aggregate before gelation and so mostly remained outside the cell and were the least cytotoxic. The rhodamine-functionalised LMWGs were found both inside and outside the cell, but were the most cytotoxic. Not only does this show that LMWGs can be used to image different parts of the cell using different fluorophore-functionalised gelators, but reinforces that small changes in the gelator structure strongly affect the behaviour.

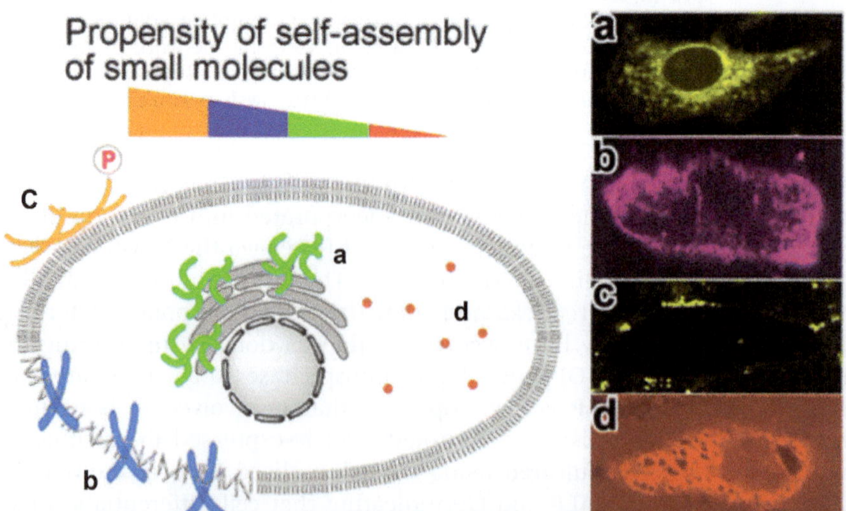

Figure 12.12 (a) NBD-gelator, (b) DNS-gelator, (c) DBD-gelator and (d) rhodamine-gelator. Left shows an illustration of the distinct spatial distribution of the small molecules in a cellular environment. Right shows fluorescent confocal microscope images of the HeLa cells incubated with 500 μM. Scale bar = 25 μm.
Reprinted from ref. 120. Copyright (2013) with permission from the American Chemical Society.

Self-assembly of peptide-based materials on surfaces can be used on metal implants such as vascular stents, bone plates and artificial joints in the body. By modifying the surface of the metal with these molecules, the surface of the metal implant has a greater stability as it prevents dissolution and degradation of the material. When these molecules are present on the surface, biological adhesion is promoted, and so the body's own tissue can grow over the metal and increase new bone or blood vessel growth. Sargeant *et al.* have used a PA gelator with an RGDS cellular adhesion sequence to self-assemble on NiTi substrates.[121] When these PA-modified NiTi implants were placed *in situ*, the tissue came into contact with the PA and initiated a response. Cells then attached to the surface and secreted proteins, forming an ECM. These cells also absorb some of the biodegradable PA. When PAs were covalently bound to the NiTi surface *via* salinization, there was greater cell proliferation. After 7 days, PAs showed no toxicity towards the cells that had grown and spread on the surface. These modified NiTi surfaces could potentially be used in implants to promote implant-specific cellular responses.

Hydrogels have also been used to immobilise enzymes for catalysis. These enzymes when immobilised show superactivity. Xu's group further investigated this to see whether these hydrogels would improve the activity of artificial enzymes.[122] Hemin was used to mimic peroxidase, an enzyme that is used to oxidise a wide range of organic and inorganic substrates. The use of a peptide-based scaffold to trap the hemin serves two purposes: (1) to aid the function of hemin as the sequence is similar to that of the active site of peroxidase, and (2) the immobilization helps with recovery of the hemin. Fmoc-Phe was used to form separate gels using Na_2CO_3 in a 7.4 pH buffer, to which hemin chloride was added. The hemin showed greater activity in these gels than when free in solution. The use of energy-dispersive X-ray spectroscopy (EDX) showed that the hemin molecules were localised on the gel fibres. This localisation also prevented dimerisation of the hemin, and hence resulted in an increase in the activity of the artificial enzyme. This hydrogel-immobilised hemin also showed high stability and reusability. This is a result of the gel providing protection from oxidative inactivation of the hemin.

12.6 Conclusions and Outlook

The use of LMWGs for biological applications has many advantages over other materials. LMWGs are often inexpensive and are able to be used on a large scale. The wide variety of molecules that can be used as gelators means that gels with different functionalities and responses can be prepared. This means that gels can be tailored to specific needs. In many cases, LMWGs have also been shown to have little or no cytotoxicity and so are biocompatible. Gelation can also be reversible and so makes the gels reusable.

LMWGs show great potential for use in biological applications due to the ease of incorporating bioactive molecules and cells without affecting the properties of the gel itself. The responsive behaviour upon a stimulus is useful for sensors that can be used in the detection of analytes for the diagnosis of illnesses. The wide variety of stimuli means that these LMWGs can be used for a broad range of sensing applications.

Cell growth and proliferation is enabled by hydrogels due to the fibrous scaffold not only mimicking the ECM but also providing a hydrated environment that can be modified to support the differentiation of different cells. The gelators are non-toxic and help adhere the cells to the gel fibres to form 2D and 3D cell growth. Cell growth and differentiation is crucial for the development of regenerative medicines.

The use of LMWGs in drug delivery and therapeutic medicine enables targeted delivery of drugs, which is important to reduce toxicity to other parts of the body, enables sustained release of the drug and reduces the amount of drug that needs to be administered to the patient. The thixotropic behaviour of gels allows them to be injected into the target site *e.g.* a tumour or a bone without changing the physical properties of the gel. They also show promise in other applications, for example in organic electronics, cell imaging and catalysis.

The next step forward is integrating gels into more complex assemblies to make them more usable; in the example of sensors, this is difficult as the gels are wet and so the longevity of a device would be comprised by the wet gel drying out and losing its function. Another problem is that designing a new gelator for a specific function is difficult and often unpredictable, so more research into every step of self-assembly needs to be done to fully understand what makes a molecule able to gel. Overall, LMWGs provide excellent materials for use in a wide range of biological and medicinal applications and have the promise of becoming more widely used in the future.

References

1. M. de Loos, B. L. Feringa and J. H. van Esch, *Eur. J. Org. Chem.*, 2005, **17**, 3615–3631.
2. P. Terech and R. G. Weiss, *Chem. Rev.*, 1997, **97**, 3133–3160.
3. V. Jenning, A. Gysler, M. Schäfer-Korting and S. H. Gohla, *Eur. J. Pharm. Biopharm.*, 2000, **49**, 211–218.
4. B. Xu, *Langmuir*, 2009, **25**, 8375–8377.
5. W. Pilnik and F. M. Rombouts, *Carbohydr. Res.*, 1985, **142**, 93–105.
6. B. D. Wall and J. D. Tovar, *Pure Appl. Chem.*, 2012, **84**, 1039–1045.
7. K. J. Skilling, F. Citossi, T. D. Bradshaw, M. Ashford, B. Kellam and M. Marlow, *Soft Matter*, 2014, **10**, 237–256.
8. S. Maude, E. Ingham and A. Aggeli, *Nanomedicine*, 2013, **8**, 823–847.
9. N. Stephanopoulos, J. H. Ortony and S. I. Stupp, *Acta Mater.*, 2013, **61**, 912–930.
10. G. Fichman and E. Gazit, *Acta Biomater.*, 2014, **10**, 1671–1682.
11. R. Tian, J. Chen and R. Niu, *Nanoscale*, 2014, **6**, 3474–3482.
12. S. J. Buwalda, K. W. M. Boere, P. J. Dijkstra, J. Feijen, T. Vermonden and W. E. Hennink, *J. Controlled Release*, 2014, **190**, 254–273.

13. V. Javvaji, A. G. Baradwaj, G. F. Payne and S. R. Raghavan, *Langmuir*, 2011, **27**, 12591–12596.
14. J. Jagur-Grodzinski, *Polym. Adv. Technol.*, 2010, **21**, 27–47.
15. V. B. R. Prashant, P. Kalshetti, D. N. Dixit and P. P. Parekh, *J. Pharm. Pharm. Sci.*, 2012, **4**, 1–7.
16. B. Guo and P. Ma, *Sci. China: Chem.*, 2014, **57**, 490–500.
17. *Polymeric and Self-Assembled Hydrogels*, ed. X. J. Loh and O. A. Sherman, Royal Society of Chemistry, Cambridge, 2013, vol. 1, pp. 1–262.
18. S. Zhang, *Nat. Biotechnol.*, 2003, **21**, 1171–1178.
19. J. Cui, Z. Shen and X. Wan, *Langmuir*, 2009, **26**, 97–103.
20. L. A. Estroff and A. D. Hamilton, *Chem. Rev.*, 2004, **104**, 1201–1218.
21. R. V. Ulijn, N. Bibi, V. Jayawarna, P. D. Thornton, S. J. Todd, R. J. Mart, A. M. Smith and J. E. Gough, *Mater. Today*, 2007, **10**, 40–48.
22. Y.-H. Chen, J. T. Yang and K. H. Chau, *Biochemistry*, 1974, **13**, 3350–3359.
23. L. A. Estroff, L. Leiserowitz, L. Addadi, S. Weiner and A. D. Hamilton, *Adv. Mater.*, 2003, **15**, 38–42.
24. C. Yan and D. J. Pochan, *Chem. Soc. Rev.*, 2010, **39**, 3528–3540.
25. J. A. Foster and J. W. Steed, *Angew. Chem., Int. Ed.*, 2010, **49**, 6718–6724.
26. H. Wang and Z. Yang, *Nanoscale*, 2012, **4**, 5259–5267.
27. P. Dastidar, *Chem. Soc. Rev.*, 2008, **37**, 2699–2715.
28. R. Luboradzki, O. Gronwald, M. Ikeda, S. Shinkai and D. N. Reinhoudt, *Tetrahedron*, 2000, **56**, 9595–9599.
29. K. A. Houton, K. L. Morris, L. Chen, M. Schmidtmann, J. T. A. Jones, L. C. Serpell, G. O. Lloyd and D. J. Adams, *Langmuir*, 2012, **28**, 9797–9806.
30. A. Aufderhorst-Roberts, W. J. Frith, M. Kirkland and A. M. Donald, *Langmuir*, 2014, **30**, 4483–4492.
31. M. M. Smith and D. K. Smith, *Soft Matter*, 2011, **7**, 4856–4860.
32. J. H. van Esch, *Langmuir*, 2009, **25**, 8392–8394.
33. M. George and R. G. Weiss, *Acc. Chem. Res.*, 2006, **39**, 489–497.
34. D. Mandal, A. Nasrolahi Shirazi and K. Parang, *Org. Biomol. Chem.*, 2014, **12**, 3544–3561.
35. R. A. Gortner and W. F. Hoffman, *J. Am. Chem. Soc.*, 1921, **43**, 2199–2202.
36. F. M. Menger, Y. Yamasaki, K. K. Catlin and T. Nishimi, *Angew. Chem., Int. Ed.*, 1995, **34**, 585–586.
37. C. Wang, D. Zhang, J. Xiang and D. Zhu, *Langmuir*, 2007, **23**, 9195–9200.
38. L. Chen, S. Revel, K. Morris, L. C. Serpell and D. J. Adams, *Langmuir*, 2010, **26**, 13466–13471.
39. R. Orbach, L. Adler-Abramovich, S. Zigerson, I. Mironi-Harpaz, D. Seliktar and E. Gazit, *Biomacromolecules*, 2009, **10**, 2646–2651.
40. J. Shi, Y. Gao, Z. Yang and B. Xu, *Beilstein J. Org. Chem.*, 2011, **7**, 167–172.
41. Y. Ogawa, C. Yoshiyama and T. Kitaoka, *Langmuir*, 2012, **28**, 4404–4412.

42. L. Chen, S. Revel, K. Morris and D. J. Adams, *Chem. Commun.*, 2010, **46**, 4267–4269.
43. Z. Yang, G. Liang and B. Xu, *Acc. Chem. Res.*, 2008, **41**, 315–326.
44. P. Berndt, G. B. Fields and M. Tirrell, *J. Am. Chem. Soc.*, 1995, **117**, 9515–9522.
45. J. D. Hartgerink, E. Beniash and S. I. Stupp, *Proc. Natl. Acad. Sci.*, 2002, **99**, 5133–5138.
46. S. Khan, S. Sur, C. J. Newcomb, E. A. Appelt and S. I. Stupp, *Acta Biomater.*, 2012, **8**, 1685–1692.
47. J. P. Schneider, D. J. Pochan, B. Ozbas, K. Rajagopal, L. Pakstis and J. Kretsinger, *J. Am. Chem. Soc.*, 2002, **124**, 15030–15037.
48. A. Aggeli, I. A. Nyrkova, M. Bell, R. Harding, L. Carrick, T. C. B. McLeish, A. N. Semenov and N. Boden, *Proc. Natl. Acad. Sci.*, 2001, **98**, 11857–11862.
49. J. K. Kretsinger, L. A. Haines, B. Ozbas, D. J. Pochan and J. P. Schneider, *Biomaterials*, 2005, **26**, 5177–5186.
50. Z. Yang, G. Liang, M. Ma, Y. Gao and B. Xu, *J. Mater. Chem.*, 2007, **17**, 850–854.
51. C. Yu, M. Xue, K. Liu, G. Wang and Y. Fang, *Langmuir*, 2014, **30**, 1257–1265.
52. E. R. Draper, J. J. Walsh, T. O. McDonald, M. A. Zwijnenburg, P. J. Cameron, A. J. Cowan and D. J. Adams, *J. Mater. Chem. C*, 2014, **2**, 5570–5575.
53. S. Roy, D. K. Maiti, S. Panigrahi, D. Basak and A. Banerjee, *RSC Adv.*, 2012, **2**, 11053–11060.
54. S. Basak, J. Nanda and A. Banerjee, *Chem. Commun.*, 2013, **49**, 6891–6893.
55. C. Li and H. Wonneberger, *Adv. Mater.*, 2012, **24**, 613–636.
56. J. Raeburn, G. Pont, L. Chen, Y. Cesbron, R. Levy and D. J. Adams, *Soft Matter*, 2012, **8**, 1168–1174.
57. A. Mahler, M. Reches, M. Rechter, S. Cohen and E. Gazit, *Adv. Mater.*, 2006, **18**, 1365–1370.
58. C. Tang, R. Ulijn and A. Saiani, *Eur. Phys. J. E: Soft Matter Biol. Phys.*, 2013, **36**, 1–11.
59. R. Vegners, I. Shestakova, I. Kalvinsh, R. M. Ezzell and P. A. Janmey, *J. Pept. Sci.*, 1995, **1**, 371–378.
60. Z. Xie, A. Zhang, L. Ye, X. Wang and Z.-g. Feng, *J. Mater. Chem.*, 2009, **19**, 6100–6102.
61. L. Chen, T. O. McDonald and D. J. Adams, *RSC Adv.*, 2013, **3**, 8714–8720.
62. Y. Gao, Z. Yang, Y. Kuang, M.-L. Ma, J. Li, F. Zhao and B. Xu, *Pept. Sci.*, 2010, **94**, 19–31.
63. L. Chen, J. Raeburn, S. Sutton, D. G. Spiller, J. Williams, J. S. Sharp, P. C. Griffiths, R. K. Heenan, S. M. King, A. Paul, S. Furzeland, D. Atkins and D. J. Adams, *Soft Matter*, 2011, 7, 9721–9727.
64. X. Wang and M. Liu, *Chem. - Eur. J.*, 2014, **20**, 10110–10116.

65. P. A. Korevaar, C. J. Newcomb, E. W. Meijer and S. I. Stupp, *J. Am. Chem. Soc.*, 2014, **136**, 8540–8543.
66. J. Nanda, A. Biswas and A. Banerjee, *Soft Matter*, 2013, **9**, 4198–4208.
67. A. M. Smith, R. J. Williams, C. Tang, P. Coppo, R. F. Collins, M. L. Turner, A. Saiani and R. V. Ulijn, *Adv. Mater.*, 2008, **20**, 37–41.
68. M. Zhou, A. M. Smith, A. K. Das, N. W. Hodson, R. F. Collins, R. V. Ulijn and J. E. Gough, *Biomaterials*, 2009, **30**, 2523–2530.
69. W. Helen, P. de Leonardis, R. V. Ulijn, J. Gough and N. Tirelli, *Soft Matter*, 2011, **7**, 1732–1740.
70. K. J. C. van Bommel, M. C. A. Stuart, B. L. Feringa and J. van Esch, *Org. Biomol. Chem.*, 2005, **3**, 2917–2920.
71. A. Z. Cardoso, A. E. Alvarez Alvarez, B. N. Cattoz, P. C. Griffiths, S. M. King, W. J. Frith and D. J. Adams, *Faraday Discuss.*, 2013, **166**, 101–116.
72. D. J. Adams, M. F. Butler, W. J. Frith, M. Kirkland, L. Mullen and P. Sanderson, *Soft Matter*, 2009, **5**, 1856–1862.
73. J. Raeburn, T. O. McDonald and D. J. Adams, *Chem. Commun.*, 2012, **48**, 9355–9357.
74. J. Raeburn, B. Alston, J. Kroeger, T. O. McDonald, J. R. Howse, P. J. Cameron and D. J. Adams, *Mater. Horiz.*, 2014, **1**, 241–246.
75. E. K. Johnson, L. Chen, P. S. Kubiak, S. F. McDonald, D. J. Adams and P. J. Cameron, *Chem. Commun.*, 2013, **49**, 8698–8700.
76. S. C. Bremmer, J. Chen, A. J. McNeil and M. B. Soellner, *Chem. Commun.*, 2012, **48**, 5482–5484.
77. M. J. Webber, C. J. Newcomb, R. Bitton and S. I. Stupp, *Soft Matter*, 2011, **7**, 9665–9672.
78. Z. Yang, H. Gu, D. Fu, P. Gao, J. K. Lam and B. Xu, *Adv. Mater.*, 2004, **16**, 1440–1444.
79. K. Thornton, A. M. Smith, C. L. R. Merry and R. V. Ulijn, *Biochem. Soc. Trans.*, 2009, **37**, 660–664.
80. H. Wang, C. Ren, Z. Song, L. Wang, X. Chen and Z. Yang, *Nanotechnology*, 2010, **21**, 225606.
81. H. Liu, Z. Lv, K. Ding, X. Liu, L. Yuan, H. Chen and X. Li, *J. Mater. Chem. B*, 2013, **1**, 5550–5556.
82. J.-B. Guilbaud, E. Vey, S. Boothroyd, A. M. Smith, R. V. Ulijn, A. Saiani and A. F. Miller, *Langmuir*, 2010, **26**, 11297–11303.
83. L. Szkolar, J.-B. Guilbaud, A. F. Miller, J. E. Gough and A. Saiani, *J. Pept. Sci.*, 2014, **20**, 578–584.
84. M. Hughes, P. W. J. M. Frederix, J. Raeburn, L. S. Birchall, J. Sadownik, F. C. Coomer, I. H. Lin, E. J. Cussen, N. T. Hunt, T. Tuttle, S. J. Webb, D. J. Adams and R. V. Ulijn, *Soft Matter*, 2012, **8**, 5595–5602.
85. R. J. Williams, A. M. Smith, R. Collins, N. Hodson, A. K. Das and R. V. Ulijn, *Nat. Nanotechnol.*, 2009, **4**, 19–24.
86. M. Ikeda, T. Tanida, T. Yoshii, K. Kurotani, S. Onogi, K. Urayama and I. Hamachi, *Nat. Chem.*, 2014, **6**, 511–518.
87. J.-B. Guilbaud, C. Rochas, A. F. Miller and A. Saiani, *Biomacromolecules*, 2013, **14**, 1403–1411.

88. Z. Yang, G. Liang, L. Wang and B. Xu, *J. Am. Chem. Soc.*, 2006, **128**, 3038–3043.
89. J. Dong, J. M. Canfield, A. K. Mehta, J. E. Shokes, B. Tian, W. S. Childers, J. A. Simmons, Z. Mao, R. A. Scott, K. Warncke and D. G. Lynn, *Proc. Natl. Acad. Sci.*, 2007, **104**, 13313–13318.
90. A. B. G. Lansdown, U. Mirastschijski, N. Stubbs, E. Scanlon and M. S. Ågren, *Wound Repair Regen.*, 2007, **15**, 2–16.
91. P. J. Knerr, M. C. Branco, R. Nagarkar, D. J. Pochan and J. P. Schneider, *J. Mater. Chem.*, 2012, **22**, 1352–1357.
92. C. M. Micklitsch, P. J. Knerr, M. C. Branco, R. Nagarkar, D. J. Pochan and J. P. Schneider, *Angew. Chem., Int. Ed.*, 2011, **50**, 1577–1579.
93. M. A. Greenfield, J. R. Hoffman, M. Olvera de la Cruz and S. I. Stupp, *Langmuir*, 2009, **26**, 3641–3647.
94. J. Shi, Y. Gao, Y. Zhang, Y. Pan and B. Xu, *Langmuir*, 2011, **27**, 14425–14431.
95. L. Chen, G. Pont, K. Morris, G. Lotze, A. Squires, L. C. Serpell and D. J. Adams, *Chem. Commun.*, 2011, **47**, 12071–12073.
96. Y. Zhang, H. Gu, Z. Yang and B. Xu, *J. Am. Chem. Soc.*, 2003, **125**, 13680–13681.
97. X. Zhang, X. Chu, L. Wang, H. Wang, G. Liang, J. Zhang, J. Long and Z. Yang, *Angew. Chem., Int. Ed.*, 2012, **51**, 4388–4392.
98. X.-D. Xu, B.-B. Lin, J. Feng, Y. Wang, S.-X. Cheng, X.-Z. Zhang and R.-X. Zhuo, *Macromol. Rapid Commun.*, 2012, **33**, 426–431.
99. Z. A. C. Schnepp, R. Gonzalez-McQuire and S. Mann, *Adv. Mater.*, 2006, **18**, 1869–1872.
100. J. Raeburn, A. Z. Cardoso and D. J. Adams, *Chem. Soc. Rev.*, 2013, **42**, 5143–5156.
101. H. Cui, A. G. Cheetham, E. T. Pashuck and S. I. Stupp, *J. Am. Chem. Soc.*, 2014, **136**, 12461–12468.
102. A. Baral, S. Roy, A. Dehsorkhi, I. W. Hamley, S. Mohapatra, S. Ghosh and A. Banerjee, *Langmuir*, 2014, **30**, 929–936.
103. V. Jayawarna, M. Ali, T. A. Jowitt, A. F. Miller, A. Saiani, J. E. Gough and R. V. Ulijn, *Adv. Mater.*, 2006, **18**, 611–614.
104. A. Richter, G. Paschew, S. Klatt, J. Lienig, K.-F. Arndt and H.-J. Adler, *Sensors*, 2008, **8**, 561–581.
105. D. Buenger, F. Topuz and J. Groll, *Prog. Polym. Sci.*, 2012, **37**, 1678–1719.
106. S. Kiyonaka, K. Sada, I. Yoshimura, S. Shinkai, N. Kato and I. Hamachi, *Nat. Mater.*, 2004, **3**, 58–64.
107. J. H. Kim, S. Y. Lim, D. H. Nam, J. Ryu, S. H. Ku and C. B. Park, *Biosens. Bioelectron.*, 2011, **26**, 1860–1865.
108. P. K. Vemula, J. E. Kohler, A. Blass, M. Williams, C. Xu, L. Chen, S. R. Jadhav, G. John, D. I. Soybel and J. M. Karp, *Sci. Rep.*, 2014, **4**, 4645.
109. K. Fan, J. Yang, X. Wang and J. Song, *Soft Matter*, 2014, **10**, 8370–8375.
110. *Smart Materials for Drug Delivery: Volume 1*, ed. C. Lorenzo-Alvarez and A. Concheiro, 2013, Royal Society of Chemistry, Cambridge, vol. 1, pp. 1–393.

111. *Smart Materials for Drug Delivery: Volume 2*, ed. C. Lorenzo-Alvarez and A. Concheiro, 2013, Royal Society of Chemistry, Cambridge, vol. 1, pp. 1–393.
112. Z. M. Yang, K. M. Xu, Z. F. Guo, Z. H. Guo and B. Xu, *Adv. Mater.*, 2007, **19**, 3152–3156.
113. S. P. Patil, S.-H. Kim, J. R. Jadhav, J.-h. Lee, E. M. Jeon, K.-T. Kim and B. H. Kim, *Bioconjugate Chem.*, 2014, **25**, 1517–1525.
114. H. Wang, J. Wei, C. Yang, H. Zhao, D. Li, Z. Yin and Z. Yang, *Biomaterials*, 2012, **33**, 5848–5853.
115. D. Tadic and M. Epple, *Biomaterials*, 2004, **25**, 987–994.
116. N. Amosi, S. Zarzhitsky, E. Gilsohn, O. Salnikov, E. Monsonego-Ornan, R. Shahar and H. Rapaport, *Acta Biomater.*, 2012, **8**, 2466–2475.
117. J. P. Jung, J. V. Moyano and J. H. Collier, *Integr. Biol.*, 2011, **3**, 185–196.
118. S. Zhang, M. A. Greenfield, A. Mata, L. C. Palmer, R. Bitton, J. R. Mantei, C. Aparicio, M. O. de la Cruz and S. I. Stupp, *Nat. Mater.*, 2010, **9**, 594–601.
119. T. D. Sargeant, C. Aparicio, J. E. Goldberger, H. Cui and S. I. Stupp, *Acta Biomater.*, 2012, **8**, 2456–2465.
120. Y. Gao, Y. Kuang, X. Du, J. Zhou, P. Chandran, F. Horkay and B. Xu, *Langmuir*, 2013, **29**, 15191–15200.
121. T. D. Sargeant, M. S. Rao, C.-Y. Koh and S. I. Stupp, *Biomaterials*, 2008, **29**, 1085–1098.
122. Q. Wang, Z. Yang, X. Zhang, X. Xiao, C. K. Chang and B. Xu, *Angew. Chem., Int. Ed.*, 2007, **46**, 4285–4289.

CHAPTER 13

Stimuli-responsive Molecularly Imprinted Polymers

ORTENSIA ILARIA PARISI[a,b] AND FRANCESCO PUOCI*[a]

[a] Department of Pharmacy, Health and Nutritional Sciences, University of Calabria, 87036 Rende (CS), Italy; [b] Department of Informatics, Modeling, Electronics and Systems Engineering, University of Calabria, 87036 Rende (CS), Italy
*Email: francesco.puoci@unical.it

13.1 Introduction

Molecularly imprinted polymers (MIPs) are synthetic materials able to selectively rebind a target molecule in preference to other closely related compounds. Molecular imprinting technology (MIT) involves the synthesis of MIPs in the presence of a template analyte (including ions, molecules, complexes or molecular, ionic or macromolecular assemblies, and microorganisms) with complementary and selective binding sites into the polymeric matrix characterized by the memory of the template.

The idea of molecular imprinting was first introduced by Polyakov in 1931 as unusual adsorption properties of silica particles prepared in the presence of an additive using a novel synthetic procedure.[1] The observed selectivity toward the additive molecule present during silica drying was explained in terms of a template effect due to changes in the silica pore structure reflecting the nature of the additive itself.

In 1949, Dickey[2] reported the polymerization of sodium silicate in the presence of methyl orange (and other alkyl orange dyes) as a template employing a synthetic strategy which was very similar to that of Polyakov's.

RSC Smart Materials No. 14
Chemoresponsive Materials: Stimulation by Chemical and Biological Signals
Edited by Hans-Jörg Schneider

Published by the Royal Society of Chemistry, www.rsc.org

The silica gel showed an increased selectivity for the dye template, which was present during the polymerization process, as related to the other dyes. During the subsequent years, researchers adopted Dickey's method in order to prepare specific silica-based sorbents, however, the interest in imprinted silica decreased due to the limited stability and reproducibility of these materials.

The decline of molecular imprinting in silica coincided with the introduction of this innovative technology in synthetic organic polymers by Wulff and Sarhan in 1972.[3] The developed approach was based on reversible covalent bonds between imprinted molecules and functional monomers, therefore, this technique was called the covalent approach.

However, the most important development was the introduction of the non-covalent approach by Mosbach in the 1980s,[4,5] which involves the formation of non-covalent interactions to create an imprint of the template molecule within the polymeric matrix. This method is used more widely because it is based on the observation that most of the interactions between biological molecules consist of non-covalent forces.

Many researchers have explored novel strategies to prepare these kinds of materials that can grant new potential functions in many fields of chemistry and biology as sensors,[6–8] artificial antibodies,[9,10] adsorbents for solid phase extraction,[11–13] chromatographic stationary phases[14,15] and drug delivery systems (DDS).[16–18]

During the last decades, DDS have receive considerable attention and a special chapter by Peppas *et al.* in this book is devoted to smart hydrogels for this application.

Currently, the combination of molecular imprinting with stimuli-sensitivity is attracting significant research interest. Stimuli-responsive MIPs combine the selective recognition properties for a template molecule with the ability to respond to specific external stimuli (such as the presence of another molecule, pH, temperature, magnetic field and irradiation) with consequent changes in their properties including the affinity of the polymeric matrix for the template.

Therefore, this chapter aims to be an overview of molecular imprinting technology, the design of MIPs and synthetic approaches with particular attention devoted to the field of stimuli-responsive MIPs.

13.2 Molecular Imprinting Technology

Molecularly imprinted polymers (MIPs) are synthesized by polymerizing functional monomers and crosslinking agents around a print molecule, called a template, to obtain a three-dimensional polymeric network.

During the reaction, the pre-polymerization complex formed between the template and the functional monomers is stabilized within the resulting rigid polymer and once polymerization has taken place, the template is removed to give a porous and highly crosslinked polymer matrix (Figure 13.1).

The resulting MIPs are stable polymers characterized by significant molecular recognition abilities due to the presence of binding sites that are

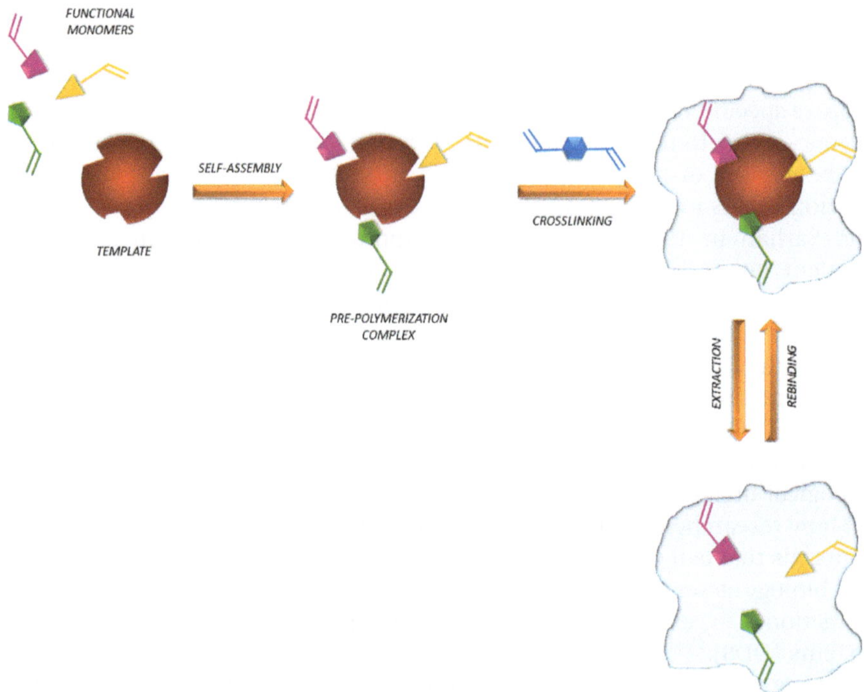

Figure 13.1 Representation of the molecular imprinting process.

complementary to the target molecule in terms of size, shape and chemical functionality arrangement. The high degree of crosslinking allows the microcavities to maintain their shape after the extraction of the target molecule, thus, the functional groups are held in an optimal configuration for selective rebinding of the template in preference to other closely related structures.[19] This behaviour emulates the recognition and binding properties of natural biomolecules, such as antibodies and enzymes, but without the associated stability limitations. Molecularly imprinted polymers are resistant to a wide range of chemical and physical conditions such as pH, organic solvents, temperature and pressure and, at the same time, possess several advantages over their biological counterparts including low cost, easy preparation, storage stability and excellent reusability without loss of activity.

Three different main methodologies to prepare MIPs have been developed based on the formation of covalent or non-covalent interactions between the template and the functional monomers during the pre-polymerization step and the following rebinding phase, respectively.

13.2.1 The Covalent Approach

The covalent approach for the synthesis of MIPs involves the formation of reversible covalent interactions between the template molecule and one or more functional monomers before the polymerization process. Then, the template is removed from the obtained polymeric network by chemical cleavage of the corresponding covalent bonds, which are re-formed upon rebinding of the target analyte.

The covalent imprinting strategy was introduced by Wulff[20,21] and co-workers who prepared MIPs by the synthesis of specific sugar or amino acid derivatives that contained a polymerizable function such as vinylphenylboronate. After the polymerization reaction, the sugar moiety was hydrolyzed obtaining a polymeric material able to selectively bind the template.

The advantage of this approach is the high stability of the template–monomer interaction, which leads to a more homogeneous binding site distribution, minimising the existence of non-specific sites. However, the design of an appropriate template–monomer complex in which covalent bond formation and cleavage are rapidly reversible under mild conditions makes this strategy less flexible. An acid hydrolysis procedure is often required to cleave the covalent bonds between the template and the functional monomer,[19] therefore, the achievement of a thermodynamic equilibrium is difficult due to the strong nature of the covalent interactions and the consequent slow binding and dissociation.[22] Moreover, only a slight number of molecules (including alcohols, aldehydes, ketones, amines and carboxylic acids) are suitable templates for covalent imprinting, thus, this method becomes rather restrictive.

Shea *et al.*[23] reported the imprinting of carboxylic acids by using carboxylic esters linkages, but due to the slow rebinding kinetics and the need for activated intermediates, carboxylic acids have limited potential when used in the covalent approach.

The use of Schiff's bases represents a useful strategy for the imprinting of amine and aldehyde templates and it is reversible without additional activation. However, the reformation of the covalent bonds during the rebinding phase is too slow for many potential applications including chromatography.

Diols and carbonyl compounds are able to form reversible ketal bonds, which have been used in order to imprint mono- and di-ketones.[24]

The requirements of covalent imprinting are quite different to those for a non-covalent approach, particularly with respect to ratios of template, functional monomer and crosslinking agent, and of the two methodologies, the non-covalent one has been used more widely.

13.2.2 The Non-covalent Approach

The non-covalent approach for the synthesis of MIPs was first introduced by Arshady and Mosbach.[4]

The strategy is based on the formation of relatively weak non-covalent interactions between the template and selected functional monomers, which are the same before the polymerization process and during the rebinding step. In an appropriate solvent, a template–monomer complex is formed by the establishment of interactions such as hydrogen bonds, ionic interactions, van der Waals forces and dipole–dipole interactions and, after the polymerization reaction and template extraction, the obtained polymeric matrix can rebind the target molecule through the same non-covalent interactions.[22]

Nowadays, non-covalent imprinting represents the most widely used method to prepare MIPs due to several advantages including the simplicity of the experimental procedure, high flexibility in terms of functionalities on a template that can be targeted and the available wide range of functional monomers able to interact with almost any kind of template.

The non-covalent protocol is rather simple because it avoids the tedious synthesis of a pre-polymerization complex, which is formed by the self-assembly between the template and functional monomers in a suitable solvent, followed by a crosslinked co-polymerization.[25,26] Moreover, the removal of the template is generally much easier and accomplished by continuous extraction.

Furthermore, there is a great variety of functionalities that can be introduced into the binding sites using non-covalent strategies, which impart high affinity and selectivity to the site itself. For this purpose, the most common functional monomer employed is methacrylic acid, first reported by Mosbach,[27] but many monomers are commercially available and a number of custom-designed functional monomers have also been reported. Other acid monomers used include 4-vinylbenzoic acid,[28] acrylic acid[29] and 2-acrylamido-2-methyl-1-propanesulfonic acid,[30] while the most commonly employed basic monomer is 4-vinylpyridine.[31] Among the neutral monomers, acrylamide,[32] vinyl pyrrolidone[33] and 2-hydroxyethyl methacrylate[34] are widely used. Also the combination of different functional monomers have been used to improve the binding and selectivity properties of MIPs.[24]

In addition, the range of attracting compounds that can be imprinted is significantly expanded including templates characterized by the presence of carboxyl, keto or amino groups, since many weak interactions dominate the recognition of the template molecules by MIPs.[22] On the other hand, compounds with a single interacting group lead to imprinted polymers with restricted recognition properties due to the need of more functionalities suitable for the formation of non-covalent bonds.[19]

The template–functional monomer interactions are governed by an equilibrium process, therefore, a high amount of monomer is used in order to displace the equilibrium towards the formation of the pre-polymerization complex. However, the excess of free monomers is randomly incorporated into the polymeric matrix leading to the formation of non-selective binding sites.

Based on these considerations, the non-covalent imprinting approach represents the most promising method due to the wide number of compounds, including biological molecules, which are able to interact in a non-covalent way with functional monomers.[35]

13.2.3 The Semi-covalent Approach

The semi-covalent approach represents an intermediate option between the two synthetic strategies mentioned and it is attributed to Whitcombe *et al.*[36]

During the pre-polymerization stage, the template is covalently bound to a functional monomer as in the covalent approach, but the rebinding step is entirely non-covalent in nature.

The method is aimed at combining the advantages of the covalent and non-covalent approaches such as the high affinity of covalent binding and mild operation conditions of non-covalent rebinding which have no kinetic restrictions except diffusion.[24]

13.3 The Design of Molecularly Imprinted Polymers

The design and synthesis of MIPs are very complex due to the influence of several factors including the nature and levels of template, type and concentration of functional monomers, crosslinking agents and initiators, the reaction solvent, the method of initiation and the time of polymerization process. All these parameters should be optimized to prepare an ideal imprinted polymer to preserve the structure of the binding cavities after template extraction and, at the same time, characterized by a high flexibility in order to facilitate a fast equilibrium between release and reuptake of the template.[22]

The *template* should be chemically inert and stable under the reaction conditions, thus, it should not contain functionalities that are involved in or prevent polymerization, but should be able to interact with the monomers.

Monomers provide the functional groups for the formation of covalent or non-covalent interactions with the template molecule (Figure 13.2).

The strength of the interactions affects the affinity and selectivity of MIP recognition sites.[37,38] In addition, the imprinting efficiency can be influenced by the adopted template to functional monomer molar ratio. Lower molar ratios lead to fewer binding sites in the polymer matrix due to a low number of template–monomer complexes; on the other hand, higher ones determine high non-specific binding capacity. Therefore, the molar ratio of template to monomer should be optimized to achieve high imprinting efficiency.[22] In the covalent approach, the template dictates the number of functional monomers that can be covalently attached in a stoichiometric way, while in the non-covalent method, monomers are used in excess relative to the number of template moles in order to favour the formation of the pre-polymerization complex.

Another parameter that influences MIP selectivity and binding capacity for the template is represented by the type and the amount of *crosslinking agent* employed during the polymerization process. The crosslinker fixes monomer functional groups around the imprinted molecule forming a highly crosslinked rigid polymer able to retain an optimal shape of the binding cavities (also after the template removal), characterized by stable configurations.

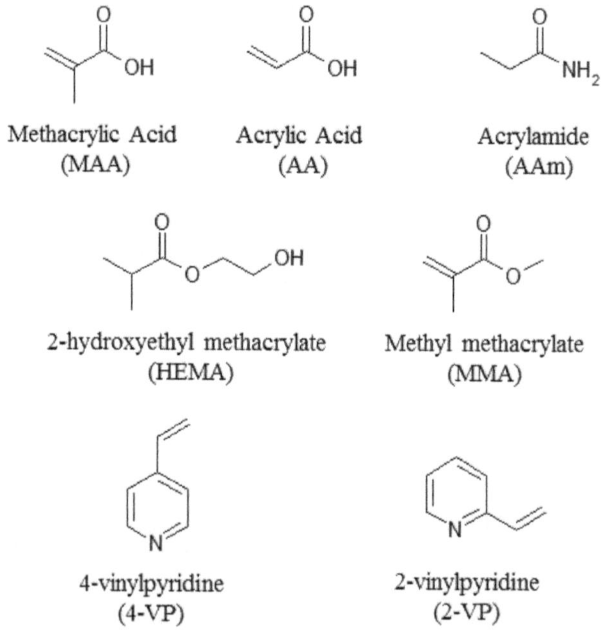

Figure 13.2 Chemical structures of some of the most common functional monomers.

Furthermore, it controls the morphology of the polymer matrix (gel-type, macroporous or microgel powder) and imparts mechanical stability.[19] An optimal crosslinking degree is important in order to obtain stable binding sites after template removal, adequate mechanical stability and accessible porous materials. The cavities have to maintain a three-dimensional structure complementary in both shape and functionalities to that of the template for rebinding the target analyte. The most commonly used crosslinkers include ethylene glycol dimethacrylate (EGDMA), divinylbenzene (DVB), trimethylolpropane trimethacrylate (TRIM) and *N*,*N*′-methylenebisacrylamide (MBAA) (Figure 13.3).

Different *initiators* can be used as the free radical source in the synthesis of molecularly imprinted polymers and, for this purpose, they are employed at low levels compared to the functional monomer. One of the most widely employed initiators is 2,2′-azobisisobutyronitrile (AIBN) (Figure 13.4), which can be decomposed by thermolysis or photolysis (UV) to produce stabilised carbon-centred radicals able to initiate the reaction.[19]

Other thermal initiators include water-soluble inorganic compounds, such as ammonium or potassium persulfate,[39,40] used either alone or in combination with *N*,*N*,*N*′,*N*′-tetramethylenediamine (TEMED),[41] and organic peroxides, such as benzoyl peroxide and lauroyl peroxide.[42,43] Photochemical decomposition of initiators (such as benzophenone[44] and 2,2′-dimethoxy-2-phenylacetophenone[45]) allows the preparation of MIPs at low

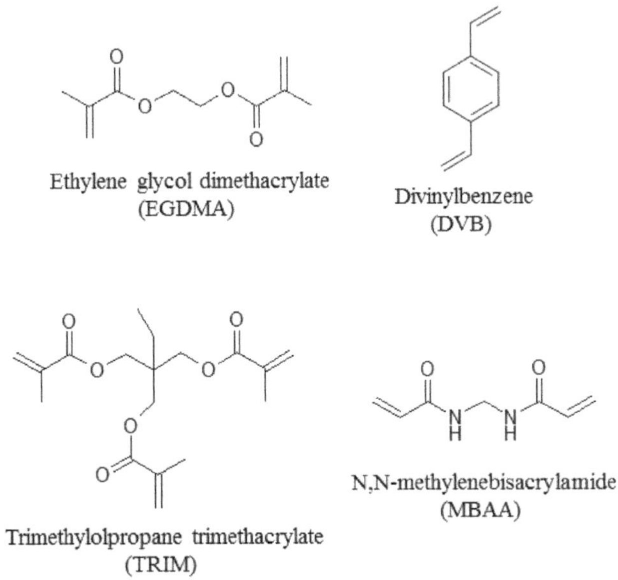

Figure 13.3 Chemical structures of some of the most common crosslinking agents.

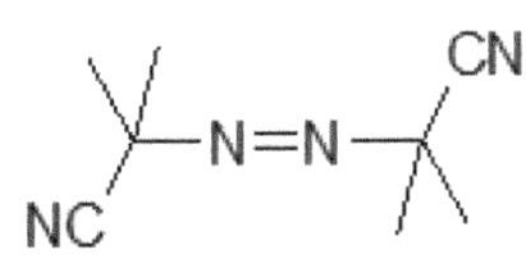

Figure 13.4 Chemical structure of 2,2′-azobisisobutyronitrile (AIBN).

temperature, avoiding template degradation in the case of thermo-sensitive compounds, and resulting in a higher stability of the pre-polymerization complex, which leads to an increased number and quality of binding sites. Therefore, the obtained MIPs are characterized by a better selectivity compared to the same polymers polymerized thermally.

Porogenic solvent plays a key role in the polymerization because it acts as not only a solvent but also a porogen influencing both the imprinting process and the physical state and morphology of the synthesized MIP.[46] Template, functional monomers, crosslinking agent and initiator have to be soluble in the chosen porogenic solvent, which is also responsible for the formation of the porous structure of the MIP. The porogen also influences the bonding strength between template and functional monomers, especially in the non-covalent approach. Aprotic and low polar organic solvents, such as toluene, acetonitrile and chloroform, are often used in non-covalent imprinting in order to stabilize hydrogen bonds and reduce interference during template–monomer complex formation. However, MIPs prepared in organic solvents work poorly in aqueous media, thus, the development of water-compatible imprinted polymers is attracting considerable research interest.

13.4 The Synthesis of Molecularly Imprinted Polymers

Molecularly imprinted polymers can be synthesized by employing different methods based on the desired final application.

The most common method to prepare MIPs is represented by *bulk polymerization* due to its rapidity and simplicity. However, the obtained monolithic polymer has to be crushed, ground and sieved to achieve particles characterized by appropriate size ranges. These operations lead to a reduction of the polymer yield. In addition, mechanical grinding leads to highly irregular particles in shape and size and some recognition sites are destroyed, resulting in a lower affinity for the template molecule.

To overcome these drawbacks, several polymerization strategies have been reported including suspension polymerization,[47] emulsion polymerization,[48] multi-step swelling polymerization[49] and precipitation polymerization.

Suspension polymerization is a simple method for the preparation of MIPs that do not require the grinding step. Water is employed as a continuous phase to suspend droplets of the pre-polymerization mixture in the presence of a suspension stabilizer or surfactant.[50] However, the obtained polymers are polydisperse in size and show poor recognition properties due to two main reasons: the presence of water, which can weaken the non-covalent interactions between template and functional monomers, and the presence of stabilizers or surfactants that also interfere with the chemistry of non-covalent imprinting. In order to overcome these limitations, suspension polymerization in perfluorocarbon solvents, used as a more compatible suspension medium, has been studied.[51]

Emulsion polymerization is another synthetic approach used to prepare high-yield monodispersed polymeric particles and it is successfully employed for protein imprinting.[48] However, this method is also characterized by the presence of surfactants.

The *multi-step swelling method* allows the synthesis of spherical imprinted particles monodisperse in size and shape by the step-wise swelling of a seed polymer with a mixture of fresh monomer and solvent.[52] Non crosslinked seed particles prepared by emulsion polymerization can be swollen with porogen, template, functional monomer and crosslinker in the second stage for the preparation of MIP particles.[24] This strategy is also employed to perform *in situ* modification and beads synthesized in this way can be rendered magnetic through inclusion of iron oxide particles. However, the use of an aqueous suspension as a continuous phase could interfere with the imprinting process and template–monomer interactions leading to a decrease in selectivity.

Precipitation polymerization is a relatively simple and promising technique to synthesize uniform and spherical imprinted polymeric particles in a single-step reaction and with excellent control over the particle size. In this method, more porogenic solvent is employed than in bulk polymerization and the growing polymer chains continue to grow individually by capturing newly formed oligomers and monomers from the surrounding solution. Due to the high dilution factor, a higher amount of template molecules is needed for the preparation of MIPs.

13.5 Stimuli-responsive Molecularly Imprinted Polymers

Stimuli-responsive polymers are a class of high-performance materials able to respond to specific external stimuli, including the presence of another molecule, temperature, pH, magnetic field and irradiation, with changes in their properties.

By combining molecular imprinting, and thus the selective recognition capacity for a template molecule, with the ability to respond to specific environmental stimuli with consequent changes in polymer properties, stimuli-responsive molecularly imprinted polymers have been developed.

The affinity of MIPs for the template molecule depends on the maintenance of the three-dimensional structure of the binding cavities inside the polymeric matrix. Therefore, the ability to reversibly rebind or release the template is affected by the change in the polymeric structure in response to an external stimulus. Due to this behavior, stimuli-responsive MIPs represent innovative smart materials for potential applications in several fields including drug delivery, bio- and chemosensors, artificial antibodies and separation sciences.

13.5.1 Thermo-responsive Molecularly Imprinted Polymers

Thermo-responsive polymers are able to respond in a controlled way to temperature, changing the dimension of their structure. This behavior is due to the presence of both hydrophilic and hydrophobic groups in the polymeric network.

These kinds of polymeric materials can exhibit either a lower critical solution temperature (LCST), below which they swell in deionized water, or an upper critical solution temperature (UCST), above which they swell.

The responsive behavior of polymers with LCST properties is characterized by the formation of hydrogen bond interactions between the hydrophilic groups in the polymer chains with water molecules at low temperatures. When the temperature increases and reaches values higher than the LCST, the hydrophilic interactions are destroyed while hydrophobic interactions between functionalities, such as methyl, ethyl and propyl groups, become stronger resulting in the aggregation of polymer chains and the subsequent contraction of the matrix. On the contrary, polymers characterized by UCST properties swell at high temperatures and shrink at low temperatures.

Poly(*N*-isopropylacrylamide) (PNIPAAm) is the most widely studied thermo-responsive polymer due to its LCST which is around 32 °C and, thus, close to human body temperature.[53]

In recent years, MIPs exhibiting thermo-responsive behavior have also been studied to modulate the affinity for the template as a function of the temperature, which involves changes in polymer structure. The swollen state leads to a release of the template; on the other hand, the contracted polymer can rebind the target molecule (Figure 13.5).

During the last years, *N*-isopropylacrylamide (NIPAAm) was largely employed as a functional monomer in order to synthesize thermo-responsive MIPs for different templates including proteins. The stimuli-sensitive recognition obtained with these systems is very similar to the recognition of proteins observed in nature.

One of the first studies to report thermo-responsive imprinted polymers was based on NIPAAm, acrylic acid and *N*,*N*′-methylene-bis(acrylamide), which were synthesized in the presence of DL-norephedrine hydrochloride or DL-adrenaline hydrochloride as a template.[54] This research represents one of the first examples of synthetic polymer gels that can undergo a conformational change (swelling change), but retain selective recognition properties in the shrunken state.

In 2009, Qin *et al.*[55] prepared a thermo-sensitive macroporous imprinted hydrogel showing selectivity for lysozyme by adopting an imprinting procedure that is based on metal coordinate interactions. For this purpose, *N*-(4-vinyl)-benzyl iminodiacetic acid, *N*-isopropylacrylamide, acrylamide and *N*,*N*′-methylene-bis(acrylamide) were used as metal chelate monomer, functional monomers and crosslinking agent, respectively. The synthesized imprinted hydrogel showed high selectivity in the purification of lysozyme from a mixture of proteins and a real sample.

In another study, a dibenzothiophene sulfone (DBTS) imprinted chitosan hydrogel was synthesized by the crosslinking of chitosan with glutaraldehyde in the presence of DBTS as the template.[56] The prepared hydrogel was found to be stimuli-responsive with temperature and showed a LCST between the swollen and the collapsed phases at 50 °C.

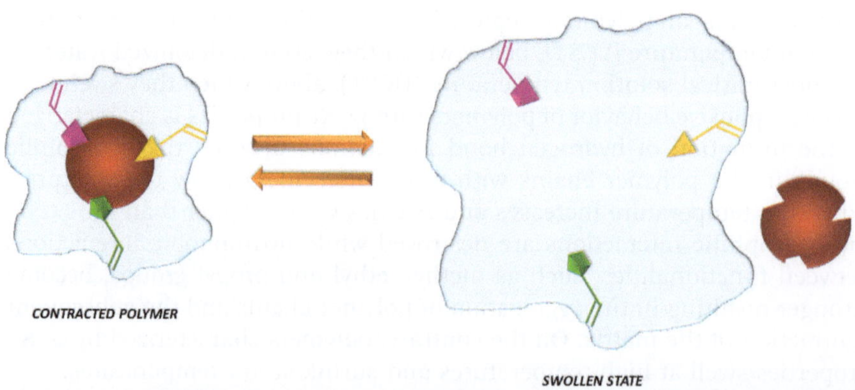

Figure 13.5 Schematic representation in which the contracted polymer can rebind the template, while the swollen state results in the release of the target molecule.

Recently, Zhao *et al.*[57] prepared a novel MIP-based thermo-sensitive electrochemical sensor by an electropolymerization method using *N*-isopropylacrylamide and theophylline as monomer and template, respectively. The resultant sensor showed excellent thermo-sensitive behavior and a good recognition capacity for template molecules from its analogues. This approach may allow the application of this type of electrochemical sensor for the stimulus-responsive recognition of target analytes.

Another strategy to prepare thermo-responsive MIPs is the grafting of PNIPAAm brushes onto the surface of preformed imprinted microspheres *via* surface-initiated reversible addition-fragmentation chain transfer (RAFT) polymerization of *N*-isopropylacrylamide.[58] The introduction of PNIPAAm brushes significantly improves the surface hydrophilicity and imparts stimuli-sensitivity resulting in water-compatibility and thermo-responsive binding properties.

13.5.2 pH-Responsive Molecularly Imprinted Polymers

pH-Responsive polymers contain ionizable groups (including carboxyl, sulfonic acid and amino groups) able to accept or donate protons with changes of environmental pH leading to swelling/shrinking structural changes in the polymer matrix.[59]

pH-sensitive hydrogels may be divided into two main categories depending on the type of ionizable groups: cationic and anionic hydrogels.[60] The pH-sensitivity of cationic hydrogels is due to the protonation of basic functionalities in the polymer chain. At low pH, the basic groups are protonated resulting in charge repulsion that leads to swelling of the polymer. On the contrary, at higher pH values, the groups become less ionized and the charge repulsion is reduced. Therefore, the polymer–polymer interactions increase with a contraction of the polymeric matrix. Regarding anionic hydrogels, the most common pH-responsive functional groups are carboxyl groups. At low pH, carboxyl groups are protonated and hydrophobic interactions dominate, resulting in a volume shrinkage. At high pH, carboxyl groups are ionized causing the swelling of the material.

In addition to the repulsion of charges, the increased number of ionic groups leads to the swelling of the pH-sensitive polymer also due to the presence of water of solvation. Water molecules are oriented around the ions forming a hydration shell, which results in an increase of the hydrogel volume.

Recently, pH-responsive MIPs have been developed by incorporating pH-sensitive functional groups into the imprinted polymers.

Methacrylic acid, acrylic amide and 4-vinylpyridine are the most widely used functional monomers in molecular imprinting, however, the presence of these monomers is not enough to impart pH sensitivity due to the high crosslinked structure of MIPs. Therefore, imprinted hydrogels with a low degree of crosslinking have been synthesized for pH response.

Kanekiyo *et al.*[61] developed a novel molecular imprinting strategy, using amylose as a host matrix, consisting of helical inclusion-complex formation between amylose modified with acryloyl groups (acryloylamylose) and the template bisphenol A in aqueous solution. The formation of the complex was followed by a radical copolymerization in the presence of a monomer with ionizable units (acrylic acid) and a crosslinker (*N*,*N*′-methylene-bisacrylamide). The resultant amylose-based polymers are characterized by binding sites that reversibly change their structure depending on the pH of the solution.

A pH-sensitive imprinted nanospheres/hydrogel composite exhibiting controlled release of dexamethasone-21 phosphate disodium (DXP) was developed as a potential coating for biocompatible sensing and DXP delivery.[62] The use of implantable glucose sensors is limited due to their short-term reliability in the human body, which is caused by tissue reactions including inflammation. The local release of dexamethasone allows control of inflammation, improving the biocompatibility and the lifetime of these kinds of sensors. The imprinted pH-sensitive nanospheres were prepared by UV-initiated precipitation polymerization using DXP, 2-hydroxyethyl methacrylate (HEMA), 2-(diethylamino)ethyl methacrylate (DEAEMA, an acid-swellable functional monomer) and EGDMA as template, functional monomers and crosslinker, respectively. The synthesized nanospheres showed faster DXP release rate at a lower pH value within the tested pH range (6.0–7.4), which is desirable for suppressing inflammation because inflammation induces an acidic microenvironment. pH-Sensitive MIP nanosphere/hydrogel composites were then prepared by dispersing imprinted nanospheres into the optimized hydrogel monomer solution, containing 2-hydroxyethyl methacrylate, *N*-vinyl-2-pyrrolidinone and 2-methacryloyloxyethyl phosphorylcholine, followed by UV polymerization.

In another research study, new networks based on star polymers were designed to be pH-responsive and recognitive for D-glucose.[63] The star polymer networks were prepared using molecular imprinting with D-glucose and crosslinking with poly(ethylene glycol) dimethacrylate with an ethylene glycol chain. The synthesized materials exhibited over 300% more uptake for D-glucose compared to a structurally similar sugar such as D-fructose. Then, the star polymer networks with pH-sensitivity were prepared by copolymerization with methacrylic acid and the resultant hydrogels showed a sharp transition in swelling around a pH of 4.5.

pH-Responsive molecularly imprinted particles were successfully prepared by pore-filling poly (acrylic acid) (PAA) gels into bisphenol A imprinted polyethersulfone particles by Zhao *et al.*[64] In this study, it was confirmed that changing the acidity of the solution reversibly controls the rebinding ability toward BPA. Furthermore, the BPA uptake of the pore-filled particles showed chemical valve behavior at a pH in the range of 3–6. The pH-sensitivity is due to the incorporation of PAA gel into the imprinted cavities, which swelled at higher pH (pH 7.5) leading to a lower binding capacity and, on the other hand, shrank in acidic solution (pH 2.5), resulting in an improved template affinity (Figure 13.6).

This approach is useful for the preparation of pH-responsive MIPs and is applicable to the imprinting of other hydrophobic molecules. Moreover,

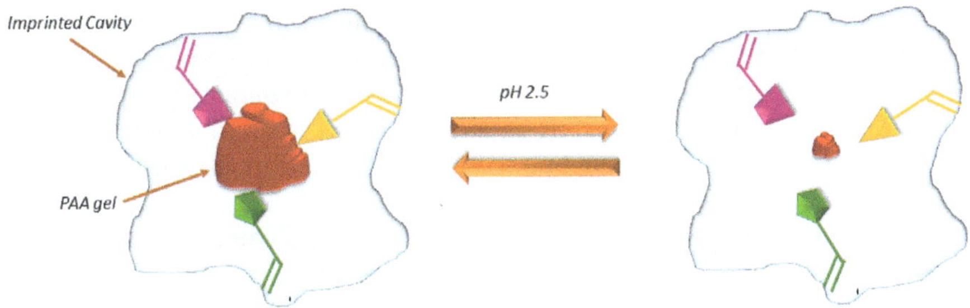

Figure 13.6 Switching of the substrate affinity of imprinted cavities due to PAA gel swelling.

the obtained results highlighted the potential of these kinds of stimuli-responsive molecularly imprinted polymers as smart chemicals and as drug delivery systems.

13.5.3 Photo-responsive Molecularly Imprinted Polymers

Photo-responsive polymers are a class of smart materials able to respond to external changes of light by a physical or chemical change and, usually, group isomerization or dissociation occurs due to the light stimulus.

The synthesis of photo-responsive polymers involves the introduction of a photosensitive group, such as azobenzene or triphenylmethane, into the polymer backbone, therefore, the preparation of photo-responsive MIPs involves replacing the conventional functional monomers by photo-responsive monomers.

Generally, a photo-responsive functional monomer consists of three main groups: the photo-responsive group, the recognition group and the polymerizable group.[60]

The most commonly investigated photo-responsive monomers contain azobenzene as a photosensitive group. The *trans–cis* photo-isomerization of the azobenzene functional monomer leads to modifications of the geometry and dipole moment of the chromophore with a relevant alteration of shape, size and chemical functionalities of the imprinted cavities. This results in a reduced capacity to recognize the template molecule when the polymer is irradiated by UV light. On the contrary, when irradiated by visible light, the recognition site returns to its original state and selectively rebinds the template. Therefore, the interaction between light and a photo-sensitive polymeric material may be used to modulate drug delivery.

A photo-responsive molecularly imprinted polymer was prepared by bulk polymerization from an azobenzene-based functional monomer, such as 4-[(4-methacryloyloxy)phenylazo]benzoic acid (MPABA), and using caffeine as a template.[65] It was observed that the substrate affinity of the synthesized MIP was photo-switchable due to the photo-isomerization of azobenzene, resulting in an alteration of the binding sites' geometry. Upon irradiation at 365 nm, 58.3% of receptor-bound caffeine was released from the MIP material. Subsequent irradiation at 440 nm caused 96.4% of the released caffeine to be rebound by the MIP material. However, although the photo-regulated substrate release and uptake processes were generally repeatable, a gradual reduction in the extent of substrate release and rebinding was observed. This may be caused by the slow deformation of MIP receptors during the course of repeated photo-switching.

Acetonitrile is largely employed for precipitation polymerization in order to synthesize spherical imprinted particles. Therefore, a methacrylate azo-benzene functional monomer with a pyridine group (4-((4-methacryloyloxy)-phenyl-azo)pyridine) with good solubility in acetonitrile was designed and synthesized in order to prepare 2,4-dichlorophenoxyacetic acid (2,4-D) imprinted microspheres.[66] The binding affinity was found to be photo-responsive towards the template and decreased upon UV light irradiation, whereas template could be recovered during the visible light-induced back-isomerization.

Gong *et al.*[67] synthesized photo-responsive MIPs, by precipitation polymerization, for the photo-regulated release and uptake of pharmaceuticals in aqueous media. Paracetamol, a common analgesic and antipyretic drug, 4-[(4-methacryloyloxy)phenylazo] benzenesulfonic acid and *N,N′*-hexylenebismethacrylamide were employed as template, water-soluble azo-benzene-containing monomer and crosslinking agent, respectively. It was observed that the affinity of the prepared hydrogel for paracetamol could be photo-regulated. Upon irradiation at 353 nm, 83.6% of paracetamol was released, while the subsequent irradiation at 440 nm caused 94.1% of the released paracetamol to be rebound.

13.5.4 Magnetic-responsive Molecularly Imprinted Polymers

Magnetic-responsive polymers are materials that show the ability to quickly respond to an external magnetic field resulting in a directional movement. These kinds of systems can find applications in several fields such as bio-engineering, separation sciences and biomedicine. Magnetically-guided delivery strategies have great potential in enhancing the therapeutic profile of drugs, promoting their specific accumulation at the pathological site. Moreover, magnetic polymers are also able to act as diagnostic molecular imaging agents. The combination of disease diagnosis with therapy leads to a newly emerging concept called theragnosis, therefore, magnetic-responsive MIPs could also play a key role in this area due to their magnetic and template recognition properties.

The general procedure to prepare magnetic-responsive MIPs involves three consecutive steps: the synthesis of Fe_3O_4 magnetic nanoparticles by a co-precipitation or a solvothermal reduction method; the surface modification of the Fe_3O_4 magnetic nanoparticles to promote surface polymerization, which is the ideal method to improve binding capacity; the synthesis of magnetic-responsive MIPs through a sol–gel process or a free radical polymerization.

Core–shell structural magnetic molecularly imprinted polymers, characterized by combined properties of molecular recognition and controlled release, were synthesized by the co-polymerization of methacrylic acid and trimethylolpropane trimethacrylate around aspirin (ASP) at the surface of double-bond-functionalized Fe_3O_4 nanoparticles.[68] The obtained spherical magnetic nanoparticles showed superparamagnetism and could be separated quickly by an external magnetic field. Binding experiments were carried out demonstrating high adsorption capacity and selectivity to ASP. Furthermore, ASP release profiles indicated that the magnetic MIPs could find potential applications in the controlled release of drugs.

In another study, Parisi *et al.*[69] described the synthesis of an innovative delivery system for targeted cancer therapy, which combines the drug controlled release ability of molecularly imprinted polymers with the magnetic properties of magnetite, by an easy and smart synthetic approach. The developed strategy involves a new engineered precipitation photopolymerization using 1,4-dimethyl-6-hydroxy-9*H*-carbazole as a template. The synthesized polymeric materials exhibited not only good selective recognition and controlled release properties, but also high magnetic responding capacity (Figure 13.7).

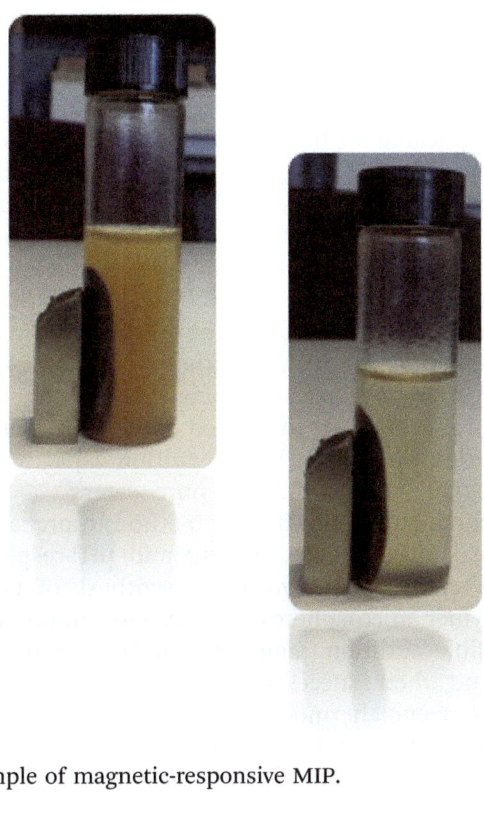

Figure 13.7 Example of magnetic-responsive MIP.

Both *in vitro* drug release and cytotoxicity studies on different cancer cell lines, such as HeLa and MCF-7, were performed, showing high inhibitory activity against the tested cell lines. Therefore, the obtained results indicated the potential application of synthesized MIPs as magnetic targeted drug delivery nanodevices.

13.5.5 Dual- and Multi-responsive Molecularly Imprinted Polymers

Nowadays, dual- and multi-stimuli responsive molecularly imprinted polymers are relatively less explored and there are only a few reports on these kinds of polymeric materials involving magnetic and photo-responsive MIPs, photo- and thermo-responsive MIPs, photo-, thermo- and pH-responsive MIPs.

Photonic and magnetic dual responsive molecularly imprinted polymers were prepared by a combination of stimuli-responsive polymers and a molecular imprinting technique.[70] The resultant materials exhibited specific affinity for caffeine, while photo-isomerization induced reversible uptake and release of this molecule upon alternate UV and visible light irradiation. With irradiation at 365 nm, 62.5% of the caffeine was released from the dual-responsive MIPs, while subsequent irradiation with visible light caused 93.6% of the released caffeine to be rebound. The novel dual-responsive MIPs were used as sorbents for the enrichment of caffeine from real water and beverage samples. In addition, the magnetic properties of these polymeric systems provided fast and simple separation, thus, the obtained materials could be applicable for trace caffeine analysis in complex samples.

Fang *et al.*[71] described an easy and efficient approach to obtain azobenzene (azo)-containing imprinted microspheres with both photo- and thermo-responsive template binding properties in pure aqueous media. The developed synthetic strategy involves the first synthesis of "living" azo-containing MIP microspheres with surface-immobilized alkyl halide groups *via* atom transfer radical precipitation polymerization (ATRPP) and their subsequent modification *via* surface-initiated atom transfer radical polymerization (ATRP) of *N*-isopropylacrylamide. The introduction of PNIPAAm brushes onto the azo-containing imprinted microspheres imparted thermo-responsive properties. Moreover, the recognition capacity was found to be photo-responsive toward the template in pure water.

In another research study, an efficient method to prepare water-compatible and multiple stimuli (photo, thermo and pH)-responsive imprinted microspheres was reported.[72] It involves the synthesis of narrowly dispersed "living" core polymer microspheres with surface-immobilized dithioester groups *via* reversible addition-fragmentation chain transfer precipitation polymerization and their successive grafting of an azobenzene (azo)-containing MIP layer and thermo- and pH-responsive hydrophilic polymer brushes. The introduction of the polymeric brushes onto the core–shell imprinted microspheres improved the surface hydrophilicity and provided a responsive layer with controllable collapse and dissolution of polymer brushes towards external stimuli. This resulted in stimuli-responsive template binding (or releasing) properties.

13.6 Conclusions

Molecularly imprinted polymers can grant new potential functions in many fields of chemistry and biology as drug delivery systems, sensors, artificial antibodies, adsorbents for solid phase extraction and chromatographic stationary phases.

During the last decades, novel strategies to prepare MIPs were explored and the combination of molecular imprinting with stimuli-sensitivity is attracting significant research interest. Stimuli-responsive MIPs combine the selective recognition properties for a template molecule with the ability to respond to specific external stimuli (such as the presence of another molecule, pH, temperature, magnetic field and irradiation) with consequent changes in their properties including the affinity of the polymeric matrix for the template. Due to these properties and their versatility, stimuli-responsive molecularly imprinted polymers have great potential in several applications including drug delivery and environmental protection.

References

1. M. V. Polyakov, *Zh. Fiz. Khim.*, 1931, **2**, 799.
2. F. H. Dickey, *Proc. Natl. Acad. Sci. U. S. A.*, 1949, **35**, 227.
3. G. Wulff and A. Sarhan, *Angew. Chem., Int. Ed. Engl.*, 1972, **11**, 341.
4. R. Arshady and K. Mosbach, *Makromol. Chem.*, 1981, **182**, 687.
5. L. Andersson, B. Sellergren and K. Mosbach, *Tetrahedron Lett.*, 1984, **25**, 5211.
6. F. L. Dickert, P. Lieberzeit and M. Tortschanoff, *Sens. Actuators, B*, 2000, **65**, 186.
7. K. Hirayama, Y. Sakai, K. Kameoka, K. Noda and R. Naganawa, *Sens. Actuators, B*, 2002, **86**, 20.
8. K. Haupt and K. Mosbach, *Chem. Rev.*, 2000, **100**, 2495.
9. N. Lavignac, C. J. Allender and K. R. Brain, *Anal. Chim. Acta*, 2004, **510**, 139.
10. L. Ye and K. Mosbach, *React. Funct. Polym.*, 2001, **48**, 149.
11. A. Molinelli, R. Weiss and B. Mizaikoff, *J. Agric. Food Chem.*, 2002, **50**, 1804.
12. R. Weiss, A. Molinelli, M. Jakusch and B. Mizaikoff, *Bioseparation*, 2001, **10**, 379.
13. A. Bereczki, A. Tolokan, G. Horvai, V. Horvath, F. Lanza, A. J. Hall and B. Sellergren, *J. Chromatogr. A*, 2001, **930**, 31.
14. C. C. Hwang and W. C. Lee, *J. Chromatogr. B*, 2001, **765**, 45.
15. S. Peter, L. Schweitz and S. Nilsson, *Electrophoresis*, 2003, **24**, 3892.

16. D. Cunliffe, A. Kirby and C. Alexander, *Adv. Drug Delivery Rev.*, 2005, **57**, 1836.
17. F. Puoci, G. Cirillo, M. Curcio, O. I. Parisi, F. Iemma and N. Picci, *Expert Opin. Drug Delivery*, 2011, **8**, 1379.
18. R. Suedee, *Int. J. Pharm. Sci. Rev. Res.*, 2013, **20**, 235.
19. H. Yan and K. H. Row, *Int. J. Mol. Sci.*, 2006, 7, 155.
20. G. Wulff, A. Sharhan and K. Zabrocki, *Tetrahedron Lett.*, 1973, **14**, 4329.
21. G. Wulff, R. Vesper, E. Grobe and A. Sarhan, *Makromol. Chem.*, 1977, **178**, 2799.
22. L. Chen, S. Xu and J. Li, *Chem. Soc. Rev.*, 2011, **40**, 2922.
23. K. J. Shea and E. A. Thompson, *J. Org. Chem.*, 1978, **43**, 4253.
24. C. Alexander, H. S. Andersson, L. I. Andersson, R. J. Ansell, N. Kirsch, I. A. Nicholls, J. O'Mahony and M. J. Whitcombe, *J. Mol. Recognit.*, 2006, **19**, 106.
25. J. Svenson, J. G. Karlsson and I. A. Nicholls, *J. Chromatogr. A*, 2004, **1024**, 39.
26. B. Ekberg and K. Mosbach, *Trends Biotechnol.*, 1989, 7, 92.
27. G. Vlatakis, L. I. Andersson, R. Müller and K. Mosbach, *Nature*, 1993, **361**, 645.
28. L. Andersson, B. Sellergren and K. Mosbach, *Tetrahedron Lett.*, 1984, **25**, 5211.
29. B. Sellergren, B. Ekberg and K. Mosbach, *J. Chromatogr.*, 1985, **347**, 1.
30. I. R. Dunkin, J. Lenfeld and D. C. Sherrington, *Polymer*, 1993, **34**, 77.
31. W. Chen, F. Liu, K. A. Li, Y. H. Yang and S. Y. Tong, *Anal. Lett.*, 2000, **33**, 809.
32. J. Zhou, X. W. He and Y. J. Li, *Anal. Commun.*, 1999, **36**, 243.
33. T. Takagishi, A. Hayashi and N. Kuroki, *J. Polym. Sci., Polym. Chem. Ed.*, 1982, **20**, 1533.
34. K. Sreenivasan, *Talanta*, 1997, **44**, 1137.
35. B. Sellergren and J. S. Kenneth, *J. Chromatogr.*, 1993, **635**, 31.
36. M. J. Whitcombe, M. E. Rodriguez, P. Villar and E. Vulfson, *J. Am. Chem. Soc.*, 1995, **117**, 7105.
37. A. R. Koohpaei, S. J. Shahtaheri, M. R. Ganjali, A. Rahimi Forushani and F. Golbabaei, *Talanta*, 2008, 75, 978.
38. H. T. Zhang, T. Song, F. L. Zong, T. C. Chen and C. P. Pan, *Int. J. Mol. Sci.*, 2008, **9**, 98.
39. N. Pérez, M. J. Whitcombe and E. N. Vulfson, *J. Appl. Polym. Sci.*, 2000, 77, 1851.
40. H. Asanuma, T. Akiyama, K. Kajiya, T. Hishiya and M. Komiyama, *Anal. Chim. Acta*, 2001, **435**, 25.
41. N. Sallacan, M. Zayats, T. Bourenko, A. B. Kharitonov and I. Willner, *Anal. Chem.*, 2002, **74**, 702.
42. J. Damen and D. C. Neckers, *J. Org. Chem.*, 1980, **45**, 1382.
43. K. Hirayama, Y. Sakai, K. Kameoka, K. Noda and R. Naganawa, *Sens. Actuators, B*, 2002, **86**, 20.

44. S. A. Piletsky, H. Matuschewski, U. Schedler, A. Wilpert, E. V. Piletska, T. A. Thiele and M. Ulbricht, *Macromolecules*, 2000, **33**, 3092.
45. S. Striegler, *Bioseparation*, 2002, **10**, 307.
46. B. Sellergren and K. J. Shea, *J. Chromatogr.*, 1993, **635**, 31.
47. T. Jing, X. Gao, P. Wang, Y. Wang, Y. Lin, X. Hu, Q. Hao, Y. Zhou and S. Mei, *Anal. Bioanal. Chem.*, 2009, **393**, 2009.
48. C. J. Tan and Y. W. Tong, *Anal. Chem.*, 2007, **79**, 299.
49. K. Hoshina, S. Horiyama, H. Matsunaga and J. Haginaka, *J. Chromatogr. A*, 2009, **1216**, 4957.
50. N. Pérez-Moral and A. G. Mayes, *Bioseparation*, 2001, **10**, 287.
51. A. G. Mayes and K. Mosbach, *Anal. Chem.*, 1996, **68**, 3769.
52. M. Nakamura, M. Ono, T. Nakajima, Y. Ito, T. Aketo and J. Haginaka, *J. Pharm. Biomed. Anal.*, 2005, **37**, 231.
53. H. G. Schild, *Prog. Polym. Sci.*, 1992, **17**, 163.
54. M. Watanabe, T. Akahoshi, Y. Tabata and D. Nakayama, *J. Am. Chem. Soc.*, 1998, **120**, 5577.
55. L. Qin, X. He, W. Zhang, W. Li and Y. Zhang, *Anal. Chem.*, 2009, **81**, 7206.
56. J. Aburto and S. L. Borgne, *Macromolecules*, 2004, **37**, 2938.
57. Z. Zhao, Y. Teng, G. Xu, T. Zhang and X. Kan, *Anal. Lett.*, 2013, **46**, 2180.
58. G. Pan, Y. Zhang, X. Guo, C. Li and H. Zhang, *Biosensor. Bioelectron.*, 2010, **26**, 976.
59. S. R. Tonge and B. J. Tighe, *Adv. Drug Delivery Rev.*, 2001, **53**, 109.
60. S. Xu, H. Lu, X. Zheng and L. Chen, *J. Mater. Chem. C*, 2013, **1**, 4406.
61. Y. Kanekiyo, R. Naganawa and H. Tao, *Angew. Chem., Int. Ed.*, 2003, **42**, 3014.
62. C. Wang, A. Javadi, M. Ghaffari and S. Gong, *Biomaterials*, 2010, **31**, 4944.
63. E. Oral and N. A. Peppas, *J. Biomed. Mater. Res., Part A*, 2004, **68A**, 439.
64. W. F. Zhao, B. H. Fang, N. Li, S. Q. Nie, Q. Wei and C. S. Zhao, *J. Appl. Polym. Sci.*, 2009, **113**, 916.
65. G. gong, M. H. Lam and H. Yu, *Adv. Funct. Mater.*, 2006, **16**, 1759.
66. L. Fang, S. Chen, Y. Zhang and H. Zhang, *J. Mater. Chem.*, 2001, **21**, 2320.
67. C. Gong, K. wong and M. H. Lam, *Chem. Mater.*, 2008, **20**, 1353.
68. X. Kan, Z. Geng, Y. Zhao, Z. Wang and J. J. Zhu, *Nanotechnology*, 2009, **20**, 165601.
69. O. I. Parisi, C. Morelli, F. Puoci, C. Saturnino, A. Caruso, D. Sisci, G. Elvi Trombino, N. Picci and M. S. Sinicropi, *J. Mater. Chem. B*, 2014, **2**, 6619.
70. S. Xu, J. Li, X. Song, J. Liu, H. Lu and L. Chen, *Anal. Methods*, 2013, **5**, 124.
71. L. Fang, S. Chen, X. Guo, Y. Zhang and H. Zhang, *Langmuir*, 2012, **28**, 9767.
72. Y. Ma, Y. Zhang, M. Zhao, X. Guo and H. Zhang, *Chem. Commun.*, 2012, **48**, 6217.

CHAPTER 14

Responsive Peptide-based Materials: Stimulation by Chemical and Biological Signals

BRITTA E. I. RAMAKERS,[a] HUI GUO[a,b] AND DENNIS W. P. M. LÖWIK*[a]

[a] Radboud University Nijmegen, Institute for Molecules and Materials, Bio-organic Chemistry, Heyendaalseweg 135, 6525 AJ, Nijmegen, The Netherlands; [b] State Key Laboratory of Molecular Engineering of Polymers, Advanced Materials Laboratory, Department of Macromolecular Science, Fudan University, Shanghai 200433, People's Republic of China
*Email: d.lowik@science.ru.nl

14.1 Introduction

Many biomolecules in nature display stimuli-responsive behaviour often accompanied with a conformational change. And as one of the most important group of biomolecules, proteins are also able to switch fold.[1] The response can be initiated by a series of stimuli, such as pH, ionic strength, redox potential, temperature, light, enzymes and so on, controlling a great number of biological pathways. Peptides, with simpler and synthetically more accessible structures compared to proteins, have been examined extensively as building blocks for stimuli-responsive materials.[2–4] These resulting responsive materials are believed to have potential applications in

RSC Smart Materials No. 14
Chemoresponsive Materials: Stimulation by Chemical and Biological Signals
Edited by Hans-Jörg Schneider

Published by the Royal Society of Chemistry, www.rsc.org

areas such as drug delivery, tissue engineering and bio-sensing; for example, the material may be applied through a syringe, and undergo a transition when triggered by temperature, pH, ionic strength, oxidative species or enzymes at the site of an injury to act as a scaffold for tissue regrowth or, alternatively, to release drugs or genes in response.[5,6]

As mentioned, often, the response of these peptide-based materials is reflected in a conformational transition. Here we will focus on peptide-based materials that can be switched between the main conformational structures of peptides: α-helix, β-sheet and random coil. A transition between these can often be ascribed to a change in intramolecular hydrogen bond interactions that stabilise a particular fold.

Among the types of local structures in proteins and peptides, the α-helix is the most regular and the most predictable from sequence, as well as the most prevalent. Linus Pauling was the first to predict the existence of α-helices.[7] An α-helix is a rod-like structure and the right-handed coil that the polypeptide chain folds into is held together by hydrogen bonds between the backbone carbonyl oxygens and the amide hydrogens.[8] These hydrogen bonds are made between a residue and one situated four residues ahead. The resulting helix makes a complete turn approximately every 3.6 amino acid residues. The stability of the helix is further influenced by the amino acid side chains and the total dipole moment of the entire helix. A tertiary structure that can arise from α-helices is the coiled coil, a structural motif in which α-helices intertwine like the strands of a rope. Peptides that adopt a coiled coil structure usually consist of repeats of seven residues, the so-called heptad repeat.

The residues in this heptad repeat, labelled a–g, are usually hydrophobic at positions a and d (see Figure 14.1) providing most of the stability by forming a hydrophobic interface. Incorporation of charged residues at positions e and g stabilizes the structure further.[9]

Another important secondary structural feature is that of the β-sheet; this is often a key structural motif in naturally occurring proteins and peptides.[8] In β-sheets the peptide chain is found in an almost extended conformation forming hydrogen bonds with neighbouring strands through their backbone giving rise to a so-called β-pleated sheet. Typically, the side chains of subsequent residues in the amino acid chain are positioned on opposite sides. The formation of β-sheets can give rise to protein aggregates and fibrils, which have been associated with diseases such as Alzheimer's but also protein-based materials such as silk owe their strength to it. When there are just two short connected anti-parallel β-strands, the secondary structure motif is called a β-hairpin.[8] The loop between the two β-strands is called an α, β or γ-turn depending on the number of amino acids in the turn. Such turns play an important role in many protein–protein recognition events.

In general, small peptides do not have the tendency to form secondary structures. One of the ways to design small peptides such that they form a secondary structure is by introducing amphiphilicity. The resulting so-called peptide amphiphiles are peptide-based molecules that have the tendency to

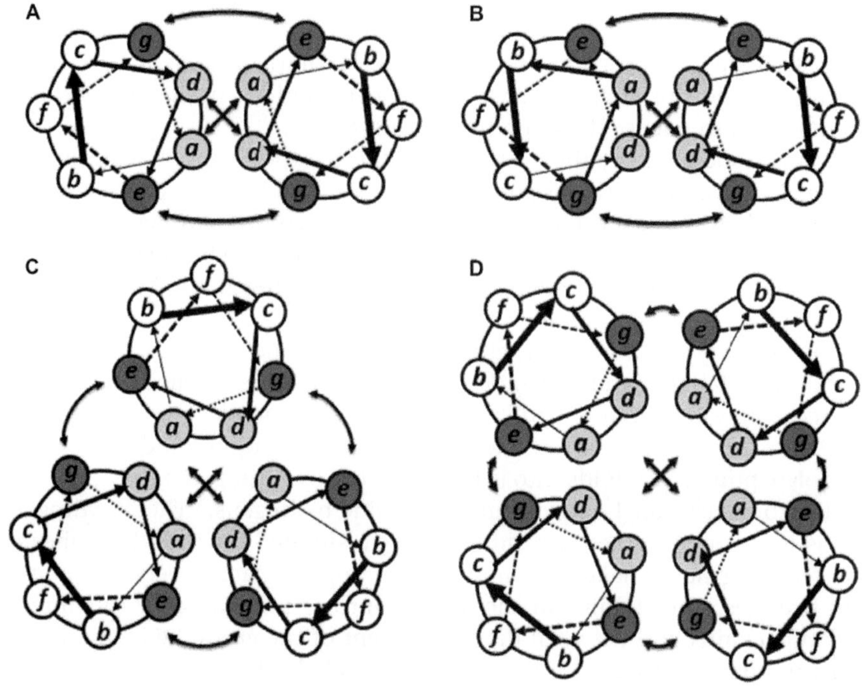

Figure 14.1 Helical wheel diagrams of a variety of coiled coil assemblies: (A) parallel dimeric coiled coil, (B) antiparallel dimeric coiled coil, (C) parallel trimeric coiled coil, (D) parallel tetrameric coiled coil. The curved arrows indicate salt bridges and the crossed arrows indicate the hydrophobic interactions.
Reproduced from ref. 9 with permission from the Royal Society of Chemistry.

self-assemble.[10–12] They have drawn a lot of attention over the past decade because of their ability to self-assemble into a range of nanostructures, such as nano-fibres, nanotubes, ribbons, and vesicles, controlled by the interplay between the hydrophobic forces on the one hand and the polar and hydrogen bonding interactions on the other. Three classes of amphiphilic peptides can be identified: (i) peptides that consist of both polar and non-polar residues, giving rise to both hydrophobic and hydrophilic properties, (ii) hydrophilic peptides attached to hydrophobic lipid alkyl chains and (iii) peptide-based copolymers.[13,14]

The topic of switchable peptide-based materials has been reviewed quite extensively over the past years.[2–5,9,13,15–18] Here, we will mainly focus on the most important and most recent progress of peptide-based stimuli-responsive materials. Furthermore, we will not discuss peptide–polymer hybrids; many examples of these switchable systems have also been reported.[13,14] Instead, we will mainly focus on lower molecular weight compounds. Moreover, while the switchable behaviour of the peptides can be under the control of many factors, in the following sections, we will only discuss stimulation by chemical (pH, ions and reduction–oxidation) and biological signals.

14.2 pH-Induced Changes

pH, as an easily controllable stimulus, plays a major role in the assembly process of most peptide-based materials. Many peptides can respond to switches in pH due to protonation or deprotonation of basic and acidic amino acids. In nature, different organs, tissues, subcellular compartments, and pathophysiological conditions have different pH levels and gradients, providing an opportunity to utilize the pH-responsive behaviour in medical applications.[5]

14.2.1 α-Helix

Amino acids with an ionisable side chain would be the most obvious choice to be employed in designing pH-switchable peptides. Hence, amino acids such as Lys, Glu and His are most typically used in designing pH-responsive peptides.[19–22]

Conticello's group[23] reported the design and characterization of a 41-residue peptide containing three histidines, TZ1H (Figure 14.2). This peptide was based on a trimeric coiled coil motif that is able to reversibly assemble into long-aspect ratio helical fibrils. The protonation and deprotonation of the histidines was believed to lead to a conformational transition which in turn was associated with the self-assembly of the peptide. Below pH 5.8, the peptides existed in a random coil, which upon increasing the

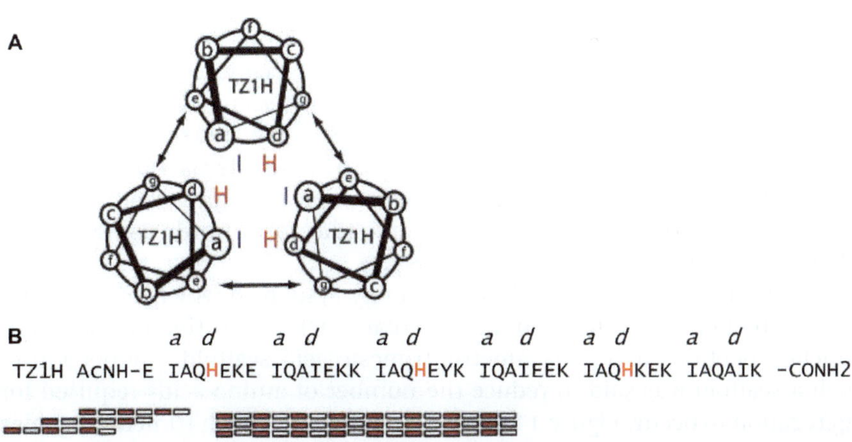

Figure 14.2 (A) Helical wheel diagram corresponding to a cross section of the trimeric bundle resulting from self-assembly of TZ1H into a helical fibril. (B) Amino acid sequence of TZ1H.

pH changed conformation and associated into three-stranded helical bundles. This transition was reversible. The authors claim that 'this peptide engineering strategy represents a promising approach to the design of intelligent materials that combine well-defined structures with environmentally responsive self-assembly'.

Peptide-based hydrogels, with biodegradability, biocompatibility and cell-matrix recognition, have many advantages in biomedical applications. Fletcher *et al.* reported a 21-residue α-helical peptide with the sequence LKELAKVLHELAKLVSEALHA,[24] to form self-healing hydrogels at weight fractions below 0.1%. The peptide was designed such that, depending on the pH and thus charge of the peptide, the helices associated either in-register to form structures of defined size or in an offset fashion to give fibrils of indefinite length. It was found that highly-charged peptides gave low-viscosity solutions. When the charge of the peptide was reduced close to either +1 or −1 they formed thermo-stable gels. However, in a neutralised state, the peptide formed insoluble fibrils. All the transitions between the various types of aggregation were shown to be reversible by a simple addition of acid or base to change pH.

14.2.2 β-Sheet

β-Sheet motifs have also been studied extensively as the basis for pH-triggered peptide switches. The β-sheets usually give rise to fibrous assemblies that, due to the presence of interactions between these fibres, can form crosslinks, which can result in the formation of a three-dimensional structured hydrogel.[25] These β-sheet based peptide hydrogels have found applications such as pH-triggered drug delivery systems or self-healing materials.[26]

In pioneering work by Aggeli *et al.*,[27–29] it was shown how responsive gels could be obtained by tailoring their peptides such that interactions between the peptides was mediated by attractive forces between side-chains to give lateral interactions between adjacent β-sheets. They stated that a precise control over assembly at the molecular level will provide the opportunity to gain control over the mesoscopic and macroscopic structures that form. And further, they also claim that the structures that form fibrils are stabilised when the net charge is around +1 or −1.

The group of Besenius developed a binary system in which cationic and anionic monomers formed supramolecular stacks.[30] To this purpose they attached peptides to a C_3-symmetric trimesic acid scaffold. Connection to such a scaffold was said to reduce the number of amino acids required for aggregation to occur. Figure 14.3 shows how a pair of such trimeric peptides can form supramolecular polymers when mixed in a 1:1 ratio. The structures are stabilised by a combination of Coulomb attractive forces and hydrogen-bonding interactions. Depending on the amino acids incorporated, the pH window in which assembly occurred could be changed from between 2 and 12 to 4 and 8. Outside this pH window, disassembly was

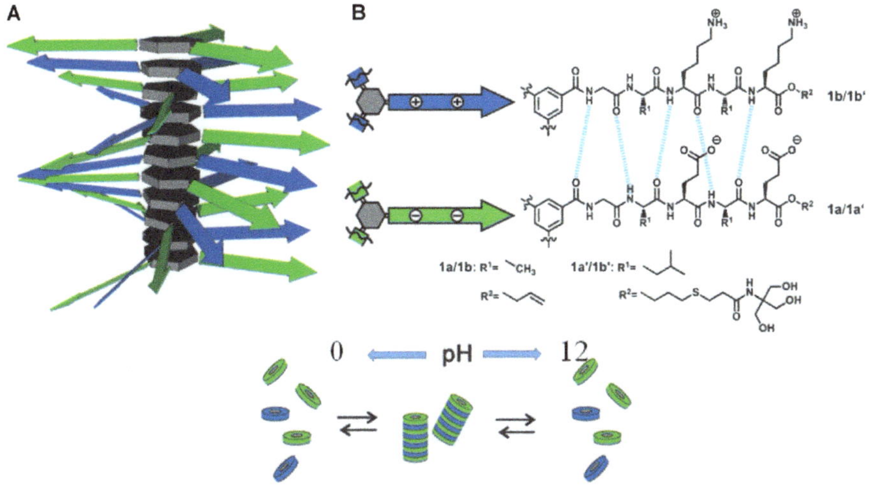

Figure 14.3 (A) Schematic representation of the supramolecular copolymer and (B) the anionic/cationic β-sheet-encoded dendritic co-monomers that assemble only in a particular pH window.
Reprinted with permission from ref. 30.

observed through a rapid depolymerisation due to loss of salt bridges between the peptides.

A well-known peptide, derived from the amyloid β-peptide (Aβ), to form β-sheet structures is the so-called Tjernberg peptide with the sequence KLVFF.[31] Kumaraswamy *et al.* showed that this peptide can spontaneously form a variety of aggregates controlled by pH leading to a stacking of micelles. As expected, these structures were rich in β-sheets (Figure 14.4).[32] At low pH, repulsive forces between the peptides prevented higher order structures from forming. This only happened at neutral or basic pH promoted by attractive longitudinal and lateral stacking interactions. Moreover, the authors show that the Tjernberg peptide might be employed to block the formation of amyloid fibrils in an *in vitro* study. It was shown that cells could be protected against the stress of Aβ aggregates at concentrations between 100 and 500 μM.

14.2.3 Peptide Amphiphiles

A great variety of pH-responsive peptide amphiphiles (PAs) has been developed over the last two decades. Pioneering research on this topic was performed by the Stupp group. Throughout the years, they have developed a series of peptide amphiphiles constructed from designed peptide sequences functionalised with lipid alkyl chains. In their initial ground-breaking work, they demonstrated how such lipidated peptides could be made to self-assemble reversibly into nano-fibre networks. The formation of the resulting aqueous gels could be controlled through changes in pH (Figure 14.5).[33,34]

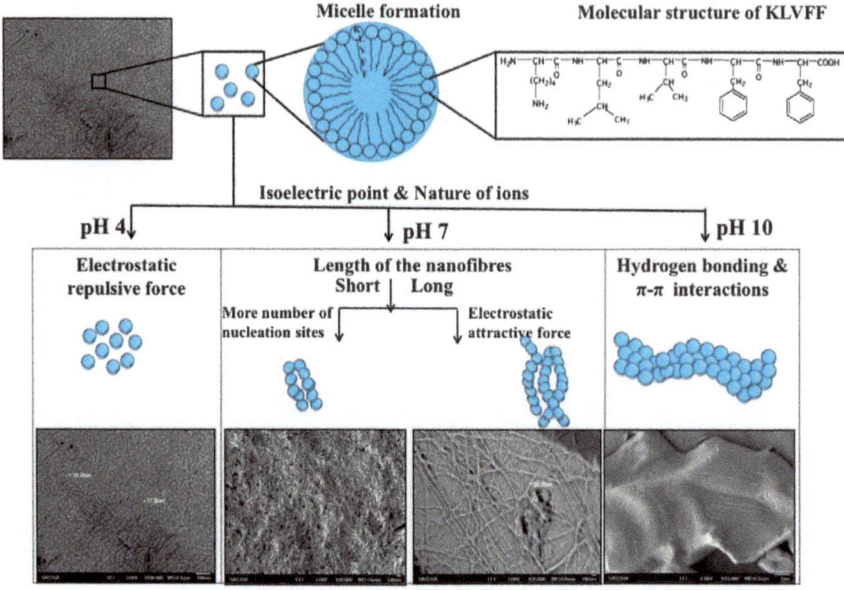

Figure 14.4 Micelle stacking mechanism for the formation of higher order structures. Reproduced from ref. 32 with permission from the Royal Society of Chemistry.

At high pH, the peptides were in a monomeric state, while at lower pH, they self-assembled into cylindrical micelles. The incorporation of a few cysteine residues allowed the group to polymerise the assembled structures by means of oxidation. This stabilised the assemblies such that raising the pH did not lead to disassembly.

An example of the group's more recent research on this topic[35] showed how these peptide amphiphiles can be employed to prepare aligned monodomain gels. It was found that thermal treatment (heating to 80 °C) of an aqueous solution of peptide amphiphiles with the sequence $V_3A_3E_3$ connected to a palmitoyl chain with its N-terminus gave gels up to fourfold stiffer upon adding $CaCl_2$ or lowering the pH than without such treatment. Materials with a noodle-like texture were obtained and these could be easily aligned by simply drawing a pipette with the PA solution into a $CaCl_2$ or low pH solution. This allowed mixtures of these aligned assemblies to be prepared with cells which consequently also got aligned. It is expected that such materials might have potential for the *in situ* construction of 'cellular wires' that can be employed, for example, to study processes in which directional cell migration and growth is important.

Another group of peptide amphiphiles is constituted by peptides that consist of both polar and nonpolar residues, either in blocks or alternating in sequence, introducing the hydrophobic and hydrophilic domains. For example, Uesaka *et al.* prepared peptides with a hydrophobic $(\text{Leu-Aib})_6$ peptide tail connected to three hydrophilic sarcosine (*N*-methyl glycine)

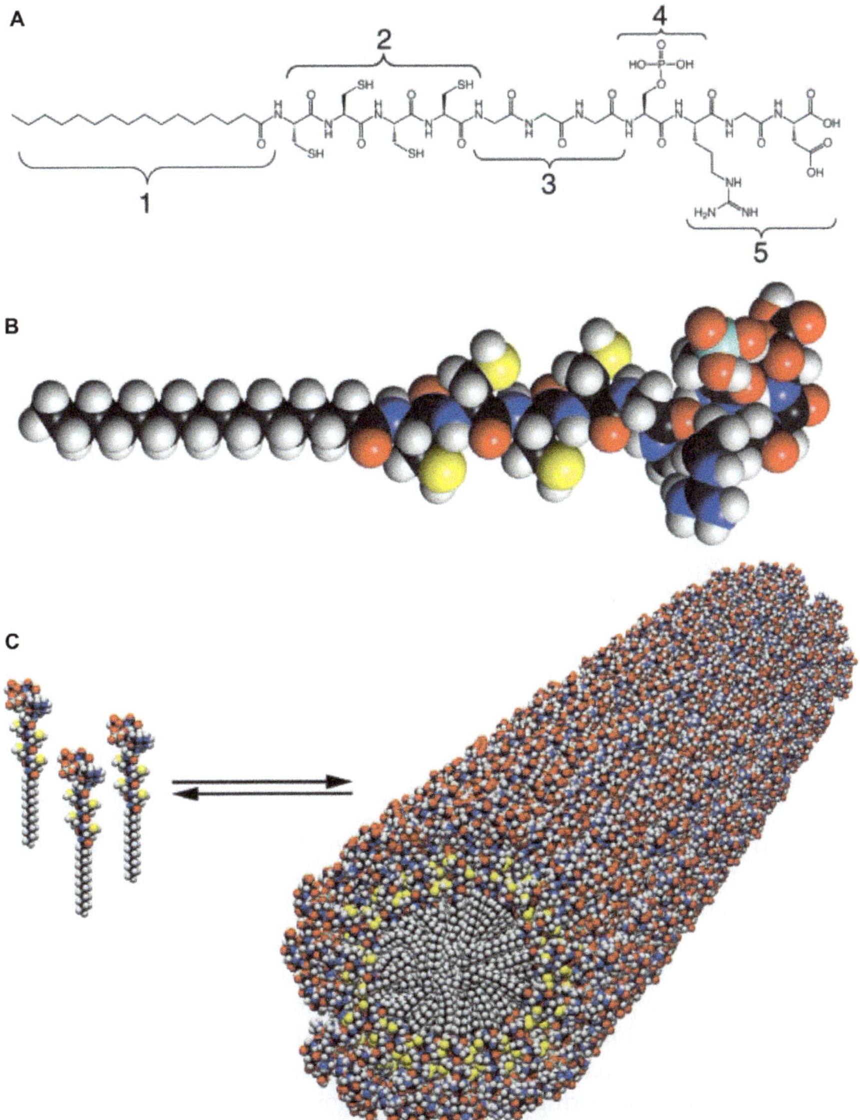

Figure 14.5 (A) Chemical structure and (B) molecular model of the (C) PA assembling reported by Hartgerink *et al.*
Reprinted with permission from ref. 34.

blocks.[36] The expected pH dependence of these PAs was observed using electron microscopy and was correlated with the number of histidine residues (1 or 2) that connected the peptide parts. The assemblies that were observed ranged in morphology from a twisted ribbon, a helical ribbon, to a nanotube (Figure 14.6). The assemblies of the PA with two histidine residues changed from a twisted ribbon at pH 3.0, to a helical ribbon at pH 5.0, up to

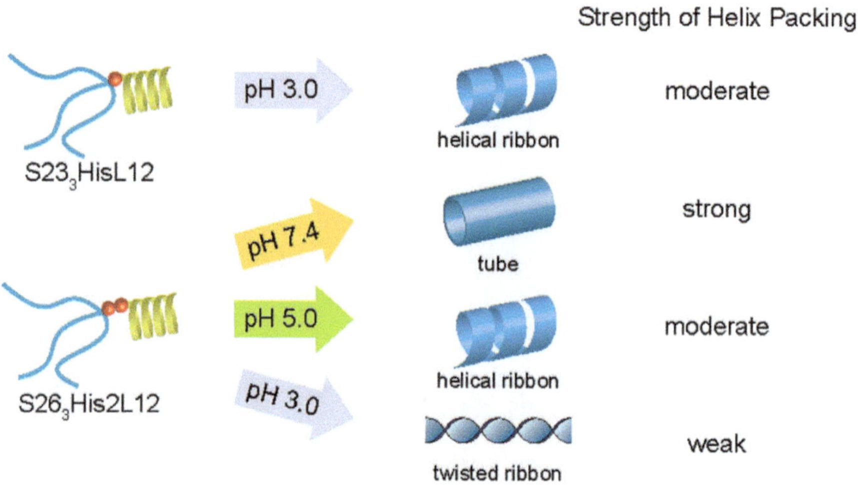

Figure 14.6 pH-Responsive assembly of peptides of the form A_3B as described by Uesaka *et al.*
Reprinted with permission from ref. 36. Copyright (2014) American Chemical Society.

a nanotube at pH 7.4. In contrast, the other PA with only one histidine residue could only change from helical ribbons to planar sheets within the same pH range.

14.3 Ion-induced Changes

Similar to pH, ions other than protons can also play a crucial role in peptide assembly. In nature, interactions of metal ions with residues such as histidine, methionine, cysteine, aspartic acid, and glutamic acid can induce conformational changes in proteins to result in a variety of biological responses but can also initiate the formation of supramolecular structures. Inspired by this, peptides have been developed to take advantage of these naturally occurring interactions that enhance structural stability and promote self-assembly upon binding of a variety of metal cations.[37] In addition to these naturally occurring amino acids, non-natural metal-ligand containing residues have been synthesized that allow triggering of higher order assembly.[18]

14.3.1 α-Helix

The group of Conticello reported the synthesis of a peptide named TZ1C2.[38] It is a modified version of an peptide sequence reported earlier called TZ1 that was shown to form trimeric coiled coils, which in itself was shown to be responsive to silver(I) cations.[39] In the 41-residue TZ1C2 sequence, two cysteine residues replaced two isoleucines thus introducing ligands that might coordinate metal ions when positioned in a proper orientation. It was shown that TZ1C2 could adopt two distinct orientations dictated by the presence or absence of cadmium ions. As shown in Figure 14.7, the binding of cadmium ions resulted in a so-called registry shift of the heptad repeats. The alignment of the helices in the presence of Cd^{2+} changes such that instead of well-defined trimeric coiled coils, long fibrillar structures form. It was shown that this transformation was reversible by the addition of EDTA scavenging the cadmium cations. This system can be compared with the reversible coiled coil based systems reported earlier by the Woolfson group, which has developed a variety of peptides with switchable secondary structures.[40,41]

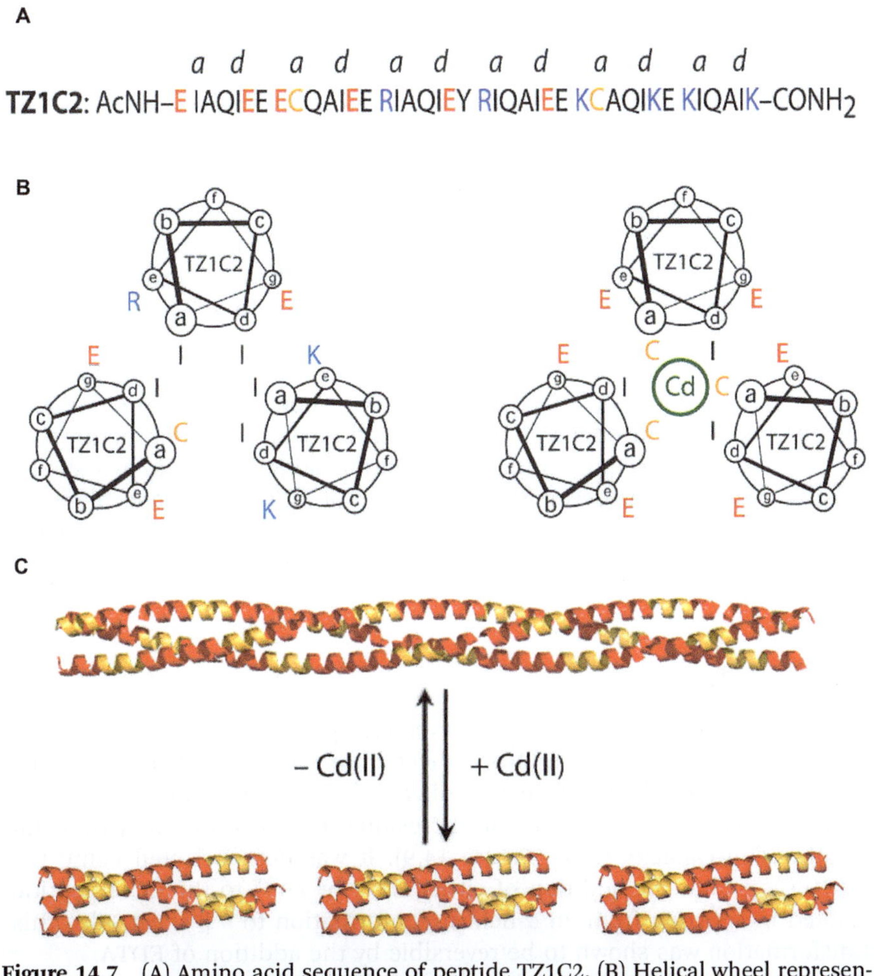

Figure 14.7 (A) Amino acid sequence of peptide TZ1C2. (B) Helical wheel representation of TZ1C2 in the staggered (left) and aligned (right) orientations. (C) Schematic representation of the registry shift that results from Cd(II) coordination to the cysteine residues of TZ1C2.
Reprinted with permission from ref. 38. Copyright (2013) American Chemical Society.

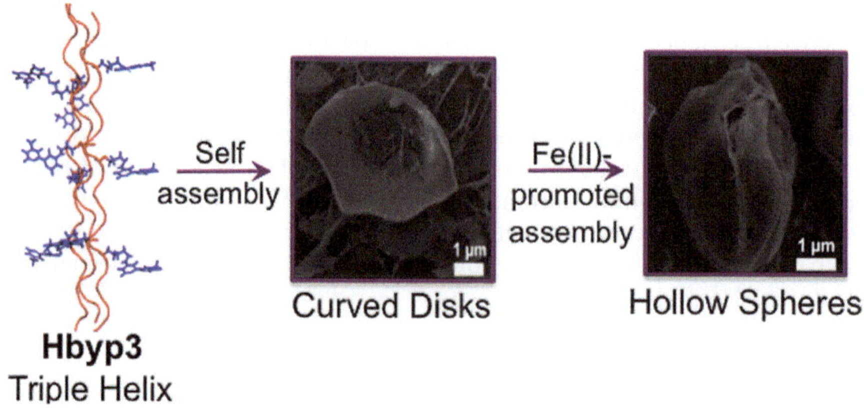

Figure 14.8 Metal ion-induced morphology change of assemblies of Hbyp3. Reprinted with permission from ref. 42. Copyright (2013) American Chemical Society.

The Chmielewski group took inspiration from natural collagen to design a 27-residue peptide, named Hbyp3, that in addition to a number of Pro-Hyp-Gly collagen type repeats, also contained three bipyridine functionalised residues.[42] This Hbyp3 peptide was shown to self-assemble into curved disks of micrometre size. The addition of metal ions such as Fe(II), Co(II), Zn(II) and Cu(II) resulted in coordination of the metal ions, initiating further assembly of the disks into hollow spheres (Figure 14.8). SAXS measurements on the spheres revealed that the triple helical structure was preserved during the morphological transition. Finally, it was also shown that this transition was reversible by addition of EDTA to chelate the metal ions to give back disk-like structures. The authors suggest that these spheres possess the potential to encapsulate materials comparable to viral capsid based strategies.

14.3.2 β-Sheet

The group of Koksch[43] reported a reversible peptide secondary structure switch between an α-helix and a β-sheet induced by metal ions. To this purpose they started with a weakly coiled coil forming peptide CC and replaced four amino acids with histidine residues to give a peptide with metal complexation capacity CCM (Figure 14.9). It was demonstrated using CD spectroscopy that the addition of either Cu^{2+} or Zn^{2+} to the CCM peptide resulted in a switch from an α-helical conformation to a β-sheet. Also this transformation was shown to be reversible by the addition of EDTA.

Also Hoernke *et al.* designed a number of coiled coil forming peptides that could switch from an α-helical state to a β-sheet conformation.[44] This propensity was introduced by the incorporation of three valine residues. Moreover, metal binding histidine residues were introduced at either i, $i+1$ and i, $i+7$ or and i, $i+4$ positions. These peptides were studied at the

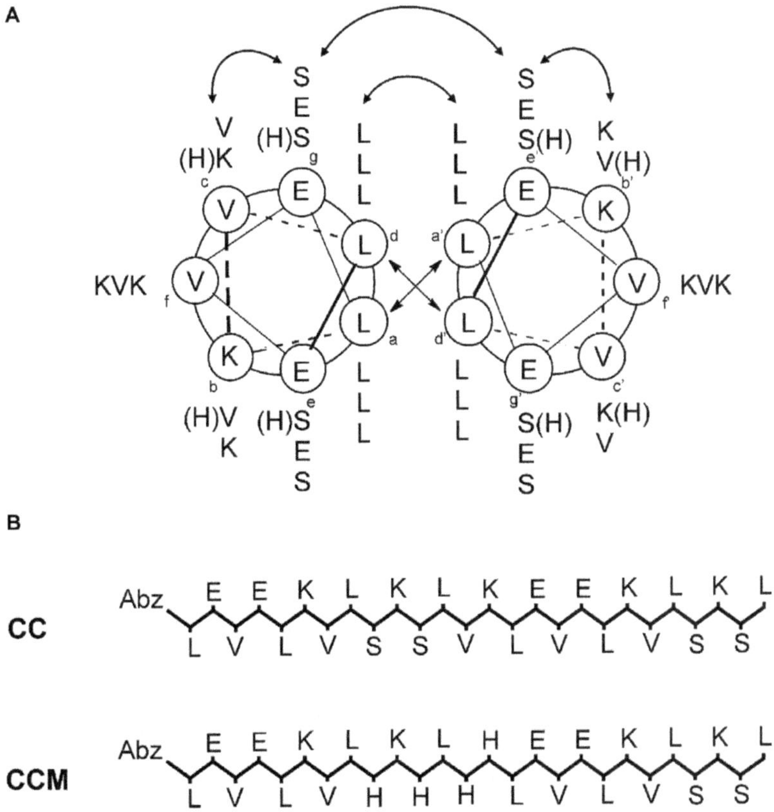

Figure 14.9 (A) Helical wheel representation and (B) primary sequences of coiled coil peptide CC and a metal complexing analogue CCM and their amino acid sequences.
Reproduced from ref. 43 with permission from the Royal Society of Chemistry.

air–water interface of a Langmuir trough. Once at the interface, the peptide was found to adopt an α-helical structure. Depending on the peptide sequence, the conformation could be influenced by the addition of metal ions such Cu^{2+} and Zn^{2+} initiating a transformation into a β-sheet. Further, it was found that the interaction with the interface was predominant and that metal ions were able to prevent aggregation in bulk but not at the interface.

Instead of inducing assembly with metal ions, it is also possible that the ions prevent aggregation. For example, Mold *et al.* demonstrated that copper ions could prevent amyloid-β_{1-42} from forming amyloid fibrils under near-physiological conditions *in vitro*.[45] Using fluorimetry and transmission electron microscopy, it was shown that Cu^{2+} was able to prevent the assembly of the Aβ peptide into β-sheet rich amyloid fibrils. Metal to peptide ratios of 1 : 10 up to 10 : 1 were evaluated and it was found that copper ions consistently suppressed β-sheet formation.

14.3.3 Peptide Amphiphiles

Stoichiometric amounts of salts have also been shown to have a marked effect on the structure of cationic amphiphilic peptides such as Ala_9Lys and Ile_3Lys.[46] A series of different salts were added to these peptides and it was found that the addition of ClO^{3-}, NO^{3-} and Br^- prevented the formation of elongated nanorods. Instead, addition of these sodium salts resulted in the formation of nanostacks, nanospheres and short nanorods (Table 14.1). In contrast to this, the addition of anions such as Cl^-, $SO_4{}^{2-}$, $HPO_4{}^{2-}$, $PO_4{}^{3-}$, and $C_6H_5O_7{}^{3-}$ led to extended axial nanofibrils being formed. The authors also found that these anions increased the diameter, surface roughness and inclination for nanostructuring of these peptides. In keeping with the Hofmeister anion sequence, the anions can be ordered from $ClO^{3-} < NO^{3-} \leq Br^- < Cl^- < SO_4{}^{2-} < HPO_4{}^{2-} < PO_4{}^{3-} < C_6H_5O_7{}^{3-}$ starting with the least efficient ion for promoting nanofibril growth. The exact mechanisms behind the different assembly processes remain unclear; however, it is likely that the specific ion effects are due to complex interactions between electrostatic and non-electrostatic forces.

The Schneider group reported the design of a metal-responsive peptide-based hydrogelation system triggered by zinc ion binding (Figure 14.10).[47] In an aqueous environment and in the absence of metal ions, the peptide was unfolded and soluble. The addition of Zn^{2+} ions resulted in chelation of the metal to a negatively charged, non-natural amino acid in the peptide, which in turn triggered peptide folding and subsequent self-assembly into a β-sheet-rich fibrillar hydrogel.

Azuma *et al.* have also studied metal induced switching by means of introducing non-natural amino acids.[48] They provide a convenient synthesis of dipicolylamine-containing peptides and show how these peptides respond

Table 14.1 Heights and lengths of the nanostructures formed in the solutions of Ala_9Lys containing sodium salts with different anions. Reprinted with permission from ref. 46. Copyright (2011) American Chemical Society.

	Nanostructure	Height (nm)	Length (nm)
A_9K	Spheres	3.2 ± 0.6	
	Rods	3.0 ± 0.5	<100
$A_9K/NaClO_3$	Spheres	3.7 ± 0.5	
$A_9K/NaNO_3$	Spheres	4.0 ± 0.6	
$A_9K/NaBr$	Spheres	3.8 ± 0.8	
$A_9K/NaCl$	Spheres	4.3 ± 0.7	
	Entangled fibrils	3.8 ± 0.5	>500
A_9K/Na_2SO_4	Spheres	4.5 ± 0.8	
	Entangled fibrils	4.0 ± 0.5	>500
A_9K/Na_2HPO_4	Spheres	5.0 ± 0.5	
	Entangled fibrils	4.0 ± 1.0	>500
A_9K/Na_3PO_4	Spheres		
	Closely packed fibrils	5.5 ± 0.7	>500
$A_9K/Na_3C_6H_5O_7$	Spheres		
	Closely packed fibril bundles		>500

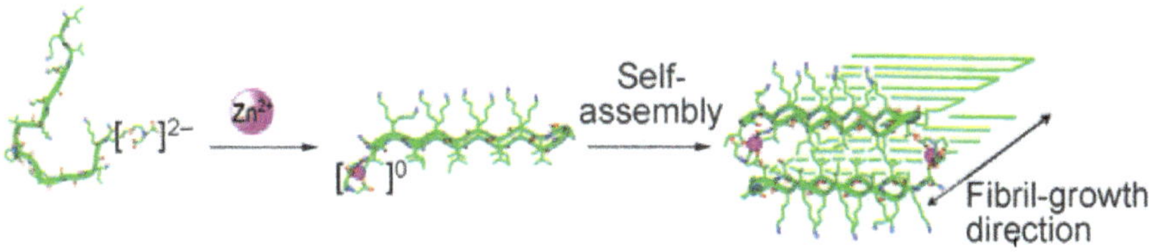

Figure 14.10 A schematic of metal-triggered folding and self-assembly of a zinc binding peptide reported by Micklitsch *et al.*
Reprinted with permission from ref. 47.

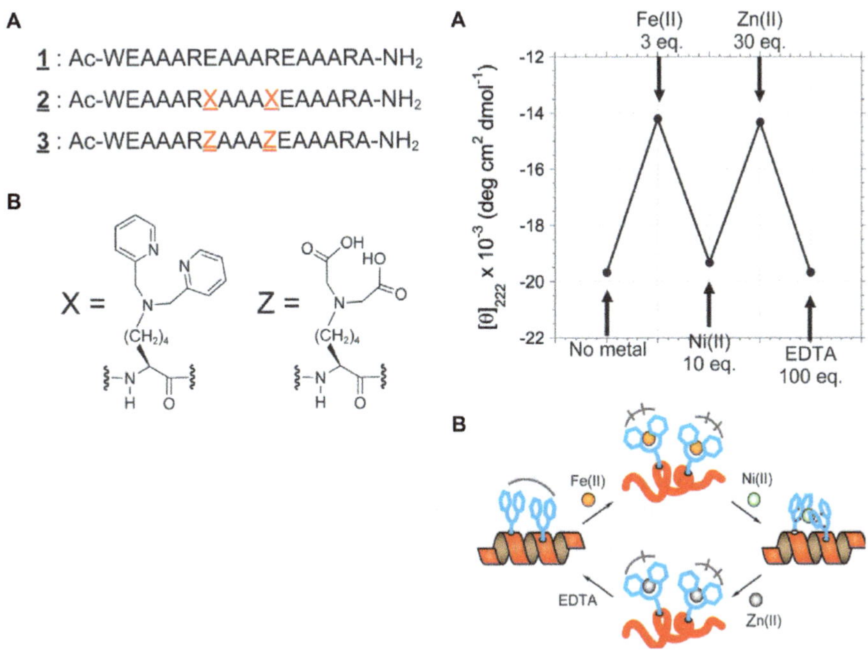

Figure 14.11 (Left A + B) Structures of dipicolyl and diacetic acid containing peptides that (right A + B) can be conformationally switched.
Reproduced from ref. 48 with permission from the Royal Society of Chemistry.

to the addition of metal ions (Figure 14.11). The authors found, using circular dichroism spectrometry and isothermal titration calorimetry, that Fe(II) and Cu(II) gave rise to the formation of a complex of dipicolylamine containing 1 : 1 peptide to metal ion. In contrast to this, Ni(II) gave cross-linked structures by forming a 2 : 1 complex of peptide to metal ion. In addition, the authors showed that reversible switching between the peptide structures was possible through the careful selection of metal ions.

Lu *et al.* reported the development of two peptide-based chemosensors derived from a known β-hairpin forming 12-residue peptide HP7.[49] The peptide included histidine residues for metal binding and tryptophan residues as reporter moieties. The authors examined the effect of a variety of metal ions and found that Cu(II) and Hg(II) had the most pronounced fluorescence quenching. They determined that these peptides had a 1 micromolar detection limit for both copper and mercury ions and that the dissociation constant for Cu(II) was smaller than for Hg(II) meaning that the peptides had a larger affinity for copper ions. Moreover, CD spectroscopy showed that both these ions have a distinctive effect on the secondary structure of the peptide, thus providing an additional means of distinguishing between these two ions.

14.4 Redox-induced Changes

Only a limited number of studies have concentrated on redox (reduction–oxidation) sensitive peptides. Considering their potential application as controlled drug carriers in oxidative stress environments, the investigation of polypeptide materials that undergo a change in conformation or solubility upon oxidation is important. Typically, disulfide linkages are incorporated into the main chain and side chain of the peptides, as a cross-linker or for the attachment of functional moieties. Redox potentials can be utilized for very precise and well controlled intracellular delivery of therapeutics.

14.4.1 α-Helix

Kramer and Deming reported the preparation of sugar-functionalized poly-L-cysteine, which can be oxidized using hydrogen peroxide to induce a helix-to-coil transition without loss of water solubility.[50] They also synthesised block co-polypeptides containing poly(L-leucine) as the hydrophobic segment and either galactosylated poly-L-cysteine sulfone (poly(α-gal-C^{O2})) or galactosylated poly-L-lysine (poly(α-gal-K)) as the hydrophilic segment.[51] The poly-(α-gal-C^{O2}) segment is known to be disordered in water whereas the poly-(α-gal-K) segment is known to be an α-helix. Likewise, the block co-polypeptides containing these hydrophilic segments adopt different morphologies, with the disordered, poly-(α-gal-C^{O2}) containing block co-polypeptides forming vesicles in which the sugar groups are on the surface and so can bind biological targets.

Another interesting example of a peptide that undergoes a redox induced conformational change is reported by Wang *et al.*[52] They adapted the temperature switchable peptide with the sequence SIRELEAKIRELELRIG described by Kammerer and Steinmetz[53,54] so that is was able to bind Cu^{2+} ions on one face of the β-sheet. Subsequently, they were able to switch reversibly between an α-helical monomer and a Cu^{2+}-bound β-sheet aggregate, by reduction and oxidation of the copper ions (Figure 14.12). The Cu^2 induced β-sheet aggregation of this peptide could provide an interesting model to study prion assembly, in which Cu^{2+} is thought to coordinate to histidine residues.

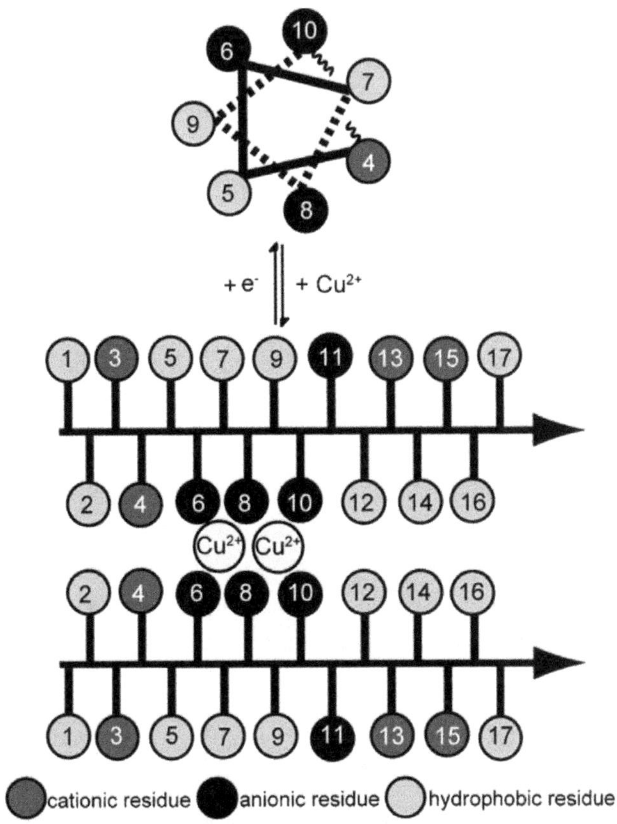

Figure 14.12 A schematic of how an α-helical peptide converts into a β-sheet assembly upon addition of Cu(II). Reduction of the metal ion causes reversal of the protein conformation.
Reprinted with permission from ref. 52.

14.4.2 β-Sheet

Oxidation-responsive PEG-based polypeptide micelles that disassemble upon the addition of an oxidising substance such as H_2O_2 have also been reported (Figure 14.13).[55] These micelles were assembled from PEG_{45}-*b*-poly(L-EG_2MA-C)$_{22}$ block copolymers and were subject to oxidation-induced disassembly due to the transformation of the thioethers to sulfones, which are more polar. Furthermore, the authors showed that the *in vitro* release of the model drug DOX could be triggered through the addition of an oxidant.

Miao *et al.* prepared selenium-containing peptides that could undergo reversible self-assembly.[56] This self-assembly process could be controlled using biocompatible redox triggers and they were also able to form vitamin C induced hydrogels.

Bowerman and Nilsson designed a peptide, C(FKFE)$_2$CG, that could be cyclised through the formation of a disulfide bond, which consequently

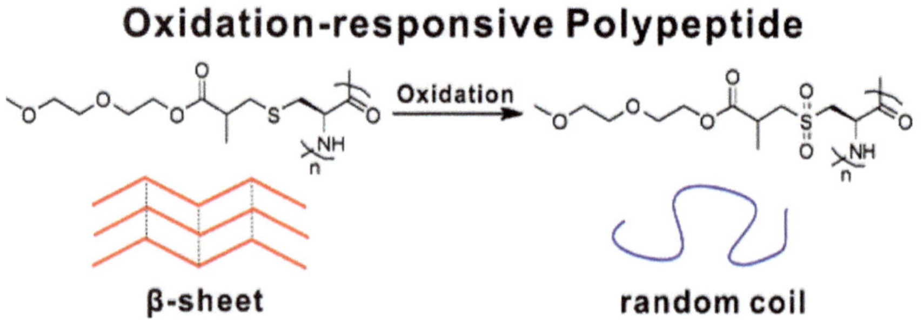

Figure 14.13 Oxidation-responsive polypeptide.
Reprinted with permission from ref. 55. Copyright (2014) American Chemical Society.

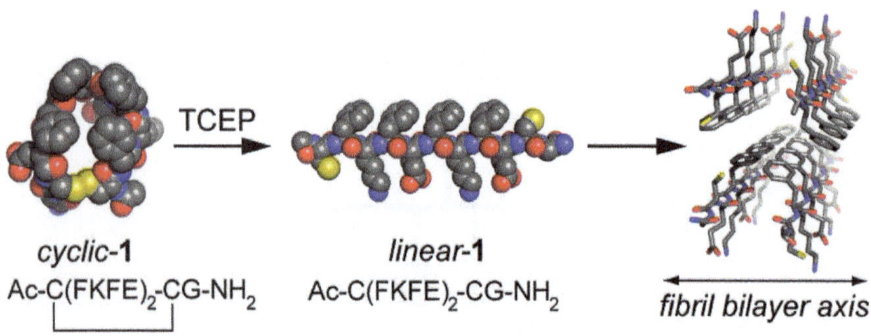

Figure 14.14 Cyclic to linear peptide conformational switch using a reductive trigger.
Reprinted with permission from ref. 57. Copyright (2010) American Chemical Society.

prevented the peptide from adopting a β-sheet conformation.[57] The reduction of this disulfide bond induced a conformational change that resulted in the formation of fibrillar structures (Figure 14.14). At high peptide concentrations, this self-assembly process resulted in the formation of rigid viscoelastic hydrogels.

14.4.3 Peptide Amphiphiles

Cysteine residues are often incorporated into redox-responsive peptide amphiphiles. An excellent example includes work by Hartgerink *et al.* in which cysteine-containing peptides were functionalised with an alkyl tail.[33] These molecules were shown to self-assemble upon acidification, giving rise to nano-fibres, and disassembled at neutral or basic pH. It was found that the oxidation of these molecules prevented their assembly at acidic pH, whereas the oxidation of the self-assembled fibres made them more pH stable (Figure 14.15).

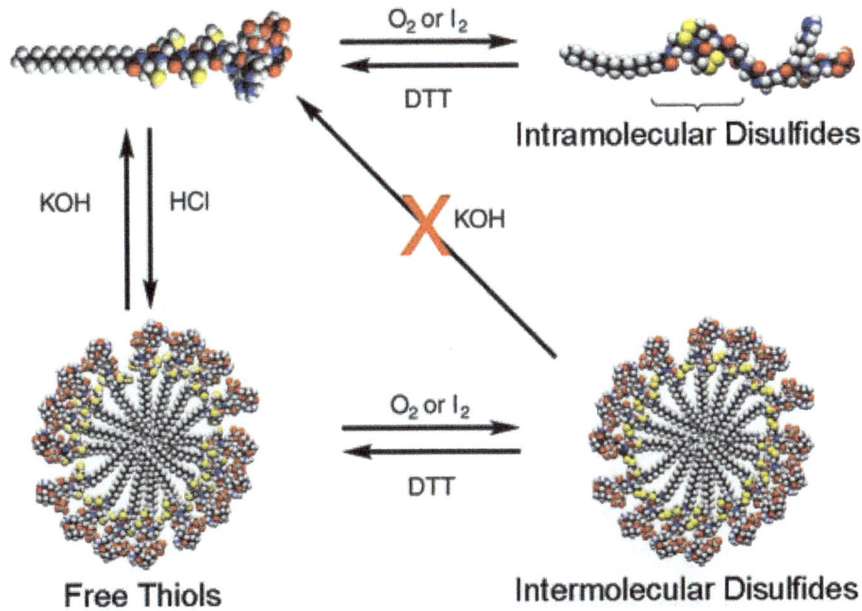

Figure 14.15 Schematic illustrating the self-assembly and covalent capture of the PAs based on pH and oxidation state.
Reprinted with permission from ref. 33. Copyright (2002) National Academy of Sciences, USA.

14.5 Biological Stimuli

Recently, peptide-based systems that respond to biological signals were also developed. Such materials will not only help in the understanding of certain biological processes, for example aggregate formation, and how these can be controlled, but they can also be the basis for a variety of (bio)-materials that respond to the presence or absence of particular biological signals such as enzymes.[58,59] Typically, most research in this area involves mainly peptides that adopt a β-sheet conformation. Therefore, in this part, we will not distinguish between α-helix, β-sheet and PA based systems as was done earlier.

Camus *et al.* have developed peptides that could be induced to self-assemble upon modification by an enzyme (Figure 14.16).[60] The design of the peptides is based on an O > S acyl shift of the backbone as developed by Mutter.[61] It was claimed that such peptides could be very useful in the study of folding and self-assembling peptides in a so-called *in statu nascendi* manner. Camus *et al.* showed that enzymatic cleavage by the dipeptidyl peptidase IV (DPPIV), expressed on several immune cells, resulted in the initiation of the formation of amyloid fibrils. Moreover, they demonstrated that these switch peptides could be used to identify molecules, such as tannic acid, myricentin and dopamine, that interfere with the fibrillisation

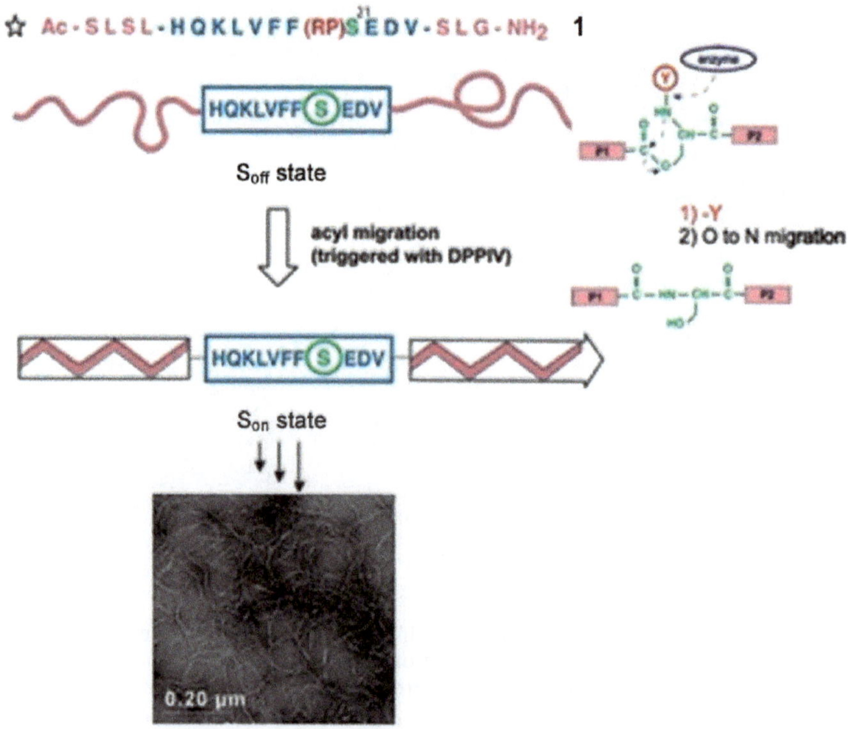

Figure 14.16 Schematic showing DPPIV triggered assembly and fibrillisation. Reprinted with permission from ref. 60.

process. Nepomniaschiy *et al.* have employed this O > S acyl shift strategy to construct a redox-sensitive switch peptide *via* a Staudinger reaction.[62]

Voss *et al.* constructed a lipopolysaccharide sensor based on the putative ligand binding domain of the lipopolysaccharide (LPS) binding protein CD14.[63] In their study, they functionalised this peptide with two different fluorescent labels at the termini that were known to give fluorescence quenching when the peptide was in an unconstrained state. They showed that in the presence of LPS, fluorescence could be restored most likely due to a conformational change of the peptide that separates the chromophores. Detection of LPS at a concentration as low as 150 nM was established. A variant of this peptide could even be employed to detect LPS in the presence of fetal calf serum.

In a similar fashion, the Schmuck group has prepared a peptide beacon that can detect nucleic acids.[64] For this they attached two pyrene moieties at the N-termini of a dimeric peptide. The pyrene units can form a π-stacked pair when it is in an unbound folded form, leading to excimer emission. Upon binding to ds-DNS, the fold changes to an extended conformation changing the emission to that of monomeric pyrene (Figure 14.17). They show in addition that their peptide-based beacon is taken up by HeLa cells and is hardly cytotoxic.

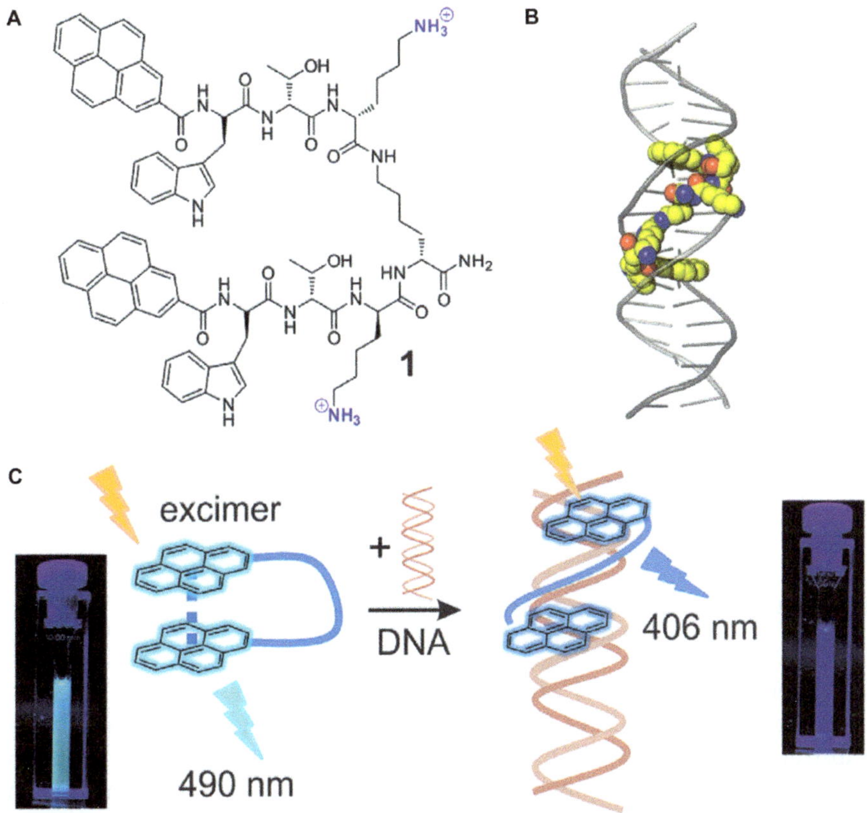

Figure 14.17 (A) The structure of a pyrene-functionalised peptide beacon, (B) a possible binding mode to ds-DNA and (C) a schematic of its mode of action.
Reprinted with permission from ref. 64. Copyright (2012) American Chemical Society.

The Ulijn group showed how their amphiphilic Fmoc-dipeptide system could be placed under enzymatic control by capping the C-terminus as a methyl ester.[65] Self-assembly was initiated by the action of subtilisin, an enzyme that hydrolysed the ester. Changing the amounts of enzyme led to different types of higher order structures. They claim the catalytic activity and mobility of the catalytic clusters control the order of the assemblies.

Taking inspiration from nature where post-translational modifications are an important signal for a great variety of processes, Riemen and Water investigated the structural consequences of methylation and phosphorylation of a β-hairpin peptide.[66] They showed that methylation of the peptide resulted in increased stability of the β-hairpin conformation. When the same peptide was phosphorylated using protein kinase A, the reverse was observed—a less stable hairpin. Wang *et al.* constructed an electrochemical-based system for the detection of the same enzyme.[67] They constructed a

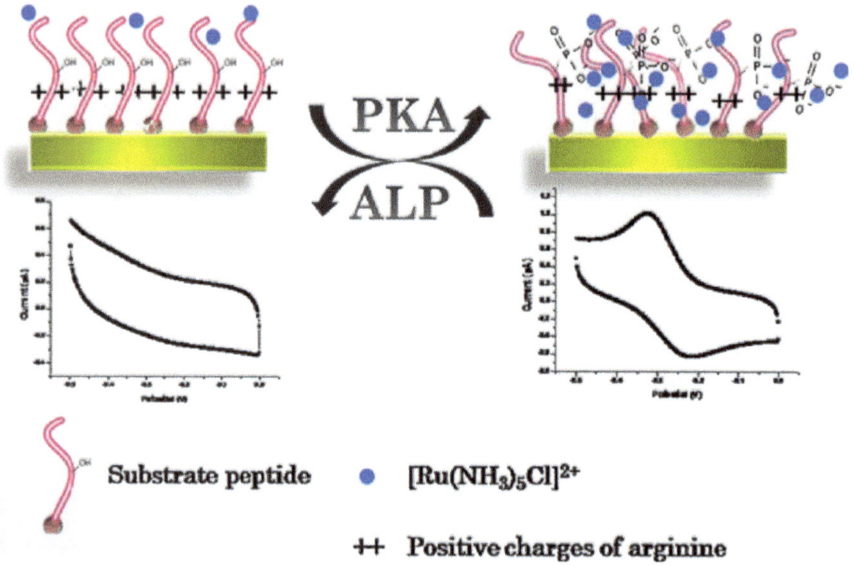

Figure 14.18 An electrochemical sensor for phosphorylation. Reprinted with permission from ref. 67.

self-assembled monolayer of peptides on a surface blocking the electrochemical probe $[Ru(NH_3)_5Cl]^{2+}$ from the electrode. Phosphorylation of the substrate peptide led to disturbance of the monolayer resulting in an electrochemical signal. Using alkaline phosphatase, this process was shown to be reversible (Figure 14.18).

References

1. P. N. Bryan and J. Orban, *Curr. Opin. Struct. Biol.*, 2010, **20**, 482–488.
2. D. W. P. M. Löwik, E. H. P. Leunissen, M. van den Heuvel, M. B. Hansen and J. C. M. van Hest, *Chem. Soc. Rev.*, 2010, **39**, 3394–3412.
3. M. B. Hansen, L. Ruizendaal, D. W. P. M. Löwik and J. C. M. van Hest, *Drug Discovery Today: Technol.*, 2009, **6**, e1–e40.
4. R. J. Mart, R. D. Osborne, M. M. Stevens and R. V. Ulijn, *Soft Matter*, 2006, 2, 822–835.
5. S. Wong, M. S. Shim and Y. J. Kwon, *J. Mater. Chem. B*, 2014, **2**, 595–615.
6. Z. Luo and S. Zhang, *Chem. Soc. Rev.*, 2012, **41**, 4736–4754.
7. L. Pauling, R. B. Corey and H. R. Branson, *Proc. Natl. Acad. Sci. U. S. A.*, 1951, **37**, 205–211.
8. L. Stryer, *Biochemistry*, W.H. Freeman and Co., New York, 1998.
9. B. Apostolovic, M. Danial and H. A. Klok, *Chem. Soc. Rev.*, 2010, **39**, 3541–3575.
10. H. Cui, M. J. Webber and S. I. Stupp, *Pept. Sci.*, 2010, **94**, 1–18.
11. A. Dehsorkhi, V. Castelletto and I. W. Hamley, *J. Pept. Sci.*, 2014, **20**, 453–467.

12. D. W. P. M. Löwik and J. C. M. van Hest, *Chem. Soc. Rev.*, 2004, **33**, 234–245.
13. L. A. Canalle, D. W. P. M. Löwik and J. C. M. van Hest, *Chem. Soc. Rev.*, 2010, **39**, 329–353.
14. J. Y. Shu, B. Panganiban and T. Xu, *Annu. Rev. Phys. Chem.*, 2013, **64**, 631–657.
15. X. I. Ambroggio and B. Kuhlman, *Curr. Opin. Struct. Biol.*, 2006, **16**, 525–530.
16. K. Chockalingam, M. Blenner and S. Banta, *Protein Eng., Des. Sel.*, 2007, **20**, 155–161.
17. K. Pagel and B. Koksch, *Curr. Opin. Chem. Biol.*, 2008, **12**, 730–739.
18. S. Tashiro and M. Shionoya, *Chem. Lett.*, 2013, **42**, 456–462.
19. B. Apostolovic and H. A. Klok, *Biomacromolecules*, 2008, **9**, 3173–3180.
20. M. Altman, P. Lee, A. Rich and S. G. Zhang, *Protein Sci.*, 2000, **9**, 1095–1105.
21. K. Pagel, S. C. Wagner, K. Samedov, H. von Berlepsch, C. Bottcher and B. Koksch, *J. Am. Chem. Soc.*, 2006, **128**, 2196–2197.
22. S. C. Wagner, M. Roskamp, H. Coelfen, C. Boettcher, S. Schlecht and B. Koksch, *Org. Biomol. Chem.*, 2009, **7**, 46–51.
23. Y. Zimenkov, S. N. Dublin, R. Ni, R. S. Tu, V. Breedveld, R. P. Apkarian and V. P. Conticello, *J. Am. Chem. Soc.*, 2006, **128**, 6770–6771.
24. N. L. Fletcher, C. V. Lockett and A. F. Dexter, *Soft Matter*, 2011, **7**, 10210–10218.
25. A. M. Jonker, D. W. P. M. Löwik and J. C. M. van Hest, *Chem. Mat.*, 2012, **24**, 759–773.
26. A. Altunbas and D. Pochan, in *Peptide-Based Materials*, ed. T. Deming, Springer, Berlin Heidelberg, 2012, ch. 206, vol. 310, pp. 135–167.
27. A. Aggeli, M. Bell, N. Boden, L. M. Carrick and A. E. Strong, *Angew. Chem., Int. Ed.*, 2003, **42**, 5603–5606.
28. A. Aggeli, M. Bell, N. Boden, J. N. Keen, P. F. Knowles, T. C. B. McLeish, M. Pitkeathly and S. E. Radford, *Nature*, 1997, **386**, 259–262.
29. A. Aggeli, M. Bell, L. M. Carrick, C. W. G. Fishwick, R. Harding, P. J. Mawer, S. E. Radford, A. E. Strong and N. Boden, *J. Am. Chem. Soc.*, 2003, **125**, 9619–9628.
30. H. Frisch, J. P. Unsleber, D. Ludeker, M. Peterlechner, G. Brunklaus, M. Waller and P. Besenius, *Angew. Chem., Int. Ed.*, 2013, **52**, 10097–10101.
31. L. O. Tjernberg, J. Näslund, F. Lindqvist, J. Johansson, A. R. Karlström, J. Thyberg, L. Terenius and C. Nordstedt, *J. Biol. Chem.*, 1996, **271**, 8545–8548.
32. P. Kumaraswamy, S. Sethuraman and U. M. Krishnan, *Soft Matter*, 2013, **9**, 2684–2694.
33. J. D. Hartgerink, E. Beniash and S. I. Stupp, *Proc. Natl. Acad. Sci. U. S. A.*, 2002, **99**, 5133–5138.
34. J. D. Hartgerink, E. Beniash and S. I. Stupp, *Science*, 2001, **294**, 1684–1688.

35. S. Zhang, M. A. Greenfield, A. Mata, L. C. Palmer, R. Bitton, J. R. Mantei, C. Aparicio, M. O. de la Cruz and S. I. Stupp, *Nat. Mater.*, 2010, **9**, 594–601.
36. A. Uesaka, M. Ueda, A. Makino, T. Imai, J. Sugiyama and S. Kimura, *Langmuir*, 2014, **30**, 1022–1028.
37. T. Tanaka, T. Mizuno, S. Fukui, H. Hiroaki, J.-i. Oku, K. Kanaori, K. Tajima and M. Shirakawa, *J. Am. Chem. Soc.*, 2004, **126**, 14023–14028.
38. P. Anzini, C. Xu, S. Hughes, E. Magnotti, T. Jiang, L. Hemmingsen, B. Demeler and V. P. Conticello, *J. Am. Chem. Soc.*, 2013, **135**, 10278–10281.
39. S. N. Dublin and V. P. Conticello, *J. Am. Chem. Soc.*, 2008, **130**, 49–51.
40. M. J. Pandya, E. Cerasoli, A. Joseph, R. G. Stoneman, E. Waite and D. N. Woolfson, *J. Am. Chem. Soc.*, 2004, **126**, 17016–17024.
41. E. Cerasoli, B. K. Sharpe and D. N. Woolfson, *J. Am. Chem. Soc.*, 2005, **127**, 15008–15009.
42. D. E. Przybyla, C. M. R. Perez, J. Gleaton, V. Nandwana and J. Chmielewski, *J. Am. Chem. Soc.*, 2013, **135**, 3418–3422.
43. K. Pagel, T. Vagt, T. Kohajda and B. Koksch, *Org. Biomol. Chem.*, 2005, **3**, 2500–2502.
44. M. Hoernke, J. A. Falenski, C. Schwieger, B. Koksch and G. Brezesinski, *Langmuir*, 2011, **27**, 14218–14231.
45. M. Mold, L. Ouro-Gnao, B. M. Wieckowski and C. Exley, *Nat. Sci. Rep.*, 2013, **3**, 1256.
46. M. W. Cao, Y. M. Wang, X. Ge, C. H. Cao, J. Wang, H. Xu, D. H. Xia, X. B. Zhao and J. R. Lu, *J. Phys. Chem. B*, 2011, **115**, 11862–11871.
47. C. M. Micklitsch, P. J. Knerr, M. C. Branco, R. Nagarkar, D. J. Pochan and J. P. Schneider, *Angew. Chem., Int. Ed.*, 2011, **50**, 1577–1579.
48. Y. Azuma, H. Imai, T. Yoshimura, T. Kawabata, M. Imanishi and S. Futaki, *Org. Biomol. Chem.*, 2012, **10**, 6062–6068.
49. M.-C. Lu, L.-Y. Chiu, L.-Y. Chiu, C.-Y. Lin and J.-C. Horng, *Anal. Methods*, 2013, **5**, 1702–1707.
50. J. R. Kramer and T. J. Deming, *J. Am. Chem. Soc.*, 2012, **134**, 4112–4115.
51. J. R. Kramer, A. R. Rodriguez, U.-J. Choe, D. T. Kamei and T. J. Deming, *Soft Matter*, 2013, **9**, 3389–3395.
52. X. J. Wang, I. Bergenfeld, P. S. Arora and J. W. Canary, *Angew. Chem., Int. Ed.*, 2012, **51**, 12099–12101.
53. R. A. Kammerer, D. Kostrewa, J. Zurdo, A. Detken, C. Garcia-Echeverria, J. D. Green, S. A. Muller, B. H. Meier, F. K. Winkler, C. M. Dobson and M. O. Steinmetz, *PNAS*, 2004, **101**, 4435–4440.
54. R. A. Kammerer and M. O. Steinmetz, *J. Struct. Biol.*, 2006, **155**, 146–153.
55. X. Fu, Y. Ma, Y. Shen, W. Fu and Z. Li, *Biomacromolecules*, 2014, **15**, 1055–1061.
56. X. Miao, W. Cao, W. Zheng, J. Wang, X. Zhang, J. Gao, C. Yang, D. Kong, H. Xu, L. Wang and Z. Yang, *Angew. Chem., Int. Ed.*, 2013, **52**, 7781–7785.
57. C. J. Bowerman and B. L. Nilsson, *J. Am. Chem. Soc.*, 2010, **132**, 9526–9527.

58. M. E. Hahn and N. C. Gianneschi, *Chem. Commun.*, 2011, **47**, 11814–11821.
59. M. Zelzer, S. J. Todd, A. R. Hirst, T. O. McDonald and R. V. Ulijn, *Biomater. Sci.*, 2013, **1**, 11–39.
60. M. S. Camus, S. Dos Santos, A. Chandravarkar, B. Mandal, A. W. Schmid, G. Tuchscherer, M. Mutter and H. A. Lashuel, *ChemBioChem*, 2008, **9**, 2104–2112.
61. M. Mutter, A. Chandravarkar, C. Boyat, J. Lopez, S. Dos Santos, B. Mandal, R. Mimna, K. Murat, L. Patiny, L. Saucede and G. Tuchscherer, *Angew. Chem., Int. Ed.*, 2004, **43**, 4172–4178.
62. N. Nepomniaschiy, V. Grimminger, A. Cohen, S. DiGiovanni, H. A. Lashuel and A. V. Brik, *Org. Lett.*, 2008, **10**, 5243–5246.
63. S. Voss, R. D. Fischer, G. Jung, K.-H. Wiesmüller and R. Brock, *J. Am. Chem. Soc.*, 2007, **129**, 554–561.
64. J. Wu, Y. Zou, C. Li, W. Sicking, I. Piantanida, T. Yi and C. Schmuck, *J. Am. Chem. Soc.*, 2012, **134**, 1958–1961.
65. A. R. Hirst, S. Roy, M. Arora, A. K. Das, N. Hodson, P. Murray, S. Marshall, N. Javid, J. Sefcik, J. Boekhoven, J. H. van Esch, S. Santabarbara, N. T. Hunt and R. V. Ulijn, *Nat. Chem.*, 2010, **2**, 1089–1094.
66. A. J. Riemen and M. L. Waters, *J. Am. Chem. Soc.*, 2010, **132**, 9007–9013.
67. J. Wang, M. Shen, Y. Cao and G. Li, *Biosens. Bioelectron.*, 2010, **26**, 638–642.

CHAPTER 15

Target Biomolecule-responsive Hydrogels

TAKASHI MIYATA

Department of Chemistry and Materials Engineering, Kansai University, 3-3-35, Yamate-cho, Suita, Osaka 564-8680, Japan
Email: tmiyata@kansai-u.ac.jp

15.1 Introduction

Hydrogels are physically or chemically crosslinked hydrophilic polymer networks swollen in an aqueous solution. They have already been utilized as adsorbents, chromatography columns, contact lenses, foods and biomaterials because of their various fascinating properties such as swelling, permeability, mechanical properties, adsorption and biocompatibility.[1,2] Since the volume phase transition of hydrogels was discovered by Tanaka in 1978, many researchers have focused on a unique property in that some hydrogels undergo changes in volume in response to environmental changes such as pH, temperature and so on. Such unique hydrogels are named stimuli-responsive hydrogels or smart hydrogels as they can sense a stimulus as a signal and induce volume changes by themselves. A variety of stimuli-responsive hydrogels that exhibit reversible swelling/shrinking behavior in response to changes in environmental parameters such as temperature,[3–6] pH,[7,8] electric field[9,10] and light[11] have been prepared for the development of sensors, actuators, separation systems, drug delivery systems (DDSs) and cell culture systems.

The responsive swelling/shrinking behavior of stimuli-responsive hydrogels is mainly caused by changes in hydrophilicity of the polymer chains or

RSC Smart Materials No. 14
Chemoresponsive Materials: Stimulation by Chemical and Biological Signals
Edited by Hans-Jörg Schneider

Published by the Royal Society of Chemistry, www.rsc.org

in osmotic pressure by charged groups produced. For example, polymers with carboxy or amino groups that are ionized by pH changes have been generally used in synthesizing pH-responsive hydrogels.[12–16] Temperature-responsive hydrogels that exhibit great shrinkage with rising temperature have normally been prepared using polymers with a low critical solution temperature (LCST), at which the polymers undergo soluble/insoluble changes in an aqueous solution: poly(*N*-isopropyl acrylamide) (PNIPAAm) has been widely utilized for designing temperature-responsive hydrogels because its polymer chains change from hydrophilic to hydrophobic with rising temperatures above the LCST (32 °C).[17–20] A lot of pH- and temperature-responsive hydrogels have been developed for fabricating smart biomaterials and biomedical systems as pH and temperature changes are important signals to monitor several physiological changes.

Natural feedback systems such as homeostasis are the most important biosystems to maintain lives. Such natural feedback systems perceive not only pH and temperature changes but also specific ions and biological molecules such as hormones, and induce conformational changes or rearrangements of their constitutional biomolecules to elicit biological functions. For example, metabolizing glucose is controlled by insulin, which is a hormone secreted from the Wrangell Hans Island of the pancreas in response to blood glucose concentration. In addition, specific biomolecules such as tumor markers are useful signals for monitoring the biological conditions of the human body and diseases. Therefore, biologically stimuli-responsive hydrogels that undergo changes in volume in response to a target biomolecule can be useful tools for developing self-regulated DDSs and molecular diagnostics. Although most researchers focused on only pH- and temperature-responsive hydrogels for constructing smart systems several years ago, recently, several researchers have synthesized biologically stimuli-responsive hydrogels that exhibit swelling/shrinking changes in response to a target biomolecule, so-called biomolecule-responsive hydrogels.[21,22] Such biomolecule-responsive hydrogels have many potential applications in biomedical fields because they can provide a material platform that can detect a target biomolecule such as biomarkers and induce their structural changes that regulate drug release or transduce a signal to mechanical properties.

Target biomolecule-responsive swelling/shrinking behavior involves both biomolecular recognition events and responsive network changes. In designing biomolecule-responsive hydrogels, therefore, the molecular recognition ability of their ligands must be combined with a structural change in their networks. There has not been so much research on target biomolecule-responsive hydrogels, in spite of their many potential applications, because of difficulties in the combinations. Strategies for designing biomolecule-responsive hydrogels can be classified into three categories: (i) combination of molecular recognition with changes in charged groups, (ii) combination of molecular recognition with changes in hydrophilicity of polymer chains, (iii) combination of molecular recognition with crosslinked structures.

In biomolecule-responsive hydrogels designed by strategy (i), positively or negatively charged groups produced by molecular recognition events resulted in network swelling owing to increasing osmotic pressure. In strategy (ii), molecular recognition events induce drastic changes in the hydrophilicity of polymer chains such as PNIPAAm, followed by responsive swelling or shrinking behavior. In strategy (iii), using molecular complexes as dynamic crosslinks, molecular recognition events cause a decrease or increase in crosslinking density, followed by responsive swelling or shrinking. Most biomolecule-responsive hydrogels have been prepared based on the three strategies. This chapter provides an overview of important and historic research in the development of target biomolecule-responsive hydrogels and their applications.

15.2 Saccharide-responsive Hydrogels

15.2.1 Saccharide-responsive Hydrogels using pH-Responsive Polymers

The first biomolecule-responsive hydrogels were glucose-responsive hydrogels for an autonomous treatment of insulin-dependent diabetes mellitus (IDDM). Insulin is a hormone secreted from the Wrangell Hans Island of the pancreas and controls glucose metabolism. Diabetes is caused by the inability of the pancreas to control blood glucose concentration. In the treatment of diabetes, a necessary amount of insulin is administered while monitoring the blood glucose concentration. Glucose-responsive hydrogels are promising biomaterials for constructing an artificial pancreas for administering a necessary amount of insulin in response to the blood glucose concentration. Therefore, some researchers have reported glucose-responsive hydrogels for developing self-regulated insulin release systems, as described in Chapter 2 by Peppas *et al.*

There have been mainly three strategies for designing glucose-responsive hydrogels: (i) combination of enzymatic reaction with pH-responsive polymers, (ii) introduction of glucose-binding sites into temperature-responsive polymers, (iii) utilization of saccharide–lectin complexes as dynamic crosslinks. Of the three strategies, combining glucose oxidase (GOD) with pH-responsive polymers is the most traditional method for preparing glucose-responsive hydrogels. The present section focuses on glucose-responsive hydrogels using GOD and the next sections will describe that using phenylboronic acids and lectins.

Glucose-responsive hydrogels that can control insulin release in response to glucose concentration have been prepared by loading GOD within pH-responsive networks. The glucose-responsive hydrogels and self-regulated insulin release system are illustrated in Figure 15.1. GOD and pH-responsive polymers play important roles in glucose recognition and responsive network changes, respectively. When glucose diffuses into the GOD-loaded hydrogel, GOD catalyzes the hydrolysis of glucose to gluconic acid.

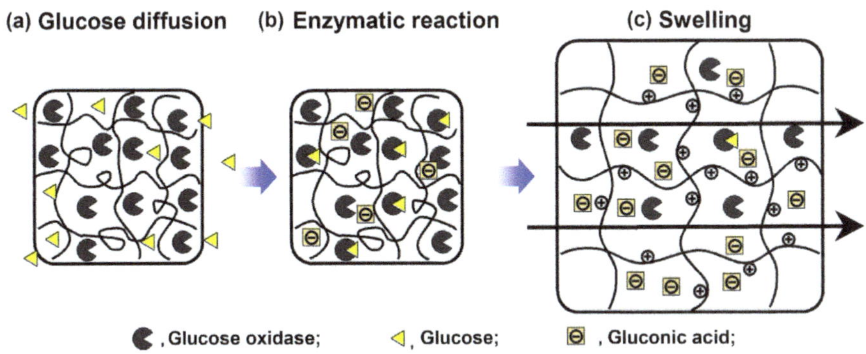

Figure 15.1 Schematic representation of a glucose-responsive hydrogel consisting of glucose oxidase and pH-responsive networks.

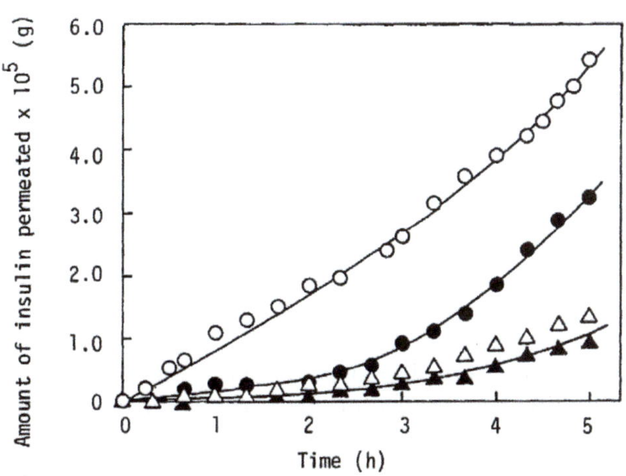

Figure 15.2 Permeation profile of insulin through a glucose-responsive polymer membrane consisting of a poly(amine) and glucose oxidase-immobilized hydrogel. Glucose concentration: (▲) 0 M; (●) 0.1 M; (○) 0.2 M; (△) 0.2 M without glucose oxidase.[23]

The resulting gluconic acid lowers the pH within the pH-responsive networks, followed by their network expansion. As a result, insulin is permeated through the expanded networks in response to the glucose concentration. Thus, glucose-responsive hydrogels and self-regulated insulin releases can be strategically designed by the combination of GOD with pH-responsive polymers. For example, glucose-responsive insulin release systems were fabricated by loading GOD to sense glucose within pH-responsive copolymer membranes from *N*,*N*-diethylaminoethyl methacrylate (DEAEMA) and 2-hydroxypropyl methacrylate (HPMA).[23] Insulin permeability through the glucose oxidase-loaded DEAEMA–HPMA copolymer membranes was effectively enhanced with increasing glucose concentration (Figure 15.2). In addition, glucose-responsive polymer capsules containing insulin were also prepared by a conventional interfacial precipitation method.[24] Insulin release from the capsule membranes was inhibited in the absence of glucose but was enhanced in its presence. The steady-state behavior of glucose-responsive hydrogel membranes prepared by entrapping glucose oxidase within a DEAEMA–hydroxyethyl methacrylate copolymer was investigated theoretically and experimentally.[25,26]

Complex formation between methacrylic acid (MAAc) and ethylene glycol (EG) was utilized as a pH-responsive network for designing glucose-responsive hydrogels.[27,28] Poly(MAAc-EG) hydrogel underwent drastic changes in volume in response to pH changes because the complexes between the carboxy groups of MAAc and etheric groups of EG formed at low pH but dissociated at high pH. The swelling rate of the GOD-loaded hydrogels at the high glucose concentration of a hyperglycemic condition was slower than that at the lower glucose concentration of normal blood glucose. Thus, using a combination of GOD with pH-responsive polymers, a variety of glucose-responsive hydrogels have been designed for fabricating self-regulated insulin release systems.

15.2.2 Saccharide-responsive Hydrogels using Temperature-responsive Polymers

Lectins are carbohydrate-binding proteins that form complexes with carbohydrate chains of glycoproteins and glycolipids on a cell's surface. Lectins have been widely used as carbohydrate recognition tools for fabricating biosensors and biochemical systems. For example, glucose-responsive insulin release systems were developed by using the competitive and complementary binding properties of glycosylated insulin and glucose to lectins.[29–31] The unique carbohydrate-binding properties of lectins have been employed to design saccharide-responsive hydrogels. For example, saccharide-responsive hydrogels were prepared by combining the carbohydrate-binding properties of lectins with the temperature-responsive property of PNIPAAm.[32] Concanavalin A (ConA) used in this study is a tetramer lectin that has four binding sites for specific saccharides such as glucose and mannose derivatives. ConA-loaded PNIPAAm hydrogels exhibited a drastic shift in LCST by complex formation between ConA and ionic saccharide dextran sulfate. Therefore, their swelling ratio increased dramatically in the presence of ionic saccharide dextran sulfate because their LCST increased by complex formation of ConA with the ionized saccharide. The hydrogel was collapsed to its native volume by replacing the ionic saccharide dextran sulfate with the non-ionic saccharide.

A smart strategy for synthesizing glucose-responsive hydrogels uses the combination of the glucose-binding property of phenylboronic acids with the LCST of temperature-responsive polymers. Phenylboronic acid and its derivatives reversibly bind with 1, 2- or 1, 3-*cis*-diols such as glucose. The phenylboronic acid groups in water exist in equilibrium between uncharged and negatively charged forms (Figure 15.3). As glucose forms a complex

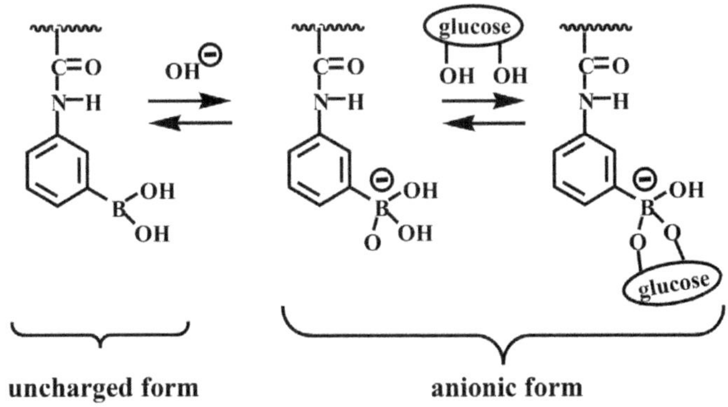

Figure 15.3 Complex formation of phenylboronic acid and glucose.[34]

with the charged phenylboronic acid group more preferentially than with uncharged groups, the presence of glucose leads to an increase in charged groups and a decrease in uncharged groups due to a shift of the equilibrium of the phenylboronic acid moiety. Therefore, the solubility of temperature-responsive polymers with phenylboronic acid groups is drastically enhanced by the presence of glucose because its LCST is shifted by increasing charged groups based on the complex formation of phenylboronic acid groups and glucose. For example, the copolymer of *N*-isopropyl acrylamide (NIPAAm) and 3-(acrylamido)phenylboronic acid (APBA) exhibited a drastic shift in the LCST to a higher temperature in the presence of free glucose, based on a shift in the equilibrium between the uncharged and charged phenylboronic acid groups through their complex formation with glucose.[33] Therefore, totally synthetic hydrogels showing glucose-responsive swelling/shrinking behavior were prepared by combining the LCST of temperature-responsive polymers with the shift in the equilibrium of the phenylboronic acid due to its complex formation with glucose.[34] The LCST of the NIPAAm–APBA copolymer hydrogels in the presence of free glucose was higher than that in its absence (Figure 15.4). Therefore, at a constant temperature between LCSTs in the presence and absence of glucose, the NIPAAm–APBA copolymer hydrogels swelled greatly in response to free glucose. In addition, the NIPAAm–APBA copolymer hydrogels effectively released insulin in a buffer solution with glucose while insulin release was depressed for the hydrogel immersed in a solution without glucose. The reversibly glucose-responsive swelling/shrinking behavior of the NIPAAm–APBA copolymer hydrogels enabled repeated on–off release of insulin on changing the concentration of external glucose.

Recently, the chemical structure of the glucose-responsive hydrogels prepared by introducing glucose-recognition sites to temperature-responsive networks was optimized to exert glucose-sensitivity under physiological pH and temperature, *i.e.* pH 7.4 and 37 °C.[35] To decrease the pK_a of phenylboronic acid, strongly electron-withdrawing substituents were introduced

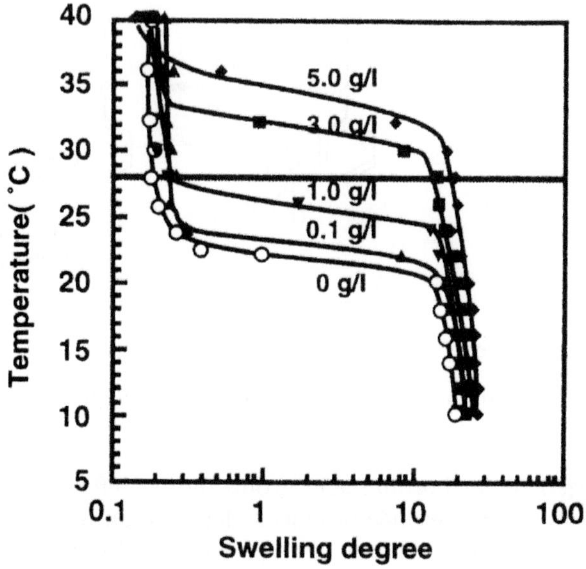

Figure 15.4 Temperature dependence of swelling curves for the PNIPAAm copolymer hydrogel with phenylboronic acid moieties at different glucose concentrations.[34]

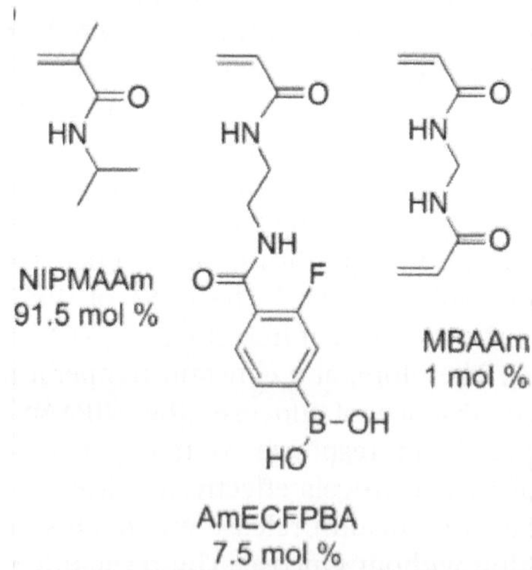

Figure 15.5 Structures of the monomers contained in the gel and their optimized molar amounts in the feed solution so as to obtain glucose sensitivity under physiological pH and temperature (pH 7.4 and 37.8 °C).[35]

to the phenyl ring of the phenylboronic acid. For example, 4-(2-acrylamidoethylcarbamoyl)-3-fluorophenylboronic acid (AmECFPBA) with a pK_a value of 7.2 was synthesized by introducing *para*-carbamoyl and *meta*-fluoro substituents (Figure 15.5). The copolymer hydrogel prepared from *N*-isopropyl

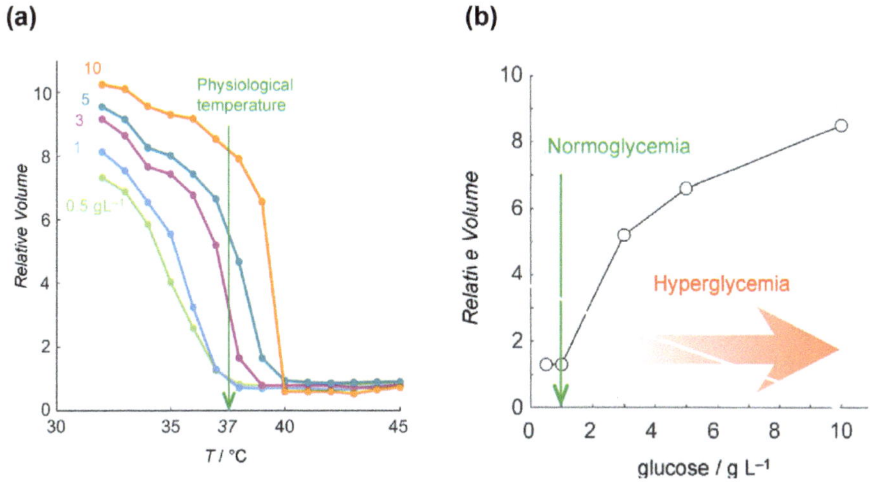

Figure 15.6 (a) Temperature dependence of swelling curves for the PNIPMAAm copolymer hydrogel with AmECFPBA at various glucose concentrations at pH 7.4. (b) Effect of the glucose concentration on the swelling ratio of the PNIPMAAm copolymer hydrogel with AmECFPBA at 37 °C.[35]

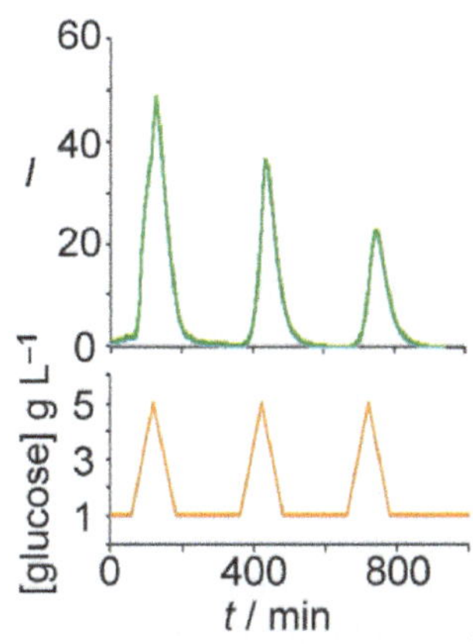

Figure 15.7 Repeated on–off release of FITC-insulin from the PNIPMAAm copolymer hydrogel with AmECFPBA under physiological conditions (pH 7.4, $I = 0.15$, 37 °C).[35]

methacrylamide (NIPMAAm) and AmECFPBA remained shrunken in a buffer solution with a glucose concentration of up to 1 g L^{-1}, whereas a swelling of the hydrogel was observed with any further increase in glucose concentration (Figure 15.6). The repeated on–off release of insulin on changing glucose concentration at a physiological condition was achieved using the NIPMAAm–AmECFPBA copolymer hydrogel (Figure 15.7). These glucose-responsive hydrogels provide a unique platform for self-regulated insulin release systems.

15.2.3 Saccharide-responsive Hydrogels using Dynamic Crosslinks

A unique strategy for designing glucose-responsive hydrogels has been proposed: the strategy uses lectin–saccharide complexes as dynamic crosslinks of the hydrogel networks. Synthetic glycopolymers with pendant glucose are very useful in preparing glucose-responsive hydrogels using the strategy. Glycopolymers with pendant saccharide form complexes with lectins but the complexes are dissociated by the presence of a target saccharide that is recognized by the lectin.[36] For example, the addition of lectin ConA into a buffer solution containing poly(2-glucosyloxyethyl methacrylate) (PGEMA) as a glycopolymer with pendant glucose resulted in rapid precipitation by their complex formation, but the resulting PGEMA–ConA complexes were dissociated by the addition of free glucose and mannose owing to the competitive and complementary binding properties.[37] As such, formation and dissociation of the PGEMA–ConA complex can be employed as glucose-responsive crosslinks of hydrogel networks; glucose-responsive hydrogels that underwent changes in volume in response to glucose concentration were developed using PGEMA–ConA complexes as dynamic crosslinks in their networks. The bioconjugated hydrogels with PGEMA–ConA complexes as dynamic crosslinks were prepared by copolymerization of a monomer with a pendant glucose (GEMA) and *N,N′*-methylenebisacrylamide (MBAA) after the formation of GEMA–ConA complex.[38] The resulting hydrogels (ConA-entrapped PGEMA hydrogels) swelled immediately in a buffer solution with free glucose and mannose, but did not change their volume in a solution with galactose (Figure 15.8). From compressive modulus measurements, it was clear that the crosslinking density of the ConA-entrapped PGEMA hydrogels decreased with an increase in glucose concentration. This indicates that the glucose-responsive swelling behavior of the ConA-entrapped PGEMA hydrogels was attributed to the dissociation of the complexes between ConA and pendant glucose of GEMA by a complex exchange with free glucose (Figure 15.9).

To obtain reversibly glucose-responsive hydrogels that swell in the presence of free glucose and shrink in its absence, the ConA-copolymerized PGEMA hydrogels were also prepared by copolymerization of GEMA with chemically modified ConA with polymerizable groups.[39] The ConA-copolymerized PGEMA hydrogels underwent reversible changes in volume in response to stepwise changes in glucose concentration but the ConA-entrapped hydrogel did not. These were attributed to the fact that ConA leaked out of the ConA-entrapped hydrogel but not out of the ConA-copolymerized hydrogel.

Complex formation between a polymer with pendant glucose (glycopolymer) and ConA is also useful in achieving a sol–gel phase transition by changes in environmental glucose concentration. For example, an aqueous solution containing vinylpyrrolidinone–allylglucose or acrylamide–allylglucose copolymers changed to a gel state with the addition of ConA.[40,41] The sol–gel

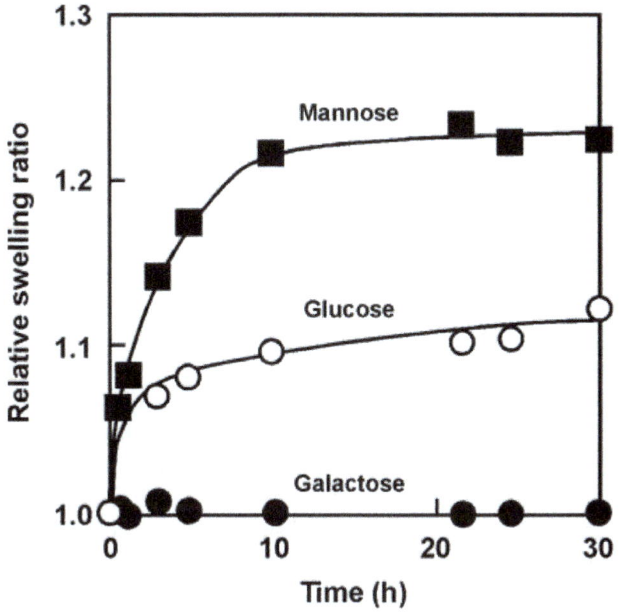

Figure 15.8 Swelling ratio changes of the PGEMA–ConA hydrogel as a function of time when the hydrogel was immersed in a buffer solution containing 1 wt% of monosaccharide: (○), glucose; (■), mannose; (●) galactose.[38]

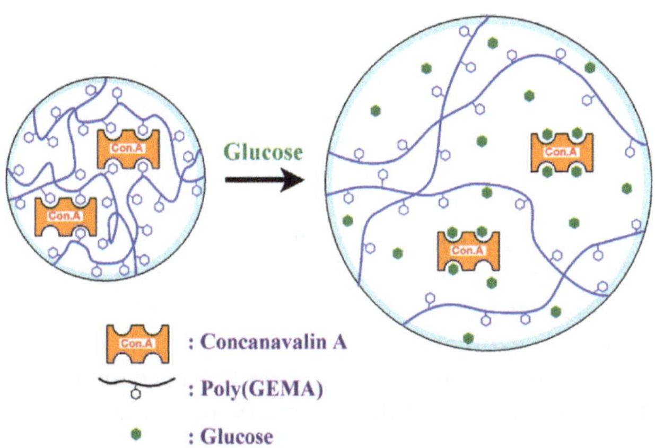

Figure 15.9 Schematic representation of the glucose-responsive swelling of ConA-entrapped hydrogel.[38]

phase transition was strongly dependent upon the glucose concentration because of the formation and dissociation of the complex between the glycopolymers and ConA in response to free glucose. Regulation of the release of lysozyme and insulin in response to the glucose concentration was achieved using the glucose-responsive sol–gel phase transition.[42] Thus, the reversibly glucose-responsive swelling/shrinking behavior and sol–gel phase transition using glycopolymer–ConA complexes as dynamic crosslinks show great potential as smart tools for fabricating self-regulated DDSs in which insulin is autonomously released in response to repeated changes in glucose concentration.

15.3 Protein-responsive Hydrogels

15.3.1 Protein-responsive Hydrogels using Enzymatic Reactions

Some enzymes provide valuable disease signals for monitoring several physiological changes. Enzymes located in a specific place in the body play important roles as useful marks for site-specific drug delivery. Therefore, enzymes have been widely used as target proteins to monitor physiological changes or to guide drugs to a specific site. Enzyme-responsive hydrogels that undergo structural changes in response to a target enzyme have been designed using polymer chains degraded by enzymatic activity. For example, natural polymers such as gelatin and dextran have been utilized as enzymatically degradable network chains. This section focuses on protein-responsive hydrogels that can be degraded by enzymatic activity.

Dextranase is a microbial enzyme predominantly presenting in the colon and can degrade dextran enzymatically. Dextranase-responsive hydrogels synthesized by crosslinking dextran with diisocyanate were degraded *in vitro* by dextranase, *in vivo* in rats and in a human colonic fermentation model.[43] In the presence of dextranase, drug release from the dextranase-responsive hydrogels was enhanced by enzymatic degradation of dextran networks. As azoreductase, an enzyme produced by the microbial flora of the colon, is also a useful mark, azoreductase-responsive hydrogels were prepared by copolymerization of acrylamide derivatives, acrylic acid and crosslinkers with azoaromatic bonds that can be degraded by azoreductase.[44–46] As the pH-responsive copolymer hydrogels with azoaromatic bonds swell at high pH and shrink at low pH, protein drugs loaded in the hydrogel are protected against digestion by proteolytic enzymes in the stomach (low pH) but azoreductase becomes accessible to the azoaromatic crosslinks in the colon. As a result, the protein drugs can be predominantly released in the colon because of the enzymatic degradation of azoaromatic crosslinks.

Matrix metalloproteinase (MMP)-responsive hydrogels were rationally designed by crosslinking multi-armed PEGs with oligopeptide cleavable by MMPs, which are a protease family extensively involved in tissue development and remodeling (Figure 15.10).[47–49] The MMP-responsive hydrogels were useful as extracellular matrix (ECM)-mimicking scaffolds because they were finally degraded by enzymatic activity of MMPs. When the MMP-responsive hydrogels containing recombinant human bone morphogenetic protein-2 (rhBMP-2) were implanted to the site of critical-sized defects in rat crania, they were completely invaded by cells and rhBMP-2 was released

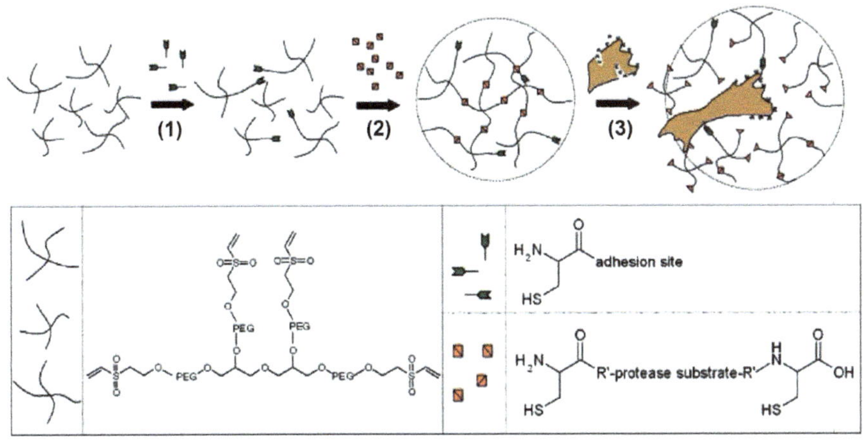

Figure 15.10 Schematic illustration of the ECM-mimicking hydrogels composed of a PEG-based hydrogel crosslinked by the oligopeptides which are cleavable by the MMPs.[49]

from degraded hydrogels, followed by regeneration of bone tissue.[47,50] Enzyme-responsive hydrogels were also prepared by introducing tetrapeptide (Cys-Tyr-Lys-Cys), which is degraded by α-chymotrypsin, as a crosslinker into poly(acrylamide).[51] The presence of α-chymotrypsin resulted in the degradation of hydrogel with the Cys-Tyr-Lys-Cys tetrapeptide crosslink by hydrolysis of the tetrapeptide but had no effect on a Cys-Ser-Lys-Cys crosslink. Thus, the strategy using enzyme-cleavable peptides as dynamic crosslinks is a fascinating way to design enzyme-responsive hydrogels for constructing self-regulated DDSs and cell culture systems. The most important and most recent progress of peptide-based stimuli responsive materials are reviewed in Chapter 14 of this book by Löwik *et al.*

By monitoring two or more proteins simultaneously, physiological changes from several diseases can be detected at the same time. Dual-enzyme-responsive hydrogels that were degraded in the presence of two enzymes were strategically designed by the formation of an interpenetrating polymer network (IPN) of oligopeptide-terminated PEG and dextran that were specifically decomposed by papain and dextranase, respectively.[52,53] The IPN hydrogels were degraded in the presence of both papain and dextranase, but not in the presence of either enzyme alone. The presence of both α-chymotrypsin and dextranase resulted in the release of lipid microspheres from the gelatin-dextran IPN hydrogels, but not in the presence of either enzyme alone (Figure 15.11). In addition, dual-stimuli-responsive hydrogels that were degraded in response to a target enzyme within a certain temperature were prepared by combining enzymatic biodegradation with the temperature-responsive property of PNIPAAm.[54,55] To design temperature-responsive dextran hydrogels with two different LCSTs, PNIPAAm-grafted dextran with polymerizable groups were copolymerized with NIPAAm and *N*,*N*-dimethylacrylamide (DMAAm). The temperature-responsive dextran

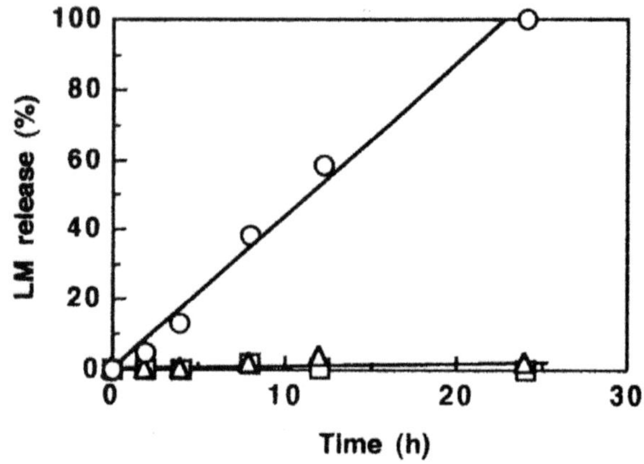

Figure 15.11 Lipid microsphere release from gelatin/dextran IPN hydrogel in phosphate buffer at 37 °C. ○, 5 U mL^{-1} α-chymotrypsin + 0.5 U mL^{-1} dextranase; △, 5 U mL^{-1} α-chymotrypsin; □, 0.5 U mL^{-1} dextranase.[53]

hydrogels were degraded in the presence of an enzyme at a lower temperature, but not at a higher temperature. The temperature-responsive biodegradation within a specific temperature range is due to the fact that the formation of an enzyme–substrate complex was influenced by structural changes of the polymer chains in response to temperature changes. Thus, these dual-stimuli-responsive hydrogels have a fail-safe mechanism for guaranteed drug delivery to a certain disease.

In contrast to enzyme-responsive hydrogels that undergo degradation, a few researchers reported enzyme-responsive hydrogels that were formed in the presence of a specific enzyme. For example, enzyme-mediated hydrogel formation within a few minutes was achieved by crosslinking functionalized poly(ethylene glycol) (PEG) and a lysine-containing polypeptide through the enzymatic reaction of transglutaminase (TGase) that catalyzes an acyl-transfer reaction between the γ-carboxamide group of protein-bound glutaminyl residues and the ε-amino group of lysine residues.[56,57]

Differing from the enzyme-responsive hydrogels that were formed or degraded in the presence of a target enzyme, enzyme-responsive hydrogels that undergo changes in volume in response to a specific protease were developed by combining enzymatic cleavage of peptide linkers with producing or eliminating charged groups. For example, PEG-based hydrogels with an enzyme-cleavable peptide linker, Fmoc-Asp-Ala-Ala-Arg, which is hydrolyzed by thermolysin, swelled specifically in the presence of thermolysin because of the production of doubly charged peptide fragments by enzymatic hydrolysis (Figure 15.12).[58] On the other hand, when the PEG-based hydrogels with positively charged peptide chains were exposed to the protease, they collapsed owing to the elimination of the charged peptides from the hydrogel networks by enzymatic hydrolysis.[59] Such enzyme-responsive hydrogels have some applications in the selective release of therapeutic agents at specific sites at which the target enzyme is located.

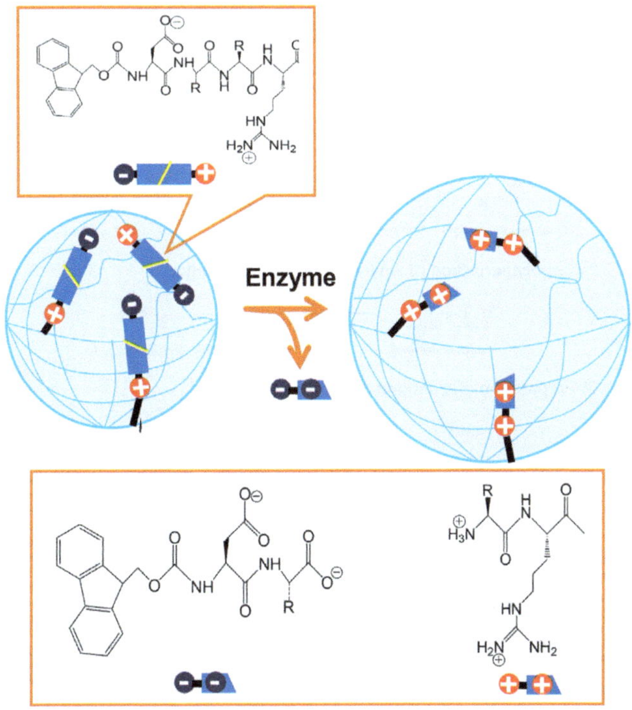

Figure 15.12 Schematic representation of selective enzyme-triggered charge-induced swelling of enzyme-responsive hydrogels with the zwitterionic peptide linkers that are hydrolyzed by a specific enzyme.[58]

15.3.2 Protein-responsive Hydrogels using Dynamic Crosslinks

An antibody recognizes a specific antigen and forms an antigen–antibody complex through multiple interactions. Formation of the antigen–antibody complexes plays important roles in immune responses to protect an organism from infection. To detect a signal biomolecule for monitoring physiological changes, a variety of highly sensitive diagnostic systems such as an enzyme-linked immunosorbent assay (ELISA) have been fabricated using the specificity and versatility of the antigen-recognition functions of antibodies.[60] Antigen–antibody complexes are very useful tools as dynamic crosslinks for designing protein-responsive hydrogels that undergo changes in volume in response to a specific antigen as a target protein. This section focuses on protein-responsive hydrogels prepared using antigen–antibody complexes as dynamic crosslinks.[61,62]

A typical method to prepare antigen-responsive hydrogels with antigen–antibody complexes as dynamic crosslinks is shown in Figure 15.13(a). First of all, rabbit immunoglobulin G (rabbit IgG) as an antigen and an antibody (goat anti-rabbit IgG) were modified using *N*-succinimidylacrylate (NSA) to introduce vinyl groups. After the synthesis of a poly(acrylamide)

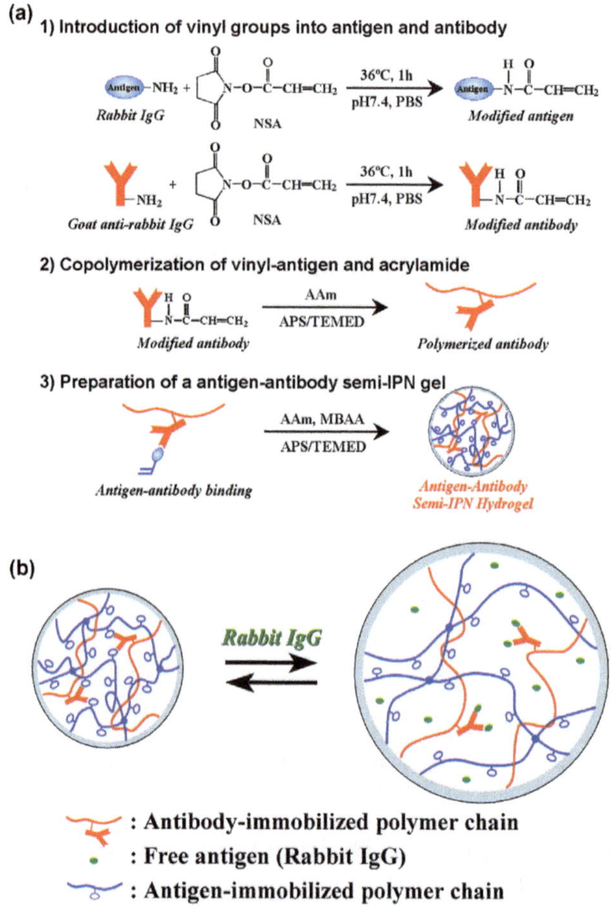

Figure 15.13 Schematic representation of the preparation (a) and responsive behavior (b) of antigen–antibody-conjugated hydrogels with a semi-IPN structure.[61]

(PAAm)-grafted antibody by copolymerization of acrylamide (AAm) and acryloyl-antibody, antigen–antibody bioconjugated hydrogels with a semi-interpenetrating polymer network (semi-IPN) structure were obtained by the copolymerization of the vinyl-antigen, AAm and *N,N′*-methylenebisacrylamide in the presence of the PAAm-grafted antibody. Antigen–antibody entrapment hydrogels without a semi-IPN structure were also synthesized by using a native antibody in place of the PAAm-grafted antibody.

The resulting antigen–antibody entrapment and semi-IPN hydrogels with antigen–antibody complexes as dynamic crosslinks swelled immediately when they were immersed in a buffer solution with rabbit IgG as a target antigen. Their equilibrium swelling ratio strongly depended on the antigen concentration of the buffer solution. However, the antigen–antibody hydrogels showed no change in volume in the presence of free goat IgG that was not recognized by the antibody linked with hydrogel networks. These demonstrate that the hydrogels with antigen–antibody complexes as dynamic crosslinks can recognize a target rabbit IgG and induce structural changes. The surface plasmon resonance (SPR) measurements revealed that the affinity constant of the antibody for a native antigen was much higher than that for modified antigen. This means that the antigen–antibody complexes as dynamic crosslinks linked with PAAm networks are dissociated by the addition of a free native antigen. Furthermore, the compressive modulus measurements showed that the crosslinking density of the antigen–antibody hydrogels decreased gradually in proportion to the increasing free antigen concentration in a buffer solution. Consequently, the antigen-responsive swelling of the antigen–antibody hydrogels is explained by the complex exchange mechanism as follows (Figure 15.13(b)): in the antigen–antibody hydrogels prepared by the method shown in Figure 15.13(a), the antigen–antibody complexes acted as dynamic crosslinks. Therefore, the hydrogels swelled gradually in a buffer solution with a target antigen because their crosslinking density decreased by the dissociation of antigen–antibody complexes linked with the networks.

The antigen–antibody semi-IPN hydrogels underwent reversible changes in volume in response to the antigen concentration when they were alternately immersed in a buffer solution with and without a free antigen (Figure 15.14). On the other hand, antigen–antibody entrapment hydrogels without IPN structures exhibited no shrinkage when the hydrogels swollen in a buffer solution with an antigen were transferred to a buffer solution without an antigen.[63] This suggests that only semi-IPN hydrogels can show reversible swelling/shrinking changes in response to stepwise changes in the antigen concentration. In the case of the antigen–antibody entrapment hydrogels, the antibody entrapped within their networks leaked out of the gels while they swelled in a buffer solution with a free antigen. However, the PAAm-grafted antibody in the semi-IPN hydrogels did not leak out of the gel networks as it was entangled with a network with grafted antigens. As a result, semi-IPN structures in hydrogels with biomolecular complexes as dynamic crosslinks played important roles in reversible changes in volume in response to a target biomolecule.

Many researchers have utilized stimuli-responsive hydrogels to regulate drug release in response to signal changes such as pH and temperature. To apply the antigen–antibody semi-IPN hydrogels to self-regulated drug delivery systems, the permeation of a model drug through the hydrogels was investigated in the presence and absence of a free antigen. The antigen–antibody semi-IPN hydrogels permeated hemoglobin with a molecular weight of 64.5 kDa in the presence of a target antigen, but not in its absence (Figure 15.14).[61] The antigen–antibody semi-IPN hydrogels enabled more sensitive on–off regulation of the hemoglobin permeation in response to stepwise changes in the antigen concentration than the antigen–antibody entrapment hydrogels. On the other hand, vitamin B_{12} with a molecular weight of 1.4 kDa permeated slightly through the antigen–antibody semi-IPN hydrogel in the absence of the

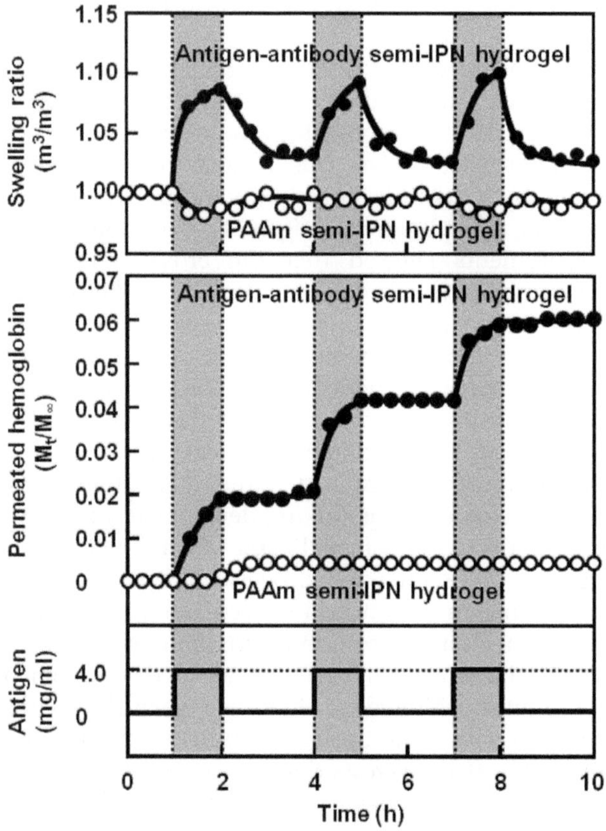

Figure 15.14 Reversible swelling changes and antigen-responsive permeation profiles of hemoglobin through the PAAm semi-IPN hydrogel (○) and the antigen–antibody-conjugated semi-IPN gel (●) in response to stepwise changes in the antigen concentration between 0 and 4 mg ml^{-1}.[61]

target antigen and its permeation was accelerated in its presence.[64] These results imply that the network size of the antigen–antibody semi-IPN hydrogel in a buffer solution without the target antigen is smaller than the molecular size of hemoglobin and larger than vitamin B_{12}. However, the network size of the antigen–antibody semi-IPN hydrogel swollen in a buffer solution with the target antigen became larger than the size of hemoglobin. Thus, the reversibly antigen-responsive hydrogels are promising candidates for constructing smart drug delivery systems that regulate drug release in response to a specific protein such as rabbit IgG.

15.4 Molecularly Imprinted Hydrogels

Molecular imprinting is an attractive technique to create synthetic hosts with molecular cavities for molecular recognition,[65–69] as described in Chapter 13 by Puoci *et al.* In standard molecular imprinting, after functional

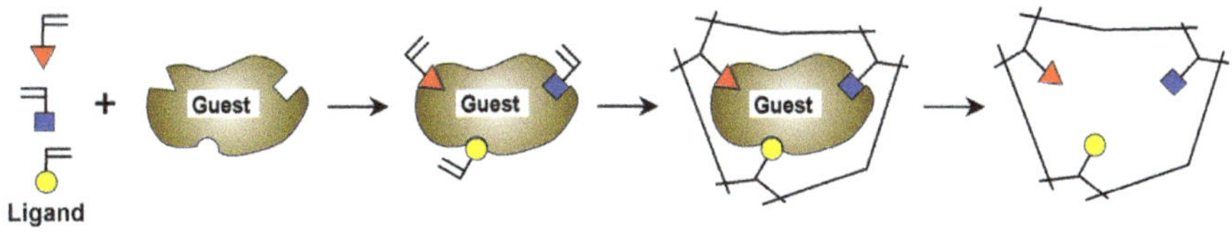

Figure 15.15 Schematic illustration of molecular imprinting.

monomers are prearranged around template molecules by noncovalent interactions, they are copolymerized with the crosslinker for forming polymer networks. Then the molecularly imprinted polymers are obtained by extracting the template molecule from the resultant polymer networks to form molecular cavities as molecular recognition sites (Figure 15.15). The molecularly imprinted polymers with the molecular recognition sites can recognize the guest molecule (template molecule) by a combination of reversible binding and shape complementarity. Standard molecular imprinting has utilized a large amount of crosslinker to fix the structure of cavities for molecular recognition, resulting in the formation of highly crosslinked polymer networks. Recently, some researchers reported that molecular imprinting is a useful technique for creating molecular recognition sites within swollen hydrogels slightly crosslinked with a minute amount of crosslinker. The molecular recognition ability of the molecularly imprinted polymer prepared using stimuli-responsive polymers can be regulated by environmental stimuli, as described in Chapter 13 by Puoci *et al.*

To design stimuli-responsive hydrogels with molecular recognition abilities, molecular imprinting was combined with temperature-responsive polymers.[70,71] Temperature-responsive hydrogels with molecular recognition sites were synthesized by copolymerization of NIPAAm, acrylic acid (AAc) as a ligand monomer and a minute amount of crosslinker in the presence of norephedrine as a template molecule in molecular imprinting.[70] The norephedrine-imprinted NIPAAm–AAc copolymer hydrogels swollen at a low temperature below its LCST exhibited no change in volume in the presence of norephedrine. However the hydrogels shrunken at a high temperature above its LCST swelled gradually with increasing norephedrine concentration but not with increasing adrenaline concentration (Figure 15.16). Their norephedrine-responsive swelling was attributed to a shift of LCST by the binding of norephedrine with its recognition site created by molecular imprinting. These results reveal that temperature-responsive hydrogels designed by molecular imprinting memorize a target molecule in their shrunken state, but cannot memorize it in their swollen state.

Molecular imprinting is also effective in preparing molecule-responsive hydrogels by using molecular complexes as dynamic crosslinks. For example, target molecule-responsive hydrogels with β-cyclodextrin (CD) as a ligand were prepared *via* molecular imprinting using bisphenol A (BPA), which could act as an endocrine disruptor, a template molecule and a minute amount of crosslinker (Figure 15.17).[72] After complex formation between acryloyl-CD and template BPA with a stoichiometry of 2 : 1, the

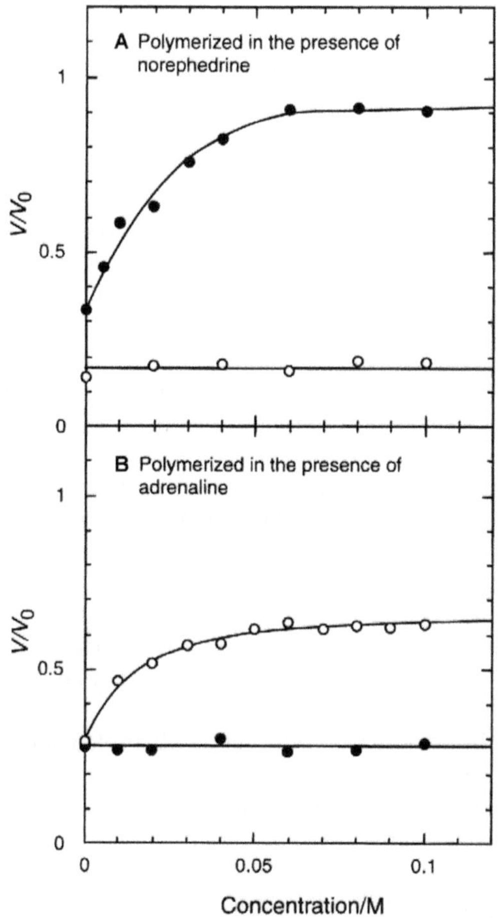

Figure 15.16 Equilibrium swelling ratios at 50 °C as a function of concentration of either norephedrine (●) or adrenaline (○) in water for molecular recognition hydrogels prepared in the presence of norephedrine (A) and adrenaline (B).[70]

resulting sandwich-like CD–BPA–CD complexes with acryloyl groups were copolymerized with AAm and MBAA. BPA-imprinted hydrogels with molecular recognition sites for BPA were obtained by extracting the template BPA from the resultant networks. The BPA-imprinted and non-imprinted hydrogels showed a gradual decrease in swelling ratio after their immersion in aqueous BPA solution (Figure 15.18). It should be noted that the BPA-imprinted hydrogel showed greater shrinkage than the non-imprinted hydrogel in response to target BPA. The compressive modulus measurements revealed that the apparent crosslinking density of the BPA-imprinted and non-imprinted hydrogels increased with increasing BPA concentration. Furthermore, the change in apparent crosslinking density of the BPA-imprinted hydrogel was greater than that of the non-imprinted hydrogel. The BPA-imprinted hydrogel also showed more effective adsorption of BPA than the

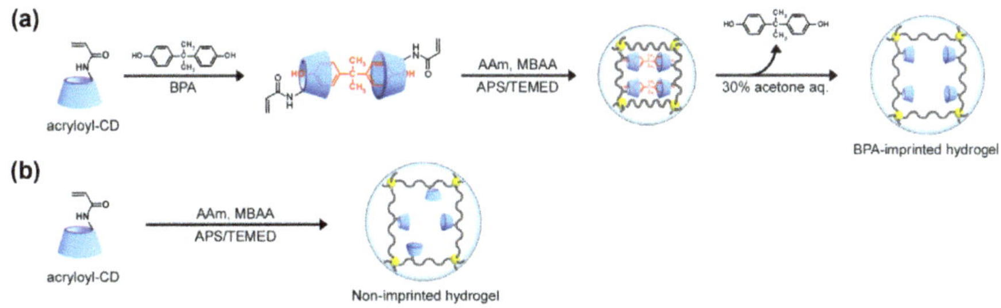

Figure 15.17 Synthesis of BPA-imprinted hydrogel (a) and non-imprinted hydrogel (b).[72]

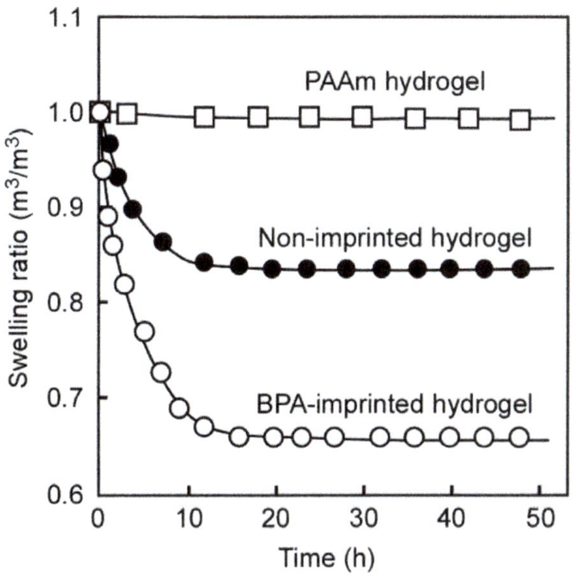

Figure 15.18 Changes in the swelling ratios of BPA-imprinted, non-imprinted, and PAAm hydrogels in an aqueous BPA solution as a function of time.[72]

non-imprinted hydrogel. The highly BPA-responsive shrinkage of the BPA-imprinted hydrogel can be explained by the tentative model schematically illustrated in Figure 15.19. In the BPA-imprinted hydrogel, CDs as ligands for BPA were arranged at suitable positions for the formation of 2 : 1 complexes between CD and BPA because gel networks were formed by molecular imprinting using BPA as a template. Therefore, BPA was more effectively adsorbed into the BPA-imprinted hydrogel than the non-imprinted hydrogel. The formation of the CD–BPA–CD complexes led to an increase in apparent crosslinking density of the BPA-imprinted and non-imprinted hydrogels, followed by shrinkage in response to BPA. The more drastic BPA-responsive shrinkage of the BPA-imprinted hydrogel than that of the non-imprinted hydrogel was attributed to a greater change in the crosslinking density.

Standard molecular imprinting uses low-molecular-weight monomers with a functional group as ligands that interact with a template molecule. Recently,

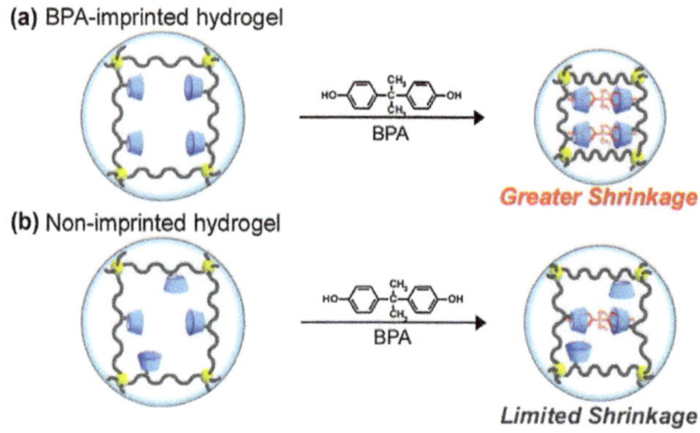

Figure 15.19 Schematic of swelling behaviors of BPA-imprinted hydrogel (a) and non-imprinted hydrogel (b).[72]

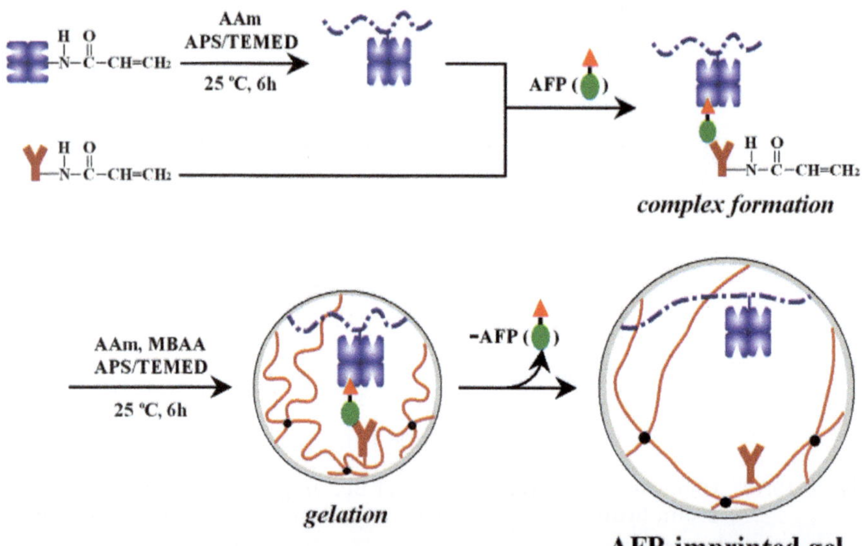

Figure 15.20 Synthesis of AFP-imprinted hydrogels with lectins and antibodies as ligands for template AFP molecules.[73]

biomolecules such as lectins and antibodies were utilized as high-molecular-weight ligands in developing biomolecule-responsive hydrogels by molecular imprinting. For example, glycoprotein-responsive hydrogels were strategically prepared by biomolecular imprinting that used lectin and an antibody as biomolecular ligands. α-Fetoprotein (AFP) is widely used as a tumor-specific marker glycoprotein for the diagnosis of hepatocellular carcinoma, germ cell tumors and metastatic cancers of the liver. In preparing the AFP-responsive hydrogels that underwent changes in volume in response to target AFP, lectin ConA and anti-AFP antibody were used as ligands for the saccharide and peptide chains of target AFP, respectively[73,74] (Figure 15.20). After synthesis of

ConA-conjugated PAAm (ConA–PAAm) by copolymerization of acryloyl-modified ConA with AAm and the formation of lectin–AFP–antibody complexes among the resulting ConA–PAAm, template AFP and acryloyl anti-AFP antibody, acryloyl anti-AFP antibody was copolymerized with MBAA and AAm, followed by the formation of semi-IPN structures. The AFP-imprinted hydrogel with ConA and anti-AFP antibody was obtained by the extraction of the template AFP from the resulting semi-IPN hydrogel network.

The AFP-imprinted hydrogel shrank gradually but non-imprinted hydrogel prepared without template AFP swelled slightly when they were immersed in a buffer solution with target AFP (Figure 15.21). On the other hand, both the AFP-imprinted and non-imprinted hydrogels exhibited slight swelling in the presence of glycoprotein ovalbumin that is recognized by ConA but not recognized by the anti-AFP antibody. Only AFP-imprinted hydrogel accurately recognized AFP as a target tumor marker and shrank gradually with increasing AFP concentration in a buffer solution. The apparent crosslinking

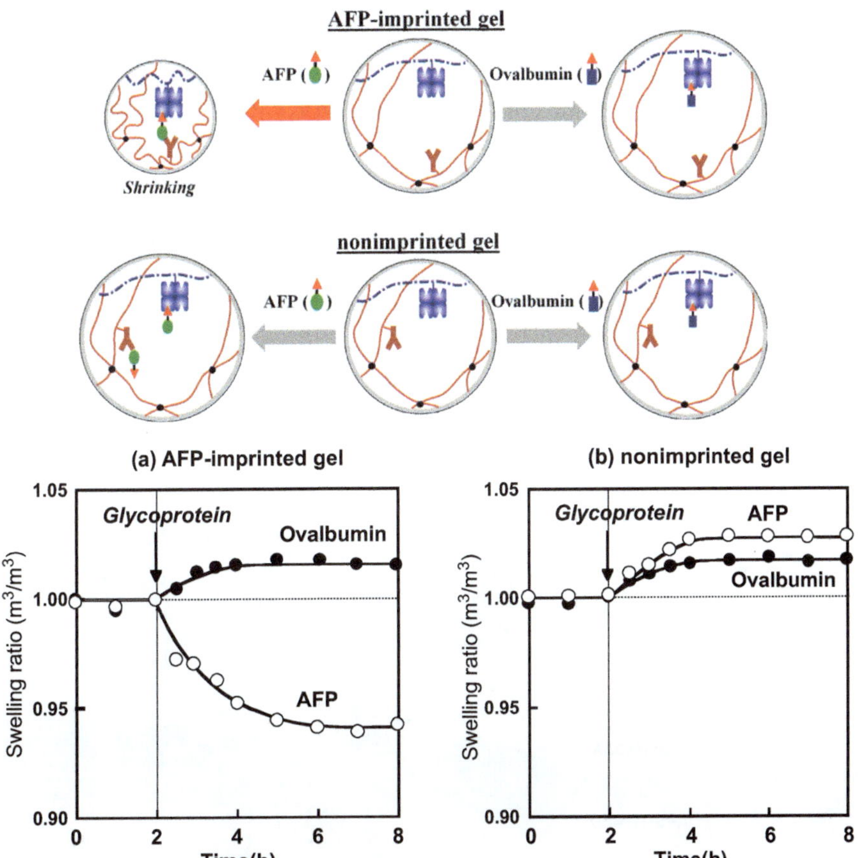

Figure 15.21 Swelling ratio changes of AFP-imprinted gel (a) and non-imprinted gel (b) in a buffer solution with AFP (○) and ovalbumin (●).[73]

density of the AFP-imprinted hydrogels, which was determined by compressive modulus measurements, increased with an increase in the AFP concentration, but that of non-imprinted hydrogel remained unchanged. Using molecular imprinting, lectins and antibodies as ligands were arranged at optimal positions for the simultaneous recognition of saccharide and peptide chains in the AFP molecule. Therefore, the AFP-imprinted hydrogel exhibited AFP-responsive shrinkage because its crosslinking density increased by the formation of sandwich-like lectin–AFP–antibody complexes that act as dynamic crosslinks. Such biomolecule-imprinted hydrogels have many potential applications as smart devices for constructing novel sensor systems and molecular diagnostics because their responsive behavior enables accurate detection and recognition of tumor marker glycoprotein.

Biomolecule-responsive hydrogels were also developed using aptamers as ligands in molecular imprinting. For example, PDGF-ββ and its aptamer were utilized as a target biomolecule and its ligand in preparing biomolecule-responsive hydrogels by molecular imprinting (Figure 15.22).[75] PDGF-ββ, a dimeric protein, is released during blood clotting and is an important biomarker for angiogenesis and arteriosclerosis. The PDGF-ββ-imprinted hydrogels exhibited responsive shrinkage that is visible to the naked eye down to femtomolar concentrations of PDGF-ββ. The PDGF-ββ recognition behavior of the aptamer ligands was amplified by the formation of the PDGF-ββ–aptamer complexes as dynamic crosslinks and the consequential reduction of excluded volume in the hydrogel. Thus, molecular imprinting that uses biomolecular ligands and a minute amount of crosslinker is a promising technique to combine molecular recognition with responsive network changes in tailor made designs of biomolecule-responsive hydrogels.

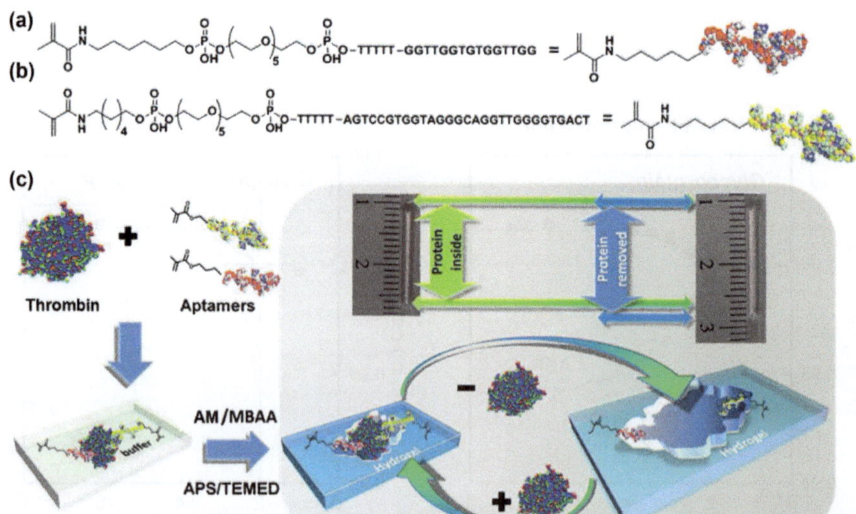

Figure 15.22 Schematic illustration of the reversible swelling/shrinking behavior of PDGF-ββ imprinted aptamer hydrogel in response to PDGF-ββ.[75]

15.5 Micro- and Nano-sized Biomolecule-responsive Hydrogels

15.5.1 Biomolecule-responsive Hydrogel Particles

Rapid response is very important in fabricating smart biomaterials that sense an environmental change and induce structural changes. Therefore, rapidly responsive changes are required in stimuli-responsive hydrogels. In general, the kinetics of hydrogel swelling/shrinking depends on several parameters, such as size, surface area, and porosity. Microscale and nanoscale stimuli-responsive hydrogels undergo rapid volume changes in response to environmental factors because of their large surface/volume ratio, as described in Chapter 3 by Schneider *et al.* In addition, the relaxation time for gel swelling is proportional to the square of the gel radius and the inverse of the collective diffusion coefficient.[76,77] Stimuli-responsive gel particles with nanoscale sizes have many advantages since nanosized stimuli-responsive hydrogels can respond to pH and temperature changes much more rapidly than conventional stimuli-responsive bulk hydrogels. These suggest that the kinetics of the biomolecule-responsive hydrogels can be improved by their miniaturization. For example, rapid response to a change in the glucose concentration was achieved by synthesizing submicron-sized gel particles with glycopolymer–lectin complexes as dynamic crosslinks.

The submicron-sized gel particles that exhibited glucose-responsive size changes were synthesized by surfactant free emulsion copolymerization of *N,N*-diethylaminoethyl methacrylate (DEAEMA), poly(ethylene glycol) dimethacrylate (PEGDMA), 2-glucosyloxyethyl methacrylate (GEMA) and modified lectin (acryloyl ConA) (Figure 15.23).[78] General emulsion polymerizations utilize surfactants to stabilize and disperse monomer emulsions and produce particles in water. Such general emulsion polymerizations with surfactants have disadvantages in synthesizing biomolecule-responsive gel particles because the surfactants adsorbed on the resulting particle surfaces often affect the properties and functions of the particles. The ConA-copolymerized gel particles synthesized by surfactant-free emulsion

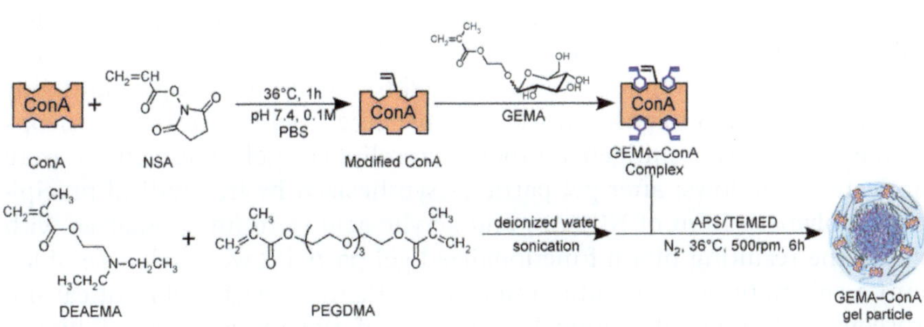

Figure 15.23 Synthesis of bioconjugated gel particles with GEMA–ConA complexes by surfactant-free emulsion polymerization.[78]

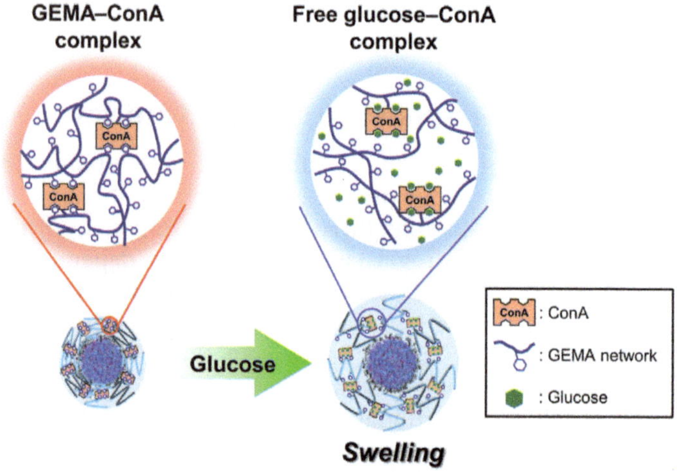

Figure 15.24 Schematic of the glucose-responsive behavior of GEMA–ConA gel particles.[78]

polymerization were colloidally stable in a buffer solution. XPS surface analysis revealed that the ConA-copolymerized gel particles had a heterogeneous structure in which a relatively hydrophobic DEAEMA was localized inside the particles and hydrophilic GEMA–ConA complexes and PEGDMA were localized near the particle surfaces. The ConA-copolymerized gel particles swelled rapidly in a buffer solution with free glucose but did not change at all with free galactose (Figure 15.24). The swelling ratio of the ConA-copolymerized gel particles strongly depended on the glucose concentration of the buffer solution. The more rapid glucose-responsive swelling of the ConA-copolymerized gel particles compared to the macrosized PGEMA–ConA complex hydrogels described in Section 15.2.3 demonstrates that the response rate of biomolecule-responsive hydrogels can be effectively enhanced by their miniaturization. The rapid glucose-responsive function of the ConA-copolymerized gel particles can provide a useful tool for self-regulated insulin release in which insulin is autonomously released in response to blood glucose concentration.

Biomolecule-responsive gel particles were utilized to fabricate a dynamically tunable microlens array.[79] A label-free biosensing technique as a new protein detection technology was developed by combining antigen–antibody crosslinked gel microlenses with a simple brightfield optical microscopy technique.[80,81] The antigen–antibody crosslinked gel microlenses were fabricated as follows: after gel particles synthesized by free radical precipitation polymerization of NIPAAm and acrylic acid were functionalized with biotin, the resulting biotin-functionalized gel particles were self-assembled onto a substrate by Coulombic interaction. Using a brightfield optical microscopy technique, the optical properties of the microlenses of biotin-functionalized gel particles were investigated in the presence of avidin and polyclonal anti-biotin. The microlenses showed a difference in the appearance of the differential interference contrast (DIC) images as a function of the avidin and anti-biotin concentrations because the crosslinking density changed by the complex formation of biotin in gel particles with avidin and anti-biotin. The unique characteristics of the biomolecule-responsive microlens assay enable the detection of a target protein by monitoring the focal length of the microlens.

15.5.2 Biomolecule-responsive Hydrogel Layers

The surface plasmon resonance (SPR) sensor has been widely used in the biochemical and biomedical fields because it enables us to selectively and quantitatively measure kinetics of association and dissociation of biomolecular complexes and to determine affinity constant.[82–84] When analytes bind to ligands immobilized on the SPR sensor chip surfaces, the permittivity in the proximity of the surface changes sensitively, resulting in resonance angle shifts. In general, proteins, saccharides and DNA have been covalently immobilized on the SPR sensor chip surfaces by standard amine coupling methods. Recently, hydrogel layers with antibodies as ligands were formed by surface-initiated atom transfer radical polymerization (SI-ATRP).[85] The formation of gel layers by SI-ATRP enables us to three-dimensionally immobilize a large amount of ligands and to control the layer thickness. The SPR signals were dramatically enhanced by the formation of gel layers with a large amount of antibodies.

In addition, the formation of molecularly imprinted hydrogel layers with molecular recognition sites on the gold chip surfaces *via* SI-ATRP was reported as a novel approach to develop a SPR sensing system to detect a target protein.[86] Protein-imprinted hydrogel layers with recognition sites for a target protein were prepared on the SPR sensor chips by molecular imprinting with a minute amount of crosslinker. Recognition sites for a target protein in the hydrogel thin layer were created by molecular imprinting that used a lectin (ConA) as a template protein and a monomer with pendant glucose (GEMA) as its ligand (Figure 15.25). The ConA-imprinted hydrogel layers were prepared by grafting from SPR sensor chips *via* SI-ATRP combined with molecular imprinting. After GEMA as a ligand monomer was mixed with ConA in a phosphate buffer solution for their complex formation, it was copolymerized with AAm as a main monomer and MBAA as a crosslinker by SI-ATRP. Then, ConA-imprinted hydrogel layer chips were obtained by extracting template ConA from the resulting networks in an aqueous glucose solution. Using a SPR sensor, the resonance angle shifts of the ConA-imprinted and non-imprinted hydrogel layer chips in response to a target ConA were investigated. The resonance angle shift of the ConA-imprinted hydrogel layer chip in response to target ConA was much larger than that of the non-imprinted hydrogel layer chip. Analysis of the resonance angle shift concluded that the affinity constants of ConA to the ConA-imprinted and non-imprinted hydrogel layer chips were 2.02×10^{7} and 5.26×10^{5} (M^{-1}), respectively. The larger resonance angle shift and affinity

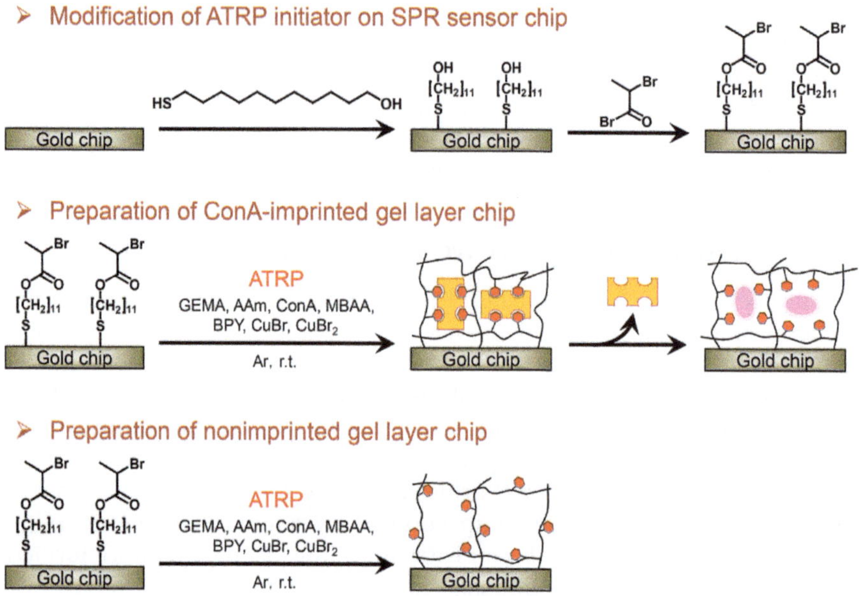

Figure 15.25 Preparation of ConA-imprinted and non-imprinted hydrogel layer chips *via* SI-ATRP.[86]

constant to ConA of the ConA-imprinted hydrogel layer chip was attributed to the fact that molecular imprinting with a minute amount of crosslinker enabled GEMA as a ligand to be arranged at optimal positions for binding target ConA. These indicate that hydrogels composed of flexible polymer networks with a minute amount of crosslinker and large open spaces are very suitable to memorize macromolecules with a large size such as proteins in molecular imprinting. In addition, the resonance angle shift of the ConA-imprinted hydrogel layer chip strongly depended on the layer thickness, which was controllable by the conditions during the preparation of SI-ATRP such as polymerization time and monomer concentrations. Thus, the sensitivity of the SPR sensor was effectively enhanced by the combination of molecular imprinting and SI-ATRP. Controlling the thickness of molecularly imprinted hydrogel layers by SI-ATRP contributes significantly to the development of sensors and separating systems in the biochemical and biomedical fields.

15.6 Other Biomolecule-responsive Hydrogels

A variety of fascinating biomolecule-responsive hydrogels that undergo structural changes in response to cell surface receptors, antibiotic drugs, DNA and so on have been reported. For example, noncovalently associated hydrogels were produced by the complex formation of a low-molecular-weight heparin-modified PEG star polymer (PEG-LMWH) and a dimeric heparin-binding growth factor (VEGF) (Figure 15.26).[87] The presence of

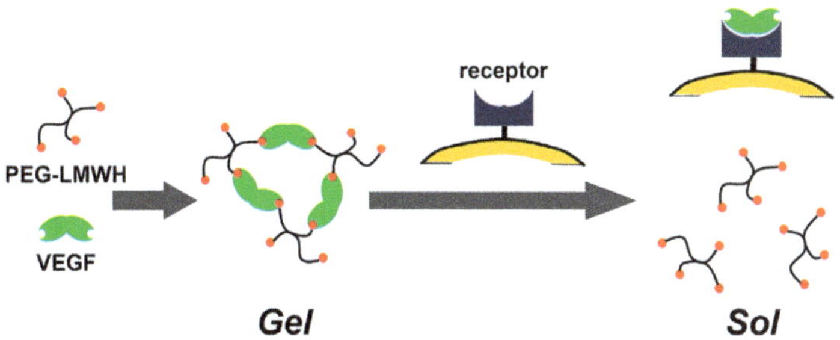

Figure 15.26 Cell-responsive sol–gel transition systems by gelation of multi-armed PEG with an adhesion receptor-binding motif by the addition of a bis-cysteine peptide crosslinker.[87]

VEGF receptors caused the dissociation of the complexes by selective removal of VEGF, followed by receptor-mediated erosion of the PEG-LMWH/VEGF hydrogels. Using the hydrogels that underwent the receptor-mediated erosion, VEGF release in response to cell surface receptors was achieved, indicating high potential for use in vascular therapies.

To fabricate the trigger-inducible release of human vascular endothelial growth factor, antibiotic drug-responsive hydrogels were formed by the addition of the aminocoumarin antibiotic coumermycin to polyacrylamide conjugated with genetically engineered bacterial gyrase subunit B (GyrB) that undergoes dimerization with coumermycin.[88] By the addition of clinically validated novobiocin (albamycin), the hydrogel changed to a sol state owing to the dissociation of the complexes between the GyrB subunits and coumermycin. Then the human vascular endothelial growth factor 121 (VEGF121) entrapped within the hydrogel networks was rapidly released. Optimal administration of the rapidly growing number of protein-based biopharmaceuticals can be achieved using such antibiotic-inducible release using drug-responsive hydrogels.

Substantial conformational changes of well-defined folding motifs of proteins provided useful tools for preparing stimuli-responsive hydrogels that undergo volume changes or sol–gel phase transition in response to environmental changes such as pH and temperature.[89,90] Focusing on the protein calmodulin (CaM), which undergoes two conformational changes from an extended conformation in the presence of Ca^{2+} to a collapsed conformation upon the binding of phenotiazine as a ligand, the Ca^{2+}-responsive hydrogel was prepared by incorporating genetically engineered CaM and phenothiazine derivatives as pendant moieties within the hydrogel network.[91] Noncovalent crosslinking of the hydrogel network was formed in the presence of Ca^{2+} because the phenothiazine binding site of CaM was accessible and immobilized phenothiazine derivative was bound to CaM. However, when Ca^{2+} was removed from the hydrogel in the presence of a calcium chelator, it swelled owing to the dissociation of CaM-phenothiazine derivative complexes as dynamic crosslinks. Using CaM-phenothiazine hydrogel membranes, the permeation of small molecules was regulated by their reversible swelling/shrinking cycles in response to Ca^{2+} concentration. The CaM-phenothiazine hydrogel also acted as a gate for controlling flow from a reservoir in microfluidics.

Allosteric enzyme and transporter protein undergo conformational changes in response to a given stimulus and regulate their functions such as catalytic activity and binding affinity. Such conformational changes of proteins are very useful in designing biomolecule-responsive hydrogels that exhibit swelling/shrinking in response to a target biomolecule. For example, ATP-responsive hydrogels were strategically synthesized using bacterial enzyme adenylate kinase (AKe), which catalyzes the phosphoryl transfer reaction, as conformationally changeable crosslinks.[92] When ATP binds to the active site of AKe, the distance in Cα-atoms between A55 and V169 in AKe decreases from 29.5 Å to 12.4 Å by its large conformational change. Using a triple mutant of AKe, AKtm (C77S, A55C and V169C), which were prepared by a genetic engineering procedure, as macrocrosslinkers, the ATP-responsive hydrogels were synthesized by crosslinking poly-(hydroxypropyl methacrylamide) (polyHPMA) having pendant maleimide groups with AKtm *via* a thiol-maleimide coupling reaction. The AKtm-crosslinked hydrogels shrank immediately in a buffer solution with ATP because of the conformational transition of AKtm from an "open" to a "closed" conformation in response to ATP (Figure 15.27). Thus, the responsive behavior of the AKtm-crosslinked hydrogels transduced the nanoscale conformational changes induced by the substrate binding to a macroscale mechanical motion.

In addition to saccharides and proteins, DNA and RNA are important signals to monitor physiological conditions because a large number of specific DNA and RNA are closely related to diseases such as cancer. To detect a tiny amount of target DNA and RNA sensitively and accurately, diagnosis tools such as DNA chips and quantitative real-time reverse-

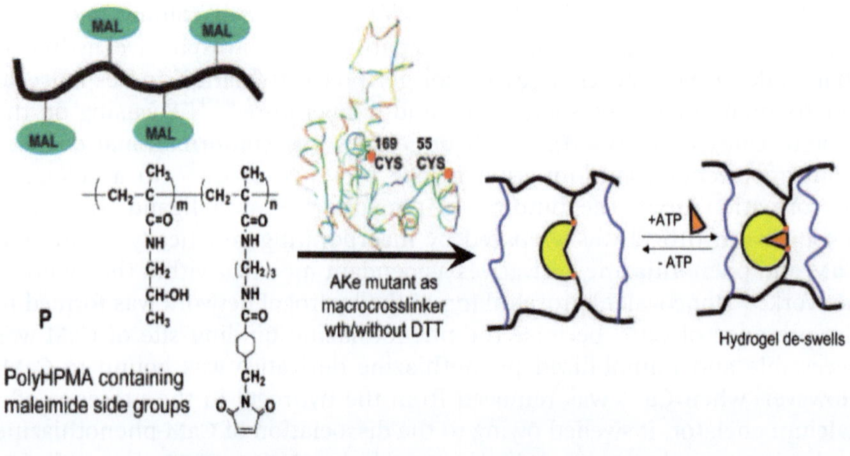

Figure 15.27 Schematic illustration of the reversible swelling/shrinking behaviors of AKtm-crosslinked hydrogels in response to ATP concentration.[92]

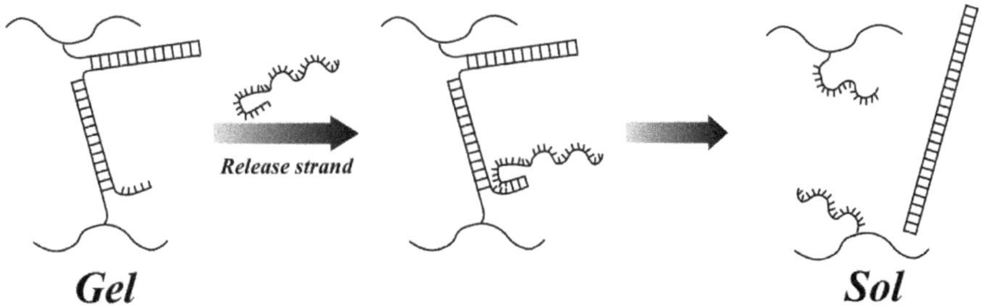

Figure 15.28 Schematic illustration of the sol–gel transition behavior of bioconjugated hydrogels with DNA as dynamic crosslinks.[93]

transcription PCR have been developed using the hybridization between complementary strands of DNA or RNA. Similarly, utilization of the hybridization of DNA and RNA enables the designs of DNA- and RNA-responsive hydrogels that undergo sol–gel phase transition or changes in volume in response to target DNA and RNA. For example, two different single-stranded (ss) DNA–PAAm conjugates were synthesized by copolymerization of AAm and two ssDNAs with polymerizable acryloyl groups.[93] The addition of DNA (crosslinker strand) with a sequence complementary to both ssDNA–PAAm conjugates resulted in the formation of a bioconjugated hydrogel with double-stranded (ds) DNA as dynamic crosslinks. The addition of another DNA (release strand) fully complementary to the crosslinker strand induced the dissolution of the hydrogel by dissociation of the duplex between the ssDNA–PAAm conjugates and crosslinker strand that acted as dynamic crosslinks (Figure 15.28). Trapping and release of fluorescent semiconductor quantum dots (QDs) from DNA-responsive hydrogels were visualized to investigate the controllable macroscopic rheological properties of DNA-responsive sol–gel transitions.[94]

An interesting approach for engineering highly selective target molecule-responsive hydrogels uses DNA aptamers as dynamic crosslinks that selectively recognize various target molecule.[95] Aptamers are ssDNAs that bind to a specific target molecule (for example, proteins), and are usually created by an *in vitro* selection process. After complex formation between a thrombin-specific aptamer and thrombin, PAAm with pendant ssDNAs complementary to a part of the thrombin-specific aptamer was mixed with the resultant aptamer–thrombin complexes, followed by formation of the hydrogel.[96] The addition of release strands with a sequence fully complementary to the thrombin-specific aptamer resulted in the dissolution of the hydrogel and release of thrombin. Target biomolecule-responsive hydrogels that underwent sol–gel phase transition in response to target adenosine triphosphate (ATP) were also designed using duplex formation between the ssDNA–PAAm conjugate and aptamer as a crosslink.[97] The hydrogel formed from ssDNA–PAAm and ATP was dissociated in the presence of ATP because of the dissociation of DNA duplex crosslinks by competitive binding of target ATP to the aptamer.

In contrast to DNA-responsive sol–gel phase transitions, DNA-responsive hydrogels that swell or shrink by the recognition of the DNA sequences have been reported.[98,99] Two types of DNA, which were a ssDNA crosslink without an intramolecular base pair and a DNA crosslink with a stem-loop structure, were utilized as conformationally changeable crosslinks for designing DNA-responsive hydrogels.[98] In a solution with a target DNA with a complementary sequence, the PAAm hydrogel with ssDNA as a changeable crosslink shrank but the PAAm hydrogel with stem-loop structured DNA swelled. On the other hand, their hydrogels did not change the volume at all in the presence of noncomplementary DNA. Their DNA-responsive shrinking and swelling were attributed to a decrease and increase in the chain length of DNA crosslinks induced by the duplex formation of the conformationally changeable DNA crosslinks and stem-loop DNA with target DNA, respectively. Such DNA-responsive swelling/shrinking behavior of the hydrogels provide many advantages in sensing systems and the selective release of therapeutic agents in response to target DNA and proteins related to disease.

15.7 Conclusions

Target biomolecule-responsive hydrogels are unique biomaterials that undergo changes in volume in response to a target biomolecule. There have been a variety of biomolecule-responsive hydrogels that can sense saccharides and proteins and that induce volume changes or a sol–gel phase transition. The design of such biomolecule-responsive hydrogels requires the cooperative combination of biomolecular recognition events with responsive structure changes of their networks. Standard strategies for designing biomolecule-responsive hydrogels use a combination of biomolecular recognition events with pH- or temperature-responsive polymers. Recently, another strategy using biomolecular complexes as dynamic crosslinks has become increasingly important because it enables us to prepare a variety of biomolecule-responsive hydrogels. The reversible association/dissociation behavior of biomolecular complexes can provide tools for creating smart biomaterials with a wide variety of uses. Responsive swelling/shrinking changes and sol–gel phase transitions of biomolecule-responsive hydrogels have contributed significantly to the development of self-regulated DDSs, sensors, microdevices and cell culture systems. The knowledge gained from the successful design of biomolecule-responsive hydrogels provides the basis for the development of innovative smart biomaterials in various biomedical applications. Even though biomolecule-responsive hydrogels require further research work into potential applications, they are likely to become quite innovative and important smart biomaterials in the future.

Abbreviations

AAc	acrylic acid
AAm	acrylamide

AFP	α-fetoprotein
AKe	adenylate kinase
Ala	alanine
AmECFPBA	4-(2-acrylamidoethylcarbamoyl)-3-fluorophenylboronic acid
APBA	3-(acrylamido)phenylboronic acid
Arg	arginine
Asp	aspartic acid
ATP	adenosine triphosphate
BPA	bisphenol A
CaM	calmodulin
CD	cyclodextrin
ConA	concanavalin A
Cys	cysteine
DDS	drug delivery system
DEAEMA	*N,N*-diethylaminoethyl methacrylate
DIC	differential interference contrast
DMAAm	*N,N*-dimethylacrylamide
ECM	extracellular matrix
EG	ethylene glycol
ELISA	enzyme-linked immunosorbent assay
GEMA	2-glucosyloxyethyl methacrylate
GOD	glucose oxidase
GyrB	gyrase subunit B
HPMA	2-hydroxypropyl methacrylate
IgG	immunoglobulin G
IPN	interpenetrating polymer network
LCST	low critical solution temperature
Lys	lysine
MAAc	methacrylic acid
MBAA	*N,N′*-methylenebisacrylamide
MMP	matrix metalloproteinase
NIPAAm	*N*-isopropyl acrylamide
NIPMAAm	*N*-isopropyl methacrylamide
NSA	*N*-succinimidylacrylate
PAAm	poly(acrylamide)
PEG	poly(ethylene glycol)
PEGDMA	poly(ethylene glycol) dimethacrylate
PEG-LMWH	low-molecular-weight heparin-modified PEG star polymer
PGEMA	poly(2-glucosyloxyethyl methacrylate)
PNIPAAm	poly(*N*-isopropyl acrylamide)
QD	quantum dot
rhBMP-2	recombinant human bone morphogenetic protein-2
SI-ATRP	surface-initiated atom transfer radical polymerization
SPR	surface plasmon resonance
TGase	transglutaminase
Tyr	tyrosine
VEGF	vascular endothelial growth factor

References

1. N. A. Peppas, *Hydrogels in Medicine and Pharmacy*, CRC, Boca Raton, 1987.
2. T. Miyata, in *Supramolecular Design for Biological Applications*, ed. N. Yui, CRC, Boca Raton, USA, 2002, pp. 95–136.
3. T. Tanaka, *Phys. Rev. Lett.*, 1978, **40**, 820–823.
4. Y. Hirokawa and T. Tanaka, *J. Chem. Phys.*, 1984, **81**, 6379–6380.
5. G. Chen and A. S. Hoffman, *Nature*, 1995, **373**, 49–52.
6. R. Yoshida, K. Uchida, T. Kaneko, K. Sakai, A. Kikuchi, Y. Sakurai and T. Okano, *Nature*, 1995, **374**, 240–242.
7. T. Tanaka, D. Fillmore, S.-T. Sun, I. Nishio, G. Swislow and A. Shah, *Phys. Rev. Lett.*, 1980, **45**, 1636–1639.
8. M. Annaka and T. Tanaka, *Nature*, 1992, **335**, 430–432.
9. T. Tanaka, I. Nishio, S.-T. Sun and S. Ueno-Nishio, *Science*, 1982, **218**, 467–469.
10. Y. Osada, H. Okuzaki and H. Hori, *Nature*, 1992, **355**, 242–244.
11. A. Suzuki and T. Tanaka, *Nature*, 1990, **346**, 345–347.
12. R. A. Siegel, *Adv. Polym. Sci.*, 1993, **109**, 233–267.
13. C. S. Brazel and N. A. Peppas, *Macromolecules*, 1995, **28**, 8016–8020.
14. L.-C. Dong and A. S. Hoffman, *J. Controlled Release*, 1991, **5**, 141–152.
15. T. Miyata, K. Nakamae, A. S. Hoffman and Y. Kanzaki, *Macromol. Chem. Phys.*, 1994, **195**, 1111–1120.
16. K. Nakamae, T. Nizuka, T. Miyata, M. Furukawa, T. Nishio, K. Kato, T. Inoue, A. S. Hoffman and Y. Kanzaki, *J. Biomater. Sci. Polym. Ed.*, 1997, **9**, 43–53.
17. A. S. Hoffman, *J. Controlled Release*, 1987, **6**, 297–305.
18. T. Okano, *Adv. Polym. Sci.*, 1993, **110**, 179–197.
19. L.-C. Dong and A. S. Hoffman, *J. Controlled Release*, 1990, **13**, 21–31.
20. T. Okano, Y. H. Bae, H. Jacobs and S. W. Kim, *J. Controlled Release*, 1990, **11**, 255–265.
21. T. Miyata, T. Uragami and K. Nakamae, *Adv. Drug Delivery Rev.*, 2002, **54**, 79–98.
22. T. Miyata and T. Uragami, in *Biomedical Applications of Hydrogels Handbook*, ed. R. M. Ottenbrite, K. Park and T. Okano, Springer, New York, USA, 2010, pp. 65–86.
23. K. Ishihara, M. Kobayashi, N. Ishimaru and I. Shinohara, *Polym. J.*, 1984, **16**, 625–631.
24. K. Ishihara and K. Matsui, *J. Polym. Sci. Polym. Lett. Ed.*, 1986, **24**, 413–417.
25. G. Albin, T. A. Horbett and B. D. Ratner, *J. Controlled Release*, 1985, **2**, 153–164.
26. S. Cartier, T. A. Horbett and B. D. Ratner, *J. Membr. Sci.*, 1995, **106**, 17–24.
27. C. M. Hassan, F. J. Doyle III and N. A. Peppas, *Macromolecules*, 1997, **30**, 6166–6173.
28. R. S. Parker, F. J. Doyle III and N. A. Peppas, *IEEE Trans. Biomed. Eng.*, 1999, **46**, 148–157.

29. M. Brownlee and A. Cerami, *Science*, 1979, **206**, 1190–1191.
30. L. A. Seminoff, G. B. Olsen and S. W. Kim, *Int. J. Pharm.*, 1989, **54**, 241–249.
31. K. Makino, E. J. Mack, T. Okano and S. W. Kim, *J. Controlled Release*, 1990, **12**, 235–239.
32. E. Kokufuta, Y.-Q. Zhang and T. Tanaka, *Nature*, 1991, **351**, 302–304.
33. T. Aoki, Y. Nagao, K. Sanui, N. Ogata, A. Kikuchi, Y. Sakurai, K. Kataoka and Y. Okano, *Polym. J.*, 1996, **28**, 371–374.
34. K. Kataoka, H. Miyazaki, M. Bunya, T. Okano and Y. Sakurai, *J. Am. Chem. Soc.*, 1998, **120**, 12694–12695.
35. A. Matsumoto, T. Ishii, J. Nishida, H. Matsumoto, K. Kataoka and Y. Miyahara, *Angew. Chem., Int. Ed.*, 2012, **51**, 2124–2128.
36. T. Miyata and K. Nakamae, *Trends Polym. Sci.*, 1997, **5**, 198–206.
37. K. Nakamae, T. Miyata, A. Jikihara and A. S. Hoffman, *J. Biomater. Sci., Polym. Ed.*, 1994, **6**, 79–90.
38. T. Miyata, A. Jikihara, K. Nakamae and A. S. Hoffman, *Macromol. Chem. Phys.*, 1996, **197**, 1135–1146.
39. T. Miyata, A. Jikihara, K. Nakamae and A. S. Hoffman, *J. Biomater. Sci. Polym. Ed.*, 2004, **15**, 1085–1098.
40. S. J. Lee and K. Park, *J. Mol. Recognit.*, 1996, **9**, 549–557.
41. A. A. Obaidat and K. Park, *Pharm. Res.*, 1996, **13**, 989–995.
42. A. A. Obaidat and K. Park, *Biomaterials*, 1997, **18**, 801–806.
43. L. Hovgaard and H. Brøndsted, *J. Controlled Release*, 1995, **36**, 159–166.
44. M. Saffran, G. S. Kumar, C. Savariar, J. C. Burnham, F. Williams and D. C. Neckers, *Science*, 1986, **233**, 1081–1086.
45. H. Ghandehari, P. Kopeckova and J. Kopecek, *Biomaterials*, 1997, **18**, 861–872.
46. E. O. Akala, P. Kopeckova and J. Kopecek, *Biomaterials*, 1998, **19**, 1037–1047.
47. M. P. Lutolf, J. L. Lauer-Fields, H. G. Schmoekel, A. T. Metters, F. E. Weber, G. B. Fields and J. A. Hubbell, *Proc. Natl. Acad. Sci. U. S. A.*, 2003, **100**, 5413–5418.
48. M. P. Lutolf and J. A. Hubbell, *Biomacromolecules*, 2003, **4**, 713–722.
49. M. P. Lutolf, G. P. Raeber, A. H. Zisch, N. Tirelli and J. A. Hubbell, *Adv. Matter.*, 2003, **15**, 888–892.
50. M. P. Lutolf, F. E. Weber, H. G. Schmoekel, J. C. Schense, T. Kohler, R. Müller and J. A. Hubbel, *Nat. Biotechnol.*, 2003, **21**, 513–518.
51. K. N. Plunkett, K. L. Berkowski and J. S. Moore, *Biomacromolecules*, 2005, **6**, 632–637.
52. N. Yamamoto, M. Kurisawa and N. Yui, *Macromol. Rapid Commun.*, 1996, **17**, 313–318.
53. M. Kurisawa and N. Yui, *J. Controlled Release*, 1998, **54**, 191–200.
54. K. M. Huh, J. Hashi, T. Ooya and N. Yui, *Macromol. Chem. Phys.*, 2000, **201**, 613–619.
55. Y. Kumashiro, T. Ooya and N. Yui, *Macromol. Rapid Commun.*, 2004, **25**, 867–872.
56. J. J. Sperinde and L. G. Griffith, *Macromolecules*, 1997, **30**, 5255–5264.

57. B.-H. Hu and P. B. Messersmith, *J. Am. Chem. Soc.*, 2003, **125**, 14296–14299.
58. P. D. Thornton, R. J. Mart and R. V. Ulijn, *Adv. Mater.*, 2007, **19**, 1252–1256.
59. P. D. Thornton, G. McConnell and R. V. Ulijn, *Chem. Commun.*, 2005, 5913–5915.
60. E. P. Diamandis and T. K. Christopoulos, *Immunoassay*, Academic Press, New York, USA, 1996.
61. T. Miyata, N. Asami and T. Uragami, *Nature*, 1999, **399**, 766–769.
62. T. Miyata, N. Asami and T. Uragami, *Macromolecules*, 1999, **32**, 2082–2084.
63. T. Miyata, N. Asami and T. Uragami, *J. Polym. Sci., Polym. Phys.*, 2009, **47**, 2144–2157.
64. T. Miyata, N. Asami, Y. Okita and T. Uragami, *Polym. J.*, 2010, **42**, 834–837.
65. K. Mosbach, *Trends Biochem. Sci.*, 1994, **19**, 9–14.
66. K. J. Shea, *Trends Polym. Sci.*, 1994, **2**, 166–173.
67. G. Wulff, *Angew. Chem., Int. Ed. Engl.*, 1995, **34**, 1812–1832.
68. M. Byrne, K. Park and N. A. Peppas, *Adv. Drug Delivery*, 2002, **54**, 149–161.
69. N. M. Bergmann and N. A. Peppas, *Prog. Polym. Sci.*, 2008, **33**, 271–288.
70. M. Watanabe, T. Akahoshi, Y. Tabata and D. Nakayama, *J. Am. Chem. Soc.*, 1998, **120**, 5577–5578.
71. T. Oya, T. Enoki, A. U. Grosberg, S. Masamune, T. Sakiyama, Y. Takeoka, K. Tanaka, G. Wang, Y. Yilmaz, M. S. Feld, R. Dasari and T. Tanaka, *Science*, 1999, **286**, 1543–1545.
72. A. Kawamura, T. Kiguchi, T. Nishihata, T. Uragami and T. Miyata, *Chem. Commun.*, 2014, **50**, 11101–11103.
73. T. Miyata, M. Jige, T. Nakaminami and T. Uragami, *Proc. Natl. Acad. Sci. U. S. A.*, 2006, **103**, 1190–1193.
74. T. Miyata, T. Hayashi, Y. Kuriu and T. Uragami, *J. Mol. Recognit.*, 2012, **25**, 336–343.
75. W. Bai, N. A. Gariano and D. A. Spivak, *J. Am. Chem. Soc.*, 2013, **135**, 6977–6984.
76. T. Tanaka and D. J. Fillmore, *J. Chem. Phys.*, 1979, **70**, 1214–1218.
77. T. Tanaka, E. Sato, Y. Hirokawa, S. Hirotsu and J. Peetermans, *Phys. Rev. Lett.*, 1985, **55**, 2455–2458.
78. A. Kawamura, Y. Hata, T. Miyata and T. Uragami, *Colloid Surf. B*, 2012, **99**, 74–81.
79. S. Nayak and L. A. Lyon, *Angew. Chem., Int. Ed.*, 2005, **44**, 7686–7780.
80. J. Kim, S. Nayak and L. A. Lyon, *J. Am. Chem. Soc.*, 2005, **127**, 9588–9592.
81. J. Kim, N. Singh and L. A. Lyon, *Angew. Chem., Int. Ed.*, 2006, **45**, 1446–1449.
82. J. Homola, *Chem. Rev.*, 2008, **108**, 462–493.
83. N. Calakos, M. K. Bennett, K. E. Peterson and R. H. Scheller, *Science*, 1994, **263**, 1146–1149.

84. S. C. Schuster, R. V. Swanson, L. A. Alex, R. B. Bourret and M. I. Simon, *Nature*, 1993, **365**, 343–347.
85. Y. Kuriu, M. Ishikawa, A. Kawamura, T. Uragami and T. Miyata, *Chem. Lett.*, 2012, **41**, 1660–1662.
86. Y. Kuriu, A. Kawamura, T. Uragami and T. Miyata, *Chem. Lett.*, 2014, **43**, 825–827.
87. N. Yamaguchi, L. Zhang, B.-S. Chae, C. S. Palla, E. M. Furst and K. L. Kiick, *J. Am. Chem. Soc.*, 2007, **129**, 3040–3041.
88. M. Ehrbar, R. Schoenmakers, E. H. Christen, M. Fussenegger and W. Weber, *Nat. Mater.*, 2008, **7**, 800–804.
89. W. A. Petka, J. L. Harden, K. P. McGrath, D. Wirtz and D. A. Tirrell, *Science*, 1998, **281**, 389–392.
90. C. Wang, R. J. Stewart and J. Kopecek, *Nature*, 1999, **397**, 417–420.
91. J. D. Ehrick, S. K. Deo, T. W. Browning, L. G. Bachas, M. J. Madou and S. Daunert, *Nat. Mater.*, 2005, **4**, 298–302.
92. W. Yuan, J. Yang, P. Kopečková and J. Kopeček, *J. Am. Chem. Soc.*, 2008, **130**, 15760–15761.
93. D. C. Lin, B. Yurke and N. A. Langrana, *J. Biomech. Eng.*, 2004, **126**, 104–110.
94. T. Liedl, H. Dietz, B. Yurke and F. Simmel, *Small*, 2007, **3**, 1688–1693.
95. H. Yang, H. Liu, H. Kang and W. Tan, *J. Am. Chem. Soc.*, 2008, **130**, 6320–6321.
96. B. Wei, I. Cheng, K. Q. Luo and Y. Mi, *Angew. Chem., Int. Ed.*, 2008, **47**, 331–333.
97. H. Yang, H. Liu, H. Kang and W. Tan, *J. Am. Chem. Soc.*, 2008, **130**, 6320–6321.
98. Y. Murakami and M. Maeda, *Biomacromolecules*, 2005, **6**, 2927–2929.
99. Y. Murakami and M. Maeda, *Macromolecules*, 2005, **38**, 1535–1537.

CHAPTER 16

Polymer Brushes with Chemical Responses

ELZA CHU AND ALEXANDER SIDORENKO*

Department of Chemistry & Biochemistry, University of the Sciences, Philadelphia, PA, USA
*Email: a.sidorenko@usciences.edu

16.1 Introduction: Polymer Brushes

Polymer brushes (PBs) have emerged as a unique interfacial design which can be sensitive to different physical and chemical stimuli. By the term "polymer brush" most researchers mean a layer of polymer chains permanently tethered to a solid substrate with one end; the density of chains is high enough to ensure stretching of the polymer coils in a direction orthogonal to the substrate plane. Some deviations may include a liquid–air interface and physical adsorption instead of covalent bonds; these systems are also referred as PBs or "brush-like" polymer layers (Figure 16.1).

This review is focused on PB responsiveness to chemical stimuli. However, it is difficult, if not impossible, to distinguish between the chemical and physical nature of different mechanisms involved in the complex response of various PBs. Our review begins with a brief introduction to PBs (Section 16.1) and concise mechanistic summary of stimuli and responses in PB (Section 16.2). In the main part, we cover three of the most important cases of "chemically" stimulated PBs: polyelectrolyte PBs (Section 16.3), brushes bearing RedOx groups (Section 16.4), and photochemical response in PBs (Section 16.5). In Section 16.6 we will scan through hybrid and complex

RSC Smart Materials No. 14
Chemoresponsive Materials: Stimulation by Chemical and Biological Signals
Edited by Hans-Jörg Schneider

Published by the Royal Society of Chemistry, www.rsc.org

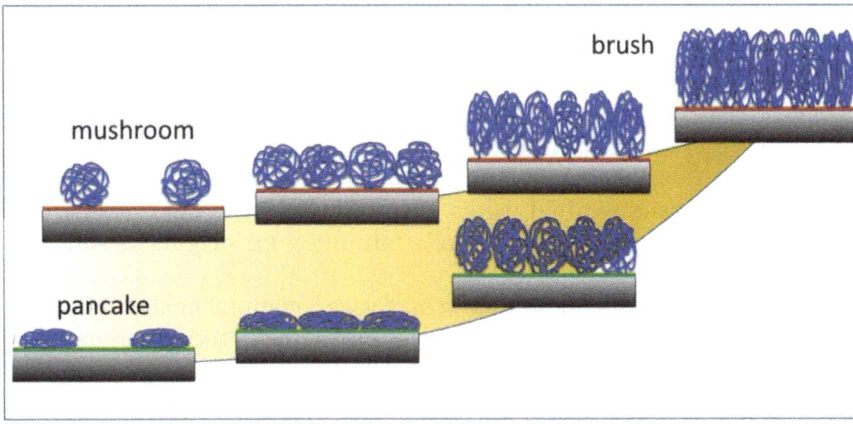

Figure 16.1 The concept of a polymer brush as surface-tethered polymer coils with dense packing.

brushes and in Section 16.7 we provide some remarks on applications and a brief summary.

16.1.1 Early Studies and Models

The singular behavior of grafted polymer chains was acknowledged for some colloidal dispersions back in 1952.[1,2] The need to study the practical aspects of stability and miscibility of composite materials, adhesives and paints inspired fundamental research in polymer adsorption on and grafting to solid surfaces.[3–7] Alexander[8] and de Gennes[9] applied the classical Flory's theory of polymer solutions to surface-tethered polymer chains in the approximation of strong adsorption to the surface (Alexander) and no adsorption (grafting only, de Gennes) using scaling methods and predicted several regimes of grafted polymers: (i) low grafting density, mushroom-like (no adsorption) or pancake-like (adsorption) conformations of the coils when average coil-to-coil distance d is larger than coil diameter D ($d>D$); (ii) the overlap occurs at $d=D$, after which coils adopt (iii) a stretched conformation, *i.e.* regime of brush. In the regime of overlap, the height of the coils h scales with the grafting density σ as

$$h \propto N\sigma^{\alpha} \tag{16.1}$$

where N is the polymerization degree and the parameter α reflects the interactions between polymer and solvent: $\alpha=2/3$ in a theta-solvent, $\alpha=1/3$ in a good solvent, and approaches unity in a non-solvent.[10] The scaling eqn (16.1) simply states that for a given grafting density (grafted amount) and an average polymerization index (molecular weight), the thickness of the grafted polymer layers in a stretched state depends on the quality of the solvent, *i.e.* has a *response* to the solvent.

Early experiments with a surface forces apparatus[11] decisively demonstrated that PBs have a complex density profile depending on the thermodynamic quality of the solvent they swell in.

Using a self-consistent field (SCF) approach, Milner *et al.* have shown theoretically that the concentration profile of a swollen polymer brush has a parabolic form.[12] In the same publication, they coined the term "polymer brush" *per se*. These seminal works have pioneered the field of PBs and shown that the height of polymer coils forming a PB may be sensitive to different stimuli.

More recently, Brittain and Minko *et al.*[13] gave a comprehensive definition for PBs. As a universal measure of grafting compactness, they recommended the parameter of reduced tethered density Σ. It can be interpreted as "the number of chains that occupy an area that a free non-overlapping polymer chain would normally fill under the same experimental conditions". Σ combines two parameters: the gyration radius of a polymer coil R_g (at experimental conditions, depending on the polymerization degree) and grafting density σ:

$$\Sigma = \pi R_g{}^2 \sigma. \tag{16.2}$$

Grafting density is determined as the number of chains per area unit, usually expressed in units of nm^{-2}, and calculated from brush thickness h as follows:

$$\sigma = \frac{h \rho N_A}{M_n}, \tag{16.3}$$

where ρ is bulk polymer density, N_A is Avogadro's number, and M_n is the number averaged molecular weight of the polymer. Thus, Σ defines the stretching of polymer coils and may serve as a quantitative criterion for a PB.

Minko and Brittain recommend distinguishing PBs from broadly defined "end-grafted polymers" using the "Σ"-criterion: the grafted polymer is in the PB regime if $\Sigma > 5$, and in a transition regime at $1 < \Sigma < 5$. However, in most practical cases, the thickness of a grafted polymer layer above 3–5 nm (dry) in a good solvent corresponds to the regime of stretched coils, *i.e.* PBs.[13]

16.1.2 Synthesis of PBs

Most commonly, PBs are "grafted to", "grafted from" or "grafted through" a pre-modified anchoring layer on a substrate (Figure 16.2). The "grafting to" method (Figure 16.2a) consists of a chemical attachment or physical adsorption of end-functionalized polymers $(P)_nF$ to surface S. As a rule, click-chemistry reactions are utilized in order to ensure reliable grafting of the polymer chains onto the surface $(P)_nF_S$. Usually the substrate surface requires additional functionalization. Some examples of such surface anchors include self-assembled monolayers of thiols (on gold), alkoxysilanes (Si/SiO_2

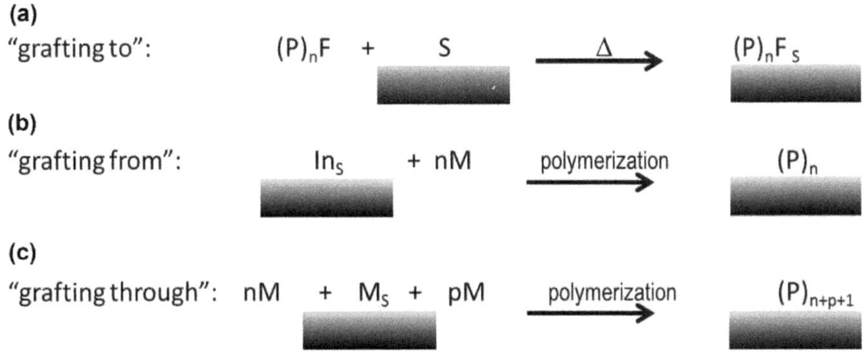

Figure 16.2 Overview of techniques for producing grafted polymer layers: "grafting to" (a), "grafting from" (b), and "grafting through" (c).

wafers, ITO glass), and chemisorbed functional polymers. The PBs obtained by the "grafting to" approach rarely reach 10 nm in thickness due to geometric constrains. Grafting of the first layer of polymer coils efficiently screens out the substrate surface which prevents the grafting of chains of the second echelon.[14–17] This approach is advantegous for modelling fundamental investigations of PB behavior where narrow distributions of grafted polymers are required. The use of well-defined polymer chains also allows for grafting of PBs that require specific conditions of polymer synthesis, such as, polyethylene glycol (PEG).

The "grafting from" method consists of two steps. The attachment of a surface initiator In_s is followed by radical polymerization (Figure 16.2b).[18–21] Relatively thick PBs (up to 100 nm or thicker) can be conventionally obtained in this case. The relatively broad distribution of polymer molecular weight typical for classic radical polymerization can sometimes be a drawback. This issue was resolved by introducing "living"/controlled radical polymerization (CRP)[22,23] to the fabrication of PBs.[24–26] In this case, the "grafting from" occurs in the regime of "living" radicals when all initiator groups are activated within a short initial time and (ideally) all chains are growing parallel during the entire process.

The "grafting through" approach for PB synthesis has not received much attention yet. Instead of introducing an initiator in "grafting from", a polymerizable –C=C group (either vinyl or (meth-) acrylate) M_S is introduced on the substrate surface (Figure 16.2c). Radical polymerization is initiated in bulk and propagates traditionally, eventually involving surface-attached M_S groups in the growing chains.[27–30] The resulting PBs are similar to those obtained by the conventional "grafting from" approach, but with two important differences. First, the surface anchoring of vinyls or acrylates is more feasible compared to most initiators sensitive to temperature and light. Second, and more importantly, a wide variety of initiators may be used in bulk including photoinitiators. The latter opens an avenue to photo-lithographically patterned PBs as well as PBs with gradient thickness.[28,31]

In some cases, post-grafting modification (PGM) of PBs is a favorable, at times even unavoidable, step to introduce chemical functions which are not miscible with the process of grafting/polymerization. A comprehensive review on post-grafting/post-polymerization modifications was recently published by Galvin and Genzer.[32] They provide several reasons for PGM, the most common of which address the following issues and drawbacks of direct synthesis: (i) the monomers are incompatible with the polymerization conditions and/or the substrate; (ii) the polymerization of bulky monomers gives a low yield or low molecular weight; (iii) the entropic constraints associated with the interfacial/lateral confinement. Usually, click-reactions are used for PGM as listed in Figure 16.3. In the context of this review, two groups of PBs are worth mentioning specifically as those that require PGM. First, poly(acrylic) (PAA) and poly(methacrylic) acid (PMAA) PBs are difficult

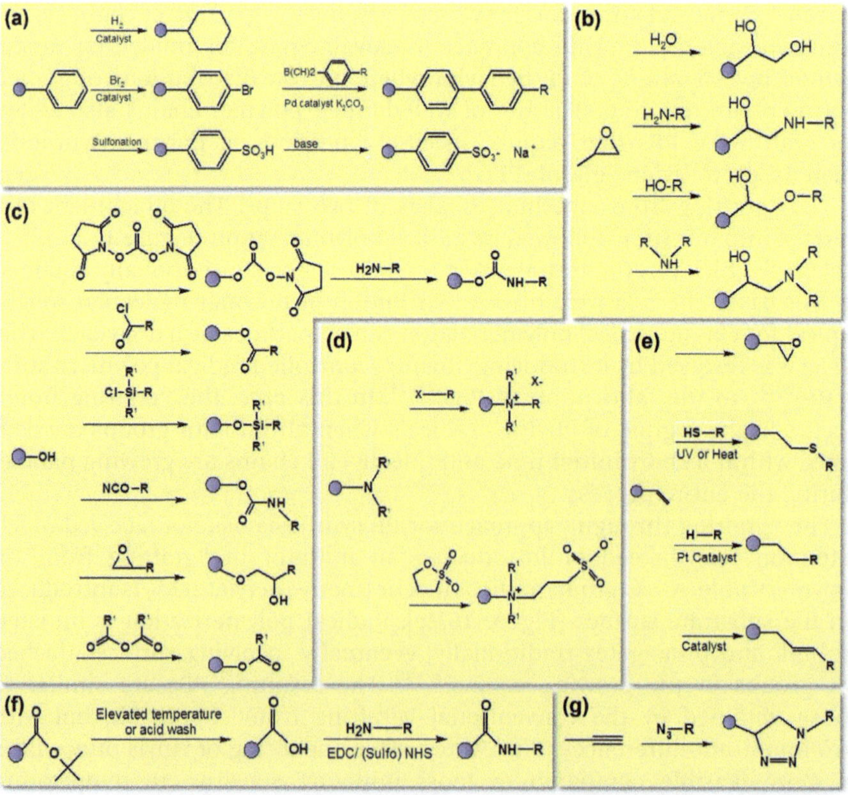

Figure 16.3 Selected click-reactions used in the post-grafting modification approach.
Reprinted from ref. 32. Copyright 2015, with permission from Elsevier.

to synthesize directly from the corresponding monomers. In this case, "grafting from" using *tert*-butyl (meth)acrylate (*t*BMA) is followed by hydrolysis resulting in the corresponding brush. Another example is modification of poly(2-vinyl pyridine) (P2VP) and poly(4-vinyl pyridine) (P4VP). The reaction of quaternization of the PB composed of these polymers generates cations along the chains with high yield. PGM also eliminates the need for secondary purification requirements of PB systems from possible by-products and excess reagents.

16.1.3 PBs of Complex Architecture

The strong application potential of PBs has stimulated research in hybrid, complex, and brush-like polymer layers. *Binary PBs* composed of two different polymers were prepared by a step-by-step "grafting from" approach using polystyrene (PS) and P2VP as reported in 1999.[33] Another approach to binary PBs is "grafting to". Minko *et al.* used a sequential two-step grafting of end-tethered PS and P2VP.[34] Tsukruk *et al.* simplified the procedure to one-step grafting using Y-shaped block copolymers with the central point tethered to a complementary functional site on the surface.[35] *Block copolymer PBs* consisting of blocks of PS and poly(methyl methacrylate) (PMMA) were obtained by sequential surface-initiated CRP (SI-CRP).[36–38] In both cases, the PBs revealed a lateral microphase separation of nanometre-scale dimension similar to block copolymers in bulk.[39–41] More complex structures can be obtained by grafting pre-synthesized polymer constructs with specific architectures. For example, *bottle-brush polymers* consisting of a backbone chain and different side-chains can be either adsorbed or tethered to solid substrates by click-reactions between the functional groups of the backbone and complementary functions of the surface.[42–44] Other brush-like architectures include a loop-brush,[45,46] grafted hyperbranched polymers,[47–53] dendrons,[54,55] and star-like polymers.[16]

The sensitivity of PBs to different stimuli makes them a very interesting subject for fundamental research in the fields of polymers and interfaces. Such responses are important elements of nanotechnology and other hi-tech applications.

16.2 Responses in Polymer Brushes

As commonly accepted by several scholastic disciplines, a stimulus-response model deals with the reaction of a system, usually a complex one, to an external action. By applying this concept to PBs, one may consider two kinds of responses: permanent (usually chemical) modifications and reversible reactions to a chemical or physical stimulus. Permanent chemical modification is an important topic with many attractive applications, such as the synthesis of PBs with required physical characteristics, introduction of chemical functions, or the use of PBs as nanoscopically-sized chemical reactors with specific interiors for synthetic and catalytic purposes. Each of these topics is rather broad and rapidly developing, and merits a separate review. However, we usually perceive stimuli-responsive PBs as systems with reversible reactions—often these reactions are changes in the physical state of the interface governed by thermodynamics. Such responsiveness is an intrinsic property of PBs as stable interphases between a solid–liquid or solid–gas that are comprised of soft matter. Inspired by this behavior, responsive PBs are often referred to as "smart" or "intelligent".[56]

The physical state of a polymer coil is a result of its interactions with the environment or its elements (*e.g.* solvent, solutes, pH, *etc.*). The equilibrium state of a coil corresponds to its molecular conformation and is expressed thermodynamically through the gyration radius R_g. Naturally, PBs consisting of stretched polymer coils are more susceptible to changes in the environment. In most cases, such changes shift the equilibrium and result in swelling or de-swelling in order to minimize free energy and adopt a new conformation of polymer molecules. The chains are confined within two dimensions, *i.e.* are vertically stretched normal to the surface plane. Therefore, swelling or de-swelling results in an increase or decrease in brush thickness, h, as depicted in eqn (16.2) and (16.3).[57]

Another way to change interactions between the polymer coils of PBs and the environment is to make alterations in the chemical nature or physical state of the coils. A change of temperature may cause such an alteration if weak interactions between polymer units are involved, for example hydrogen bonds or van der Waals forces. Another example is the reversible chemical modification of polymers, for example photoinduced *trans–cis* isomerization, or the presence of counter-ions in polyelectrolyte PBs. In all those cases, PBs reversibly react to changes of the environment. We will refer to this phenomenon as a *reversible response*.

In the process of transitioning between two states, the PB can be *locked* in a metastable state. It usually occurs in PBs with complex structures. As an example, let us consider binary PBs composed of non-polar (PS) and polar (P2VP) chains (Figure 16.4).[33] If placed into a non-selective solvent good for both counterparts, all chains are swollen and present at the surface. The reconstruction of the PB occurs in a selective solvent good for non-polar PS, namely toluene. The chains of PS remain swollen, while polar P2VP collapses in non-polar toluene. In a dry state they form two strata: the top one is composed of PS, and the stratum of P2VP is adjacent to the substrate (Figure 16.4, top). If placed directly into a selective solvent good for P2VP and poor for PS (*e.g.* methanol), the glassy layer of PS prohibits the chains of P2VP from reaching the top, even if small molecules of solvent may penetrate the PS layer. The system is in a metastable state. In order to release it, a trigger can be used. The exposure of a PS–P2VP brush to a common solvent such as tetrahydrofuran (THF) allows the response (Figure 16.4, middle to bottom). The class of responses where a stimulus triggers such an outcome is further referred to as *triggered responses*. The control over lock and release

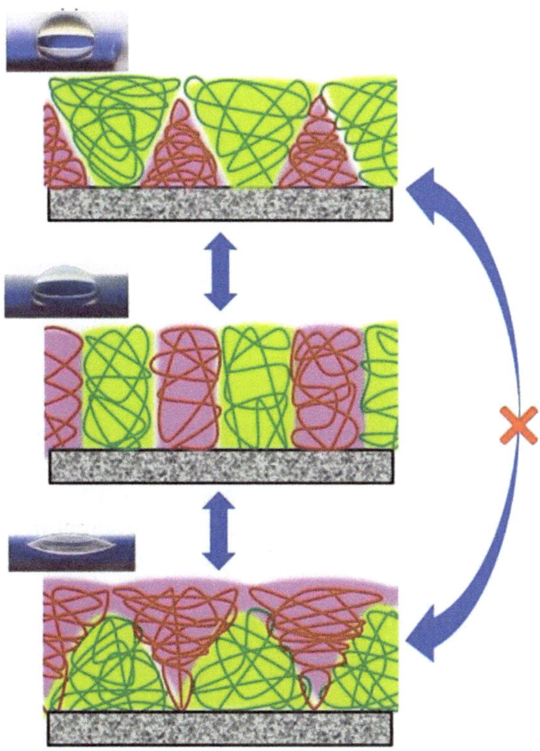

Figure 16.4 Conformation of the grafted layer in different environments and principle of switching properties: structure in a nonselective solvent (middle), in a selective solvent for "green" polymer (top), and in a selective solvent for "red" polymer (bottom). The polymer that occupies the top stratum governs the surface properties, for example wetting (inserts).

using different stimuli is a new and exciting way to complement systems with reversible responses.

16.3 Responsive Polyelectrolyte Brushes

The incorporation of ions along the polymer chains of PBs results in the substantial alteration of the properties of polyelectrolyte brushes (PEBs), as compared to the neutral ones. It is easy to foresee that electrostatic repulsion between the charged sites and osmotic pressure caused by the high intrinsic ionic strength of PEBs will increase the volume of the polymer coils. Due to lateral confinement, the only direction that PEBs may swell is normal to grafting plane. Apparently, this qualitative prediction requires a quantitative background. Almost immediately after successful application of the SCF approach to neutral PBs, two groups independently performed theoretical studies of responses of the PEBs using the same method: Miklavic and Marčelja[58] and Misra *et al.*[59] Ref. 59 in particular highlights that the total one-dimensional potential $U(x)$ of the brush is the sum of electrostatic contribution:

$$-\frac{U_{\mathrm{el}}}{kT}=\frac{fZe\psi}{kT} \tag{16.4}$$

and a neutral brush U_{PB} according to Milner's application of Flory polymer theory for the free energy:

$$-\frac{U_{\mathrm{PB}}}{kT}=-\ln(1-\Phi)-2\chi\Phi, \tag{16.5}$$

where ψ is the electrostatic potential, f is the degree of ionization, Ze is the charge of the ions, Φ the monomer fraction in the brush, and χ the polymer–solvent interaction parameter.

The results of the computations were expressed as density profiles $\Phi(x)$ of PEBs depending on the degree of ionization f, surface charge σ_0, and ionic strength I (Figure 16.5a–c). The surface charge affects only the stratum of PEBs adjacent to the substrate because the Debye screening length is small compared to the brush height. The other two responses are more robust.

16.3.1 Basic Stimuli and Responses in PEBs

The *pH response* in PEBs is the most straightforward one. However, it is important to remember that the ionization degree of polyelectrolytes is a function of pH only in the instance of weak electrolytes (see Chapters 3 and 10 of this book). Therefore, two kinds of PEBs are to be distinguished: strong electrolyte PEBs (sometimes called "quenched", they include, for example, polystyrene sulfonic acid (PSS) and quaternized polyvinyl pyridines qP2VP and qP4VP) that are virtually independent from the pH, and weak electrolyte PEBs (sometimes called "annealed"). The latter include PMAA, P2VP, *etc.*; capable of (usually) reversible changes of the ionization degree in response to the pH of the medium:

$$\frac{f}{1-f}=\frac{K_{\mathrm{a}}}{C_{\mathrm{H^+}}} \tag{16.6}$$

where K_a is the dissociation constant of polyacid and f is the fraction of dissociated sites. A similar equation can be obtained for polybases. The scaling gives the following connection between height and ionization degree for "osmotic brushes" (in the absence of added salt).[10]

$$h\propto Nf^{1/2} \tag{16.7}$$

The mechanistic description of an "osmotic brush" regime originates from the electroneutrality of PEBs. The presence of fixed ions and the counterions c_i in a concentration that exceeds ionic strength in the bulk c_i^{b} causes the osmotic pressure $\Delta\Pi/k_BT=\Sigma_i c_i-\Sigma_i c_i^{\mathrm{b}}$ to balance the entropic elasticity of the chains and the brush to swell.

The *concentration of added salt* in the medium affects both weak and strong PEBs due to charge screening. According to both experimental

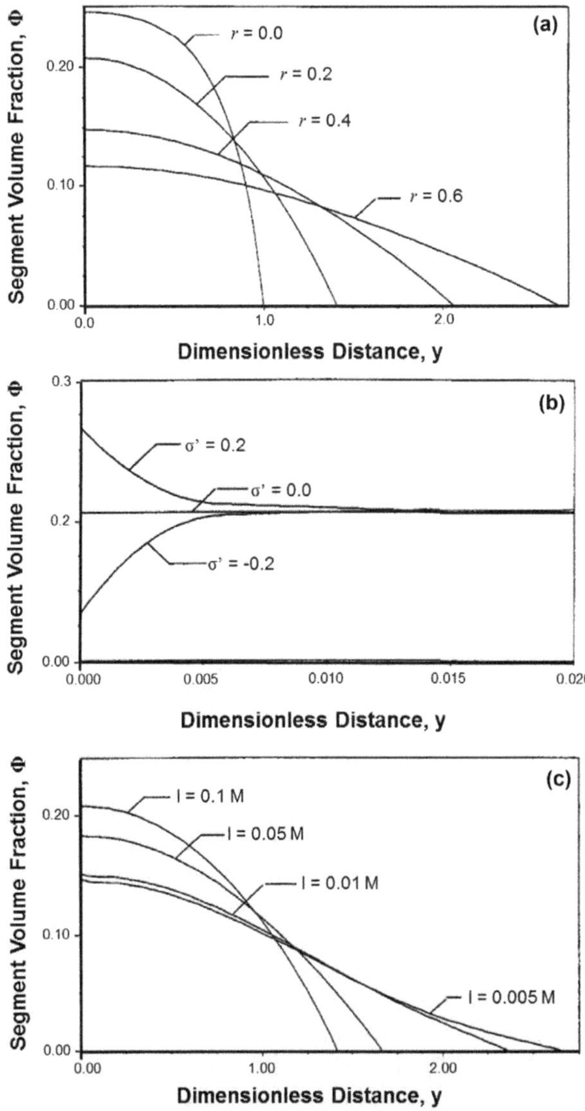

Figure 16.5 Effect of brush charging r (a), surface charge σ_0 (b), and ionic strength I (c) on segment density distribution.
Reprinted (adapted) with permission from ref. 59. Copyright 2015 American Chemical Society.[59]

results[60,61] and model calculations,[62] effective screening occurs when the concentration of added salt exceeds the concentration of counterions trapped within the PEB ("salt brush" regime).[63] As a result, the brush thickness decreases as a function of salt concentration c^{b}:

$$h \propto Nf^{2/3}(c^{\mathrm{b}})^{-1/3} \tag{16.8}$$

The *nature of counterions* may also critically affect the behavior of the PEB. All the theoretical considerations and experimental observations mentioned were made for monovalent counterions. What happens to PEBs in the presence of di- or multivalent ions? Multivalent ions build bridges between the charged sites of PEBs. Both intra- and inter-chain interactions contribute to the *ionic collapse* of the brush.[64–66] Moreover, multiple charges of the counterions prevent them from leaving the PEB continuum, akin to multivalent complexes with chelating agents.[67] In terms of stimuli-response behavior, PEBs demonstrate locked or irreversible collapse.[68,69]

To summarize, the concentration and nature of ions (including protons, *i.e.* pH) in the environment stimulate basic responses in PEBs, such as swelling or collapse (and height changes, respectively). The methods of spectroscopic and IR ellipsometry,[70,71] neutron reflectivity,[72–74] SFA[63,75] and AFM[76,77] were widely used to investigate this response and associated processes. Next, we consider secondary responses, *i.e.* changes in the macroscopic properties of the PEBs that are derivatives of the basic microscopic morphology.

16.3.2 Mechanics of PEBs

The phenomenon of swelling of polymer films is usually associated with changes in mechanical properties, such as compressibility of the films, adhesion and friction between the films and the probes. Soon after the commercialization of AFM technology in the mid-1990s, it was used to measure mechanical responses in PEBs.[76–79] In particular, work by Moya *et al.* explores the mechanical responses of highly charged poly(sulfopropyl methacrylate) (PSPM) PEBs using an indentation technique of AFM.[80] They introduced the parameter of compressibility and observed an adverse response to the ionic strength. At low salt concentrations, the electrostatic repulsion of the charges reduces the compressibility as compared with high ionic strength. On the contrary, PEBs are highly swollen and therefore more compressible at low salt concentrations, *i.e.* this trend works in the opposite direction. According to Stamm *et al.*, similar behavior is demonstrated by two weak PEBs: PAA and P2VP.[76] However, the dissociation of weak PEBs[81] makes the picture rather complicated and the response becomes multidimensional, as it is an interplay and superposition of several parameters, *i.e.* grafting density, concentration of the supporting salt/electrolyte, dissociation degree, complex density profile, *etc.*

The swelling/collapse of polymer chains in PEBs induced by pH and ionic strength may be used for mechanical actuation. If deposited on an AFM cantilever, PEB transforms ionic signal (either pH or concentration of electrolyte) into a macroscopic mechanical response. Huck *et al.* describe an elegant method of visualization and mechanosensoric measurements of the response to pH and ionic strength.[82] The brushes of poly(methacryloyl ethylene) phosphate (PMEP) were grown on one side of an Au-coated AFM cantilever by the "grafting from" method. The response of these anionic

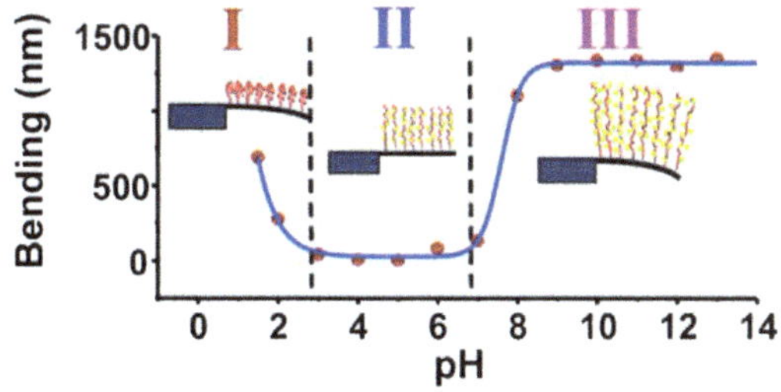

Figure 16.6 Observed bending of single-side PMEP brush-modified cantilever with changing pH and schematic illustration of the brush conformation in different regimes.
Reprinted (adapted) with permission from ref. 82. Copyright 2015 American Chemical Society.

PEBs on the conformational level to changes in pH and salt concentration is shown in Figure 16.6.

There are three different ionic states of the PEB: (i) fully protonated at $pH<2$, not soluble in water and collapsed, (ii) mono-protonated $-HPO_4^-$, relaxed, and (iii) di-protonated PO_4^{-2} at $pH>8$, strongly swollen. The cantilever bending occurs when compressive stress is high: a strong swollen PEB in state (iii) and collapsed state (i) when the brush is too dense to accommodate the collapsed coils. Similarly, the compressive stress takes place in collapsed PEBs at high ionic strength. Interestingly, the local changes in pH may be generated by an applied electric bias, resulting in the reversible bending of the cantilever.[83]

16.3.3 Surface Properties of PEBs: Adhesion and Friction

The ability to switch properties of one or both opposing surfaces in adhesion and friction may lead to numerous applications, ranging from colloid stabilization and MEMS technology to controlled adhesive joints. The physico–chemical methods of surface control of PEBs have received significant attention during the last decade. Theoretical work by Pincus,[84] Borisov,[85] Birshtein and Zhulina[86,87] laid the foundation for the theory of adhesion and friction between PEBs. The majority of experimental work is based on the SFA technique and colloidal probe AFM. The latter uses microscopic lateral resolution and spherical microprobes, with well-defined geometries, to investigate both normal (adhesion as energy of interaction on retraction branch of the force–distance circle) and shear forces (friction). Direct force measurements were reported for oppositely charged strong PEBs in ref. 88. The effect of a monovalent (1 : 1) electrolyte on adhesion between polycations (poly (2-(methacryloyloxy)ethyltrimethylammonium chloride), PMETA chloride) and polyanions (PSPM) was studied. A very weak adhesion force at high KCl concentration (2.0 M) increased rapidly from 0.1–1.0 nN up to 140 nN with dilution of KCl down to 1.0 M. High adhesion between PMETA and PSPM was associated with polyelectrolyte complex formation, while KCl effectively screens out the ionized sites of the PEB. Drechsler *et al.* have probed the adhesion between weak polycationic PEBs of P2VP and the surface of silica particles at different pH values and concentrations of added monovalent anion (Cl^-).[77] The system revealed very responsive yet complex and non-linear behavior which indicated a superposition of several effects: osmotic repulsion by highly swollen PEBs at low pH, van der Waals interactions and Coulombic attraction/repulsion at different pH values and salt concentrations. Figure 16.7a shows that the average values of adhesion energy increase by two orders of magnitude from pH 2.5 (highly swollen state) to 3.75 (partly protonated brush) and then decrease to about half reaching a plateau at pH 5.0 and with a neutral electrolyte. The effect of added electrolyte is more predictable: the decrease in adhesion matches the screening effect of KCl concentration (Figure 16.7b). A surprisingly small number of direct force measurements (either SFA or AFM) for the effects of divalent ions on PEB interactions are available in the literature. Klein *et al.* have investigated the effect of divalent cations Mg^{2+} on adhesion between a PEB of hyaluronic acid (HA) and a silica colloidal probe.[65] The surfaces of both HA brush and silica are charged similarly, thus revealing repulsive forces, both electrostatic and osmotic by nature. However, in the presence of divalent ions, the adhesion prevails as a result of a change in the HA brush charge from negative to positive. The effect of a trivalent counterion $[Ru(NH_3)_6]^{3+}$ was evaluated in ref. 64 where SFA measurements of brush-to-brush adhesion were combined with electrochemical analysis. The effect of ionic bridging in the case of trivalent ions is more pronounced compared to divalent ions, and the threshold ion concentration is on the scale of micromolar compared to millimolar for divalent ions.[65]

The friction between two PEBs or a PEB and a bare probe was the subject of recent publications, both theoretical and experimental. For neutral brushes, the Alexander–de Gennes scaling model was used to elucidate shear-force induced deformation.[89] Zhulina and Rubinstein *et al.* recently

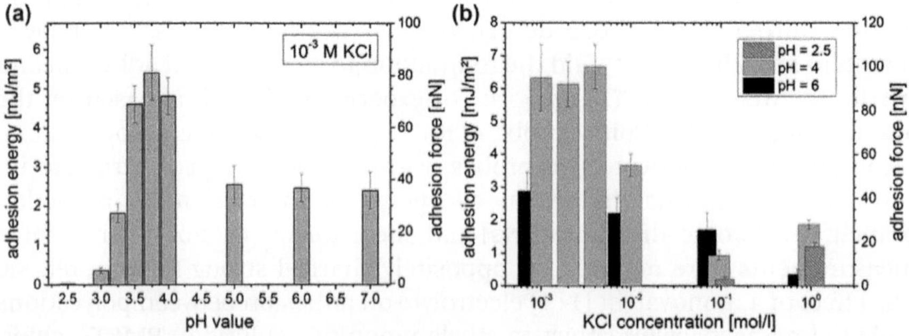

Figure 16.7 Average values of the adhesion energy and force measured between a silica colloidal probe and a PEB of P2VP at various pH values (a, the concentration of added electrolyte KCl was constant 1×10^{-3} M) and concentration of electrolyte (b).
Reprinted (adapted) with permission from ref. 77. Copyright 2015 American Chemical Society.

developed a scaling model for friction forces between two PEBs sliding one over the other. They demonstrated the enhanced lubricating properties compared to both neutral brush surfaces and bare charged surfaces.[87] Liberelle and Giasson *et al.* found no significant effect of monovalent electrolyte (NaCl) concentration on the friction between two PAA brushes, although grafting density and separation distance were important factors for lubrication.[90] The effect of lubrication by PEBs of PMETA was observed in experiments by Takahara[91] and Zhou *et al.*[92] using a colloidal probe (bare) AFM technique. In the former work, the general effect of lubrication was demonstrated while no significant effect of added electrolyte NaCl on the friction coefficient was observed. In contrast, the effect of the counterion nature on the tribological properties of the PEB was demonstrated and elucidated in the latter work. In particular, the friction coefficient increased by about two orders of magnitude when Cl^- ions were substituted by bulky PF_6^- and was increased ten-fold in the case of bis(trifluoromethanesulfonimide) ($TFSI^-$). According to the authors, the general trend is as follows: swollen PEBs show superior lubrication, partially collapsed chains reveal moderate friction, and collapsed PEBs have high friction. This collapse caused by hydrophobic counterions, such as $TFSI^-$ and PF_6^-, has been referred as "hydrophobic collapse" and has been investigated in several works.[61,74,93–95]

16.3.4 Electrochemical Gating of PEBs

As we summarized in Section 16.3.1, PEB collapse caused by an external electrolyte (nature and concentration of counterions, pH) is a basic response to chemical stimuli. This transformation occurs at the molecular scale. However, if deposited on an electrode, PEBs are capable of (reversible) collapse revealing macroscopic electrochemical sensitivity, *i.e.* a secondary response. This curious phenomenon was studied by several groups.

The diffusion of electrochemical labels (usually ions) through PEBs depends on the state of PEB. If collapsed, the PEB become virtually impenetrable which can be conveniently used to monitor the PEB's state. This property was employed by Tirrell *et al.* where the authors monitored the state of a strong PEB (PSS) in the presence of $[Ru(NH_3)_6]^{3+}$ by SFA and cyclic voltammetry (CVA) in parallel.[64] The oxidation and reduction peaks of $[Ru(NH_3)_6]^{3+}$ trapped by the brush show distinctively different characteristics compared to "free" $[Ru(NH_3)_6]^{3+}$ ions. Huck *et al.* have shown that strong PEBs composed of quaternized poly(dimethylamino)ethyl methacrylate (qPDMAEMA) and PMETA may have switchable electron-transfer resistance. This depends on either the swollen or collapsed state of the PEB as caused by the concentration and nature of the counterions (charge screening or hydrophobic collapse).[96–98]

In the case of the weak PEB, P2VP, the electrochemical gating was probed by Katz and Minko *et al.*[99] Using Faradaic impedance spectroscopy, they found that the pH of the electrolyte solution has a strong effect on electron

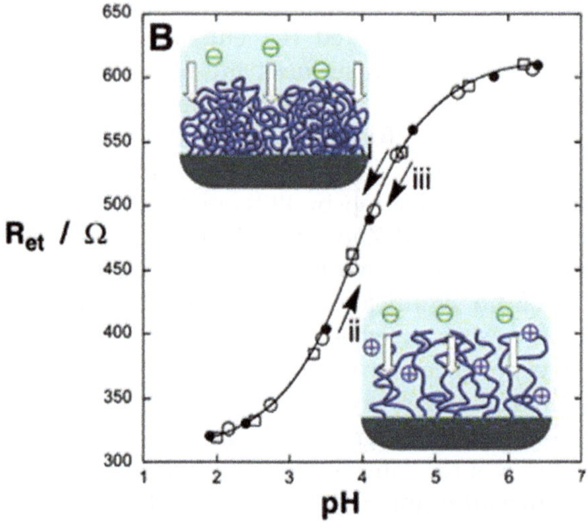

Figure 16.8 Titration curve showing changes of the electron-transfer resistance derived from the impedance spectra upon variation of pH values: (i) the first run of pH values in the acidic direction (●); (ii) the second run of pH values in the basic direction (○); and (iii) the third run of pH values in the acidic direction (□).[99]
Reprinted (adapted) with permission from ref. 99. Copyright 2015 American Chemical Society.

transfer: the collapsed cationic brush at high pH is about two times more resistant compared to swollen PEBs at low pH as probed by $K_3[Fe(CN)_6]$ (Figure 16.8). A similar effect was observed by Li *et al.*[100] when PAA grafted onto an Au electrode served as a pH-controlled gate for reduction/oxidation of cytochrome C.

This effect was explored in a series of recent papers by Katz *et al.*[101–104] The pH-mediated electrochemical response to many complex biochemical stimuli was successfully demonstrated when an electrode with a grafted PEB was used. The Boolean-type response based on bioelectrocatalytic conversion of glucose and ethyl butyrate in the presence of three hydrolytic enzymes—esterase, urease, and glucose oxidase—is only one example of a number of sophisticated systems proposed by this group. A bias applied to the electrode bearing a PEB creates a local (interfacial) pH that differs from the bulk. This feature has been used to control reversible swelling/collapse and thus permeability of the PEB, akin to electronic transistors. The term "electrochemical" or "chemical gating" used in these papers originates from this similarity.[105]

16.4 Polymer Brushes with Electrochemical Response

This section focuses on responsive PBs directly involved in electrochemical transformations, *i.e.* bearing RedOx moieties, in contrast to the PEBs with responses mediated by changes in pH/ion concentration considered in the previous section.

16.4.1 Non-covalent Modifications of PEBs with RedOx Species

RedOx-active responsive PBs is a relatively new field of study. However, the main synthetic approaches have been developed for several different systems. The most straightforward one is the modification of a ready PB with RedOx probes. Huck *et al.* have studied the stability and RedOx behavior of the polycationic PEB of PMETA bearing quaternary trimethyl ammonium groups with immobilized RedOx counterions of $[Fe(CN)_6]^{3-}/[Fe(CN)_6]^{4-}$.[97] They used a simple ion-exchange procedure to introduce the RedOx species of $[Fe(CN)_6]^{3-}$. They found that quaternized ammonium ions form stable ion pairs with ferricyanide ions. As a result, the brush is relatively stable, easily switchable between the reduced and oxidized state, and reveals a fast electron transfer. Moreover, depending on the RedOx state, the PB reveals responses in surface properties: upon reduction of ferricyanide, the PB become more hydrophilic. The water contact angle (CA) drops 20–25° depending on the thickness of the PB within 10–30 seconds of applied potential. The response is fully reversible and repeatable within the stability range of the PMETA-$[Fe(CN)_6]^{3-}$ ion pair. The authors hypothesize that the shift in wettability is a result of weaker ion-pairing between the reduced form of ferricyanide $[Fe(CN)_6]^{4-}$ and cations of PMETA compared to the oxidized form $[Fe(CN)_6]^{4-}$. The kinetics of the wetting response is restricted by the transport of electroactive ions within the PB layer.[98]

A similar approach, *i.e.* non-covalent immobilization of RedOx species onto a functional PB, was explored by Minko and Katz.[103] They functionalized the PB of P4VP with pendant RedOx groups of $Os(dmo\text{-}bpy)_2Cl_2$. The RedOx behavior of the Os-P4VP brush was studied by CVA. The experiments showed that the PB was very stable: it revealed a reversible, $E^\circ = 0.29$ V, electrochemical response with a small peak separation. Being polycationic by nature, their PBs are sensitive to pH. They swell at low pH (3.0) and collapse at neutral conditions (pH 7.0). This behavior can be easily traced with CVA: clear RedOx activity at acidic conditions diminishes when the pH reaches a value of 6.0 (collapsed PB).

16.4.2 Covalently Attached Pendant RedOx Sites

Another approach has been used by Huck *et al.* in an attempt to prepare PBs terminated with a RedOx group.[106] They used click-chemistry reactions between a Br-terminated poly[1-ethyl 3-(2-methacryloyloxyethyl)imidazolium chloride], PEMEImCl) brush and propargylferrocencarboxylate resulting in ferrocenyl (Fc) groups at the free ends of the polycationic PEB. Both CVA and electrochemical impedance spectroscopy were used to characterize the brush.

The electrochemical response of the Fc-terminated PEB was affected by both the polymer conformation (as a function of nature and concentration of counterions in the external environment) and distance between RedOx probes and the electrode surface. The diffusion limitation in electron transfer was very strong in collapsed brushes, which resulted in a decrease or full loss of electrochemical activity. However, even more important in the context of this review was the mediating effect of the Fc group. The electrochemical oxidation of ascorbic acid was mediated by Fc. It resulted in a very low (approx. 200nM) concentration detection limit of ascorbic acid.

The method of direct synthesis of Fc functionalized PBs on indium tin oxide (ITO) electrodes by surface-initiated controlled radical polymerization (SI-CPR) has been developed by Pyun.[107] They synthesized 2-(methacryloyloxy)ethyl ferrocencarboxylate (MOEFc) and grafted PMOEFc from a sacrificial initiator attached to an ITO electrode by phosphonic acid functionality. The PB of PMOEFc revealed some electrochemical activity. The parameters of the RedOx process strongly depend on the length of the chains and conformation of the PB. In contrast to PEBs with pendant or terminated RedOx groups,[106] the PB of PMOEFc reveals a cyclic voltammogram with a high peak current and small peak-to-peak separation in the collapsed state while the swollen PB shows shallow and low RedOx peaks (Figure 16.9). The authors attribute this observation to fast electron exchange between Fc groups in the dense (collapsed) conformation of the PB.

An alternative method of grafting of PFcMA onto ITO silane chemistry by SI-CPR was proposed by Kang *et al.*[108] They prepared enzyme-mediated an amperometric biosensor and showed its high efficiency for glucose determination by immobilization of glucose oxidase onto the chains of the PB.

Another interesting design of RedOx responsive brushes was realized in ref. 109. The authors used the RedOx-controlled reversible coupling of Fc and β-cyclodextrin (β-CD) to test whether the hydrophobic PB bearing Fc can be reversibly loaded/unloaded with the hydrophilic β-CD depending on the oxidation state of the Fc. In the oxidized state the positively charged Fc^+ sites do not bind effectively to the β-CD cap. As a result, the wettability of the PB varies within a wide range, from $\sim 26°$ up to $\sim 98°$ and back as revealed by the water CA. This electrochemically controlled coupling of Fc and β-CD species was used by Liu *et al.* for reversible cross-linking: free bis-β-CD moieties cross-linked two sites of the Fc of the brush, while electrical treatment results in dissociation of the cross-links due to de-bonding of the β-CD and Fc^+.[110]

Recently, Gutmann *et al.*[111] explored the RedOx chemistry of phenothiazine immobilized as a pendant group and electrochemical probe on PBs. The PBs bearing phenothiazines were prepared in two ways: by direct grafting using 10-(4-vinylbenzyl)-10*H*-phenothiazine monomer and by click-chemistry modification. The results of both approaches were compared and the electrochemical responses were discussed.

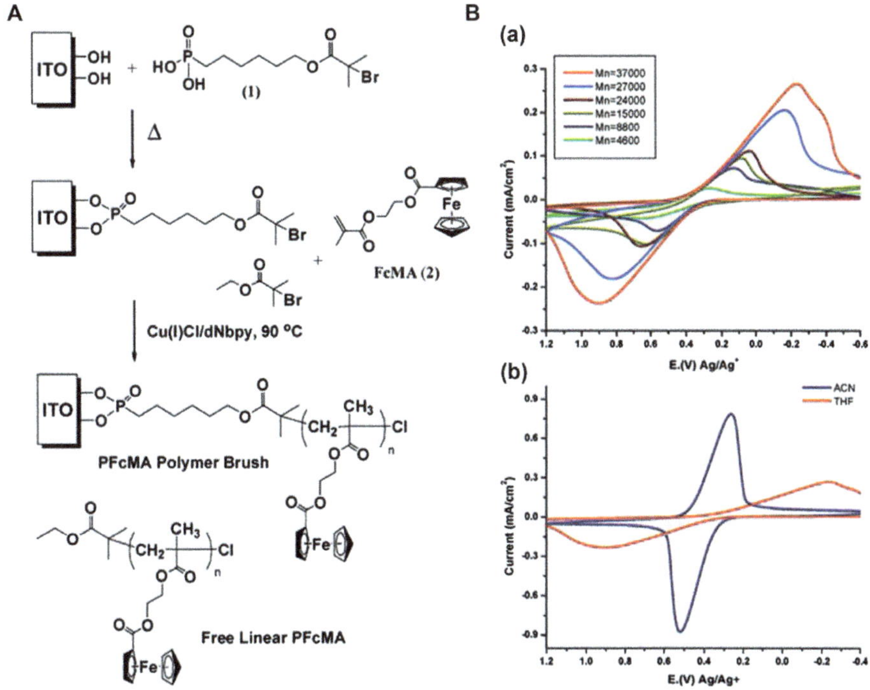

Figure 16.9 (A) Synthetic scheme for the modification of ITO and phosphonic acid 1 and SI-ATRP with FcMA (2) in the presence of ethyl 2-bromoisobutyrate as the "sacrificial" initiator. (B) cyclic voltammogram of PFcMA brushes of varying molar mass on ITO in THF (a); comparative cyclic voltammogram of high molar mass PFcMA brush ($M_n = 37\,000$ g mol^{-1}) on ITO in acetonytrile or THF at the scan rate = 100 mV s^{-1} (b).
Reprinted (adapted) with permission from ref. 107. Copyright 2015 American Chemical Society.

16.5 Photosensitive PBs and Photochemistry

The manipulation of surface properties with light is very attractive for many applications as it can be fast, distant, and non-destructive. Another important benefit of the photo-stimulated response is a possibility to generate lateral gradients and/or patterns, *i.e.* photolithography. Yet, only a few studies related to the control of PB behavior have been published. Here we consider two basic options: *irreversible locking*/coupling/decoupling of PBs using photo-induced chemical reactions and *reversible switching* based on several chromophores.

16.5.1 Photo-induced Coupling for Surface Modification and Locking in PBs

The method of lateral patterning of PBs was explored in a recent report by Locklin and Popik *et al.*[112] The starting material was poly(3-hydroxymethyl)naphthalene-2-ol 4-vinylbenzoate PNQMP-4VB. Quantitatively, each monomer of this 50 nm thick brush contained a functional group of 3-(hydroxymethyl)naphthalene-2-ol capable of Diels–Alder ligation with vinyloxy moieties upon short (2 min) irradiation with UV light (Figure 16.10a). The authors demonstrated their approach to be feasible by laterally resolved attachment of rhodamine B and fluorescein dyes using a 12 μm pitch TEM grid (Figure 16.10b).

The opposite effect of photo-induced decoupling was applied for light-responsive PBs by del Campo *et al.*[113] They prepared PBs of poly(4,5-dimethoxy-2-nitrobenzyl methacrylate) (PNVOCMA). The photocleavage of the NVOCMA monomers resulted in removal of the NVOC "cages" and the release of methacrylate moieties. The process is relatively slow (full conversion was reached in approx. 20 h) and therefore the approach was suggested for gradual "tuning" of the PB instead of a fast "click". The photocleavage that resulted in the formation of PEBs permits pH sensitivity. The formation of a pattern upon UV light exposure was probed by AFM.

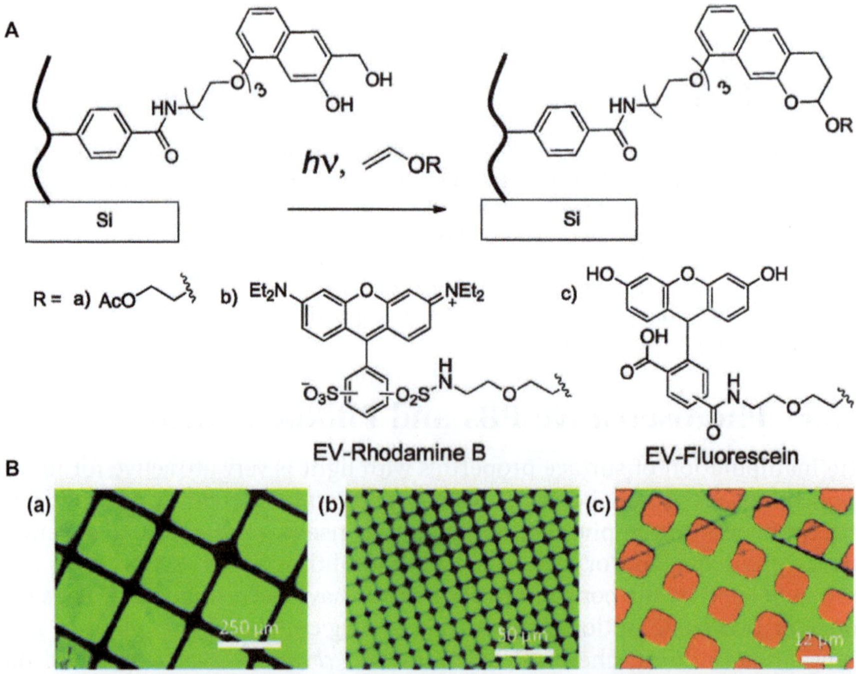

Figure 16.10 Photo-Diels–Alder functionalization of a poly(NQMP-4VB)-coated silicon wafer (A); fluorescence microscopy images of poly(NQMP-4VB)-coated wafers (a, b) photopatterned in 0.1 mM EV-fluorescein solution using (a) 250 and (b) 12 μm pitch TEM grids and (c) upon sequential immobilization of rhodamine B and fluorescein.[112]
Reprinted (adapted) with permission from ref. 112. Copyright 2015 American Chemical Society.

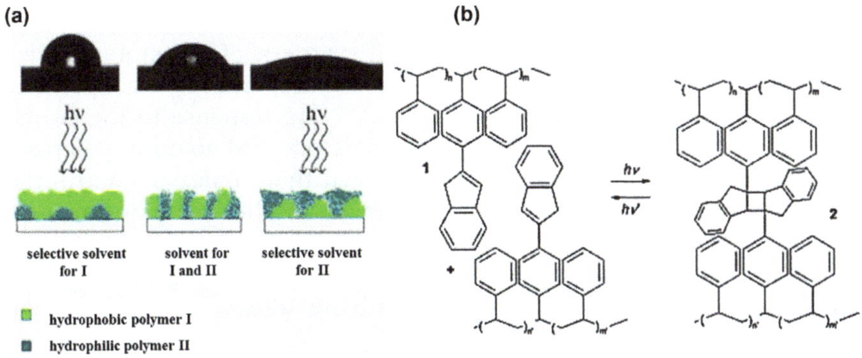

Figure 16.11 Illustration of switchable polymer brushes that can be fixed in one state *via* photo-crosslinking of one component and the corresponding water contact angle measurements' drop profiles (a). Formation, $h\gamma$ (and potential cleaving, $h\gamma'$), of a photo-crosslink *via* photodimerization of two phenylindene moieties in PSC (poly(styrene-*co*-2-(4′-styryl)-indene) 1)(b).
Reprinted from ref. 114. Copyright 2015 with permission from Elsevier.

The irreversible cross-linking of some elements of the responsive PB may result in the locking of PB properties. This concept was realized by Wolff *et al.*[114] They introduced the photodimerazable moiety phenylindene as a pendant to the PS components of several binary brushes. The cross-link occurs *via* $2\pi + 2\pi$ coupling induced by UV light (wavelength 320–400 nm) (Figure 16.11). The binary PB of PS–P4VP (1 : 1) demonstrated good switching in surface composition as revealed by wetting (water CA): hydrophobic PS on top (approx. 82°) after exposure to toluene (selective solvent for PS) gave way to P4VP on top (approx. 51°) after exposure to acidic water (selective/ionizing solvent for P4VP and non-solvent for PS). The cross-linking of the swollen PS layer (exposed to toluene) resulted in a modest fixation—a switching in wettability from approx. 95° to 77° respectively. In contrast, the cross-linking of collapsed PS led to a very prominent preservation: approx. 67° to 66° after toluene and water respectively. These important findings were supported by comprehensive analyses of the surface morphology.

16.5.2 Reversible Photosensitive PBs

Some of the most studied photosensitive organic compounds are derivatives of azobenzene (AB).[115] The reversible isomerization of AB from *trans* to *cis* and *vice versa* by irradiation with light of different wavelengths has been used in many applications.[116,117] In most cases, AB is attached to polymer chains as a pendant, either by post-modification or is readily incorporated into the monomer followed by polymerization. The latter approach was used by Camorani *et al.* in their synthesis of a photoresponsive PB: poly(4-ω-methacryloyloxyhexyloxy-4′-ethoxyazobenzene), a polymer known for its liquid crystalline properties.[118] The response to the switch was recorded using polarized Raman spectroscopy and confocal polarized Raman microscopy. The authors demonstrated their ability to control the molecular order within the PB using polarized irradiation. Unfortunately, other characteristics of the PB photoresponse are absent in the paper and there is no follow-up available. A recent review by Seki[119] addresses specific issues related to liquid crystalline PBs with photoresponsive azobenzene pendant groups.

The method of synthesis and stimulus-response characterization of photosensitive PBs was reported by Locklin *et al.*[120,121] They used spiropyran (SP) as a photoresponsive element (Figure 16.12a). The intermolecular transformation of SP (colorless) into merocyanine (MC, purple) occurred upon irradiation with UV light, while the reverse conversion was caused by visible light. The brushes obtained by ring-opening metathesis polymerization of the norbornyl functionalized spiropyran monomer contained the photoresponsive element of spiropyran as a pendant group (Figure 16.12b) and were 10 to 120 nm in thickness.[120] In the other approach, the monomer of spiropyrano-methacrylate (SPMA) was copolymerized with MMA by SI-CPR to produce a PB with a thickness of up to 80 nm.[121] The spiropyran-PB demonstrated reversible photochromism and changes in wettability due to the formation of MC, the highly polar and colorful isomer of the SP (Figure 16.12c). The dimeric complexes of MC with different metals (Cu^{2+}, Co^{2+}, Ni^{2+}, Fe^{2+}, Hg^{2+}, Zn^{2+}, Mg^{2+}) are colorful and stable, yet decomplexation is reversible and occurs upon irradiation with visible light.[122] The presence of the metal complexes substantially improves switching in wettability: a PB of 60 nm in thickness indicated 104° in SP form (colorless) and approx. 35° in MC form complexed with Fe^{2+} upon irradiation with UV light. Lateral patterns were demonstrated on the PB upon irradiation through a stencil mask, as depicted by imaging ellipsometry.

16.6 Complex Chemoresponsive PBs

In the previous sections we summarized the studies of responses in PBs caused by various chemical stimuli. In many cases, however, it is very difficult to differentiate and/or identify the nature of the primary cause of the response. Now we will briefly outline some of the most outstanding examples of such complex systems that may have a strong impact on future developments in the fundamental understanding and practical applications of PBs. Each of these topics has gained a lot of attention in recent years that has resulted in massive publication records. Throughout the course of this section we will direct interested readers to specific reviews related to these subjects.

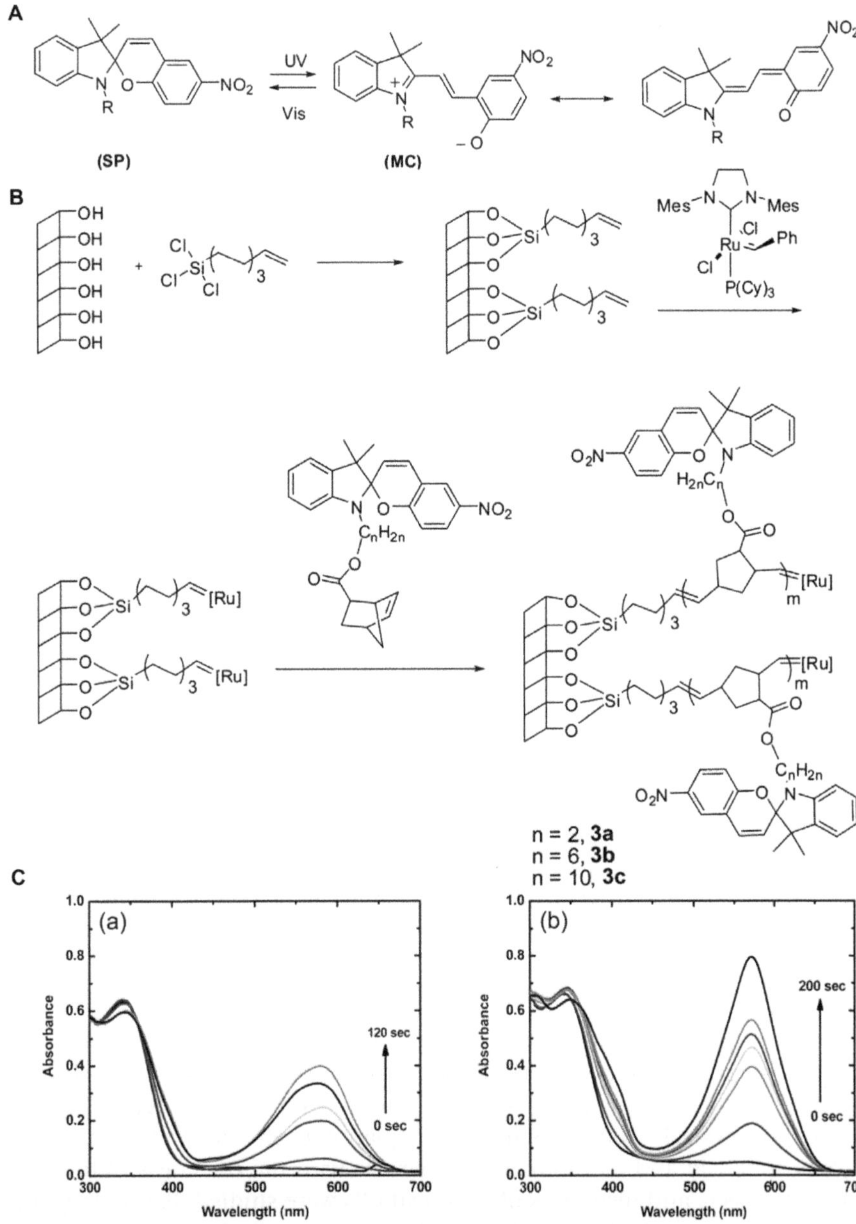

Figure 16.12 (A) Isomeric molecular structure of a spiropyran irradiated with light, spiropyran (left), merocyanine (center) and quinoidal canonical form (right). (B) Schematic of procedure used to form spiropyran-based polymer brushes using SI-ROMP from a surface oxide. (C) Change in absorption spectra of polymer brush (120 nm thick) with different UV exposure time, (a) in CH_2Cl_2 (time intervals: 0, 10, 20, 30, 60 and 120 s) and (b) in DMF (time intervals: 0, 10, 20, 30, 60, 120 and 200 s).[120] Reprinted (adapted) with permission from ref. 120. Copyright 2015 American Chemical Society.

16.6.1 Mixed Brushes and Polyampholytes

The complex and multifaceted responses often require combined or hybrid approaches to the principles of design of responsive systems. In the case of PBs, these can be devised using binary (or mixed, in general) brushes. In particular, independent (orthogonal) responses to two different stimuli can be achieved if two different polymers are grafted to the same surface, for example polyelectrolyte (either anionic or cationic) and a thermo-responsive polymer.[123–128] Another reported combination is a copolymer brush composed of both polyelectrolyte and photosensitive (spiropyran) monomers.[123]

An interesting response can be achieved for changes in pH if two PEBs are grafted in parallel: a polycationic and polyanionic PEB. Such mixed PEBs are called polyampholyte PBs. In the case of asymmetric polyampholyte PBs, the properties and particularly the pH responses are guided by the major component. However, in the case of nearly symmetric PBs with molar ratios close to 1 : 1, two oppositely charged polymers form the polyelectrolyte complex when the charges are efficiently compensated. It results in a low swellability and domination of hydrophobic forces (water non-wetting) at neutral pH values.[129–133]

In some instances, very profound effects can be obtained if only one of the binary PB components swells/collapses in response to an external stimulus. It is usually achieved through dynamic reconfiguration in mixed PBs.[134] Interplay between surface energy, free energy of the polymer chains confined at the surface, and the solvent causes microphase separation within the mixed PBs. They generates a complex microscopic structure where lateral phase separation and normal segregation occur in the form of ripples and dimples. If one of the components of the PB swells in response to an external stimulus, it controls the surface properties. In the collapsed state, the surface properties are dominated by the inactive "matrix" component. Several systems based on this active–inactive mixed PB sensitive to pH were developed and studied.[135–137]

16.6.2 PBs on Non-planar Substrates

Most of the systems listed above are planar PBs. They are convenient models to study, but are limited in applications. On the other hand, both primary (swelling, changes in thickness, and nature of the surface) and secondary responses to physical chemical stimuli are quantitatively different if the substrate is spherical and its radius is comparable with the thickness of the PB. Different aspects and responses of spherical PBs were studied. Particularly, in a series of reviews by Ballauff *et al.*, the behavior of spherical PBs stimulated by changes in pH, ionic strength and nature, temperature, and adsorption of proteins are summarized.[138–143] Other examples of non-planar PBs with strong application potential are membranes and porous systems. Stimuli-responsive PBs deposited inside sub-microscopic pores can control the flow of a fluid (or even a gas) depending on the nature and characteristics of the fluid or dissolved species, *i.e.* ions including H^+, RedOx species, biomolecules, *etc.* Such developments in "smart" membranes can provide far-reaching advancements in prospective directions of modern technology dealing with filtration, chromatography, and microfluidics. Recent advances in this field are summarized in several reviews[144–147] as well as in Chapter 10 of this book.

16.6.3 Bioadhesion: Cells and Proteins

Ability to adhere is an evolutionary property of living cells. Adhesion is a common feature in the life of most tissue cells, bacteria, and organisms. Cell adhesion is a complex process of binding a cell to a surface or extracellular matrix formed by other cells. Interactions of a cell with a solid surface are mediated by specific proteins, so-called cell adhesion molecules. In general, proteins readily adsorb onto most of the solid surfaces. It is due to a variety of functions and a broad range of interactions with the surfaces: ionic, van der Waals, hydrophobic, *etc.*[148] The ability of PBs to control or modify cell adhesion and protein adsorption makes them very attractive for a wide range of medical applications. In particular, some PBs are anti-biofouling, *i.e.* they can repel cells and resist adsorption of proteins.[149] The most widely used model of an anti-biofouling surface is PBs of poly(ethylene glycol) (or poly(ethylene oxide), PEG).[150] Yet the origins of reduced protein adsorption on PEG brushes are still debated. Recently, in pursuit of higher efficiency and convenient preparation, many more anti-biofouling PBs have been designed: poly(oligo-ethylene glycol)methacrylate (POEGMA),[151–153] polyacrylamide (PAAm),[28,154–156] and poly(*N*-isopropylacrylamide) (PNIPAAm). The latter polymer is particularly interesting in the context of this review since it is capable of conformational changes as a function of hydrogen bond formation and dissociation depending on the temperature. More importantly, these changes occur at the physiologically important temperature range of 35–40 °C. Above the lower critical solution temperature of PNIPAAm, the chains are desolvated and collapsed. This facilitates protein adsorption and cell adhesion,[157] the cells are attached and proliferate.

The idea of responsive PBs to trigger adhesion or repulsion of living cells was earlier expressed by Nagasaki and Kataoka in 1996: "By utilizing this system, cultured cells will be obtained easily without any damage from trypsin".[56] Unfortunately, in many cases the response to either physical (temperature, as in PNIPAAm) or chemical (pH, ionic strength) stimuli occurs at relatively harsh conditions outside the physiological range. This conundrum motivates further pursuit of chemically-stimulated PBs with anti-biofouling responses. Often, complex and hybrid systems that incorporate several PBs and other elements are promising. Several of the most recent examples of such complex approaches are listed below to demonstrate the breadth and variety of the designs.

- Orthogonal click functionalization of multicomponent PEG brushes for *in situ* surface modifications during cell culture.[158]
- Two types of switchable mixed brushes were compared for efficiency in protein adsorption suppression and bacterial adhesion reduction: polycationic PEB + zwitterionic PB *versus* polyanioinic PEB + hydrophilic POEGMA + hydrophobic PMMA.[159]

- Dual-responsive (pH and glucose concentration) PBs containing phenylboronic acid for reversible capture and release of cancer cells.[160]
- Silver nanoparticles embedded into zwitterionic PBs as an anti-biofouling platform for the prevention of microbial colonization.[161]
- Switchable biocompatible mixed PBs with self-sterilizing and anti-biofouling properties.[162]

16.7 Brief Summary and Remarks on Applications

The PB is a unique responsive system that enables the control of surface properties in a dynamic mode. The existing methods of PB fabrication allow us to generate "smart" PBs with complex responses to different physical and chemical stimuli. PBs have strong application potential in many fields. However, wide industrial application of PBs is hindered by poor reproducibility, expensive preparation (especially on non-planar surfaces) and poor stability against wear and contamination of the readily available systems. We can foresee further directions of research in the field of responsive PBs that will address these issues:

— a reversible, quick (within seconds or faster) and high contrast response to a stimulus or a combination of stimuli of different physical and/or chemical nature;
— laterally resolved PBs, *i.e.* patterned and/or gradient PBs;
— very simple and inexpensive preparation;
— self-cleaning and wear-resistant PBs;
— bio-related PBs: biochemical response and miscibility with biological objects, bioadhesion control.

The development of responsive brushes, brush-like systems, and methods of synthesis/fabrication that satisfy these conditions is a challenge. In a recent review dedicated to responsive surfaces for life science applications, Minko *et al.* made the prognosis: "Future developments will be based on multicomponent and multifunctional, heretically organized biointerfaces that combine properties of functional materials and devices".[163] Is this a solution? The future will show. An interested reader is referred to this and other reviews dedicated to medical and life-science applications of PBs.[164–168]

Abbreviations and Acronyms

AB	azobenzene
AFM	atomic force microscopy
CA	contact angle
CRP	controlled radical polymerization
CVA	cyclic voltammetry
dmo-bpy	4,4′-dimethoxy-2,2′-bipyridine
Fc	ferrocenyl
HA	hyaluronic acid

ITO	indium tin oxide
MC	merocyanine
MEMS	microelectromechanical systems
PB	polymer brush
PDMAEMA	poly(dimethylamino)ethyl methacrylate
PEB	polyelectrolyte brush
PFcMA	poly(ferrocenyl methacrylate)
PGM	post-grafting modification
qP2VP	quaternized poly(2-vinyl pyridine)
qP4VP	quaternized poly(4-vinyl pyridine)
qPDMAEMA	quaternized poly(dimethylamino)ethyl methacrylate
SCF	self-consistent field
SFA	surface force apparatus
SP	spiropyran
*t*BMA	*tert*-butyl(meth)acrylate
$TFSI^-$	bis(trifluoromethanesulfonimide)
THF	tetrahydrofuran
β-CD	β-cyclodextrin

Polymer	Acronym	Structure
Poly((3-hydroxymethyl) naphthalene-2-ol-4 vinylbenzoate)	PNQMP-4VB	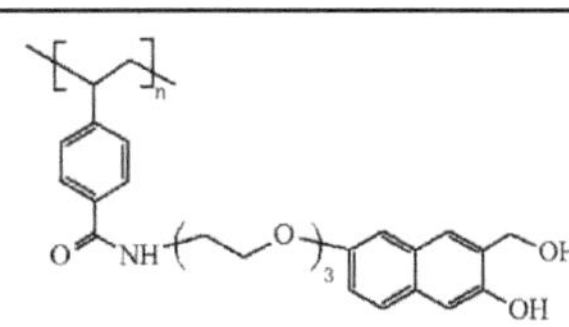
Poly(1-ethyl 3-(2-methacryloyl oxyethyl)imidazolium chloride	PEMEImCl	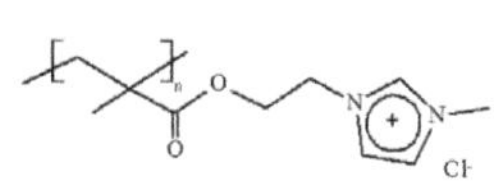
Poly(2-(methacryloyl-oxy)ethyl ferrocen-carboxylate)	PMOEFc	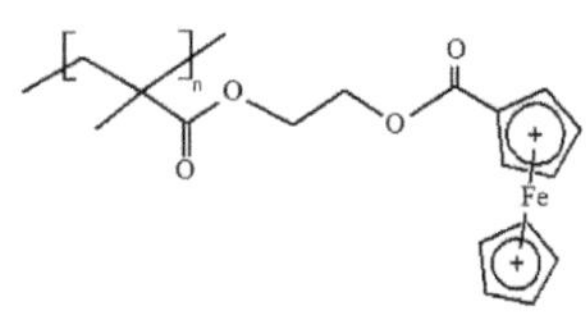
Poly(ferrocenylmethyl methacrylate)	PFcMMA	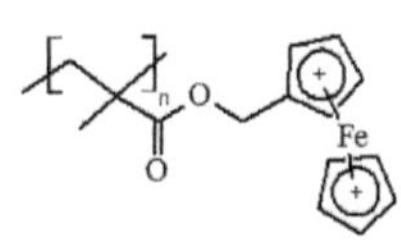
Poly(2-(methacryloyloxy) ethyltrimethyl ammonium) chloride	PMETA Cl^-	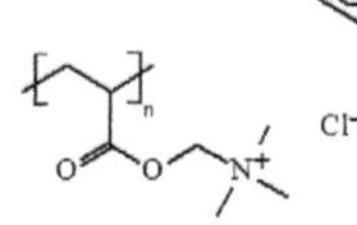
[1] Poly(2-vinyl pyridine) [2] Poly(4-vinyl pyridine)	[1] P2VP [2] P4VP	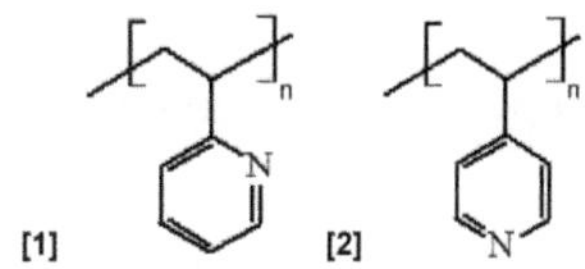

Polymer	Acronym	Structure
Poly(4,5-dimethoxy-2-nitro benzyl methacrylate)	PNVOCMA	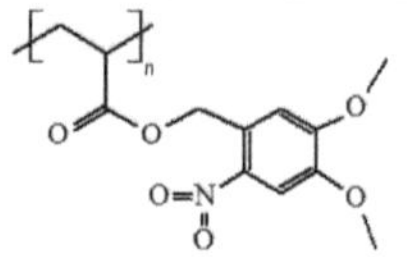
Poly(4-ω-methacryloyloxy-hexyloxy-4′-ethoxy-azobenzene)	PM6A	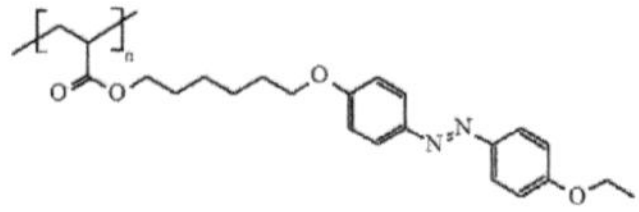
[1] Poly(acrylic acid) [2] Poly(methacrylic acid)	[1] PAA [2] PMeAAc	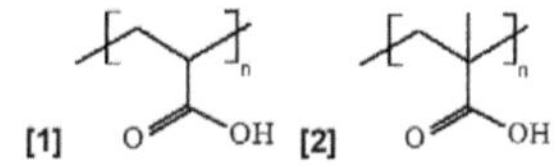
Poly(ethylene glycol)	PEG	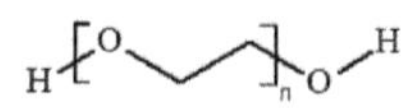
Poly(methacryloyl ethylene) phosphate	PMEP	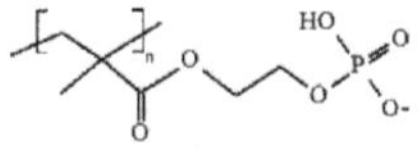
Poly(methyl methacrylate)	PMMA	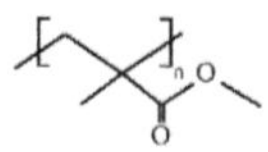
Poly(oligo-ethylene glycol) methacrylate	POEGMA	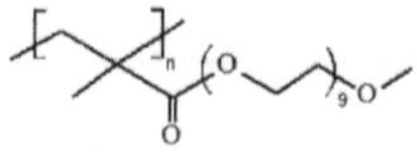
Poly(sulfopropyl methacrylate)	PSPM	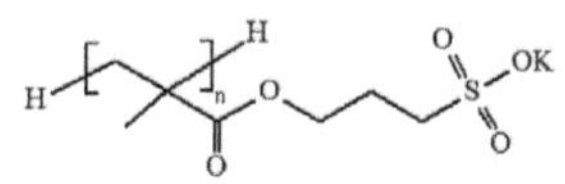
[1] Polyacrylamide [2] Poly(*N*-isopropylacrylamide)	[1] PAAm [2] PNIPAAm	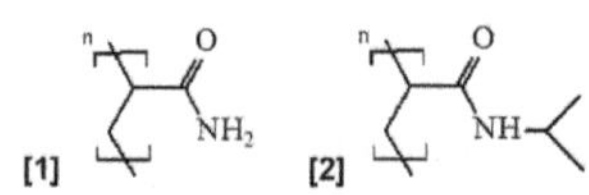
Polystyrene	PS	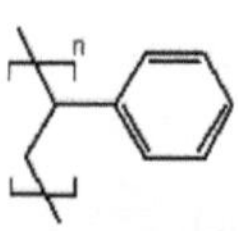
Poly(styrenesulfonate)	PSS	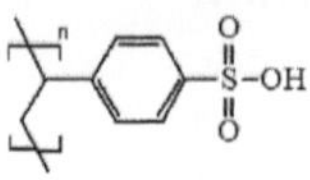
[1] Poly(dimethylamino)ethyl methacrylate [2] Quaternized poly(dimethyl amino)ethyl methacrylate	[1] PDMAEMA [2] qPDMAEMA	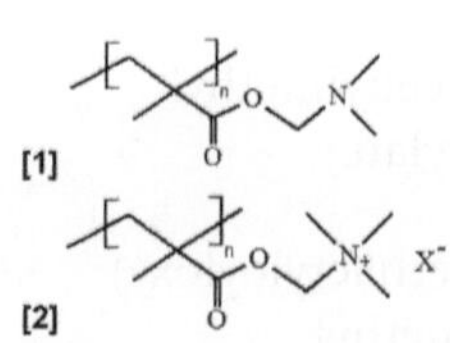
Spiropyranomethacrylate	SPMA	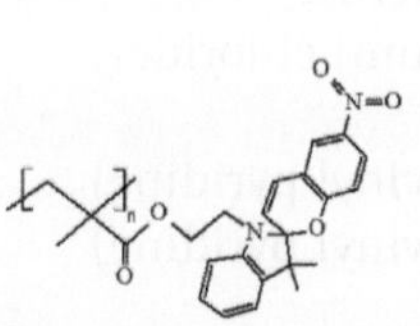

References

1. E. L. Mackor and J. H. Van der Waals, The Statistics of the Adsorption of Rod-Shaped Molecules in Connection with the Stability of Certain Colloidal Dispersions, *J. Colloid Sci.*, 1952, **7**, 535–550.
2. E. J. Clayfield and E. C. Lumb, Theoretical Approach for Polymeric Dispersant Action. II. Calculation of the Dimensions of Terminally Adsorbed Macromolecules, *J. Colloid Interface Sci.*, 1966, **22**(3), 285–293.
3. V. B. Tikhomirov, Effect of Adhesion on the Strength of Bonded, Fiber-Reinforced Materials, *Dokl. Akad. Nauk SSSR*, 1966, **167**(4), 867–868.
4. G. J. Howard and P. McConnell, Adsorption of Polymers at the Solution-Solid Interface. I. Polyethers on Silica, *J. Phys. Chem.*, 1967, **71**(9), 2974–2981.
5. K. Thinius and B. Hoesselbarth, Behavior of Inorganic Fillers in Thermoplastics. Surface-Modified Inorganic Fillers for Thermoplastics, *Plaste Kautsch.*, 1970, **17**(7), 474–477.
6. H. Schonhorn, Wetting Phenomena Pertaining to Adhesion, *Prog. Surf. Membr. Sci.*, 1972, **5**, 121–137.
7. L. L. Miller and M. R. Van de Mark, Electrode Surface Modification via Polymer Adsorption, *J. Am. Chem. Soc.*, 1978, **100**(2), 639–640.
8. S. Alexander, Adsorption of Chain Molecules with a Polar Head a Scaling Description, *J. Phys.*, 1977, **38**(8), 983–987.
9. P. G. De Gennes, Conformations of Polymers Attached to an Interface, *Macromolecules*, 1980, **13**(5), 1069–1075.
10. R. Toomey and M. Tirrell, Functional Polymer Brushes in Aqueous Media from Self-Assembled and Surface-Initiated Polymers, *Annu. Rev. Phys. Chem.*, 2008, **59**, 493–517.
11. G. Hadziioannou, S. Patel, S. Granick and M. Tirrell, Forces Between Surfaces of Block Copolymers Adsorbed on Mica, *J. Am. Chem. Soc.*, 1986, **108**(11), 2869–2876.
12. S. T. Milner, T. A. Witten and M. E. Cates, Theory of Grafted Polymer Brush, *Macromolecules*, 1988, **21**, 2610–2619.
13. W. J. Brittain and S. Minko, A Structural Definition of Polymer Brushes, *J. Polym. Sci., Part A: Polym. Chem.*, 2007, **45**(16), 3505–3512.
14. B. Zdyrko, V. Klep and I. Luzinov, Synthesis and Surface Morphology of High-Density Poly(ethylene glycol) Grafted Layers, *Langmuir*, 2003, **19**(24), 10179–10187.
15. H. Huang, S. E. Rankin, L. S. Penn, R. P. Quirk and T. H. Cheong, Transition from Mushroom to Brush during Formation of a Tethered Layer, *Langmuir*, 2004, **20**(14), 5770–5775.
16. K. Viswanathan, T. E. Long and T. C. Ward, Silicon Surface Modification with Trialkoxy Silyl-Functionalized Star-Shaped Polymers, *J. Polym. Sci., Part A: Polym. Chem.*, 2005, **43**(16), 3655–3666.
17. H. Huang and L. S. Penn, Dense Tethered Layers by the "Grafting-To" Approach, *Macromolecules*, 2005, **38**(11), 4837–4843.

18. S. Minko, G. Gafijchuk, A. Sidorenko and S. Voronov, Radical Polymerization Initiated from a Solid Substrate. 1. Theoretical Background, *Macromolecules*, 1999, **32**(14), 4525–4531.
19. S. Minko, A. Sidorenko, M. Stamm, G. Gafijchuk, V. Senkovsky and S. Voronov, Radical Polymerization Initiated from a Solid Substrate. 2. Study of the Grafting Layer Growth on the Silica Surface by in situ Ellipsometry, *Macromolecules*, 1999, **32**(14), 4532–4538.
20. J. Ruhe and W. Knoll, Functional Polymer Brushes, in *Supramolecular Polymers*, ed. A. Ciferri, 2000, pp. 565–613.
21. S. Edmondson and S. P. Armes, Synthesis of Surface-Initiated Polymer Brushes Using Macro-Initiators, *Polym. Int.*, 2009, **58**(3), 307–316.
22. K. Matyjaszewski, Controlled Radical Polymerization, in *ACS Symposium Series 685*, ed. K. Matyjaszewski, Washington D.C., 1997.
23. K. Matyjaszewski, Controlled Radical Polymerization, *Curr. Opin. Solid State Mater. Sci.*, 1996, **1**(6), 769–776.
24. J. Pyun and K. Matyjaszewski, Synthesis of Nanocomposite Organic/Inorganic Hybrid Materials Using Controlled/"Living" Radical Polymerization, *Chem. Mater.*, 2001, **13**(10), 3436–3448.
25. R. Barbey, L. Lavanant, D. Paripovic, N. Schuwer, C. Sugnaux, S. Tugulu and H.-A. Klok, Polymer Brushes via Surface-Initiated Controlled Radical Polymerization: Synthesis, Characterization, Properties, and Applications, *Chem. Rev.*, 2009, **109**(11), 5437–5527.
26. K. Matyjaszewski and N. V. Tsarevsky, Macromolecular Engineering by Atom Transfer Radical Polymerization, *J. Am. Chem. Soc.*, 2014, **136**(18), 6513–6533.
27. M. Bialk, O. Prucker and J. Ruhe, Grafting of Polymers to Solid Surfaces by Using Immobilized Methacrylates, *Colloids Surf., A*, 2002, **198–200**, 543–549.
28. T. P. Enright, D. Hagaman, M. Kokoruz, N. Coleman and A. Sidorenko, Gradient and Patterned Polymer Brushes by Photoinitiated "Grafting Through" Approach, *J. Polym. Sci., Part B: Polym. Phys.*, 2010, **48**(14), 1616–1622.
29. P. Kim, L. D. Zarzar, X. Zhao, A. Sidorenko and J. Aizenberg, Microbristle in Gels: Toward All-Polymer Reconfigurable Hybrid Surfaces, *Soft Matter*, 2010, **6**(4), 750–755.
30. E. Chu and A. Sidorenko, Surface Reconstruction by a "Grafting Through" Approach: Polyacrylamide Grafted onto Chitosan Film, *Langmuir*, 2013, **29**(40), 12585–12592.
31. J. Genzer, Surface-Bound Gradients for Studies of Soft Materials Behavior, *Annu. Rev. Mater. Res.*, 2012, **42**, 435–468.
32. C. J. Galvin and J. Genzer, Applications of Surface-Grafted Macromolecules Derived from Post-Polymerization Modification Reactions, *Prog. Polym. Sci.*, 2012, **37**(7), 871–906.
33. A. Sidorenko, S. Minko, K. Schenk-Meuser, H. Duschner and M. Stamm, Switching of Polymer Brushes, *Langmuir*, 1999, **15**(24), 8349–8355.

34. S. Minko, S. Patil, V. Datsyuk, F. Simon, K.-J. Eichhorn, M. Motornov, D. Usov, I. Tokarev and M. Stamm, Synthesis of Adaptive Polymer Brushes via "Grafting To" Approach from Melt, *Langmuir*, 2002, **18**(1), 289–296.
35. M. C. LeMieux, Y.-H. Lin, P. D. Cuong, H.-S. Ahn, E. R. Zubarev and V. V. Tsukruk, Microtribological and Nanomechanical Properties of Switchable Y-shaped Amphiphilic Polymer Brushes, *Adv. Funct. Mater.*, 2005, **15**(9), 1529–1540.
36. A block copolymer PB is a brush formed by two blocks of which only the end of one block is tethered to the substrate surface. This is to be distinguished from a block copolymer with one block fully adsorbed on the surface. In the latter case only "free" (non-adsorbed) block forms a PB. A narrow molecular weight distribution of polymer blocks is required in order to observe well-defined microphase separartion and nanometer scale patterns.
37. M. Husseman, E. E. Malmstroem, M. McNamara, M. Mate, D. Mecerreyes, D. G. Benoit, J. L. Hedrick, P. Mansky, E. Huang, T. P. Russell and C. J. Hawker, Controlled Synthesis of Polymer Brushes by "Living" Free Radical Polymerization Techniques, *Macromolecules*, 1999, **32**(5), 1424–1431.
38. R. A. Sedjo, B. K. Mirous and W. J. Brittain, Synthesis of Polystyrene-Block-Poly(Methyl Methacrylate) Brushes by Reverse Atom Transfer Radical Polymerization, *Macromolecules*, 2000, **33**(5), 1492–1493.
39. B. Zhao, W. J. Brittain, W. Zhou and S. Z. D. Cheng, Nanopattern Formation from Tethered PS-b-PMMA Brushes upon Treatment with Selective Solvents, *J. Am. Chem. Soc.*, 2000, **122**(10), 2407–2408.
40. S. Minko, M. Mueller, M. Motornov, M. Nitschke, K. Grundke and M. Stamm, Two-Level Structured Self-Adaptive Surfaces with Reversibly Tunable Properties, *J. Am. Chem. Soc.*, 2003, **125**(13), 3896–3900.
41. S. Minko, M. Mueller, V. Luchnikov, M. Motornov, D. Usov, L. Ionov and M. Stamm, Mixed Polymer Brushes: Switching of Surface Behavior and Chemical Patterning at the Nanoscale, in *Polymer Brushes*, ed. R. C. Advincula, 2004, pp. 403–425.
42. D. L. Hall-Edgefield, T. Shi, K. Nguyen and A. Sidorenko, Hybrid Molecular Brushes with Chitosan Backbone:Facile Synthesis and Surface Grafting, *ACS Appl. Mater. Interfaces*, 2014, **6**(24), 22026–22033.
43. X. Li, S. L. Prukop, S. L. Biswal and R. Verduzco, Surface Properties of Bottlebrush Polymer Thin Films, *Macromolecules*, 2012, **45**(17), 7118–7127.
44. Y. Li, J. Zou, B. P. Das, M. Tsianou and C. Cheng, Well-Defined Amphiphilic Double-Brush Copolymers and Their Performance as Emulsion Surfactants, *Macromolecules*, 2012, **45**(11), 4623–4629.
45. J. Alonzo, Z. Huang, M. Liu, J. W. Mays, R. G. Toomey, M. D. Dadmun and S. M. Kilbey, Looped Polymer Brushes Formed by Self-Assembly of Poly(2-vinylpyridine)-Polystyrene-Poly(2-vinylpyridine) Triblock Copolymers at the Solid-Fluid Interface. Kinetics of Preferential Adsorption, *Macromolecules*, 2006, **39**(24), 8434–8439.

46. S. Sakurai, H. Watanabe and A. Takahara, Preparation and Characterization of Looped Polydimethylsiloxane Brushes, *Polym. J.*, 2014, **46**(2), 117–122.
47. A. Sidorenko, X. W. Zhai, S. Peleshanko, A. Greco, V. V. Shevchenko and V. V. Tsukruk, Hyperbranched Polyesters on Solid Surfaces, *Langmuir*, 2001, **17**(19), 5924–5931.
48. A. Sidorenko, X. W. Zhai, A. Greco and V. V. Tsukruk, Hyperbranched Polymer Layers as Multifunctional Interfaces, *Langmuir*, 2002, **18**(9), 3408–3412.
49. H. Mori, A. Boeker, G. Krausch and A. H. E. Mueller, Surface-Grafted Hyperbranched Polymers via Self-Condensing Atom Transfer Radical Polymerization from Silicon Surfaces, *Macromolecules*, 2001, **34**(20), 6871–6882.
50. Y. Liu, M. L. Bruening, D. E. Bergbreiter and R. M. Crooks, Multilayer Dendrimer-Polyanhydride Composite Films on Glass, Silicon, and Gold Wafers, *Angew. Chem., Int. Ed. Engl.*, 1997, **36**(19), 2114–2116.
51. H. Mori and A. H. E. Mueller, Surface-Grafted Hyperbranched Polymers, in *Polymer Brushes*, ed. R. C. Advincula, 2004, pp. 167–186.
52. A. Sidorenko, C. Houphouet-Boigny, O. Villavicencio, M. Hashemzadeh, D. V. McGrath and V. V. Tsukruk, Photoresponsive Langmuir monolayers from azobenzene-containing dendrons, *Langmuir*, 2000, **16**(26), 10569–10572.
53. A. Sidorenko, C. Houphouet-Boigny, O. Villavicencio, D. V. McGrath and V. V. Tsukruk, Low generation photochromic monodendrons on a solid surface, *Thin Solid Films*, 2002, **410**(1–2), 147–158.
54. W. Cui, C.-F. Su, H. Merlitz, C.-X. Wu and J.-U. Sommer, Structure of Dendrimer Brushes: Mean-Field Theory and MD Simulations, *Macromolecules*, 2014, **47**(11), 3645–3653.
55. O. V. Borisov, A. A. Polotsky, O. V. Rud, E. B. Zhulina, F. A. M. Leermakers and T. M. Birshtein, Dendron Brushes and Dendronized Polymers: A Theoretical Outlook, *Soft Matter*, 2014, **10**(13), 2093–2101.
56. Y. Nagasaki and K. Kataoka, An Intelligent Polymer Brush, *Trends Polym. Sci.*, 1996, **4**(2), 59–64.
57. S. Minko, Responsive Polymer Brushes, *Polym. Rev.*, 2006, **46**(4), 397–420.
58. S. J. Miklavic and S. Marčelja, Interaction of Surfaces Carrying Grafted Polyelectrolytes, *J. Phys. Chem.*, 1988, **92**(23), 6718–6722.
59. S. Misra, S. Varanasi and P. P. Varanasi, A Polyelectrolyte Brush Theory, *Macromolecules*, 1989, **22**(11), 4173–4179.
60. P. Guenoun, A. Schlachli, D. Sentenac, J. W. Mays and J. J. Benattar, Free-Standing Black Films of Polymers: A Model of Charged Brushes in Interaction, *Phys. Rev. Lett.*, 1995, **74**(18), 3628–3631.
61. M. Biesalski, D. Johannsmann and J. Ruhe, Electrolyte-Induced Collapse of a Polyelectrolyte Brush, *J. Chem. Phys.*, 2004, **120**(18), 8807–8814.

62. E. B. Zhulina and O. V. Borisov, Poisson-Boltzmann Theory of pH-Sensitive (Annealing) Polyelectrolyte Brush, *Langmuir*, 2011, **27**(17), 10615–10633.
63. M. Balastre, F. Li, P. Schorr, J. Yang, J. W. Mays and M. V. Tirrell, A Study of Polyelectrolyte Brushes Formed from Adsorption of Amphiphilic Diblock Copolymers Using the Surface Forces Apparatus, *Macromolecules*, 2002, **35**(25), 9480–9486.
64. R. Farina, N. Laugel, P. Pincus and M. Tirrell, Brushes of Strong Polyelectrolytes in Mixed Mono- and Tri-valent Ionic Media at Fixed Total Ionic Strengths, *Soft Matter*, 2013, **9**(44), 10458–10472.
65. L. Jiang, S. Titmuss and J. Klein, Interactions of Hyaluronan Layers with Similarly Charged Surfaces: The Effect of Divalent Cations, *Langmuir*, 2013, **29**(39), 12194–12202.
66. V. S. Guptha and P.-Y. Hsiao, Polyelectrolyte Brushes in Monovalent and Multivalent Salt Solutions, *Polymer*, 2014, **55**(12), 2900–2912.
67. M. J. Roman, E. A. Decker and J. M. Goddard, Fourier Transform Infrared Studies on the Dissociation Behavior of Metal-Chelating Polyelectrolyte Brushes, *ACS Appl. Mater. Interfaces*, 2014, **6**(8), 5383–5387.
68. S. Moya, O. Azzaroni, T. Farhan, V. L. Osborne and W. T. Huck, Locking and Unlocking of Polyelectrolyte Brushes: Toward the Fabrication of Chemically Controlled Nanoactuators, *Angew. Chem., Int. Ed.*, 2005, **44**(29), 4578–4581.
69. Y. Tran and P. Auroy, Complexation and Distribution of Counterions in a Grafted Polyelectrolyte Layer, *Eur. Phys. J. E: Soft Matter Biol. Phys.*, 2001, **5**(1), 65–79.
70. D. Aulich, O. Hoy, I. Luzinov, K.-J. Eichhorn, M. Stamm, M. Gensch, U. Schade, N. Esser and K. Hinrichs, In-situ IR Synchrotron Mapping Ellipsometry on Stimuli-Responsive PAA-b-PS/PEG Mixed Polymer Brushes, *Phys. Status Solidi C*, 2010, 7(2), 197–199.
71. K. Hinrichs, D. Aulich, L. Ionov, N. Esser, K.-J. Eichhorn, M. Motornov, M. Stamm and S. Minko, Chemical and Structural Changes in a pH-Responsive Mixed Polyelectrolyte Brush Studied by Infrared Ellipsometry, *Langmuir*, 2009, **25**(18), 10987–10991.
72. S. M. Kilbey and J. F. Ankner, Neutron Reflectivity as a Tool to Understand Polyelectrolyte Brushes, *Curr. Opin. Colloid Interface Sci.*, 2012, **17**(2), 83–89.
73. G. Sudre, D. Hourdet, C. Creton, F. Cousin and Y. Tran, pH-Responsive Swelling of Poly(acrylic acid) Brushes Synthesized by the Grafting Onto Route, *Macromol. Chem. Phys.*, 2013, **214**(24), 2882–2890.
74. I. E. Dunlop, R. K. Thomas, S. Titmus, V. Osborne, S. Edmondson, W. T. S. Huck and J. Klein, Structure and Collapse of a Surface-Grown Strong Polyelectrolyte Brush on Sapphire, *Langmuir*, 2012, **28**(6), 3187–3193.
75. F. Li, M. Balastre, P. Schorr, J.-F. Argillier, J. Yang, J. W. Mays and M. Tirrell, Differences between Tethered Polyelectrolyte Chains on Bare

Mica and Hydrophobically Modified Mica, *Langmuir*, 2006, **22**(9), 4084–4091.
76. A. Drechsler, A. Synytska, P. Uhlmann, M. M. Elmahdy, M. Stamm and F. Kremer, Interaction Forces between Microsized Silica Particles and Weak Polyelectrolyte Brushes at Varying pH and Salt Concentration, *Langmuir*, 2010, **26**(9), 6400–6410.
77. A. Drechsler, A. Synytska, P. Uhlmann, M. Stamm and F. Kremer, Tuning the Adhesion of Silica Microparticles to a Poly(2-vinylpyridine) Brush: An AFM Force Measurement Study, *Langmuir*, 2012, **28**, 15555–15565.
78. C. Prinz, P. Muller and M. Maaloum, Annealed Polyelectrolyte Brushes under Normal and Lateral Compressions, *Macromolecules*, 2000, **33**(13), 4896–4902.
79. T. Farhan, O. Azzaroni and W. T. S. Huck, AFM Study of Cationically Charged Polymer Brushes: Switching Between Soft and Hard Matter, *Soft Matter*, 2005, **1**(1), 66–68.
80. J. L. Cuellar, I. Llarena, S. E. Moya and E. Donath, Indentation of Highly Charged PSPM Brushes Measured by Force Spectroscopy: Application of a Compressible Fluid Model, *Macromolecules*, 2013, **46**(6), 2323–2330.
81. R. Dong, M. Lindau and C. K. Ober, Dissociation Behavior of Weak Polyelectrolyte Brushes on a Planar Surface, *Langmuir*, 2009, **25**(8), 4774–4779.
82. F. Zhou, W. Shu, M. E. Welland and W. T. S. Huck, Highly Reversible and Multi-Stage Cantilever Actuation Driven by Polyelectrolyte Brushes, *J. Am. Chem. Soc.*, 2006, **128**(16), 5326–5327.
83. F. Zhou, P. M. Biesheuvel, E.-Y. Choi, W. Shu, R. Poetes, U. Steiner and W. T. S. Huck, Polyelectrolyte Brush Amplified Electroactuation of Microcantilevers, *Nano Lett.*, 2008, **8**(2), 725–730.
84. P. Pincus, Colloid Stabilization with Grafted Polyelectrolytes, *Macromolecules*, 1991, **24**(10), 2912–2919.
85. J. L. Harden, O. V. Borisov and M. E. Cates, Deformation of Polyelectrolyte Brushes in Strong Flows: Good Solvent Regime, *Macromolecules*, 1997, **30**(4), 1179–1187.
86. T. M. Birshtein and E. B. Zhulina, Shear of a Planar Polyelectrolyte Brush, *Makromol. Chem., Theory Simul.*, 1992, **1**(4), 193–204.
87. E. B. Zhulina and M. Rubinstein, Lubrication by Polyelectrolyte Brushes, *Macromolecules*, 2014, **47**(16), 5825–5838.
88. E. Spruijt, S. Cohen, A. Martien and J. van der Gucht, Dynamic Force Spectroscopy of Oppositely Charged Polyelectrolyte Brushes, *Macromolecules*, 2010, **43**(3), 1543–1550.
89. A. Galuschko, L. Spirin, T. Kreer, A. Johner, C. Pastorino, J. Wittmer and J. Baschnagel, Frictional Forces between Strongly Compressed, Nonentangled Polymer Brushes: Molecular Dynamics Simulations and Scaling Theory, *Langmuir*, 2010, **26**(9), 6418–6429.
90. B. Liberelle and S. Giasson, Friction and Normal Interaction Forces between Irreversibly Attached Weakly Charged Polymer Brushes, *Langmuir*, 2008, **24**(4), 1550–1559.

91. M. Kobayashi, H. Yamaguchi, Y. Terayama, Z. Wang, K. Ishihara, M. Hino and A. Takahara, Structure and Surface Properties of High-Density Polyelectrolyte Brushes at the Interface of Aqueous Solution, *Macromol. Symp.*, 2009, **279** (Polymers at Frontiers of Science and Technology–MACRO 2008), 79–87.
92. Q. Wei, M. Cai, F. Zhou and W. Liu, Dramatically Tuning Friction Using Responsive Polyelectrolyte Brushes, *Macromolecules*, 2013, **46**(23), 9368–9379.
93. O. Azzaroni, S. Moya, T. Farhan, A. A. Brown and W. T. S. Huck, Switching the Properties of Polyelectrolyte Brushes via "Hydrophobic Collapse", *Macromolecules*, 2005, **38**(24), 10192–10199.
94. X. Wei and T. Ngai, Ion-Induced Hydrophobic Collapse of Surface-Confined Polyelectrolyte Brushes Measured by Total Internal Reflection Microscopy, *Polym. Chem.*, 2012, **3**(8), 2121–2128.
95. G.-L. He, H. Merlitz and J.-U. Sommer, Molecular Dynamics Simulations of Polyelectrolyte Brushes Under Poor Solvent Conditions: Origins of Bundle Formation, *J. Chem. Phys.*, 2014, **140**(10), 104911.
96. F. Zhou, H. Hu, B. Yu, V. L. Osborne, W. T. S. Huck and W. Liu, Probing the Responsive Behavior of Polyelectrolyte Brushes Using Electrochemical Impedance Spectroscopy, *Anal. Chem.*, 2007, **79**(1), 176–182.
97. E.-Y. Choi, O. Azzaroni, N. Cheng, F. Zhou, T. Kelby and W. T. S. Huck, Electrochemical Characteristics of Polyelectrolyte Brushes with Electroactive Counterions, *Langmuir*, 2007, **23**(20), 10389–10394.
98. E. Spruijt, E.-Y. Choi and W. T. S. Huck, Reversible Electrochemical Switching of Polyelectrolyte Brush Surface Energy Using Electroactive Counterions, *Langmuir*, 2008, **24**(19), 11253–11260.
99. M. Motornov, R. Sheparovych, E. Katz and S. Minko, Chemical Gating with Nanostructured Responsive Polymer Brushes: Mixed Brush versus Homopolymer Brush, *ACS Nano*, 2008, **2**(1), 41–52.
100. J. Zhou, X. Lu, J. Hu and J. Li, Reversible Immobilization and Direct Electron Transfer of Cytochrome C on a pH-Sensitive Polymer Interface, *Chem. - Eur. J*, 2007, **13**(10), 2847–2853.
101. V. Bocharova, T. K. Tam, J. Halamek, M. Pita and E. Katz, Reversible Gating Controlled by Enzymes at Nanostructured Interface, *Chem. Commun.*, 2010, **46**(12), 2088–2090.
102. T. K. Tam, J. Zhou, M. Pita, M. Ornatska, S. Minko and E. Katz, Biochemically Controlled Bioelectrocatalytic Interface, *J. Am. Chem. Soc.*, 2008, **130**(33), 10888–10889.
103. T. K. Tam, M. Ornatska, M. Pita, S. Minko and E. Katz, Polymer Brush-Modified Electrode with Switchable and Tunable Redox Activity for Bioelectronic Applications, *J. Phys. Chem. C*, 2008, **112**(22), 8438–8445.
104. V. Privman, S. Domanskyi, S. Mailloux, Y. Holade and E. Katz, Kinetic Model for a Threshold Filter in an Enzymatic System for Bioanalytical and Biocomputing Applications, *J Phys. Chem. B*, 2014, **118**(43), 12435–12443.

105. T. K. Tam, M. Pita, M. Motornov, I. Tokarev, S. Minko and E. Katz, Electrochemical Nanotransistor from Mixed-Polymer Brushes, *Adv. Mater.*, 2010, **22**(16), 1863–1866.
106. B. Yu, H. Hu, D. Wang, W. T. S. Huck, F. Zhou and W. Liu, Electrolyte-Modulated Electrochemistry and Electrocatalysis on Ferrocene-Terminated Polyelectrolyte Brushes, *J. Mater. Chem.*, 2009, **19**(43), 8129–8134.
107. B.-Y. Kim, E. L. Ratcliff, N. R. Armstrong, T. Kowalewski and J. Pyun, Ferrocene Functional Polymer Brushes on Indium Tin Oxide via Surface-Initiated Atom Transfer Radical Polymerization, *Langmuir*, 2010, **26**(3), 2083–2092.
108. Z. B. Zhang, S. J. Yuan, X. L. Zhu, K. G. Neoh and E. T. Kang, Enzyme-Mediated Amperometric Biosensors Prepared via Successive Surface-Initiated Atom-Transfer Radical Polymerization, *Biosens. Bioelectron.*, 2010, **25**(5), 1102–1108.
109. L. Q. Xu, D. Wan, H. F. Gong, K.-G. Neoh, E.-T. Kang and G. D. Fu, One-Pot Preparation of Ferrocene-Functionalized Polymer Brushes on Gold Substrates by Combined Surface-Initiated Atom Transfer Radical Polymerization and "Click Chemistry", *Langmuir*, 2010, **26**(19), 15376–15382.
110. T. W. Chuo, T. C. Wei, Y. Chang and Y. L. Liu, Electrically Driven Biofouling Release of a Poly(Tetrafluoroethylene) Membrane Modified with an Electrically Induced Reversibly Cross-Linked Polymer, *ACS Appl. Mater. Interfaces*, 2013, **5**(20), 9918–9925.
111. A. A. Golriz, T. Kaule, M. B. Untch, K. Kolman, R. Berger and J. S. Gutmann, Redox Active Polymer Brushes with Phenothiazine Moieties, *ACS Appl. Mater. Interfaces*, 2013, **5**, 2485–2494.
112. S. Arumugam, S. V. Orski, J. Locklin and V. V. Popik, Photoreactive Polymer Brushes for High-Density Patterned Surface Derivatization Using a Diels-Alder Photoclick Reaction, *J. Am. Chem. Soc.*, 2012, **134**(1), 179–182.
113. J. Cui, T.-H. Nguyen, M. Ceolín, R. Berger, O. Azzaroni and A. del Campo, Phototunable Response in Caged Polymer Brushes, *Macromolecules*, 2012, **45**(7), 3213–3220.
114. F. Hoffmann and T. Wolff, Photochemical Fixation of Structures in Binary Polymer Brushes-Influence of Layer Thickness and Grafting Method, *J. Colloid Interface Sci.*, 2008, **322**(2), 434–447.
115. A. Noble, III. Zur Geschichte des Azobenzols und des Benzidins, *Liebigs Ann*, 1856, **98**(2), 253–256.
116. A. Natansohn and P. Rochon, Photoinduced Motions in Azobenzene-Based Amorphous Polymers. Possible Photonic Devices, *Adv. Mater.*, 1999, **11**(16), 1387–1391.
117. A. Natansohn and P. Rochon, Photoinduced Motions in Azobenzene-Based Polymers, in *Photoreactive Organic Thin Films*, ed. Z. Sekkat and W. Knoll, 2002, pp. 399–427.

118. P. Camorani, L. Cristofolini, M. P. Fontana, L. Angiolini, L. Giorgini and F. Paris, Azo-Containing Polymer Brushes: Photoalignment and Application as Command Surfaces, *Mol. Cryst. Liq. Cryst.*, 2009, **502**, 56–64.
119. T. Seki, Meso- and Microscopic Motions in Photoresponsive Liquid Crystalline Polymer Films, *Macromol. Rapid Commun.*, 2014, **35**(3), 271–290.
120. S. Samanta and J. Locklin, Formation of Photochromic Spiropyran Polymer Brushes via Surface-Initiated, Ring-Opening Metathesis Polymerization: Reversible Photocontrol of Wetting Behavior and Solvent Dependent Morphology Changes, *Langmuir*, 2008, **24**(17), 9558–9565.
121. K. Fries, S. Samanta, S. Orski and J. Locklin, Reversible Colorimetric Ion Sensors Based on Surface Initiated Polymerization of Photochromic Polymers, *Chem. Commun.*, 2008, **47**, 6288–6290.
122. S. Samanta, K. Fries, S. Orski and J. Locklin, Formation of Photo-Responsive Surfaces by Surface-Initiated Ring Opening Metathesis Polymerization and Atom Transfer Radical Polymerization: Reversible Optodes for Metal Ion Sensors, *ACS Symp. Ser.*, 2010, **1050** (Smart Coatings III), 73–85.
123. D. S. Achilleos, T. A. Hatton and M. Vamvakaki, Light-Regulated Supramolecular Engineering of Polymeric Nanocapsules, *J. Am. Chem. Soc.*, 2012, **134**(13), 5726–5729.
124. B.-Y. Zhang, W.-D. He, W.-T. Li, L.-Y. Li, K.-R. Zhang and H. Zhang, Preparation of Block-Brush PEG-b-P(NIPAM-g-DMAEMA) and its Dual Stimulus-Response, *Polymer*, 2010, **51**(14), 3039–3046.
125. Y. Stetsyshyn, J. Zemla, O. Zolobko, K. Fornal, A. Budkowski, A. Kostruba, V. Donchak, K. Harhay, K. Awsiuk, J. Rysz, A. Bernasik and S. Voronov, Temperature and pH Dual-Responsive Coatings of Oligoperoxide-Graft-Poly(N-isopropylacrylamide): Wettability, Morphology, and Protein Adsorption, *J. Colloid Interface Sci.*, 2012, **387**(1), 95–105.
126. S.-i. Yamamoto, J. Pietrasik and K. Matyjaszewski, Temperature- and pH-Responsive Dense Copolymer Brushes Prepared by ATRP, *Macromolecules*, 2008, **41**(19), 7013–7020.
127. J. Lindqvist, D. Nystroem, E. Oestmark, P. Antoni, A. Carlmark, M. Johansson, A. Hult and E. Malmstroem, Intelligent Dual-Responsive Cellulose Surfaces via Surface-Initiated ATRP, *Biomacromolecules*, 2008, **9**(8), 2139–2145.
128. Y. Lu, A. Zhuk, L. Xu, X. Liang, E. Kharlampieva and S. A. Sukhishvili, Tunable pH and Temperature Response of Weak Polyelectrolyte Brushes: Role of Hydrogen Bonding and Monomer Hydrophobicity, *Soft Matter*, 2013, **9**(22), 5464–5472.
129. N. Houbenov, S. Minko and M. Stamm, Mixed Polyelectrolyte Brush from Oppositely Charged Polymers for Switching of Surface Charge and Composition in Aqueous Environment, *Macromolecules*, 2003, **36**(16), 5897–5901.

130. L. Ionov, N. Houbenov, A. Sidorenko, M. Stamm, I. Luzinov and S. Minko, Inverse and Reversible Switching Gradient Surfaces from Mixed Polyelectrolyte Brushes, *Langmuir*, 2004, **20**(23), 9916–9919.
131. Y. Mikhaylova, L. Ionov, J. Rappich, M. Gensch, N. Esser, S. Minko, K.-J. Eichhorn, M. Stamm and K. Hinrichs, In Situ Infrared Ellipsometric Study of Stimuli-Responsive Mixed Polyelectrolyte Brushes, *Anal. Chem.*, 2007, **79**(20), 7676–7682.
132. L. Ionov, N. Houbenov, A. Sidorenko, M. Stamm and S. Minko, Stimuli-Responsive Command Polymer Surface for Generation of Protein Gradients, *Biointerphases*, 2009, **4**(2), FA45–FA49.
133. W. Sun, S. Zhou, B. You and L. Wu, Polymer Brush-Functionalized Surfaces with Reversible, Precisely Controllable Two-Way Responsive Wettability, *Macromolecules*, 2013, **46**(17), 7018–7026.
134. S. Santer, A. Kopyshev, J. Donges, H.-K. Yang and J. Ruehe, Dynamically Reconfigurable Polymer Films: Impact on Nanomotion, *Adv. Mater.*, 2006, **18**(18), 2359–2362.
135. M. Motornov, R. Sheparovych, R. Lupitskyy, E. MacWilliams, O. Hoy, I. Luzinov and S. Minko, Stimuli-Responsive Colloidal Systems from Mixed Brush-Coated Nanoparticles, *Adv. Funct. Mater.*, 2007, **17**(14), 2307–2314.
136. M. F. Delcroix, G. L. Huet, T. Conard, S. Demoustier-Champagne, F. E. Du Prez, J. Landoulsi and C. C. Dupont-Gillain, Design of Mixed PEO/PAA Brushes with Switchable Properties Toward Protein Adsorption, *Biomacromolecules*, 2013, **14**(1), 215–225.
137. H. Lee, V. Tsouris, Y. Lim, R. Mustafa, J. Choi, Y. H. Choi, H.-W. Park, M. Meron, B. Lin and Y.-Y. Won, Macroscopic Lateral Heterogeneity Observed in a Laterally Mobile Immiscible Mixed Polyelectrolyte-Neutral Polymer Brush, *Soft Matter*, 2014, **10**(21), 3771–3782.
138. A. Jusufi, O. Borisov and M. Ballauff, Structure Formation in Polyelectrolytes Induced by Multivalent Ions, *Polymer*, 2013, **54**(8), 2028–2035.
139. A. L. Becker, K. Henzler, N. Welsch, M. Ballauff and O. Borisov, Proteins and Polyelectrolytes: A Charged Relationship, *Curr. Opin. Colloid Interface Sci.*, 2012, **17**(2), 90–96.
140. Y. Lu and M. Ballauff, Thermosensitive Core-Shell Microgels: From Colloidal Model Systems to Nanoreactors, *Prog. Polym. Sci.*, 2011, **36**(6), 767–792.
141. M. Ballauff, Spherical Polyelectrolyte Brushes, *Prog. Polym. Sci.*, 2007, **32**(10), 1135–1151.
142. A. Wittemann and M. Ballauff, Interaction of Proteins with Linear Polyelectrolytes and Spherical Polyelectrolyte Brushes in Aqueous Solution, *Phys. Chem. Chem. Phys.*, 2006, **8**(45), 5269–5275.
143. M. Ballauff, Spherical Polyelectrolyte Brushes, in *Polymer Brushes*, ed. R. C.Advincula, 2004, pp. 231–248.
144. M. L. Bruening, D. M. Dotzauer, P. Jain, O. Lu and G. L. Baker, Creation of Functional Membranes Using Polyelectrolyte Multilayers and Polymer Brushes, *Langmuir*, 2008, **24**(15), 7663–7673.

145. I. Zharov and A. Khabibullin, Surface-Modified Silica Colloidal Crystals: Nanoporous Films and Membranes with Controlled Ionic and Molecular Transport, *Acc. Chem. Res.*, 2014, **47**(2), 440–449.
146. A. G. Koutsioubas, Formation of Polymer Brushes Inside Cylindrical Pores, in *Polymer Brushes*, ed. V. Mittal, 2012, pp. 115–131.
147. P. Jain, G. L. Baker and M. L. Bruening, Applications of Polymer Brushes in Protein Analysis and Purification, *Annu. Rev. Anal. Chem.*, 2009, **2**, 387–408.
148. T. Kawai, K. Saito and W. Lee, Protein Binding to Polymer Brush, Based on Ion-Exchange, Hydrophobic, and Affinity Interactions, *J. Chromatogr., B: Anal. Technol. Biomed. Life Sci.*, 2003, **790**(1–2), 131–142.
149. A. Hucknall, S. Rangarajan and A. Chilkoti, In Pursuit of Zero: Polymer Brushes that Resist the Adsorption of Proteins, *Adv. Mater.*, 2009, **21**(23), 2441–2446.
150. K. L. Prime and G. M. Whitesides, Adsorption of Proteins onto Surfaces Containing End-Attached Oligo(Ethylene Oxide): A Model System Using Self-Assembled Monolayers, *J. Am. Chem. Soc.*, 1993, **115**(23), 10714–10721.
151. H. Ma, J. Hyun, P. Stiller and A. Chilkoti, Non-Fouling'' Oligo(Ethylene Glycol)-Functionalized Polymer Brushes Synthesized by Surface-Initiated Atom Transfer Radical Polymerization, *Adv. Mater.*, 2004, **16**(4), 338–341.
152. H. Ma, D. Li, X. Sheng, B. Zhao and A. Chilkoti, Protein-Resistant Polymer Coatings on Silicon Oxide by Surface-Initiated Atom Transfer Radical Polymerization, *Langmuir*, 2006, **22**(8), 3751–3756.
153. A. A. Brown, N. S. Khan, L. Steinbock and W. T. S. Huck, Synthesis of Oligo(Ethylene Glycol) Methacrylate Polymer Brushes, *Eur. Polym. J.*, 2005, **41**(8), 1757–1765.
154. I. Cringus-Fundeanu, J. Luijten, H. C. Van der Mei, H. J. Busscher and A. J. Schouten, Synthesis and Characterization of Surface-Grafted Polyacrylamide Brushes and Their Inhibition of Microbial Adhesion, *Langmuir*, 2007, **23**(9), 5120–5126.
155. I. Fundeanu, D. Klee, A. J. Schouten, H. J. Busscher and H. C. van der Mei, Solvent-Free Functionalization of Silicone Rubber and Efficacy of PAAm Brushes Grafted from an Amino-PPX Layer Against Bacterial Adhesion, *Acta Biomater.*, 2010, **6**(11), 4271–4276.
156. I. Fundeanu, H. C. van der Mei, A. J. Schouten and H. J. Busscher, Microbial Adhesion to Surface-Grafted Polyacrylamide Brushes After Long-Term Exposure to PBS and Reconstituted Freeze-Dried Saliva, *J. Biomed. Mater. Res., Part A*, 2010, **94A**(3), 997–1000.
157. L. Bureau and M. Vayssade, Collapse, Compression, and Adhesion of Poly(N-Isopropylacrylamide) Brushes, in *Polymer Brushes*, ed. V. Mittal, 2012, pp. 47–63.
158. J. Sha, E. S. Lippmann, J. McNulty, Y. Ma and R. S. Ashton, Sequential Nucleophilic Substitutions Permit Orthogonal Click Functionalization of Multicomponent PEG Brushes, *Biomacromolecules*, 2013, **14**(9), 3294–3303.

159. O. Rzhepishevska, S. Hakobyan, R. Ruhal, J. Gautrot, D. Barbero and M. Ramstedt, The Surface Charge of Antibacterial Coatings Alters Motility and Biofilm Architecture, *Biomater. Sci.*, 2013, **1**(6), 589–602.
160. H. Liu, Y. Li, K. Sun, J. Fan, P. Zhang, J. Meng, S. Wang and L. Jiang, Dual-Responsive Surfaces Modified with Phenylboronic Acid-Containing Polymer Brush To Reversibly Capture and Release Cancer Cells, *J. Am. Chem. Soc.*, 2013, **135**(20), 7603–7609.
161. R. Hu, G. Li, Y. Jiang, Y. Zhang, J.-J. Zou, L. Wang and X. Zhang, Silver-Zwitterion Organic-Inorganic Nanocomposite with Antimicrobial and Antiadhesive Capabilities, *Langmuir*, 2013, **29**(11), 3773–3779.
162. G. Cheng, H. Xue, Z. Zhang, S. Chen and S. Jiang, A Switchable Biocompatible Polymer Surface with Self-Sterilizing and Nonfouling Capabilities, *Angew. Chem., Int. Ed.*, 2008, **47**(46), 8831–8834.
163. H. Kuroki, I. Tokarev and S. Minko, Responsive Surfaces for Life Science Applications, *Annu. Rev. Mater. Res.*, 2012, **42**, 343–372.
164. O. Azzaroni, Polymer Brushes Here, There, and Everywhere: Recent Advances in Their Practical Applications and Emerging Opportunities in Multiple Research Fields, *J. Polym. Sci., Part A: Polym. Chem.*, 2012, **50**(16), 3225–3258.
165. P. Uhlmann, H. Merlitz, J.-U. Sommer and M. Stamm, Polymer Brushes for Surface Tuning, *Macromol. Rapid Commun.*, 2009, **30**(9–10), 732–740.
166. E. Bittrich, S. Burkert, K.-J. Eichhorn, M. Stamm and P. Uhlmann, Control of Protein Adsorption and Cell Adhesion by Mixed Polymer Brushes Made by the "Grafting-To" Approach, *ACS Symp. Ser.*, 2012, **1120** (Proteins at Interfaces III), 179–193.
167. S. Zapotoczny, Stimuli Responsive Polymers for Nanoengineering of Biointerfaces, *Methods Mol. Biol.*, 2012, **811** (Nanotechnology in Regenerative Medicine), 51–78.
168. N. Ayres, Polymer Brushes: Applications in Biomaterials and Nanotechnology, *Polym. Chem.*, 2010, **1**(6), 769–777.

CHAPTER 17

Autonomously Moving Interface and Colloidal Objects

DAIGO YAMAMOTO, ERIKA NAWA AND AKIHISA SHIOI*

Department of Chemical Engineering & Materials Science, Doshisha University, 1-3 Tatara Miyakodani, Kyotanabe, Kyoto 610-0321, Japan
*Email: ashioi@mail.doshisha.ac.jp

17.1 Introduction

Living organisms can be distinguished from non-living organisms by their vital motions. Animals can move their bodies,[1] and flowers can move their petals. These motions are triggered by various types of chemical, physical, and environmental stimuli. Smaller-scale vital motions can also be observed in the human body. Numerous proteins and cells move autonomously and appear to be very busy as they perform their allotted roles,[2] Biological systems could not maintain their vitality without this dynamic nature.

Unlike man-made machines, vital motions are maintained by complicated sequences of chemical reactions including mass transport. Moreover, in many cases, mass transport is sustained by vital motion, and this is called active transport. The stimuli-responsive nature of vital motions originates from the linkage between the motions and the sequence of chemical reactions. Even a weak stimulus may cause a dramatic change in the resultant motion, due to a nonlinearity in the time evolution of the sequential dynamics.[3] Many scientific efforts have focused on the subtle mechanisms inherent to this linkage, and as a result, great strides have been made in our scientific understanding of biological systems.[2]

RSC Smart Materials No. 14
Chemoresponsive Materials: Stimulation by Chemical and Biological Signals
Edited by Hans-Jörg Schneider

Published by the Royal Society of Chemistry, www.rsc.org

An alternative method for understanding the linkage between chemical reactions and autonomous motions is to design man-made chemical systems with similar characteristics to those found in biological motion. These man-made systems are simpler than biological systems, making the physics and chemistry of the underlying linkage between the chemical reactions and autonomous motions easier to understand.

The sustainable motions driven by chemical reactions are accomplished under a strong nonequilibrium state,[4] without which the system would monotonously relax toward equilibrium. Monotonous relaxation never produces vital motion. A strong nonequilibrium state can only be maintained by an open system accompanying a mass transport across the system interface.[4] The thermodynamics for this scenario require a small scale, *i.e.*, a large specific area, for the moving objects. If the scale of the moving objects is large, the chemical reaction and the related mass transports in the vicinity of the interface do not influence their bulk properties, and the objects will not respond to the surrounding stimuli. Thus, a moving object that appears to have biological characteristics will be colloidal in scale.

The other significant thermodynamic point is the spontaneous generation of regulated motions from chemical reactions. In the second law of thermodynamics, Kelvin's principle states that the mechanical work cannot be obtained from a reservoir with constant temperature.[5] From a microscopic viewpoint, this demonstrates that the thermal motion of molecules cannot be transformed into regular motion.[6] The regular motion of a colloidal object is mechanical, hence it cannot be obtained from the object's thermal motion. Consider a case in which chemical reactions proceed on the surface of a colloidal object and a mechanical force for driving the object is generated from these reactions.[7] If the chemical reaction proceeds homogeneously over the surface, the mechanical force will act on the object homogeneously. In other words, no systematic force acts upon the object, and, at most, only transient random forces act upon the object. So these forces can give only random motions to the object. These random fluctuations are very similar to the colloidal object's thermal fluctuation. However, a systematic motion is a type of mechanical work and can be produced only from the random fluctuation caused by chemical reactions. In the first section of this chapter, we show several types of regulated motions from the random forces generated by the chemical reactions exhibited by a catalytic particle.[8] These regulated motions result from an anisotropic dissipation of the mechanical work generated by random forces. This is a type of ratchet effect and may be a universal method for obtaining regulated motions from the random forces acting on an object.

For sustainable motion, man-made machines require gradients of intensive variables such as temperature, pressure, and chemical potential. This is a fundamental requirement of the second law of thermodynamics. Living cells also use the same gradient, the chemical potential gradient in particular, to generate the mechanical work that propels them and to maintain their physiological functions.[2,9] For example, a pH gradient produced from ATP hydrolysis across the mitochondrial inner membrane is used for reverse ATP synthesis.[10] A white blood cell chases a moving bacterium by a process known as chemotaxis,[11] wherein a trace amount of the bacterium's physiological products generates a concentration gradient tracked by the white blood cell. Both these gradients are microscopic in scale, and no man-made machine could convert them into mechanical work. In some cases, the gradient itself is produced by a chemical reaction, such as in the pH gradient as mentioned earlier.[10] The design of colloidal objects that can convert the microscopic gradient of a chemical potential into mechanical work may be the first step in mimicking these biological functions. In the second section of this chapter, we discuss an amphiphilic molecular assembly with a sustainable cyclic motion by a pH gradient.[12] This is a rare example where mechanical work is generated from the microscopic gradient of a chemical potential.

With respect to chemical systems, stimuli responsiveness is nearly identical to a function that amplifies a molecular level event into a macroscopic response. The human body and other living systems depend on this characteristic for life maintenance. A trace amount of chemicals can sometimes cause a fatal event *via* an allergic reaction. Human physiological functions may be lost by a small change in the ion balance of sodium, potassium, and calcium. These molecules and ions act on the body *via* chemical reactions with proteins at the molecular level, while the resultant effect is macroscopic. This characteristic is quite different from that in a typical man-made machine. If we want to change the motions of a macroscopic machine, we must introduce to the system a macroscopic perturbation such as macroscopic power or a large amount of chemicals. This is because the man-made machine is designed in such a way that the system maintains stability despite various perturbations from the surroundings. This stability is necessary in any reliable man-made system. If the system were to be unstable with respect to a particular perturbation, even a weak perturbation could change the characteristics of its macroscopic motion. While a living system is dynamically stable, it also has numerous inherent instabilities.[13] When these instabilities are contained in chemical reaction rate equations, the system with these instabilities exhibits a responsiveness against chemical stimuli whereby a trace amount of chemicals induces the instability and changes the macroscopic motion. In the third section of this chapter, we describe the process by which an oil/water interface exhibits a macroscopic motion depending on the cation species.[14,15] Even the qualitative nature of the spontaneous motions of the oil/water interface changes in the presence of specific cations. This is an example where the interface must touch specific cations in order to move.

17.2 Spontaneous Regular Motion of a Particle from Random Fluctuation

In many biological systems, chemical reactions provide the energy required for vital motions. Motions assisted by chemical reactions are responsive to chemical stimuli *via* their effect on those reactions. This may be an attractive tool for developing a smart power source that has biomimetic characteristics. From this perspective, catalytic particles driven by chemical reactions are analogs of biomimetic energy conversion systems.[7] These particles have attracted much attention in recent years. A typical example is the use of a Janus particle comprising two types of materials[16–20] in which one side of the particle is made of one material, and the other side is made of the other. These two types of materials exhibit contrasting catalytic activities. The chemical reaction rate of one side of the particle is much faster than that of the other side. This produces a chemical potential gradient in the chemicals surrounding the particle, and this gradient can drive the particle *via* Marangoni and self-electrophoretic effects[21,22] or both. (Marangoni effect means that surface tension gradient causes the surface flow. In some cases, it grows to a macroscopic scale by a hydrodynamic instability that is called Marangoni instability.[23] This will be explained in Section 17.4 of this chapter. The latter operate when oxidation and reduction reactions proceed at each side.) This type of micromotor generates a driving force that propels the particle in a direction determined by the particle structure.[24]

However, a few groups (including the authors of this chapter) revealed that a Janus structure is not necessary for a catalytic particle to have regulated motions.[8,25,26] Figure 17.1 shows a trajectory of a single component Pt particle in pure water (a1, a2) and in a H_2O_2 aqueous solution (b1–b5).[8] All the particles exhibit Brownian motion without the presence of H_2O_2, but they can exhibit non-Brownian motion in an H_2O_2 aqueous solution. Platinum decomposes H_2O_2 into H_2O and O_2. However, the O_2 bubbles are invisible in the vicinity of the particle. The shapes of particles are classified into spherical (b1, b2), teardrop (b3), boomerang (b4), and windmill (b5) types. When a spherical Pt particle is put into an H_2O_2 aqueous solution, it exhibits a random motion that appears to be enhanced Brownian (a1 and a2). For the other shapes, regulated motions such as translation (b3), rotation (b4), and spin (b5) are obtained. For the random motions of a spherical particle, Figure 17.2 shows the apparent diffusion coefficients for several concentrations of H_2O_2. The apparent diffusion coefficient increases with an increase in H_2O_2 concentration.

Since all particles are made of platinum (>99%), the average rate of H_2O_2 decomposition is nearly constant over the particle surface. However, the transient rate of decomposition may depend on the surface position, and this transient inhomogeneity may provide a driving force to the particle. Such a driving force is transient, and the strength and direction of the force vary every moment. A particle under such driving forces will exhibit random motions. As a result, the trajectory of a spherical particle appears to be random, and the mean square displacement $\langle r(t)^2 \rangle$ is proportional with time. The particle's mobility increases by the transient driving force caused by the reaction. As a result, the apparent diffusion coefficient increases with increasing H_2O_2 concentration.

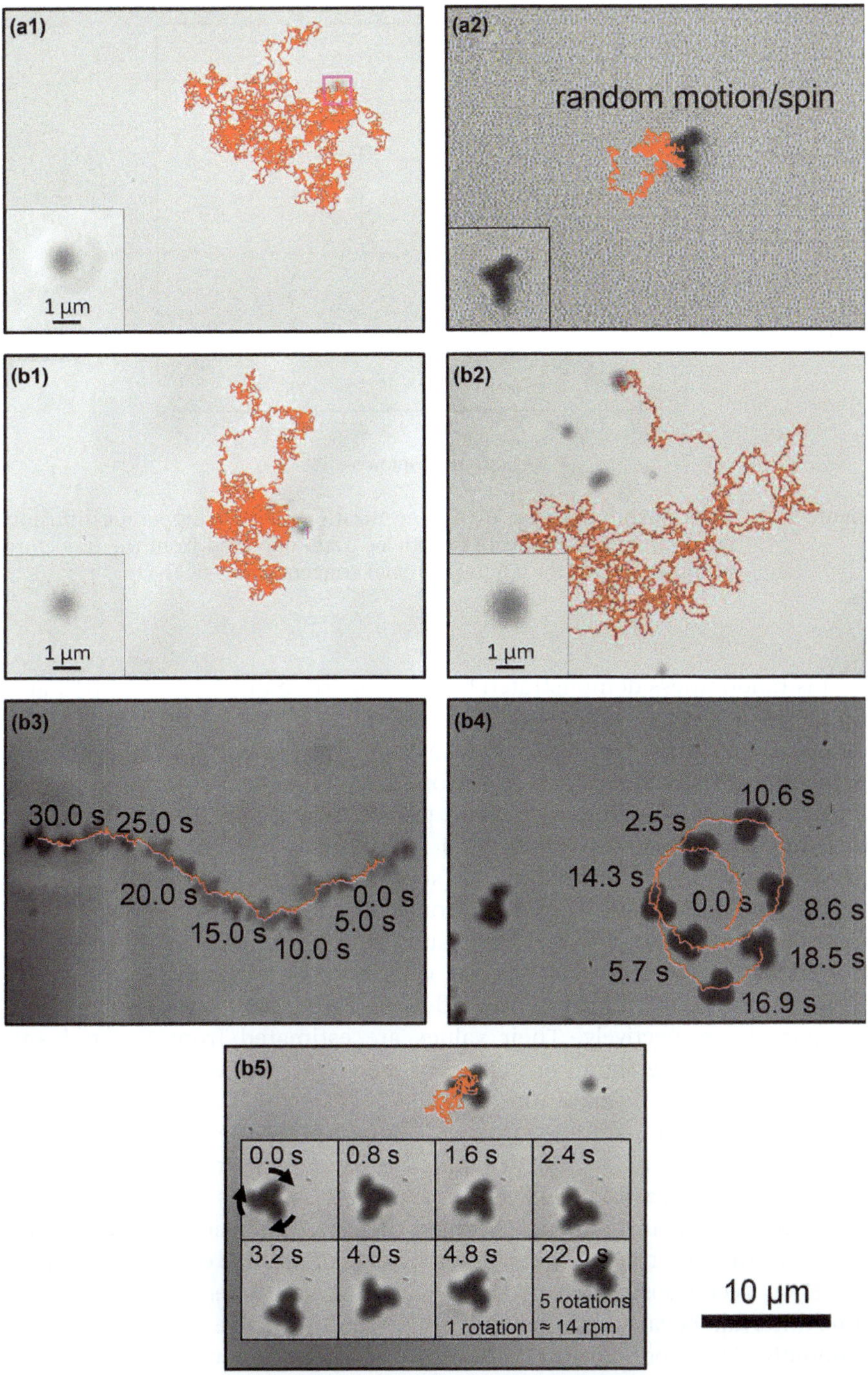

Figure 17.1 Trajectory and snapshots of various motions of Pt particles in (a) water and (b) 1% H_2O_2 water. Both spherical and non-spherical Pt particles exhibit Brownian motion in water, while they also exhibit non-Brownian motions such as (b1, b2) active random motion, (b3) translation, (b4) rotation, and (b5) spin motions, depending on their morphology. (*J. Chem. Phys.*, 2013, **139**, 034705)[8]

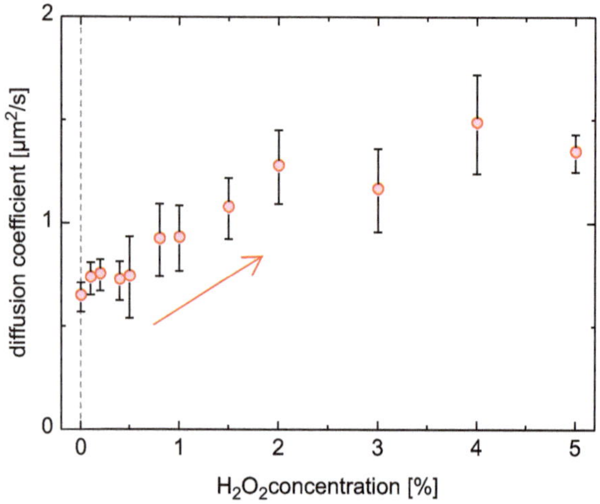

Figure 17.2 Relationship between H_2O_2 concentrations and the apparent diffusion coefficients of spherical Pt particles. Data obtained from the trajectory of ten particles ($d \approx 0.6$ μm) at each concentration of H_2O_2.

This randomness demonstrated by the proportionality between $\langle r(t)^2 \rangle$ and t appears to be lost on a very short timescale.[24] This tendency is more remarkable in larger particles, where, as shown in Figure 17.3, the $\langle r(t)^2 \rangle$ of a particle in a H_2O_2 solution is not proportional to t on a timescale shorter than 0.2 s.[8] The experimental result gives a relationship of $\langle r(t)^2 \rangle \propto t^{\alpha}$ for this time range. The exponent α (1.4 for the case in Figure 17.3) is intermediate between 1 (random) and 2 (straight motion). This result suggests that a straight motion does occur on a short timescale. However, it is unlikely that this straight motion results from the inertia of a Pt particle. If inertia dominates the motion, the velocity of Pt particle $v(t)$ is given by $v(t) = v(0)\exp(-ft/m)$ where f and m denote the viscous drag and mass of the particle, respectively. Their values are estimated from $f = 3\pi\eta d$ and $m = \rho_{Pt}(\pi d^3)/6$. Here, η and ρ_{Pt} represent the viscosity of water ($\approx 10^{-3}$ Pa s) and the density of platinum ($\approx 21.5 \times 10^3$ kg m^{-3}), respectively, and d ($\approx 10^{-6}$ m) is the particle diameter. The period for the inertial motion is approximately given by m/f, or $\rho_{Pt}d^2/18\eta$. At room temperature, this is approximately equal to 1–2 μs. This timescale is much shorter than the timescales shown in Figure 17.3. Thus, the one dimensional motion does not result from inertia but is sustained by the continuous energy input from the H_2O_2 decomposition. This straight motion continues only for a short time interval, because the driving force acting on the particle changes randomly. Therefore, the particle exhibits random motions for a longer period of time.

The straight motion of a Pt spherical particle for a short time is probably associated with the regulated motion of Pt aggregates with

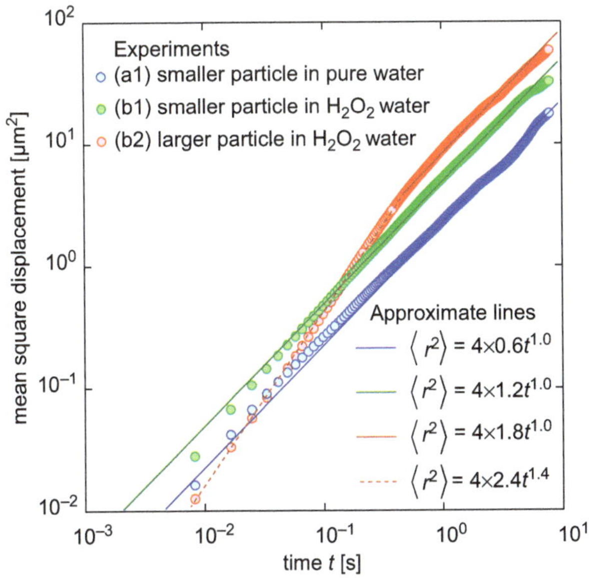

Figure 17.3 Relationship between time and mean square displacement obtained from the trajectory of particles (a1), (b1), and (b2) of Figure 17.1. The solid lines and the dotted line show approximate lines of slope 1.0 and 1.4, respectively. (*J. Chem. Phys.*, 2013, **139**, 034705).[8]

non-spherical shapes. The motion of aggregates can also be classified into (a) translation, (b) rotation, and (c) spin types. As shown in Figure 17.1, these characteristics are fully determined by the geometrical symmetry of the aggregates. The density of platinum is greater than that of water; hence the Pt aggregates sink to the bottom of the cell. Thus, the trajectory shown in (b3–b5) of Figure 17.1 is restricted at the surface of the bottom plate, so we focus on the symmetry of the aggregate shape projected onto the bottom plate. Figure 17.4 shows an example of a teardrop-type aggregate that translates in the H_2O_2 solution while maintaining this projection shape. We put this two-dimensional shape in an arbitrary direction and drew an x-axis that passes through its center of gravity (G). We then drew another axis with the azimuth ψ, where ψ is an arbitrary angle. Then we defined an angle θ from the ψ-axis as shown in Figure 17.4. Let $R(\psi;\theta)$ be the distance between point G and the periphery at angle θ. Angle θ varies from $-\pi$ to π. Then, we have a function $\delta R(\psi;\theta) = R(\psi;\theta) - R_{av}(\psi)$ where $R_{av}(\psi)$ is the average of $R(\psi;\theta)$ over all θ values. The Fourier series of $\delta R(\psi;\theta)$ is given by

$$\delta R(\psi, \theta) = \sum_{n=1}^{\infty} a_n(\psi)\cos(n\theta) + \sum_{n=1}^{\infty} b_n(\psi)\sin(n\theta) \tag{17.1}$$

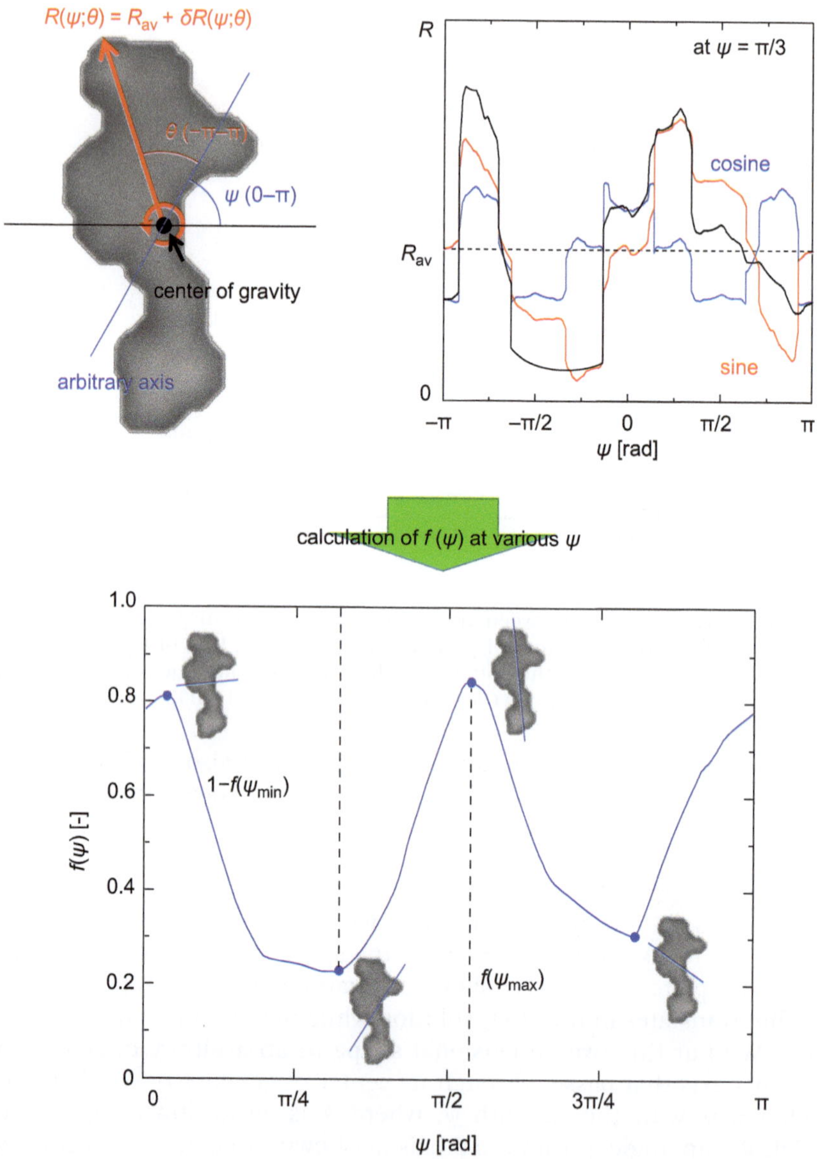

Figure 17.4 Typical example of a quantitative analysis of particle morphology using the particle from (b3) of Figure 17.1. Top: (left) schematic representation of quantitative analysis of particle morphology and (right) profiles of $R(\theta)$ $(-\pi \le \theta \le \pi)$ from an axis tilted at angle ψ $(\psi = \pi/3)$. The profile can be divided into terms of cosine (blue line) and sine (red line). Bottom: calculated result for $f(\psi)$ of an arbitrary axis at various values of angle ψ $(0 \le \psi \le \pi)$. (*J. Chem. Phys.*, 2013, **139**, 034705)[8]

The shape corresponding to the first term is symmetrical with the ψ-axis because it is an even function with respect to θ. The second term gives the departure from perfect symmetry. Thus, we may define the degree of symmetry with respect to the ψ-axis ($f(\psi)$) by

$$f(\psi) = \frac{\sum_{n=1}^{\infty} a_n(\psi)^2}{\sum_{n=1}^{\infty} a_n(\psi)^2 + \sum_{n=1}^{\infty} b_n(\psi)^2} \tag{17.2}$$

The ψ value varies from 0 to π, and the function $f(\psi)$ is shown at the bottom of Figure 17.4. Its maximum and minimum values are at $\psi_{max} = 0.53\pi$ and $\psi_{min} = 0.32\pi$, respectively. We may define an axial symmetry–asymmetry balance (SAB) by SAB$=f\ (\psi_{max})/(1-f\ (\psi_{min}))$, where the denominator may be regarded as the maximum degree of asymmetry.

The SAB values are calculated for the various aggregate shapes exhibiting regulated motions in the H_2O_2 solution. Figure 17.5 demonstrates that spin, rotation, and translation are obtained for SAB$<$1, SAB$=$1, and SAB$>$1, respectively.[27] This result implies that the projection shape determines the type of two-dimensional motion. Therefore, it is reasonable to

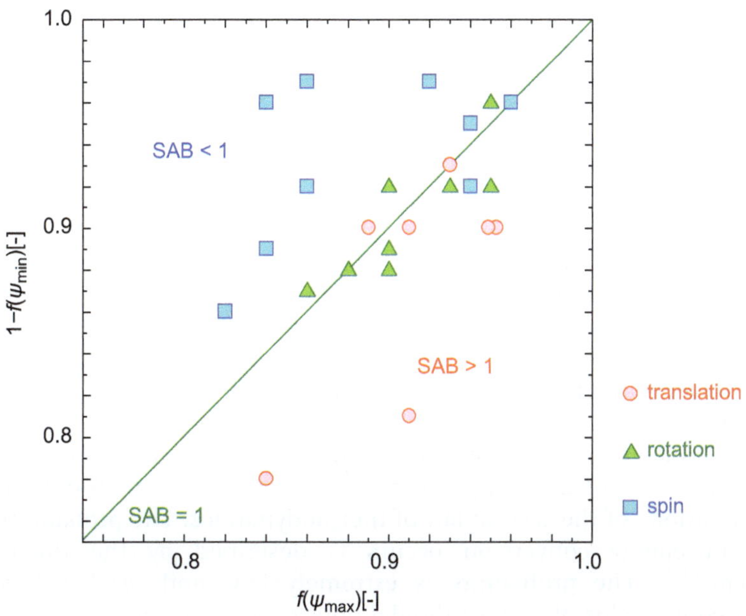

Figure 17.5 Relationship between the dynamical modes and the axial symmetry–asymmetry balance (SAB) of particles with various morphologies. (*KONA Powder Part. J.*, 2015, **32**, 2)[27]

consider that the asymmetry of the drag forces produces the regulated motion. In general, the asymmetry of the viscous drag force is not significant at a low Reynolds number, because the viscous drag force is equal to the integration of viscous stress over the surface. Unless vortices develop, this drag force is dominated by the area of an object projected onto a plane perpendicular to the flow and independent of the flow direction. However, vortices may develop even at a very low Reynolds number.[28,29]

The equation of motion of a random driving force F_r (torque N_r) and an anisotropic drag force $\eta_v v(t)$ ($\eta_\omega \omega(t)$) may be described as

$$m\frac{\mathrm{d}\vec{v}(t)}{\mathrm{d}t} = \vec{F}_r - \eta_v\,\vec{v}(t) \quad \text{for translation, and} \tag{17.3}$$

$$I\frac{\mathrm{d}\vec{\omega}(t)}{\mathrm{d}t} = \vec{N}_r - \eta_\omega\,\vec{\omega}(t) \quad \text{for rotation} \tag{17.4}$$

Here, I and ω denote the moment of inertia and angular velocity, respectively. As shown in Figure 17.6, by reducing the drag force in a particular direction or rotation by 10% less than other directions or rotations, we can reproduce trajectories very similar to those observed.

The example given here uses chemical reactions to obtain a random driving force. However, other methods may be used for generating a random driving force. For example, the random vibration of a plate provides a random fluctuation of an object on the plate. The resultant motion is random if the object shape is isotropic. However, when the object shape is anisotropic, regular motions can be obtained from random vibration.[30] An alternating current electric field or pressure field are examples of other possible methods.[31–34] A charged particle in the AC field and a particle in an alternating pressure gradient exhibit only alternating motions that cannot give a systematic translation. However, when the system is designed in such a way that the dissipation is anisotropic, a systematic translation appears by a ratchet mechanism. Ratchet mechanisms are used not only for biological systems,[35,36] but also for man-made active transport systems.[37–39] Designing asymmetry into dissipation may be a universal method for obtaining regulated motions from random fluctuations.

A spontaneous energy conversion from thermal motions to mechanical work under a constant temperature is believed to be impossible due to the limitations of the second law of thermodynamics. The probability that such an energy conversion occurs is described by the fluctuation theorem.[40–42] The probability is extremely low, and reaches zero for macroscopic scales. When a colloidal object swims in water at a constant velocity, the motion can generate mechanical work even if the time period of the constant velocity is very short and the motion appears to be random. For example, we can obtain mechanical work from a random vital

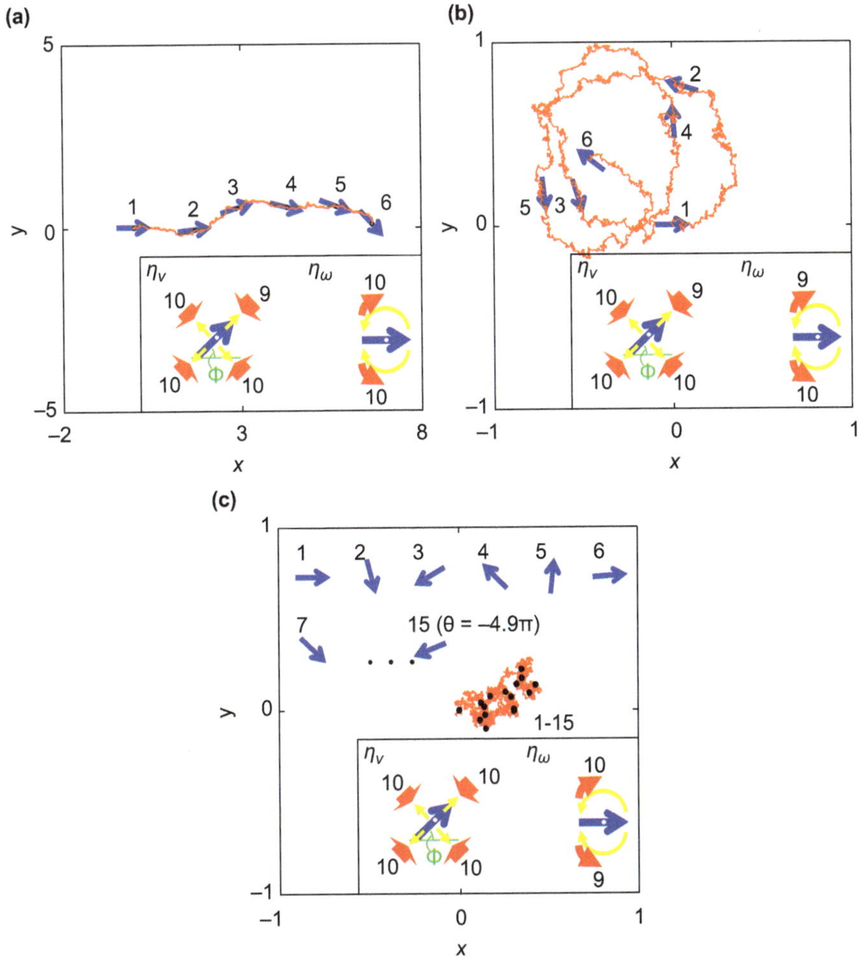

Figure 17.6 Calculated result of the trajectory ($0 \leq t \leq 1.5 \times 10^4$) for (a) translation, (b) rotation, and (c) spin motions of Pt aggregates based on eqn (17.3) and (17.4). Blue arrows represent particles with the angular displacement of rotation (ϕ) [see the insets in (a)–(c)]. All parameters, excluding drag, η_v, and η_ω, are fixed as follows: $\langle |F(t)|^2 \rangle = 1/6$, $\langle N(t)^2 \rangle = 1/3$, $x(0) = y(0) = v_x(0) = v_y(0) = \phi(0) = \omega(0) = 0$, $m = 0.5$, and $I = 1.0$. Values of direction-dependent drag are described in the insets of (a)–(c). (*J. Chem. Phys.*, 2013, **139**, 034705)[8]

motion of microorganisms[43] despite the fact that the thermal motion of dead microorganisms cannot be transformed into mechanical work. The Brownian ratchet considered by R. Feynman[6] cannot convert thermal motions into mechanical work if the second law is satisfied for molecular length scales. However, essentially the same ratchet can generate mechanical work from the random motions of colloids driven by non-thermal forces. This ratchet utilizes anisotropic dissipation to regulate the random motions.

17.3 Chemotactic Motion of a Vesicle under a pH Gradient

A chemical potential or concentration gradient produces a translating liquid droplet on a solid substrate. This type of motion has been studied since the 1990s, and various ideas for driving a droplet have been proposed.[44–50] The simplest idea is to prepare a hydrophobicity (hydrophilicity) gradient on a solid surface on which a water droplet propels itself toward the less hydrophobic area. This idea was developed as a means to generate stimuli-responsive motion. When a gradient of hydrophobicity is generated by an irradiation of light, a light-responsive motion by a droplet can be obtained. When a liquid droplet was placed on a substrate surface modified with a calix[4]resorcinarene derivative with photochromic azobenzene units, asymmetrical photoirradiation caused a gradient in surface free energy due to the photoisomerization of surface azobenzenes, leading to the directional motion of the droplet.[48] A pH-responsive droplet was also studied. An oil/water interface containing the surfactant bis-(2-ethylhexyl)phosphate has a lower interfacial tension at a higher pH. When an oil droplet containing this surfactant is put into a buffer solution, the droplet moves around by spontaneous Marangoni instability.[51] A liquid droplet with this characteristic can navigate a maze.[52] Oil droplets that spontaneously generate a chemical potential gradient have also been studied.[53] An oil droplet containing an anion exhibits self-propulsion on a glass surface in a water phase with a dissolving cationic surfactant.[54,55] Adsorption of the cationic surfactant on the glass surface makes it hydrophobic. However, a chemical reaction between the adsorbed surfactant and the anion in the oil droplet produces fluctuation in the hydrophobicity around the droplet. In this case, the droplet moves away from less hydrophobic areas that occasionally form. Once the droplet moves, the glass surface behind the moving droplet is always less hydrophobic than the surface at the front of the droplet, because the chemical reaction fully proceeds at its rear surface due to the sufficient contact between the glass surface and the droplet. This effect maintains droplet motion.[56–60]

All these studies have been performed with liquid droplets. This is because the Marangoni instability and contact angle change can both be used to drive a liquid droplet. Marangoni instability causes a convection from a weak fluctuation of interfacial tension which drives the droplet. Convection and contact angle change are characteristics of an oil/water and a liquid/solid interface. These interfaces are quite different from biological cells, in which both sides of the interface (membrane) are essentially in aqueous solutions. In this case, it appears to be difficult to use Marangoni instability and contact angle change for self-motion. Actually, biological motions use the reversible sol/gel transition of an amoeboid motion responding to stimuli[61] and the oscillatory conformational change in proteins that drive molecular motors.[62] Marangoni instability and contact angle change are not considered to be associated with these biological motions.

Vesicles (liposomes) have been studied as model cells that perform chemical reactions and material separations that exhibit biomimetic characteristics.[63,64] Both sides of a vesicle membrane are in the aqueous phase. Thus, if we are able to design a self-moving vesicle, this motion may involve a mechanism quite different from that used in droplet motions. The shape change of vesicles responding to physical and chemical stimuli has been the focus of both experimental and theoretical study.[65,66] These studies have addressed the vesicle transformation to minimum membrane energy. Thus, the dynamics of the shape change may be described on the basis of a monotonous relaxation to the minimum energy state. In contrast, sustainable motions such as vital motions require activation of the energy state in addition to relaxation dynamics. In a biological system, a chemical reaction and a chemical potential gradient provide the energy for this activation.[2]

From these viewpoints, making a man-made system with a periodic shape change of vesicles has become a matter of interest.[67–69] Recently, a stepwise dissolution of vesicles was reported, wherein the vesicle decreases its size by an addition of chemicals or an irradiation of light. However, this dissolution process is not a simple relaxation; it accompanies an oscillatory shape change. One example is a periodic holing of a vesicle.[69] With light irradiation or addition of chemicals, a hole opens on the vesicle membrane and closes after a while, and this cycle repeats itself. Another example involves a more complicated shape change.[67,68] After an initial shape change from spherical to other more complicated shapes, the vesicle returns to a spherical shape, and this cycle repeats itself. These motions have periodic characteristics. However, the vesicle sizes in both systems decrease with time, and hence they can be regarded as the stepwise dissolution of vesicles. We have also reported vesicle self-motion *via* chemical reaction, however, in our study the vesicle size decreased with the reaction.[70] Self-propulsion of a vesicle that maintains its size has been studied using biomolecules,[71] whereby the motion mechanism mimics that of *Listeria monocytogenes*, and the vesicle moves by consuming an actine filament in its surroundings. As far as we know, this is the only example that has been reported of a sustainable motion by a vesicle.

Recently, our group found a vesicle that exhibits an oscillatory shape change by a pH gradient.[12] The vesicle maintains its size during the motion, and hence a continuous energy conversion from the local pH gradient to mechanical work is realized in a man-made chemical system. Figure 17.7(A) shows the vesicle's oscillatory motion. The vesicle is formed with oleic acid and oleate in a bicine buffer. The vesicle appears to be double spherical in shape and is similar to a stomatocyte.[66] On one side, there is a space filled with an aqueous phase between the inner and the outer spheres, denoted as the swollen part in this figure. The confocal microscopy images in

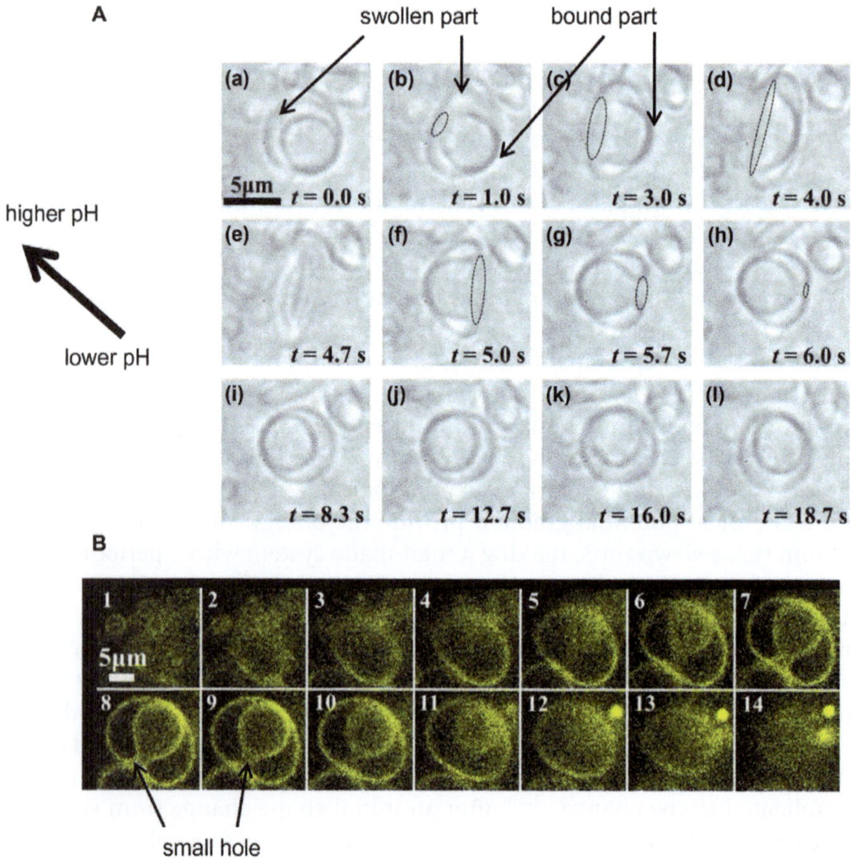

Figure 17.7 One cycle of a vesicle shape change and confocal microscopy images are shown. The vesicle is composed of oleate in a bicine buffer ($pH \cong 8.1$). (A) Vesicle transformation. The direction of the pH gradient is shown by the arrow. (B) Confocal microscopy images at various cross sections of a double spherical vesicle. Photographs were taken at a depth of approximately 1 μm. Rhodamine 6G was used as the fluorescent dye. (*Soft Matter*, 2013, **9**, 7832)[12]

Figure 17.7(B) indicate the presence of a small hole in the swollen part, connecting the external phase with the inner water pool. On the other side, there is no space between the spheres so these two membrane layers are considered to be tightly bound. This is denoted as the bound part in the figure. A pH gradient forms by a diffusion of sodium hydroxide or a carbonate buffer with $pH \approx 10$. After diffusion, every vesicle rotates in such a way that the swollen part faces the higher-pH side. This rotation can be seen in Figure 17.7(A)i → 7(A)l. After this rotation, a small hole in the swollen part begins to open and becomes larger, as shown in 7(A)b → 7(A)d. This expansion of the hole makes the vesicle shape nearly flat, as seen in 7(A)e.

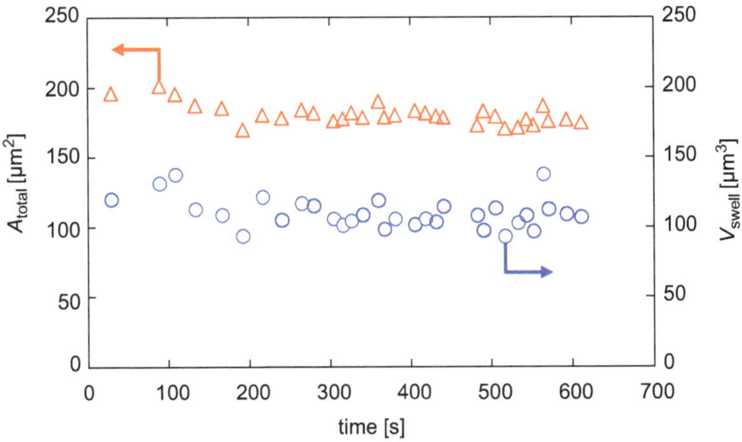

Figure 17.8 Membrane area (triangle) and volume of water pool (circle) of a vesicle. Both values were evaluated at a double spherical state for every cycle of the shape change. The high-pH source was 1 M NaOH. (*Soft Matter*, 2013, **9**, 7832)[12]

This motion continues until the vesicle closes and is facing the direction opposite to the pH gradient, as seen in 7(A)h. This type of motion (from 7(A)a → 7(A)h) is denoted as reversal. After reversal, the swollen part faces the lower-pH side (7(A)i). However, this direction appears to be unstable, and the vesicle begins to rotate to return to its initial direction (7(a)i → 7(A)l). This cycle is repeated.

Figure 17.8 shows the vesicle membrane area and the volume of the swollen part as a function of time during 30 cycles. Using confocal microscopy, we observe that the vesicle structure is a cross section of a solid of revolution. The revolution axis is parallel to the pH gradient and passes through the center of the vesicle. We evaluated the membrane area and the volume of the swollen part based on this geometry. The vesicle size is nearly constant despite the complicated shape change, although the results are a little scattered due to the uncertainty of the photograph and the vesicle's departure from a perfect solid of revolution. The oscillatory shape change enables a continuous energy conversion from a local pH gradient to a local mechanical force. So this vesicle can carry an object along a pH gradient by this cyclic motion.[12]

This oscillatory shape change is composed of two types of dynamical modes. The first is reversal, and the other is the rotation of the vesicle such that the swollen part faces the higher-pH side. We have previously proposed a mathematical model for these shape changes, however, an intuitive physical interpretation derived from the model is explained here.

In the absence of a pH gradient, the directions taken by the double spherical vesicle are random. As a simple model, we allow only the two directions shown in Figure 17.9(a), denoted as parallel and antiparallel. Here, the parallel direction indicates that the swollen part faces the higher-pH

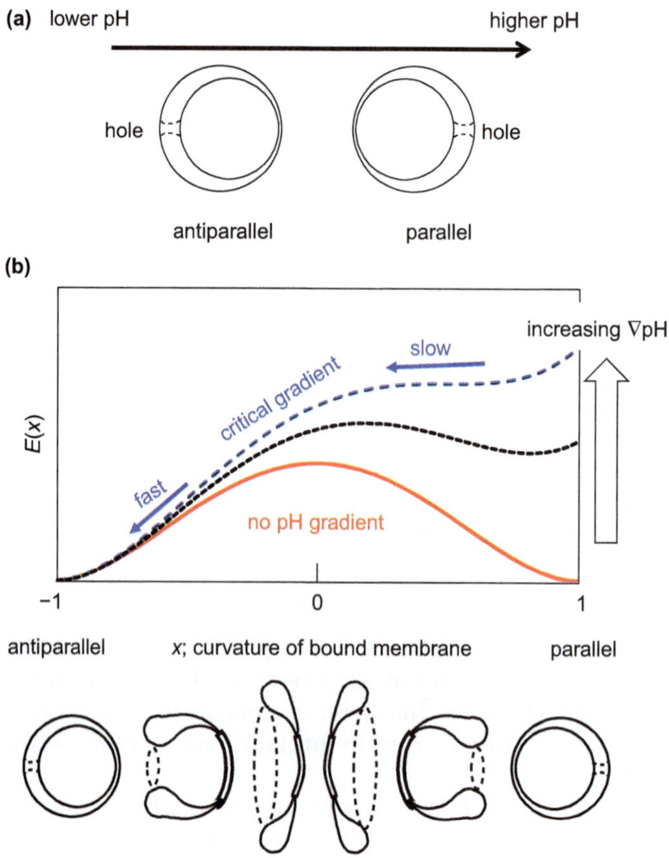

Figure 17.9 Schematic representation of the energy landscape of a vesicle membrane. (a) Definition of parallel and antiparallel vesicles. (b) Energy landscape against a reduced curvature x. The x represents the state of the vesicle depicted underneath the curve. The minimum energy is present at both sides. A higher pH increases the energy of the small hole in the swollen part. Thus, the minimum energy of the parallel vesicle increases with increasing pH.

side, and the antiparallel direction is the one doing the opposite. Without a pH gradient, parallel and antiparallel directions are each taken by half of the total vesicle populations. This indicates that there are two equivalent minima in the vesicle membrane's energy landscape. This energy landscape is illustrated in Figure 17.9(b), and is denoted as $E(x)$. Here, x represents the reduced curvature of the bound part shown in Figure 17.7, which may be a vesicle shape parameter.

A small hole is present in the center of the swollen part. When a pH gradient is applied, this hole opens. The vesicle flattens and closes toward the antiparallel direction. This suggests that the hole in a lower-pH environment is more stable than that in a higher-pH environment. The hole is surrounded by numerous head groups of oleic acid and its ionized form (oleate). The ratio of oleate to acid increases with an increase in pH.[72] This results in stronger electrostatic repulsions between the head groups, which may increase the diameter of the holes. Thus, a higher pH increases the energy minimum for parallel vesicles in the energy landscape, as shown in Figure 17.9(b), and the resultant energy landscape becomes asymmetric. The difference in the two minima increases with an increase in the degree of the pH gradient. With a pH gradient, the two energy minima are no longer the same.

As the pH gradient increases, the minimum energy of the parallel vesicle also increases until this minimum disappears. At this critical pH gradient, the vesicle changes to adopt the antiparallel direction. In the early stages of this transformation, the vesicle shape changes very slowly, because the gradient of the membrane energy (dE/dx) is extremely small. However, the rate of shape change accelerates as the gradient becomes larger during the transformation.

Figure 17.10 shows the experimentally reduced curvature of the bound state membrane as a function of time. The curvature change is very slow at the early stages and then accelerates during the transformation. A mathematical model that reproduces the curve of Figure 17.10 along with appropriate parameters has been proposed.[12]

After the reversal, the vesicle is oriented in the antiparallel direction with the swollen part facing the lower-pH side. We occasionally observed a vesicle resembling a tadpole, with a spherical head and a tail which formed after numerous repetitions of oscillatory shape change. Then, the vesicle exhibits a translational motion toward the higher-pH side which suggests that a

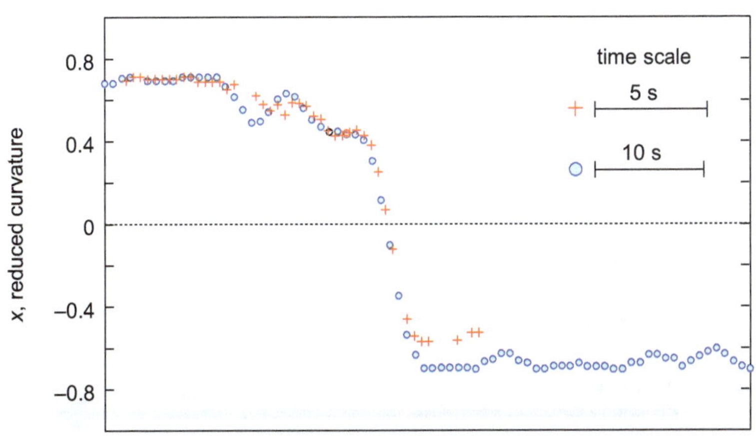

Figure 17.10 Reduced curvature of a vesicle. The definition of the curvature is shown elsewhere.[12] The high-pH source was either 1 M NaOH (plus) or a carbonate buffer solution (circle). The time is scaled so that the results overlap. (*Soft Matter*, 2013, **9**, 7832)[12]

driving force (phoretic force) is operating. The membrane of the swollen part of the double spherical vesicle is more hydrophilic than that of the bound part, because the swollen part contains a bulk water phase. This means that the concentration of the ionized form of the surfactant (oleate) is probably higher in the swollen part than in the bound part. As such, a thermodynamic force to drive the swollen part toward the higher-pH side may appear. In actual experiments, numerous vesicles and fragments of vesicles appear to restrict the translation. However, this thermodynamic force may cause a rotation of the vesicle.

Consider a balancing toy under gravity, as shown in Figure 17.11(a). The two weights are at the lowest position in Figure 17.11(a2). This state is stable

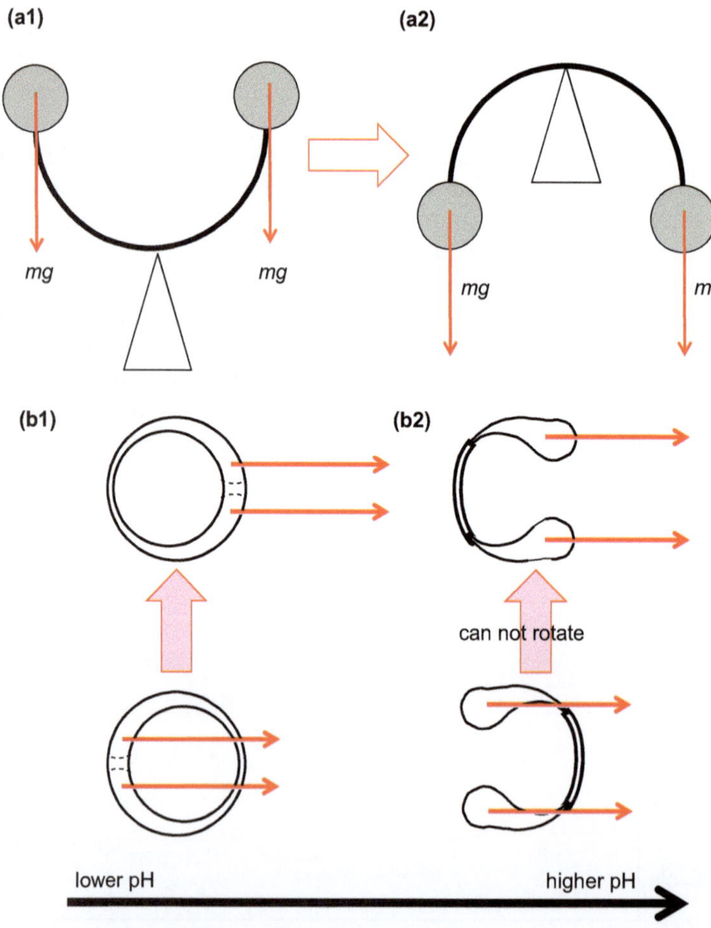

Figure 17.11 Schematic explanation of rotational instability. (A) Balancing a toy under gravity. The arrangement of a1 is unstable and changes to a2 *via* rotation. The latter arrangement is stable against weak perturbations. A phoretic force toward the higher-pH side plays the role of gravity. The arrangements at the bottom of b1 and b2 are unstable and change to the corresponding upper arrangements.

against weak rotational perturbations. However, there is another balancing point (Figure 17.11(a1)) at which the two weights are at the highest positions. This state is unstable against rotational perturbations. In this case, the two weights go to the lowest positions by rotating around the fulcrum of the toy. As shown in Figure 17.11(b1), the swollen part of the double spherical vesicle upon which the driving force acts corresponds to the weight of the toy. The thermodynamic force plays the role of gravity. When the translation is inhibited due to the surrounding vesicles and their fragments or due to their adsorption onto a solid substrate, the vesicle may rotate from a state of rotational instability to the most stable state, *i.e.*, with the swollen part on the higher-pH side.

As shown in Figure 17.11(b2), rotation may occur during reversal, when the vesicle does not close completely. However, this would be very difficult to achieve, as Figure 17.10 indicates that the period from having a flat shape to having a completely closed form is much shorter than the following period of the rotation. Thus, considering that there is insufficient time for rotation during reversal, we conclude that rotation begins after the vesicle completely closes.

17.4 Ion-sensitive Motion of an Oil/Water Interface

Liquid with a higher surface tension pulls the surrounding liquid more strongly. Thus, a gradient in surface tension causes liquid flow in the surface. In some cases, this surface flow is enhanced by an instability based on hydrodynamics, leading to an occurrence of macroscopic convection. This is called Marangoni instability, which is probably one of the oldest examples of spontaneously moving chemical objects that has long intrigued scientists. The well-known "tears of wine", which has been studied since the 19th century, is explained by the Marangoni effect. In a Benard–Marangoni instability, numerous hexagonal cells in fluid have natural convection, and this is one of the most well studied examples of nonlinear pattern formation in physics.[23]

When a drop of detergent falls onto a water surface, it spreads over the surface spontaneously by the Marangoni effect, causing a surface flow. However, this flow occurs only once immediately after the droplet falls. In contrast, two immiscible liquid phases, such as oil and water, can exhibit successive occurrences of Marangoni instability, typically when surfactants are dissolved in one of the two phases. This type of Marangoni instability has been studied since the mid-20th century[73–76] and is referred to as interfacial turbulence. This interfacial flow accelerates the mass transfer rate across the interface, and the stagnant film theory, which explains the rate-determining step of mass transport, cannot be applied to this type of interface. Moreover, spontaneous emulsifications at an oil/water interface may be associated with successive Marangoni instabilities.

The successive occurrence of Marangoni instabilities is theoretically explained by hydrodynamics and a boundary condition at the interface. In this theory, several dimensionless parameters govern the instability, and the Marangoni number (Ma) is the most crucial of these.[74,76] For heat transport across a fluid interface such as air/water, the Marangoni number is defined as

$$\mathrm{Ma} = \frac{(-\partial\sigma/\partial T)(\Delta T/d)d}{\kappa\eta} \tag{17.5}$$

where σ and η, respectively, denote the interfacial tension at the surface and the viscosity of the liquid phase. T and κ are the temperature and thermal diffusivity of the liquid phase, respectively. The temperature gradient is $\Delta T/d$. When the Marangoni number exceeds a critical value, a hydrodynamic instability develops and causes spontaneous convections. For two immiscible liquids with mass transports, the situation is more complicated than the thermal Marangoni instability. Diffusion coefficients for the two immiscible liquids appear, instead of thermal diffusivity, in the governing equations. The viscosities of the two immiscible fluids also affect the occurrence of the instability, and concentration gradients form on each side of the interface. In this case, the ratio of the diffusion coefficients, viscosities, and concentration gradients play crucial roles.[76]

Although the phenomenon of two immiscible liquid phases with mass transport of the solutes appears to be complicated, the governing equations are based on the hydrodynamics. Specifically, the interfacial chemistry effects of the chemical species on the thermodynamic properties of the interface focuses on the parameter $\partial\sigma/\partial C_i$ which is the effect of the concentration of the i-th adsorbate on the interfacial tension. When chemical reactions between oil-soluble and water-soluble chemicals proceed and affect the interfacial tension, several types of chemical species are adsorbed. Then, their concentrations at the interface vary in complicated ways with time. A typical approach for this complicated case is to combine the reaction rate equations with the hydrodynamic equations.[77,78] In the orthodox description of Marangoni instability, very small spontaneously formed fluctuations make up the hydrodynamic instability aspect. Although the interfacial chemistry affects this growth mechanism, the chemistry does not change the essential physics of the instability growth. This appears to hold true unless the reaction rate equations contain an instability mechanism such as the Belousov–Zhabotinsky (BZ) reaction.[79,80] Numerous types of interfacial reactions that have been studied in connection with Marangoni instability are not oscillatory, as are BZ reactions,[81] hence the orthodox framework mentioned appears to be quite useful.

In some cases, however, small changes in the chemical nature of the solutes at the interface or in the interfacial potential cause successive occurrences of Marangoni instability. These small changes do not strongly affect the hydrodynamic properties such as the viscosity and density of the solution and the diffusivity of the solute molecules. Many experiments related to these studies have dealt with the simple mass transport of surfactants between oil and water phases. No oscillatory characteristics seem to appear in the adsorption and desorption kinetics. Several researchers have taken an in-depth orthodox approach to this case and have proposed a thorough study of the hydrodynamic considerations,[82] while others have considered the effect of the interfacial chemistry and thermodynamics on instability.[14,15,83–86] In the latter studies, an extraordinary adsorption/desorption or an electrochemical instability was considered. The extraordinary adsorption and desorption concept means that a critical adsorption density (concentration of solutes) is required for desorption (adsorption).[83,86] The electrochemical instability theory is applicable to ionic surfactants.[84,85] The adsorption of ionic surfactants forms an electrical double layer at the interface. The capacitance can become negative within a restricted range of adsorption density. When the adsorption proceeds and enters this restricted range, the interface must undergo a transition to avoid a negative capacitance.

Our group has studied ion-sensitive Marangoni instability at the oil/water interface.[14,15,87–89] We consider this to be one of the most useful systems to study the role of interfacial chemistry in successive occurrences of Marangoni instability, because it is dramatically affected by cation species.[90] When the oil phase containing bis-(2-ethylhexyl)phosphate (DEHPA), an anionic surfactant, comes in contact with a water phase containing Ca^{2+} or Fe^{3+}, a Marangoni instability continuously occurs at the interface. However, the instability is not observed in the presence of Mg^{2+}, Ba^{2+}, Sr^{2+}, Cu^{2+}, and Co^{2+}. The difference in these cations is in their atomic scale. Hence our results indicate that such small scale differences are amplified to the macroscopic level by instabilities. Elucidation of these instabilities is interesting from the viewpoint of the stimuli-responsive interface. Ca^{2+} and Fe^{3+} may be regarded as irritant chemicals to the interface. Moreover, since Ca^{2+}, Mg^{2+}, Ba^{2+} and Sr^{2+} belong to the same group in the periodic table of the elements, they may exhibit similar characteristics in various situations. Thus, the marked difference between Ca^{2+} and the other cations (Mg^{2+}, Ba^{2+}, Sr^{2+}) in the resulting instabilities may provide significant insights into the role of interfacial chemistry that characterizes instability.

Figure 17.12 shows the dynamic interfacial tension of the oil/water interface containing Ca^{2+}. At the interface, a continuous spontaneous interfacial flow appears. Essentially the same result is obtained for Fe^{3+}. In this experiment, the interfacial tension was measured just after contact between an oil phase containing di-(2-ethylhexyl)phosphoric acid (DEHPA) and the water phase. At this point, the distribution of the solutes between the two phases is in a nonequilibrium state which generates the driving force for spontaneous convection. Figure 17.13(a–c) show snapshots of the interface over which trace silica particles are spread. We can observe an interfacial flow that is a radial propagation of flow from a source. (The white circles shown in Figure 17.13(a) and (c) indicate the periphery of the radial

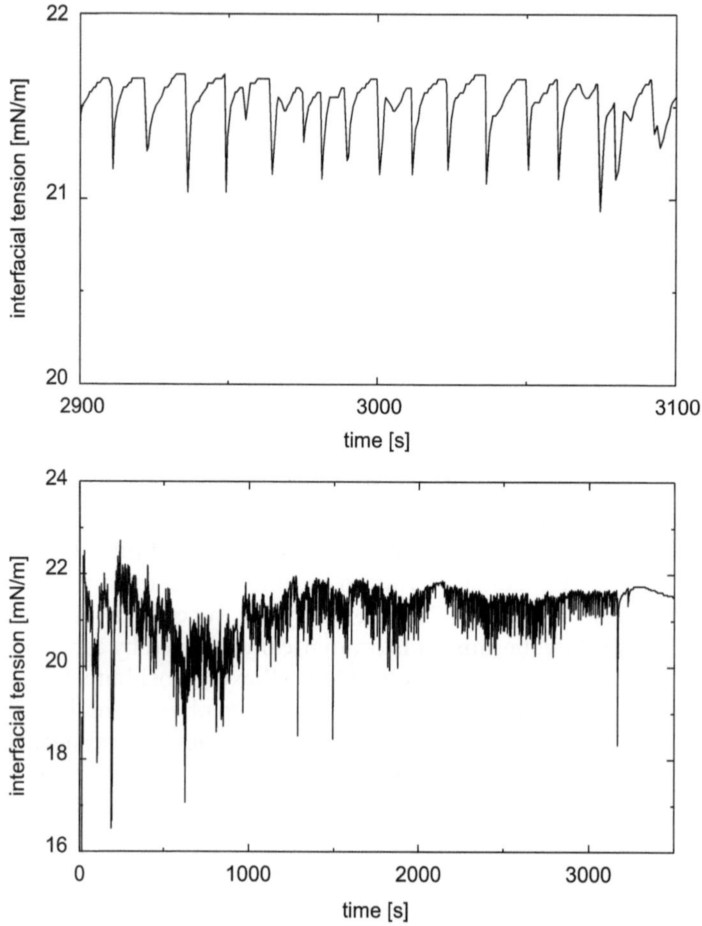

Figure 17.12 Oscillation of interfacial tension for a flat oil/water interface. Oil and water phases contain 5 mM DEHPA and 5 M $CaCl_2$. The upper panel expands the result from 2900–3100 s in the lower panel. (*Langmuir*, 2000, **16**, 8383)[88]

propagation.) Numerous radial propagations were observed to occur simultaneously.

Essentially the same instability is observed for an interface that is nearly at a distribution equilibrium. One way to cause instability is to provide mechanical compression at the interface. We used a pendant droplet for this purpose, changing the droplet size by increasing or decreasing the droplet volume[15,91] as shown in the photographs in Figure 17.14. As a result, the interfacial area increases or decreases. When the area is decreased, some of the surfactant molecules adsorbed at the interface have to be desorbed, and *vice versa*. Figure 17.14 shows an evaluation of the interfacial tension from a transient droplet shape under periodic change in the interfacial area. When the water phase contains Ca^{2+} and Fe^{3+}, the interfacial tension oscillates or

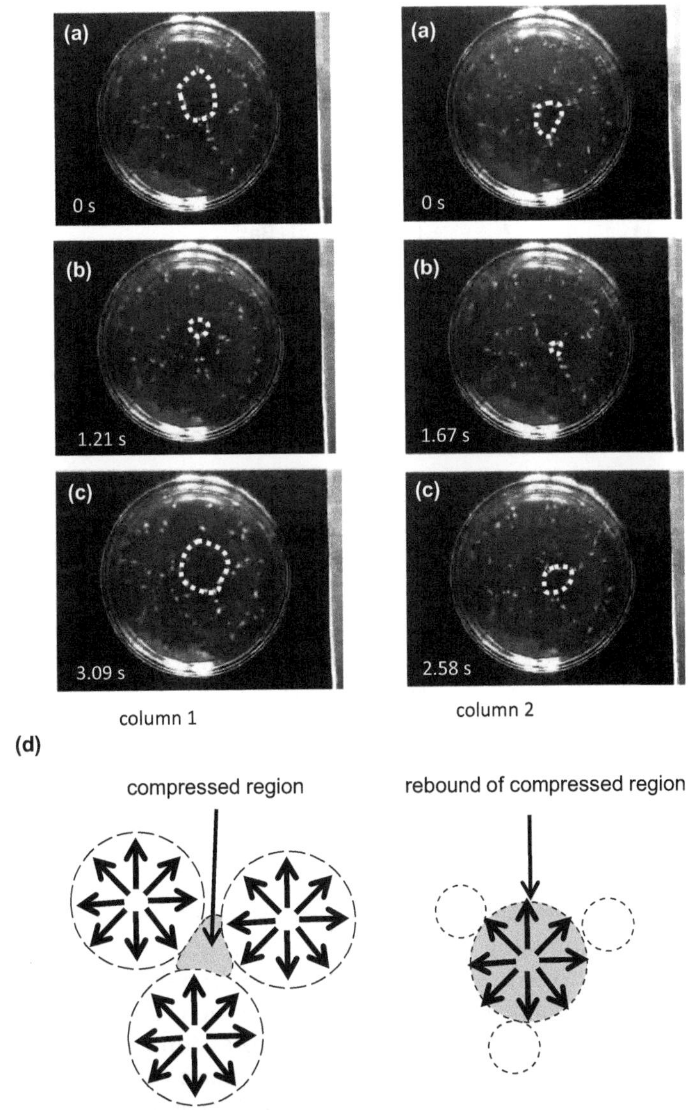

Figure 17.13 Motion of silica powder spread over the oil/water interface. $CaCl_2$ (5 M) and DEHPA (5 mM) are dissolved in water and *n*-heptane, respectively. (a) Silica powder is moved by a spontaneous interfacial flow that propagates radially, and expands to form a circular pattern. (b) The circular domain is compressed by another flow born nearby, so the pattern collapsed. (c) Radial flow appears from the compressed region. (d) Illustration of the sequence of events. Diameter of the petri dish is 6 cm. (*Chaos*, 2012, **22**, 037111)[15]

greatly fluctuates after compression of the interface.[15] However, the interfacial tension does not oscillate after an expansion of the interface. For the other cations, significant oscillations are not observed even after compression.

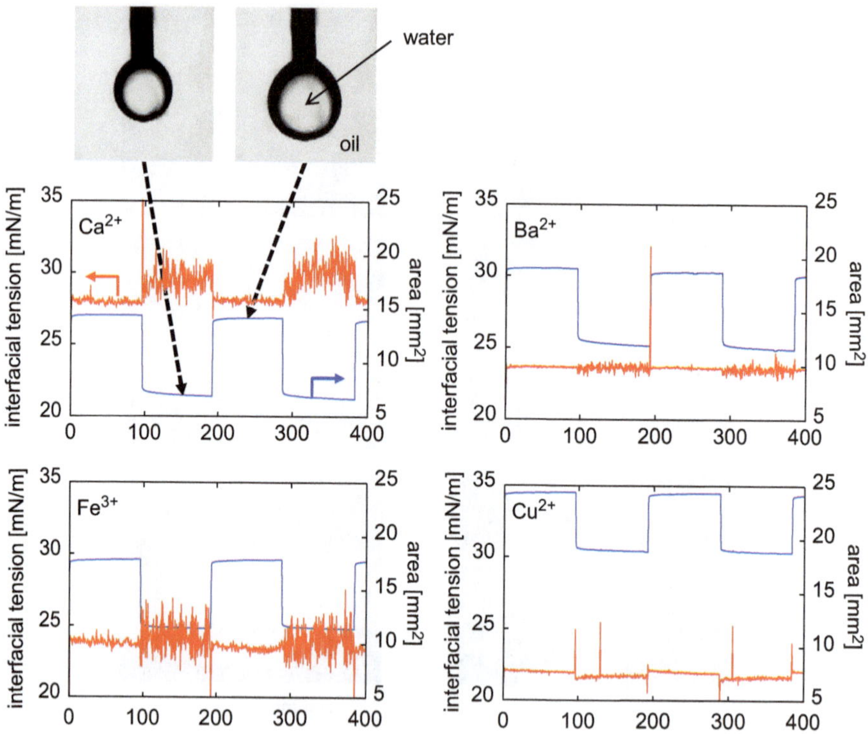

Figure 17.14 Interfacial tension of the pendant drop experiment under a cyclic volume change shown in the photograph. The rectangular shape represents the area change, and the red line is the interfacial tension. (*Chaos*, 2012, **22**, 037111)[15]

Compression is associated with the occurrence of instability. This suggests that the compression rate may be a decisive parameter for the occurrence of instability. We performed the pendant drop experiment shown in Figure 17.14 at various compression rates. When the droplet area is reduced from A_1 to A_2 within time Δt, the compression rate is expressed by $(1/A_1)(\Delta A/\Delta t)$, where $\Delta A = A_1 - A_2$. Figure 17.15 shows the amplitude of the interfacial tension oscillation as a function of the compression rate.[15] The ordinate is the mean value for numerous sets of amplitudes between a maximum and the neighboring minimum. The mean amplitude becomes approximately 0.2 mN m^{-1} without oscillation, because an irregular fluctuation of interfacial tension is always observed. For Ca^{2+}, the mean amplitude begins to increase at approximately 0.5–0.6 s^{-1}. For Mg^{2+}, Sr^{2+}, and Ba^{2+}, no such increase is observed.

For a flat oil/water interface experiment (Figures 17.12 and 17.13), compression of the interface is achieved by spontaneous convection itself. When an area of the interface is located between two radial propagations, it is compressed, as shown in Figure 17.13(d). Figure 17.13(b)→(c) show that a new interfacial flow appears at this compressed area. The spontaneous

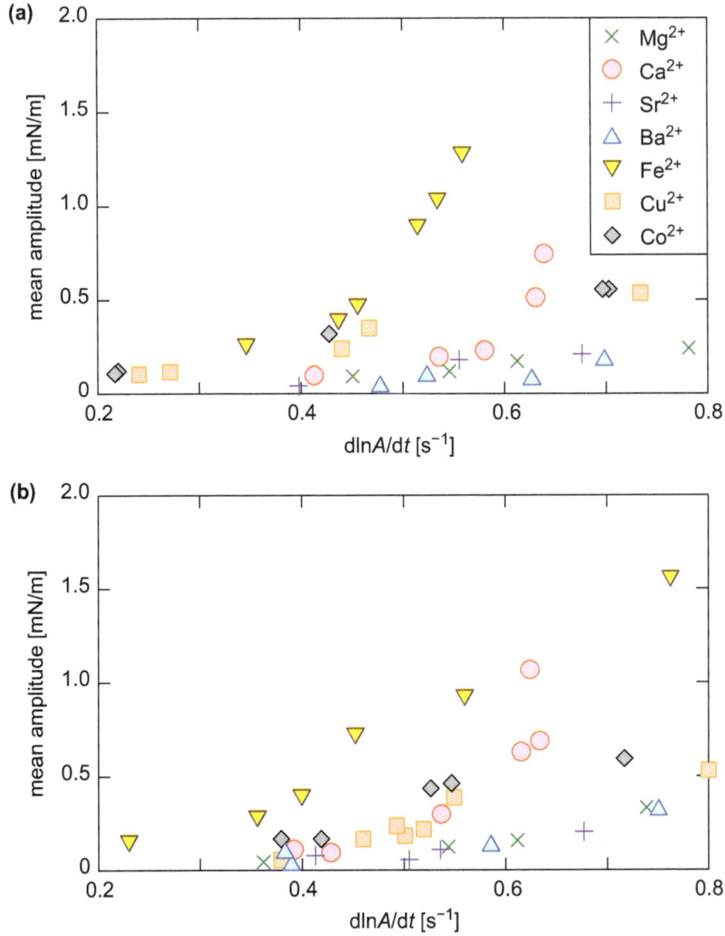

Figure 17.15 Dependency of the mean amplitude of the tension fluctuation on the rate of area change, dln*A*/d*t*, in an oil phase containing (a) 10 mM and (b) 100 mM DEHPA. The cations dissolved in water are Mg^{2+} (×), Ca^{2+} (circle), Sr^{2+} (+), Ba^{2+} (triangle), Fe^{3+} (inverted triangle), Cu^{2+} (square), and Co^{2+} (diamond). The cation concentrations are set to 1 M for $MgCl_2$, $CaCl_2$, $SrCl_2$, and $BaCl_2$, and 5 mM for $FeCl_3$, $CuCl_2$, and $CoCl_2$. (*Chaos*, 2012, **22**, 037111)[15]

compression rate determined from Figure 17.13 was approximately 0.4–0.5 s^{-1} that is sufficient to cause the oscillation following the results shown in Figure 17.15.[15] This result suggests that compression induces subsequent interfacial flow. Thus, an occurrence of interfacial flow continues as a chain process.

The significant role of compression may be demonstrated by another experimental method using an electric field across the interface.[91] When electric voltage is applied to a pendant drop, the droplet volume changes according to variations in the interfacial tension. When the electric potential

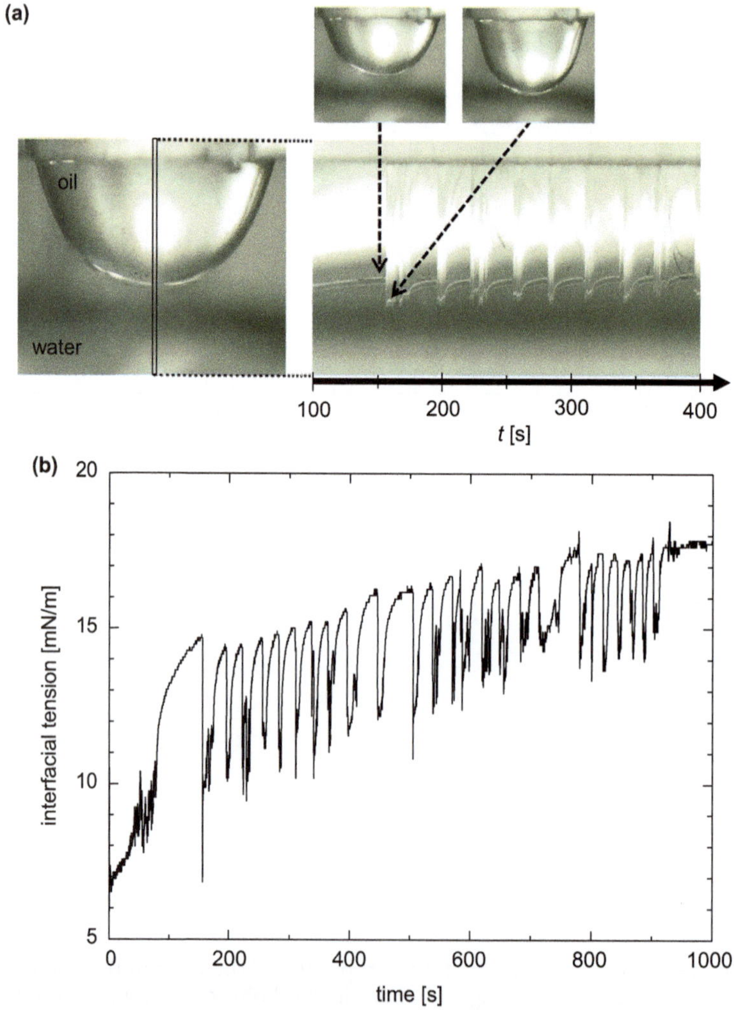

Figure 17.16 Dynamics of a hemispherical oil droplet at 600 mV. The concentrations of DEHPA and $Ca(OH)_2$ are 2 and 5 mM, respectively. (a) Spatiotemporal plot of a hemispherical droplet. The part of the photograph denoted by the rectangle is cut and arranged along a time sequence. (b) Interfacial tension change. Nitrobenzene is used as the oil. (*Bull. Chem. Soc. Jpn.*, 2014, **87**, 863)[91]

of the water phase is higher than that of the oil phase, the droplet shrinks after voltage is applied. After this shrinkage, the droplet exhibits a quasi-periodic pulsation. This experiment's spatiotemporal plot and interfacial tension are shown in Figure 17.16. The droplet length rapidly increases after spontaneous compression of the interface. Interestingly, such an instability was not observed for a flat oil/water interface despite using the same voltage and the same oil/water phases. The flat interface cannot shrink because the interfacial area is determined by the sample container. Thus, this result suggests that compression of the interface causes the instability.

Considering mechanical balance, the interfacial tension of the source should be lower than that of the surroundings. Thus, we must consider that an abrupt decrease in interfacial tension occurs at the compressed region. Since the interfacial tension depends on the adsorbed chemicals and their adsorption densities, we must consider the chemistry at the interface. To do so, we measured the extraction equilibrium of a cation with a DEHPA-containing oil phase and the electrical conductivity of the oil phase. As a result, we obtained following extraction equilibria.[14]

$$Ca^{2+} + 2(RH)_{2,org} \rightleftarrows (CaR_2 \cdot (RH)_2)_{org} + 2H^+ \tag{17.6}$$

$$Fe^{3+} + 2(RH)_{2,org} \rightleftarrows (FeR_2 \cdot (R_2H))_{org} + 3H^+ \tag{17.7}$$

$$M^{2+} + (RH)_{2,org} \rightleftarrows MR_2 + 2H^+, \qquad MR_2 \rightleftarrows MR^+ + R^- \tag{17.8}$$

Eqn (17.8) is applicable for Mg^{2+}, Sr^{2+}, and Ba^{2+}. Moreover, the complexes $CaR_2 \cdot (RH)_2$ and $FeR_2 \cdot (R_2H)$ tend to aggregate at the interface where the complexes form.[14] These species are oil-soluble. However, the aggregation retards the desorption process. Thus, the aggregate remains at the interface and its desorption requires energy to overcome the aggregate desorption. Compression of the interface provides a mechanical energy that can be used for this purpose. Once desorption has occurred in a restricted area of the interface, the tension must decrease dramatically, because the DEHPA molecules are quickly adsorbed at the interface. DEHPA is a more effective surfactant than the complex. Then, a radial flow develops by a Marangoni instability in the contrasting interfacial tensions. In this scenario, the complex formation and the aggregation provide a nonlinear characteristic in the desorption kinetics. That is, desorption requires the interfacial compression provided by the neighboring instability in the typical biphasic experiment (Figure 17.13). Therefore, this type of instability continues as a chain process.

17.5 Summary

In this chapter, we have reviewed self-moving colloidal objects with a focus on the authors' studies. First, we discussed the motion of a platinum particle by the decomposition of hydrogen peroxide. In contrast to studies using Janus particles, we use a single component Pt particle. In this case, the chemical reaction rate is macroscopically homogeneous over the Pt surface. This yields only random fluctuations by the particle. However, the particle shape produces anisotropic viscous drag. The asymmetry of the viscous dissipation transforms these random fluctuations into regular motions such as translation, spin, and rotation. Anisotropic viscous dissipation may be used as a universal mechanism for obtaining a regular motion from a random driving force and for designing a new self-moving object. Next, we discussed a vesicle that exhibits an oscillatory shape change under a pH gradient. pH gradients are widely used by living organisms to generate vital motions. However, as far as we know, a man-made amphiphilic molecular assembly with this characteristic has not been found. The vesicle presented in this chapter exhibits a continuous motion without any decrease in size. The oscillatory motion comprises two types of dynamics with quite different timescales. One is the reversal of the vesicle membrane, arising from the membrane energy's dependence on the pH gradient. The other is a rotation of the vesicle caused by a rotational instability produced by a phoretic force toward its higher-pH side. The former reversal motion is much faster than the latter rotation, and hence these two types of motions appear alternately and form an oscillatory cycle. Finally, we addressed ion-sensitive motion at an oil/water interface. The oil/water interface introduced in this chapter exhibits a spontaneous agitation when the water phase contains Ca^{2+} or Fe^{3+}, which is a type of Marangoni instability. For this instability to occur, a compression of the interface is necessary. The compression may be given by either mechanical or electrochemical methods. However, the compression may also occur by spontaneous agitation of the oil/water interface. Once the spontaneous agitation occurs, a radial propagation of the interfacial flow appears and a local area of the interface between the two radial flows is spontaneously compressed. This compression provides the mechanical energy required for the next instability. The surfactant molecules used in this experiment can form a complex with the cations. The complex molecules are oil-soluble but tend to aggregate at the interface. This aggregation forms an energy wall impeding desorption to the oil phase, but the compression provides sufficient energy to overcome the energy wall. This is an example where interfacial chemistry is fundamentally associated with the spontaneous growth of macroscopic convections. A difference in the molecular length scale causes a difference in the macroscopic response *via* this type of instability.

Abbreviation List

A_{total}	area of vesicle membrane
$A(t)$	surface area of pendant droplet
$\Delta A(t)$	change in surface area of pendant droplet
BZ	Belousov–Zhabotinsky (reaction)
$CaR_2(RH)_2$	complex molecule of Ca^{2+} and DEHPA
DEHPA	di-(2-ethylhexyl)phosphoric acid
$E(x)$	sum of the elastic energy of a vesicle membrane and the energy stored at the hole in swollen part
F_r	force acting on particle

$FeR_2 \cdot (R_2H)$	complex molecule of Fe^{2+} and DEHPA
M^{2+}	Mg^{2+}, Sr^{2+}, Ba^{2+}, Cu^{2+} or Co^{2+}
Ma	Marangoni number
MR_2	reaction product of bis-(2-ethylhexyl)phosphate anion and M^{2+}
R	bis-(2-ethylhexyl)phosphate group
RH	pronated form of bis-(2-ethylhexyl)phosphate (di-(2-ethylhexyl)-phosphoric acid)
$R(\psi,\theta)$	distance between the center of gravity and the periphery at the azimuth θ measured from an axis with an angle ψ
SAB	symmetry–asymmetry balance
T	temperature
V_{swell}	volume of water pool in swollen part of a vesicle
$a_n(\psi)$	the n-th Fourier coefficient for the cosine terms
$b_n(\psi)$	the n-th Fourier coefficient for the sine terms
d	diameter of Pt particle or distance between two plates with different temperature
$f(\psi)$	fraction of the power of the cosine components (degree of symmetry)
g	gravitational acceleration
I	moment of inertia
m	mass of a particle
N_r	torque acting on a particle
$\langle r(t)^2 \rangle$	mean square distance of moving particle
v	velocity
x	curvature of vesicle membrane of the bound state or x-component of a moving particle
$x(t)$	x-component of the position vector of a moving particle at time t
$y(t)$	y-component of the position vector of a moving particle at time t
$\delta R(\psi,\theta)$	deviation of $R(\psi,\theta)$ from the average
Δt	time increment
η	coefficient of viscous drag force
η_v	coefficient of viscous drag force for translational motion
η_ω	coefficient of viscous drag force for rotational motion
κ	thermal diffusivity
θ	azimuth
ρ	density
σ	surface tension
ω	angular velocity
ψ	angle of an appropriately selected direction
ψ_{max}	ψ value giving the maximum degree of symmetry
ψ_{min}	ψ value giving the minimum degree of symmetry
∇pH	strength of pH gradient

References

1. A. A. Biewener, *Animal Locomotion*, OUP Oxford, 2003.
2. R. Phillips, J. Kondev, J. Theriot and N. Orme, *Physical Biology of the Cell*, Garland Science, 2013.
3. S. H. Strogatz, *Nonlinear Dynamics and Chaos: With Applications to Physics, Biology, Chemistry, and Engineering*, Westview Press, 1994.
4. D. K. Kondepudi and I. Prigogine, *Modern thermodynamics: from heat engines to dissipative structures*, John Wiley, 1998.
5. J. Dunning-Davies, *Nature*, 1965, **208**, 576–577.
6. R. P. Feynman, R. B. Leighton and M. L. Sands, *The Feynman Lectures on Physics*, Addison-Wesley Publishing Company, 1964.
7. M. Guix, C. C. Mayorga-Martinez and A. Merkoçi, *Chem. Rev.*, 2014, **114**, 6285–6322.
8. D. Yamamoto, A. Mukai, N. Okita, K. Yoshikawa and A. Shioi, *J. Chem. Phys.*, 2013, **139**, 034705(1–5).
9. G. Pocock and C. D. Richards, *Human Physiology: The basis of medicine*, OUP Oxford, 2006.
10. J. Santo-Domingo and N. Demaurex, *J. Gen. Physiol.*, 2012, **139**, 415–423.
11. R. I. Gamow, B. Böttger and F. S. Barnes, *Biophys. J.*, 1971, **11**, 860–867.
12. E. Nawa, Y. Nishigaki, D. Yamamoto and A. Shioi, *Soft Matter*, 2013, **9**, 7832–7842.
13. J. P. Keener and J. Sneyd, *Mathematical Physiology: I: Cellular Physiology*, Springer, 2009.
14. T. Hosohama, K. Megumi, S. Terakawa, J. Nishimura, Y. Iida, T. Ban and A. Shioi, *Langmuir*, 2011, **27**, 14131–14142.
15. T. Miyaoka, J. Nishimura, Y. Iida, S. Maki and A. Shioi, *Chaos*, 2012, **22**, 037111(1–12).
16. L. Baraban, M. Tasinkevych, M. N. Popescu, S. Sanchez, S. Dietrich and O. G. Schmidt, *Soft Matter*, 2012, **8**, 48–52.
17. A. Brown and W. Poon, *Soft Matter*, 2014, **10**, 4016–4027.
18. W. Gao, X. Feng, A. Pei, Y. Gu, J. Li and J. Wang, *Nanoscale*, 2013, **5**, 4696–4700.
19. W. Gao, A. Pei, R. Dong and J. Wang, *J. Am. Chem. Soc.*, 2014, **136**, 2276–2279.
20. W. F. Paxton, K. C. Kistler, C. C. Olmeda, A. Sen, S. K. St. Angelo, Y. Cao, T. E. Mallouk, P. E. Lammert and V. H. Crespi, *J. Am. Chem. Soc.*, 2004, **126**, 13424–13431.
21. W. F. Paxton, P. T. Baker, T. R. Kline, Y. Wang, T. E. Mallouk and A. Sen, *J. Am. Chem. Soc.*, 2006, **128**, 14881–14888.
22. Y. Wang, R. M. Hernandez, D. J. Bartlett, J. M. Bingham, T. R. Kline, A. Sen and T. E. Mallouk, *Langmuir*, 2006, **22**, 10451–10456.
23. P. Colinet, J. C. Legros and M. G. Velarde, *Nonlinear Dynamics of Surface-Tension-Driven Instabilities: With a Foreword by I. Prigogine*, Wiley, 2001.
24. B. t. Hagen, S. v. Teeffelen and H. Löwen, *J. Phys.: Condens. Matter*, 2011, **23**, 194119.

25. B. Jingjing, Y. Zhan, M. Nakajima, S. Yajing, M. Takeuchi, H. Qiang and T. Fukuda, *Rob., IEEE Trans.*, 2014, **30**, 33–39.
26. N. I. Kovtyukhova, *J. Phys. Chem. C*, 2008, **112**, 6049–6056.
27. D. Yamamoto and A. Shioi, *KONA Powder Part. J.*, 2015, **32**, 2–21.
28. J. Soulages, M. S. N. Oliveira, P. C. Sousa, M. A. Alves and G. H. McKinley, *J. Non-Newtonian Fluid Mech.*, 2009, **163**, 9–24.
29. S. Taneda, *J. Phys. Soc. Jpn.*, 1979, **46**, 1935–1942.
30. W. Wang, L. A. Castro, M. Hoyos and T. E. Mallouk, *ACS Nano*, 2012, **6**, 6122–6132.
31. J. S. Bader, R. W. Hammond, S. A. Henck, M. W. Deem, G. A. McDermott, J. M. Bustillo, J. W. Simpson, G. T. Mulhern and J. M. Rothberg, *Proc. Natl. Acad. Sci. U. S. A.*, 1999, **96**, 13165–13169.
32. R. Eichhorn, J. Regtmeier, D. Anselmetti and P. Reimann, *Soft Matter*, 2010, **6**, 1858–1862.
33. J. Ok, E. Lopez-Oña, D. Nikitopoulos, H. Wong and S. Park, *Microfluid. Nanofluid.*, 2011, **10**, 1045–1054.
34. J. Regtmeier, S. Grauwin, R. Eichhorn, P. Reimann, D. Anselmetti and A. Ros, *J. Sep. Sci.*, 2007, **30**, 1461–1467.
35. R. Ait-Haddou and W. Herzog, *Cell Biochem. Biophys.*, 2003, **38**, 191–214.
36. M. Bier, *Contemp. Phys.*, 1997, **38**, 371–379.
37. S. Maeda, Y. Hara, T. Sakai, R. Yoshida and S. Hashimoto, *Adv. Mater.*, 2007, **19**, 3480–3484.
38. L. Mahadevan, S. Daniel and M. K. Chaudhury, *Proc. Natl. Acad. Sci. U. S. A.*, 2004, **101**, 23–26.
39. R. Yoshida and Y. Murase, *Colloids Surf., B*, 2012, **99**, 60–66.
40. G. E. Crooks, *Phys. Rev. E: Stat. Phys., Plasmas, Fluids, Relat. Interdiscip. Top.*, 2000, **61**, 2361–2366.
41. G. Gallavotti and E. G. D. Cohen, *Phys. Rev. Lett.*, 1995, **74**, 2694–2697.
42. C. Maes, *J. Stat. Phys.*, 1999, **95**, 367–392.
43. R. Di Leonardo, L. Angelani, D. Dell'Arciprete, G. Ruocco, V. Iebba, S. Schippa, M. P. Conte, F. Mecarini, F. De Angelis and E. Di Fabrizio, *Proc. Natl. Acad. Sci. U. S. A.*, 2010, **107**, 9541–9545.
44. C. D. Bain, *ChemPhysChem*, 2001, **2**, 580–582.
45. M. K. Chaudhury and G. M. Whitesides, *Science*, 1992, **256**, 1539–1541.
46. S. Daniel, M. K. Chaudhury and J. C. Chen, *Science*, 2001, **291**, 633–636.
47. F. D. Dos Santos and T. Ondarçuhu, *Phys. Rev. Lett.*, 1995, **75**, 2972–2975.
48. K. Ichimura, S.-K. Oh and M. Nakagawa, *Science*, 2000, **288**, 1624–1626.
49. S.-W. Lee and P. E. Laibinis, *J. Am. Chem. Soc.*, 2000, **122**, 5395–5396.
50. D. T. Wasan, A. D. Nikolov and H. Brenner, *Science*, 2001, **291**, 605–606.
51. T. Ban, T. Yamagami, H. Nakata and Y. Okano, *Langmuir*, 2013, **29**, 2554–2561.
52. I. Lagzi, S. Soh, P. J. Wesson, K. P. Browne and B. A. Grzybowski, *J. Am. Chem. Soc.*, 2010, **132**, 1198–1199.
53. P. G. de Gennes, *Physica A*, 1998, **249**, 196–205.
54. E. Nakache and M. Dupeyrat, *Bioelectrochem. Bioenerg.*, 1982, **9**, 583–590.

55. E. Nakache, M. Dupeyrat and M. Vignes-Adler, *Faraday Discuss. Chem. Soc.*, 1984, 77, 189–196.
56. A. Shioi, T. Ban and S. Suzuki, *Phys. Rev. E: Stat., Nonlinear, Soft Matter Phys.*, 2008, 77, 036208.
57. A. Shioi, T. Ban and S. Suzuki, *Chem. Phys. Lett.*, 2008, **467**, 210–215.
58. A. Shioi, K. Katano and Y. Onodera, *J. Colloid Interface Sci.*, 2003, **266**, 415–421.
59. Y. Sumino, H. Kitahata, K. Yoshikawa, M. Nagayama, S.-i. M. Nomura, N. Magome and Y. Mori, *Phys. Rev. E: Stat., Nonlinear, Soft Matter Phys.*, 2005, **72**, 041603.
60. Y. Sumino, N. Magome, T. Hamada and K. Yoshikawa, *Phys. Rev. Lett.*, 2005, **94**, 068301.
61. D. Marsland, in *Int. Rev. Cytol.*, ed. G. H. Bourne and J. F. Danielli, Academic Press, 1956, vol. 5, pp. 199–227.
62. A. Rajendran, M. Endo and H. Sugiyama, *Chem. Rev.*, 2013, **114**, 1493–1520.
63. S. M. Christensen and D. Stamou, *Soft Matter*, 2007, **3**, 828–836.
64. P. Stano and P. L. Luisi, *Chem. Commun.*, 2010, **46**, 3639–3653.
65. D. Abreu, M. Levant, V. Steinberg and U. Seifert, *Adv. Colloid Interface Sci.*, 2014, **208**, 129–141.
66. U. Seifert, *Adv. Phys.*, 1997, **46**, 13–137.
67. T. Hamada, Y. Hirabayashi, T. Ohta and M. Takagi, *Phys. Rev. E: Stat., Nonlinear, Soft Matter Phys.*, 2009, **80**, 051921.
68. F. Nomura, M. Nagata, T. Inaba, H. Hiramatsu, H. Hotani and K. Takiguchi, *Proc. Natl. Acad. Sci. U. S. A.*, 2001, **98**, 2340–2345.
69. O. Sandre, L. Moreaux and F. Brochard-Wyart, *Proc. Natl. Acad. Sci. U. S. A.*, 1999, **96**, 10591–10596.
70. T. Miura, H. Oosawa, M. Sakai, Y. Syundou, T. Ban and A. Shioi, *Langmuir*, 2010, **26**, 1610–1618.
71. P. A. Giardini, D. A. Fletcher and J. A. Theriot, *Proc. Natl. Acad. Sci. U. S. A.*, 2003, **100**, 6493–6498.
72. R. D. Kulkarni and P. Somasundaran, *Colloids Surf.*, 1980, **1**, 387–405.
73. P. L. T. Brian and K. A. Smith, *AIChE J.*, 1972, **18**, 231–233.
74. J. R. A. Pearson, *J. Fluid Mech.*, 1958, **4**, 489–500.
75. T. K. Sherwood and J. C. Wei, *AIChE J.*, 1955, **1**, 522–527.
76. C. V. Sternling and L. E. Scriven, *AIChE J.*, 1959, **5**, 514–523.
77. K. Warmuziński and J. Buzek, *Chem. Eng. Sci.*, 1990, **45**, 243–254.
78. K. Warmuziński and M. Tańczyk, *Chem. Eng. Sci.*, 1991, **46**, 2031–2039.
79. R. J. Field, E. Koros and R. M. Noyes, *J. Am. Chem. Soc.*, 1972, **94**, 8649–8664.
80. A. M. Zhabotinsky, *Chaos*, 1991, **1**, 379–386.
81. A. De Wit, K. Eckert and S. Kalliadasis, *Chaos*, 2012, **22**, 037101(1–5).
82. N. M. Kovalchuk and D. Vollhardt, *Adv. Colloid Interface Sci.*, 2006, **120**, 1–31.
83. Y. Ikezoe, S. Ishizaki, H. Yui, M. Fujinami and T. Sawada, *Anal. Sci.*, 2004, **20**, 435–440.

84. T. Kakiuchi, *J. Electroanal. Chem.*, 2002, **536**, 63–69.
85. Y. Kitazumi and T. Kakiuchi, *Bull. Chem. Soc. Jpn.*, 2011, **84**, 1312–1320.
86. K. Yoshikawa and Y. Matsubara, *J. Am. Chem. Soc.*, 1983, **105**, 5967–5969.
87. A. Shioi, H. Kumagai, Y. Sugiura and Y. Kitayama, *Langmuir*, 2002, **18**, 5516–5522.
88. A. Shioi, Y. Sugiura and R. Nagaoka, *Langmuir*, 2000, **16**, 8383–8389.
89. A. Shioi, K. Yamada, R. Shiota, F. Hase and T. Ban, *Bull. Chem. Soc. Jpn.*, 2006, **79**, 1696–1703.
90. T. Ban, T. Fujii, K. Kurisaka and A. Shioi, *Chem. Lett.*, 2006, **35**, 1134–1135.
91. S. Maki, R. Yamamoto, D. Yamamoto and A. Shioi, *Bull. Chem. Soc. Jpn.*, 2014, **87**, 863–870.

Subject Index

Note: Page numbers in *italic* refer to figures and tables.